Chemische Elementarprozesse

Herausgegeben von

Hermann Hartmann

in Zusammenarbeit mit

J. Heidberg · H. Heydtmann · G. H. Kohlmaier

Mit 86 Abbildungen

Springer-Verlag Berlin Heidelberg GmbH

Hermann Hartmann, O. Professor, Dr. rer. nat., Direktor des Instituts für physikalische Chemie, 6000 Frankfurt (Main), Robert-Mayer-Str. 11

Joachim Heidberg, Dr. phil. nat., Dipl. Chem.,
Horst Heydtmann, Dr. phil. nat., Dipl. Chem.,
Gundolf H. Kohlmaier, Ph. D., M. Sc.,
Institut für physikalische Chemie der Universität, 6000 Frankfurt (Main),
Robert-Mayer-Str. 11

ISBN 978-3-642-85890-1   ISBN 978-3-642-85889-5 (eBook)
DOI 10.1007/978-3-642-85889-5

Vorwort

Die Theorie der chemischen Phänomene kann man systematisch in die Statik und in die Kinetik der Materie unterteilen. Von der Statik, welche sich im molaren Bereich als die Theorie der Gleichgewichte und im molekularen Bereich als die Theorie der stationären Zustände molekularer Gebilde präsentiert, wissen wir schon wesentlich mehr als von der Kinetik. Das liegt natürlich daran, daß Vorgänge in sehr differenzierter Weise durch submikroskopische Umstände bestimmt werden, so daß die Kinetik von vornherein schwieriger als die Statik ist. Bei der zukünftigen Entwicklung der theoretischen Chemie wird sich dementsprechend das Schwergewicht der Forschung mehr und mehr nach der Kinetik verlagern. In dem engeren Bereich der eigentlichen chemischen Kinetik ist das deutlich zu sehen. In der jüngst vergangenen Zeit sind wesentliche Fortschritte in der Theorie der Reaktionen und speziell in der Theorie der Elementarreaktionen oder chemischen Elementarprozesse erzielt worden. Parallel dazu sind neue experimentelle Techniken und Methoden entwickelt worden, die die Aussagen der Theorien besser zu überprüfen gestatten und die neuartige Informationen über chemische Vorgänge liefern können.

Im Jahre 1965 sind im Rahmen der Konstanzer Internationalen Ferienkurse für Theoretische Chemie Vorträge und Vorlesungen zum Thema „Chemische Elementarprozesse" gehalten worden. Während dieses Kurses entstand der Plan zur Veröffentlichung eines Buches unter dem gleichen Titel. Das nunmehr vorliegende Buch, das nicht als Handbuch oder als eine Sammlung zusammenfassender Berichte gemeint ist, beabsichtigt, dem Studenten höherer Semester und dem ferner stehenden Wissenschaftler den Zugang zu dem vielgestaltigen, faszinierenden Gebiet der neueren Kinetik zu eröffnen. Diesem Ziel steht nicht entgegen, daß an einzelnen Stellen Neuentwicklungen dargelegt werden, die der Sachverständige erkennen wird und die das Buch auch für ihn interessant machen dürften.

Um dem Leser eine Orientierung zu erleichtern, soll an dieser Stelle die Anordnung der Kapitel des Buches erläutert werden. Am Anfang stehen zwei Kapitel, die als allgemeine Einführung in die theoretische Behandlung chemischer Elementarprozesse angesehen werden können. Die folgenden 6 Kapitel sind alle theoretischer Art. In diesem Abschnitt werden die wichtigsten modernen Aspekte der Theorie der Elementarprozesse behandelt. Dann folgen vier Kapitel über die Anwendung moderner experimenteller Methoden, welche geeignet sind, Ergebnisse der theoretischen Betrachtung zu überprüfen. Die hier behandelten experimentellen Methoden beziehen sich auf Prozesse in der Gasphase. Nicht nur diejenigen, welche sich mit chemischen Elementarprozessen unter möglichst reinen Bedingungen beschäftigen, sind an Elementarprozessen überhaupt interessiert. Vor allem im Bereich der organischen Chemie stellt sich dem Experimentator sehr oft die Frage, ob es nicht theoretische Vorstellungen gibt, die eine sicher vorläufige, aber doch wertvolle Ordnung seines Beobachtungsmaterials über Reaktivitäten möglich machen. Diesem Aspekt sind die folgenden zwei Kapitel gewidmet. In dem letzten Abschnitt dieses Buches ist schließlich in gleicher Weise von theoretischen Überlegungen und experimentellen Methoden die Rede. Das gemeinsame Charakteristikum ist dabei, daß es sich in allen Fällen um Prozesse in flüssiger Phase handelt.

Ich möchte an dieser Stelle allen Autoren, die Mühe und Zeit geopfert haben, damit dieses Buch zustande komme, herzlich Dank sagen. Dank gebührt auch meinen Mitarbeitern im Frankfurter Institut für physikalische Chemie, vor allem den Herren Dr. Heidberg, Dr. Heydtmann und Dr. Kohlmaier, die viel Zeit geopfert und mit einer nie erlahmenden Begeisterung für die Sache alle Schwierigkeiten überwunden haben. Ohne ihre Hilfe wäre dieses Buch nicht zustande gekommen.

H. Hartmann

Inhalt

Reaction Cross Sections, Rate Coefficients and Nonequilibrium Kinetics. By K. E. SHULER 1

1. Introduction 1
2. Reaction Cross Sections and Rate Coefficients 4
3. Examples of Relation between σ_R and k_σ 9
   3.1. Complete Equilibrium Distributions 9
   3.2. The Model of Reactive Hard Spheres 10
   3.3. A Generalization of the Hard Sphere Model 11
   3.4. Nearly Constant Cross Sections 12
   3.5. Ion-Molecule Reactions $(A^+ + B \to C^+ + D)$ 12
   3.6. Reactants Having Velocity Distributions at Two Temperatures ... 14
   3.7. Vibration-Translation Energy Exchange 16
4. Nonequilibrium Kinetics 18
   4.1. Reactant Nonequilibrium Distributions 18
   4.2. Product Nonequilibrium Distributions 19
   4.3. Perturbation of Distribution Functions by Chemical Reactions 19
   4.4. Phenomenological and Statistical Rate Coefficients 20
References 21

Remarks on the Generalization of Activated Complex Theory. By R. A. MARCUS 23
References 25

Wellenmechanische Aspekte in der Theorie der Elementarreaktionen. Von H. PREUSS. Mit 2 Abbildungen 26

1. Ausgangspunkte 26
2. Grundriß des wellenmechanischen Verfahrens 34
3. Die Energiehyperflächen 37
Literatur 41

Theory of Non-Adiabatic Transitions. Recent Development of the Landau-Zener-(Linear) Model. By E. E. NIKITIN. With 12 Figures . 43

1. Adiabatic Approximation and Non-Adiabatic Coupling 43
2. Two-State Approximation for a One-Dimensional Linear Model 45
3. The Landau-Zener Formula 49
4. Effect of the Turning Point. Weak Adiabatic Coupling. Strong Adiabatic Coupling. The Landau-Zener Formula 52
5. Non-Adiabatic Reflection from and Transmission through a Potential Barrier. Small Adiabatic Splitting. Overbarrier Non-Adiabatic Transmission and Reflection. Underbarrier Non-Adiabatic Transmission (Tunneling) and Reflection. Adiabatic Reflection from a Potential Hump Corresponding to the Lower Electronic Term. 59
6. Mean Transition Probability 64
7. Extension of the Landau-Zener Model to Multistate Systems 67

8. Two-Dimensional Extension of the Landau-Zener Model 70
9. Application of the Landau-Zener Model to Elementary Processes 75
References 76

Statistische Quantenmechanik chemischer Reaktionen. Von G. L.
 Hofacker. Mit 1 Abbildung 78
1. Einleitende Bemerkungen 78
 1.1. Die Dichtematrix 78
 1.2. Die Bewegungsgleichung 79
 1.3. Formale Störungsrechnung für die von Neumann-Gleichung 80
2. Intramolekulare Relaxation 84
3. Die Molekel gekoppelt an einen Thermostaten 95
Literatur 106

Unimolecular Reaction Rate Theory. By R. A. Marcus. With 1 Figure 109
1. Introduction 109
2. The Specific Rate Constant 109
3. The First Order Rate Expression 112
4. The Density of Energy States 113
5. The Influence of Centrifugal Potential 114
6. Concluding Remarks 114
Appendix 115
References 116

Zwischenmolekulare Energieübertragung bei Stößen unter beson-
 derer Berücksichtigung von mehratomigen Molekülen in hoch-
 angeregten Schwingungszuständen. Von G. H. Kohlmaier. Mit 3
 Abbildungen 117
1. Einführung 117
2. Phänomenologische Beschreibung der Energieübertragung 118
 2.1. Ultraschalltechnik 118
  2.1.1. Relaxationszeit für ein 2-Niveau-Schema 119
  2.1.2. Relaxationszeit für ein n-Niveau-Schema 120
  2.1.3. Relaxationszeit τ für mehratomige Moleküle................. 120
 2.2. Chemische Aktivierung 121
  2.2.1. Einstufige Desaktivierung 123
  2.2.2. Mehrstufige Desaktivierung mit T Schritten................. 123
  2.2.3. Experimentelles Beispiel 124
3. Mechanik der Stöße 124
 3.1. Mechanische Grundlagen 124
 3.2. Erhaltungssätze bei Stößen 129
 3.3. Bewegungsablauf bei Stößen 131
  3.3.1. Beispiel: Eindimensionaler elastischer Stoß zwischen zwei
   Atomen.. 132
  3.3.2. Beispiel: Eindimensionaler unelastischer Stoß zwischen einem
   Molekül und einem Atom..................................... 132
4. Theorie der Energieübertragung bei Stößen...................... 134
 4.1. Vibration-Translationsübergänge und Vibration-Vibrationsüber-
  gänge bei zweiatomigen Molekülen in niedrigen Energiezuständen 134
 4.2. Zweiatomige Moleküle mit Energien in der Nähe der Dissoziationsgrenze 136
 4.3. Erweiterung des Landau-Teller-Modells auf mehratomige Moleküle
  in niedrigen Energiezuständen 136

4.4. Statistische Theorie für hochangeregte mehratomige Moleküle .. 136
5. Vergleich der experimentellen Resultate mit der Theorie 138
   5.1. Resultate für Moleküle bei niedrigen Energien: Ergebnisse von
      Ultraschallmessungen 138
   5.2. Resultate für mehratomige Moleküle bei hohen Energien: Ergebnisse
      der chemischen Aktivierung und der Fluoreszenzausbeutemessungen .. 140
Literatur 142

Monte-Carlo-Rechnungen in der chemischen Kinetik. Von H. HEYDT-
   MANN. Mit 4 Abbildungen 143
1. Einleitung 143
2. Bimolekulare Reaktionen 144
3. Unimolekulare Reaktionen 153
Literatur 156

Molecular Beam Studies of Chemical Reactions. By J. P. TOENNIES.
   With 34 Figures 157
1. Introduction 157
   1.1. General 157
   1.2. Basic Technique and Definitions 158
2. The Essential Components of a Molecular Beam Apparatus 161
   2.1. Effusive Beam Sources 162
   2.2. Beam Sources with Energies Greater than 10 kcal/mole 165
   2.3. Velocity Selectors 167
   2.4. State Selectors 168
   2.5. Beam Detectors 170
   2.6. Some Typical Values Needed for Estimating Product Intensities ... 172
3. Elastic Scattering and Chemical Reactions 173
   3.1. The Intermolecular Potential 173
   3.2. Classical Theory of Elastic Scattering 174
   3.3. Quantum Mechanical Theory of Elastic Scattering................ 179
   3.4. The Central Field Impact Theory of Chemical Reactions 180
   3.5. The Transformation of Cross Sections from the Center of Mass Coordi-
      nate System to the Laboratory Coordinate System 184
4. Some Recent Results on Chemical Reactions 188
   4.1. Summary of Data 188
   4.2. Rebound Reactions 189
      4.2.1. The Reaction $K + HBr \rightarrow H + KBr$ 189
      4.2.2. The Reaction $K + CH_3I(Br) \rightarrow CH_3 + KI(Br)$ 197
      4.2.3. Discussion 200
   4.3. Stripping Reactions 205
      4.3.1. The Reaction $K + Br_2 \rightarrow Br + KBr$ 205
      4.3.2. Discussion 210
References 216

**Einige Beispiele für die Anwendung von Stoßwellen zur Untersu-
chung chemischer Reaktionen.** Von W. JOST und H. GG. WAGNER.
   Mit 2 Abbildungen 219
1. Beschreibung von Stoßwellen 219

2. Erzeugung von Stoßwellen und Messungen im Stoßwellenrohr zur
   Untersuchung chemischer Reaktionen 222
3. Beispiele für Untersuchungen an unimolekularen Reaktionen 227
   3.1. Theoretische Beschreibung unimolekularer Reaktionen 227
   3.2. Ergebnisse für den Zerfall vier- und mehratomiger Moleküle 229
   3.3. Ergebnisse für den Zerfall dreiatomiger Moleküle 230
      3.3.1. Der Zerfall von CO_2 233
      3.3.2. Der Zerfall von H_2O 235
      3.3.3. Der Zerfall von SO_2 236
      3.3.4. Der Zerfall von N_2O 238
Literatur 240

Atom Reactions in Discharge-Flow Systems. By B. A. THRUSH 243
1. Experimental Methods 243
2. Applications 246
3. Chemiluminescent Reactions 248
References 250

Flash Photolysis By B. A. THRUSH 252
1. Introduction 252
2. The Apparatus 253
3. Spectroscopic Studies 255
4. The Recombination of Atoms 256
5. Radical Reactions in Gases 257
6. Reactions in Solution 258
References 259

Molekülstruktur und chemische Reaktivität. Von F. BECKER. Mit
   2 Abbildungen 261
1. Reaktionstypen und Reaktionsmechanismen in der organischen Chemie . 261
   1.1. Klassifizierung organischer Reagenzien und Reaktionen 261
   1.2. Beispiele für Reaktionsmechanismen 266
2. Empirische Struktur-Reaktivitäts-Beziehungen 273
   2.1. Thermodynamische Beziehungen 274
   2.2. Klassifizierung der strukturellen Effekte 275
   2.3. Lineare ΔG-Beziehungen 277
   2.4. Die Trennung von induktiven, mesomeren und sterischen Effekten 284
Literatur 289

**Abschätzung relativer freier Aktivierungsenthalpien mittels der HMO-
   Methode** Von O. E. POLANSKY und P. SCHUSTER. Mit 10 Abbildungen 290
1. Einleitung 290
2. Reaktionstheoretische Grundlagen 292
3. Konkurrenzreaktionen 302
4. Abschätzung der Beiträge $\Delta H_\pi^{\ddagger}$ mittels der HMO-Methode 305
5. Anwendungsbeispiel: Reaktionen von Diazoalkanen mit Olefinen 308
   5.1. Konkurrenz I 311

5.2. Konkurrenz II 318
5.3. Konkurrenz III 323
5.4. Reaktionen mit weiteren Molekülen Diazoalkan 324
6. Diskussion 325
Literatur 329

Reaktionen in Lösungen unter erhöhten statischen Drucken. Von
    H. Heydtmann. Mit 5 Abbildungen 331
1. Einleitung 331
2. Deutung des Aktivierungsvolumens 334
   2.1. Der Einfluß der Änderung von Abständen zwischen im Anfangs- oder
      Endzustand gebundenen Atomen 335
   2.2. Der Einfluß sterischer Behinderung in der Umgebung des Reaktions-
      zentrums 335
   2.3. Der Einfluß der Solvatation 338
3. Interpretation von Reaktionsmechanismen mit Hilfe der Druckabhängigkeit von Reaktionsgeschwindigkeiten 344
Literatur 346

Electron Transfer Reactions. By R. A. Marcus. With 2 Figures 348
1. Introduction 348
2. Potential Energy Surfaces and Mechanism of Electron Transfer 348
3. Assumptions 350
4. Deductions 353
References 354

**Chemische Elementarprozesse und kernmagnetische Resonanz in
    Flüssigkeiten.** Von J. Heidberg. Mit 5 Abbildungen 357
1. Einleitung 357
2. Die Umlagerung von Spins in adiabatischen Theorien 357
   2.1. Blochsche Gleichungen 358
   2.2. Blochsche Gleichungen mit Berücksichtigung chemischer Umlagerun-
      gen 360
3. Quantenstatistische Beschreibung eines Ensembles gekoppelter Kerne
   mit dem Spin 1/2. Die Boltzmanngleichung 364
   3.1. Reine Zustände 364
   3.2. Gemischte Zustände. Die Kernspindichtematrix 365
   3.3. Zeitabhängige Störungsrechnung mit der Kernspindichtematrix. Die
      Boltzmanngleichung 370
4. Der Umlagerungsterm in der Boltzmanngleichung 374
   4.1. Klassifizierung chemischer Umlagerungen 374
   4.2. Die symmetrische intramolekulare Umlagerung. Der Diracsche Spin-
      austauschoperator 376
   4.3. Verallgemeinerung für asymmetrische intramolekulare Umlagerungen.
      Hinweis auf die Behandlung des Relaxationsproblems 383
5. Eine Lösung der generalisierten Boltzmanngleichung 384
6. Kinetische Größen aus der kernmagnetischen Resonanz. Die Kinetik von
   Prozessen im chemischen Gleichgewicht 388
7. Anwendungsbeispiele 390

XII Inhalt

7.1. Intermolekulare Protonenübertragung 390
    in wäßrigen Lösungen von Ammoniumionen. Eine Anwendung der
    adiabatischen Theorie in Gegenwart von Kern-Spin-Spin-Kopplung
7.2. Ligandenaustausch in Komplexen. 394
    Austausch von Atomen mit elektrischem Kernquadrupolmoment.
    Ligandenaustausch in paramagnetischen Komplexen
7.3. Anwendung des Dichtematrixformalismus. 400
    Inversion von Ammoniakderivaten. Ringinversion
8. Grenzen der Anwendbarkeit. Ergänzende Methoden und Entwicklungen 402
9. Anhang; Der Spin und die Transformationseigenschaften von Spin-
    funktionen bei räumlichen Drehungen. 406
Literatur 414

Kinetik schneller Reaktionen in Lösung und chemische Relaxation.
    Von M. Eigen und L. De Maeyer. Ausgearbeitet von J. Heidberg
    und G. H. Kohlmaier. Mit 3 Abbildungen 417

1. Einleitung 417
2. Übersicht der Meßmethoden 417
3. Theoretische Basis der Relaxationsmethoden 418
    3.1. Störung des chemischen Gleichgewichts 418
    3.2. Linearisierte Geschwindigkeitsgleichungen und Relaxationszeiten . 419
        3.2.1. Einstufige Reaktionen 419
        3.2.2. Mehrstufige Relaxation. Normalvariable 421
        3.2.3. Quasikontinuierliche Relaxationsspektren. Mittlere Relaxations-
            zeiten 422
    3.3. Reaktionseffekte bezüglich Normalvariablen 424
4. Relaxationsverfahren 424
    4.1. Sprung- und Impuls-Verfahren (Einschwingverfahren) ... 424
        4.1.1. Mathematische Formulierung 424
        4.1.2. Experimentelle Technik 425
    4.2. Stationäre Verfahren (Periodische Störungen) 426
        4.2.1. Mathematische Formulierung 426
        4.2.2. Experimentelle Technik 427
5. Anwendungsbeispiele 428
    5.1. Metallkomplexreaktionen in wäßriger Lösung 428
    5.2. Die Protolyse des Wassers. Ein diffusionsbestimmter Prozeß 430
    5.3. Kooperative Umwandlungen von Biopolymeren 435
        5.3.1. Einführung 435
        5.3.2. Thermodynamik auf der Grundlage des linearen Ising-Modells 435
        5.3.3. Ein Modell für die Kinetik der Umwandlung. Experimentelle Er-
            gebnisse 442
    5.4. Reaktionen eines allosterischen Enzyms 447
        5.4.1. Zur Deutung sigmoider Bindungsisothermen 447
        5.4.2. Gleichgewichtseigenschaften eines allosterischen Systems 447
        5.4.3. Relaxation eines allosterischen Systems. Die Meßbarkeit von
            Relaxationszeiten und Reaktionseffekte bezüglich Normalvariablen 448
        5.4.4. Die Bindung von Nikotinamid-adenin-dinucleotid an Hefe-D-
            Glycer-aldehyd-3-phosphatdehydrogenase 450
Anhang: Ein Beispiel für die mathematische Behandlung mehrstufiger
    Reaktionssysteme 452
Literatur 457

Reaction Cross Sections, Rate Coefficients and Nonequilibrium Kinetics

Kurt E. Shuler

1. Introduction

The observed rate of a chemical reaction is usually expressed through an empirical equation involving the concentration of reactants and/or products and a proportionality constant k, the *rate coefficient*. This rate coefficient is determined experimentally from measurements of the change of concentration of reactants and/or products with time. If the empirical rate equation correctly describes, *in all aspects*, the reaction under investigation, the rate coefficient should be independent of the concentration of the reactants and/or products and of the time and should depend only upon the temperature of the (isothermal) reaction system [1].

Gas phase chemical reactions of all types — in particular those which are „bi-molecular" — whether they involve atoms, molecules, photons, electrons, etc., proceed by collisions of the reactant species. Not all collisions, however, are effective in giving rise to product molecules. In some collisions between reactants there may be an unfavorable geometrical orientation for product formation, others may not be sufficiently energetic for the formation or rupture of a bond, and still others may be too energetic to permit the necessary rearrangements leading to the desired product formation. These collisions will lead to an energy transfer (internal and/or translational) between the colliding species rather than to product formation. The empirical rate coefficient k clearly is related to the number of successful collisions per unit time, i.e., the frequency of collisions which lead to product formation. The various theories of gas phase chemical kinetics have as one of their objectives the calculation of this rate coefficient k from the dynamics and energetics of the collision processes. The successes and failures of some of the theoretical approaches are discussed in detail in a recent review article [2].

Recent advances in experimental techniques permit the measurement for some reactions of the "efficiency" of elementary chemical collision processes leading to product formation as a function of the energy of the reactants (kinetic and internal) and the geometry of the encounter. The results of these measurements are expressed as *reaction cross sections*, σ_R. It is the purpose of these lectures to explore the relation between the macroscopic rate coefficient k and the microscopic reaction cross section σ_R.

We shall now put in a more quantitative form the concepts discussed above. Consider the gas phase reaction

$$A + B \leftrightharpoons C + D \tag{I}$$

where the species $S = A, B, C, D$ may be atoms, radicals, molecules, ions, electrons or photons. The above "bi-molecular" reaction scheme also includes dissociation reactions

$$X + A_2 \rightarrow X + A + A$$

and isomerizations

$$X + A \leftrightharpoons X + A'.$$

The empirical rate expression for the rate of change of concentration of species A, for instance, is

$$-\frac{dn_A}{dt} = k_f n_A n_B - k_r n_C n_D \tag{1.1}$$

where n_A is concentration of species A in some chosen set of units (number density, partial pressure, etc.), and k_f and k_r are the rate coefficients for the forward and reverse reactions. Since reaction (I) as written represents an elementary process, i.e., a single step involving only the species A, B, C, and D, the order with respect to each species equals one.

The rate equation for a gas phase reaction as formulated in Eq. (1.1) implies:

a) The density of the system is sufficiently low and the forces between species are sufficiently short-range that there are no correlations and that binary collisions suffice to describe the elementary processes.

b) There exists a "rate constant" k which is independent of the concentrations n_S and of the time t.

c) Since the rate of a chemical reaction is demonstrably affected by the temperature T of the reaction system, the rate constant k must be a function of the temperature, i.e., $k = k(T) \neq k(n_S, t)$.

On the basis of purely equilibrium thermodynamic arguments involving the variation of the equilibrium constant with temperature and using the hypothesis of a dynamic equilibrium, it can be shown [1] that the rate constant should be of the form (Arrhenius equation)

$$k = A(T)e^{-E/kT} \tag{1.2}$$

where $A(T)$ is the *frequency factor* and E is the *activation energy*. The empirical activation energy E in the above equation is usually interpreted as the minimum energy required for a collision between species A and B to lead to the formation of reaction products C and D. It is an *empirical* activation energy in that it is determined experimentally from the variation of the rate coefficient k with the temperature T. The appearance and use of a "temperature" T in Eq. (1.2), rests on the tacit assumption that a "temperature" can be measured and defined in the reaction systems throughout the course of the reaction. This holds only for reactions in which the translational and internal degrees of freedom are very close to their equilibrium distribution at all times.

This formulation of chemical rate processes in which the empirical rate depends upon the equilibrium, thermodynamic temperature T greatly limits its application to real systems. For example, reactions in which one of the species A or B are (at time $t = 0$) mono-energetic electrons or monochromatic photons and in which the rate of reaction is more rapid than the rates of translational and internal equilibration cannot be encompassed within the above macroscopic equilibrium theory. Additional examples are furnished by reactions behind shock fronts, in expansion nozzles, in high temperature arcs and plasmas, and a number of other systems where there is no such parameter as an "equilibrium temperature". A different approach is required to discuss the rates of chemical reactions in such systems [3].

The above formulation of reaction rates is a *macroscopic* one involving the concepts of a macroscopic rate constant k and a thermodynamic temperature T. We shall now discuss briefly the *microscopic* (i.e., elementary event) approach. The species $S \equiv (A, B, C, D)$ which enter into reaction I are not point masses but have quantized internal degrees of freedom. Reaction I should really be written

$$A(i) + B(j) \rightleftarrows C(l) + D(m) \tag{II}$$

where the letter in the parentheses specifies the internal quantum states (rotational, vibrational, electronic) of each species. Furthermore, each species S has a velocity v_S relative to some coordinate system. The collision between species $A(i, v_A)$ and $B(j\ v_B)$ must in addition be specified by their orbital angular momenta. When a collision between $A(i, v_A)$ and $B(j, v_B)$ leads to products $C(l, v_C)$ and $D(m, v_D)$ as in Reaction II, the rate process can be specified by a *reaction cross section* σ_R which is a function of the various parameters listed above. This cross section can be determined experimentally from molecular beam experiments or can be calculated, in principle, from quantum scattering theory.

Many collisions lead only to energy transfer rather than product formation, i.e.,

$$A(i, v_A) + B(j, v_B) \rightleftarrows A(i', v_A') + B(j', v_B') \tag{III}$$

when the primes indicate the quantum states and velocities after the collision. Such processes can be specified by an energy transfer cross

1*

section. The product species of the elementary processes II and III may be distributed over the whole set of accessible states l, m, i', j', v_C, v_D, v_A', v_B' compatible with the collision kinematics.

One can now define a *statistical rate coefficient* k_σ as the weighted average of the cross section σ_R from each reactant state to all product states. This averaging is carried out over the velocity and internal quantum state distributions of the reactants. The explicit appearance of the distribution functions of the reactants in the expression for the rate coefficient k now permits one to evaluate k explicitly for nonequilibrium reaction systems where a "temperature" cannot be uniquely specified.

2. Reaction Cross Sections and Rate Coefficients

The relation between the bimolecular statistical rate coefficient k_σ and the reaction cross sections σ_R can be obtained by considering the following idealized molecular beam experiment. A molecular beam is a highly attenuated, uni-directional flow of molecules. We form a molecular beam of reactant A in the internal quantum state i where the molecules $A(i)$ move with the velocity v_A relative to a coordinate system fixed in the laboratory. This may be extremely difficult in practice, but can be done in principle. The molecular beam of species $A(i)$ collides with a second molecular beam of species B in quantum state j, in which the molecules have velocity v_B. We thus consider the elementary reaction

$$A(i, v_A) + B(j, v_B) \rightarrow C(l) + D(m) . \tag{II}$$

We restrict our attention here to single binary collisions clearly separated in time. The products $C(l)$ and $D(m)$ formed in the binary collision remain in the specified quantum states, undergoing no further collisions prior to observation. This presupposes sufficiently attenuated beams or sufficiently short-range forces for which only binary interactions of A with B need be considered.

To define a cross section we denote the volume of intersection of the two beams by τ, the velocity distribution of the molecules in the two beams by $n_{A(i)} f_{A(i)}(v_A) dv_A$ and $n_{B(j)} f_{B(j)}(v_B) dv_B$, the relative speed by $v = |v_A - v_B|$ and let $N_{C(l)}(\Omega')$ stand for the number of product molecules C in internal quantum state l arriving per unit time at a detector situated at Ω' and subtending a differential solid angle element $d\Omega'$. The number density of species A in the internal quantum state i is denoted by $n_{A(i)}$, and $f_{A(i)}(v_A)$ is the normalized velocity distribution of $A(i)$. The *differential* chemical reaction cross section $\sigma_R(m l \,|\, ij; \Omega'v)$ for the reaction II is then defined operationally by the equation

$$dN_{C(l)}(\Omega')$$
$$= \sigma_R(m\,l\,|\,ij;\, \Omega'\,v)\, v\, n_{A(i)} f_{A(i)}(v_A)\, n_{B(j)} f_{B(j)}(v_B)\, \tau\, d\Omega'\, dv_A\, dv_B \tag{2.1}$$

where $D(m)$ is observed in coincidence with $C(l)$. The independent variables of the cross section are, as they should be, *all* observables of the experiment. Thus product C is observed in quantum state l and product D in quantum state m. It is important to note and appreciate the fact that the cross section is a function of the product quantum states m and l as wel as of the reactant quantum states i and j. Specification of the scattering angle Ω' for $C(l)$ is sufficient to determine the scattering angle for $D(m)$ as well because of conservation of linear momentum. It is frequently more convenient to work with scattering angles in relative coordinates, where we now write $d\Omega$ for the solid angle element. Since the number of particles scattered into a given area element is independent of the coordinate system used to describe the scattering, we must have the relation

$$\sigma_R(m\,l\,|\,i\,j;\ \Omega'v)\,d\Omega' = \sigma_R(m\,l\,|\,i\,j;\ \Omega v)\,d\Omega\ . \tag{2.2}$$

The dimension of the cross section for binary collisions is (length)^2.

The *total* reaction cross section is defined by

$$\sigma_R(m\,l\,|\,i\,j;\ v) = \int \sigma_R(m\,l\,|\,i\,j;\ \Omega v)\,d\Omega\ . \tag{2.3}$$

An alternative expression for the total reaction cross section can be obtained whenever the initial relative translational motion of the colliding molecules can be described by classical mechanics. In that case the flux of the relative motion can be specified for a given impact parameter b. The total reaction cross section then is

$$\sigma_R(m\,l\,|\,i\,j,\ v) = 2\pi \int P(m\,l\,|\,i\,j;\ v,\,b)\,b\,d\,b \tag{2.4}$$

where $P(m\,l\,|\,i\,j;\ v,\,b)$ is the probability of reaction for the set of variables $(m\,l\,|\,i\,j;\ v,\,b)$.

The macroscopic rate, $-dn_A/dt$, of the forward reaction I is a sum of the rates of the elementary reactions II and is obtained by summing, or integrating, Eq. (2.1) over all possible quantum states of reactants and products, all possible scattering angles, and all possible velocities $\mathbf{v}_A$ and $\mathbf{v}_B$. If one carries out this procedure, first of all on the left hand side of Eq. (2.1), one obtains the total number of molecules C produced per unit time in the volume element τ. Division by the volume τ then yields dn_C/dt, the total rate of formation of species C in the absence of back reaction. In terms of the total cross section one then has

$$\frac{dn_C}{dt} = -\frac{dn_A}{dt}$$
$$= \sum_{i\,j\,l\,m} \iint v\,\sigma_R(m\,l\,|\,i\,j;\ v)\,n_{A(i)}\,f_{A(i)}(\mathbf{v}_A)\,n_{B(j)}\,f_{B(j)}(\mathbf{v}_B)\,d\mathbf{v}_A\,d\mathbf{v}_B\ . \tag{2.5}$$

If one now writes for the number density of $A(i)$ the equation

$$n_{A(i)} = n_A\,x_{A(i)} \tag{2.6}$$

6   Kurt E. Shuler

where n_A is the number density of species A, $x_{A(i)}$ is the mole fraction of A in quantum state i, then one obtains the macroscopic rate expression for the forward reaction (in the absence of back reaction)

$$-\frac{dn_A}{dt} = k_\sigma n_A n_B \tag{2.7}$$

with the statistical rate coefficient k_σ given by

$$k_\sigma = \sum_{ijlm} x_{A(i)} x_{B(j)} \int \int v\, \sigma_R\,(ml\,|\,ij;\,v)\, f_{A(i)}\,(v_A)\, f_{B(j)}\,(v_B)\, dv_A\, dv_B. \tag{2.8}$$

Equation (2.8) relating the binary reaction cross section σ_R to the statistical rate coefficient k_σ has been obtained above by arguments based on a simple idealized experiment. A more formal derivation can be carried out from the Boltzmann transport equation [4]. This latter derivation, which is briefly sketched below, is particularly instructive in that it points up clearly the range of validity of relation (2.8).

It should be noted that in both derivations the assumption of *statistical independence* of the reactants prior to collision is required. This is assured in the idealized attenuated molecular beam experiment discussed above since the beam sources are independent. In the Boltzmann equation, the statistical independence of the reactants is related to the molecular chaos assumption. Without this assumption, or its equivalent, the reaction rate is not simply proportional to the product of the densities of reactants, but also involves a correlation function which describes the deviation from statistical independence. Whether a correlation function is necessary or not depends on the initial conditions and on the ratio of the range of interaction to the average distance between the reactants. For neutral species in a dilute gas the range of interaction is small compared to the average distance between molecules and the assumption of statistical independence of reactants applies well to the macroscopic system. If we denote the range of molecular interaction by a, the mean free path by λ, and the total number density by N, then the above conditions for statistical independence can be expressed as $a/\lambda \simeq Na^3 \ll 1$.

Equation (2.8) shows clearly that the statistical rate coefficient k_σ depends on the distribution functions of the reactants for both the velocity and the internal degrees of freedom. In the idealized molecular beam experiment one has control over these distributions; in a macroscopic reaction system one usually does not. In that case the distribution functions must be obtained as the solution of the relevant transport equation or by direct experiment. When a rate process takes place the internal and velocity distributions are perturbed. The deviation of these distributions from their equilibrium forms depends on the rate of the reaction compared to the rate of relaxation of the internal and velocity distributions. When the rate of reaction is slow compared to the rates of rotational, vibrational, electronic and velocity relaxation, the distribution functions $x_{A(i)}$, $f_{A(i)}(v_A)$, etc., deviate only slightly from their equilibrium forms and can thus be expressed to a good approximation as functions of temperature only. The macroscopic rate coefficient k then depends only on the temperature. If, however, the rate of the reaction is not negligibly small compared to the rates of internal or velocity relaxation, then the distribution functions $x_{A(i)}$, $f_{A(i)}(v_A)$, etc., may be functions of time and of the concentrations of reactants and products. In this case, the macroscopic rate coefficient will also be a function of these variables [3, 4].

To put this discussion on a more quantitative basis we consider a spatially uniform bimolecular gas phase reaction which consists of the elementary reactions of type II,

$$A(i) + B(j) \rightleftharpoons C(l) + D(m) \tag{2.9}$$

for all the internal states of the species A, B, C and D. We assume that the rate of change with time of the density of molecules of species A in internal quantum state i and with speeds in the range v_A to $v_A + dv_A$, i.e. the distribution function $n_{A(i)} f_{A(i)}(v_A)$, is determined by the spatially homogeneous Boltzmann equation extended to include chemical reaction [4, 5],

$$\frac{d[n_{A(i)} f_{A(i)}(v_A)]}{dt} = \sum_{j,S} n_{A(i)} n_{S(j)} \int \int \frac{p_{AS}}{\mu_{AS}} \sigma^e_{AS}(ij; p_{AS}\Omega) \times$$

$$\times \left[f'_{A(i)}(v_A) f'_{S(j)}(v_S) - f_{A(i)}(v_A) f_{S(j)}(v_S) \right] dp_S d\Omega \; +$$

$$+ \sum_{jlmS} \int \int \frac{p_{AS}}{\mu_{AS}} \sigma^i_{AS}(lm|ij; p_{AS}\Omega) \times$$

$$\times \left[n_{A(l)} n_{S(m)} f'_{A(l)}(v_A) f'_{S(m)}(v_S) - n_{A(i)} n_{S(j)} f_{A(i)}(v_A) f_{S(j)}(v_S) \right] dp_S d\Omega \; +$$

$$+ \sum_{jlm} \int \int \frac{p_{AB}}{\mu_{AB}} \sigma^*_{AB}(lm|ij; p_{AB}\Omega) \times$$

$$\times \left[n_{C(l)} n_{D(m)} f'_{C(l)}(v_C) f'_{D(m)}(v_D) - n_{A(i)} n_{B(j)} f_{A(i)}(v_A) f_{B(j)}(v_B) \right] dp_B d\Omega \tag{2.10}$$

where S denotes all species present, including reactants, products and inert gases and where p stands for the relative momentum and μ for the reduced mass. The first term on the right hand side of the equation gives the rate of change with time of the distribution function due to elastic collisions, the second term that due to inelastic collisions, and the third term that due to reactive collisions. The cross sections appearing in these three terms are those for elastic, inelastic, and reactive collisions respectively. The solution of Eq. (2.10) for the velocity and internal distribution function of each species as a function of time and/or concentration constitutes the complete solution for the rate of the set of reactions (2.9).

This equation may be reduced formally to a rate equation by integration over velocities and summation over internal states. The velocity distribution function when integrated over all velocities yields the number density of that species in a specified internal state, i.e.

$$n_{A(i)} = \int_{\mathbf{v}_A} n_{A(i)} f(\mathbf{v}_A)\, d\mathbf{v}_A \tag{2.11}$$

Hence the equation for the rate of change with time of $n_{A(i)}$ is obtained by integrating the Boltzmann Eq. (2.10) over all velocities $\mathbf{v}_A$ (or momenta p_A). In this integration the elastic collision integral must vanish since elastic collisions cannot change the number density of $A(i)$. Thus we obtain

$$\frac{dn_{A(i)}}{dt} = \sum_{jlmS} \iiint \frac{p_{AS}}{\mu_{AS}}\, \sigma^i_{AS}(lm|ij; p_{AS}\Omega) \times$$

$$\times \left[n_{A(l)}\, n_{S(m)} f'_{A(l)}(\mathbf{v}_A) f'_{S(m)}(\mathbf{v}_S) - n_{A(i)}\, n_{S(j)} f_{A(i)}(\mathbf{v}_A) f_{S(j)}(\mathbf{v}_S) \right] dp_A\, dp_S\, d\Omega\; + \tag{2.12}$$

$$+ \sum_{jlm} \iiint \frac{p_{AB}}{\mu_{AB}}\, \sigma^*_{AB}(lm|ij; p_{AB}\Omega) \times$$

$$\times \left[n_{C(l)}\, n_{D(m)} f'_{C(l)}(\mathbf{v}_C) f'_{D(m)}(\mathbf{v}_D) - n_{A(i)}\, n_{B(j)} f_{A(i)}(\mathbf{v}_A) f_{B(j)}(\mathbf{v}_B) \right] dp_A\, dp_B\, d\Omega.$$

Next we note that the number density n_A of species A, is the sum of $n_{A(i)}$ over all internal states, i.e.

$$n_A = \sum_i n_{A(i)}\ . \tag{2.13}$$

The rate equation for dn_A/dt is thus obtained by summing Eq. (2.12) over all internal states i. In this summation the inelastic collision integral vanishes because inelastic collisions do not change the density of species, n_A. We then obtain the rate equation

$$-\frac{dn_A}{dt} = k_{f\sigma}\, n_A n_B - k_{r\sigma}\, n_C n_D \tag{2.14}$$

with

$$k_{f\sigma} = \sum_{ijlm} \frac{n_{A(i)}}{n_A}\, \frac{n_{B(j)}}{n_B} \iiint \frac{p_{AB}}{\mu_{AB}}$$

$$\sigma^*_{AB}(lm|ij; p_{AB}\Omega)\, f_{A(i)}(\mathbf{v}_A)\, f_{B(j)}(\mathbf{v}_B)\, dp_A\, dp_B\, d\Omega$$

$$k_{r\sigma} = \sum_{ijlm} \frac{n_{C(l)}}{n_C}\, \frac{n_{D(m)}}{n_D} \iiint \frac{p_{AB}}{\mu_{AB}} \tag{2.15}$$

$$\sigma^*_{AB}(lm|ij; p_{AB}\Omega)\, f'_{C(l)}(\mathbf{v}_C)\, f'_{D(m)}(\mathbf{v}_D)\, dp_A\, dp_B\, d\Omega\ .$$

Eq. (2.15) for the rate coefficients is equivalent to Eq. (2.8) and thus represents a derivation of the statistical rate coefficient from the Boltzmann equation.

3. Examples of Relation between σ_R and k_σ

3.1. Complete Equilibrium Distributions

We proceed now to develop a number of specific expressions for the statistical rate coefficient k_σ, Eq. (2.8), for various forms of the distribution functions and for several different forms of cross sections. First we introduce the average cross section for the chemical reaction II

$$\sigma_R(v) \equiv \sum_{ijlm} x_{A(i)} x_{B(j)} \sigma_R(ml|ij; v) , \tag{3.1}$$

which depends both on the relative speed of the reactants and on the parameters characterizing their internal energy distribution. If the velocity distributions $f_{A(i)}(v_A)$ and $f_{B(j)}(v_B)$ have their equilibrium form, and are independent of the internal energy states, they then depend only on temperature

$$f_A(v_A, T) = \left(\frac{m_A}{2\pi kT}\right)^{3/2} \exp\left(-\frac{m_A v_A^2}{2kT}\right). \tag{3.2}$$

The statistical rate coefficient can then be rewritten more simply as

$$k_\sigma(T) = \iint v\sigma_R(v) f_A(v_A) f_B(v_B) \, dv_A \, dv_B . \tag{3.3}$$

If the distributions $f_A(v_A)$ and $f_B(v_B)$ are characterized by one temperature T, then the change of variables from laboratory velocities to relative and center-of-mass velocities is made with the use of the equations

$$v = v_A - v_B$$
$$(m_A + m_B) V = m_A v_A + m_B v_B , \tag{3.4}$$

where v is the relative velocity and V is the center-of-mass velocity. For this transformation the differential velocity elements are

$$dv_A \, dv_B = dV \, dv \tag{3.5}$$

and the total initial kinetic energy is

$$\tfrac{1}{2} m_A v_A^2 + \tfrac{1}{2} m_B v_B^2 = \tfrac{1}{2}\mu v^2 + \tfrac{1}{2}(m_A + m_B) V^2 , \tag{3.6}$$

where μ is the reduced mass, $\mu = m_A m_B/(m_A + m_B)$. Substitution of Eqs. (3.5, 3.6) into Eq. (3.3) allows integration over the center-of-mass velocity to yield

$$k_\sigma(T) = \left(\frac{\mu}{2\pi kT}\right)^{3/2} \int v\sigma_R(v) e^{-\frac{\mu v^2}{2kT}} \, dv . \tag{3.7}$$

The differential relative velocity element in spherical coordinates can be written as $d\mathbf{v} = 4\pi v^2 dv$ since the integrand depends only on the relative speed v. A further change of variables to the relative kinetic energy, $E = \frac{1}{2}\mu v^2$, then leads to

$$k_\sigma(T) = \frac{1}{kT}\left(\frac{8}{\pi\mu kT}\right)^{1/2} \int_0^\infty \sigma_R(E)\, E\, e^{-E/kT}\, dE. \tag{3.8}$$

For reactants with complete equilibrium Maxwell-Boltzmann distribution functions the statistical rate coefficient $k_\sigma(T)$ is thus a Laplace transform of $E\sigma_R(E)$ [5].

3.2. The Model of Reactive Hard Spheres

The simplest collision theory of chemical kinetics is based on the model of hard sphere interaction. To obtain the cross section $\sigma_R(E)$ for this model it is most convenient to consider the reactants and products each to be in a single quantum state only. The total elastic scattering cross section for hard spheres of diameter d is $\pi b^2_{\max} = \pi d^2$, where $b_{\max}$ is the largest impact parameter leading to elastic scattering. Reactive scattering is said to occur only if the impact parameter and initial kinetic energy are such that on impact of the hard spheres, the energy along the line of centers exceeds a given threshold value E^*. The total reaction cross section is again $\pi b^2_{\max}$, but now this upper limit on b must be determined from the conditions just cited. If E is the initial relative energy and E_c the energy along the line of centers on impact, then the relation between b, d, E, and E_c can readily be seen to be [5]

$$\frac{b^2}{d^2} = 1 - \left(\frac{E_c}{E}\right). \tag{3.9}$$

For a given b, d, and E, the maximum value of b for which reaction occurs is $b_{\max}$, which corresponds to $E_c = E^*$. Hence the total reaction cross section for this model is

$$\sigma_R(E) = 0; \qquad\qquad\qquad E < E^*$$
$$\sigma_R(E) = \pi d^2\left[1 - \frac{E^*}{E}\right]; \qquad E \geqslant E^* . \tag{3.10}$$

Substitution of Eq. (3.10) into the equation for the rate coefficient, Eq. (3.8), yields

$$k_\sigma(T) = \frac{1}{kT}\left(\frac{8}{\pi\mu kT}\right)^{1/2} \int_{E^*}^\infty \pi d^2\left[1 - \frac{E^*}{E}\right] E\, e^{-\frac{E}{kT}}\, dE \tag{3.11}$$

or

$$k_\sigma(T) = \pi d^2\left(\frac{8kT}{\pi\mu}\right)^{1/2} e^{-\frac{E^*}{kT}} \tag{3.12}$$

which is an Arrhenius equation with $A(T) = \pi\, d^2\left(\frac{8kT}{\pi\mu}\right)^{1/2}$. This model can be extended to include internal degrees of freedom [6].

3.3. A Generalization of the Hard Sphere Model

We postulate the following general form for the total reaction cross section,

$$\sigma_R(E) = 0\,; \qquad\qquad\qquad\qquad\qquad E < E^*$$
$$\sigma_R(E) = \pi\, d^2\left[1 - \frac{E^*}{E}\right]\alpha(E - E^*)\,; \quad E \geqslant E^*\,. \qquad (3.13)$$

In this model, the cross section again has a threshold energy E^*. The function $\alpha(E-E^*)$ takes account of the deviation of $\sigma_R(E)$ from its hard sphere value (3.10) for $E > E^*$. If the reactants and products have more than one internal quantum state, then the parameters E^* and d, as well as the function α itself may vary with the quantum state. Omitting these complications we substitute the expression (3.13) for the cross section into Eq. (3.8) to find

$$k_\sigma(T) = \pi\, d^2\left(\frac{8\,kT}{\pi\mu}\right)^{1/2}\chi(T)\,e^{-E^*/kT} = k_\sigma^{H.S.}(T)\,\chi(T) \quad (3.14)$$

where the function $\chi(T)$ is given by

$$\chi(T) = \left(\frac{1}{kT}\right)^2\int_0^\infty \xi\,\alpha(\xi)\,e^{-\xi/kT}d\xi\,. \qquad (3.15)$$

From the above equations we note two important points. First, if the threshold energy E^* of the cross section is large compared to kT, then the rate coefficient $k_\sigma(T)$ is determined predominantly by the behavior of $\sigma_R(E)$ near E^* [see either Eq. (3.12) or (3.14)]. Secondly, deviations of the rate coefficient from the hard sphere model appear only in the temperature dependent pre-exponential factor $\chi(T)$.

As far as is known to date, chemical reaction cross sections for reactions with activation energies $(E^* > 0)$ are not strongly varying functions of the energy E past the threshold energy E^*, i.e., for $E > E^*$. The function $\chi(T)$ thus is quite insensitive to the form of $\sigma(E)$ for $E > E^*$ and the temperature dependence of the rate coefficient for reactions with $E^* > kT$ is determined primarily by the exponential factor, $\exp(-E^*/kT)$ and not by the form of the cross section $\sigma(E)$. *This indicates how difficult it is to obtain any information about the energy dependence of cross sections from the temperature dependence of rate coefficients.* From measurements of the phenomenological rate coefficient k as a function of T, and a plot of $\log k$ vs. T^{-1}, empirical activation energies can be evaluated. The accuracy of such

12   KURT E. SHULER

measurements is however seldom sufficient to detect any pre-exponential
temperature dependence at all. Even in those cases where estimates of the
pre-exponential temperature dependence have been made, the accuracy
has not been sufficient to permit any conclusion about the energy dependence
of the cross section above threshold.

3.4. Nearly Constant Cross Sections

If the total reaction cross section is nearly constant over the range of
relative energies up to a few kT, which from the arguments presented above
implies $E^* < kT$, then the rate coefficient takes on a particularly simple
form. We see from Eq. (3.7) that we can now approximate $k_\sigma(T)$ by

$$k_\sigma(T) \simeq \left(\frac{\mu}{2\pi kT}\right)^{3/2} \sigma_R \, 4\pi \int\limits_0^\infty v^3 e^{-\frac{\mu v^2}{2kT}} dv \tag{3.16}$$

or

$$k_\sigma(T) \simeq 2\left(\frac{2kT}{\pi\mu}\right)^{1/2} \sigma_R \tag{3.17}$$

with $\sigma_R \neq f(v)$. Since the mean velocity $\bar{v}$ equals $(8kT/\pi\mu)^{1/2}$ the last equation takes the simple form

$$k(T) \simeq \bar{v}\,\sigma_R. \tag{3.18}$$

The simple expression (3.18) has been employed frequently in the literature
for the approximate evaluation of the rate constant k. The above "derivation" clearly shows its limited range of validity. The error involved in
using this approximation when σ_R is not really constant with energy
should be evaluated whenever possible to gain some feeling for the useful-
ness (or uselessness) of this approximation (see e.g. section 3.7.).

3.5. Ion-Molecule Reactions $(A^+ + B \to C^+ + D)$

Ion-molecule reactions, as a class, are distinguished in several ways.
Most notably, they have rate constants which are usually temperature
independent. In addition, they appear to proceed with no activation energy
(above that required in endothermic reactions), and they have, in general,
rather large cross sections (up to 10^{-14} cm^2). These effects all stem from the
strong long range attractive forces between ions and polarizable molecules
or atoms.

Briefly, the *long range* potential for radial motion of an ion and a molecule
is given by

$$V(r) = -\frac{\alpha e^2}{2r^4} + \frac{Eb^2}{r^2} \tag{3.19}$$

where α is the polarizability of the neutral species, e is the unit of electrical charge, E the initial relative translational energy, b the impact parameter and r the distance between the ion and neutral particle. Any ion-molecule pair which can pass over the potential barrier of the long range potential $V(r)$ will stand a good chance of reacting if the reaction is exothermic. Thus it is easy to estimate the cross section for the reaction by calculating the cross section for passage over the potential barrier.

The potential $V(r)$ has a maximum at

$$r^* = (\alpha\, e^2 / E\, b^2)^{1/2} \tag{3.20}$$

of height

$$V(r^*) = E^2\, b^4 / 2\, \alpha\, e^2 . \tag{3.21}$$

In order to pass over this potential barrier, the relative translational energy must be larger than $V(r^*)$, i.e.,

$$E > E^2\, b^4 / 2\, \alpha\, e^2 . \tag{3.22}$$

Equivalently, for a given E the impact parameter b must be less than $b_{\max}$,

$$b_{\max} = \left(\frac{2\, \alpha\, e^2}{E} \right)^{1/4} = \left(\frac{4\, \alpha\, e^2}{\mu\, v^2} \right)^{1/4} . \tag{3.23}$$

If we assume that a constant fraction, g, of the ion-molecule pairs entering the reaction region actually do react, the cross section for exothermic ion molecule reactions is found to be

$$\sigma_r = 2\pi \int_0^{b_{\max}} g\, b\, d\, b = \pi\, g\, b^2_{\max} = \frac{2\pi g}{v} \left(\frac{\alpha\, e^2}{\mu} \right)^{1/2} . \tag{3.24}$$

The factor g has usually been taken as 1 for an exothermic reaction and 0 for an endothermic reaction.

Since the cross section according to Eq. (3.24) is proportional to $1/v$, where v is the relative velocity, the statistical rate coefficient k_σ will be independent of T. From Eqs. (3.7) and (3.24) one obtains (with $d\mathbf{v} = 4\pi v^2 dv$)

$$k_\sigma = 4\pi \left(\frac{\mu}{2\pi k T} \right)^{3/2} \int_0^\infty dv\, v^3\, \sigma_R(v) \exp\left(-\mu\, v^2 / 2\, k\, T \right) = 2\pi g \left(\frac{\alpha\, e^2}{\mu} \right)^{1/2} . \tag{3.25}$$

Since $v\, \sigma_R$ is a constant in this approximation, the statistical rate coefficient is independent of the velocity distribution.

This result, first given by EYRING, HIRSCHFELDER, and TAYLOR [7] and later by GIOUMOUSIS and STEVENSON [8], has provided the basis for most theoretical interpretation of ion-molecule rate data. It should be noted that exceptions do occur (the most important, perhaps, being the $(NNO)^+$ systems) and that the formulae above are valid only for relatively low kinetic energies (0—10 eV).

3.6. Reactants Having Velocity Distributions at Two Temperatures

If the two reactant species in a bimolecular reaction each have a Maxwell-Boltzmann distribution of velocities corresponding, respectively, to temperatures T_1 and T_2, the problem of converting the velocity dependent cross section to a statistical rate coefficient is somewhat more difficult. Such a situation is most likely to occur if the masses of the reactant species are very different. The collisional momentum transfer between the different species is then reduced, and the inter-species translational relaxation may be very slow. While the definition of the rate constant of Eq. (2.8) is still valid, the transformations to relative and center-of-mass coordinates must be modified in order to carry out the required integration.

We assume the velocity distributions of the reactants to be given by

$$f_i(v) = C_i \exp\left\{ -\frac{m_i v_i^2}{2 k T_i} \right\} \tag{3.26}$$

with C_i, the normalization constant, equal to

$$C_i = [m_i / 2\pi k T_i]^{3/2} . \tag{3.27}$$

For reactions of type II, i.e. reactions involving only a single internal quantum state of reactants, the statistical rate coefficient can then be written as

$$k_\sigma = \iint d\mathbf{v}_1 \, d\mathbf{v}_2 \, v \sigma_R(v) \, f(\mathbf{v}_1) \, f(\mathbf{v}_2) \tag{3.28}$$

where $v = |\mathbf{v}_1 - \mathbf{v}_2|$ is the relative velocity. Eq. (3.28) is valid, for instance, for electron-atom systems with the atoms initially in their ground electronic state. If one now sets

$$\mathbf{v}' = \mathbf{v} = \mathbf{v}_1 - \mathbf{v}_2$$
$$\mathbf{V}' = [q_1 \mathbf{v}_1 + q_2 \mathbf{v}_2]/[q_1 + q_2] \tag{3.29}$$
$$q_i = \frac{m_i}{T_i}$$

one finds that

$$f_1 f_2 = C_1 C_2 \exp\left\{ -\frac{\theta v'^2}{2 k} - \frac{\Theta V'^2}{2 k} \right\} \tag{3.30}$$

where

$$\Theta = q_1 + q_2 = \frac{m_1}{T_1} + \frac{m_2}{T_2} \tag{3.31}$$

$$\theta = \frac{q_1 q_2}{\Theta} = \frac{\mu}{T^*} \tag{3.32}$$

with the reduced mass $\mu = m_1 m_2/(m_1 + m_2)$ and the effective "temperature"

$$T^* = \frac{m_1 T_2 + m_2 T_1}{m_1 + m_2} . \tag{3.33}$$

Since the Jacobian of this transformation is unity, we have $d\mathbf{v}_1 d\mathbf{v}_2 = d\mathbf{v}' d\mathbf{V}'$. The expression for the statistical rate coefficient then becomes, after integration over $d\mathbf{V}'$ and over the angular coordinates of $d\mathbf{v}'$,

$$k_\sigma(T) = 4\pi \left(\frac{\mu}{2\pi k T^*}\right)^{3/2} \int_0^\infty v^3 \sigma(v)\, e^{-\frac{\mu v^2}{2kT^*}} \, dv \; . \tag{3.34}$$

This expression is identical in form to that obtained for a *single* temperature system [Eq. (3.7)] with the effective temperature T^* [Eq. (3.33)] in place of the equilibrium temperature T.

As an example, we may consider the ionization of neutral atoms in a plasma characterized by an electron temperature, T_{el}, and a neutral atom temperature, T_N with $T_{el} \neq T_N$. Assume the cross section for single ionization is of the threshold form

$$\sigma(v) = \begin{cases} 0 & ; \; \dfrac{\mu v^2}{2} < I \\[2ex] a\left(\dfrac{\mu v^2}{2} - I\right); & \dfrac{\mu v^2}{2} \geqslant I \end{cases}$$

where I is the ionization potential, $\mu = \dfrac{m_{el}\, m_N}{m_{el} + m_N} \simeq m_{el}$ is the reduced mass, and a is the cross section parameter with dimensions cm²/erg with $a = (d\sigma/dE)_{E=I}$. Then we have, in the above notation,

$$k_\sigma = 4\pi \left(\frac{\mu}{2\pi k T^*}\right)^{3/2} a \int_{\left[\frac{2I}{\mu}\right]^{1/2}}^\infty dv\, v^3 \left(\frac{\mu v^2}{2} - I\right) \exp\left[-\frac{\mu v^2}{2k T^*}\right] \tag{3.35}$$

Performing the integration, we find

$$k_\sigma(T^*) = 2a \left(\frac{2k T^*}{\mu}\right)^{1/2} (I + 2k T^*) \exp\left[-\frac{I}{k T^*}\right] \tag{3.36}$$

where

$$T^* = \frac{m_{el} T_N + m_N T_{el}}{m_{el} + m_N} \simeq \left(\frac{m_{el}}{m_N}\right) T_N + T_{el} \; . \tag{3.37}$$

For an atom-electron plasma system with $T_{el} > T_N$, one then has $T^* \simeq T_{el}$. When $I \gg k T_{el}$, i.e., in a low temperature plasma, Eq. (3.36) reduces to

$$k_\sigma(T_{el}) \simeq 2a \left(\frac{2k T_{el}}{m_{el}}\right)^{1/2} I e^{-\frac{I}{k T_{el}}} \; . \tag{3.38}$$

with k_σ small in magnitude owing to the exponential term. It will be noted that to within the above approximations, the rate coefficient for ionization is independent of the neutral species temperature. Eq. (3.36) is probably not applicable to high temperature plasmas with $I \ll kT_{el}$ since the above threshold law is not valid in this region.

3.7. Vibration-Translation Energy Exchange

We give here a brief discussion of the calculation of the cross section for translational-vibrational energy transfer in inelastic collisions and of the rate constant corresponding to this cross section. The point of this example is not so much to inform about the method of calculating energy transfer cross sections and rate coefficients but rather to show clearly and quantitatively the real danger of over simplified formulae for rate constants such as $k(T) = \bar{v}\sigma(\bar{v})$.

The most widely accepted theory of energy exchange between vibrational and translational degrees of freedom is generally based on the work of Landau and Teller [9], and Herzfeld [10] and co-workers. Although recent work has cast some doubt on the validity of two key assumptions [11], the qualitative nature of the velocity dependence of the cross sections remains unchanged. Since several excellent review articles [12] are readily available in the literature only an outline of the procedure is presented here.

In the simplest version of the theory a harmonic oscillator is forced by a time dependent perturbation representing the effect of the collision with the incident atom or molecule. In first order time dependent perturbation theory, the probability of excitation of the oscillator from the ground state to the first vibrational level is given by

$$P_{0\to1}(v) = \frac{1}{\hbar^2} \left| \int_0^\infty dt \, \langle \varphi_1^* | V(x,t) | \varphi_0 \rangle \, e^{-i\omega_{01}t} \right|^2 \tag{3.39}$$

where $\omega_{01} = \dfrac{E_1 - E_0}{\hbar}$, $\varphi_i(x)$ is the wave function of the ith harmonic oscillator level, and $V(x,t)$ is the time (and thus velocity) dependent interaction. With the approximations of a one-dimensional classical trajectory determined by an exponential repulsive potential, and retaining in the interaction only the linear term in the oscillator coordinate, $V(x,t)$ becomes [13]

$$V(x,t) = \frac{xE}{L} \operatorname{sech}\left(\frac{\pi v t}{4L}\right) \tag{3.40}$$

where L is a parameter of the potential, E is the translational energy, and v is the velocity of the incident atom or molecule. The integral of Eq. (3.39) may easily be evaluated by contour integration to yield

$$P_{0\to 1}(v) \propto v^2 \operatorname{sech}^2\left(\frac{2\omega L}{v}\right) . \tag{3.41}$$

This one dimensional result can be extended to three dimensions in order to obtain a cross section by using the modified wave number approximation or some equivalent approximation [12]. The velocity dependence remains the same in all approximations which have been employed and one finds

$$\sigma_{0\to 1}(v) \propto v^2 \operatorname{sech}^2\left(\frac{2\omega L}{v}\right) \simeq v^2\, e^{-4\omega L/v} \tag{3.42}$$

where only the leading term of $\operatorname{sech}^2\left(\dfrac{2\omega L}{v}\right)$ for $\omega L/v \gg 1$ has been retained. For velocities at which first order perturbation theory is valid, this is a very reasonable expansion.

To find the thermal rate coefficient for vibrational excitation, we must now integrate Eq. (3.42) over the (3 dimensional) Maxwellian velocity distribution, i.e.,

$$k_\sigma^{0\to 1}(T) \propto \int_0^\infty e^{-\frac{\mu v^2}{2kT}} v^3\left[v^2 e^{-\frac{4\omega L}{v}}\right] dv . \tag{3.43}$$

This integration cannot be carried out analytically, but a good approximation can be obtained by using the method of steepest descent [10]. Noting that the combined exponential terms have a rather sharp maximum at the minimum of $\dfrac{\mu v^2}{2kT} + \dfrac{4\omega L}{v}$, we expand the exponential about this point and integrate to obtain

$$k_\sigma^{0\to 1}(T) \simeq \left[\frac{4\omega L kT}{\mu}\right]^{5/3} \left(\frac{\pi kT}{6\mu}\right)^{1/2} \exp\left\{-\frac{3}{2}\left[\frac{16\mu\omega^2 L^2}{kT}\right]^{1/3}\right\} \tag{3.44}$$

so that

$$k_\sigma^{0\to 1}(T) \propto A\,T^2 \exp\left[-B/T^{1/3}\right]. \tag{3.45}$$

Note that the major temperature dependence of the rate coefficient (for reasonably large values of $\mu\omega^2 L^2/kT$) is governed by the exponential term.

In this example, the difference between the properly defined statistical rate coefficient $k_\sigma = \int v\,\sigma(v)f(v)\,dv$ of Eqs. (2.8) or (3.3) and an approximate rate coefficient $\bar{k}_\sigma$ of the form $\bar{k}_\sigma \simeq \bar{v}\,\sigma(\bar{v})$ is striking. Using $\bar{v} = (8kT/\pi\mu)^{1/2}$ for the mean velocity, one finds

$$\bar{k}_\sigma^{0\to 1}(T) \propto C T^{3/2} \exp\left[-D/T^{1/2}\right] \tag{3.46}$$

where C and D are easily evaluated numerical constants. The principal difference between the expressions (3.45) and (3.46) is in the magnitude of the exponential terms.

In the first place it should be noted that the $T^{-1/3}$ exponential dependence of $k_\sigma(T)$ in Eq. (3.45) has been repeatedly confirmed experimentally over a wide range of temperatures for a number of species for which extensive data are available. The approximate formula (3.46) for the rate coefficient thus has the wrong temperature dependence. In addition, there is a difference of several orders of magnitude between $\bar{k}_\sigma$ and k_σ. For instance, for $\mu = 14$ A.u., $\omega = 4\times10^{14}$ sec^{-1}, $L = 2\times10^{-9}$ cm corresponding to vibrational-translational energy transfer in N_2 at 10,000 °K, one obtains approximately $\exp(-25)$ for the exponential term in the approximate expression (3.46) and $\exp(-10)$ for the correct expression (3.45), a ratio of about 10^6. This tremendous difference is due to the fact that a very large contribution to the integral (3.43) comes from velocities $v \gg \bar{v}$ for which the product of the cross section and the Maxwell distribution is much larger than for the mean velocity $\bar{v}$. The approximate expression (3.46) thus gives a value for the rate coefficient for energy transfer which is too small by several orders of magnitude.

4. Nonequilibrium Kinetics

We shall consider now briefly and qualitatively the implication of the above results for chemical rate processes with nonequilibrium distributions of reactants and/or products.

4.1. Reactant Nonequilibrium Distributions

There are a large number of examples of chemical reactions with non-equilibrium distributions of reactants. In some cases, these nonequilibrium distributions are an integral feature of the preparation of the system and are more or less outside of the control of the experimenter; in other cases, the reactants are intentionally prepared in nonequilibrium distributions. A sudden input of energy into a system initially in chemical and thermal equilibrium, with a pulse duration short compared to the "relaxation time" for energy transfer and chemical reactions is one of the techniques leading to non-thermal distributions of reactants. Specific examples of such a preparation of a reaction system are: plasmas (input of electrical energy), shocked gases (input of kinetic energy), and flash photolysis (photon pulse). Deliberate preparation of reactants in nonequilibrium distributions is exemplified by mono-energetic electron beams, monochromatic photon beams, and velocity or internal state selected atomic and molecular beams.

4.2. Product Nonequilibrium Distributions

Nonequilibrium distributions of products of elementary chemical reactions seem to be the rule rather than the exception. This is now well established. The reason for the delay in recognizing this stemmed from the need to develop specialized detection techniques to measure the distribution of products over the internal quantum states and velocities.

Several recent review articles [14] give an excellent survey of the status of the field, both experimental and theoretical. Nonequilibrium excitation of products of chemical reactions has been observed for all internal degrees of freedom (electronic, vibrational, and rotational) as well as for the velocity distributions. Among the many types of reactions which give rise to products in nonequilibrium distributions are:

1. Recombinative Excitation:

$$A + B + M \rightarrow AB^* + M$$
$$A + B + M \rightarrow AB \; + M^*$$

2. Dissociative Energy Transfer:

$$AB + M^* \rightarrow A^* + B + M$$

3. Exothermic Exchange Reactions:

$$A + BC \rightarrow AB^* + C$$
$$A + BC \rightarrow AB \; + C^*$$

4. Photodissociation:

$$ABC + h\nu \rightarrow AB^* + C$$

In these reactions A, B, C and M represent atoms or molecules and the symbol * indicates excitation to rotational, vibrational or electronic nonequilibrium states. Nonequilibrium excitation of products has also been observed in molecular beam studies [15].

4.3. Perturbation of Distribution Functions by Chemical Reactions

A chemical reaction, like any other transport process, will perturb the velocity and internal distribution function of the reactant species. The nature and extent of this perturbation have been studied by a number of authors [3, 4]. These studies have shown that this perturbation is significant whenever the activation energy is less than or of the same order of magnitude as kT, i.e., whenever $E_{\mathrm{act}} < 10\,kT$. This result is in accord with the discussion presented in Section 2, where the criterion for the (approximate) maintenance of the equilibrium distribution of reactants and products has been shown to be the more rapid rate of equilibration of the velocity and internal distributions compared to the rate of chemical reaction.

2*

In a macroscopic rate experiment then, where $E_{act} \leq kT$ and/or where the inelastic cross sections σ_i (momentum and internal energy transfer) are small compared to the reactive cross section σ_R, one must expect a significant perturbation of an initial Maxwell-Boltzmann distribution of reactants. In a steady state (i.e. flow) experiment, this will manifest itself as a steady-state nonequilibrium distribution of reactants; in a non-steady state experiment, one will find time dependent nonequilibrium distributions. For details as to the type of distributions which might be obtained, the reader is referred to the literature quoted above and the references contained therein.

4.4. *Phenomenological and Statistical Rate Coefficients*

We now discuss briefly the implication and effects of nonequilibrium chemical rate processes on the phenomenological rate coefficient k [Eq. (1.1)] and the statistical rate coefficient k_σ [Eq. (2.8)].

Let us first consider the statistical rate coefficient k_σ. In principle, k_σ can be evaluated for any and all nonequilibrium situations although, to be sure, it may not be possible to carry out such a program in practice. However, and this is the important point, there is a clear and unequivocal prescription for the calculation of the statistical rate coefficient, i.e., Eq. (2.15). If the elastic, inelastic and reactive cross sections are known, either through calculations or experiments, the distribution functions, both for velocity and for the internal degrees of freedom, can be calculated from the Boltzmann equation (2.10). Alternatively, these distribution functions may be determined through appropriate experiments. The knowledge of the distribution functions and of the reactive cross section $\sigma_R\,(\equiv \sigma_{AB}^*)$ then permits the evaluation of the statistical rate coefficient k_σ from Eq. (2.15). Thus regardless whether one is dealing with initial nonequilibrium distributions, final nonequilibrium distributions and/or perturbations of the distribution functions by the chemical reaction, the statistical rate coefficient can, in principle, be calculated by a clear and unique prescription. For nonequilibrium rate processes, the statistical rate coefficient will depend on the concentration of reactants and products and will be an implicit function of time through the time dependence of the distribution functions. It is, however, a perfectly well defined transport coefficient.

It is important to note that the statistical rate coefficient is an implicit function of the elastic and inelastic cross sections. Only for "equilibrium chemical reactions", i. e., for reactions where the distribution functions of reactants and products maintain a Maxwell-Boltzmann distribution at the equilibrium temperature $T_{eq.}$ at all times t during the course of the reaction, will this dependence vanish since then the first two terms on the right hand side of Eq. (2.10) are identically equal to zero.

The problem with the phenomenological rate coefficients k_f and k_r of Eq. (1.1) is that they do not really provide a clear and well defined specification of a chemical rate process except for the case of "equilibrium" chemical reactions as defined above. It is only in that case, or for conditions very close to that case, that the rate coefficient k is indeed a rate *constant*, independent of time and concentration, and dependent only on temperature. To be sure, a "rate coefficient" k can be determined experimentally from Eq. (1.1) for nonequilibrium chemical reactions. If these measurements are carried out with sufficient accuracy, they will show that the rate coefficient k is a function of the initial concentrations, $n_A(0)$ and $n_B(0)$, and of the time t. Furthermore, under these nonequilibrium conditions, the phenomenological rate coefficient no longer gives a unique specification of the reaction system but may well depend upon the experimental technique employed to determine the rate. Under conditions far from equilibrium, and particularly under conditions where the form of the departure from equilibrium is not known or well understood, the phenomenological rate coefficient is at best an ill defined measure of the chemical dynamics of the reaction system. At worst, it may even be a meaningless or misleading measure.

The above statement may be considered a harsh, unfair and even erroneous judgment. At the present time, however, the author is not aware of any experiments carried out in conjunction with the appropriate detailed calculations which would permit a comparison of a phenomenological rate coefficient with the corresponding statistical rate coefficient *for reaction systems far from equilibrium*. In the absence of such a comparison, the issue must be considered open for discussion and, preferably, open for some much needed experimental and theoretical work.

Acknowledgements

These lectures are based to a large extent on some recent work carried out jointly with JOHN C. LIGHT, Department of Chemistry, University of Chicago and JOHN ROSS, Department of Chemistry, Brown University. I wish to thank Drs. LIGHT and ROSS for their permission to use parts of our joint work as the basis for these lectures.

References

1. For a more detailed discussion see e.g. LAIDLER, K. J.: Chemical Kinetics. New York: McGraw-Hill Book Co. 1950, or KONDRATIEV, V. N.: Chemical Kinetics of Gas Reactions. Reading, Mass.: Addison-Wesley Publishing Co. Inc., 1964.
2. LAIDLER, K. J., and J. C. POLANYI: Theories of the Kinetics of Bimolecular, Reactions, in: Progress in Reaction Kinetics **3**, 1 (1965). Pergamon Press, Oxford; G. Porter, Editor.

3. See e.g., Shuler, K. E., in: Fifth Symposium (International) on Combustion, pp. 56 et seq. New York: Reinhold Publ. Co. 1955; Montroll, E. W., and K. E. Shuler, in: Adv. Chem. Phys. 1, 361 (1958).

4. Ross, J., and P. Mazur: J. Chem. Phys. 35, 19 (1961).

5. Eliason, M. A., and J. O. Hirschfelder: J. Chem. Phys. 30, 1426 (1959).

6. Fowler, R., and E. A. Guggenheim: Statistical Thermodynamics, p. 497 et seq. Cambridge: University Press 1949.

7. Eyring, H., J. O. Hirschfelder, and H. S. Taylor: J. Chem. Phys. 4, 479 (1936).

8. Gioumousis, G., and D. P. Stevenson: J. Chem. Phys. 29, 294 (1958).

9. Landau, L., and E. Teller, Phys. Z. Sowjet. 10, 34 (1936).

10. Schwartz, R. N., Z. I. Slawsky, and K. F. Herzfeld: J. Chem. Phys. 20, 1591 (1952); Schwartz, R. N., and K. F. Herzfeld: J. Chem. Phys. 22, 767 (1954).

11. Mies, F. H.: J. Chem. Phys. 42, 2709 (1965).

12. Takayanagi, K.: Progs. Theoret. Phys. (Kyoto) Suppl. 25 (1963); Herzfeld, K. F., and T. A. Litovitz: Absorption and Dispersion of Ultrasonic Waves. New York: Academic Press 1959.

13. Rapp, D., and T. E. Sharp: J. Chem. Phys. 38, 2641 (1963).

14. See e.g., the articles by Broida, Polanyi, Callear, Zare, and Herschbach, and by Shuler, Carrington, and Broida, in: Chemical Lasers, Applied Optics Supplement No. 2, Optical Society of America, Washington, D. C. (1965), Edited by W. R. Bennett and K. E. Shuler.

15. Advances in Chemical Physics, John Ross, Editor, Interscience Publishers, New York, Vol. X, (1965).

Kurt E. Shuler
National Bureau of Standards
Washington, D. C. USA

Remarks on the Generalization of Activated Complex Theory

R. A. MARCUS

Activated complex theory has been very useful to chemists in their treatment of chemical reaction rates: For any given potential energy surface one can calculate the rate constant of a reaction if the assumptions of activated complex theory are correct. In effect, activated complex theory is principally used nowadays to calculate the pre-exponential factor A in the Arrhenius expression for rate constant, $k = A \exp(-E/kT)$. The activation energy E is usually too difficult to calculate *a priori* except for a few reactions.

The assumptions of activated complex theory are now being examined more rigorously than before and one should have a good idea of their correctness or error in the not too distant future, partly as a result of electronic computer calculations.

The principal assumptions of activated complex theory are (1) an equilibrium between reactants and those activated complexes that are moving in the forward direction along the reaction coordinate q, and (2) a treatment of the motion along q as being rectilinear. Several additional, usually minor approximations are also usually made (e.g., neglect of centrifugal potential on the course of motion along the reaction coordinate).

Assumption (2) has been made largely for convenience. It is an unnecessary one and has recently been removed [1], by using a curvilinear reaction coordinate and translating the usual derivation into one for which a curvilinear metric is appropriate at every stage of the derivation.

Assumption (1) is under current active investigation. Some years ago several authors [2] pointed out that if, somehow, the reacting system remained in the same quantum state for the vibrational-rotational motion, as the system moved along the reaction coordinate, a striking consequence would occur: If the distribution of quantum states of the reactants were one of thermal equilibrium then the distribution of quantum states of activated complexes moving *in the forward direction* along the reaction coordinate q would also be an equilibrium one. Since the former type of thermal equilibrium is expected to be achieved in typical experiments, the latter type of equilibrium would also be achieved then, and so assumption (1) would be justified. The condition that the system remain in the same vibrational-rotational quantum state during the motion along q might well be called the "vibrationally-adiabatic" condition: The process would be quantum mechanically adiabatic with respect to the vibration-rotation degrees of freedom.

Clearly, the investigation of the correctness of this condition must be one of analytical mechanics — quantum mechanical or classical. The entire concept of "remaining in the same vibrational-rotational state" must be more sharply defined, as must the nature of the reaction coordinate curves throughout the configuration space. To this end, the analytical mechanics of linear collisions was investigated recently [3], with the aid of a coordinate system which passed smoothly from one suited to reactants to one suited to products during the reactive collision. The coordinate system is not one of the usual ones, that is, it is not tied to "separation of variables", but has been used in treatments of orbits of particles in accelerators.

More recently, this analytical mechanics was extended from linear collisions to collisions in ordinary space [3b]. In this case it was necessary to study in some detail the evolution of the various degrees of freedom during the course of the collision. A "vibrationally-adiabatic" approximation was used to solve the various equations of motion. This evolution may be described as follows:

In a reaction $A + BC \rightarrow AB + C$ the BC vibration evolves into an ABC symmetric stretching vibration of the activated complex and finally into an AB vibration. The initial radial relative translational coordinate of A with respect to BC evolves into the usual asymmetric translational coordinate in the activated complex (reaction coordinate) and finally into the radial relative translational coordinate of the products. The two orbital relative translational coordinates of A with respect to BC evolve into two rotations of the activated complex and finally into the two orbital or relative translational coordinates of C with respect to AB. The two rotations of BC evolve, after some coupling with the orbital motions, into the two bending modes of a linear activated complex, and finally into the two rotations of AB.

Extensive numerical integrations of the classical mechanical equations of motion of the atoms during the collision have been performed by several research groups, particularly by those of KARPLUS and those of BUNKER [4]. (Pioneering work was done by WALL and his collaborators [5].) An analysis [6] of the numerical studies [4a] for the $H + H_2 \rightarrow H_2 + H$ reaction provides encouraging results for assumption (1) of activated complex theory, for this reaction at least. (There are also reactions for which one would not expect the vibrationally adiabatic assumption to be valid.) Reasoning in this direction has also led to a statistical dynamical theory of chemical reaction cross-sections [7].

Nowadays, very detailed results on chemically reactive collisions are being obtained experimentally, due to the use of molecular beam techniques, to the study of light emission from exothermic reactions in flow systems and in flash photolyses, to the investigations of ion-molecule reactions, and to others. Thus, theories are needed which are much more detailed than those used for rate constants of reactions at thermal equilibrium. Indeed, these new experimental sources of information represent the main motivation for the numerical integrations and for the analytical mechanics studies.

Hopefully, from such studies one can learn about a variety of problems including (1) contribution of vibrational excitation to the reaction rate, (2) accuracy of the activated complex theory, (3) extent of vibrational excitation of the reaction products, (4) angular distribution of reactant and product molecules in molecular scattering, (5) extent of quantization of the various degrees of freedom in the actual complex theory, (6) calculation of atom tunneling contribution to the reaction rate, (7) nonadiabatic contributions (vibrational, rotational excitation), (8) relation between classical and quantum mechanical cross-sections, and (9) relation of threshold energy [8] of a reaction to other properties (e.g., to activation energy).

The potentialities for future investigators in this field are clearly considerable.

References

1. Marcus R. A.: J. Chem. Phys. **41**, 2614, 2624 (1964); **43**, 1598 (1965).

2. Hirschfelder, J. O., and E. Wigner: J. Chem. Phys. **7**, 616 (1939); Eliason, M. A., and J. O. Hirschfelder: **30**, 1426 (1959); Hofacker, L.: Z. Naturforsch. **18a**, 607 (1963).

3. a) Marcus, R. A.: J. Chem. Phys. **45**, 4493, 4500 (1966);
   b) ibid (to the published);
   c) For another approach see Micha, D.: Arkiv Fysik **30**, 411, 425, 437 (1965).

4. E.g.:
   a) Karplus, M., R. N. Porter, and R. D. Sharma: J. Chem. Phys. **43**, 3259 (1965), and others;
   b) Bunker, D. L., and N. C. Blais: J. Chem. Phys. **41**, 2377 (1964) and others, as well as papers by Polanyi, by Raff and by Wall and their collaborators.

5. Wall, F. T., L. A. Hiller, jr., and J. Mazur: J. Chem. Phys. **29**, 255 (1958).

6. Marcus, R. A.: J. Chem. Phys. **45**, 2138 (1966).

7. Marcus, R. A.: J. Chem. Phys. **45**, 2630 (1966); **46**, 959 (1967).

8. Kuppermann, A., and J. M. White: J. Chem. Phys. **44**, 4352 (1966).

R. A. Marcus

University of Illinois
Urbana, Illinois/USA

Wellenmechanische Aspekte in der Theorie der Elementarreaktionen

H. Preuss

Mit 2 Abbildungen

1. Ausgangspunkte

Jedes chemische Element E ist eindeutig durch die Ladung Z und die Masse M [näherungsweise in Protonenmassen (Neutronen) angegeben] seines Atomkerns gekennzeichnet:

$$E \leftrightarrow E^{ZM} \tag{1}$$

In der Gasphase ergeben sich alle denkbaren molekularen Gebilde, Radikale, Moleküle (Ionen) und Atome (Ionen) X_k zwischen diesen Elementen zu

$$X_k = \left[\prod_{Z,M} E_{k_{ZM}}^{ZM} \right]^{\delta}, \tag{2}$$

wenn die Zahlen

$$k_{ZM} = 0, 1, 2, \ldots ; \quad k = \{k_{ZM}\} \tag{2a}$$

jedem Element E^{ZM} zugeordnet werden und k für die jeweilige Gesamtheit aller k_{ZM} steht. Für δ, die Ladungszahl des Systems X_k, gilt

$$\delta = 0, \pm 1, \pm 2, \ldots . \tag{2b}$$

Die zu einer bestimmten Anzahl von Elementen gehörenden X_k, welche miteinander in Wechselwirkung getreten sind, bezeichnen wir als die *Komponenten einer Reaktionsgruppe*, wobei wir unter X_0 in jedem Falle ein freies Elektron verstehen wollen ($X_0 \equiv e$).

Für $E^{2,4} = He$ ergibt sich zum Beispiel nach den derzeitigen Kenntnissen die folgende Reaktionsgruppe mit sechs Komponenten:

$$
\begin{matrix}
 & e & \\
He & & He_2^{++} \\
He^+ & & He_2^{+} \\
 & He^{++} &
\end{matrix}
\tag{3}
$$

Liegen nur He und H vor, so resultiert eine Reaktionsgruppe von 14 Komponenten, wenn wir für Wasserstoff nur das häufigste Isotop betrachten[1]:

$$
\begin{array}{cccc}
 & & e & \\
 & \text{He} & & \text{H} \\
 & & & \text{H}^+ \\
 & \text{He}^+ & & \text{H}^- \\
 \text{He}^{++} & & & \text{H}_2^- \\
 & \text{He}_2^{\ddagger} & & \text{H}_2 \\
 & & & \text{H}_2^{\ddagger} \\
 & \text{He}_2^{\ddagger+} & & \text{H}_3^{\ddagger} \\
 & & \text{HeH}^+ &
\end{array}
\tag{4}
$$

Allgemein ist eine Unterscheidung der Komponenten nach (2) nicht vollständig, da auch der räumliche Aufbau berücksichtigt werden muß. Wir unterscheiden daher die einzelnen X_k auch bezüglich der relativen Lagen der Atome zueinander, die sich durch Rotation des Gesamtsystems nicht ineinander überführen lassen.

So sind auf diese Weise zum Beispiel

$$
\begin{array}{ccc}
\text{E} \quad \text{E}' & \quad\text{und}\quad & \text{E} \quad \text{E}' \\
\text{E}'' & & \text{E}''
\end{array}
$$

Abb. 1

verschieden.

Die Definition einer Komponente X_k läßt sich eindeutig nur wellenmechanisch durchführen:

Danach wird allgemein die Bewegung von N Atomkernen und n Elektronen durch die zeitabhängige Schrödingergleichung ($t = $ Zeit)

$$
\{ \mathrm{H}\,(r, \mathscr{R}) + \mathrm{K}\,(\mathscr{R}) \}\, \Psi\,(r, \mathscr{R}, \sigma, t) = i\,\frac{\partial\,\Psi\,(r, \mathscr{R}, \sigma, t)}{\partial\,t}
\tag{5}
$$

beschrieben, die hier in atomaren Einheiten (at. E.) formuliert ist. r stellt die Gesamtheit aller Vektoren r_i ($i = 1, .. n$) zu den n Elektronen dar, entsprechend ist $\mathscr{R}$ die Gesamtheit aller $\mathscr{R}_\lambda$, ($\lambda = 1, 2, .. N$) zu den N Atomkernen. σ steht für alle n Spinfunktionen σ_i ($i = 1, 2, .. n$), die je nach den beiden Spinstellungen α und β sein können. Der Gesamthamiltonoperator $\mathscr{H}$ wird durch

$$
\mathscr{H} = \mathrm{H}\,(r, \mathscr{R}) + \mathrm{K}\,(\mathscr{R})
\tag{6}
$$

gegeben. Im einzelnen ist

$$H\,(r,\mathscr{R}) = K\,(r) + V\,(r,\mathscr{R}) + W\,(\mathscr{R}) \qquad (6\,\mathrm{a})$$

und weiter

$$V\,(r,\mathscr{R}) = -\sum_{i=1}^{n}\sum_{\lambda=1}^{N}\frac{Z_\lambda}{r_{\lambda i}} + \sum_{i=1}^{n-1}\sum_{j=i+1}^{n}\frac{1}{r_{ij}}\,, \qquad (6\,\mathrm{b})$$

$$W\,(\mathscr{R}) = +\sum_{\lambda=1}^{N-1}\sum_{\mu=\lambda+1}^{N}\frac{Z_\lambda\,Z_\mu}{R_{\lambda\mu}}\,. \qquad (6\,\mathrm{c})$$

Die Gleichungen (6b) und (6c) stellen die Wechselwirkungen der Elektronen mit den N Kernen, die die Atomkernladungen Z_λ ($\lambda = 1, 2, ..$ N) haben, und untereinander dar, sowie die Atomkernwechselwirkung. Es ist

$$r_{\lambda i} = \left|\mathscr{R}_\lambda - r_i\right|\,, \qquad (7\,\mathrm{a})$$

$$R_{\lambda\mu} = \left|\mathscr{R}_\lambda - \mathscr{R}_\mu\right|\,, \qquad (7\,\mathrm{b})$$

$$r_{ij} = \left|r_i - r_j\right|\,. \qquad (7\,\mathrm{c})$$

$K\,(\mathscr{R})$ und $K\,(r)$ bedeuten die Operatoren der kinetischen Energie in Kern- und Elektronenkoordinaten. Geben wir r_i und $\mathscr{R}_\lambda$ in kartesischen Koordinaten an,

$$r_i = \{X_i, Y_i, Z_i\}\,, \qquad (8\,\mathrm{a})$$

$$\mathscr{R}_\lambda = \{R_{\lambda X}, R_{\lambda Y}, R_{\lambda Z}\}\,, \qquad (8\,\mathrm{b})$$

so ist (in at. E.)

$$K\,(\mathscr{R}) = -\frac{1}{2\times 1836}\sum_{\lambda=1}^{N}\frac{1}{M_\lambda}\Delta_\lambda\,, \qquad (9\,\mathrm{a})$$

$$K\,(r) = -\frac{1}{2}\sum_{i=1}^{n}\Delta_i\,, \qquad (9\,\mathrm{b})$$

und im einzelnen:

$$\Delta_\lambda = \frac{\partial^2}{\partial R_{\lambda X}^2} + \frac{\partial^2}{\partial R_{\lambda Y}^2} + \frac{\partial^2}{\partial R_{\lambda Z}^2}\,, \qquad (10\,\mathrm{a})$$

$$\Delta_i = \frac{\partial^2}{\partial X_i^2} + \frac{\partial^2}{\partial Y_i^2} + \frac{\partial^2}{\partial Z_i^2}\,. \qquad (10\,\mathrm{b})$$

M_λ bedeutet, wie in (1), die Atomkernmasse des λ-ten Atoms in Vielfachen der Protonen(Neutronen)masse, die in atomaren Einheiten ungefähr 1836 Elektronenmassen beträgt.

Werden die Kerne festgehalten gedacht und die Vektoren $\mathcal{R}_\lambda$ als Parameter aufgefaßt, so tritt $K(\mathcal{R})$ in (5) nicht auf, und wir haben die Gleichung

$$H(r, \mathcal{R})\, \Psi(r, \sigma, \mathcal{R}, t) = i\, \frac{\partial \Psi(r, \sigma, \mathcal{R}, t)}{\partial t} \qquad (5a)$$

zu behandeln. Betrachten wir weiter stationäre Zustände, also Zustände des Systems, bei welchem die Wahrscheinlichkeitsdichte

$$\overline{W} = \Psi^* \Psi \qquad (11)$$

von der Zeit nicht abhängt, so muß Ψ von der Form

$$\Psi = \psi(r, \mathcal{R}, \sigma)\, e^{-i\varepsilon t} \qquad (12)$$

sein, und (5a) geht über in die zeitunabhängige Schrödingergleichung für ψ

$$H(r, \mathcal{R})\, \psi = \varepsilon \psi\,, \qquad (5b)$$

wobei wegen (11)

$$\int \psi^* \psi\, dr\, d\sigma = \int \overline{W}\, d\tau = 1 \qquad (11a)$$

sein muß.

Da die Lösungen von (5b) noch (11a) erfüllen müssen (Eigenfunktionen), existieren in der Regel nur für bestimmte ε-Werte (ε_p) Lösungsfunktionen von (5b),

$$H(r, \mathcal{R})\, \psi_p = \varepsilon_p\, \psi_p\,, \qquad (5b')$$

wobei es vorkommen kann, daß für einige ε-Werte mehrere (g_p) ψ-Funktionen $(\psi_{pn};\ n = 1, 2 .. g_p)$ erhalten werden (Entartung). Während zwei ψ-Funktionen zu verschiedenen ε-Werten $(\varepsilon_p, \varepsilon_{p'})$ zueinander orthogonal sind,

$$\int \psi_p^* \psi_{p'}\, d\tau = \delta_{pp'} = \begin{cases} 1 & p = p' \\ 0 & p \neq p' \end{cases},$$

braucht dies für die Funktionen, die zu einem g-fach entarteten ε-Wert gehören, nicht von vornherein der Fall zu sein.

Die sogenannten Eigenwerte ε_p in (5b') sind Funktionen der Atomlagen im Raum

$$\varepsilon_p = \varepsilon_p(\mathcal{R})\,, \qquad (13)$$

und stellen nach (6b), (6c) und (9b) die Gesamtenergie des Elektronensystems im Coulombfeld der Atomkerne dar. Neben dem diskreten Spektrum der ε_p-Werte gibt es auch ε-Bereiche, in denen für jeden ε-Wert eine ψ-Funktion existiert (Kontinuum).

Die $\mathcal{R}$-Abhängigkeit von ε in (13) läßt sich detaillierter betrachten. Drehen wir das Gesamtsystem der Atome starr um eine beliebige Achse, so muß ε unverändert bleiben, ebenso, wenn wir das System starr parallel verschieben. Diese beiden Bewegungsformen sind durch 2×3 Bestimmungsgrößen gegeben (Vektoren der Rotation und Translation), so daß ε nicht von den N Vektoren $\mathscr{R}_\lambda$ abhängt, die durch $3\,N$ Zahlen (Parameter) repräsentiert werden, sondern nur von $3\,N - 6 = F$ Größen, die wir im folgenden mit $R_1 R_2 \ldots R_F$ bezeichnen wollen. ε hängt also von F unabhängigen Parametern ab

$$\varepsilon_p(\mathscr{R}) \Longrightarrow \varepsilon'_p(\mathscr{R}'') \tag{14}$$

wobei $\mathscr{R}''$ für die Gesamtheit der R_j $(j = 1 \ldots F)$ steht. Liegen die N Atome auf einer Geraden, so beträgt $F = N - 1$. Wir haben also

Tab. 1

N	F (beliebige Anordnung)	F (lineare Anordnung)
2	1	1
3	3	2
4	6	3
5	9	4

Wir bezeichnen $\varepsilon'_p(\mathscr{R}'')$ in (14) als Energiehyperfläche des aus N Atomen bestehenden Systems. Nur für $N = 2$ und $N = 3$ (lineare Anordnung der Kerne) ist eine graphische Darstellung von ε', der Gesamt-Molekül-Energie des Systems, möglich.

Betrachten wir nun nach (2) je k_{ZM} Atome vom Element E^{ZM}, so haben wir ein System von

$$N = \sum k_{ZM} \tag{15}$$

Atomen vor uns, deren Elektronenzahl n sich als die Summe der Elektronenzahlen n_{ZM} der einzelnen neutralen Atome ergibt, abzüglich δ nach (2) und (2b)

$$n = \sum n_{ZM} - \delta = \sum_{\lambda=1}^{N} Z_\lambda - \delta . \tag{16}$$

Diese N Atome werden nun diejenigen Lagen $\mathscr{R}^{(0)}$ einnehmen wollen, in denen ε' ein Minimum hat. In diesen gilt

$$\tag{17}$$
$$\varepsilon'_p(R_1^{(0)}, \ldots R_j^{(0)} + \varDelta R_j^{(0)}, \ldots R_F^{(0)}) - \varepsilon'_p(R_1^{(0)}, \ldots R_j^{(0)}, \ldots R_F^{(0)}) > 0; (j = 1 \cdots F)$$

für alle $j = 1 \ldots F$. Damit ist aber noch nicht gesagt, ob sich das System wirklich in diesem Minimum aufhalten wird. Dazu müssen wir noch untersuchen, ob an der Stelle $\mathscr{R}^{(0)}$ Kernschwingungen möglich sind, denn es könnte sein, daß das Minimum so flach ist, daß solche Kernbewegungen nicht möglich sind, ohne daß das System auseinanderfliegt.

Wir erhalten die Rotations- und Schwingungszustände eines N-atomigen Systems aus der Schrödingergleichung der Kernbewegung

$$\left\{ K\left(\mathscr{R}\right) + \varepsilon_p'\left(\mathscr{R}''\right) \right\} X\left(\mathscr{R}, t\right) = i\,\frac{\partial X\left(\mathscr{R}, t\right)}{\partial t}\,, \tag{18}$$

in welcher X die Wellenfunktion der Kerne bedeutet. Fragen wir nach stationären Zuständen, so ist entsprechend (12) der Ansatz

$$X\left(\mathscr{R}, t\right) = \chi\left(\mathscr{R}\right) e^{-i\bar{\varepsilon}t} \tag{19}$$

zu verwenden, der, in (18) eingesetzt, diese Wellengleichung in

$$\left\{ K\left(\mathscr{R}\right) + \varepsilon_p'\left(\mathscr{R}''\right) \right\} \chi_{pq}\left(\mathscr{R}\right) = \bar{\varepsilon}_{pq}\,\chi_{pq}\left(\mathscr{R}\right) \tag{20}$$

überführt. $\bar{\varepsilon}_{pq}$ bedeutet die Energien, die das Kernsystem im Felde ε_p' annehmen kann. Die Quantenzahlen q zählen diese Zustände, zu denen auch ein Kontinuum gehört.

In (20) sind auch die nicht quantisierten Translationsbewegungen des starren Systems enthalten, was wir am Auftreten von K $(\mathscr{R})$ erkennen. Man kann K $(\mathscr{R})$ zerlegen

$$K\left(\mathscr{R}\right) \equiv K\left(\mathscr{R}_S\right) + K'\left(\mathscr{R}'\right)\,, \tag{21}$$

wobei $\mathscr{R}_S$ der Vektor zum Massenschwerpunkt bedeutet. Damit ist (20) in die Schwerpunktsbewegungen und in die sogenannten inneren Bewegungen separierbar geworden, denn mit

$$\chi_{pq}\left(\mathscr{R}\right) = \chi_S\left(\mathscr{R}_S\right)\overline{\chi}_{pq}\left(\mathscr{R}'\right) \tag{22}$$

zerfällt (20) in zwei Wellengleichungen

$$K\left(\mathscr{R}_S\right)\chi_S\left(\mathscr{R}_S\right) = \varepsilon^{(S)}\chi_S\left(\mathscr{R}_S\right) \tag{23a}$$

$$\left\{ K'\left(\mathscr{R}'\right) + \varepsilon_p'\left(\mathscr{R}''\right) \right\}\overline{\chi}_{pq}\left(\mathscr{R}'\right) = \bar{\bar{\varepsilon}}_{pq}\,\overline{\chi}_{pq}\left(\mathscr{R}'\right)\,, \tag{23b}$$

wobei

$$\bar{\varepsilon}_{pq} = \varepsilon^{(S)} + \bar{\bar{\varepsilon}}_{pq}\,; \tag{23c}$$

$\varepsilon^{(S)}$ stellt dann die Translationsbewegungsenergie des Schwerpunktes dar

$$\varepsilon^{(S)} = \frac{1}{2}\left(\sum_{\lambda=1}^{N} M_\lambda\right) 1836\, v^2\,, \tag{24}$$

mit v der Geschwindigkeit in at. E. Die Rotations- und Schwingungszustände der Energie sind durch $\bar{\bar{\varepsilon}}_{pq}$ gegeben, wobei auch hier u. a. ein Kontinuum auftritt. Zu jeder Energiehyperfläche ε_p' gehören dann ganz bestimmte diskrete und kontinuierliche $\bar{\bar{\varepsilon}}$-Werte. Die Komponenten X_k einer Reaktionsgruppe

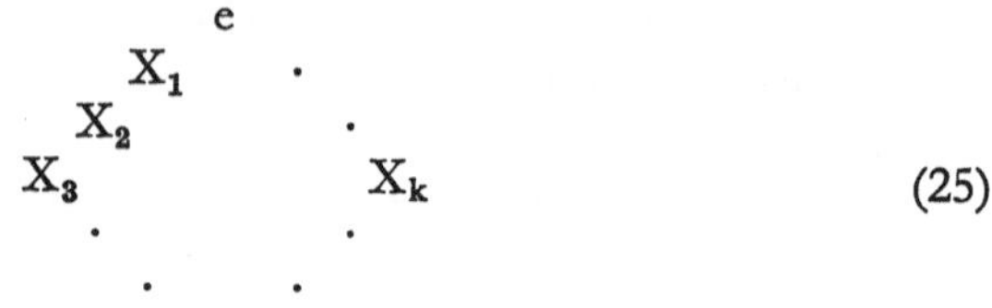

$$\begin{array}{c} \overset{e}{} \\ X_1 \quad\; \cdot \\ X_2 \quad\quad\; \cdot \\ X_3 \quad\quad\quad\quad X_k \\ \cdot \quad\quad\; \cdot \\ \cdot \end{array} \qquad (25)$$

nach (2), (15) und (16) definieren wir so, daß nur dann ein X_k vorliegt, wenn sich für dieses System nach (23 b) mindestens ein diskreter Zustand ergibt.

Liegt ein System vor, bei dem sich eine Reihe von verschiedenen räumlich endlichen Lagen der Atome ergeben, für die nach (17) die Energie $\varepsilon'_p(\mathscr{R}'')$ ein Minimum ist, so bezeichnen wir die jeweiligen Minima als zu verschiedenen Komponenten X_k gehörig, wenn sich die daraus ergebenden Atomkonstellationen, entsprechend dem Beispiel der Abb. 1, nicht durch Rotation des Gesamtsystems ineinander überführen lassen und das ursprüngliche Gesamtsystem im Sinne nach (23 b) mindestens einen diskreten $\overline{\overline{\varepsilon}}$-Wert annehmen kann.

Treten die Komponenten X_k miteinander in Wechselwirkung, so kann es vorkommen, daß sie sich ineinander umwandeln, soweit dies unter Beibehaltung der Gesamtmasse möglich ist. Einen solchen Vorgang wollen wir ganz allgemein Reaktion nennen.

Wir können daher eine Reihe von Fällen unterscheiden, die wir im folgenden angeben wollen, zusammen mit der für jeden Vorgang typischen chemischen und physikalischen Interpretation [2]:

$$X_k \rightleftharpoons X_{k'} \qquad\qquad \text{Umwandlung} \qquad\qquad\qquad (26a)$$

$$X_k \rightleftharpoons X_{k'} + X_{k''} \qquad \text{Zerfall, Rekombination, Addition} \quad (26b)$$

$$X_k + X_{k'} \rightleftharpoons X_{k''} + X_{k'''} \;\text{Substitution, Austausch, Umbau} \quad (26c)$$

Für jede Reaktion läßt sich eine Energiehyperfläche angeben, auf der die Reaktion verläuft, gegebenenfalls sind mehr Flächen notwendig zur Beschreibung der Vorgänge (26), wenn beim Übergang von einer Seite zur anderen das Gesamtsystem auch Übergänge zu anderen Energiehyperflächen vornehmen kann, was dann besonders wahrscheinlich ist, wenn sich die Energieflächen im für die Reaktion maßgeblichen Raumgebiet sehr nahe kommen.

Treten von außen Wechselwirkungen auf, etwa Einstrahlungen $(h\nu)$ so kann ein Übergang zu einer anderen Energiehyperfläche auftreten

$$X_k + h\nu \rightleftharpoons X_k^* , \qquad\qquad\qquad (27)$$

wobei X_k^* die elektronisch angeregte Komponente von X_k bedeutet. Entsprechende Vorgänge können auch in (26) eintreten, indem einige der Komponenten angeregt sind. So zum Beispiel in

$$X_k + X_{k'} \rightleftharpoons X_k + X_{k'}^* \,, \tag{26c'}$$

wo es sich um eine Stoßanregung an der gleichen Komponente handeln kann. Unter Stoßionisation verstehen wir den Übergang

$$X_k + X_{k'} \rightarrow X_{k''} + X_0 \quad (X_0 = e) \,. \tag{28}$$

Auch in (26b) und (26c) können Ionen auftreten. Schließlich kann eine Einstrahlung nach (27) zu Ionen führen

$$X_k + h\nu \rightarrow X_{k'} + X_0 \,. \tag{29}$$

Auch in diesen Fällen liegt ein Übergang zu anderen Energiehyperflächen vor, wobei die für den Übergang notwendige Energie aus dem Strahlungsfeld gewonnen wird, beziehungsweise wieder als Strahlung abgegeben werden kann, wenn der Vorgang rückläufig ist. Gegebenenfalls können die Energiebeträge auch, wie dies unter Umständen bei (26) der Fall sein könnte, aus der inneren Energie der Systeme oder aus Translationsenergien entnommen werden.

Es sei die Verabredung getroffen, daß in einer Reaktionsgruppe vorerst nur diejenigen Komponenten aufgenommen werden, die bei den jeweiligen Reaktionen zur Energiehyperfläche des Grundzustandes gehören, also zur tiefsten Energiefläche. Andernfalls, wenn die Gruppe auf angeregte Zustände erweitert werden soll, kann zum Beispiel nach (5b') und (14) auch He_2 als Komponente auftreten.

Was die Existenz von Komponenten anbetrifft, so ist es bemerkenswert, daß hier Absolutrechnungen bedeutsam sein können, da gewisse Komponenten möglicherweise präparativ sehr schwer zugänglich sind, die bei den Wechselwirkungen innerhalb einer Reaktionsgruppe eine große Rolle spielen. Jede Reaktion enthält nämlich die Möglichkeit, wesentlich auf das nach hinreichender Zeit sich einstellende Gleichgewicht zwischen den Komponenten einer Gruppe einzuwirken [3]. Leider ist zum Beispiel im Hinblick auf die Gruppe (4) die Existenz oder Nichtexistenz von He_2H^+, HeH_2^+, H_4^{++}, H_4^+ und He_3^+ nicht gesichert. Vorläufige Rechnungen an He_2H^+ zeigen, daß dieses System möglicherweise als metastabiles Molekül existiert (Protonenbrücke zwischen zwei He-Atomen) [4].

Mit Hilfe der statistischen Mechanik (statistischen Thermodynamik) lassen sich die Gleichgewichte bestimmen, die neben der Temperatur noch von der Konzentration einiger vorzugebender Komponenten abhängen können. In diesem Rahmen ist die Berechnung der Zustandssummen erforderlich, so daß auf diese Weise die verschiedenen Energien $\bar{\varepsilon}$ nach (23b) erforderlich sind. Erst die wellenmechanischen Rechnungen ermöglichen daher die Bestimmung der einzelnen Terme in den Zustandssummen. Wenn die Temperatur nicht allzu hoch ist, genügt es, den elektronischen Grundzustand zu betrachten, so daß die hier angegebene Definition der Reaktionsgruppe ausreicht. Bei sehr hohen Temperaturen dagegen müssen auch angeregte Elektronenzustände, die man aus (5b′) erhalten kann, berücksichtigt werden [5].

Die Thermodynamik erfaßt also nur Anfangs- und Endzustand eines Systems, kann also allein Aussagen über Gleichgewichte machen. Welche Übergänge (Reaktionen) im einzelnen auftreten, und ob sie überhaupt zu erwarten sind, ist eine Frage der Kinetik. Thermodynamik und Kinetik müssen letzten Endes auf die Wellenmechanik zurückgreifen, da die einzelnen Energiezustände der Komponenten und der damit im Zusammenhang stehenden Energiehyperflächen zur Diskussion des Reaktionsvorganges notwendig sind [6].

Die angegebenen Reaktionen in (26), (26c′) und (28) finden in einem Zusammenstoß der beteiligten Komponenten statt. Man bezeichnet solche Vorgänge als Elementarreaktionen oder Elementarschritte (Reaktionsschritte). In der Regel laufen mehrere solcher Elementarschritte gleichzeitig (Parallel- oder Nebenreaktion) oder in zeitlicher Folge ab (Folgereaktion) und geben in ihrer Gesamtheit das Bild einer komplexen Reaktion, die durch ihre Bruttoumsatzgleichung angegeben werden kann. Es ist im einzelnen fast immer sehr schwierig, bei einer vorgegebenen komplexen Reaktion die einzelnen Reaktionsschritte zu erkennen. Diese Aufklärung des Reaktionsmechanismus führt dann auf die Elementarreaktionen und auf ihr Verhältnis zueinander. Neben den Reaktionsmechanismen ist die Erfassung der Elementarreaktionen das Hauptziel der Theorie der Reaktionen.

2. Grundriß des wellenmechanischen Verfahrens [7]

Wir gehen dazu von Gleichung (5) aus, wobei der Hamiltonoperator nach (6) und (6a) sowie nach (6b) und (6c) erklärt ist. Die Lösung von (5) wird in der Form einer Reihenentwicklung

$$\Psi(r, \sigma, \mathscr{R}, t) = \sum_m \Phi_m(\mathscr{R}, t)\, \psi_m(r, \sigma, \mathscr{R}) \tag{30}$$

angesetzt, worin die ψ_m ein orthonomiertes, vollständiges Funktionensystem darstellen sollen. Mit diesem Ansatz erfassen wir auch nichtstationäre Zustände.

Wegen

$$K(\mathscr{R})\, \Phi_m \psi_m = \psi_m K(\mathscr{R})\, \Phi_m + \Phi_m K(\mathscr{R})\, \psi_m$$

$$- \frac{1}{1836} \sum_{\lambda=1}^{N} \frac{1}{M_\lambda} \nabla_\lambda \Phi_m \nabla_\lambda \psi_m \tag{31}$$

mit

$$\nabla_\lambda X = \frac{\partial X}{\partial R_{\lambda x}}\, e_x + \frac{\partial X}{\partial R_{\lambda y}}\, e_y + \frac{\partial X}{\partial R_{\lambda z}}\, e_z \tag{31a}$$

$$(e_x,\, e_y,\, e_z,\ \text{Einheitsvektoren})\ (X = \Phi_m, \psi_m)$$

ergibt sich

$$\sum_m \left\{ \Phi_m K(r)\psi_m + (V(r,\mathscr{R}) + W(\mathscr{R}))\Phi_m\psi_m + \psi_m K(\mathscr{R})\Phi_m \right.$$
$$\left. + \Phi_m K(\mathscr{R})\psi_m - \frac{1}{1836}\sum_{\lambda=1}^{N}\frac{1}{M_\lambda}\nabla_\lambda\Phi_m\nabla_\lambda\psi_m \right\} = i\sum_m\psi_m\frac{\partial\Phi_m}{\partial t}, \tag{32}$$

wenn (30) in (5) eingesetzt wird. Über ψ_m in (30) haben wir bisher noch nicht verfügt. Wir wollen jetzt annehmen, daß ψ_m die Lösungen der Schrödingergleichung für das Elektronensystem darstellen, wenn die Lagen der Kerne im Raum als Parameter angesehen werden. Damit nehmen wir (5b), (5b′) an und erhalten

$$H(r,\mathscr{R})\,\psi_m(r,\sigma,\mathscr{R}) = \varepsilon_m(\mathscr{R})\,\psi_m(r,\sigma,\mathscr{R}), \tag{33}$$

wobei

$$H(r,\mathscr{R}) = K(r) + V(r,\mathscr{R}) + W(\mathscr{R}) \tag{33a}$$

nach (6a). Unter dieser Annahme geht (32) über in

$$\sum_m\left\{\psi_m[K(\mathscr{R}) + \varepsilon(\mathscr{R})]\Phi_m + \Phi_m K(\mathscr{R})\psi_m - \frac{1}{1836}\sum_{\lambda=1}^{N}\frac{1}{M_\lambda}\nabla_\lambda\Phi_m\nabla_\lambda\psi_m\right\}$$
$$= i\sum_m\psi_m\frac{\partial\Phi_m}{\partial t}. \tag{34}$$

Nehmen wir die letzten beiden Terme in der Klammer als klein an und vernachlässigen sie, so geht (34) in die Gleichung der adiabatischen Näherung über, denn wir erhalten Gleichungen für Φ_m in der Form

$$[K(\mathscr{R}) + \varepsilon(\mathscr{R})]\Phi_m = i\frac{\partial\Phi_m}{\partial t}, \tag{35}$$

die mit (18) identisch sind, wenn Φ_m in X_m übergeht. Der stationäre Fall ergibt sich dann wieder zu (20). Da wir jetzt nicht den adiabatischen und stationären Fall behandeln wollen, dürfen wir diese beiden Terme nicht weglassen. Wir müssen daher Φ_m in der Form

$$\Phi_m = \sum_j C_{mj}(t)\chi_{mj}(\mathscr{R})e^{-i\bar\varepsilon_{mj}t} \tag{36}$$

ansetzen, wobei die χ_{mj} die Gleichungen nach (20) erfüllen. Wegen (20), (36) und

$$i\frac{\partial\Phi_m}{\partial t} = i\sum_j\left\{\frac{\partial C_{mj}}{\partial t}\chi_{mj}e^{-i\bar\varepsilon_{mj}t} - i\,C_{mj}\bar\varepsilon_{mj}\chi_{mj}e^{-i\bar\varepsilon_{mj}t}\right\} \tag{37}$$

geht (34), wenn man zuerst die linke Seite nach rechts bringt

$$\sum_m\left\{\psi_m\left[K(\mathscr{R}) + \varepsilon(\mathscr{R}) - i\frac{\partial}{\partial t}\right]\Phi_m + \Phi_m K(\mathscr{R})\psi_m\right.$$
$$\left. - \frac{1}{1836}\sum_{\lambda=1}^{N}\frac{1}{M_\lambda}\nabla_\lambda\Phi_m\nabla_\lambda\psi_m\right\} = 0 \tag{38}$$

3*

und von links mit ψ_n^* multipliziert und integriert

$$\sum_m \left\{ \delta_{nm} \left[K(\mathscr{R}) + \varepsilon(\mathscr{R}) - i\frac{\partial}{\partial t} \right] \Phi_m + \Phi_m [K(\mathscr{R})]_{nm} \right.$$
$$\left. - \frac{1}{1836} \sum_{\lambda=1}^{N} \frac{1}{M_\lambda} [\nabla_\lambda]_{nm} \nabla_\lambda \Phi_m \right\} = 0 \qquad (39)$$

in die Form über

$$\sum_m \sum_j \left\{ -i\,\delta_{nm} \frac{dC_{mj}}{dt} \chi_{mj}(\mathscr{R}) + [K(\mathscr{R})]_{nm} C_{mj} \chi_{mj}(\mathscr{R}) \right.$$
$$\left. - \frac{C_{mj}}{1836} \sum_{\lambda=1}^{N} \frac{1}{M_\lambda} [\nabla_\lambda]_{nm} \nabla_\lambda \chi_{mj}(\mathscr{R}) \right\} e^{-i\bar{\varepsilon}_{mj}t} = 0 \quad (n = 0,1,2\ldots). \qquad (40)$$

Dabei ist im einzelnen

$$[K(\mathscr{R})]_{nm} = \int \psi_n^*(r,\sigma,\mathscr{R}) \, K(\mathscr{R}) \, \psi_m(r,\sigma,\mathscr{R}) \, dr \, d\sigma \qquad (41\,a)$$

$$[\nabla_\lambda]_{nm} = \int \psi_n^*(r,\sigma,\mathscr{R}) \, \nabla_\lambda \psi_m(r,\sigma,\mathscr{R}) \, dr \, d\sigma . \qquad (41\,b)$$

Um die Bedingungsgleichungen für die Koeffizienten in (36) zu erhalten, multiplizieren wir (40) von links mit $\chi_{nj'}$ und integrieren über den ganzen $\mathscr{R}$-Raum. Es ergibt sich danach

$$i\frac{dC_{nj'}}{dt} = \sum_m \sum_j C_{mj} \left\{ [[K(\mathscr{R})]_{nm}]_{nj'\,mj} \right.$$
$$\left. - \frac{1}{1836} \sum_{\lambda=1}^{N} \frac{1}{M_\lambda} [[\nabla_\lambda]_{nm} \nabla_\lambda]_{nj'\,mj} \right\} e^{-i(\bar{\varepsilon}_{mj} - \bar{\varepsilon}_{nj'})t} . \qquad (42)$$
$$(n = 0,1,2\ldots) \qquad\qquad\qquad (j' = 0,1,2\ldots)$$

Führen wir für die geschweifte Klammer noch die Abkürzung

$$\Omega_{nj'\,mj} = [[K(\mathscr{R})]_{nm}]_{nj'\,mj} - \frac{1}{1836} \sum_{\lambda=1}^{N} \frac{1}{M_\lambda} [[\nabla_\lambda]_{nm} \nabla_\lambda]_{nj'\,mj} \qquad (43)$$

ein, wobei

$$[[K(\mathscr{R})]_{nm}]_{nj'\,mj} = \int \chi_{nj'}^* [K(\mathscr{R})]_{nm} \chi_{mj} \, d\mathscr{R} \qquad (44\,a)$$

$$[[\nabla_\lambda]_{nm} \nabla_\lambda]_{nj'\,mj} = \int \chi_{nj'}^* [\nabla_\lambda]_{nm} \nabla_\lambda \chi_{mj} \, d\mathscr{R} , \qquad (44\,b)$$

so erhält (42) die Form

$$i\frac{dC_{nj'}}{dt} = \sum_m \sum_j C_{mj} \Omega_{nj'\,mj} \, e^{-i(\bar{\varepsilon}_{mj} - \bar{\varepsilon}_{nj'})t} \qquad (45)$$
$$(m = 0,1,2\ldots; \, j' = 0,1,2\ldots),$$

die mit den Gleichungen der zeitabhängigen Störungsrechnung (Dirac'sche Störungsrechnung) übereinstimmt. Damit kann $|\Omega|^2$ als ein Maß für die Wahrscheinlichkeit des Übergangs des Systems von dem j'-ten Kerngerüstzustand der n-ten Energiehyperfläche nach dem j-ten Kerngerüstzustand der m-ten Energiehyperfläche angesehen werden. Damit sind die Grundgleichungen der wellenmechanischen Behandlung angegeben worden. Aus ihnen lassen sich dann, wenn Ω bekannt, oder annähernd bekannt ist, Informationen über Reaktionsvorgänge gewinnen, wobei die Gleichungen (45) unter den jeweils vorliegenden Anfangsbedingungen behandelt werden müssen. Zusammen mit statistischen Verfahren können dann die einzelnen Elementarreaktionen zwischen den Komponenten einer Reaktionsgruppe erfaßt werden.

3. Die Energiehyperflächen

Wie man aus den Gleichungen des vorherigen Abschnittes ersieht, spielt die Energie $\varepsilon(\mathscr{R})$ des Moleküls als Funktion der Kernlagen im Raum eine große Rolle. Um also die wellenmechanische Behandlung des Reaktionsvorganges durchführen zu können, müssen diese Energiehyperflächen bekannt sein, wobei in jedem Falle der analytischen Darstellung der Vorzug zu geben ist. Bis vor wenigen Jahren war eine analytische Form, die ε approximativ angibt, wobei mehr als zwei Atome vorliegen sollen, nicht bekannt. 1963 gab der Verfasser [8] Approximationen von Energiehyperflächen mehratomiger Systeme an, die dann später verbessert wurden [9]. Darin wurde $\varepsilon(\mathscr{R})$ in der Form

$$\varepsilon = E + W \tag{46}$$

mit

$$E = \frac{\displaystyle\sum_{f_1\ldots f_F}^{M_1\ldots M_F} \alpha_{f_1\ldots f_F} R_1^{f_1}\ldots R_F^{f_F}}{\displaystyle\sum_{f_1\ldots f_F}^{M_1\ldots M_F} \alpha'_{f_1\ldots f_F} R_1^{f_1}\ldots R_F^{f_F}} \tag{46a}$$

und

$$W = \sum_{\lambda=1}^{N-1} \sum_{\mu=\lambda+1}^{N} \frac{Z_\lambda Z_\mu}{R_{\lambda\mu}} \tag{46b}$$

angesetzt; die Größen R_j sind oben schon erklärt worden. Die Wahl von M_j hängt von der gewünschten Genauigkeit ab, sowie von der Kenntnis, die wir gegebenenfalls schon teilweise (punktweise) von der Energiehyperfläche besitzen. Die Parameter α und α' können dann durch Forderungen an die Energiefunktion bestimmt werden, wobei folgende Möglichkeiten in Anspruch genommen werden können:

1. Die Verwendung der Vorstellung der Atomassoziationen [10].

38 H. Preuss

2. Die störungstheoretische Berechnung des Energieverlaufs für kleine oder große Kernabstände.

3. Die Berechnung bestimmter ε-Werte (Punkte) mit Hilfe von ausreichend genauen wellenmechanischen Rechnungen (SCF- bzw. CI-Verfahren) [3].

4. Die Verwendung spektroskopischer Daten bei stabilen Konstellationen der Kerne.

Alle vier Wege wurden bisher beschritten, wobei es besonders günstig ist, alle Möglichkeiten gleichzeitig auszunutzen.

Bevor (46) auf mehrzentrige Systeme angewendet wurde, waren Rechnungen an zweiatomigen Molekülen vorausgegangen, bei denen dann nur mit einem $f_1 = f$ und mit einem entsprechenden $M_1 = M$ ($0 \leqslant f \leqslant M$) gerechnet wurde [11]. Auf diese Weise wurde mit $M = 7$ die bisher beste Potentialkurve des H_2-Moleküls erhalten. Testrechnungen an H_2^+, HeH^+ und He^+ zeigen ebenfalls, daß mit $M \geqslant 4$ ausreichende Genauigkeiten für die entsprechenden Potentialkurven zu erwarten sind. Schließlich sind in letzter Zeit noch weitere zweiatomige Moleküle mit diesem Ansatz behandelt und die dazugehörigen Spektralgrößen ausgerechnet worden, die eine gute Übereinstimmung mit den zum Vergleich vorliegenden Werten zeigen [12].

Dabei waren die Forderungen nach den Punkten 3 und 4 (oben) in der folgenden Form angesetzt:

$$\text{a)} \qquad \left.\frac{\partial \varepsilon}{\partial R}\right|_{R_0} = 0 \, , \qquad\qquad (47\,a)$$

$$\text{b)} \qquad \left.\frac{\partial^2 \varepsilon}{\partial R^2}\right|_{R_0} = k \, , \qquad\qquad (47\,b)$$

$$\text{c)} \qquad \varepsilon(R_0) = B + \varepsilon(\infty) \, . \qquad\qquad (47\,c)$$

Dabei bedeutet R_0 den Gleichgewichtsabstand und B ist die dazugehörige Bindungsenergie. k ist die Kraftkonstante. Nach Punkt 1 ist zu verlangen, daß

$$\text{d)} \qquad \lim_{R \to \infty} \varepsilon(R) = \varepsilon(\infty) \, , \qquad\qquad (47\,d)$$

$$\text{e)} \qquad \lim_{R \to 0} E(R) = E(0) \, , \qquad\qquad (47\,e)$$

so daß $\varepsilon(\infty)$ die Energie der getrennten Atome (Assoziation der getr. Atome) und $E(0)$ die Energie des vereinigten Atoms (Assoziation des vereinigten Atoms) bedeutet. Schließlich folgten nach Punkt 2 die Forderungen

$$E(R) = E(0) + E_2 R^2 + E_3 R^3 + \cdots \qquad (R \ll 1) \qquad (47\,f)$$

$$\varepsilon(R) = \varepsilon(\infty) + \frac{e_1}{R} + \frac{e_2}{R^2} + \cdots \qquad (R \gg 1) \, , \qquad (47\,g)$$

die nur dann benutzt werden können, wenn einige E_j oder e_j bekannt sind [13], was immer der Fall ist, da

$$E_1 \equiv 0 \tag{48}$$

ist und die ersten $e_j \equiv 0$ sind, wenn keine Dissoziation in Ionen stattfindet.

Die Forderungen bei mehratomigen Molekülen und Atomsystemen entsprechen den obigen bei zwei Atomen, so daß sich jetzt ergibt [9]

$$\text{a)} \qquad \left. \frac{\partial \varepsilon}{\partial R_j} \right|_{R_j^{(0)}} = 0 \,, \tag{49a}$$

$$\text{b)} \qquad \left. \frac{\partial^2 \varepsilon}{\partial R_j^2} \right|_{R_j^{(0)}} = k_j \,, \tag{49b}$$

$$\text{c)} \qquad \varepsilon(\mathscr{R}^{(0)}) = B + \varepsilon(a|b|c|..|N) \,, \tag{49c}$$

wobei $\mathscr{R}^{(0)}$ für die Gesamtheit aller Gleichgewichtsabstände $R_j^{(0)}$ steht. Die Kraftkonstante k_j ist bezüglich der Koordinate R_j definiert. B bedeutet die Energie, die das System im Gleichgewicht gegenüber der Energie der N getrennten Atome $\varepsilon(a|b|c|..|N)$ gewinnt.

Für (47d) und (47e) tritt jetzt

$$\lim_{[K]} E(\mathscr{R}) = E(K) \,, \qquad (K = 1, 2, \ldots) \tag{49d}$$

wobei [K] eine der möglichen Atomassoziationen [10] der N Atome mit der dazugehörigen Energie $E(K)$ bedeutet.

Bei drei Atomen a, b und c

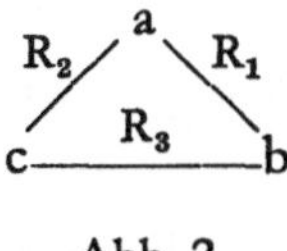

Abb. 2

liegen fünf Atomassoziationen vor

$$\text{1. } [a|b|c] \quad \text{2. } [a|b\,c] \quad \text{3. } [b|a\,c] \quad \text{4. } [c|ab] \quad \text{5. } [a\,b\,c] \,, \tag{50}$$

wobei zum Beispiel in 3. nach (49d) gilt (s. Abb. 2)

$$\lim_{\substack{R_2 \to 0 \\ R_1 \to \infty \\ R_3 \to \infty}} E(R_1, R_2, R_3) = E(b|a\,c) \,. \tag{51}$$

Die Forderungen (47f) und (47g) können wiederum für die jeweiligen Atomabstände übernommen werden.

Betrachten wir nochmals den Fall dreier Atome wie in Abb. 2, so geht (46) speziell über in

$$\varepsilon = \frac{\displaystyle\sum_{l=0}^{M_1} \sum_{k=0}^{M_2} \sum_{m=0}^{M_3} \alpha_{klm}\, R_1^l\, R_2^k\, R_3^m}{\displaystyle\sum_{l=0}^{M_1} \sum_{k=0}^{M_2} \sum_{m=0}^{M_3} \alpha'_{klm}\, R_1^l\, R_2^k\, R_3^m} + W \tag{52}$$

mit

$$W = +\frac{Z_a\,Z_b}{R_1} + \frac{Z_a\,Z_c}{R_2} + \frac{Z_b\,Z_c}{R_3}. \tag{53}$$

Sind einige der Atome gleich, so gelten für die α und α' Symmetriebeziehungen, die die Anzahl der freien Parameter α und α' einschränken. Für H_3^+ gilt zum Beispiel

$$\alpha_{klm} \equiv \alpha_{lkm} \equiv \alpha_{kml} \equiv \alpha_{mlk} \equiv \alpha_{lmk} \equiv \alpha_{mkl}. \tag{54}$$

Entsprechendes ist für α'_{klm} erfüllt. Setzen wir noch $M_1 = M_2 = M_3 = M = 1$ in (52), so geht damit (52) über in

$$\varepsilon = \frac{\alpha_{000} + \alpha_{100}\,(R_1 + R_2 + R_3) + \alpha_{110}\,(R_1 R_2 + R_1 R_3 + R_2 R_3) + \alpha_{111}\, R_1 R_2 R_3}{1 + \alpha'_{100}\,(R_1 + R_2 + R_3) + \alpha'_{110}\,(R_1 R_2 + R_1 R_3 + R_2 R_3) + \alpha'_{111}\, R_1 R_2 R_3}$$

$$+ W, \tag{55}$$

wobei

$$W = \frac{1}{R_1} + \frac{1}{R_2} + \frac{1}{R_3}. \tag{56}$$

Da $M = 1$ ist, lassen sich nur einige Forderungen an E und ε erfüllen. Wenn

$$\varepsilon\,(R_0) = B + \varepsilon\,(\infty)$$

$$\left.\frac{\partial \varepsilon}{\partial R}\right|_{R_0} = 0 \tag{57}$$

$$\lim_{R \to \infty} \varepsilon\,(R) = \varepsilon\,(\infty)$$

erfüllt ist, nachdem ein H^+ nach Unendlich gegangen ist und ferner

$$\varepsilon\,(\overset{\triangle}{R}_0) = \varepsilon\,(\infty) + \overset{\triangle}{B}$$

$$\left.\frac{\partial \varepsilon}{\partial \overset{\triangle}{R}}\right|_{\overset{\triangle}{R}_0} = 0 \tag{58}$$

$$\varepsilon\,(\overline{R}_0) = \varepsilon\,(\infty) + \overline{B}$$

erfüllt ist, wobei $\overset{\triangle}{B}$ und $\overline{B}$ die Bildungsenergien der dreieckigen und gestreckten Form mit den entsprechenden H—H-Abständen $\overset{\triangle}{R_0}$ und $\overline{R}_0$ bedeuten, so erhält man schließlich (in at.E.):

$$
\begin{aligned}
\alpha_{000} &= -22,648 & \alpha'_{000} &= 1,000 \\
\alpha_{100} &= -\ 3,619 & \alpha'_{100} &= 1,817 \\
\alpha_{110} &= -\ 5,528 & \alpha'_{110} &= 1,000 \\
\alpha_{111} &= -\ 2,925 & \alpha'_{111} &= 2,925 \ .
\end{aligned}
\tag{59}
$$

Ein Vergleich dieser Näherung, die erwartungsgemäß noch sehr grob sein muß, mit sehr genauen punktweisen Rechnungen [*14*] zeigt, daß für (in at.E.)

$$
1,0 \leqslant \overline{R} \leqslant 4,0 \qquad 1,0 \leqslant \overset{\triangle}{R} \leqslant 4,0
\tag{60}
$$

der Fehler nicht über 12% steigt. Im Durchschnitt beträgt die Abweichung ungefähr 3—5%.

Entsprechende Rechnungen sind bisher noch am HeHH-System durchgeführt worden [*9*], bei dem zur Zeit noch kein Vergleich mit genauen Rechnungen möglich ist.

Die bisherigen Rechnungen zeigen aber, daß mit (46) ein Ansatz vorliegt, mit dem die Energiehyperflächen mit ausreichender Genauigkeit approximiert werden können. Es werden jedoch noch viele Untersuchungen notwendig sein, bis genügend Erfahrungen und Ergebnisse vorliegen, um bei Kenntnis der Gleichungen für die wellenmechanische Behandlung von Atomkernbewegungen (Abschnitt 2) und von Energiehyperflächen in analytischer Form (Abschnitt 3) den Übergang zur rein theoretischen Behandlung der Reaktionsvorgänge vornehmen zu können.

Literatur

1. Eine Behandlung dieser Reaktionsgruppe bezüglich der Gleichgewichte als Funktion der Temperatur und des Druckes ist vor einiger Zeit vom Verfasser (bisher unpubliziert) durchgeführt worden.

2. man vgl.: PREUSS, H.: Mol. Phys. **8**, 233 (1964).

3. Als zur Zeit vorliegendes Absolutrechnungsverfahren kann u. a. die SCF—MO—LC (LCGO)-Methode genannt werden, bei der bisher die behandelten Moleküle am größten waren und eine vergleichsweise kurze Rechenzeit notwendig ist (PREUSS, H.: Z. f. Naturforschung **19a**, 1335 (1964); **20a**, 17 (1965); **20a**, 27 (1965); **20a**, 1290 (1965); PREUSS, H., G. DIERCKSEN: Int. Journal f. Quantum Chem. (im Druck, 1967)), sowie eine kurze Notiz: Z. f. Naturforschung **21a**, 863 (1966).

4. PREUSS, H. (unpubliziert).

5. Bezüglich der Berechnung von Zustandssummen sei auf das vorzügliche Buch von I. N. Godnew, „Berechnung thermodynamischer Funktionen aus Moleküldaten" hingewiesen (Deutscher Verlag der Wissenschaft Berlin 1963).
6. Man vgl. etwa Frost, A. A. and R. G. Pearson: Kinetik und Mechanismen homogener chemischer Reaktionen (Deutsche Übersetzung von F. Helffe-rich und H. Schindewolf, Verlag Chemie 1964).
7. Man vgl. Hellmann, H.: Einführung in die Quantenchemie. Wien: Deuticke 1936 (Kap. VIII).
8. Preuss, H.: Rev. Mod. Phys. 35, 646 (1963), s. auch Habilitationsschrift des Verfassers (1963).
9. Preuss H.: Theoret. Chim. Acta 2, 344 (1964); 2, 362 (1964); weitere Arbeiten in Druck.
10. Preuss, H.: Z. f. Naturforschung 12a, 599 (1957); 13a, 364 (1958), Na-turwissenschaften 11, 241 (1961); Theoret. Chim. Acta 1, 42 (1962); Z. f. Naturforschung 18a, 489 (1963); Mol. Phys. 8, 441 (1964) sowie Literatur-zitat [8] und [9].
11. Preuss, H.: Theoret. Chim. Acta 2, 102 (1964).
12. Kraemer, W.: Diplomarbeit (1966, Universität Frankfurt/M.).
13. Bingel, W. A.: J. Chem. Phys. 30, 1250, 1254 (1959); Z. f. Naturforschung 16a, 668 (1961); Morgenau, H.: Phys. Rev. 38, 747 (1931); Pauling, L., and J. Y. Beach: Phys. Rev. 47, 686 (1935).
14. Conroy, H.: J. Chem. Phys. 40, 609 (1964).

H. Preuß

Max-Planck-Institut
für Physik und Astrophysik
München

Theory of Non-Adiabatic Transitions. Recent Development of the Landau-Zener (Linear) Model

E. E. NIKITIN

With 12 Figures

1. Adiabatic Approximation and Non-Adiabatic Coupling

A system of electrons and nuclei is described by the wave equation

$$i\hbar \frac{\partial \Psi}{\partial t} = H\Psi = \left[\sum_{\alpha} T_{\alpha} + H_{\text{el}}\right]\Psi \tag{1.1}$$

where $\sum_{\alpha} T_{\alpha}$ is the operator of the kinetic energy of the nuclei, and H_{el} is the electronic Hamiltonian for fixed nuclei. It may be written as follows:

$$H_{\text{el}} = \sum_{i} T_i + \sum_{i \leq k} V_{ik} + \sum_{i,\alpha} V_{i\alpha} + \sum_{\alpha \leq \beta} V_{\alpha\beta} + V_{SO} . \tag{1.2}$$

Here $\sum T_i$ is the kinetic energy of electrons, $\sum V_{ik}$ is the interaction between pairs of electrons, $\sum V_{i\alpha}$ is the potential energy of electron-nuclei interaction, $\sum V_{\alpha\beta}$ is the electrostatic interaction of nuclei (Roman letters refer to electrons and Greek refer to nuclei). The spin-orbital interaction V_{SO} also should be included in H_{el}, as this interaction is usually stronger than the non-adiabatic coupling due to nuclear motion.

By adiabatic approximation for an electronic wave function we mean the wave function found as a solution of the Schrödinger equation with fixed nuclei

$$H_{\text{el}}(r, R)\,\varphi_l(r, R) = E_l(R)\,\varphi_l(r, R) \tag{1.3}$$

where r stands for the set of electronic coordinates, R denotes the set of nuclear coordinates. Here R is considered as a parameter, because H_{el} depends on R parametrically. A set $\varphi_l(r, R)$ is the commonly one used in quantum chemical calculations of electronic structure.

There are two ways for taking into account the nuclear motion. First we may consider nuclei to be particles subject to wave-mechanical treatment. Then an adiabatic basic set is constructed:

$$\Phi_{ln}(r, R, t) = \varphi_l(r, R)\,\chi_{ln}(R)\exp\left(-\frac{i}{\hbar} E_{ln} t\right) . \tag{1.4}$$

Here χ_{ln} is a wave function for nuclei in a state n moving on a potential energy surface l. Accordingly we have for χ_{ln} a set of uncoupled equations

$$\left[\sum_{\alpha} T_{\alpha} + E_l(\boldsymbol{R})\right] \chi_{ln}(\boldsymbol{R}) = E_{ln}\chi_{ln}(\boldsymbol{R}) \; . \tag{1.5}$$

Adiabatic functions Φ_{ln} are coupled by the operator $\sum T_{\alpha}$ acting on functions $\varphi_l(\boldsymbol{r}, \boldsymbol{R})$. If the coupling is taken into account, equation (1.5) is to be complemented by non-diagonal terms, and then $E_l(\boldsymbol{R})$ cannot be interpreted as the potential energy of nuclei. This quantum-mechanical approach to non-adiabatic coupling has been discussed recently by LONGUET-HIGGINS [1]. We shall not consider it here in detail and shall confine ourselves to the more simple semiclassical approach. It consists essentially in dividing the general problem of coupled electronic-nuclear motion into two parts: wave-mechanical treatment of electrons, and classical treatment of nuclei. Accordingly, for nuclei we introduce a trajectory $\boldsymbol{R} = \boldsymbol{R}(t)$ and consider the time dependent electronic Hamiltonian

$$H_{\text{el}}(\boldsymbol{r}, \boldsymbol{R})\,\Psi(\boldsymbol{r}, t) = i\hbar\,\frac{\partial\Psi(\boldsymbol{r}, t)}{\partial t} \; . \tag{1.6}$$

The Hamiltonian $H_{\text{el}}(\boldsymbol{r}, \boldsymbol{R})$ depends on t through $\boldsymbol{R} = \boldsymbol{R}(t)$ and the choice of the trajectory represents in itself a problem, independent of eq. (1.6).

In the semiclassical approach we use the basic set $\varphi_l(\boldsymbol{r}, \boldsymbol{R})$ with $\boldsymbol{R} = \boldsymbol{R}(t)$ to expand the wave function. Writing

$$\Psi = \sum_l a_l(t)\,\varphi_l(\boldsymbol{r}, \boldsymbol{R}(t))\,\exp\left[-\frac{i}{\hbar}\int^t E_l(\boldsymbol{R})\,dt\right] \; , \tag{1.7}$$

we try to satisfy eq. (1.6). Putting (1.7) in (1.6) and using the identity which follows from (1.3):

$$H_{\text{el}}(\boldsymbol{r}, \boldsymbol{R}(t))\,\varphi_l = E_l(\boldsymbol{R}(t))\,\varphi_l \; , \tag{1.8}$$

we come to the following set of equations:

$$i\hbar\,\dot{a}_l = \sum_{l'} a_{l'}\langle\varphi_l^*\left(-i\hbar\frac{\partial}{\partial t}\right)\varphi_{l'}\rangle\exp\left[-\frac{i}{\hbar}\int^t (E_{l'} - E_l)\,dt\right] \; . \tag{1.9}$$

Comparing this with a standard form of equations for time-dependent perturbation theory [see *1, 2, 3*] we find the following expression for an operator W which is responsible for non-adiabatic coupling

$$W = -i\hbar\frac{\partial}{\partial t} \; . \tag{1.10}$$

To estimate the coupling efficiency we rewrite (1.10) in the form

$$W_{ll'} = \left(-i\hbar\frac{\partial}{\partial t}\right)_{ll'} = -i\hbar\,\boldsymbol{v}\langle\varphi_l^*\frac{\partial\varphi_{l'}}{\partial\boldsymbol{R}}\rangle \; . \tag{1.11}$$

Denoting $|\langle \varphi_l^* \frac{\partial \varphi_{l'}}{\partial \boldsymbol{R}} \rangle|^{-1}$ by a characteristic length δR we find $W_{ll'} \approx \hbar v / \delta R$.

Non-adiabatic coupling is small when the perturbation treatment is valid

$$|W_{ll'}| \ll |E_l - E_{l'}| = \Delta E_{ll'} \ . \tag{1.12}$$

Thus we obtain the condition of small coupling

$$\Delta E_{ll'} \cdot \delta R / \hbar v \gg 1 \ . \tag{1.13}$$

The ratio $\Delta E \, \delta R / \hbar v$ is called the Massey parameter. It is useful for classification of non-adiabatic transitions. From (1.13) we see that non-adiabatic coupling is strong when two electronic terms cross at some point $R = R_c$ or come close together at $R = R_p$ $(\Delta E (R_p) < \hbar v / \delta R)$.

Thus, the condition for adiabatic approximation can be put as follows. For regions of the phase-space of nuclei, where condition (1.13) is fulfilled for a given electronic state l and all states l', the adiabatic electronic energy E_l may be interpreted as the potential energy of nuclei. The velocity v in (1.13) corresponds to the classical nuclear motion on the potential energy surface $E_l (\boldsymbol{R})$.

When a wave function Ψ is expressed as a linear combination of type (1.7) with coefficients a_l *independent* of time, Ψ will also be an adiabatic function. The main feature of the non-adiabaticity originates from the kinematic coupling of electrons and nuclei and is expressed in the time dependence of the contributions from different adiabatic electronic states to the total function Ψ for the motion over some classical trajectory $\boldsymbol{R} = \boldsymbol{R} (t)$.

When condition (1.13) is not fulfilled in some regions of the phase space, the electronic energy $E_l (\boldsymbol{R})$ ceases having the sense of nuclear potential energy, and treatment must be based on the solution of coupled equations that include all electronic terms.

A difficulty arises here: how to choose a trajectory simulating nuclear motion on several electronic terms? It is clear that any particular trajectory chosen will involve certain not well defined parameters. However, it appears possible to establish a correspondence between quantum and semiclassical approaches in an important case which would allow the definition of the precise sense of a classical trajectory. This question is discussed below in connection with the Landau-Zener model.

2. Two-State Approximation for a One-Dimensional Linear Model

As we have seen earlier, strong non-adiabatic coupling seems important only for terms which approach one another closely. We concentrate now on a case, where we consider only *two* electronic states. The close spacing of three and more terms is not a very improbable event, but still *two-state approximation* will be considered for simplicity.

Supposing we have two adiabatic functions φ_1 and φ_2 corresponding to adiabatic terms E_1, E_2 that are shown in Fig. 1 by full lines marked with indices 1 and 2. If there is a non-adiabatic transition between these terms, functions φ_1 and φ_2 must depend quite markedly on R. We now

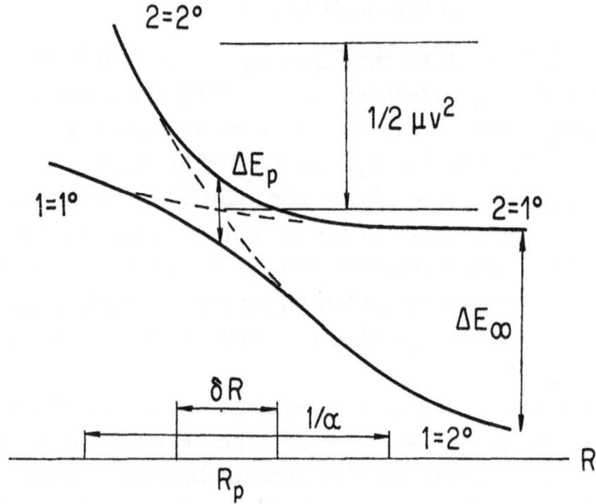

Fig. 1. Adiabatic and zero-order terms of a linear model (full and dashed lines, respectively)

want to represent φ_l as a linear combination of another pair of functions φ_l^0, the dependence of which on R is very weak. Thus we put

$$\begin{aligned}
\varphi_1 &= \quad \varphi_1^0 \cos\chi + \varphi_2^0 \sin\chi \\
\varphi_2 &= -\varphi_1^0 \sin\chi + \varphi_2^0 \cos\chi ,
\end{aligned}$$
(2.1)

where φ_1 and φ_2, and φ_1^0 and φ_2^0 are considered to be orthonormal functions. In the basic set φ_1 and φ_2 the two-state Hamiltonian H_{el} is diagonal

$$H_{el}(\varphi) = \begin{pmatrix} E_1(R) & 0 \\ 0 & E_2(R) \end{pmatrix}$$
(2.2)

with matrix elements E_1 and E_2 strongly varying with R. Moreover, $\Delta E_p = (E_1 - E_2)_{\min}$ ist small enough for the non-adiabatic coupling to be efficient. In the basic set φ_1^0 and φ_2^0 the Hamiltonian is *not* diagonal[1]

$$H_{el}(\varphi^0) = \begin{pmatrix} H_{11} & H_{12} \\ H_{21} & H_{22} \end{pmatrix} ,$$
(2.3)

---

[1] We may refer to the φ_l^0 states as being adiabatically coupled, because there exists a coupling between φ_1^0 and φ_2^0 due to the non-diagonal element H_{12} of the adiabatic Hamiltonian.

but the matrix elements H_{ik} depend very weakly on R due to the slight dependence of φ_l^0 on R. Now we want to define φ_l^0 in such a manner that asymptotically when R is far from the region $R \sim R_p$ of strong coupling φ_l and φ_l^0 coincide. It means that when $|R - R_p| \to \infty$ we must have

$$H_{12}(R)/[H_{11}(R) - H_{22}(R)] \to 0 \ . \tag{2.4}$$

The only step remaining to be carried out is to choose an approximate expression for the matrix elements. Restricting ourselves to a one-dimensional case, we develop H_{12} and $H_{11} - H_{22}$ in series of $R - R_p$ and retain only the first two terms

$$
\begin{aligned}
H_{12}(R) &= H_{12}(R_p) + H'_{12}(R_p)(R - R_p) + \cdots \ , \\
H_{11} - H_{22} &= \Delta H(R) = \Delta H(R_p) + \Delta H'(R_p)(R - R_p) + \cdots
\end{aligned} \tag{2.5}
$$

The importance of higher terms has to be investigated later. Now we define R_p from equation

$$\Delta H(R_p) = 0 \ . \tag{2.6}$$

Condition (2.4) means that $H'_{12}(R_p)$ must be zero at $R = R_p$. Thus, with our choice of φ_l^0 we have the following matrix for H_{el}:

$$
H_{\text{el}}(\varphi^0) = \begin{pmatrix} E_0 + \dfrac{\Delta F}{2} x & a \\[2ex] a & E_0 - \dfrac{\Delta F}{2} x \end{pmatrix}, \quad x = R - R_p \tag{2.7}
$$

where

$$E_0 = H_{11}(R_p) = H_{22}(R_p), a = H_{12}(R_p) \text{ and } \Delta F = -\frac{\partial}{\partial R}(H_{11} - H_{22})\big|_{R=R_p} \ .$$

Diagonal matrix elements $H_{11}(R)$ and $H_{22}(R)$ are shown in Fig. 1 by dashed lines bearing indices 1^0 and 2^0. In terms of these parameters the eigenvalues E_i are

$$E_{1,2} = E_0 \mp \Delta E, \quad \Delta E = \frac{1}{2}[(\Delta F x)^2 + 4a^2]^{1/2} \tag{2.8}$$

and

$$\chi = \frac{1}{2}\operatorname{arctg}\frac{2a}{\Delta F x} \ . \tag{2.9}$$

Here $x(t)$ is a function of time which still has to be defined.

The non-adiabatic wave function Ψ can be expanded in either set

$$\Psi(t) = a_1(t)\varphi_1 \exp\left[-\frac{i}{\hbar}\int^t E_1 dt\right] + a_2(t)\varphi_2 \exp\left[-\frac{i}{\hbar}\int^t E_2 dt\right] , \tag{2.10}$$

$$\Psi(t) = b_1(t)\varphi_1^0 \exp\left[-\frac{i}{\hbar}\int^t H_{11} dt\right] + b_2(t)\varphi_2^0 \exp\left[-\frac{i}{\hbar}\int^t H_{22} dt\right] . \tag{2.11}$$

From (1.9) and (2.1) we have the following equations for a_i or b_i:

$$i \dot{a}_1 = \quad i \dot{\chi} \exp\left[-\frac{i}{\hbar} \int^t (E_2 - E_1)\, dt\right] a_2 \qquad (2.12)$$

$$i \dot{a}_2 = - i \dot{\chi} \exp\left[\frac{i}{\hbar} \int^t (E_2 - E_1)\, dt\right] a_1$$

$$\hbar i \dot{b}_1 = a \exp\left[-\frac{i}{\hbar} \int^t (H_{22} - H_{11})\, dt\right] b_1 \qquad (2.13)$$

$$\hbar i \dot{b}_2 = a \exp\left[\frac{i}{\hbar} \int^t (H_{22} - H_{11})\, dt\right] b_1$$

where we neglected the very weak dependence of φ_i^0 on R. However, it should be taken into account if H_{12} turns out to be zero, for some reason or other. This point will not be discussed here.

Each pair of equations decouples when $|x| \to \infty$. In (2.12) $|\dot{\chi}| \to 0$ at this limit, and in (2.13) very fast oscillations of exponentials in this limit will bring about decoupling.

Near $x = 0$ the coupling, generally speaking, is strong over the range δR, which has to be found later. Thus, in this transition region we have either:

adiabatic (non-crossing) potentials E_1 and E_2 plus non-adiabatic coupling
or
crossing zero-order potentials H_{11} and H_{22} plus adiabatic coupling.

Neither of these can be described as a usual potential, so that inside the range δR we can say as well that there is no *potential* at all. Instead we have the problem of coupling, represented by (2.12) and (2.13) in various bases. Because asymptotically φ_i^0 goes in φ_i these systems are completely equivalent, and we may use either of these if it proves to be convenient.

Consider now limitations which are imposed on the range of validity of the theory by linear approximation (2.5).

Let us first take the low velocity region. The decoupling of (2.12) will be due mostly to adiabatic decoupling, the condition of which is just (2.4). For a rough estimation we may conceive that the smoothly varying functions φ_i^0 change essentially at a length $1/\alpha$ of the order of a Bohr radius. Thus we have $\Delta F \sim \Delta E_\infty\, \alpha$ where ΔE_∞ is a difference in electronic energies of separated atoms. From (2.4) we see that decoupling occurs when x is higher than δR

$$\delta R \sim \frac{a}{\Delta F} \sim \frac{a}{\alpha \Delta E_\infty}. \qquad (2.14)$$

On the other hand, δR should be much less than $1/\alpha$, so that higher order terms in (2.5) could be neglected.

To meet this condition we have the first limitation of the theory

$$\delta R \alpha \approx \frac{2a}{\Delta E_\infty} = \frac{\Delta E_p}{\Delta E_\infty} \ll 1 \ . \tag{2.15}$$

A situation where the maximum splitting is much smaller than the asymptotic one is called pseudocrossing of terms, which occurs at $R = R_p$.

Consider now the high velocity limit. The exponential term in (2.13) has the form

$$\exp\left[-\frac{i}{\hbar} \int^t (H_{11} - H_{22})\, dt\right] = \exp\left[-\frac{i}{\hbar} \Delta F \int^t x\, dt\right] \ . \tag{2.16}$$

Near $R = R_p$ let us put

$$x(t) \approx v\, t \tag{2.17}$$

where v is the velocity of nuclei near the crossing point of zero-order terms. Then we have

$$\exp\left[-\frac{i}{\hbar} \Delta F \int^t x\, dt\right] = \exp\left[-\frac{i}{\hbar} \frac{\Delta F v\, t^2}{2}\right] \ . \tag{2.18}$$

Oscillations that decouple (2.13) become fast after a period δt which is

$$\frac{\Delta F v}{\hbar} (\delta t)^2 \approx 1 \ . \tag{2.19}$$

Thus, for δR we have $\delta R \approx v\, \delta t$, and the condition $\delta R \alpha \ll 1$ means

$$\delta R \alpha \approx \left(\frac{\hbar v}{\Delta F}\right)^{1/2} \alpha \ll 1 \ . \tag{2.20}$$

Putting here $\Delta F \approx \Delta E_\infty \alpha$, we have

$$\Delta E_\infty / \hbar v \alpha \gg 1 \ . \tag{2.21}$$

This is another limitation of the theory.

For velocities of chemical interest and for energy gaps ΔE_∞ of the order of e.v. conditions (2.15) and (2.20) do not limit in any way the applicability of the theory. The only trouble is the choice of trajectory $x = x(t)$, and now we wish to show how it has been overcome in the last 30 years.

3. The Landau-Zener Formula

It were first LANDAU [4] and ZENER [5] who attacked the problem. They suggested that higher terms in (2.17) could be neglected. Thus, two trajectories were approximated by *one* in the near vicinity of the crossing point

$$x = v\, t, \qquad v = \frac{dx}{dt}\bigg|_{R=R_p} \tag{3.1}$$

since the velocity at this point is the same for both electronic terms. We shall discuss their results now for two limiting cases.

Consider the case where a is so small that a perturbation treatment of (2.13) is valid. Denoting by P_{12}^0 the probability for a transition from φ_1^0 to φ_2^0 and by P_{12} the probability for a transition from φ_1 to φ_2, we can see from fig. 1 that $P_{12}^0 = 1 - P_{12}$, as adiabatic and zero-order terms reverse after going through the coupling region.

We start with $t = -\infty$ which corresponds to $R \ll R_p$, and end at $t = +\infty$ which corresponds to $R \gg R_p$. At the beginning we have

$$b_1(-\infty) = 1 , \quad b_2(-\infty) = 0 . \tag{3.2}$$

At the end $|b_2(+\infty)|^2$ gives just P_{12}^0. To the first order we have

$$b_2(+\infty) = \int_{-\infty}^{\infty} \frac{a}{i\hbar} \exp\left[-\frac{i\Delta F}{2\hbar} v t^2\right] dt = \frac{a}{\hbar i}\left[\pi \Big/ -\frac{i\Delta F v}{2\hbar}\right]^{1/2} . \tag{3.3}$$

Thus

$$P_{12}^0 = 2\pi a^2 / \Delta F \hbar v , \quad if \ P_{12}^0 \ll 1 . \tag{3.4}$$

When a is sufficiently high, P_{12}^0 can be of the order of unity and perturbation treatment fails.

It was LANDAU [4] who pointed out how to calculate the main factor in the expression for P_{12} for strong adiabatic coupling (low v). The method consists essentially in solving eq. (2.12) at a low velocity limit. As the non-adiabatic coupling is weak, one would try to solve (2.12) using the perturbation method

$$a_2(+\infty) = -\int_{-\infty}^{\infty} \dot{\chi} \exp\left[\frac{i}{\hbar} \int^t [(\Delta F v t)^2 + 4 a^2]^{1/2} dt\right] dt . \tag{3.5}$$

If we calculated the integral (3.5) the result would be exponentially low

$$a_2(+\infty) \sim \exp(-v_0/v) \tag{3.6}$$

where v_0 is a constant independent of velocity v.

Now, the non-adiabatic coupling parameter v/v_0, which is low for low velocities, enters this expression in such a way that the result of integration, although low, cannot be presented as a series in powers of (v/v_0). It means that (2.12) *cannot be solved* by a conventional perturbation method, even when the probability of transition is small. This fact became clearly understood only recently [6].

Nevertheless the exponential term in (3.5) still can be calculated by the Landau method (see sect. 52 and sect. 53 of ref. 3; and also ref. 4). Considering the complex variable t we displace the integration path in (3.5) up to the branching point of $\Delta E(t)$, and separate the exponent

$$a_2(+\infty) \sim \exp\left[\frac{i}{\hbar} \int_0^{t_c} [\Delta F^2 x(t)^2 + 4 a^2]^{1/2} dt\right] \tag{3.7}$$

where t_c is a solution of eq. (3.8) with $Im\,(t_c) > 0$:

$$[\varDelta E(t_c)]^2 = [\varDelta F x(t_c)]^2 + 4\,a^2 = 0 \ . \tag{3.8}$$

Thus t_c is the time when *adiabatic terms cross.*

It is a purely imaginary number $t_c = 2\,a\,i/\varDelta F v$ that is in line with the usual opinion that at real distances (or times) two adiabatic terms of a same symmetry *do not cross* (non-crossing rule).

The pre-exponential factor in (3.7) is so far unknown.

Thus, we have from (3.7) and (3.8)

$$P_{12} = |a_2(+\infty)|^2 \sim \exp\left[\frac{i}{\hbar}\int\limits_0^{t_c}\varDelta E\,dt\right] = \exp\left[-\frac{2\,\pi\,a^2}{\varDelta F v\,\hbar}\right] \ . \tag{3.9}$$

The correct pre-exponential factor in this expression was calculated by ZENER [5] who solved exactly eq. (2.13) with a trajectory (3.1). It was found that not only for large but also for couplings of any strength the probability is given by

$$P_{12} = \exp\left[-\frac{2\,\pi\,a^2}{\varDelta F\,\hbar\,v}\right] = 1 - P_{12}^0 \ , \tag{3.10}$$

If $2\,\pi\,a^2/\varDelta F\,\hbar\,v \ll 1$, (3.10) goes over into (3.4).

In atomic collisions nuclei usually go through the coupling region twice. Then the net probability can be calculated by adding fluxes to each term

$$P = 2\,P_{12}(1 - P_{12}) = 2\,(1 - P_{12}^0)\,P_{12}^0 \ . \tag{3.11}$$

The procedure of adding fluxes neglects the interference between probability amplitudes, so that (3.11) is to be understood as a result of averaging over oscillations. From (3.11) we have

$$P = 2\exp\left(-\frac{2\,\pi\,a^2}{\varDelta F\,\hbar\,v}\right)\left[1 - \exp\left(-\frac{2\,\pi\,a^2}{\varDelta F\,\hbar\,v}\right)\right] \ , \tag{3.12}$$

which is known as the Landau-Zener formula. The plot of P *vs* dimensionless velocity $v\,\varDelta F\,\hbar/2\,\pi\,a^2$ is shown in Fig. 2. This result has been obtained also by STUECKELBERG [7] who considered two coupled wave equations under the same limiting conditions.

Conditions of validity of (3.12) are, together with (2.15) and (2.20), also those which make it possible to neglect higher terms in expansion (3.1). The velocity does not change appreciably when the potential energy change at a length δR is small compared to kinetic energy

$$F_i\delta R \ll \frac{\mu v^2}{2}, \quad F_i = -\frac{\partial H_{ii}}{\partial R}\bigg|_{R=R_p} \ . \tag{3.13}$$

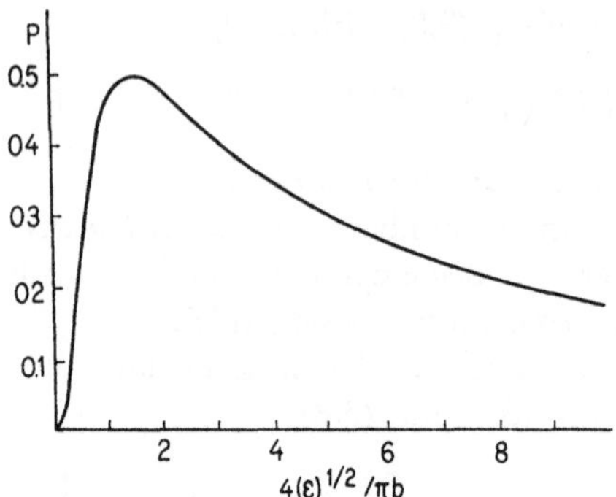

Fig. 2. The Landau-Zener transition probability P *vs* dimensionless velocity
$$4\varepsilon^{1/2}/\pi b = v\,\Delta F\,\hbar/2\,\pi\,a^2$$

This, together with the condition under which interference may be neglected can be written as

$$S_i = \frac{1}{\hbar}\int_{R_t}^{R_p}\sqrt{2\mu}\left[\frac{\mu v^2}{2} - E_i(R)\right]^{1/2}dR \gg 1\,, \quad \Delta S = |S_1 - S_2| \gg 1 \quad (3.14)$$

where R_t are turning points on each adiabatic term. It is these conditions that severely limit the applicability of the Landau-Zener formula to non-adiabatic transitions at a low energy. In what follows we shall concentrate on latest developments of the theory dealing with removal of conditions (3.14). At higher atomic energies (several hundreds of e.v.) other limitations of the Landau-Zener model will emerge and these have been discussed extensively by Bates [8].

4. Effect of the Turning Point

We try now to take into account the effect of an external field on the motion of nuclei. To proceed this way, we write down the next term in (3.1) [9]

$$x = v\,t + \frac{F}{2\mu}\,t^2\,. \tag{4.1}$$

By shifting the time scale we obtain

$$x = \frac{F}{2\mu}\,t^2 - \frac{\mu v^2}{2F}\,, \tag{4.2}$$

where F is an effective force, *if it proves to be possible* to introduce F as such.

In reality we have two forces F_1 and F_2 and we would expect the relation $F \approx (F_1 + F_2)/2 \approx (F_1 F_2)^{1/2}$, if $\Delta F \ll F$.

The semiclassical theory based on this idea has been elaborated in [9]. To find out what is meant by F when F_1 and F_2 are quite different, we should consider quantum equations for the same linear model. The quantum mechanical problem is formulated by equations:

$$\frac{\hbar^2}{2\mu}\frac{d^2\chi_1^0}{dR^2} + [E + F_1(R - R_p)]\,\chi_1^0 = a\,\chi_2^0$$

$$\frac{\hbar^2}{2\mu}\frac{d^2\chi_2^0}{dR^2} + [E + F_2(R - R_p)]\,\chi_2^0 = a\,\chi_1^0 \tag{4.3a}$$

where $E = \mu v^2/2$ is the total energy referred to the pseudo-crossing point. Here χ_k^0 are functions which are used in the wave-mechanical expansion similar to (2.11). Namely, in the two-state approximation the non-adiabatic wave function $\Psi\,(r,\,R,\,t)$ is expressed in the form:

$$\Psi(r, R, t) = [\chi_1^0(R)\,\varphi_1^0(r, R) + \chi_2^0(R)\,\varphi_2^0(r, R)]\,\exp(-iEt/\hbar). \tag{4.3b}$$

By applying the Laplace-transform to (4.3) it has been proved [10] that the wave equations (4.3) are exactly equivalent to the semiclassical equations (2.13) with the trajectory (4.2), if the effective force F is defined by

$$F = (F_1 F_2)^{1/2} \tag{4.4}$$

and if F_1 and F_2 have the same sign (positive or negative); the case of different signs is considered in section 5. This means also that initial conditions (3.2) are equivalent to the boundary conditions appropriate for (4.3): ingoing and outgoing waves in channel 1, and outgoing wave in channel 2. We shall not discuss further the reduction of (4.3) to (2.13) and the trajectory (4.2) will be dealt with bearing in mind that with definition (4.4) the semiclassical approximation is identical with an exact quantum solution. We consider now various limiting cases, where the transition probability can be found in the closed form.

In this connection we introduce two dimensionless parameters

$$\varepsilon = \frac{E}{\Delta E_p}\frac{\Delta F}{F}, \qquad b = \left(\frac{\Delta E_p}{\varepsilon_0}\frac{F}{\Delta F}\right)^{3/2} \tag{4.5}$$

where

$$E = \frac{\mu v^2}{2}, \quad \Delta E_p = |2a|, \quad \varepsilon_0 = [\hbar^2 F^4/2\mu\,\Delta F^2]^{1/3}. \tag{4.6}$$

Weak adiabatic coupling.

Consider first the case, where perturbation treatment of (2.13) may be used. To the first order we have

$$P = \left|\int_{-\infty}^{\infty}\frac{a}{i\hbar}\exp\left[\frac{i}{\hbar}\Delta F\int^t x(t)\,dt\right]dt\right|^2 = \pi^2 b^{4/3}[Ai(-\varepsilon b^{2/3})]^2 \tag{4.7}$$

where $Ai(z)$ denotes the Airy function defined by

$$Ai(z) = \frac{1}{2\pi} \int_{-\infty}^{\infty} \exp\left[iz u + \frac{i}{3} u^3\right] du \; . \qquad (4.8)$$

Using asymptotic expansions of $Ai(z)^2$ we get the following expressions for P for two limiting cases

$$P = \begin{cases} = \dfrac{\pi b}{\sqrt{\varepsilon}} \sin^2\left[\dfrac{2}{3}\varepsilon^{3/2}b + \dfrac{\pi}{4}\right] = \dfrac{8\pi a^2}{\Delta F \hbar v}\sin^2\left[\dfrac{2}{3}\left(\dfrac{E}{\varepsilon_0}\right)^{3/2} + \dfrac{\pi}{4}\right], \\[4mm] \qquad\qquad\qquad\qquad\qquad\qquad\qquad \varepsilon\, b^{2/3} = \dfrac{E}{\varepsilon_0} \gg 1 \\[4mm] = \dfrac{\pi b}{4\sqrt{|\varepsilon|}} \exp\left[-\dfrac{4}{3}|\varepsilon|^{3/2}b\right] = \dfrac{2\pi a^2}{\Delta F \hbar v}\exp\left[-\dfrac{4}{3}\left|\dfrac{E}{\varepsilon_0}\right|^{3/2}\right], \\[4mm] \qquad\qquad\qquad\qquad\qquad\qquad\qquad \varepsilon\, b^{2/3} = \dfrac{E}{\varepsilon_0} \ll -1 \end{cases} \qquad (4.9)$$

In the intermediate region of $\varepsilon\, b^{2/3}$ the dependence of P on $\varepsilon\, b^{2/3}$ is shown in Fig. 3 (curve 1).

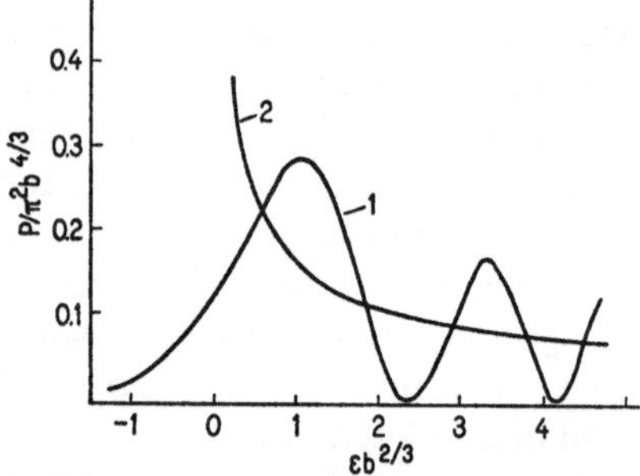

Fig. 3. Transition probability at weak adiabatic coupling influenced by the turning point (curve 1), eq. (4.7). Curve 2 is the Landau-Zener approximation (3.12)

Two features of P deserve discussion. When the nuclear energy E at the point of crossing of zero-order terms (or at the point of pseudo-crossing of adiabatic terms) is positive, P oscillates. These oscillations are due to the interference between two waves on *each* term, one representing the initial motion of nuclei, and another the motion after non-adiabatic transition in the coupling region. And the trajectory (4.2) does manage to reproduce the motion before and after turning points are reached. Averaging over oscillation represents the replacement

$$\sin^2\left[\frac{2}{3}\left(\frac{E}{\varepsilon_0}\right)^{3/2} + \frac{\pi}{4}\right] \rightarrow \frac{1}{2} \, , \qquad (4.10)$$

---

[2] The function $\Phi(z)$ introduced in the book of ref. 3 is $\pi^{1/2}$ times $Ai(z)$.

and then we come back to eq. (3.12) at the limit of small a. Curve 2 in Fig. 3 shows this averaged transition probability extrapolated to small values of $\varepsilon\, b^{2/3}$.

When the nuclear energy E at $R = R_p$ is negative, P diminishes exponentially. The non-zero value of P at negative energies is due to quantum-mechanical tunnelling.

The first-order perturbation treatment is valid under condition $P \ll 1$ which means that

$$b \ll 1 . \tag{4.11}$$

The region of parameters b and ε where eq. (4.7) and (4.9) are valid is denoted by I in Fig. 4. Shown are also positions of complex roots t_c of eq. (3.8) for the trajectory (4.2) which are of particular importance in the case of strong adiabatic coupling.

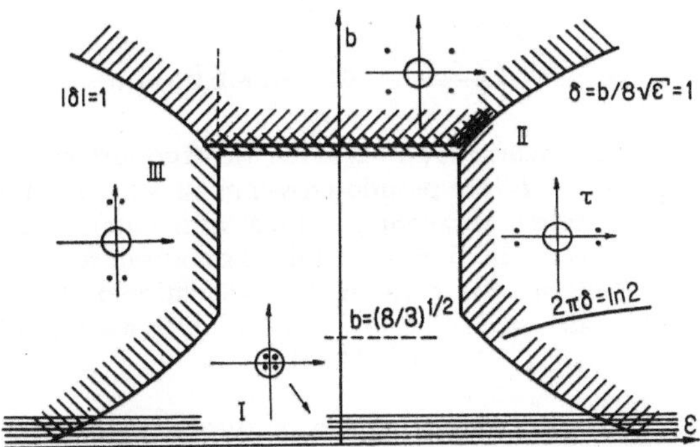

Fig. 4. Regions of applicability of various limiting formulae for the probability of a non-adiabatic transition. The Landau-Zener transition probability on curve $2\pi\delta = \ln 2$ attains a maximum value $P = {}^1/_2$. The dashed line corresponds to the value that was used for calculation of P in fig. 7. Shown are also branching points τ_c at the Riemann surface of $\Delta E(\tau)$ for each region considered. The circle embraces regions where $|s| = \left| \int\limits_0^{t_c} \Delta E(t)\, \dfrac{dt}{\hbar} \right|$ is lower than unity

Strong adiabatic coupling.

Similarly to (3.9) by using the Landau method we obtain

$$P = B \exp\left[-\frac{2}{\hbar}\, Im \int\limits_0^{t_c} [[\Delta F \varkappa(t)]^2 + 4\, a^2]^{1/2}\, dt \right] \tag{4.12}$$

where $\varkappa(t)$ is given by (4.2). In each half-plane of the complex variable t there are two branching points corresponding to double crossing of the region $R \sim R_p$. Introducing a dimensionless time $\tau = 2\,a\,t/\hbar\,b$ the exponent in (4.12) becomes

$$\frac{2}{\hbar} Im \int_0^{tc} \Delta E(t)\,dt = b\,\Delta(\varepsilon) = b\,Im\,2\int_0^{\tau_{c1}} [1 + (\tau^2 - \varepsilon)^2]^{1/2}\,d\tau \ . \quad (4.13)$$

Possible branching points are (see Fig. 5)

$$\left.\begin{array}{c} \tau_{c1} \\ \tau_{c2} \\ \tau_{c3} \\ \tau_{c4} \end{array}\right\} = \pm\,(\varepsilon \pm i)^{1/2} = \left\{\begin{array}{c} \tau' + i\,\tau'' \\ -\tau' + i\,\tau'' \\ -\tau' - i\,\tau'' \\ \tau' - i\,\tau'' \end{array}\right. \quad \tau', \tau'' > 0 \ . \quad (4.14)$$

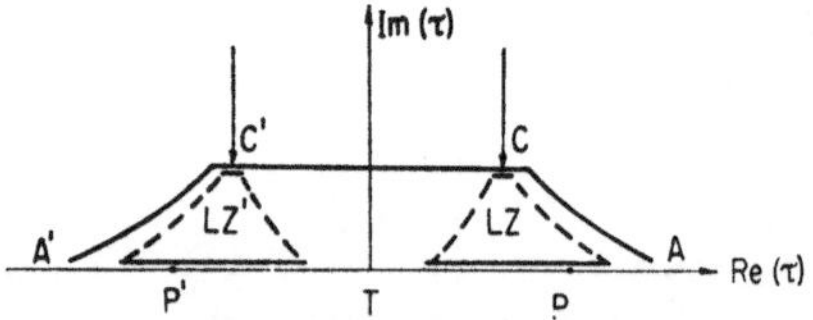

Fig. 5. Specification of characteristic points of the Riemann surface of $\Delta E = \Delta E(\tau)$: T is the turning point, P, P' are pseudo-crossings of adiabatic terms, C, C' are crossings of adiabatic terms (branching points with cuts), LZ (dashed) is a contour used to calculate P_{12} in (3.9), LZ (full) is a contour for calculation of P_{12} (3.10); $LZ + LZ'$ (full) are two contours for calculation of P (3.12) and (4.24), $Re\,(\tau)$ is a contour for calculation of P (4.7), $ACC'A'$ is a contour for calculation of P (4.16)

We may anticipate that the existence of two branching points in each half-plane will have an effect on the B factor in (4.9) because of interference, and B will be different from 2. To take interference into account (2.12) should be integrated over a path passing near τ_{c1} and τ_{c2}. This choice of a path leads to simplification of the general problem, as near to τ_{c1} and τ_{c2}, where the non-adiabatic coupling is most efficient, the following reduction of the fourth degree polynom may be made

$$1 + (\tau^2 - \varepsilon)^2 \approx (\tau - \tau_{c1})(\tau - \tau_{c2})(\tau_{c1} - \tau_{c4})(\tau_{c2} - \tau_{c3})$$
$$= -4\,(\tau'')^2\,[(\tau - i\,\tau'')^2 - (\tau')^2] \ . \quad (4.15)$$

Proper calculation gives [10]

$$P = 2\pi\,\delta\,\frac{\sin^2 \pi\,\delta}{\pi^2}\,[\Gamma(\delta)]^2\,\delta^{-2\delta}\exp(2\delta)\cdot\exp(-b\,\Delta(\varepsilon)) \quad (4.16)$$

where $\delta = b/8\,\sqrt{|\varepsilon|}$.

The conditions for the strong adiabatic coupling and the expansion to be valid are

$$\varepsilon \ll -1 , \quad \varepsilon\,b^{2/3} \ll -1 \ . \quad (4.17)$$

For this region of ε and b we find from (4.16)

$$P = \begin{cases} 2\pi\delta \exp\left[-b\,\Delta(\varepsilon)\right] = \dfrac{\pi b}{4\sqrt{|\varepsilon|}} \exp\left[-\dfrac{4}{3}|\varepsilon|^{3/2}b\right], & \delta \ll 1 \qquad (4.18) \\[2em] 4\sin^2\pi\delta \exp\left[-b\,\Delta(\varepsilon)\right] = 4\sin^2\dfrac{\pi b}{8\sqrt{|\varepsilon|}} \exp\left[-\dfrac{4}{3}|\varepsilon|^{3/2}b\right], & \delta \gg 1\;. \\[0.5em] & \qquad (4.19) \end{cases}$$

If we are not interested in the interference pattern which becomes very oscillating at $\varepsilon \to 0$, we may extrapolate (4.16) to the region $|\varepsilon| \ll 1$ and even to $\varepsilon > 0$ with $B = 2$, provided P still remains exponentially low. Thus we obtain

$$P = 2\exp\left[-b\,\Delta(\varepsilon)\right] \qquad (4.20)$$

at ε and b such that

$$b/8\sqrt{|\varepsilon|} \gg 1, \quad b \gg 1\;. \qquad (4.21)$$

Thus, in the whole region III (Fig. 4) P can be calculated making use of eq. (4.16), where the pre-exponential factor is different from 2, when $\varepsilon < -1$ and $b/8\sqrt{|\varepsilon|} \gg 1$.

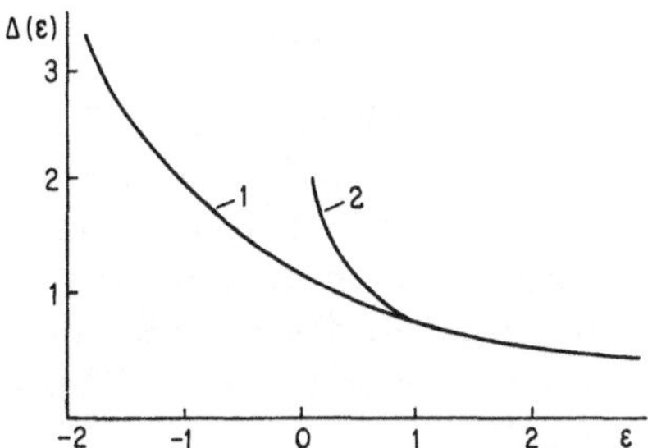

Fig. 6. Effect of the turning point on P. Curve 1: the function $\Delta(\varepsilon)$ that determines the transition probability for the case of exponentially low P (eq. 4.16, 4.18, 4.20). Curve 2: result of the Landau-Zener approximation (eq. 3.9), when the turning point is not taken into account

The function $\Delta(\varepsilon)$ defined by (4.13) is shown in Fig. 6 by curve 1. It is characterized by expansions

$$b\,\Delta(\varepsilon) = \begin{cases} \dfrac{b\pi}{4\sqrt{|\varepsilon|}}\left(1 - \dfrac{3}{32}\varepsilon^{-2} + \cdots\right), & \varepsilon \gg 1 \qquad (4.22\mathrm{a}) \\[1.5em] b\,(1.22 - 0.59\,\varepsilon + \cdots), & |\varepsilon| \ll 1 \qquad (4.22\mathrm{b}) \\[1.5em] \dfrac{4}{3}b\,|\varepsilon|^{3/2} + \cdots, & \varepsilon \ll -1\;. \qquad (4.22\mathrm{c}) \end{cases}$$

The Landau-Zener approximation (3.10) which is the first term in (4.22a) corresponds to curve 2.

The Landau-Zener formula

Let us now trace through the way the Landau-Zener formula emerges. We suppose that τ_{c1} and τ_{c4} (τ_{c2} and τ_{c3}) are closer than τ_{c1} and τ_{c2} (τ_{c3} and τ_{c4}). In this case, when integrating along the real axis, the following approximation may be used

$$
1 + (\tau^2 - \varepsilon)^2 =
\begin{cases}
(\tau')^2 \left[(\tau - \tau_{c2})(\tau - \tau_{c3}) \right] = (\tau')^2 \left[(\tau + \tau')^2 + (\tau'')^2 \right], \\
\qquad\qquad\qquad\qquad\qquad\qquad \tau \sim -\tau' \\
(\tau')^2 \left[(\tau - \tau_{c1})(\tau - \tau_{c4}) \right] = (\tau')^2 \left[(\tau - \tau')^2 + (\tau'')^2 \right], \\
\qquad\qquad\qquad\qquad\qquad\qquad \tau \sim \tau' .
\end{cases}
\tag{4.23}
$$

Then the system (2.13) will be equivalent to a twice repeated transit through the $R \sim R_p$ region with trajectory (3.1). Thus, neglecting the interference and referring to sect. 3, we may write

$$
P = 2 \exp(-2\pi\delta)\left[1 - \exp(-2\pi\delta)\right], \quad \delta = \frac{b}{8\sqrt{\varepsilon}} .
\tag{4.24}
$$

The conditions requiring a large phase difference and allowing for approximation (4.23) have the form

$$
\varepsilon \gg 1, \qquad \varepsilon\, b^{2/3} \gg 1 .
\tag{4.25}
$$

These conditions are considerably broader than (3.14). Derivation of the Landau-Zener formula given in sect. 3 assumes that the turning points on both adiabatic terms are far removed from the pseudo-crossing. However, (4.24) is valid also for cases when the conditions for the quasiclassical motion are violated. In particular (4.24) is valid for all energies ε, when one of the linear terms is of a zero slope ($F_2 = 0$). This has been studied in detail by Ovchinnikova [11].

Under condition (4.25) eq. (4.24) may be rewritten as

$$
P = 2 \exp\left[-b\,\Delta(\varepsilon)\right]\left\{1 - \exp\left[-b\,\Delta(\varepsilon)\right]\right\} .
\tag{4.26}
$$

This allows extrapolation to the region $|\varepsilon| \sim 1$, provided P remains exponentially low, as may be seen from comparison of (4.26) with (4.20). Thus in region II (Fig. 4) the transition probability may be calculated according to (4.26).

This completes the derivation of $P(\varepsilon, b)$ as a function of two parameters ε and b. It is instructive to note that the quantum equations (4.3) contain three dimensionless independent parameters, but only two of these are essential for the *transition probability* which is a squared modulus of the non-diagonal element of the scattering matrix $|S_{12}|^2$. The third parameter will enter the phases of matrix elements S_{ik}. For instance, phases S_i, defined by (3.14) include besides ε and b an additional parameter $\Delta F/F$. Various regions, where P is given in the closed form, overlap (Fig. 4).

This provides a fairly general outlook on $P(\varepsilon, b)$ at various conditions of interest from the standpoint of physics. The only region for which there is no analytical expression for P is that defined by

$$|\varepsilon| \sim 1, \quad b \sim 1 . \tag{4.27}$$

For these values of ε and b the system (2.13) has to be integrated numerically [10]. Fig. 7 gives an example of such a calculation for $b = (^8/_3)^{1/2}$. The oscillating curve 1 reproduces the effect of interference and tunnelling, and

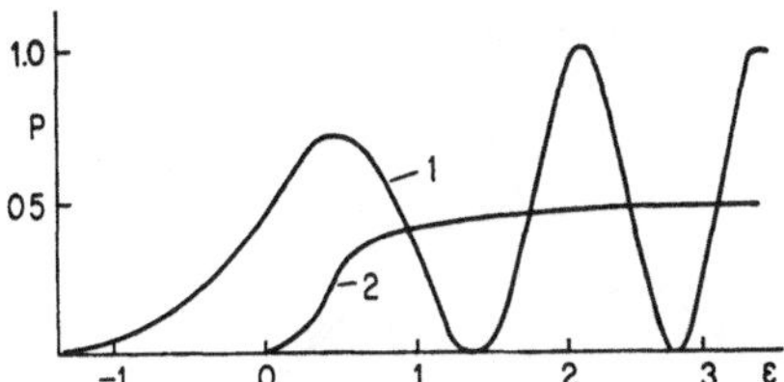

Fig. 7. Curve 1: Transition probability P at strong coupling and appreciable tunnelling $(b = (^8/_3)^{1/2})$. Curve 2: the Landau-Zener transition probability calculated for $b = (^8/_3)^{1/2}$

the monotonic curve 2 corresponds to the Landau-Zener approximation (3.12) with $b = (^8/_3)^{1/2}$. It may be seen that at large ε, when we leave the blank area of Fig. 4 and enter region II, the transition probability is approximated, in the average, by (3.12).

5. Non-Adiabatic Reflection from and Transmission through a Potential Barrier

If F_1 and F_2 are of a different sign, we have a case of transition over, or under, a potential barrier. A simple example is shown in Fig. 8. The upper adiabatic electronic term supports vibrational states that decompose due to non-adiabatic coupling with the lower term. If the coupling is neglected, the stationary states may be found from the Bohr rule:

$$J(E_n) = 2\pi \hbar (n + {}^1/_2), \quad J(E) = 2 \int_{x_1}^{x_2} (p(x)/\hbar)\, dx . \tag{5.1}$$

The period of vibration is defined as

$$T(E_n) = T_n = 2\pi / \omega(E_n) = \hbar \partial J / \partial E |_{E = E_n} . \tag{5.2}$$

From (5.2) and (5.1) we find

$$T_n = |F_1 F_2| (2\mu E_n)^{-1/2} \Delta F^{-1}, \qquad \varepsilon_n \gg 1 \tag{5.3a}$$

$$T_n = 2\pi (4a\mu)^{1/2} \Delta F^{-1}, \qquad \varepsilon_n - 1 \ll 1 . \tag{5.3b}$$

If the coupling is taken into account, the vibrational energy levels broaden. This manifests itself in the temporary capture of a particle *moving over the lower potential hump* by the upper potential well.

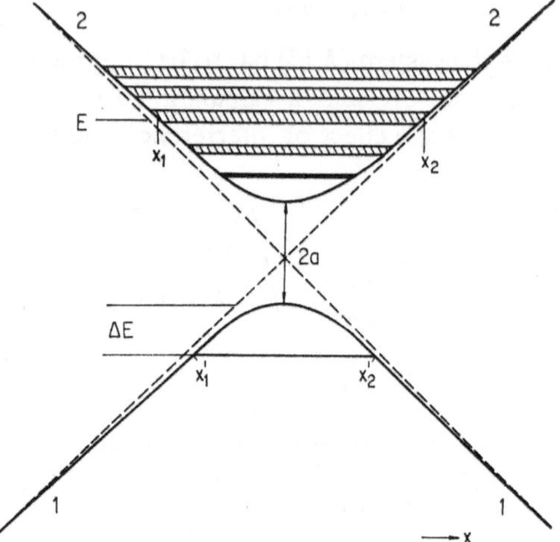

Fig. 8. Adiabatic and zero order terms operative in non-adiabatic transmission and reflection. The upper adiabatic term 2 supports vibrational states with complex eigenvalues $E_n - i\Gamma_n/2$

If definition of F in accordance with (4.4) is still retained for $F_1 F_2 < 0$ then F will be imaginary and the system (2.13) non-Hermitian. Thus, non-adiabatic transition of a particle over a potential well cannot be interpreted in terms of semiclassical approximation. It is in line with the impossibility of introducing a single trajectory describing both transmission and reflection. Consequently, we turn to quantum equations and look for a solution in the form of (1.4).

Let us introduce the integral transformation

$$\chi_m^0(x) = \frac{1}{\sqrt{\pi}} \int_C \exp\left[ik\left(x + \frac{F_1 + F_2}{2 F_1 F_2} E \right) + \right.$$

$$\left. + i\frac{k^3}{3} \cdot \frac{F_1 + F_2}{2 F_1 F_2} \frac{\hbar^2}{2\mu} \right] \cdot \frac{A_m(k)\,dk}{|F_m|^{1/2}} . \tag{5.4}$$

where C is a contour for which the integrand is zero at both ends.

Introducing a new variable $\xi = \left[|F_1 F_2| 2\mu / \Delta F \hbar^2 \right]^{1/3} k$ we get a system of equations

$$i\frac{dA_1}{d\xi} = -\frac{1}{2}\left(\xi^2 - \varepsilon b^{2/3} \right) A_1 - \frac{1}{2} b^{2/3} A_2$$

$$i\frac{dA_2}{d\xi} = \frac{1}{2}\left(\xi^2 - \varepsilon b^{2/3} \right) A_2 + \frac{1}{2} b^{2/3} A_1 , \tag{5.5}$$

which replace now eq. (2.13). Here ε and b are defined by (4.5) with $F = |F_1 F_2|^{1/2}$. The coefficients $A_m(\xi)$ satisfy the condition

$$|A_1(\xi)|^2 - |A_2(\xi)|^2 = \text{const} \tag{5.6}$$

that corresponds to flux conservation. The minus sign in (5.6) stems from a rather inconvenient representation: nuclear functions χ_m^0 refer to zero order states φ_m^0 rather than to adiabatic electronic states. It can be shown [11] that the boundary conditions

$$|A_1(-\infty)| = 1, \quad A_2(-\infty) = 0, \tag{5.7}$$

which are equivalent to the initial conditions (3.2), lead to the following expressions for the transmission coefficient P and the reflection coefficient Q:

$$P = |A_2(+\infty)/A_1(+\infty)|^2 \tag{5.8a}$$

$$Q = |1/A_1(+\infty)|^2, \tag{5.8b}$$

so that $P + Q = 1$.

Consider now various limiting cases for which P and Q can be calculated in the closed form.

Small Adiabatic Splitting

If $b \ll 1$, the system (5.5) can be solved to the first order, giving

$$P = \pi^2 b^{4/3} [Ai(-\varepsilon b^{2/3})]^2, \tag{5.9}$$

which is identical to (4.7).

Overbarrier Non-Adiabatic Transmission and Reflection

At $\varepsilon \gg 1$ there are two regions which induce reflections and which correspond to $\xi_{1,2} = \pm \varepsilon^{1/2} b^{1/3}$.

Solving system (5.5) close to these points and matching the solution with boundary conditions (5.7) we find [11]

$$P = \frac{1 + \cos J(E)}{1 + \lambda + \cos J(E)}, \quad Q = \frac{\lambda}{1 + \lambda + \cos J(E)} \tag{5.10}$$

$$\lambda = \exp[-4\pi\delta(E)]/2[1 - \exp[-2\pi\delta(E)]]$$

where, as above, $\delta(E) = b/8\varepsilon^{1/2}$ and $J(E)$ is defined by integral (5.1). It is seen from (5.10) that Q attains its maximum, $Q = 1$, when $\cos J = -1$, i.e. when $J = 2\pi(n + 1/2)$. This corresponds to resonance reflection of a particle captured by the upper potential well. To find out the shape of the resonance near its maximum at $\lambda \ll 1$, we develop $\cos(J)$ in the series

$$\cos J(E)|_{E\sim E_n} = -1 + \frac{1}{2}\left[\frac{\partial J}{\partial E}\Big|_{E=E_n}(E-E_n)\right]^2 + \cdots$$
$$= -1 + \frac{1}{2}\left[\frac{T_n}{\hbar}(E-E_n)\right]^2 . \tag{5.11}$$

Thus, for Q close to its maximum we obtain the anticipated Breit-Wigner formula

$$Q(E) = \frac{\Gamma_n^2/4}{(E-E_n)^2 + \Gamma_n^2/4}, \qquad \Gamma_n = \frac{2\hbar}{T_n}\exp\left[-2\pi\delta_n\right] . \tag{5.12}$$

Resonances are extremely sharp for E close to energy E_n of one of the lowest vibrational states in the upper well, where $\delta_n \gg 1$.

If the resonance structure of P and Q is of no interest, we may average (5.10) over an interval $\Delta J = 2\pi N$ comprising N resonances. The result is

$$\bar{P} = \int_{J-\pi N}^{J+\pi N} P(J)\,\frac{dJ}{2\pi N} = \left[1 - \exp(-2\pi\bar\delta)\right]\Big/\left[1 - \frac{1}{2}\exp(-2\pi\bar\delta)\right] \tag{5.13a}$$

$$\bar{Q} = \int_{J-\pi N}^{J+\pi N} Q(J)\,\frac{dJ}{2\pi N} = \frac{1}{2}\exp(-2\pi\bar\delta)\Big/\left[1 - \frac{1}{2}\exp(-2\pi\bar\delta)\right] \tag{5.13b}$$

where $\bar\delta$ refers to some average energy $\bar{E}$. The average value $\bar{P}$ coincides with that found by mere summation of fluxes calculated with a probability P_{12} for down-up transitions. Taking into account multiple reflections of a particle from turning points x_1 and x_2, and non-adiabatic transitions at $x = 0$, we find

$$\bar{P} = (1 - P_{12}) + P_{12}^2(1 - P_{12})\sum_{n=0}^{\infty}(1 - P_{12})^{2n} = (1 - P_{12})\Big/\left(1 - \frac{1}{2}P_{12}\right) . \tag{5.14}$$

Introducing the Landau-Zener expression (3.10) for P_{12}, we arrive at (5.13a). It is instructive to compare (5.14) with an expression for the transmission coefficient χ in chapter 16 of the book [2]. If there is a probability ϱ of reflection before achieving the activated state, and the same probability ϱ when leaving it, χ is found to be [2]

$$\chi = (1 - \varrho)/(1 + \varrho) , \tag{5.15}$$

which is different from (5.14).

It may be seen from (5.13) that at $E \to \infty$ the transmission decreases and the reflection increases. Such a behaviour is just opposite to what is anticipated for P and Q for ordinary (one-term) potential barrier. This conclusion may be of interest for chemical reactions, where the potential hump representing the activation energy always occurs together with an upper potential well overhanging it.

The usual nature of transmission and reflection is displayed for rather small energies and large adiabatic splitting, when non-adiabatic coupling is unimportant.

Underbarrier Non-Adiabatic Transmission (Tunnelling) and Reflection

If the top of a potential hump is not accessible classically, P will be different from zero only due to tunnelling. For one electronic term and an exponentially low transmission coefficient P is given by [3]

$$P = \exp\left[-2 \int_{x_1'}^{x_2'} [\,|p(x)|/\hbar]\, dx\right], \qquad (5.16)$$

where the integral is taken over a classically inaccessible region (Fig. 8). In order to take into account the effect of non-adiabatic coupling with the upper term, let us consider eigenvalues $v(\xi)$ of the matrix at the r.h.s. of (5.5). These are given by

$$v(\xi) = \pm \frac{1}{2} \left[(\xi^2 - \varepsilon\, b^{2/3})^2 - b^{4/3}\right]^{1/2}. \qquad (5.17)$$

There are four branching points of $v(\xi)$:

$$\xi = \pm\, b^{1/3} (\varepsilon \pm 1)^{1/2} \qquad (5.18)$$

that lie on real or imaginary axes of the complex variable ξ. When b is small, the four branching points are taken into account to the first order perturbation theory (5.9) by integrating (5.5) along real axes. When ε is high and positive, the four branching points are taken into account by pairs that lie near $\pm\, \varepsilon^{1/2}\, b^{1/3}$ (5.10). When ε is high and negative, as for tunnelling far from the top, we may account for one pair of branching points only, lying on the imaginary axis of ξ. The integration path is displaced to the upper half-plane of ξ so that it passes near points $\xi_1 = i\, b^{1/3} (|\varepsilon| + 1)^{1/2}$ and $\xi_2 = i\, b^{1/3} (|\varepsilon| - 1)^{1/2}$. To this approximation the solution of (5.5) is expressed by the function of parabolic cylinder [11], and the result for P is

$$P = B \exp\left[-2\, b\, Im \int_0^{z_c} [(z^2 + |\varepsilon|)^2 - 1]^{1/2}\, dz\right], \qquad (5.19a)$$

$$B = \frac{2\pi}{\delta\, |\Gamma(\delta)|^2}\, (\delta)^{\delta/2} \exp(-\delta). \qquad (5.19b)$$

It may be easily shown that the exponent in (5.19a) is identical to that in (5.16). Thus, the exponential term is the same for adiabatic and non-adiabatic tunnelling. The only difference lies in the pre-exponential $B(\delta)$ for which we obtain

$$
B(\delta) = \begin{cases} 2\pi\delta = \dfrac{\pi}{4}\,b\,|\varepsilon|^{-1/2} & \text{at} \quad \delta \ll 1 & (5.20\,\mathrm{a}) \\[2ex] 1 & \text{at} \quad \delta \gg 1 \, . & (5.20\,\mathrm{b}) \end{cases}
$$

It is seen that at $\delta \ll 1$ non-adiabatic coupling is essential, and it lowers the transmission coefficient as compared to that for adiabatic (one-term) tunnelling (for which B is always unity). If another condition $b^{2/3}\,|\varepsilon| \gg 1$ is fulfilled besides $|\varepsilon| \gg 1$, (5.19) reduces to

$$
P = \frac{\pi}{4}\,b\,|\varepsilon|^{-1/2}\exp\left[-\frac{4}{3}\,b\,|\varepsilon|^{3/2}\right]\,. \tag{5.21}
$$

This result can also be obtained from (5.9) if the asymptotic expansion of the Airy function is used.

Adiabatic Reflection from a Potential Hump Corresponding to the Lower Electronic Term

Consider now the case when the energy E is close to the top of the lower potential. Introducing a new parameter $\Delta\varepsilon = \varepsilon + 1$ and assuming that $|\Delta\varepsilon| \ll 1$, we have two branching points $\xi_{2,4} = \pm\, b^{1/3}\,(\Delta\varepsilon)^{1/2}$ near the origin, and two points $\xi_{1,3} = \pm\,\sqrt{2}\,i\,b^{1/3}$ rather far from it. It is clear from the above that ξ_1 (or ξ_3) contribute only to the pre-exponential factor B and its effect vanishes when ξ_1 (or ξ_3) move out along the imaginary axis. Instead ξ_2 (or ξ_4) contribute to the exponential factor defining the value of the integral in (5.16). Thus, taking into account the combined effect of two branching points ξ_2 and ξ_4, and completely neglecting the influence of ξ_1 and ξ_3, we shall find the transmission coefficient for the lower adiabatic potential. The solution of this problem is well known [3]

$$
P = [1 - \exp(-2\pi\,\Delta E/\hbar\,\omega^*)]^{-1}, \quad |\Delta\varepsilon| \ll 1 \tag{5.22}
$$

where ω^* is a frequency of small vibrations in the potential well, obtained by reversing the hump and approximating it close to its minimum by a parabola. In our case ω^* is identical to $2\pi/T$ defined by (5.3b). Eq. (5.22) goes over to (5.19) and (5.20), when $2\pi\,\Delta E/\hbar\,\omega^* \ll -1$.

6. Mean Transition Probability

For thermal collision processes of main interest will be the evaluation of the rate constant K for non-adiabatic transition induced by thermal motion of nuclei. The rate constant $K_{12}(T)$ is defined via cross-section $\sigma_{12}(v_\infty)$ by

$$
K_{12} = \left(\frac{\mu}{2\pi kT}\right)^{3/2}\int_0^\infty \sigma_{12}(v_\infty)\,v_\infty\exp(-\mu v_\infty^2/2kT)\,4\pi v_\infty^2\,dv_\infty \tag{6.1}
$$

where integration is carried out over the relative velocity v_∞ of collision partners at infinity. In its turn the cross-section $\sigma_{12}(v_\infty)$ is defined by the probability $P(v_\infty, b)$ depending on the impact parameter b:

$$\sigma_{12}(v_\infty) = 2\pi \int_0^\infty P(v_\infty, b) b \, db \ . \qquad (6.2)$$

In atomic collisions, when the projection of electronic angular momentum on the internuclear axis in both states is zero, $P(v_\infty, b)$ depends only on radial velocity v_R (the so-called $\sigma - \sigma$ transitions), i.e. $P(v_\infty, b) = P[v(1 - b^2/R_p^2)^{1/2}]$ where v is now referred the pseudocrossing energy E_0. Inserting (6.2) into (6.1) and making use of this dependence of P on v_∞ and b, we may present K_{12} as an average one dimensional flux through the critical surface $R = R_p$:

$$K_{12} = \frac{\int_S dS \int_0^\infty P(v) v \exp[-\mu v^2/2kT] \exp[-E_0/kT] \, dv}{\int_{-\infty}^\infty \exp[-\mu v^2/2kT] \, dv} \qquad (6.3)$$

where dS is an element of the surface. Eq. (6.3) can also be deduced from the transition state method [2], if R is considered as a reaction coordinate. This suggests that the one-dimensional model may be used for many problems at a proper choice of the reaction coordinate.

The pseudocrossing energy E_0 in (6.3) is referred to energy E_i of the initial state $(R \to \infty)$. Thus it may be readily seen that rate constants for direct and reverse processes obey the principle of detailed balancing

$$\frac{K_{12}}{K_{21}} = \exp\left[\frac{-E_2(\infty)}{kT} + \frac{E_1(\infty)}{kT}\right] . \qquad (6.4)$$

It would be worthwhile to define the mean transition probability $\langle P \rangle$ referred to one gas-kinetic collision of a radius R_p. Integrating over dS we obtain $4\pi R_p^2$. Then introducing the average one-dimensional velocity $\langle v \rangle = (kT/2\pi\mu)^{1/2}$ we define $\langle P \rangle$ by equation

$$K = 4\pi R_p^2 \langle v \rangle \langle P \rangle \exp\left[\frac{-E_0}{kT}\right] = \pi R_p^2 \left(\frac{8kT}{\pi\mu}\right)^{1/2} \langle P \rangle \exp\left(\frac{-E_0}{kT}\right) . \ (6.5)$$

If $K(T)$ is known, $\langle P \rangle$ may be easily found, and $\sigma(v_\infty)$ can be calculated from (6.2) by the inverse Laplace-transform.

Now let us consider the formulae for $\langle P \rangle$ in various limiting cases discussed in sect. 4.

For the *weak-coupling* approximation the integration (6.1) with the Airy function can be accomplished in the closed form

$$\langle P \rangle = \frac{2\sqrt{\pi}\,\pi a^2}{\hbar \Delta F (2kT/\mu)^{1/2}} \exp\left[\frac{1}{12}\left(\frac{\varepsilon_0}{kT}\right)^3\right] . \qquad (6.6)$$

Here $\langle P \rangle$ consists of two factors. The first (pre-exponential) factor is the mean probability for overbarrier transition (at a relative energy $\dfrac{\mu v_\infty^2}{2} > E_0$).

The second (exponential) factor gives a correction for tunnelling $\left(\dfrac{\mu v_\infty^2}{2} < E_0\right)$. It may result in effective lowering of the activation energy. Thus at rather low temperatures, when $(\varepsilon_0/kT) \gg 1$, $\langle P \rangle$ may exceed unity. But this does not violate the obvious condition $K < \pi R_p^2 (8 \, kT/\pi \mu)^{1/2}$, as tunnelling is essential only when the activation energy is rather high and the combined exponential (the Arrhenius factor and tunnelling correction) is low, i.e. $\exp[-E_0/kT + (^1/_{12})(\varepsilon_0/kT)^3] \ll 1$.

If the matrix element a (supposed to be independent of R) depends on other coordinates, the extra temperature dependence may come out in (6.6). It can be found by averaging over the other coordinates, independently of the averaging implied by (6.1). An important example is the so-called $\sigma - \pi$ transition between two molecular terms, where non-adiabatic coupling is of the form

$$- i \hbar \frac{\partial}{\partial t} = - i \hbar \dot\varphi \left(\frac{\partial}{\partial \varphi}\right) = - \omega \hat{J}_x \ . \tag{6.7}$$

Here φ is the angle of rotation of the interatomic axis in the plane normal to the vector of relative angular momentum M, ω is the angular velocity, and $\hat{J}_x$ is the projection of electronic angular momentum on M. Then a^2 will be

$$a^2 = \omega^2 \, |(\hat{J}_x)_{12}|^2 \, ,$$
$$\langle a^2 \rangle = \langle \omega^2 \rangle |(\hat{J}_x)_{12}|^2 = \frac{2 \, kT}{\mu \, R_p^2} |(\hat{J}_x)_{12}|^2 \ . \tag{6.8}$$

At *strong* coupling the Landau-Zener formula may be used for calculation of $\langle P \rangle$ over a large region of ε and b parameters, which makes it possible to calculate $\langle P \rangle$ in a comparatively simple way.

Introducing (3.12) in (6.1) we find

$$\langle P \rangle = 2 \, F(\gamma) - 2 \, F(2\gamma) \tag{6.9}$$

where

$$F(\gamma) = \int_0^\infty \exp\left(-\gamma/x - x^2\right) 2 \, x \, dx \ ,$$
$$\gamma = \frac{2 \pi a^2}{\Delta F \hbar} \left(\frac{\mu}{2 \, kT}\right)^{1/2} \ . \tag{6.10}$$

The function $\langle P \rangle$ shown in Fig. 9 is characterized by expansions

$$\langle P \rangle = \begin{cases} 2\sqrt{\pi}\,\gamma, & \gamma \ll 1 & (6.11\,\mathrm{a}) \\[2mm] \langle P \rangle_{\max} = 0.427, & \gamma = 0.50 & (6.11\,\mathrm{b}) \\[2mm] 4\sqrt{\dfrac{\pi}{3}}\,(\gamma/2)^{1/3} \exp\left[-3\,(\gamma/2)^{2/3}\right], & \gamma \gg 1 \ . & (6.11\,\mathrm{c}) \end{cases}$$

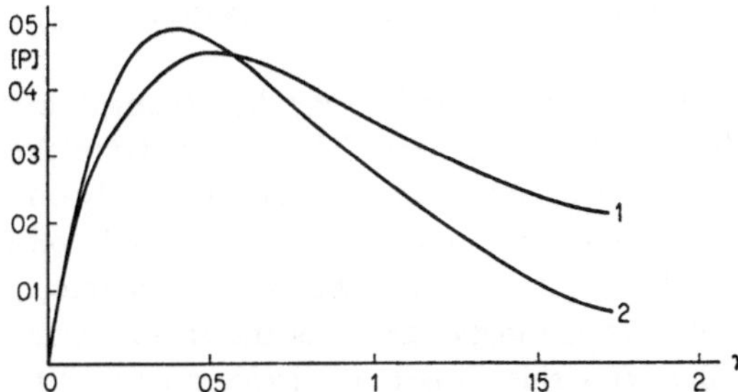

Fig. 9. Mean transition probability $\langle P \rangle$ for strong coupling versus $\gamma = (2\pi a^2/\Delta F \hbar) \cdot (\mu/2kT)^{1/2}$ (curve 1), the Landau-Zener transition probability versus $\gamma = 2\pi a^2/\Delta F \hbar v$ (curve 2)

Thus, it may be seen that (6.8) reduces to (6.6) when following conditions are satisfied

$$\gamma \ll 1 , \qquad \varepsilon_0/kT \ll 1 . \tag{6.12}$$

These imply that the first order perturbation treatment is valid and tunnelling may be neglected.

At *strong adiabatic* coupling the probability P is exponentially low and (6.1) can easily be calculated by the steepest descent method. The integrand shows a sharp maximum at a velocity v^* which is defined by

$$- b\Delta(v^*) - \mu(v^*)^2/2kT = \min . \tag{6.13}$$

Practically it is sufficient to use the first term of expansion (4.22a), so that the minimization condition takes the form

$$\frac{d}{dv}\left(\frac{2\pi a^2}{\Delta F \hbar v} + \frac{\mu v^2}{2kT} \right)\Bigg|_{v = v^*} = 0 . \tag{6.14}$$

Then v^* from (6.14) may be inserted into higher terms of expansion (4.22a) in order to calculate tunnelling corrections. This procedure gives

$$\langle P \rangle = 4\sqrt{\frac{\pi}{3}}\,(\gamma/2)^{1/3} \exp\left[- 3(\gamma/2)^{2/3} + (3/16)(\varepsilon_T)^{-2}(\gamma/2)^{-2/3} + \cdots \right], \tag{6.15}$$

$$\varepsilon_T = kT\Delta F/(\Delta E_p F) .$$

Eq. (6.15) is valid when $\gamma \gg 1$ and the second term in the exponent is low compared with the first. However, it may be higher than unity. If the tunnelling correction is neglected (the second term in the exponent of (6.15)), eq. (6.15) coincides with (6.11c).

7. Extension of the Landau-Zener Model to Multistate Systems

Consider now an extension of the Landau-Zener model on several electronic states that would allow obtainment of an analytical solution to the problem. Suppose that the zero order term H_{00} intersects a batch of zero-order terms H_{kk} $(k = 1, 2, \ldots n)$ with derivative $dH_{kk}/dx = F_k$ independent of x.

For semiclassical approximation it may be assumed without loss of generality that the slopes F_k are zero. This results from the fact that the transition probability in this approximation depends only on the difference of diagonal matrix elements. Suppose then that only off-diagonal matrix elements H_{0k} $(k = 1, 2 \ldots n)$ are different from zero, and that they are independent of x (this independence is the main feature of the linear model). Fig. 10 illustrates this situation. Circles represent the position of pseudocrossings, if adiabatic interaction (i.e. H_{0k}) is taken into account.

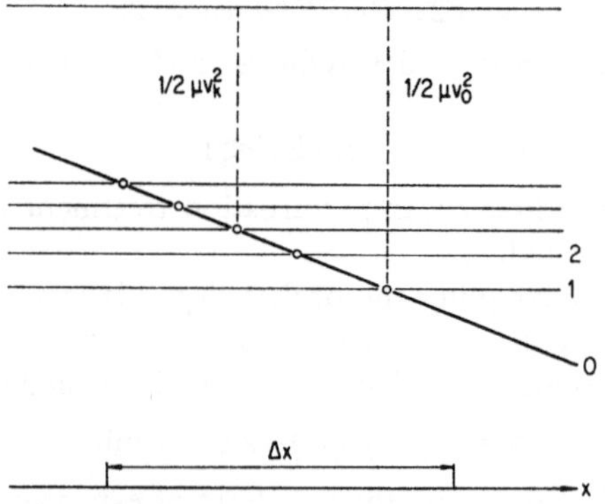

Fig. 10. Zero order terms for a multi-state model

Adiabatic wave functions φ_l are expressed as a linear combination

$$\varphi_l = \sum_{l'} C_{l'}(E_l)\,\varphi_{l'}^0 \tag{7.1}$$

where $C_{l'}$ satisfy the set of equations

$$\sum_m H_{km} C_m(E) = E C_k(E)\ . \tag{7.2}$$

Eigenvalues E_l and eigenvectors C_m for the model adopted are given by equations

$$(H_{00} - E_l) + \sum_{k=1}^{n} \frac{|H_{0k}|^2}{(E_l - H_{kk})} = 0\ , \tag{7.3a}$$

$$C_m(E_l) = \frac{H_{m0}}{E_l - H_{mm}} \left(\sum_{k=1}^{n} \frac{|H_{k0}|^2}{(E_l - H_{kk})^2} \right)^{-1}. \tag{7.3b}$$

Now let us look for a non-adiabatic wave function in the form

$$\Psi(t) = B_0(t)\varphi_0^0 + \sum_{k=1}^{n} B_k(t)\varphi_k^0 \tag{7.4}$$

taking as trajectory $x = x(t)$ the Landau-Zener approximation $x = v(t - t_0)$ where v is some average value of v_k (see Fig. 10). Substituting (7.4) into the wave equation, we obtain the following system of equations for B_k:

$$i\hbar \dot{B}_0 = -\Delta F v t B_0 + \sum_{k=1}^{n} H_{k0} B_k \qquad (7.5a)$$

$$i\hbar \dot{B}_k = H_{kk} B_k + H_{k0} B_0, \quad k = 1, 2, \ldots n \qquad (7.5b)$$

where t_0 is taken to be $t_0 = -H_{00}(x = 0)/\Delta F v$, and H_{k0} are supposed for simplicity to be real. Solutions of (7.2) have been studied by OVCHIN-NIKOVA [12] for $n = 2$, and by OSHEROV [13] and DEMKOV [14] for any n. Coefficients are expressed as countour integrals over an auxiliary variable ζ:

$$B_0(t) = A \int_C \exp\left[\frac{i\zeta^2}{2\Delta F v \hbar} + \zeta t\right] \prod_{m=1}^{n} (\zeta + i H_{mm}/\hbar)^{i\alpha_m} d\zeta, \qquad (7.6a)$$

$$B_k(t) = A \int_C \exp\left[\frac{i\zeta^2}{2\Delta F v t} + \zeta t\right] \frac{\prod\limits_{m=1}^{n} (\zeta + i H_{mm}/\hbar)^{i\alpha_m}}{(\zeta + i H_{kk}/\hbar)} d\zeta, \qquad (7.6b)$$

$$\alpha_m = H_{m0}^2/\Delta F v \hbar . \qquad (7.6c)$$

Integrals (7.6a) and (7.6b) identically satisfy (7.2b). Equation (7.5a) is satisfied if the countour C is chosen such that the integrand in (7.5a) is the same for both ends of C. The freedom still left for choosing the contour should be used to guarantee initial conditions imposed on $B_k(-\infty)$. The constant A is found from the normalization condition

$$\sum_{k=0}^{n} |B_k(t)|^2 = 1 . \qquad (7.7)$$

We shall not discuss here the details of calculation and will mention the main result only. The probability P_{lk} of transition from an initial state l to a final state k can be calculated as a probability of independent pair transitions each localized at the crossing point of terms H_{00} and H_{mm}. The probability of a pair transition between pseudocrossing terms, corresponding to the pair 0 and m is given by the Landau-Zener formula

$$p_m = \exp\left[\frac{-2\pi H_{0m}^2}{\hbar \Delta F v}\right] . \qquad (7.8)$$

Thus, for a single passing of the coupling region we have

$$P_{lk} = (1 - p_l) p_{l+1} p_{l+2} \cdots p_{k-1} (1 - p_k) \quad \text{for } \begin{array}{c} l \leqslant k \\ l \neq 0 \end{array}, \qquad (7.9a)$$

$$P_{0k} = p_1 p_2 \cdots p_{k-1} (1 - p_k) . \qquad (7.9b)$$

The probability $\mathscr{P}_0$ of transition from state φ_0^0 to all other states φ_k^0 is

$$\mathscr{P}_0 = \sum_{k=1}^{n} P_{0k} = 1 - p_1 p_2 \ldots p_n = 1 - \exp\left[-\frac{2\pi}{\Delta F \hbar v} \sum_{k=1}^{n} H_{0k}^2\right] . \quad (7.10)$$

The first order perturbation result follows from (7.7), if the exponent in (7.10) can be expanded in series

$$\mathscr{P}_0 \approx \frac{2\pi}{\hbar \Delta F v} \sum_{k=1}^{n} H_{0k}^2 . \quad (7.11)$$

The probability $P(k \to l)$ for a double passing of the coupling region, when a set of waves travelling from $x = \infty$ to the left is reflected somewhere, is given by

$$P(k \to l) = \sum_{m} P_{km} P_{ml} . \quad (7.12)$$

In (7.12) as in (3.11) all effects of interference between incoming and reflected waves have been neglected.

The semiclassical approximation applied to a linear model is valid when all the velocities v_k can be approximated by one v and the total length Δx of the coupling region is much shorter than the characteristic range $1/\alpha$. This imposes lower and upper energy limits for applicability of the model (same as in sect. 3). For low-energy processes neglecting the turning point effect is a severe limitation. To extend the applicability of a linear model to lower energies, we must consider quantum wave equations of the kind of (4.3). If F_k is taken as zero in these equations, the use of the Laplace-Transform will yield a system similar to (7.5). This suggests that quantum equations with $F_k = 0$ $(k = 1, 2, \ldots . n)$ can also be solved analytically. However, we shall not discuss it here.

8. Two-Dimensional Extension of the Landau-Zener Model

The most simple extension of the one-dimensional Landau-Zener model to many-dimensional systems is the two-dimensional case. For a two-state approximation the adiabatic electronic terms E_l are given as eigenvalues of a matrix $H_{el}(\varphi^0)$ (2.3). The same as in sect. 2 we suppose that H_{ik} are slow varying functions of two coordinates x and y, so that linear approximation is sufficient. The expansion similar to (2.5) takes now the form

$$H_{12}(x, y) = H_{12}^c + \frac{\partial H_{12}^c}{\partial x}(x - x_c) + \frac{\partial H_{12}^c}{\partial y}(y - y_c) + \cdots , \quad (8.1a)$$

$$H_{11} - H_{22} = \Delta H(x, y) = \Delta H^c + \frac{\partial \Delta H^c}{\partial x}(x - x_c) + \frac{\partial \Delta H^c}{\partial y}(y - y^c) + \cdots$$

$$(8.1b)$$

where superscript c means that the functions are calculated at fixed values $x = x_c$ and $y = y_c$ (e.g. $H_{12}^c = H_{12}(x_c, y_c)$). Bearing in mind that H_{12} is, in general, a complex quantity, and ΔH is real, we may define x_c and y_c from two real equations

$$\Delta H^c = \Delta H(x_c, y_c) = 0 \; ,$$
$$Re \, H_{12}^c = Re \, H_{12}(x_c, y_c) = 0 \; . \tag{8.2}$$

Taking x_c and y_c as the origin, we can now simplify expansion (8.1) by two linear transformations that are still at our disposal. One is a linear transformation of coordinates (rotation of the coordinate system) and another is a linear transformation of zero order functions. The latter is independent of the former, when the transformation matrix does not depend on x and y. Introducing new transformed variables X and Y referred to x_c and y_c and new transformed zero-order functions φ_1^0, φ_2^0, we achieve a simplification of (8.1) which yields the matrix

$$H_{el}(\varphi^0) = \begin{pmatrix} E_0 + \dfrac{1}{2} \Delta F_1 X & i\,A + \dfrac{1}{2} \Delta F_2 Y \\[2mm] -i\,A + \dfrac{1}{2} \Delta F_2 Y & E_0 - \dfrac{1}{2} \Delta F_1 X \end{pmatrix} , \; A = Im\,H_{12}^c \tag{8.3}$$

where ΔF_2 is supposed to be real[3].

The Landau-Zener approximation

$$X = V_1(t - t_1) \; ,$$
$$Y = V_2(t - t_2) \tag{8.4}$$

is taken for the trajectory. It is seen that on any trajectory of this type condition (2.4) is not fulfilled, so that φ_1^0 and φ_2^0 do not decouple at the limit $t \to \pm\infty$. But as the ratio $H_{12}(t)/\Delta H(t)$ at this limit tends to a constant, decoupling may be achieved by introducing a new set of zero-order functions $\overline{\varphi}_k^0$ for each trajectory determined by its own ratio V_1/V_2. For a fixed trajectory $H_{el}(t)$ can be further simplified as

$$H_{el}(t) = \begin{pmatrix} E_0 + \dfrac{1}{2} \mathcal{N} t & \mathcal{M} \\[2mm] \mathcal{M} & E_0 - \dfrac{1}{2} \mathcal{N} t \end{pmatrix} \tag{8.5}$$

---

[3] Usually the term $\dfrac{1}{2} \Delta F_2 Y$ stems from the Coulomb or exchange contribution to atomic or molecular interactions, so that this assumption may be accepted. Conversely, the $i\,A$ term is due mainly to spin-orbital coupling which may be considered as unchangeable at small displacements of nuclei.

where

$$\mathcal{N} = [(\Delta F_1 V_1)^2 + (\Delta F_2 V_2)^2]^{1/2} \,,$$

$$\mathcal{M} = \left[|A|^2 + \frac{[\Delta F_1 V_1 \Delta F_2 V_2 (t_1 - t_2)]^2}{(\Delta F_1 V_1)^2 + (\Delta F_2 V_2)^2} \right]^{1/2} . \tag{8.6}$$

The matrix (8.5) is formally equivalent to (2.7) if the trajectory (2.17) is introduced into (2.7). Then the transition probability P_{12} for this fixed trajectory can be written without actually solving equations (2.12) or (2.13)

$$P_{12} = \exp\left[-2 \pi \mathcal{M}^2 / \hbar \mathcal{N} \right] . \tag{8.7}$$

Let us interpret this expression in terms of adiabatic splitting ΔE_{12}. From (8.3) we find

$$\Delta E_{12} = [A^2 + \Delta F_1^2 X^2 + \Delta F_2^2 Y^2]^{1/2} . \tag{8.8}$$

It is seen that in the region of maximum approach of the energy surfaces these look like two paraboloids if $A \neq 0$. When $A = 0$ two surfaces intersect at one point $X = 0$, $Y = 0$ that is a vertex of a double cone. A well known theorem [3] is applicable to this case, i.e. when $Im\, H_{12} = 0$: The dimension of the multiplicity which represents the intersection of two adiabatic terms of the same symmetry is equal to $s - 2$, s being the number of degrees of freedom (see ref. 3, sect. 79).

It will be helpful to replace the time difference $t_1 - t_2$ by the distance l from the origin to the linear trajectory:

$$l = V_1 V_2 |t_1 - t_2| (V_1^2 + V_2^2)^{-1/2} . \tag{8.9}$$

If the behaviour of $\Delta E_{12}(t)$ is analyzed for this trajectory, it will be seen that $\mathcal{M}$ can be presented in the form

$$\mathcal{M} = \left[A^2 + \frac{1}{4} (\Delta F_1 V_1 \Delta F_2 V_2)^2 \frac{(V_1^2 + V_2^2)}{(\Delta F_1 V_1)^2 + (\Delta F_2 V_2)^2} \right]^{1/2}$$

$$= \min_L \Delta E(X, Y) . \tag{8.10}$$

Here the subscript L means that the minimum of $\Delta E(X, Y)$ is found on the trajectory L. Written as in (8.10) $\mathcal{M}$ is similar to ΔE_p for the one-dimensional case (sect. 2). This fact has been used by Teller [16] to calculate the mean two-dimensional transition probability for a particular case with $A = 0$ and $V_2 = 0$. Yet there is a difference for the one- and two-dimensional cases. It consists of the dependence of $\mathcal{M}$ on two additional parameters l and V_1/V_2 which define a trajectory in the coupling region.

For a two-dimensional case the representative point can pass the coupling region several times (as it does for the one-dimensional model). However, having crossed the coupling region once and moving then in two dimensions, it may fail to return by the same way to cross the coupling region again. Under these conditions it might be better to neglect completely the possibility of a double passing and to express the net probability of transition P by equation

$$P = P_{12} = \exp\left[\frac{-2\pi \mathcal{M}^2}{\hbar \mathcal{N}}\right] .$$

(8.11)

It replaces eq. (3.11) accounting for double passing by the same way. However, an approximation similar to (3.11) cannot be ruled out completely, as we shall see below.

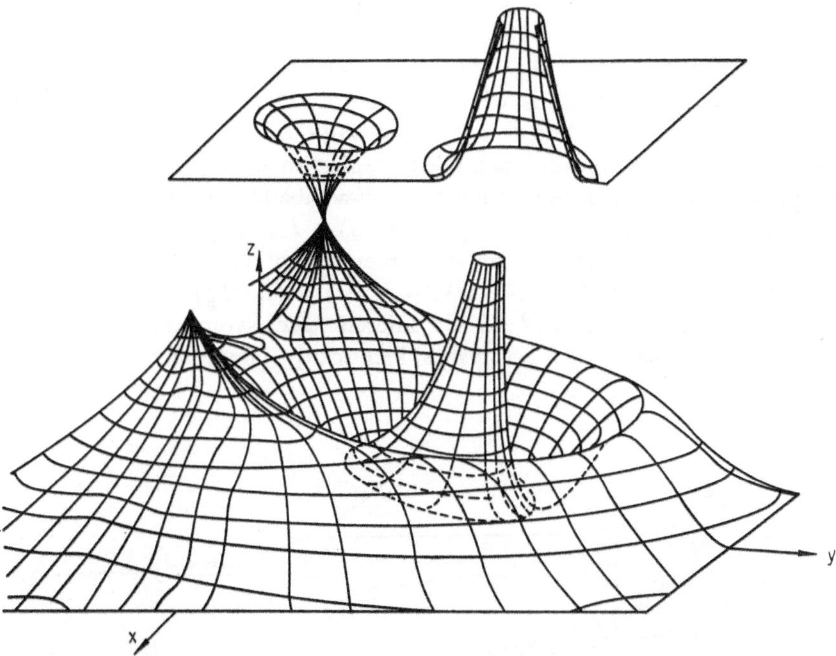

Fig. 11. Qualitative aspect of two lower adiabatic potentials for the H_3 system. Two protons are fixed at distances $\pm d$ on the y-axis. Shown are the cone intersections (effective in non-adiabatic coupling) and the saddle region (effective in exchange reactions)

Just now let us consider the validity of (8.11), taking as an example a system of three hydrogen atoms. The adiabatic electronic terms are shown schematically in Fig. 11. Two protons are fixed at distances $\pm d$ on the y-axis. The third can move in the x—y plane. When it is located on the x-axis and is separated from the other two by distances $2 d$ there will be crossing of terms at the vertex of the cone. (Here spin-orbital coupling is neglected, so that $A = 0$). The lower electronic term correlates adiabatically with $H_2(^1\Sigma_g^+) + H(^2S)$ and the upper with $H_2(^3\Sigma_u^+) + H(^2S)$;

both are of the same symmetry $^2A'$ of the point group C_S. The region of strong non-adiabatic coupling is located near the vertex and its characteristic length l^* can be estimated from equation

$$\frac{2\,\pi\,\mathcal{M}^2\,(l^*)}{\hbar\,\mathcal{N}} \approx 1 \,. \tag{8.12}$$

Taking $V_1 \sim V_2 \sim V$, $\varDelta F_1 \sim \varDelta F_2 \sim \varDelta \mathscr{E}\,\alpha$ where $1/\alpha$ is of the order of the Bohr radius, and $\varDelta\,\mathscr{E}$ is of the order of excitation $(^1\varSigma - {}^3\varSigma)$-energy, we find

$$l^* \approx \left(\frac{\hbar\,V\,\alpha}{\varDelta\,\mathscr{E}}\right)^{1/2}\left(\frac{1}{\alpha}\right) \,. \tag{8.13}$$

We see from Fig. 11 that in the $H_2 + H$ collision the probability of crossing the coupling region is of the order of l^*/d,[4] i.e. of the order $(1/\alpha\,d)\,(\hbar\,V\,\alpha/\varDelta\,\mathscr{E})^{1/2}$. Thus, it is the high value of the Massey parameter $(\varDelta\,\mathscr{E}/\hbar\,V\,\alpha \gg 1)$ that makes it possible to neglect the multiple passing of the coupling region, when calculating the non-adiabatic transition probability P in a process such as $H_2(^1\varSigma_g^+) + H(^2S) \to H_2(^3\varSigma_u^+) + H(^2S)$, or to neglect the non-adiabatic effects altogether, when calculating the rate of exchange reaction $H_2(^1\varSigma_g^+) + H(^2S) \to H(^2S) + H_2(^1\varSigma_g^+)$.

If the functions φ_1^0 and φ_2^0 are of a different space symmetry, the electrostatic contribution to H_{12} vanishes identically. Neglecting spin-orbital

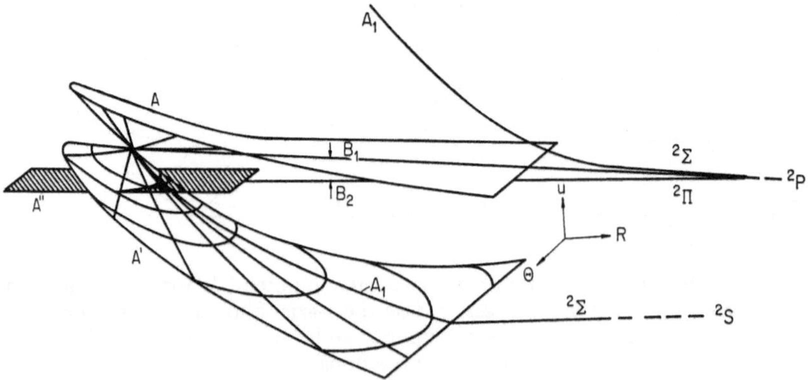

Fig. 12. Qualitative aspect of three lower adiabatic potentials for a $Na\binom{^2P}{^2S} +$ $+ N_2(^1\varSigma_g^+)$ system close to the symmetrical C_{2v} configuration. R is the intermolecular distance; the $N-N$ distance is fixed and angle $\Theta = 0$ corresponds to the C_{2v} configuration. $^2\varSigma$, $^2\varPi$ and $^2\varSigma$ are combined states arising from $^2P + {}^1\varSigma_g^+$ and $^2S + {}^1\varSigma_g^+$ for $C_{\infty v}$ symmetry; A_1, B_1, B_2 are irreducible representations of the C_{2v} point group; A' and A'' are symmetrical and antisymmetrical representations of the C_S point group. Shown are the cone and line intersections

---

[4] For thermal processes the probability is even less, as the energy at the vertex usually is higher than the mean thermal energy.

coupling, we have $H_{12} = 0$ and to the zero order two adiabatic surfaces will intersect along the line defined by

$$\Delta H(x, y) = 0 \ . \tag{8.14}$$

The expansion (8.1b) gives the tangent to this line at the point x_c, y_c. This situation is illustrated by Fig. 12 representing the qualitative behaviour of adiabatic electronic terms of a system $Na\left(^{2P}_{2S}\right) + N_2\left(^1\Sigma_g^+\right)$ close to the isosceles triangular configuration with a fixed internuclear distance R_{N-N}. There is a cone intersection for terms of the same symmetry, and line intersection for terms of different symmetry. If the spin-orbital coupling is taken into account, there will be intersection in both cases. But for the case of different space symmetry it is quite possible that in a $Na + N_2$ collision the system passes the coupling region (the pseudo-crossing line) twice, so that the approximation (3.11) rather than (8.11) would be better for this case.

In conclusion it will be emphasized that when the non-diagonal matrix element H_{12} is zero or very small, the weak dependence of φ_i^0 on x (and other coordinates), which has been neglected so far, must be taken into account. It is this dependence that manifests itself in $\sigma - \pi$ transitions in atomic collisions, or in transitions between terms of different space symmetry in molecular collisions. We shall not discuss this here, mentioning only that due to the very weak dependence of φ_i^0 on x, the corresponding transition probability may always be calculated to the first order, making use of the perturbation theory.

9. Application of the Landau-Zener Model to Elementary Processes

Since the nineteen-thirties there have appeared a number of papers discussing the application of the Landau-Zener model for the calculation of non-adiabatic transition probabilities in various elementary processes. As a rule, the authors did not pay proper attention to limitations imposed by classical approximation and by the two-state approximation, introduced originally by LANDAU [4] and ZENER [5]. Hence the results published cannot be used for direct interpretation of experimental data. We shall not discuss this point here in connection with the commonly used Landau-Zener formula. This may be found in the article by BATES [8]. A brief list of recent papers dealing with application of the extended Landau-Zener model (see [4] to [8]) to some elementary processes is given below.

List of recent papers dealing with application of the extended Landau-Zener model to some elementary processes:

Process	Specification	Ref.
1. Quenching of alkali atom fluorescence upon collisions with inert gas atoms. Alkali atoms — inert gas heat bath energy exchange	*Effect of the turning point* is taken into account *A three-term* model is used	17 18
2. Quenching of alkali atom fluorescence upon collisions with diatomic molecules	*A two-dimensional* model is used, taking into account the spin-orbital coupling in the *cone intersection* of terms	15
3. Non-adiabatic vibrational relaxation of molecules	*A two-dimensional* model is used taking into account the spin-orbital and the Coriolis coupling in *line intersection* of terms	19
4. Ionization of atoms upon collisions, and impurities in solids	A *multi-term* model is used to describe the Rydberg excitation and transition to continuum	13,14,20

Acknowledgement

The author is indebted to Dr. M. Ya. Ovchinnikova and Prof. N. D. Sokolov for helpful discussions.

References

1. Longuet-Higgins, H. C.: Advances in Spectroscopy, **2**, 429. New York: Interscience Publishers 1961.
2. Eyring, H., T. Walter, and C. E. Kimball: Quantum Chemistry, 1946.
3. Landau, L., and E. Lifschitz: Quantum Mechanics, Fizmatgiz, 1963.
4. — Phys. Z. Sow. **2**, 46 (1932).
5. Zener, C.: Proc. Roy. Soc. **A 137**, 696 (1932).
6. Pokrovskii, V. L., and I. M. Khalatnikov: Zhur. Exp. Teor. Fiz. **40**, 1713 (1961); transl. Jetp **13**, 1207 (1961).
7. Stueckelberg, E. C. G.: Helv. Phys. Acta **5**, 369 (1932).
8. Bates, D. R.: Proc. Roy. Soc. **A 257**, 22 (1960); see also Bates, D. R.: in: Atomic and Molecular Processes, ed. by D. R. Bates. New York-London: Academic Press, 1962.
9. Nikitin, E. E.: Opt. i Spektr. **11**, 452 (1961); transl. **11**, 246 (1961).
10. Bykhovskii, V. K., E. E. Nikitin, and M. Ya. Ovchinnikova: Zhur. Exp. Teor. Fiz. **47**, 750 (1964); transl. Jetp **20**, 500 (1965).
11. Ovchinnikova, M. Ya.: Opt. i Spektr. **17**, 822 (1964); transl. **17**, 447 (1964); Dokl. Akad. Nauk SSSR **161**, 641 (1965).
12. — Thesis, Inst. Chem. Phys. Ac. Sci. USSR, 1965.

13. Osherov, V. I.: Zhur. Exp. Teor. Fiz. **49**, 1157 (1965).
14. Demkov, Yu. N.: Dokl. Akad. Nauk SSSR **166**, 1076 (1966).
15. Nikitin, E. E.: Journ. Quantitative Spectroscopy and Radiative Transfer **5**, 436 (1965).
16. Teller, E.: J. Phys. Chem. **41**, 109 (1937).
17. Bykhovskii, V. K., and E. E. Nikitin: Optika i Spektroskopiya **17**, 815 (1964); transl. **17**, 444 (1964).
18. Nikitin, E. E.: Combustion & Flame, **10**, 381, (1966).
19. — Mol. Phys. **7**, 389 (1964).
20. Demkov, Yu. N., and I. V. Komarov: Zhur. Exp. Teor. Fiz. **50**, 286 (1966).

E. E. Nikitin
Institute of Chemical Physics
Academy of Sciences of the USSR
Moscow

Statistische Quantenmechanik chemischer Reaktionen

G. L. Hofacker

Mit 1 Abbildung

1. Einleitende Bemerkungen

1.1. Die Dichtematrix

Wir betrachten ein System, dessen dynamischen Zustand wir nicht vollständig bestimmen können. Es soll jedoch möglich sein, eine Wahrscheinlichkeit P_k dafür anzugeben, daß sich das System in einem von mehreren Zuständen $|1>, |2>, \ldots, |k>, \ldots$ befindet. Die Wellenfunktionen $|i>$ müssen dabei nur der Bedingung

$$< i | i > = 1 \tag{1.1}$$

genügen und für die P_k muß gelten

$$\sum_k P_k = 1 \ . \tag{1.2}$$

J. von Neumann hat gezeigt, daß man statistische Systeme mit Hilfe einer statistischen Matrix, auch Dichtematrix genannt, behandeln kann, welche gegeben ist durch

$$\varrho = \sum_k | k > P_k < k | \ . \tag{1.3}$$

ϱ hat die folgenden Eigenschaften:

$\varrho\,1)$ $\varrho^+ = \varrho$ (Hermitizität)

$\varrho\,2)$ Ist $\mathscr{L}_1, \ldots, \mathscr{L}_m$ ein vollständiger Satz kommutativer Observablen und $\left\{ |L_1, \ldots, L_m > \right\} \equiv \left\{ |L > \right\}$ ein zugehöriges vollständiges Funktionssystem, so ist

$$< L | \varrho | L > = \sum_k < L | k > P_k < k | L > = \sum_k P_k | < L | k > |^2$$

die Wahrscheinlichkeit, das System bei einer Messung in $|L_1, \ldots, L_m >$ vorzufinden.

$\varrho\,3)$ $\qquad\qquad \text{Spur}\ \varrho = \sum_L < L | \varrho | L > = \sum_k P_k = 1 \ .$

ϱ 4) Der statistische Mittelwert einer Observablen $\mathscr{A}$ ist

$$\overline{\mathscr{A}} = \mathrm{Spur}\,(\mathscr{A}\,\varrho) = \sum_{L} <L\,|\,\mathscr{A}\,\varrho\,|\,L> = \sum_{k} P_k <k\,|\,\mathscr{A}\,|\,k> \quad . \quad (1.4)$$

Es sei noch bemerkt, daß ϱ oft als Ensemble-Mittelwert eines Produktes von Wellenfunktionen aufgefaßt wird, d. h., wenn wir etwa von einer Koordinatendarstellung ausgehen,

$$\varrho\,(x'\,|\,x'') = \overline{\Psi^*\,(x'',t)\,\Psi\,(x',t)} = \sum_{e,\,m} |\,e> \overline{a_e^*\,a_m} <m\,| \quad . \quad (1.5)$$

Da die Wellenfunktion Ψ bereits auf ein klassisches Ensemble bezogen ist, bedarf es bei Einführung von ϱ jedoch nicht notwendigerweise der Mittelung über ein Ensemble.

1.2. Die Bewegungsgleichung

Ist ϱ zu einer bestimmten Zeit t_0 gegeben als

$$\varrho_0 = \varrho\,(t_0) = \sum_{k} |\,k> P_k <k\,|$$

oder in irgendeiner Darstellung, in der es nicht diagonal ist, so determiniert die Schrödingergleichung die zeitliche Evolution eines jeden in ϱ_0 enthaltenen Zustandsvektors:

$$i\,\frac{d}{dt}\,|\,k_t> = \mathscr{H}\,|\,k_t>$$

$$-i\,\frac{d}{dt}\,<k_t\,| = <k_t\,|\,\mathscr{H} \quad . \quad [1] \tag{1.6}$$

Die formale Lösung der Gleichung (1.6) ist

$$|\,k_t> = e^{-i\,\mathscr{H}\,(t-t_0)}\,|\,k> \,, \tag{1.7}$$

wo der exponentielle Operator eine Kurzschrift der folgenden Art bedeutet:

$$e^{\mathscr{A}} \equiv \sum_{l=0}^{\infty} \frac{1}{l!}\,\mathscr{A}^l \quad . \tag{1.8}$$

Damit wird

$$\varrho\,(t) = e^{-i\,\mathscr{H}\,(t-t_0)}\,\varrho_0\,e^{i\,\mathscr{H}\,(t-t_0)} \quad . \tag{1.9}$$

Durch Differenzieren finden wir die Bewegungsgleichung

$$\dot{\varrho} = -i\,\mathscr{H}\,\varrho + i\,\varrho\,\mathscr{H}$$

oder

$$i\dot{\varrho} = [\mathscr{H},\varrho] \quad . \quad [2] \tag{1.10}$$

---

[1] Wir gebrauchen im folgenden durchweg atomare Einheiten. Wo immer $\hbar$ auftritt, ist es $= 1$ zu setzen.

[2] $[\mathscr{A},\mathscr{B}] = \mathscr{A}\mathscr{B} - \mathscr{A}\mathscr{B}.$

Gleichung (1.10) heißt quantenmechanische Liouville-Gleichung oder von Neumann-Gleichung und hat für die statistische Quantenmechanik fundamentale Bedeutung [1—4].

Genau wie die Schrödingergleichung ist (1.10) eine reversible Gleichung, d. h. sie bleibt unter der Transformation

$$t \rightarrow -t \text{ und } i \rightarrow -i \text{ (ferner: } \vec{B} \rightarrow -\vec{B} \text{ für axiale Vektoren)}$$

invariant[3].

1.3. *Formale Störungsrechnung für die von Neumann-Gleichung*

Die meisten Arbeiten, betreffend die Ableitung einer master-equation, behandeln dieses Problem mit unterschiedlichen Methoden. Wir wollen hier nur den Fall eines Ensembles von Molekelpaaren betrachten, welche durch einen einzigen nichtreaktiven Stoß in Wechselwirkung treten. Dafür werden wir den Formalismus der quantenmechanischen Streutheorie heranziehen [5—8].

Wir beginnen mit der v. Neumann-Gleichung

$$i\dot{\varrho} = [\mathscr{H}, \varrho] \,,$$

deren Hamiltonoperator aus einem ungestörten Teil und aus einer Störung besteht

$$\mathscr{H} = \mathscr{H}^0 + \mathscr{V} \,.$$

Die formale Lösung der von Neumann-Gleichung ist gegeben durch (1.9), doch ist es von Vorteil, statt ϱ (t) die Matrix $\bar{\bar{\varrho}}$ (t) zu betrachten, wo

$$\bar{\bar{\varrho}}(t) = e^{i\mathscr{H}^0 t} \cdot \varrho(t) \, e^{-i\mathscr{H}^0 t} \,. \tag{1.11}$$

Diese Transformation führt uns in die sogenannte Wechselwirkungsdarstellung, welche eine Separation der Effekte von freier Bewegung und Stoß erlaubt. Im Gegensatz zur Schrödingerdarstellung (wo nur ϱ von t abhängt) und zur Heisenbergdarstellung (wo nur die Operatoren $\mathscr{A}$ von t abhängen) hängen in der Wechselwirkungsdarstellung Operatoren und Dichtematrix von der Zeit ab.

$\bar{\bar{\varrho}}$ (t) gehe aus $\bar{\bar{\varrho}}_0 = \varrho_0$ durch eine unitäre Transformation U (t, t_0) hervor,

$$\bar{\bar{\varrho}}(t) = U(t, t_0) \varrho_0 \, U^+(t, t_0) \,. \tag{1.12}$$

Wenden wir $e^{-i\mathscr{H}^0(t-t_0)}$ linksseitig und $e^{i\mathscr{H}^0(t-t_0)}$ rechtsseitig auf (1.12) an, so sagt uns der Vergleich mit (1.9), daß

$$U(t, t_0) = e^{i\mathscr{H}^0 t} e^{-i\mathscr{H}(t-t_0)} e^{-i\mathscr{H}^0 t_0}$$

$$= e^{i\mathscr{H}^0 t} e^{i\mathscr{H}(t_0-t)} e^{-i\mathscr{H}^0(t_0-t)} e^{-i\mathscr{H}^0 t}$$

oder $\hspace{11cm}$ (1.13)

$$U(t, t_0) = e^{+i\mathscr{H}^0 t} \, U(0, t_0 - t) \, e^{-i\mathscr{H}^0 t} \,.$$

---

[3] Der Übergang $i \rightarrow -i$ bei der Zeitumkehr resultiert daraus, daß $\Psi(-t)$ die Lösung der komplex konjugierten der Schrödingergleichung ist.

Die Ausdrücke in (1.13) lassen sich nicht vereinfachen, da $\mathscr{H}$ und $\mathscr{H}^0$ im allgemeinen nicht vertauschbar sind. Des weiteren hat $U(t, t_0)$ die Eigenschaften

$$
\begin{aligned}
U(t, t) &= 1 \\
U(t, t_0) &= U(t, t')\, U(t', t_0) \\
U^{-1}(t, t_0) &= U(t_0, t) = U^{+}(t, t_0) \ .
\end{aligned}
$$

Durch Differentiation von (1.13) finden wir eine Bewegungsgleichung für $U(t, t_0)$:

$$
\begin{aligned}
\frac{d}{dt} U(t, t_0) &= i\, e^{i \mathscr{H}^0 t}\, \mathscr{H}^0\, e^{-i \mathscr{H}(t - t_0)}\, e^{-i \mathscr{H}^0 t_0} \\
&\quad - i\, e^{i \mathscr{H}^0 t}\, \mathscr{H}\, e^{-i \mathscr{H}(t - t_0)}\, e^{-i \mathscr{H}^0 t_0} \\
&= - i\, \overline{\overline{\mathscr{V}}}(t)\, U(t, t_0) ;
\end{aligned}
\tag{1.14}
$$

$$
i\, \frac{dU(t, t_0)}{dt} = \overline{\overline{\mathscr{V}}}(t)\, U(t, t_0) ,
$$

wobei $\qquad \overline{\overline{\mathscr{V}}}(t) = e^{i \mathscr{H}^0 t}\, \mathscr{V}\, e^{-i \mathscr{H}^0 t} \ .$

Die Differentialgleichung (1.14) kann leicht in eine Integralgleichung umgewandelt werden

$$
U(t, t_0) = 1 - i \int_{t_0}^{t} \overline{\overline{\mathscr{V}}}(t')\, U(t', t_0)\, dt' ,
\tag{1.15}
$$

welche als Dysonsche Integralgleichung bekannt ist. Sukzessive Resubstitution unter dem Integralzeichen in (1.15) führt schließlich zu bekannten störungstheoretischen Entwicklungen. Uns interessiert hier nur der Grenzwert

$$
\lim_{\substack{t_0 \to -\infty \\ t \to 0}} \text{von (1.15)},
$$

$$
U(0, -\infty) \equiv \Omega^{\oplus} = 1 - i \lim_{\varepsilon \to 0} \int_{-\infty}^{0} e^{\varepsilon t}\, e^{i \mathscr{H} t'}\, \mathscr{V}\, e^{-i \mathscr{H}^0 t'}\, dt' \ .
\tag{1.16}
$$

Dabei haben wir in das Integral einen Konvergenzfaktor eingeführt, den wir durch einen Grenzübergang wieder entfernen[4].

Die Wirkung von $\Omega^{\oplus}$ auf die Eigenfunktionen der freien Molekeln, für die

$$
\mathscr{H}^0 |a> \; = E_a |a>
$$

---

[4] Die Existenz dieses Grenzwertes ist bewiesen in: JANCH, J. M.: Helv. Phys. Acta **31**, 127 (1958).

gilt, ist

$$\Omega^{\oplus}|a> \equiv |a^{\oplus}> = |a> - i \lim_{\varepsilon \to 0} \int_{-\infty}^{0} e^{\varepsilon t} e^{i \mathscr{H} t'} e^{-i E_a t'} \mathscr{V} |a> ,$$

oder
$$|a^{\oplus}> = |a> + \lim_{\varepsilon \to 0} \frac{1}{E_0 - \mathscr{H} + i \varepsilon} \mathscr{V} |a> , \qquad (1.17)$$

wobei wir für die letzte Umformung die symbolische Formel benutzten:

$$\int_{-\infty}^{0} e^{i \omega t'} dt' = \int_{0}^{\infty} e^{-i \omega t'} dt' = \lim_{\varepsilon \to 0} \frac{i}{\omega + i \varepsilon} . \qquad (1.18)$$

Diese Identität ist gültig unter einem ω-Integral zusammen mit einer absolut integrablen und beliebig oft differenzierbaren Funktion. Damit lautet (1.17), geschrieben als eine Operatorgleichung:

$$\Omega^{\oplus} = 1 + \lim_{\varepsilon \to 0} \frac{\mathscr{V}}{E_a - \mathscr{H} + i \varepsilon} . \qquad (1.19)$$

Andererseits können wir (1.13) in die Dysonsche Integralgleichung (1.15) einführen:

$$U(0, -\infty)|a> = |a> - i \lim_{\varepsilon \to 0} \int_{-\infty}^{0} \overline{\overline{\mathscr{V}}}(t') e^{\varepsilon t'} e^{i \mathscr{H}^0 t'} U(0, -\infty) e^{-i \mathscr{H}^0 t'} dt' |a>$$

$$= |a> - i \lim_{\varepsilon \to 0} \int_{-\infty}^{0} e^{i \mathscr{H}^0 t'} e^{\varepsilon t'} \mathscr{V} e^{-i E_a t'} dt' \, U(0, -\infty)|a> ,$$

oder
$$|a^{\oplus}> = |a> - i \lim_{\varepsilon \to 0} \int_{-\infty}^{0} e^{-i(E_a - \mathscr{H}^0 + i\varepsilon)t'} dt' \, \mathscr{V} |a^{\oplus}>$$

$$= |a> + \lim_{\varepsilon \to 0} \frac{1}{E_a - \mathscr{H}^0 + i \varepsilon} \mathscr{V} |a^{\oplus}> , \qquad (1.20)$$

wobei wir die letzte Umformung wieder vermöge der Formel (1.18) vornehmen konnten. Setzen wir in (1.20) $|a^{\oplus}> = \Omega^{\oplus}|a>$ ein, so gelangen wir zur Lippmann-Schwinger Differenzialgleichung für den Operator $\Omega^{\oplus}$:

$$\Omega_a^{\oplus} = 1 + \lim_{\varepsilon \to 0} \frac{1}{E_a - \mathscr{H}^0 + i \varepsilon} \mathscr{V} \Omega_a^{\oplus} . \qquad (1.21)$$

Beide Gleichungen, (1.19) und (1.21), sind Grenzwerte der Dysonschen Integralgleichung (1.15). Daher ist (1.19) die formale Lösung der Gleichung (1.21), genannt die Chew-Goldberger-Lösung der Lippmann-Schwinger-Differentialgleichung.

Der Operator

$$\mathscr{G}_a^{\oplus} = \lim_{\varepsilon \to 0} \frac{1}{E_a - \mathscr{H}^0 + i \varepsilon} \qquad (1.22)$$

heißt Greensche Funktion und trägt den Index des Zustandes, der zur Rechten von $\mathscr{V}\,\Omega^{\oplus}$ stehen wird. Da

$$\frac{1}{E_a - \mathscr{H}^0 + i\,\varepsilon} = \frac{1}{E_a - \mathscr{H}^0 + i\,\varepsilon} \cdot \frac{E_a - \mathscr{H}^0 - i\,\varepsilon}{E_a - \mathscr{H}^0 - i\,\varepsilon}$$

$$= \frac{E_a - \mathscr{H}^0}{(E_a - \mathscr{H}^0) + \varepsilon^2} - \frac{i\,\varepsilon}{(E_a - \mathscr{H}^0) + \varepsilon^2}$$

ist und [unter einem Integral über ω mit einer geeigneten Funktion $f(\omega)$] die einzelnen Terme geschrieben werden können als

$$\lim_{\varepsilon \to 0} \frac{\omega}{\omega^2 + \varepsilon^2} = \mathscr{P}\left(\frac{1}{\omega}\right) \qquad (\mathscr{P} \text{ bedeutet den Hauptwert des } \omega\text{-Integrals})\,[5]$$

$$\lim_{\varepsilon \to 0} \frac{i\,\varepsilon}{\omega^2 + \varepsilon^2} = i\,\pi\,\delta(\omega)\;,$$

so erhalten wir

$$\mathscr{G}_a^{\oplus} = \mathscr{P}\left(\frac{1}{\omega}\right) - i\,\pi\,\delta(E_a - \mathscr{H}^0)\;. \tag{1.23}$$

Die Lippmann-Schwinger-Differentialgleichung nimmt damit die Gestalt an:

$$\Omega^{\oplus} = 1 + \mathscr{G}^{\oplus}\,\mathscr{V}\,\Omega^{\oplus}\;. \tag{1.24}$$

Aus gutem Grund wird ein besonderer Operator

$$t^{\oplus} = \mathscr{V}\,\Omega^{\oplus} \tag{1.25}$$

eingeführt, so daß

$$\Omega^{\oplus} = 1 + \mathscr{G}^{\oplus}\,t^{\oplus}\;. \tag{1.26}$$

---

[5] Wenn $|f(\omega)| \to \infty$ für $\omega \to \omega_0$ und

$$\lim_{\varepsilon_1 \to 0} \int_a^{\omega_0 - \varepsilon_1} f(\omega)\,d\omega + \lim_{\varepsilon_2 \to 0} \int_{\omega_0 + \varepsilon_2}^b f(\omega)\,d\omega = \int_a^b f(\omega)\,d\omega$$

existiert, haben wir ein uneigentliches Integral vor uns. Existiert keiner dieser beiden Grenzwerte einzeln, sondern nur

$$\lim_{\varepsilon \to 0} \left\{ \int_a^{\omega_0 - \varepsilon} f(\omega)\,d\omega + \int_{\omega_0 + \varepsilon}^b f(\omega)\,d\omega \right\}\;,$$

so heißt dieser Grenzwert Cauchyscher Hauptwert und wird durch

$$\mathscr{P} \int_a^b f(\omega)\,d\omega$$

bezeichnet.

6*

Man kann zeigen, daß die Matrixelemente der sogenannten Streumatrix $t^{\oplus}$ folgendermaßen mit den Wirkungsquerschnitten σ zusammenhängen:

$$\sigma_{a \to b} \equiv \sigma(a\,|\,b) = (2\,\pi)^4\, M_r^2\, \frac{k_b}{k_a}\, |t_{b\,a}^{\oplus}|^2 , \qquad (1.27)$$

wo M_r die reduzierte Masse der Stoßpartner bedeutet, k_a und k_b die Beträge der Wellenzahlen der Relativimpulse zu den Zuständen a und b.

Die Kenntnis des Operators $\Omega^{\oplus}$ gibt uns die volle formale Beherrschung des Streuvorganges. Seine Berechnung ist jedoch eine der schwierigsten Aufgaben, welche die Quantenmechanik stellt. Die niederen Approximationen der Gleichung (1.19), deren nullte etwa durch

$$\Omega_0^{\oplus} = 1 + \frac{\mathscr{V}}{E - \mathscr{H}^0} \qquad \text{gegeben ist,} \qquad (1.28)$$

liefern bei den meisten Streuproblemen keine brauchbaren Streumatrizen.

2. Intramolekulare Relaxation

Die Frage, ob ein isoliertes System von molekularer Größe in einem bestimmten Sinn einen irreversiblen Prozeß durchmachen kann, ist für die chemische Reaktionskinetik von fundamentaler Bedeutung. Man mag einwenden, daß man molekulare Systeme nur selten eine so lange Zeit von allen äußeren Wechselwirkungen freihalten kann, wie das System benötigt, ins Gleichgewicht zu kommen. Die Vorstellung, daß ein Ensemble von Molekeln, das an einen Thermostaten angekoppelt ist, sich unter gewissen Bedingungen so verhalten kann, als sei der Thermostat gar nicht vorhanden, ist jedoch eine der weittragendsten und fruchtbarsten in der chemischen Kinetik. So macht die Rice-Ramsperger-Kassel-Marcus-Theorie monomolekularer Reaktionen explizit und weitgehend Gebrauch von dem Modell eines „inneren" und „äußeren" Mechanismus, wobei der innere Mechanismus meist formelhaft als „Umverteilung der inneren Energie zwischen verschiedenen inneren Freiheitsgraden", der äußere als „Stoßwechselwirkung" beschrieben wird. Jeder dieser Mechanismen kann unter gegebenen Umständen die Reaktionsgeschwindigkeit bestimmen. Wenn wir den intramolekularen Mechanismus auf quantenmechanischer Basis verstehen wollen, müssen wir ihn zunächst isoliert betrachten und uns überzeugen, daß die Stöße kein notwendiges Attribut des Relaxationsvorganges sind.

Die Aussage: „Das System geht ins Gleichgewicht" bedeutet für unendliche (oder praktisch unendliche) Systeme, daß die Rückkehr aus dem Gleichgewichtszustand während endlicher Zeit nicht möglich ist.

Bei den Systemen mit einer endlichen Anzahl von Freiheitsgraden müssen wir uns jedoch der Tatsache bewußt sein, daß endliche Wiederkehrzeiten für jeden beliebigen Zustand existieren. Wir müssen uns daher auf einen bestimmten Zeitabschnitt beschränken, der sehr kurz gegenüber der Wiederkehrzeit ist. Die Molekeln, die wir betrachten, dürfen daher eine nicht zu kleine Zahl von Freiheitsgraden besitzen. Wie wir später sehen werden, genügen aber die meisten chemisch interessanten Molekeln den notwendigen Bedingungen.

Ausgangspunkt jeder quantenstatistischen Theorie ist die v. Neumann-Gleichung (1.10). Natürlich ist es nicht möglich, eine irreversible kinetische Gleichung durch Umformung einer reversiblen Gleichung abzuleiten. Wir müssen daher unsere Fragestellung modifizieren. Unser Interesse gilt ja auch nicht der der vollständigen Lösung der v. Neumann-Gleichung, sondern nur einer in den Quantenzahlen durch Mittelung vergröberten Dichtematrix ϱ (t) auf einer Zeitskala, die sehr groß ist, verglichen mit der Dauer der Einzelprozesse, durch die das System ins Gleichgewicht hineinwandert [9—11].

Was wir daher suchen, ist eine asymptotische Lösung der v. Neumann-Gleichung für eine vergröberte Dichtematrix; nur für diese Dichtematrix können wir eine (irreversible) kinetische Gleichung finden.

Aus der statistischen Theorie irreversibler Prozesse sind eine Reihe von Methoden zur Ableitung kinetischer Gleichungen für die Diagonalelemente der Dichtematrix, genannt master-equations, bekannt [12—15].

Solche kinetische Gleichungen genügen jedoch nicht immer den Ansprüchen, die man bei der Behandlung chemischer Systeme stellt. Da chemische Reaktionen mit dem Übergang von einer Kernkonfiguration zu einer anderen verbunden sind, genügt die Kenntnis der Diagonalelemente der Dichtematrix nicht zur Charakterisierung des chemischen Zustandes zu einer gegebenen Zeit. Quantenmechanisch gesehen sind reaktionsfähige Molekeln durch Wellenpakete dargestellt und ein Ensemble reaktionsfähiger Molekeln durch eine nichtdiagonale Dichtematrix. Mit dem Problem der quantenmechanischen Beschreibung einer chemischen Spezies müssen wir uns als erstes auseinandersetzen [16, 17].

Wir können den Zustand (d. h. die Dichtematrix ϱ) eines Ensembles nur durch Messung eines vollständigen Satzes kommutativer Operatoren eindeutig festlegen. Wenn alle Mitglieder des Ensembles die chemischen Konfigurationen A1 und A2 einnehmen können, muß es wenigstens eine Methode für die Messung der chemischen Konfiguration an jedem Mitglied geben. Alle derartigen Messungen produzieren Eigenwerte der Operatoren $\mathscr{K}^{A1}$ und $\mathscr{K}^{A2}$ (genannt Operatoren der chemischen Spezies oder Konfiguration). Die relative Anzahl der Molekeln A1 und A2, nämlich N_1 und N_2, ist dann

$$\frac{N_1}{N} = \text{Spur}\,(\mathscr{K}^{A1} \varrho\; \mathscr{K}^{A1+})\,, \quad \frac{N_2}{N} = \text{Spur}\,(\mathscr{K}^{A2} \varrho\; \mathscr{K}^{A2+})\,. \tag{2.1}$$

$\mathscr{X}^{A1}$ und $\mathscr{X}^{A2}$ wollen wir durch die Messung von „lokalisierten" Energie-observablen $\mathscr{H}^{A1}$ und $\mathscr{H}^{A2}$ ergänzen (wenn notwendig durch mehr, doch würde das an unserer Überlegung nichts ändern). Diese haben die Eigenschaften

$$[\mathscr{H}^{A1}, \mathscr{X}^{A1}] = 0 \quad \text{und} \quad \mathscr{H}^{A1}|\alpha_1> = E^{A1}_{\alpha_1}|\alpha_1> \qquad (2.2)$$
$$\mathscr{X}^{A1}|\alpha_1> = |\alpha_1>$$
$$\mathscr{X}^{A1}|\alpha_2> = 0, \text{ etc. für Indizes 1 und 2 vertauscht.}$$

Wir bemerken noch, daß für die Messung von $E^{A1}_{\alpha_1}$ nur eine beschränkte Zeit, τ_m, zur Verfügung steht, die kurz sein muß im Vergleich zu der chemischen Relaxationszeit.

Die Operatoren $\mathscr{X}^{A1}$ und $\mathscr{X}^{A2}$ können wir unschwer als Projektionsoperatoren erkennen (d. h. $\mathscr{X}^2 = \mathscr{X}$ und $\mathscr{X}^+ = \mathscr{X}$). Da die Überlappung der lokalisierten Eigenfunktionen hier nur eine geringe Rolle spielt (wäre das nicht der Fall, so könnten wir nicht die chemischen Spezies durch Messung identifizieren, da das System zu schnell von einer Konfiguration zur anderen tunnelieren würde) gilt auf einer genügend kurzen Zeitskala

$$\mathscr{X}^{A1} + \mathscr{X}^{A2} = 1 \qquad (2.3)$$

und damit

$$\text{Spur}(\mathscr{X}^{A1}\varrho\,\mathscr{X}^{A1}) + \text{Spur}(\mathscr{X}^{A2}\varrho\,\mathscr{X}^{A2}) = 1 \, . \qquad (2.4)$$

Zur Zeit $t = 0$ soll unser Ensemble in der chemischen Konfiguration A1 vorliegen. Die Dichtematrix ist dann

$$\varrho_0 = \sum_{\alpha_1} |\alpha_1> P^{A1}_{\alpha_1} <\alpha_1| \qquad (2.5)$$

und, wie es sein muß, gilt

$$[\mathscr{X}^{A1}, \varrho_0] = 0 \, .$$

Der Einfachkeit halber haben wir darauf verzichtet, ein Ensemble zu betrachten, dessen Dichtematrix durch zusätzliche Observable, wie elektronischer Zustand, Rotationszustand u. ä., festgelegt werden muß. Die Situation soll am Beispiel eines Doppelminimumpotentials erläutert werden (s. Abb.).

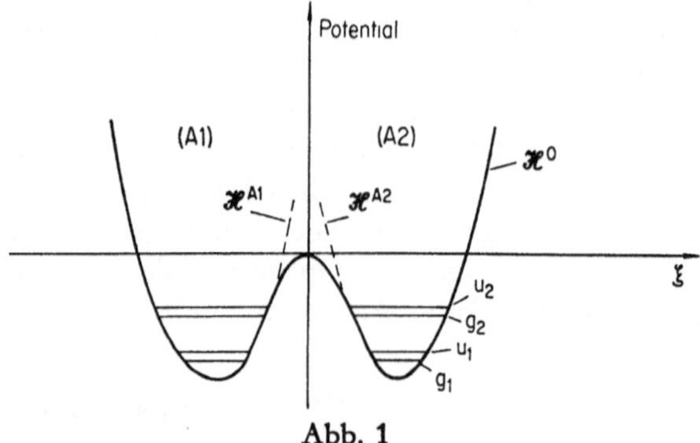

Abb. 1

ξ ist die Reaktionskoordinate; der Hamiltonoperator $\mathcal{H}^0$ des ungestörten (gesamten) Systems ist nicht konfigurationsgebunden, während $\mathcal{H}^{A1}$ und $\mathcal{H}^{A2}$ sich auf je eine der Mulden beziehen. Die Wahl von $\mathcal{H}^{A1}$ und $\mathcal{H}^{A2}$ im einzelnen ist nicht problematisch. Wir setzen

$$\mathcal{H}^0 = \mathcal{H}^{A1} + \mathcal{H}^{A2} + \mathcal{H}^{A12} \tag{2.6}$$

und nehmen an, daß die Potentiale von $\mathcal{H}^{A1}$ und $\mathcal{H}^{A2}$ im Bereich der Zustände mit negativer Energie praktisch übereinstimmen. Die Eigenzustände $|\alpha>$ von $\mathcal{H}^0$ haben die Symmetrieeigenschaften g bzw. u, in alternierender Folge. Nach Störungsrechnung 1. Ordnung ist dann

$$|\alpha_g> = \frac{1}{\sqrt{2}}(|\alpha_1> + |\alpha_2>)$$

$$|\alpha_u> = \frac{1}{\sqrt{2}}(|\alpha_1> - |\alpha_2>). \tag{2.7}$$

Dabei haben wir die Überlappung zwischen $|\alpha_1>$ und $|\alpha_2>$ vernachlässigt. Wenn das System in der Konfiguration A1 ist, führt die Messung von $\mathcal{H}^{A1}$ und $\mathcal{X}^{A1}$ auf eine Wellenfunktion

$$|\alpha_1> = \frac{\sqrt{2}}{2}(|\alpha_g> + |\alpha_u>). \tag{2.8}$$

Ein Ensemble in der Konfiguration A1 hat daher die Dichtematrix

$$\varrho_0 = \sum_{\alpha_1}|\alpha_1> P_{\alpha_1}^{A1} <\alpha_1| = \frac{1}{2}\sum_a(|\alpha_g> + |\alpha_u>) P_{\alpha_1}^{A1} (<\alpha_g| + <\alpha_u|),$$

$$\tag{2.9}$$

und es ist sofort zu sehen, daß ϱ_0 nicht mehr diagonal in der Darstellung $\{|\alpha>\}$ ist. Die Dichtematrix ϱ_0 zerfällt nun durch

1. Tunneleffekt [das Wellenpaket (2.8) hat eine Tunnelungsfrequenz $(E_{\alpha_u} - E_{\alpha_g})$]

2. Übergängen zwischen Zuständen $|\alpha>$ von etwa der gleichen Energie (dabei ist zu berücksichtigen, daß senkrecht zu ξ eine Anzahl weiterer Freiheitsgrade besteht)[6].

τ_1 sei eine charakteristische Zeit für die Änderung durch den Prozeß 1., τ_2 gebe die Zeit der Änderung von ϱ_0 durch Prozeß 2. an. Wir wissen nun, daß $\tau_m \ll \min(\tau_1, \tau_2)$ sein muß, doch haben wir es nicht in der Hand, die Meßzeit τ_m beliebig kurz zu wählen. Mit einer Meßzeit τ_m ist nämlich eine Unbestimmtheit in der Energie von der Größe τ_m^{-1} verbunden, und das kann bewirken, daß das System durch die Messung weit über die Höhe des Sattels (er habe die Höhe E_a über dem Grund der Mulde A1) hinausgehoben und zur Reaktion gebracht wird. Daher muß

---

[6] Für die explizite Konstruktion einer Basis $\{|a>\}$ nebst Operatoren $\mathcal{X}^{A1}$ und $\mathcal{X}^{A2}$ siehe [*18*].

$$E_a^{-1} \ll \tau_m \ll \min(\tau_1, \tau_2) \qquad \text{sein.}$$

Es sei noch bemerkt, daß im Falle einer Dissoziationsreaktion die Situation wesentlich einfacher ist, solange kein Tunneleffekt auftritt (dann ist $|\alpha_1> = |\alpha>$ für $E < E_0$ und $|\alpha_2> = |\alpha>$ für $E > E_0$). Kommt Dissoziation durch Tunnelieren vor, so bereitet uns die Praedissoziation die gleichen Schwierigkeiten, wie wir sie bei den Isomerisierungsreaktionen gesehen haben.

Nunmehr können wir die Ableitung der Relaxationsgleichung in Angriff nehmen. Es sei

$$\mathcal{H} = \mathcal{H}^0 + \mathcal{V},$$

$$\mathcal{H}^0 |\alpha> = E_\alpha |\alpha>,$$

$$i\dot{\varrho} = [\mathcal{H}, \varrho] \text{ und } \varrho_0 \text{ gegeben.}$$

Die v. Neumann-Gleichung bringen wir nun in eine Gestalt, die manchmal als „Dämpfungsform" bezeichnet wird, weil sie oft auf der Stelle gewisse Grenzübergänge erlaubt, welche die Gleichung irreversibel machen. Wir setzen

$$\bar{\varrho} = e^{i\mathcal{H}^0 t} \varrho(t) e^{-i\mathcal{H}^0 t},$$

differenzieren und setzen die v. Neumann-Gleichung ein

$$\dot{\bar{\varrho}}(t) = i\mathcal{H}^0 \bar{\varrho} - i\bar{\varrho}\mathcal{H}^0 + e^{i\mathcal{H}^0 t} \dot{\varrho}(t) e^{-i\mathcal{H}^0 t}$$

$$= i[\mathcal{H}^0, \bar{\varrho}(t)] - i e^{i\mathcal{H}^0 t} [\mathcal{H}, \varrho(t)] e^{-i\mathcal{H}^0 t}$$

$$= -i e^{i\mathcal{H}^0 t} [\mathcal{V}, \varrho(t)] e^{-i\mathcal{H}^0 t}.$$

Integration von 0 bis t und Anwendung des Operators $e^{-i\mathcal{H}^0 t}$ von links und $e^{i\mathcal{H}^0 t}$ von rechts ergibt

$$\varrho(t) = e^{-i\mathcal{H}^0 t} \varrho_0 e^{i\mathcal{H}^0 t} - i \int_0^t e^{-i\mathcal{H}^0 (t-\tau)} [\mathcal{V}, \varrho(\tau)] e^{i\mathcal{H}^0 (t-\tau)} d\tau. \qquad (2.10)$$

Diese Gleichung ist der v. Neumann-Gleichung äquivalent, wie man durch abermaliges Differenzieren leicht zeigen kann. Substituiert man (2.10) zurück in die v.-Neumann-Gleichung, so gelangt man zu der Dämpfungsform

$$\dot{\varrho} = -i[\mathcal{H}^0, \varrho] - [\mathcal{V}, \int_0^t e^{-i\mathcal{H}^0 \tau} [\mathcal{V}, \varrho(t-\tau)] e^{i\mathcal{H}^0 \tau} d\tau$$

$$\qquad -i[\mathcal{V}, e^{-i\mathcal{H}^0 t} \varrho_0 e^{i\mathcal{H}^0 t}]; \qquad (2.11)$$

dabei ist unter dem Integral die Integrationsvariable gegenüber (2.10) geändert worden.

Schreiben wir die Matrixelemente aus, so wird aus (2.11), mit

$$\omega_{\alpha\beta} = E_\alpha - E_\beta:$$

$$\frac{d\varrho_{\alpha\beta}}{dt} = -i\,\omega_{\alpha\beta}\varrho_{\alpha\beta} - i\sum_\gamma \left\{ V_{\alpha\gamma}\varrho_{0\gamma\beta}e^{-i\omega_{\gamma\beta}t} - \varrho_{0\alpha\gamma}V_{\gamma\beta}e^{-i\omega_{\alpha\gamma}t} \right\}$$

$$- \sum_{\gamma\varepsilon}\left\{ V_{\alpha\gamma}V_{\gamma\varepsilon}\int_0^t e^{-i\omega_{\gamma\beta}\tau}\varrho_{\varepsilon\beta}(t-\tau)d\tau - V_{\alpha\gamma}V_{\varepsilon\beta}\int_0^t e^{-i\omega_{\alpha\varepsilon}\tau}\varrho_{\gamma\varepsilon}(t-\tau)d\tau \right.$$

$$\left. - V_{\alpha\gamma}V_{\varepsilon\beta}\int_0^t e^{-i\omega_{\gamma\beta}\tau}\varrho_{\gamma\varepsilon}(t-\tau)\,d\tau + V_{\gamma\varepsilon}V_{\varepsilon\beta}\int_0^t e^{-i\omega_{\alpha\varepsilon}\tau}\varrho_{\alpha\gamma}(t-\tau)d\tau \right\}. \quad (2.12)$$

Von dieser Gleichung suchen wir eine asymptotische Lösung. Eine notwendige Bedingung für das Bestehen einer solchen Lösung ist die Existenz dreier charakteristischer Zeiten in der Ordnung

$$\tau_a \ll \tau_r \ll \tau_{tr}\,.$$

Diese Zeiten haben folgende Bedeutung:

τ_a, genannt „atomare Zeit", ist der mittlere reziproke Abstand besetzter Niveaus: $\tau_a \simeq \overline{\omega_{\alpha\beta}^{-1}}$

τ_{tr}, genannt „transversale Zeit", ist der mittlere reziproke Abstand besetzter aufeinanderfolgender Niveaus α und α', $\tau_{tr} \simeq \overline{\omega_{\alpha\alpha'}^{-1}}$

τ_r ist dann die Zeitskala, auf der möglicherweise eine kinetische Gleichung für die asymptotische Lösung der v. Neumann-Gleichung gültig ist.

Schätzen wir τ_a und τ_{tr} einmal grob ab. τ_a ist von der Größenordnung der reziproken Sattelhöhe ($E_a \simeq 1-100$ kcal/mol), also

$$\tau_a \simeq 10^{-14} - 10^{-17}\ \text{sec.}$$

τ_{tr} kann auf der Grundlage eines harmonischen Oszillatormodells abgeschätzt werden. Wir finden aus $\tau_{tr} = 1/\sigma(E_a)$, wo $\sigma(E_a)$ die Niveaudichte für die innere Energie E_a bedeutet,

$$\tau_{tr} = 10^{-2} \quad 10^9 \quad 10^{22}\ \text{sec}$$
$$f \ \ = 12 \quad\ \ 24 \quad\ 39,$$

wo f die Zahl der inneren Freiheitsgrade ist. Diese Zahlen besagen, daß wir schon für Molekeln mit einer sehr geringen Zahl von Freiheitsgraden kinetische Gleichungen für gewisse Prozesse werden ableiten können.

Im Gegensatz zu unendlichen Systemen sind in unserem Fall die Matrixelemente $V_{\alpha\beta}$ keine stetigen Funktionen von E_α und E_β. Ohne Stetigkeitseigenschaften der Summanden können wir aber die Summen in (2.12) nicht in Integrale verwandeln. Daher führen wir an Gleichung (2.12) eine Kontraktion der Quantenzahlen aus, bzw. eine Mittelung über Energieintervalle, denn die Energien E_α sind nicht entartet. Diese Methode ist der Mittelung über Zellen des Phasenraumes (coarse-graining), wie es in der klassischen statistischen Mechanik vorgenommen wird, ganz analog [19, 20]. Die Energieachse teilen wir in Intervalle der Länge ΔE, ΔE_1, ΔE_2, ..., ΔE_a, ..., ΔE_b, ... ein, mit einer Zahl von Zuständen

$$g_1, g_2, \ldots\ldots, g_a, \ldots\ldots, g_b, \ldots\ldots$$

Mittelwerte bezeichnen wir folgendermaßen:

$$V_{\bar{a}\beta} = V(E_{\bar{a}}, E_\beta) = \frac{1}{g_a} \sum_{\alpha \varepsilon a} V_{\alpha\beta}$$

$$\varrho_{a\beta} = \sum_{\alpha \varepsilon a} \varrho_{\alpha\beta} \tag{2.13}$$

$$\omega_{\bar{a}\beta} = \frac{1}{g_a} \sum_{\alpha \varepsilon a} \omega_{\alpha\beta}$$

Wenn die Zustandsdichte (oder τ_{tr}) groß genug und die Störung $\overline{\mathscr{V}}$ klein genug ist, können wir ΔE so wählen, daß die folgenden Bedingungen erfüllt sind:

$$\frac{1}{g_a} |V_{\alpha\beta}| \ll V_{\bar{a}\beta} \tag{I}$$

$$g_a/\Delta E_a \text{ variiert langsam mit } E_{\bar{a}} \tag{II}$$

$$\Delta E \ll \tau_r^{-1} \text{ (d. h. auch } \Delta E \ll \tau_m^{-1}) \tag{III}$$

$$V_{\bar{a}\bar{b}} \text{ ist eine langsam variierende Funktion von } E_{\bar{a}} \text{ und } E_{\bar{b}} \tag{IV}$$

Gleichung (2.12) kann nun über $\sum_{\alpha \varepsilon a} \sum_{\beta \varepsilon b}$ summiert werden und die Summen durch Summen über geeignete Mittelwerte ersetzt werden:

$$\frac{d\varrho_{ab}}{dt} = i\,\omega_{\bar{a}\bar{b}}\varrho_{ab} - i \sum_c \left\{ g_a V_{\bar{a}\bar{c}} \varrho_{0cb}\, e^{-i\omega_{\bar{c}\bar{b}}t} - g_b \varrho_{0ac} V_{\bar{c}\bar{b}}\, e^{-i\omega_{\bar{a}\bar{c}}t} \right\}$$

$$- \sum_{ce} \left\{ g_a g_c V_{\bar{a}\bar{c}} V_{\bar{c}\bar{e}} \int_0^t e^{-i\omega_{\bar{c}\bar{b}}\tau} \varrho_{eb}(t-\tau)\, d\tau \right.$$

$$- g_a g_b V_{\bar{a}\bar{c}} V_{\bar{e}\bar{b}} \int_0^t e^{-i\omega_{\bar{a}\bar{e}}\tau} \varrho_{ce}(t-\tau)\, d\tau$$

$$- g_a g_b V_{\bar{a}\bar{c}} V_{\bar{e}\bar{b}} \int_0^t e^{-i\omega_{\bar{c}\bar{b}}\tau} \varrho_{ce}(t-\tau)\, d\tau$$

$$\left. + g_e g_b V_{\bar{c}\bar{e}} V_{\bar{e}\bar{b}} \int_0^t e^{-i\omega_{\bar{a}\bar{e}}\tau} \varrho_{ac}(t-\tau)\, d\tau \right\}. \tag{2.14}$$

Unter dem Integralzeichen entwickeln wir $\varrho(t-\tau)$ in eine Taylor-Reihe:

$$\varrho(t-\tau) = \sum_{k=0}^{\infty} \frac{(-\tau)^k}{k!} \varrho^{(k)}(t). \tag{2.15}$$

Das gibt uns τ-Integrale der Form

$$\mathscr{I}_k = \int_0^t \frac{(-\tau)^k}{k!}\, e^{-i\omega\tau}\, d\tau \; . \tag{2.16}$$

Wir bemerken, daß man alle $\mathscr{I}_k$ durch Differenziation aus $\mathscr{I}_0$ herleiten kann

$$\mathscr{I}_k = \frac{1}{i^k\, k!} \frac{\partial^k}{\partial \omega^k} \mathscr{I}_0 \; . \tag{2.17}$$

Unter dem Summenzeichen in Gleichung (2.14) ist ω von der Größenordnung τ_a^{-1}, und wir sehen durch Ändern der Integrationsvariablen (τ/τ_a statt τ) sofort, daß für die obere Grenze des Integrals ∞ gesetzt werden darf. Dann ist

$$\mathscr{I}_0 = \int_{-0}^{\infty} e^{-i\omega\tau}\, d\tau = 2\pi\, \delta_+(\omega) \tag{2.18}$$

wobei wir $\delta_+(\omega)$ auch schreiben können:

$$\delta_+(\omega) = \frac{1}{2\pi i} \lim_{\varepsilon \to 0} \frac{1}{\omega - i\varepsilon} \; . \tag{2.19}$$

Entsprechend ist dann

$$\mathscr{I}_k = \frac{2\pi}{i^k\, k!} \frac{\partial^k}{\partial \omega^k} \delta_+(\omega) = i^{k-1} \lim_{\varepsilon \to 0} \frac{1}{(\omega - i\varepsilon)^{k+1}} \; . \tag{2.20}$$

Die in $\mathscr{V}$ quadratischen Terme sehen so aus:

$$\sum_{k=0}^{\infty} \sum_{ce} g_a\, g_c\, V_{\overline{a}\overline{c}}\, V_{\overline{c}\overline{e}}\, \varrho^{(k)}_{e\,b}\, \mathscr{I}_k(\omega_{\overline{c}\overline{b}})$$

$$= \sum_{k=0}^{\infty} g_a \int\!\!\int_0^\infty dE_{\overline{c}}\, dE_{\overline{e}}\, g(E_{\overline{c}})\, V(E_{\overline{a}}, E_{\overline{c}})\, V(E_{\overline{c}}, E_{\overline{e}})\, \varrho^{(k)}(E_{\overline{c}}, E_{\overline{b}}\,;\,t)\, \mathscr{I}_k(\omega_{\overline{c}\overline{b}})$$

$$= \sum_{k=0}^{\infty} g_a \int_0^\infty dE_{\overline{e}} \int_{-E_{\overline{b}}}^\infty d\omega\, g(E_{\overline{b}} + \omega)\, V(E_{\overline{a}}, E_{\overline{b}} + \omega)\, V(E_{\overline{b}} + \omega, E_{\overline{e}})$$

$$\times\, \varrho^{(k)}(E_{\overline{e}}, E_{\overline{b}}, t)\, \mathscr{I}_k(\omega) \; . \tag{2.21}$$

Das ω-Integral hat die Form:

$$C_k = \int_{-E_b}^\infty d\omega\, F(\omega)\, \mathscr{I}_k(\omega) \tag{2.22}$$

mit einer stetigen Funktion $F(\omega)$. Durch analytische Fortsetzung von $F(\omega)$, das genügend stark in der oberen komplexen Halbebene verschwinden soll, können wir setzen

$$\int_{-E_b}^\infty \cdots d\omega = \int_{\curvearrowright} \cdots d\omega \; ,$$

wobei der Integrationsweg von $-\infty$ bis $+\infty$ geht und die positive Halb-ebene im negativen Uhrzeigersinn umfährt. Dann wird aus (2.22):

$$C_k = F^{(k)}(0) \; . \tag{2.23}$$

Wir nehmen nun an, daß die Matrixelemente von $\varrho(t)$ exponentiell ab-fallende Funktionen sind und rechtfertigen die Annahme später, indem wir eine Lösung von (2.14) mit dieser Eigenschaft vorweisen. Das Verhältnis zweier aufeinanderfolgender Terme der k-Summe in (2.21) ist dann der Größenordnung nach:

$$\frac{\varrho^{(k)}(t)\,C_k}{\varrho^{(k-1)}(t)\,C_{k-1}} \simeq \frac{\tau_r^{-1}\,\varrho^{(k-1)}}{\varrho^{(k-1)}} \cdot \frac{F^{(k+1)}(0)}{F^{(k)}(0)} \simeq \frac{1}{\tau_r}\frac{E_a^{-1}\,F^{(k)}(0)}{F^{(k)}(0)}$$
$$= \frac{\tau_a}{\tau_r} \ll 1 \; . \tag{2.24}$$

Das bedeutet, daß für die künftige Entwicklung des Systems nur $\varrho(t)$, nicht aber die Art und Weise, wie es zu $\varrho(t)$ gekommen ist, eine Rolle spielt. Die $\delta_+(\omega)$ Funktion kann in folgender Weise zerlegt werden:

$$\delta_+(\omega) = \frac{1}{2}\left(\delta(\omega) + \frac{1}{\pi i}\,\mathscr{P}\left(\frac{1}{\omega}\right)\right) \tag{2.25}$$

wo $\mathscr{P}$ den Hauptwert eines Integrals bedeutet. Dieser Term, der die Ver-schiebung der Energieniveaus („Renormalisierung") durch die Störung $\mathscr{V}$ berücksichtigt, kann für unsere Zwecke vernachlässigt werden. Außerdem wollen wir ϱ_0 als glatt genug annehmen, so daß der inhomogene Term in (2.14) nach dem Riemann-Lebesgue-Theorem verschwindet[7-9] [19—31].

Damit sind wir zu einer kinetischen Gleichung der folgenden Art gekommen:

$$\frac{d\varrho_{ab}}{dt} = -i\,\omega_{\bar{a}\bar{b}}\,\varrho_{ab}$$
$$-\pi\,g_a\,g_b \sum_{c,\,e}\left\{V_{\bar{a}\bar{c}}\,V_{\bar{c}\bar{e}}\,\varrho_{eb}\,\delta_{cb} - V_{\bar{a}\bar{c}}\,V_{\bar{e}\bar{b}}\,\varrho_{ce}\,\delta_{ae}\right.$$
$$\left. - V_{\bar{a}\bar{c}}\,V_{\bar{e}\bar{b}}\,\varrho_{ce}\,\delta_{cb} + V_{\bar{c}\bar{e}}\,V_{\bar{e}\bar{b}}\,\varrho_{ac}\,\delta_{ae}\right\} \; . \tag{2.26}$$

Diese können wir, bezogen auf die Basis $\{|\alpha>\}$, wieder in der allgemeinen Form schreiben

$$\frac{d\varrho}{dt} = -i\,[\mathscr{H}^0, \varrho] - \pi\,[\overline{\mathscr{V}},[\overline{\mathscr{V}}, \varrho]^{(d)}] \tag{2.27}$$

---

[7] Wenn $\displaystyle\int_{-\infty}^{+\infty}|f(\omega)|\,d\omega$ existiert, dann ist
$$\lim_{t\to\infty}\int_{-\infty}^{+\infty} f(\omega)\,e^{\pm i\omega t}\,d\omega = 0$$

[8] Der allgemeine Fall ist behandelt in: Hofacker, G. L.: J. Chem. Phys. **43,** 208 (1965).

[9] Vergleiche die hier gegebene Ableitung mit E. Montroll in: Lectures in Theoretical Physics (Boulder III). New York: Interscience 1961.

$[\overline{\mathscr{V}}, \varrho]^{(d)}$ bedeutet den diagonalen Teil des Kommutators, und $\overline{\mathscr{V}}$ ist definiert durch

$$\overline{\mathscr{V}} = \sum_{\alpha, \beta} |\alpha> \overline{V}_{\alpha\beta} <\beta|$$

$$= \sum_{a, b} \sum_{\substack{\alpha \varepsilon a \\ \beta \varepsilon b}} |\alpha> V_{\overline{a}\overline{b}} <\beta| \ . \tag{2.28}$$

Wir dürfen $\overline{\mathscr{V}}$ im folgenden als reelle (und daher symmetrische) Matrix ansehen. Außerdem gilt $\overline{V}_{\alpha\beta} \neq 0$, da jedes Intervall ΔE viele Zustände von jeder Sorte enthält.

Nun können wir die Lösungen von (2.27) diskutieren. Zunächst untersuchen wir das Eigenwertproblem des linearen Gleichungssystems (2.27)

$$\mathscr{L} \varrho^{(q)} \equiv -i[\mathscr{H}^0, \varrho^{(q)}] - \pi[\overline{\mathscr{V}}[\overline{\mathscr{V}}, \varrho^{(q)}]^{(d)}] = \lambda_q \varrho^{(q)} \ , \tag{2.29}$$

wobei $\varrho^{(q)}$ eine zum Eigenwert λ_q gehörige „Eigenmatrix" des linearen Operators $\mathscr{L}$ ist. Es ist ein Vorteil, die Eigenlösungen von $\mathscr{L}$ in einem linearen Vektorraum der quadratischen Matrizen n-ter Ordnung über dem Körper der komplexen Zahlen zu suchen. Als inneres Produkt definieren wir

$$(\varrho^{(1)}, \varrho^{(2)}) = \sum_{\alpha, \beta} \varrho^{(1)*}_{\alpha\beta} \varrho^{(2)}_{\alpha\beta} \tag{2.30}$$

und wir nehmen die $\varrho^{(q)}$ als normiert an:

$$(\varrho^{(q)}, \varrho^{(q)}) = 1 \ . \tag{2.31}$$

Nun können wir zwei Theoreme beweisen:

T1) Symmetrische Eigenmatrizen von (2.29) gehören zum Eigenwert null; überdies sind sie diagonal.

Wenn $\tilde{\varrho}^{(q)}$ symmetrisch ist, verschwindet $[\overline{\mathscr{V}}, \tilde{\varrho}^{(q)}]^{(d)}$. Es bleibt

$$\tilde{\lambda}_q = -i(\tilde{\varrho}^{(q)}, [\mathscr{H}^0, \tilde{\varrho}^{(q)}])$$

oder

$$\tilde{\lambda}_q = -i \sum_{\alpha\beta} \omega_{\alpha\beta} |\tilde{\varrho}^{(q)}_{\alpha\beta}|^2 = 0 \ ,$$

da die Summe bei Vertauschung von α und β das Vorzeichen wechselt. Aus

$$-i[\mathscr{H}^0, \tilde{\varrho}^{(q)}] = 0$$

folgt

$$-i\,\omega_{\alpha\beta}\,\tilde{\varrho}^{(q)}_{\alpha\beta} = 0$$

und

$$\tilde{\varrho}^{(q)}_{\alpha\beta} = 0 \quad \text{für} \quad \alpha \neq \beta \ .$$

T2) Nichtverschwindende Eigenwerte haben einen negativen Realteil.

$\hat{\varrho}^{(q)}$ bezeichne eine nicht-symmetrische Eigenmatrix.

$$(\hat{\varrho}^{(q)}, \mathscr{L}\hat{\varrho}^{(q)}) = -i(\hat{\varrho}^{(q)}, [\mathscr{H}^0, \hat{\varrho}^{(q)}]) - \pi(\hat{\varrho}^{(q)}, [\overline{\mathscr{V}}, [\overline{\mathscr{V}}, \hat{\varrho}^{(q)}]^{(d)}])$$

$$= -i\sum_{\alpha\beta} \omega_{\alpha\beta} |\hat{\varrho}^{(q)}_{\alpha\beta}|^2 - \pi\sum_{\alpha\beta} \hat{\varrho}^{(q)}_{\alpha\beta} \overline{V}_{\alpha\beta}\{[\overline{\mathscr{V}}, \varrho]_{\beta\beta} - [\overline{\mathscr{V}}, \varrho]_{\alpha\alpha}\}$$

$$= -i\sum_{\alpha\beta} \omega_{\alpha\beta} |\hat{\varrho}^{(q)}_{\alpha\beta}|^2 - \pi\sum_{\alpha} |[\overline{\mathscr{V}}, \hat{\varrho}^{(q)}]_{\alpha\alpha}|^2 = \hat{\lambda}_q$$

d. h. $\mathrm{Re}\,\hat{\lambda}_q < 0$.

Betrachten wir zunächst den diagonalen Teil von (2.27). Es ist

$$\dot{\varrho}_{\alpha\alpha} = 0 \quad \text{oder} \quad \varrho^{(t)}_{\alpha\alpha} = \varrho_{0\alpha\alpha} = \text{const}.$$

Wie zu erwarten, bleiben auf der Zeitskala τ_r die Diagonalelemente von ϱ konstant. Sie müssen notwendigerweise gleich denen von ϱ_0 sein, weil keine äußere Wechselwirkung besteht.

Die Lösung für den nichtdiagonalen Teil von (2.27) können wir ebenfalls sofort angeben.

Ist

$$\varrho_0 = \sum_q d_q \varrho^{(q)}, \tag{2.32}$$

so haben wir

$$\varrho(t) = \sum_q \varrho^{(q)} d_q e^{+\lambda_q t}. \tag{2.33}$$

Nach T1) bleibt die Spur von $\varrho(t)$ zeitlich konstant, wie es sein muß, nach T2) verschwinden die nichtdiagonalen Anteile von ϱ_0 exponentiell. Die Dichtematrix des inneren Gleichgewichtes ist daher

$$\varrho(\infty) = \varrho_0^{(d)}. \tag{2.34}$$

Die chemische Reaktionsgeschwindigkeit R ist schließlich gegeben durch:

$$R = N \cdot \frac{d}{dt} \mathrm{Spur}\,(\mathscr{H}^{A2} \varrho \, \mathscr{H}^{A2}) \tag{2.35}$$

$$= N \cdot \mathrm{Spur}\left(\mathscr{H}^{A2} \frac{d\varrho}{dt} \mathscr{H}^{A2}\right)$$

mit $\dfrac{d\varrho}{dt} = \sum_{\hat{q}} \varrho^{(q)} \hat{\lambda}_q e^{\hat{\lambda}_q t} d_q$

kommt man auf

$$R = \sum_{\hat{q}} \hat{\lambda}_q e^{\hat{\lambda}_q t} d_q \, \mathrm{Spur}\,(\mathscr{H}^{A2} \hat{\varrho}^{(q)} \mathscr{H}^{A2}). \tag{2.36}$$

3. Die Molekel gekoppelt an einen Thermostaten

Die im vorigen Abschnitt entwickelte Methode zur Ableitung einer Transportgleichung für die Dichtematrix soll nun auf den Fall zweier in Wechselwirkung befindlicher Systeme angewandt werden. Dabei soll das System A von molekularer Ausdehnung sein, das System B jedoch sehr groß, so daß A für B nur eine kleine Störung bedeutet. Weiterhin werden wir annehmen, daß die Kopplung zwischen A und B schwach ist. Den Begriff der schwachen Kopplung wollen wir dahin gehend präzisieren, daß die Wechselwirkung von A mit B „stoßartig" sein soll und daß zwei charakteristische Zeiten τ_c und τ_0 dem Gesamtsystem eigen sind, von denen τ_c die Zeit zwischen zwei Stößen und τ_0 die Dauer eines Stoßes angibt. Über die Art und Stärke der Störung durch die Stöße, die wir durch den hermitischen Operator $\mathcal{V}^I$ darstellen wollen, müssen hier keine weiteren Annahmen gemacht werden, außer daß $\tau_c \gg \tau_0$.

Im einfachsten Fall ist A eine diatomige Molekel, die durch Stöße mit anderen Molekeln zur Dissoziation gebracht wird. Vom Standpunkt der statistischen Mechanik aus ist dieses Problem jedoch weniger schwierig, da (wenn man von der Kopplung von Schwingung und Rotation absieht) nur ein einfacher Relaxationsmechanismus, betreffend die Diagonalelemente der Dichtematrix ϱ^A, vorliegt. Wir wollen hier vor allem untersuchen, welche kinetischen Gleichungen sich im Falle zweier interferierender Relaxationsmechanismen (eines äußeren und eines inneren) ergeben und welche Bedeutung diese für die Theorie monomolekularer Reaktionen haben. Die Pauli-Gleichung (master equation) für diatomige Molekeln folgt daraus als Spezialfall. Im übrigen gilt der ganze Formalismus nicht nur für Gase. Er kann z. B. auch auf Transportvorgänge in nahezu idealen Kristallen (Massendiffusion, Exitondiffusion) Anwendung finden.

Wir betrachten ein System, das im einzelnen die folgenden Eigenschaften hat.

1. A kann zwei oder mehr chemische Konfigurationen einnehmen (durch Dissoziation oder intramolekulare Umlagerung). Außerdem hat A eine genügend große Zahl von Freiheitsgraden, um auf einer geeignet gewählten Zeitskala zu intramolekularer Relaxation zu gelangen. Das isolierte System A beschreiben wir durch eine Hamiltonfunktion

$$\mathcal{H}^A_{ges} = \mathcal{H}^A + \mathcal{V}^A, \tag{3.1}$$

wo $\mathcal{H}^A$ der ungestörten molekularen Hamiltonfunktion des vorhergehenden Paragraphen entspricht. Die Eigenfunktionen $\{|\alpha>\}$ von $\mathcal{H}^A$ werden durch griechische Buchstaben gekennzeichnet, und es ist

$$\mathcal{H}^A |\alpha> = E_\alpha |\alpha> . \tag{3.2}$$

Darüber hinaus existieren konfigurationsgebundene Hamiltonoperatoren $\mathcal{H}^{A1}$ und $\mathcal{H}^{A2}$ (eventuell mehr, wenn mehrere chemische Konfigurationen

für A möglich sind) mit Eigenfunktionen $\{|\alpha_1>\}$ resp. $\{|\alpha_2>\}$, und Eigen-
wertgleichungen

$$\mathscr{H}^{A1}|\alpha_1> = E_{\alpha_1}^{A1}|\alpha_1>$$
$$\mathscr{H}^{A2}|\alpha_2> = E_{\alpha_2}^{A2}|\alpha_2> \, .$$

$$(3.3)$$

Wir nehmen an, daß für das isolierte A die charakteristischen Zeiten die
Relation

$$\tau_a^A \ll \tau_r^A \ll \tau_{tr}^A$$

erfüllen.

2. Das System B sei dadurch gekennzeichnet, daß es groß gegen A ist
und seine Eigenzustände ein Kontinuum bilden. Ferner möge B entweder
im Gleichgewicht sein oder durch innere Kopplungen ins Gleichgewicht
gebracht werden, die so schwach sind, daß

$$\tau_r^B \gg \tau_r^A \, .$$

Wir bezeichnen die Eigenfunktionen von B mit großen lateinischen Buch-
staben, z. B. $\{|B>\}$. Das isolierte System B gehorche der Eigenwert-
gleichung

$$\mathscr{H}^B|B> = E_B|B> \, .$$

$$(3.4)$$

3. Beide Systeme sind durch den Störoperator $\mathscr{V}^I$ gekoppelt, der zu
stoßartiger Wechselwirkung mit einer Stoßdauer kurz gegen die Zeit
zwischen zwei Stößen führt. Der Hamiltonoperator des Gesamtsystems ist
dann

$$\mathscr{H} = \mathscr{H}^A + \mathscr{V}^A + \mathscr{H}^B + \mathscr{V}^I,$$

$$(3.5)$$

der des ungestörten Systems

$$\mathscr{H}^0 = \mathscr{H}^A + \mathscr{H}^B \, .$$

Sind wir nur an einer Beschreibung des Systems auf einer Zeitskala $\ll \tau_r^A$
interessiert, so mag auch

$$\tilde{\mathscr{H}}^0 = \mathscr{H}^{A1} + \mathscr{H}^{A2} + \mathscr{H}^B$$

$$(3.6)$$

als Hamiltonoperator des ungestörten Systems dienen. Je nach der interessierenden Zeitskala werden wir

$$\{|\alpha>|B>\} \quad \text{oder} \quad \{|\alpha_1>|B>, |\alpha_2>|B>\}$$

als Basisfunktionen verwenden, wobei wir sicher sein können, daß die Ab-
weichungen zwischen den $<\alpha_1|$ und $<\alpha_2|$ von der Orthogonalität auf ent-
sprechend kurzer Zeitskala keine Rolle spielen wird.

Die Dichtematrix ϱ des Gesamtsystems gehorcht der v. Neumann-
Gleichung

$$i\,\dot{\varrho} = [\mathscr{H}, \varrho] = [\mathscr{H}^0 + \mathscr{V}^A + \mathscr{V}^I, \varrho] \, ,$$

$$(3.7)$$

und wir definieren sinngemäß als Dichtematrizen der Systeme A und B

$$\varrho^A = \operatorname{Spur}_B \varrho \,,$$

$$\varrho^B = \operatorname{Spur}_A \varrho \,. \tag{3.8}$$

Unsere Aufgabe wird es sein, eine kinetische Gleichung für ϱ^A zu finden.

Aus den bisher gemachten Annahmen ergibt sich noch keine vollständige Hierarchie der Zeitskalen. Bezüglich A sollte die für die meisten chemischen Systeme gültige Relation

$$\tau_a^A \ll \tau_m^A \ll \tau_r^A \ll \tau_{tr}^A \tag{3.9}$$

bestehen. B war hinreichend charakterisiert durch

$$\tau_a^B \ll \tau_r^A \ll \tau_r^B \ll \tau_{tr}^B \,, \tag{3.10}$$

und die Wechselwirkung genügte der Ungleichung

$$\tau_0 \ll \tau_c \,. \tag{3.11}$$

τ_0 hat wiederum die Bedeutung und Größenordnung einer atomaren Zeit, was uns erlaubt, die Zeitskala nach unten durch eine einzige atomare Zeit, τ_a, zu begrenzen.

Die relativen Größenordnungen von τ_m^A, τ_c und τ_r^A bedürfen jedoch näherer Betrachtung. τ_m^A, das wir, solange es $\gg \tau_a$ und $\ll \tau_r$ ist, frei wählen können, muß auch klein gegenüber τ_c sein, da wir sonst die Möglichkeit einer Änderung der chemischen Spezies durch die Stöße ganz oder teilweise eliminieren würden (wir kämen dann wieder zu den Voraussetzungen des Kapitels 2). Bezüglich τ_c und τ_r^A können wir sagen, daß der Fall $\tau_c > \tau_r^A$ chemisch gesehen eine sehr schnelle bimolekulare Reaktion zwischen A und Teilen von B bedeutet. τ_r^A wäre dann die mittlere Zeit bis zur erstmaligen Reaktion der Molekel. Wir schließen diesen Fall hier mit ein, obgleich es natürlicher ist, ihn auf der Grundlage einer Boltzmann-Gleichung zu behandeln. Die meisten chemisch interessanten Systeme genügen der Beziehung $\tau_c \ll \tau_r^A$, d. h. es müssen im Mittel mehrere Stöße das molekulare System A treffen, bis es soweit aktiviert ist, daß der intramolekulare Mechanismus $\mathscr{V}^A$ vor der Desaktivierung die Reaktion bewerkstelligen kann. Der Fall $\tau_c \gtrsim \tau_r^A$ spielt hauptsächlich bei Desaktivierung und Quenching-Prozessen einer „heißen" bzw. angeregten Molekel A eine Rolle.

Zunächst bringen wir (3.7) in die Dämpfungsform

$$\dot{\varrho} = - i\,[\mathscr{H}^0, \varrho] - i\,[\mathscr{V}^A + \mathscr{V}^I, e^{-i\mathscr{H}^0 t}\,\varrho_0\,e^{i\mathscr{H}^0 t}]$$

$$- [\mathscr{V}^A + \mathscr{V}^I, \int_0^t e^{-i\mathscr{H}^0 \tau}[\mathscr{V}^A + \mathscr{V}^I, \varrho\,(t - \tau)]\,e^{i\mathscr{H}^0 \tau}\,d\tau] \,. \tag{3.12}$$

Wir dürfen annehmen, daß zur Zeit $t = 0$ die Systeme A und B gerade nicht wechselwirken und daher

$$\varrho_0 = \varrho_0^A \, \varrho_0^B \tag{3.13}$$

ist, mit diagonalem ϱ_0^B. Da sich A zur Anfangszeit in einer definierten chemischen Konfiguration befindet, ist ϱ_0^A nicht notwendig diagonal bezüglich $\mathcal{H}^0$; nur für einfache Dissoziationsreaktionen darf es diagonal gemacht werden. ϱ_0^A soll außerdem einem τ_m wie in (3.9) entsprechen. Das wird uns später erlauben, den inhomogenen Term in (3.12) zu vernachlässigen. Weiterhin ist es uns freigestellt, was wir als ungestörten Hamilton-operator und was als Störung $\mathscr{V}^A$ und $\mathscr{V}^I$ ansehen wollen. Aus Gründen der Einfachheit wählen wir $\mathscr{V}^A$ so, daß

$$< B \,|\, \mathscr{V}^A \,|\, C > = \delta_{BC}\, \mathscr{V}^A \tag{3.14a}$$

und $$< \alpha \,|\, \mathscr{V}^A \,|\, \alpha > = 0 \ . \tag{3.14b}$$

Desgleichen bestimmen wir für $\mathscr{V}^I$

$$< \alpha \,|\, < B \,|\, \mathscr{V}^I \,|\, B > \,|\, \alpha > = 0 \, . \tag{3.15}$$

Die Operation Spur auf die Gleichung (3.12) angewandt ergibt nun

$$\dot{\varrho}^{\,A} = -\,i\,[\mathscr{H}^A, \varrho^A] - i\,\underset{B}{\mathrm{Spur}}\,[\mathscr{V}^A + \underset{B}{\mathrm{Spur}}\,\mathscr{V}^I \varrho_0^B,\, \mathrm{e}^{-i\mathscr{H}^A t}\, \varrho_0^A\, \mathrm{e}^{i\mathscr{H}^A t}]$$
$$-\,\underset{B}{\mathrm{Spur}}\,[\mathscr{V}^A + \mathscr{V}^I,\, \int\limits_0^t \mathrm{e}^{-i\mathscr{H}^0 \tau}\,[\mathscr{V}^A + \mathscr{V}^I,\, \varrho\,(t-\tau)]\, \mathrm{e}^{i\mathscr{H}^0 \tau}\, d\tau]\, . \tag{3.16}$$

Da das System B sehr groß ist (praktisch unendlich groß), dürfen wir die Matrixelemente $\mathscr{V}^I_{A\alpha B\beta}$ als stetige Funktionen von E_A und E_B ansehen. In bezug auf E_α und E_β sind jedoch $\mathscr{V}^I_{A\alpha B\beta}$ wie auch $\mathscr{V}^A_{A\alpha B\beta}$ rapide fluktuierend, und das gleiche gilt für die Elemente der Dichtematrizen ϱ^A und ϱ. Nun können wir die gleichen Überlegungen wie in Abschnitt 2 anstellen. Wir fragen nach einer asymptotischen Lösung der Gleichung (3.16) auf der Zeitskala τ_r, und zwar für Dichtematrizen ϱ^A bzw. ϱ, die bezüglich ihrer Abhängigkeit von E_α über Intervalle der Größenordnung τ_m^{-1} gemittelt worden sind. Das verwandelt ϱ^A, ϱ, $\mathscr{V}^A$ und $\mathscr{V}^I$ in die gemittelten Größen $\bar{\varrho}^A$, $\bar{\varrho}$, $\overline{\mathscr{V}}^A$ und $\overline{\mathscr{V}}^I$. Statt $\bar{\varrho}^A$ und $\bar{\varrho}$ wollen wir jedoch weiterhin ϱ^A und ϱ schreiben. Da wir $\varrho_{0\alpha\beta}$ als eine genügend glatte Funktion von E_α und E_β angenommen haben, verschwindet der inhomogene Term in (3.16) nach dem Riemann-Lebesgue-Theorem, so daß

$$\dot{\varrho}^A = -\,i\,[\mathscr{H}^A, \varrho^A] - \underset{B}{\mathrm{Spur}}\,[\overline{\mathscr{V}}^A + \overline{\mathscr{V}}^I,\, \int\limits_0^t \mathrm{e}^{-i\mathscr{H}^0 \tau}\,[\overline{\mathscr{V}}^A + \overline{\mathscr{V}}^I,\, \varrho\,(t-\tau)]\, \mathrm{e}^{i\mathscr{H}^0 \tau}\, d\tau]\, . \tag{3.17}$$

Zu einer kinetischen Gleichung für ϱ^A können wir nur gelangen, wenn wir die genaue Gestalt von ϱ, das in dem Integralausdruck in (3.17) auftritt, kennen. Es sei bemerkt, daß ϱ wegen der Wechselwirkung nicht allgemein die Gestalt $\varrho^A \varrho^B$ hat. Da wir es mit einer stoßartigen Wechselwirkung zu tun haben, ist es jedoch sinnvoll, nach der Form von ϱ zwischen zwei Stößen und der Änderung von ϱ durch einen einzigen Stoß zu fragen.

Ist t eine Zeit zwischen zwei Stößen, so können wir zunächst unter dem Integralzeichen

$$\varrho\,(t-\tau) = \varrho^A(t-\tau)\,\varrho^B(t-\tau) \text{ für } 0 \leqslant t-\tau \leqslant t_1$$

setzen, wobei t_1 das Ende des vorausgehenden Stoßes bezeichnet. Da $\tau_c \gg \tau_0$, ist $t-t_1$ fast zu jeder Zeit von der Größenordnung τ_c, und die Entwicklung des Systems ist durch $\mathscr{V}^A$ allein bestimmt. Wir wissen aus den Überlegungen des vorigen Abschnittes, daß dann der bezüglich $\mathscr{H}^0$ nichtdiagonale Anteil von ϱ^A, den wir mit $\hat{\varrho}^A$ bezeichnen wollen, exponentiell zerfällt, während $\tilde{\varrho}^A$ konstant bleibt. Entwickeln wir $\varrho^A(t-\tau)$ wie vorher in eine Taylor-Reihe, so läßt sich nach der gleichen Methode die Kleinheit der Integrale mit höheren Potenzen von τ zeigen. Allerdings wird nun das Verhältnis aufeinanderfolgender Terme (statt $\tau_a|\tau_r$) $\tau_a|\tau_c \ll 1$ sein, da der exponentielle Zerfall von $\hat{\varrho}^A$ größenordnungsmäßig nur eine Zeit τ_c ungestört vor sich gehen kann. Der nächstfolgende Stoß bringt daher die Systeme A und B mit einem anfänglichen Dichteoperator

$$\varrho\,(t_0) = \varrho^A(t_0)\,\varrho^B(t_0)\,,$$

wo t_0 eine geeignet festgesetzte Zeit vor Beginn des Stoßes ist, in Wechselwirkung.

Ist t eine Zeit während des Stoßes, so müssen wir das τ-Integral in (3.17) bis vor den Stoß, d. h. bis etwa zur Zeit t_0 zurückverfolgen, jedoch nicht weiter, da die Wirkung von $\mathscr{V}^A$ zwischen dem betrachteten und dem vorangehenden Stoß alle unterhalb der Zeitskala τ_0 bestehenden Phasenbeziehungen (oder jegliche länger als τ_0 zurückreichende „Erinnerung" des Systems) ausgelöscht hat. Wir können also in (3.17) $\int\limits_0^t d\tau$ durch $\int\limits_{t_0}^t d\tau$ ersetzen und brauchen deshalb die v. Neumann-Gleichung (3.12) für ϱ nur auf der Zeitskala eines einzigen Stoßes zu betrachten.

Wie wir gleich sehen werden, ist es natürlicher, die Gleichung (3.17) in Beziehung zur Darstellung $\{|\alpha_1> |B>, |\alpha_2> |B>\}$ zu bringen. Da

$$\mathscr{X}^{A1} + \mathscr{X}^{A2} \approx 1$$

ist, können wir ϱ so zerlegen:

$$\varrho = (\mathscr{X}^{A1} + \mathscr{X}^{A2})\,\varrho\,(\mathscr{X}^{A1} + \mathscr{X}^{A2}) = \varrho^{A1B} + \varrho^{A2B} + \varrho^{A12B} \qquad (3.18)$$

mit

$$\varrho^{A1B} = \sum_{\alpha_1\,\beta_1\,B} |\alpha_1> |B> P_{\alpha_1\beta_1 B} <B| <\beta_1| \qquad (3.19\text{ a})$$

und

$$\varrho^{A12B} = \sum_{\alpha_1\,\alpha_2\,B} \{|\alpha_1> |B> P_{\alpha_1\alpha_2 B} <B| <\alpha_2| + |\alpha_2> |B> P_{\alpha_2\alpha_1 B} <B| <\alpha_1|\}.$$

$$(3.19\text{ b})$$

7*

Die Wirkung der Operatoren $\mathscr{V}^A$ und $\mathscr{V}^I$ auf die Dichtematrizen (3.19) ist nun physikalischer Interpretation zugänglich. Es ist praktisch, auch die chemische Relaxationszeit des Systems A ohne äußere Stöße, τ_r^{ch}, und die Relaxationszeit für die innere Energie der Systeme A 1 und A 2, τ_r^{coll}, einzuführen. τ_r ist dann etwa gleich der längeren dieser beiden Relaxationszeiten. Nun können wir im einzelnen die folgenden Annahmen formulieren:

a) $\mathscr{V}^I \varrho^{A1B}$ liegt im Teilraum $\{|\alpha_1> |B> <B| <\beta_1|\}$ des gesamten Funktionenraumes, und $[\mathscr{V}^I, \varrho^{A12B}] = 0$, d. h. die Stöße wirken auf die jeweilige chemische Spezies (Reaktant oder eine der Produktspezies), ohne Übergänge zwischen den chemischen Spezies zu bewirken.

b) Auf einer Zeitskala $\ll \tau_r^{ch}$ wird ϱ^{A12B} durch $\mathscr{V}^A$ so gut wie nicht geändert, d. h., wir dürfen unter dem Integralzeichen den Kommutator $[\mathscr{V}^A, \varrho^{A12B}] = 0$ setzen, weil $\mathscr{V}^A$ keine Übergänge zwischen Zuständen $|\alpha_1>$ und $|\alpha_2>$ während einer Zeit $\ll \tau_r^{ch}$ zustande bringt.

Die Annahmen a) und b) haben zur Folge, daß aus (3.17) zwei relevante Gleichungen für ϱ^{A1} und ϱ^{A2} hervorgehen:

$$\dot{\varrho}^{\,A1B} = -i\,[\mathscr{H}^0, \varrho^{A1B}] - [\overline{\mathscr{V}}^A + \overline{\mathscr{V}}^I, \int_{t_0}^{t} e^{-i\mathscr{H}^0\tau} [\overline{\mathscr{V}}^A + \overline{\mathscr{V}}^I, \varrho^{A1B}(t-\tau)] \times$$

$$\times\, e^{i\mathscr{H}^0\tau}\, d\tau] \qquad\qquad (3.20)$$

und eine analoge Gleichung für ϱ^{A2B}.

In (3.20) ist

$$\mathscr{H}^0 = \mathscr{H}^{A1} + \mathscr{H}^{A2} + \mathscr{H}^{A12} + \mathscr{H}^B\,.$$

$\mathscr{H}^{A12}$ bewirkt Aufspaltung oder Verbreiterung des Niveaus von A1 durch Tunnelieren zwischen den Konfigurationen A1 und A2. Dieser Vorgang spielt sich jedoch auf der Zeitskala der inneren chemischen Relaxation, τ_r^{ch}, wie sie im vorigen Abschnitt 2 besprochen wurde, ab. Auf der Zeitskala $\tau_c \ll \tau_r^{ch}$ kann diese Störung daher vernachlässigt werden. Ferner wissen wir, daß vor Einsetzen der Wechselwirkung $\mathscr{V}^I$ die innere Störung $\mathscr{V}^A$ die Tendenz hat, ϱ^{A1B} diagonal in $\{|\alpha_1> |B>\}$ zu machen. Dafür gibt es eine charakteristische Zeit, τ_r^i. Wenn wir nun t_0 weit genug in die Vergangenheit verschieben, so daß zu einer Zeit t_1, noch vor Einsetzen der Wechselwirkung $\mathscr{V}^I$, $|t_1 - t_0| \simeq \tau_r^i$, dann ist ϱ^{A1B} bei Einsetzen der Wechselwirkung diagonal in $\{|\alpha_1> |B>\}$, d. h. $\varrho^{A1B} = \tilde{\varrho}^{A1}\,\tilde{\varrho}^B$. Diese Überlegung gilt stets, solange $\tau_c > \tau_r^i$. Wird jedoch $\tau_c \ll \tau_r^i$, so fällt die Annahme a), weil zwischen aufeinanderfolgenden Stößen Phasenbeziehungen bestehen bleiben, die sich schließlich in stoßinduzierten Übergängen zwischen den chemischen Spezies bemerkbar machen.

Da ϱ^{A1B} noch vor Einsetzen der Stoßwechselwirkung diagonal geworden ist, andererseits aber die Störung $\overline{\mathscr{V}}^A$ auf der Zeitskala τ_0 neben $\overline{\mathscr{V}}^I$ unbeachtet bleiben kann, geht (3.20) über in

$$\dot{\varrho}^{A1B} = -i\,[\tilde{\mathscr{H}}^0, \varrho^{A1B}] - [\overline{\mathscr{V}}^I, \int_{t_1}^{t} e^{-i\check{\mathscr{H}}\tau}[\overline{\mathscr{V}}^I, \varrho^{A1B}(t-\tau)]\,e^{i\check{\mathscr{H}}\tau}d\tau]$$

$$(3.21)$$

$$\text{mit}\quad \tilde{\mathscr{H}}^0 = \mathscr{H}^{A1} + \mathscr{H}^{A2} + \mathscr{H}^{B}\,.$$

Die untere Grenze des Integrals, t_1, bedeutet eine Zeit vor Einsetzen der Wechselwirkung und kann daher $-\infty$ gesetzt werden. Die Lösung von (3.21) kann sofort angegeben werden. Sie ist

$$\varrho^{A1B} = \Omega^{\oplus}e^{-i\tilde{\mathscr{H}}^0(t-t_0)}\varrho^{A1B}(t_1)\,e^{i\tilde{\mathscr{H}}^0(t-t_0)}\Omega^{\oplus+}$$
$$= \Omega^{\oplus}\tilde{\varrho}^{A1}(t_1)\,\tilde{\varrho}^{B}(t_1)\,\Omega^{\oplus+}\,.$$

$$(3.22)$$

Wir können die Lösung (3.22) sogar als gültig in jedem Augenblick annehmen, da zu einer Zeit t, während der kein Stoß erfolgt, einfach

$$\Omega^{\oplus}\tilde{\varrho}^{A1}\varrho^{B}\Omega^{\oplus+} = \tilde{\varrho}^{A1}\tilde{\varrho}^{B}$$

ist.

Wir schreiben nun (3.21) in der Form

$$\Omega^{\oplus}\tilde{\varrho}^{A1}(t_1)\,\tilde{\varrho}^{B}(t_1)\,\Omega^{\oplus+} = \tilde{\varrho}^{A1}(t_1)\,\tilde{\varrho}^{B}(t_1)$$
$$-\,i\int_{-\infty}^{t}e^{-i\check{\mathscr{H}}\tau}[\overline{\mathscr{V}}^I, \Omega^{\oplus}\tilde{\varrho}^{A1}(t_1)\,\tilde{\varrho}^{B}(t_1)\,\Omega^{\oplus+}]\,e^{i\check{\mathscr{H}}\tau}\,d\tau$$

$$\text{mit}\qquad\qquad(3.23)$$

$$\tilde{\varrho}^{A1}(t_1)\,\tilde{\varrho}^{B}(t_1) = e^{-i\mathscr{H}^0(t-t_1)}\varrho^{A1}(t_0)\,\varrho^{B}(t_0)\,e^{i\mathscr{H}^0(t-t_1)}$$
$$-\,i\int_{t_0}^{t}e^{-i\mathscr{H}^0\tau}[\overline{\mathscr{V}}^A, \varrho^{A1B}(t-\tau)]\,e^{i\mathscr{H}^0\tau}\,d\tau$$

und substituieren die Gleichungen (3.23) für ϱ^{A1B} und ϱ^{A2B} zurück in (3.17). Da $|t-t_1| \ll \tau_r^{ch}$, können wir unter der Operation Spur die Zeit t_1 durch t ersetzen.

Ferner verschwindet der inhomogene Term der zweiten Gleichung (3.23) nach dem Riemann-Lebesgue-Theorem aus der substituierten Gleichung. Der Integralterm der zweiten Gleichung (3.23) gibt nach der bekannten Behandlung wieder $-i[\overline{\mathscr{V}}^A, \varrho^{A1}\varrho^{B}]^{(d)}$. Somit wird aus (3.17):

$$\dot{\varrho}^{A} = -i\,[\mathscr{H}^A, \varrho^A] - i\,\underset{B}{\text{Spur}}\,[\overline{\mathscr{V}}^A + \mathscr{V}^I, \Omega^{\oplus}\tilde{\varrho}^{A1}\varrho^{B}\Omega^{\oplus+} + \Omega^{\oplus}\tilde{\varrho}^{A2}\varrho^{B}\Omega^{\oplus+}]$$

$$-\,[\overline{\mathscr{V}}^A + \underset{B}{\text{Spur}}\,\overline{\mathscr{V}}^I\varrho^{B}, [\overline{\mathscr{V}}^A, \varrho^{A12}]^{(d)}]$$

$$-\,[\overline{\mathscr{V}}^A + \underset{B}{\text{Spur}}\,\overline{\mathscr{V}}^I\varrho^{B}, [\overline{\mathscr{V}}^A, \varrho^{A1} + \varrho^{A2}]^{(d)}]\qquad(3.24)$$

$$= -\,i\,[\mathscr{H}^A, \varrho^A] - i\,\underset{B}{\text{Spur}}\,[\overline{\mathscr{V}}^A + \mathscr{V}^I, \Omega^{\oplus}(\tilde{\varrho}^{A1} + \tilde{\varrho}^{A2})\,\varrho^{B}\,\Omega^{\oplus+}]$$

$$-\,[\overline{\mathscr{V}}^A + \underset{B}{\text{Spur}}\,\overline{\mathscr{V}}^I\varrho^{B}, [\overline{\mathscr{V}}^A, \varrho^{A}]^{(d)}]$$

oder

$$
\begin{aligned}
\dot{\varrho}^{A} = &- i\,[\mathscr{H}^{A}, \varrho^{A}] \\
&- i\,\mathop{\mathrm{Spur}}_{B}\,[\overline{\mathscr{V}}^{I}, \Omega^{\oplus}(\varrho^{A1} + \tilde{\varrho}^{A2})\,\varrho^{B}\,\Omega^{\oplus+}] \\
&- i\,[\overline{\mathscr{V}}^{A}, \mathop{\mathrm{Spur}}_{B}\,\Omega^{\oplus}(\tilde{\varrho}^{A1}+\tilde{\varrho}^{A2})\,\varrho^{B}\,\Omega^{\oplus+}] \\
&- [\overline{\mathscr{V}}^{A}, [\overline{\mathscr{V}}^{A}, \varrho^{A}]^{(d)}] \\
&- [\mathop{\mathrm{Spur}}_{B}\,\overline{\mathscr{V}}^{I}\,\varrho^{B}, [\overline{\mathscr{V}}^{A}, \varrho^{A}]^{(d)}]\ .
\end{aligned}
\tag{3.25}
$$

Dabei ist in den Termen der intramolekularen Wechselwirkung die Renormalisierung vernachlässigt worden.

Den ersten Wechselwirkungsterm auf der rechten Seite von (3.25), wir wollen ihn $\mathscr{T}$ nennen, betrachten wir uns genauer. Mit Hilfe der Lippmann-Schwinger-Gleichung (1.26) können wir $\Omega^{\oplus}$ und $\overline{\mathscr{V}}^{I}$ durch Operatoren $\mathscr{G}^{\oplus}$ und $t^{\oplus}$ ausdrücken. Das ergibt

$$
\begin{aligned}
\mathscr{T}\varrho^{A}\varrho^{B} = &- i\,\mathop{\mathrm{Spur}}_{B}\,\big\{t^{\oplus}(\tilde{\varrho}^{A1} + \tilde{\varrho}^{A2})\,\varrho^{B} - (\tilde{\varrho}^{A1} + \tilde{\varrho}^{A2})\,\varrho^{B}\,t^{\oplus+} \\
&- i\,\mathop{\mathrm{Spur}}_{B}\,\big\{t^{\oplus}(\tilde{\varrho}^{A1} + \tilde{\varrho}^{A2})\,\varrho^{B}\,t^{\oplus+}\mathscr{G}^{\oplus+} - \mathscr{G}^{\oplus}\,t^{\oplus}(\tilde{\varrho}^{A1} + \tilde{\varrho}^{A2})\,\varrho^{B}\,t^{\oplus+}\big\}\ .
\end{aligned}
\tag{3.26}
$$

Nur der zweite der Terme von (3.26), nennen wir ihn $\mathscr{T}^{0}\varrho^{A}\varrho^{B}$, hat eine nichtverschwindende Diagonale in der Darstellung $\{\,|\alpha_{1}> |B>,\ |\alpha_{2}> |B>\,\}$. Sein α_{1}, β_{1}-Matrixelement ist

$$
(\mathscr{T}^{0}\varrho^{A}\varrho^{B})_{\alpha_{1}\beta_{1}} = - i\sum_{B} 2\,Im\,(t^{\oplus}\tilde{\varrho}^{A1}\varrho^{B}\,t^{\oplus+}\mathscr{G}^{\oplus+})_{\alpha_{1}\beta_{1}}\ .
\tag{3.27}
$$

Allgemein gilt

$$
\begin{aligned}
Im\,\mathscr{G}^{\oplus\,+}_{A\alpha\,B\beta} &= \frac{1}{2}\,(\mathscr{G}^{\oplus+} - \mathscr{G}^{\oplus})_{A\alpha\,B\beta} \\
&= i\,\pi\,\delta\,(\tilde{E}_{A\alpha} - \tilde{\mathscr{H}})\,|B> |\beta> \\
&= i\,\pi\,\delta\,(\tilde{E}_{A\alpha} - \tilde{E}_{B\beta})\ ,
\end{aligned}
\tag{3.28}
$$

womit wir den Stoßterm $\mathscr{T}$ schreiben können als

$$
\begin{aligned}
\mathscr{T}\,\varrho^{A}\varrho^{B} = &- i\,\mathop{\mathrm{Spur}}_{B}\,\big\{t^{\oplus}(\tilde{\varrho}^{A1} + \tilde{\varrho}^{A2})\,\varrho^{B} - (\tilde{\varrho}^{A1} + \tilde{\varrho}^{A2})\,\varrho^{B}\,t^{\oplus+}\big\} + \\
&+ 2\pi\,\mathop{\mathrm{Spur}}_{B}\,(-\,i\,Im)\,\big\{t^{\oplus}(\varrho^{A1} + \tilde{\varrho}^{A2})\,\varrho^{B}\,t^{\oplus+}\,\delta\,(\tilde{E} - \tilde{\mathscr{H}})\big\}\ .
\end{aligned}
\tag{3.29}
$$

In der Gestalt (3.29) kann der Stoßterm $\mathscr{T}$ leicht mit dem Stoßterm einer Boltzmann-Gleichung identifiziert werden. Nimmt man an, daß das System B ein Boltzmann-Gas ist, mit

$$
\varrho^{B} = \mathscr{S}\,\varrho^{(1)}\,(1)\,\varrho^{(2)}\,(2) \ldots \varrho^{(N)}\,(N)\ ,
$$

wo die Argumente (1) bis (N) die Koordinaten der 1. bis N-ten Molekel von B darstellen, und $\mathscr{S}$ einen Symmetrisierungsoperator, so wird nach Einführung des Wirkungsquerschnittes

$$\sigma(A\,\alpha_1 \,|\, B\,\beta_1) = (2\,\pi)^4\,M_r^2\,\frac{k_{B\,\beta_1}}{k_{A\,\alpha_1}}\,|\,t^{\oplus}_{B\,\beta_1,\,A\,\alpha_1}\,|^2 \tag{3.30}$$

aus dem Diagonalteil von $\mathscr{T}^0$ nach einigen elementaren Umformungen:

$$(\mathscr{T}\,\varrho^A\,\varrho^B)_{A\alpha,\,A\alpha} = \frac{1}{(2\,\pi)^3}\,\frac{1}{M_r^2}\,\sum_{BC\gamma}\,\delta(\tilde{E}_{A\alpha} - \tilde{E}_{C\gamma})\times$$

$$\times\left\{\frac{k_{C\gamma_1}}{k_{A\alpha_1}}\,\varrho^{A1}(\gamma_1)\,\varrho^B(C)\,\sigma(C\gamma_1\,|\,A\,\alpha_1) - \frac{k_{A\alpha_1}}{k_{C\gamma_1}}\,\varrho^{A1}(\alpha_1)\,\varrho^{B2}(B)\,\sigma(A\,\alpha_1\,|\,C\gamma_1) + \right.$$

$$\left. + \frac{k_{C\gamma_2}}{k_{A\alpha_2}}\,\varrho^{A2}(\gamma_2)\,\varrho^B(C)\,\sigma(C\gamma_2\,|\,A\,\alpha_2) - \frac{k_{A\alpha_2}}{k_{C\gamma_2}}\,\varrho^{A2}(\alpha_2)\,\varrho^{B2}(B)\,\sigma(A\,\alpha_2\,|\,C\gamma_2)\right\}\,. \tag{3.31}$$

In den Gleichungen (3.30) und (3.31) bedeuten die großen lateinischen Buchstaben A, B, C, ... nun die Zustände einer individuellen Molekel des Systems B (B soll überdies nur aus einer einzigen Molekelsorte bestehen). Gleichung (3.31) ist dann nichts anderes als der Stoßterm einer Boltzmann-Gleichung. In dieser Form war der Stoßterm zuerst von UEHLING und UHLENBECK abgeleitet worden [21—27].

Wir haben in Gleichung (3.25) also zwei Haupt-Relaxationsterme, den Stoßterm $\mathscr{T}$ und den Term der inneren Übergänge

$$\mathscr{L}\,\varrho^A = [\overline{\mathscr{V}}^A,\,[\overline{\mathscr{V}}^A,\varrho^A]^{(d)}]\,. \tag{3.32}$$

Die beiden Terme, welche die Störungen $\overline{\mathscr{V}}^A$ und $\overline{\mathscr{V}}^I$ kombiniert enthalten, sind dagegen von geringerem Gewicht. Wir können das erkennen, wenn wir aus (3.25) eine Gleichung für ϱ^{A1} herausprojizieren. Wendet man $\mathscr{K}^{A1}$ auf beiden Seiten von (3.25) an, so ergibt sich:

$$\dot{\varrho}^{A1} = -\,i\,[\mathscr{H}^{A1},\varrho^{A1}] - i\,\mathscr{K}^{A1}[\mathscr{H}^{A12},\varrho^A]\,\mathscr{K}^{A1}$$

$$-\,\mathop{\mathrm{Spur}}_{B}\left\{i\,(t^{\oplus}\,\tilde{\varrho}^{A1}\,\varrho^B - \tilde{\varrho}^{A1}\,\tilde{\varrho}^B\,t^{\oplus+}) - \pi\,t^{\oplus}\,\tilde{\varrho}^{A1}\,\tilde{\varrho}^B\,t^{\oplus+}\delta(\tilde{E} - \tilde{\mathscr{H}})\right\}$$

$$-\,i\,[\mathscr{K}^{A1}\,\overline{\mathscr{V}}^A\,\mathscr{K}^{A1},\,\mathop{\mathrm{Spur}}_{B}\,\Omega^{\oplus}\,\tilde{\varrho}^{A1}\,\varrho^B\,\Omega^{\oplus+}]$$

$$-\,\mathscr{K}^{A1}\,[\overline{\mathscr{V}}^A,[\overline{\mathscr{V}}^A,\varrho^A]^{(d)}]\,\mathscr{K}^{A1}$$

$$-\,[\mathop{\mathrm{Spur}}_{B}\,\overline{\mathscr{V}}^I\,\varrho^B,\,\mathscr{K}^{A1}[\overline{\mathscr{V}}^A,\varrho^A]^{(d)}\,\mathscr{K}^{A1}]\,. \tag{3.33}$$

Wenn wir überdies die Spur bilden, so gibt die linke Seite von (3.33) $\overset{\text{A1}}{}$ die chemische Reaktionsgeschwindigkeit an. Auf der rechten Seite verschwinden die Terme in der dritten und fünften Zeile. Sie tragen also nicht *unmittelbar* zur Reaktion bei, und wir werden sie daher im folgenden außer acht lassen. Die Diskussion der kinetischen Gleichung wird dadurch wesentlich einfacher.

Schließlich können wir für den Fall, daß B ein Boltzmann-Gas ist, (3.33) als kinetische Gleichung schreiben

$$\dot{\varrho}^A = -i\,[\mathscr{H}^A, \varrho^A]$$

$$-\operatorname*{Spur}_B \left\{ i\,(t^{\oplus}\,\tilde{\varrho}^{A1}\,\varrho^B - \tilde{\varrho}^{A1}\,\varrho^B\,t^{\oplus+}) - \pi\,t^{\oplus}\,\tilde{\varrho}^{A1}\,\tilde{\varrho}^B\,t^{\oplus+}\,\delta\,(E - \tilde{\mathscr{H}}) + \right.$$

$$\left. + i\,(t^{\oplus}\,\tilde{\varrho}^{A2}\,\varrho^B - \tilde{\varrho}^{A2}\,\varrho^B\,t^{\oplus+}) - \pi\,t^{\oplus}\,\tilde{\varrho}^{A1}\,\tilde{\varrho}^B\,t^{\oplus+}\,\delta\,(E - \tilde{\mathscr{H}}) \right\}$$

$$-[\overline{\mathscr{V}}^A, [\overline{\mathscr{V}}^A, \varrho^A]^{(d)}]\;, \tag{3.34}$$

wobei
$$\tilde{\varrho}^{A1} = (\mathscr{H}^{A1}\,\varrho^A\,\mathscr{H}^{A1})^{(d)}$$

$$\tilde{\varrho}^{A2} = (\mathscr{H}^{A2}\,\varrho^A\,\mathscr{H}^{A2})^{(d)}\;.$$

Eine Lösung dieser Gleichung für kleine Abweichungen vom thermischen Gleichgewicht zwischen A und B kann man sich durch Erweiterung der Chapman-Enskog-Methode verschaffen [28—31].

Wir verzichten darauf, dies hier weiter auszuführen, und beschränken uns auf die Diskussion der Eigenschaften der abgeleiteten kinetischen Gleichungen.

Betrachten wir zunächst den Fall hohen, jedoch nicht zu hohen Druckes. Das heißt, die Bedingung (3.9) soll nicht verletzt sein, und es besteht keine Phasenbeziehung zwischen aufeinanderfolgenden Stößen, die das System A erfährt. Außerdem wollen wir das System B als so groß annehmen, daß seine zeitliche Entwicklung auf der Zeitskala, die uns für A interessiert, unerheblich ist.

Wir beginnen zur Zeit $t = 0$ mit einem Ensemble von Molekeln der chemischen Spezies A1. Für genügend hohen Druck verschwinden nun die Stoßterme in (2.34), und zwar die mit $\tilde{\varrho}^{A1}$ wie auch die mit $\tilde{\varrho}^{A2}$. Das kann man aus Gleichung (3.29) und (3.31) ablesen, doch sieht man es auch, wenn man in (3.34) $\overline{\mathscr{V}}^A = 0$ setzt. Man erhält dann Relaxation für die Diagonalelemente von ϱ^{A1} und ϱ^{A2} auf der Zeitskala τ_r^{coll}. Die Reaktionsgeschwindigkeit im Hochdruckbereich ist dann die des isolierten Ensembles, wie sie in (2.36) angegeben ist. Wir befinden uns im Hochdruckbereich, solange $\tau_r^{coll} \ll \tau_r^{ch}$ ist, und

$$\tau_0 \ll \tau_r^i < \tau_c\;.$$

Dieses Resultat begründet und erhellt die fundamentale Annahme der phänomenologischen Theorien monomolekularer Reaktionen, daß nämlich für $\tau_r^i \ll \tau_c \ll \tau_r^{ch}$ die Reaktionsgeschwindigkeit allein durch einen inneren Mechanismus der reagierenden Molekeln bestimmt ist. Die Hypothese eines von äußeren Stößen unabhängigen inneren Reaktionsmechanismus findet sich zuerst bei F. A. LINDEMANN und wurde später von RICE und KASSEL in die mikroskopische Theorie übernommen, wo sie bis heute ihren festen Platz hat [*32—39*].

An Hand der Gleichung (3.25) können wir diskutieren, wie lange die besagte Annahme gerechtfertigt ist und welche Effekte wir nach ihrem Zusammenbruch zu erwarten haben. Wir wollen uns dabei noch immer auf den Fall beschränken, wo keine reaktiven Stöße stattfinden. Die Stöße mögen aber jetzt so energiereich sein und so dicht aufeinanderfolgen, daß die innere Relaxationszeit τ_r^i in die Größenordnung der Zeit zwischen zwei Stößen kommt, d. h.

entweder $\qquad\qquad\qquad \tau_r^i \gtrsim \tau_c \qquad\qquad\qquad$ Fall (2)

oder $\qquad\qquad\qquad\quad \tau_r^i > \tau_c \;. \qquad\qquad\qquad\;$ Fall (3)

Schematisch kann man diese Fälle auf einer Zeitachse so veranschaulichen:

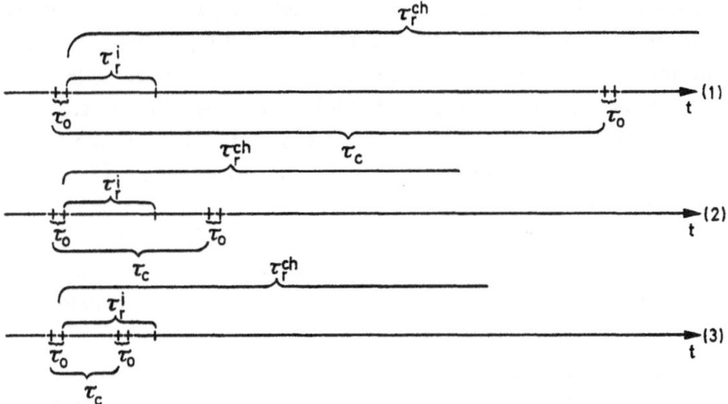

Der Fall (1) war beschrieben durch die vereinfachte Gleichung (3.34) mit dem Hochdruckgrenzwert

$$\dot{\varrho}^A = -i\,[\mathscr{H}^A, \varrho^A] - \pi\,[\mathscr{V}^A, [\mathscr{V}^A, \varrho^A]^{(d)}] \;. \qquad (3.35)$$

Die Stöße haben keinen Einfluß auf den inneren Relaxationsvorgang.

Im Fall (2) glättet $\mathscr{V}^A$ das Wellenpaket, das nach einem Stoß die reagierende Molekel durchläuft, noch vor dem nächsten Stoß aus. Wir sind also berechtigt, in den Stoßterm (3.22) die diagonalen Dichtematrizen $\tilde{\varrho}^{A1}$ und $\tilde{\varrho}^{A2}$ einzusetzen (die früher getroffenen Feststellungen a) und b) gelten weiterhin). Die gemischten Terme in (3.25) bzw. (3.33) sind jetzt aber nicht mehr zu vernachlässigen. Sie drücken die Möglichkeit reaktiver Übergänge des in der Molekel umlaufenden Wellenpakets aus. Das führt zu einer etwas höheren Reaktionsgeschwindigkeit, da die bezüglich $\tilde{\mathscr{H}}^0$ nichtdiagonalen Matrixelemente von $\Omega^{\oplus}(\tilde{\varrho}^{A1} + \tilde{\varrho}^{A2})\,\tilde{\varrho}^{B}\,\Omega^{\oplus+}$ die Zahl der reaktiven Übergänge vergrößern.

Eine zusätzliche Abweichung vom normalen Hochdruckverhalten sollte jedoch im Falle (3) zu beobachten sein. Hier erfährt eine Molekel mit umlaufendem Wellenpaket einen weiteren Stoß, welcher (vereinfacht beschrieben) die Energie des bereits vorhandenen Wellenpakets hinwegnehmen, aber auch ein zusätzliches Wellenpaket anstoßen kann. Wenn der Stoßmechanismus, wie zuerst beschrieben, eine desaktivierende Wirkung auf die in der Molekel vorhandenen Wellenpakete hat, wird die Reaktionsgeschwindigkeit absinken, möglicherweise sogar unter den normalen Hochdruckwert. Im umgekehrten Fall führen die Stöße zu größeren nichtdiagonalen Matrixelementen in $\Omega^{\oplus}(\varrho^{A1} + \varrho^{A2})\, \varrho^{B}\, \Omega^{\oplus+}$. Die Bedingung a) ist nicht mehr erfüllt, und statt (3.22) gilt jetzt

$$\varrho^{A1B} = \Omega^{\oplus}\, e^{-i\tilde{\mathscr{H}}^{0}(t-t_0)}\, \varrho^{A1}(t_1)\, \tilde{\varrho}^{B}(t_1)\, e^{i\tilde{\mathscr{H}}^{0}(t-t_0)}\, \Omega^{\oplus+} \qquad (3.36)$$

entsprechende Ausdrücke sind in (3.25) einzusetzen. Wir können aus physikalischer Anschauung heraus vermuten, wann der eine oder andere Effekt auftreten wird. Ist die Zahl der Freiheitsgrade der reagierenden Molekel klein, so ist die Wahrscheinlichkeit der Desaktivierung des inneren Wellenpaketes bei einem Stoß groß. Hat die Molekel viele Freiheitsgrade, so ist die Wahrscheinlichkeit klein, daß der Stoßpartner gerade das innere Wellenpaket „trifft". Dieser Vergleich gilt ziemlich genau bei gleicher Stoßzahl und Temperatur, da τ_r von Molekel zu Molekel nicht allzu verschieden ist [40—44].

In Flüssigkeiten ist oft sogar der Fall realisiert, daß $\tau_0 \simeq \tau_c < \tau_r^i$. Dann kann man nicht mehr von einem inneren Relaxationsprozeß sprechen, da die Kopplung mit der Umgebung zu stark geworden ist. Solche Systeme entziehen sich weitgehend theoretischer Behandlung, wenn es nicht gelingt, sie gedanklich in schwach gekoppelte Teilsysteme zu zerlegen.

Für alle oben genannten Fälle lassen sich auch in Festkörpern analoge Beispiele finden, wenn man „Stöße" als Absorption, Emission und Streuung von Phononen interpretiert. Gleichung (3.25) gilt dann noch immer wörtlich, wenn auch die Argumentation zu Gleichung (3.21) ein wenig verschieden ist. Ein Phononenmodell für das System B könnte sich auch bei Flüssigkeiten mit ausgeprägter Gitterstruktur den herkömmlichen Vorstellungen von der Stoßwechselwirkung zwischen Molekel und Thermostat als überlegen erweisen.

Literatur

1. Tolman, R. C.: The Principles of Statistical Mechanics, S. 325. Oxford: Univ. Press 1955.

2. Ter Haar, D.: Elements of Statistical Mechanics. New York: Wiley 1954.

3. Neumann, J. v.: Mathematische Grundlagen der Quantenmechanik. Berlin: Springer 1932.

4. DIRAC, P. A. M.: The Principles of Quantum Mechanics, S. 130. Oxford: Univ. Press 1958.

5. LIPPMANN, B. A., and J. SCHWINGER: Phys. Rev. **79**, 469 (1950).

6. GELL-MANN, M., and M. L. GOLDBERGER: Phys. Rev. **91**, 398 (1953).

7. GROSJEAN, C. C.: Formal Theory of Scattering Phenomena, Monographie Nr. 7. Bruxelles: Institute Interuniversitaire 1960.

8. CHEW, G. F., and M. L. GOLDBERGER: Phys. Rev. **87**, 778 (1952).

9. UHLENBECK, G. E.: Anhang in: M. KAC, Probability and Related Topics in Physical Science. New York: Interscience Publishers 1959.

10. PRIGOGINE, I.: Nonequilibrium Statistical Mechanics. New York: Interscience Publishers 1962.

11. EHRENFEST, P.: Collected Scientific Papers, S. 213. Amsterdam: North-Holland 1959.

12. HOVE, L. VAN, in: E. G. D. COHEN, Fundamental Problems in Statistical Mechanics (Nijenrode Castle 1961), S. 157. Amsterdam: North-Holland 1962.

13. KAMPEN, N. G. VAN, in: E. G. D. COHEN, Fundamental Problems in Statistical Mechanics (Nijenrode Castle 1961), S. 173. Amsterdam: North-Holland 1962; MONTROLL, E.: S. 230.

14. ZWANZIG, R. in: Lectures in Theoretical Physics (Boulder III), S. 106. New York: Interscience Publishers 1961.

15. Eine ausgezeichnete Übersicht findet sich bei ZWANZIG, R.: Physica **30**, 1109 (1964).

16. GOLDEN, S.: Suppl. Nuovo Cimento **5**, 540 (1957); **15**, 335 (1960).

17. Ausführlich ist dieses Problem diskutiert in einem im Druck befindlichen Buch: GOLDEN, S.: Quantum Mechanical Foundations of Chemical Kinetics.

18. HOFACKER, G. L.: Univ. of Florida, Quantum Theory Project Report no. 66, 1964.

19. TER HAAR, D.: Rev. Mod. Phys. **27**, 289 (1955).

20. KAMPEN, N. G. VAN in: E. G. D. COHEN, Fundamental Problems in Statistical Mechanics (Nijenrode Castle 1961), S. 173. Amsterdam: North-Holland 1963.

21. UEHLING, E. A., and G. E. UHLENBECK: Phys. Rev. **43**, 552 (1933).

22. MORI, H., and S. ONO: Prog. Theor. Phys. (Japan) **8**, 327 (1952).

23. ONO, S.: Prog. Theor. Phys. (Japan) **12**, 113 (1954).

24. NAKAJIMA, S.: Prog. Theor. Phys. (Japan) **20**, 949 (1958).

25. DAHLER, J.: J. Chem. Phys. **30**, 1447 (1959).

26. SNIDER, R. F.: J. Chem. Phys. **32**, 1051 (1960).

27. HOFACKER, G. L.: Z. Naturforsch. **18a**, 607 (1963).

28. CHAPMAN, S., and T. G. COWLING: The Mathematical Theory of Non-Uniform Gases. New York: Cambridge Univ. Press 1950.

29. HIRSCHFELDER, J. O., C. F. CURTISS, and R. B. BIRD: The Molecular Theory of Gases and Liquids. New York: John Wiley 1954.

30. ROSS, J., and P. MAZUR: J. Chem. Phys. **35**, 19 (1961).

31. PYUN, C. W., and J. ROSS: J. Chem. Phys. **40**, 2572 (1964).

32. LINDEMANN, F. A.: Trans. Faraday Soc. **17**, 599 (1922).

33. RICE, O. K., and H. C. RAMSPERGER: J. Amer. Chem. Soc. **49**, 1617 (1927).

34. KASSEL, L. S.: J. Chem. Phys. **32**, 225, 1065 (1928).

35. GOLDEN, S., and A. M. PEISER: J. Phys. and Coll. Chem. **55**, 787 (1951).

36. WILSON, D. J.: J. Phys. Chem. **64**, 323 (1960).

37. SERAUSKAS, R. V., and E. W. SCHLAG, J. Chem. Phys. **42**, 3009; **43**, 898 (1965).

38. MARCUS, R. A.: J. Chem. Phys. **20**, 359 (1952), (I); **43**, 2658 (1965), (III).

39. WIEDER, G. M., and R. A. MARCUS: J. Chem. Phys. **37**, 1835 (1962), (II).

40. BUFF, F. P., and D. J. WILSON: J. Chem. Phys. **32**, 677 (1960).

41. WILSON, D. J.: J. Chem. Phys. **64**, 322 (1960).
42. BAETZOLD, R. C., and D. J. WILSON: J. Phys. Chem. **68**, 3141 (1964).
43. BUNKER, D. L.: J. Chem. Phys. **40**, 1946 (1964).
44. PLACZEK, D. W., B. S. RABINOVITCH, and F. H. DORER: J. Chem. Phys. **44**, 279 (1966).

G. L. Hofacker
Chemistry Department
Northwestern University
Evanston, Illinois/USA

Unimolecular Reaction Rate Theory

R. A. Marcus

Rewritten by H. Heydtmann

With 1 Figure

1. Introduction

This contribution summarizes the unimolecular reaction rate theory as developed by Rice, Ramsperger, Kassel and Marcus (RRKM-theory) and especially reviews the treatment given in three earlier publications [1–3].

The RRK-theory of the 1920's was based on the hypothesis of Lindemann, that in unimolecular reactions there is a time lag for every active molecule. It was postulated that active molecules (i.e. molecules that have a critical amount of energy or more in certain internal degrees of freedom) either decompose or isomerize after a definite time (time lag) unless they are deactivated by a molecular collision. The existence of a time lag was explained by the assumption that the energy contained in the internal degrees of freedom — or some part of this energy — had to accumulate into one certain degree of freedom (critical oscillator concept). The time lag also was considered to be a function of the amount of energy surpassing the critical energy. A discussion of the RRK-theories is given in the book of Benson [4]. An excellent summary of the RRK, Slater and RRKM-theories, as well as of related topics, has been given by Rice [5]. Relevant experimental data are described in a number of articles [2, 6, 7], each of which contains many references to the literature. Of particular note also is a recent book by Bunker [8].

2. The Specific Rate Constant

The RRKM-theory gives a better description than the earlier theories of the quasi-unimolecular rate constant $k_{\mathrm{uni}} \equiv - [A]^{-1} \cdot d\,[A]/\,dt$ as a function of pressure without using an adjustable parameter. The theory is based on a reaction sequence

$$A + M \rightleftharpoons A^* + M \tag{1}$$

$$A^* \xrightarrow{\ k_{EJ}\ } A^+ \tag{2}$$

$$A^+ \longrightarrow \text{products} \tag{3}$$

where M is any third body capable of deactivating an activated molecule, A^*. (A list of notations is given in the appendix.)

The activated complex, A^+, is treated like a molecular species. It possesses $3n-4$ rotational and vibrational degrees of freedom (for a non-linear polyatomic activated complex with n atoms) and one internal translation. Reactions in which bonds are formed as well as broken are usually expected to involve rigid activated complexes (e.g. the isomerization of cyclopropane to give propene). Reactions involving only a dissociation for which the reverse reaction of recombination requires no activation energy are expected to involve loose activated complexes (e.g. the dissociation of ethane into methyl radicals). In such a loose activated complex the separating particles are assumed to rotate relatively freely, being held only by loose bonds. By contrast, a rigid complex normally has no new internal rotations, and indeed has about the same extension in space as the reactant in its vibrational and rotational ground state.

In the RRK-theory the number of internal degrees of freedom effective in transferring energy to the "critical oscillator" was an adjustable parameter. This number was found by fitting the calculated curves $\log k_{\text{uni}}$ versus $\log p$ to the experimental curves. In the RRKM-theory we distinguish from the outset "adiabatic" and "active" degrees of freedom. Only the active degrees of freedom are active in intramolecular energy transfer. All vibrations are assumed to be active, since anharmonicity effects are important and no severe restrictions should be imposed by momentum conservation laws.

Because of the increased separation distance the centrifugal potential facilitates reaction — especially for reactions with loose activated complexes — in any given rotational state of the molecule. We ignore Coriolis effects and denote by J the totality of quantum numbers that are approximately conserved on forming A^+ from A^*. Thus J is the quantum number of the adiabatic degrees of freedom which, in applications, have usually been taken to be the external rotations of the molecule. The energy for these degrees of freedom changes from E_J to E_J^+. When J refers only to rotations, the difference $E_J - E_J^+$ represents the change in centrifugal potential and we have the following energy balance (compare fig. 1):

$$E_a + E^+ + E_J^+ = E + E_J \tag{4}$$

$$E^+ = E_t^+ + E_n^+ \tag{5}$$

k_{EJg}, the specific rate constant for a dissociation (isomerization) by a particular reaction path, can be derived in the following manner: consider an energetic molecule A^* whose energy of the active modes is in an interval $(E, E + dE)$ and whose adiabatic modes are in a state J. The statistical equilibrium probability of finding such a molecule as an activated complex

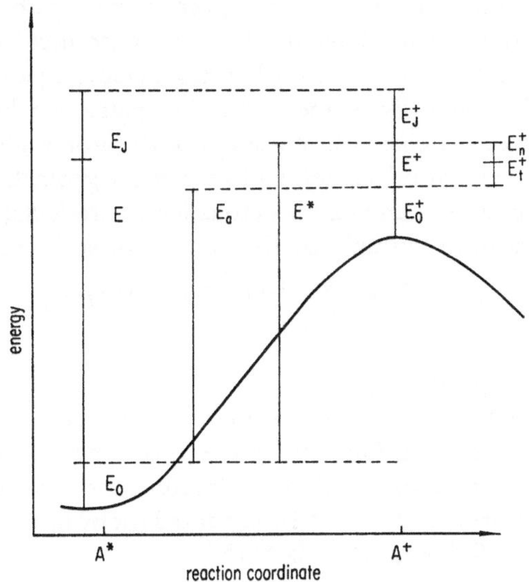

Fig. 1. Energy diagram for an unimolecular process

A^+ having an internal translational momentum in the range $(p, p + dp)$, being in a state n of the active degrees of freedom and in an interval $(q, q + dq)$ and formed by a given path, is given by the ratio of the quantum states of this A^+ and of A^* namely by

$$\frac{dq \cdot dp}{h} \frac{\Omega^+ (E_n^+)}{\Omega^* (E) \, dE} , \tag{6}$$

since $dq \cdot dp/h$ is the number of internal translational quantum states in $dq \, dp$. The corresponding probability per unit interval along q is obtained by dividing by dq. The contribution of these states of A^* to the specific unimolecular reaction rate constant, k_{EJg}, is obtained by multiplying the resulting ratios by the velocity $\dot{q}$. The coordinate q is taken to be Cartesian, so that $\dot{q}$ equals p/m where m is an effective mass. Since $d(p^2/2\,m)$ equals dE, and since k_{EJg} equals the above rate expression summed over all accessible n, we obtain

$$k_{EJg} = \sum_{E_n^+ \leqslant E^+} \Omega^+ (E_n^+) / \{h \, \Omega^* (E)\} . \tag{7}$$

A concept of equilibrium for reactants with activated complexes moving in the forward direction along the reaction coordinate is embodied in equations (2), (6) and (7). It is discussed for a quite different case — bimolecular reactions — in the "Remarks on Generalization of Activated Complex Theory" (R. A. Marcus, this book).

Equation (7) is the contribution for a given reaction path. There may be one or more reaction paths which are "geometric isomers" of each other. For each such path there may be a further degeneracy: a path may have an optical isomer [9]. Optically isomeric reaction paths can be detected by drawing a picture of the chemical migration of the atoms and seeing if the resulting figure has an optical isomer. Paths which are geometrically isomeric usually have different A^+'s and may lead to different products; so we have to sum over rather than to multiply by the number of such paths and obtain

$$k_{EJ} = \sum_g \alpha_g \sum_{E_n^+ \leqslant E^+} \Omega^+ (E_n^+) / \{h\,\Omega^* (E)\}, \tag{8}$$

where α_g and Ω^+ depend on g.

Because of symmetry restrictions some rotational states may be absent in A^*, or in A^+ or in both. On making the usual approximation employed in a classical description of rotational partition functions the absence of certain rotational states in A^* or A^+ is accounted for by introducing symmetry numbers: $\Omega^+ = W^+/\sigma^+$; $\Omega^* = W^*/\sigma^*$.

3. The First Order Rate Expression

The equilibrium probability of finding an A^* with an energy of the active modes in the range $(E, E + dE)$ and with adiabatic modes in the state J is $P_{EJ}^0\, dE$ where

$$P_{EJ}^0 = P^{-1} \Omega^* (E) \exp \left[- (E + E_J) / kT\right]. \tag{9}$$

$\Omega^*(E)$ counts all states per unit of energy and therefore includes the degeneracy factor. We denote by ω the specific collisional deactivation probability assuming a strong collision mechanism for deactivation (a single collision removes enough energy to deactivate every A^*). Considering very high pressures, where $\omega \gg k_{EJ}$, the rate of activation can be calculated. Using the equilibrium expression for A and A^* this rate equals the rate of collisional deactivation $\omega\, P_{EJ}^0\, dE[A]$. Since the rate of formation of A^* thus calculated will be assumed to hold also for lower pressures, it is implicitly taken for granted that equilibrium conditions are preserved for all molecules with energies smaller than E_a in their active modes. By using steady state arguments for A^* under conditions where $\omega \gg k_{EJ}$ no longer is true one can show that the concentration of each A^* is a fraction $\omega/(\omega + k_{EJ})$ of the equilibrium concentration. Thus the unimolecular reaction rate constant is obtained by summing $k_{EJ}\, P_{EJ}^0\, dE\, \omega/(\omega + k_{EJ})$ over all E and J:

$$k_{uni} = \int_E \sum_{J=0}^{\infty} k_{EJ}\, P_{EJ}^0\, dE\, / \,(1 + k_{EJ}/\omega). \tag{10}$$

By substituting (8) and (9) and using (4) one obtains

$$k_{\mathrm{uni}} = \frac{kT}{bP} \int\limits_{E^+=0}^{\infty} \sum\limits_{J=0}^{\infty} \frac{\sum\limits_{g} \left(\frac{\alpha_g}{\sigma^+}\right) e^{-E_a/kT} \sum\limits_{E_n^+ \leqslant E^+} W^+(E_n^+) e^{-\frac{(E^+ + E_J^+)}{kT}}}{(1 + k_{EJ}/\omega)} \frac{dE^+}{kT}.$$

$$(11)$$

The k_{uni} for formation of a particular product by a particular path g is obtained from (11) by deleting the $\sum\limits_{g}$ in (11) but not in (8). In the case of an isomerization to form B from A, B^* may reform A^* before being deactivated. A correction can be easily introduced to correct for such situations [3].

4. The Density of Energy States

For numerical calculations the evaluation of the numbers of states per unit energy becomes important. In most cases an exact calculation is only necessary for the activated complex. As for this complex internal energies are small in systems with thermal activation, quantum restrictions can not be neglected; with growing internal energies the contributions to the integral in (11) quickly become negligible. For the evaluation of $W^*(E)$, which appears in k_{EJ} and therefore also in k_{uni}, useful approximations have been found [10]. The semiclassical expression for the case that only vibrations are taken to be active is given by [11]:

$$W^*(E) = [E + E_0]^{s-1} / \{\Gamma(s) \prod_{i=1}^{s} (h\nu_i^*)\} .$$

$$(12)$$

The s normal vibrations have frequencies ν_i^*. A better approximation is attained by

$$W^*(E) = [E + aE_0]^{s-1} / \{\Gamma(s) \prod_{i=1}^{s} (h\nu_i^*)\} ,$$

$$(13)$$

where a is a quantity smaller than unity depending weakly on the energy E and approaching unity for large energies.

If there are t active rotations in A the energy E is distributed between these and the active vibrations:

$$W^*(E) = \int\limits_{x=0}^{E} W_v^*(E - x)\, W_r^*(x)\, dx .$$

$$(14)$$

This leads to an expression corresponding to equation (13):

$$W^*(E) = P_R(E + aE_0)^{s+t/2-1} / \{(RT)^{\frac{t}{2}} \Gamma\left(s + \frac{t}{2}\right) \prod_{i=1}^{s} (h\nu_i^*)\} . \quad (15)$$

Similarly active rotations in A^+ have been shown to factor the expression $1/\sigma_2^+ \sum\limits_{E_n^+ \leqslant E^+} W^+(E_n^+)$ into terms corresponding to partitioning of E^+ among the active rotations and vibrations of A^+ [1].

5. The Influence of Centrifugal Potential

We now consider the influence of changing centrifugal potential on the unimolecular reaction rate. Taking (15) as the approximation for the density of energy states we define

$$F = \frac{W^*(E^* + E_J^{\ddagger} - E_J)}{W^*(E^*)} = \left(1 - \frac{E_J - E_J^{\ddagger}}{E^* + a E_0}\right)^{s + t/2 - 1}. \tag{16}$$

For rigid activated complexes E_J and $E_J^{\ddagger}$ are normally about equal and F is very close to unity. For loose activated complexes leading to a dissociation, E_J and $E_J^{\ddagger}$ differ primarily for two rotations in which the two resulting fragments are treated as the "atoms" of a diatomic molecule. The mean value of $E_J - E_J^{\ddagger}$ can be shown to be pressure insensitive and to equal $(I^+ - I) l R T / 2 I$ where l is the number of adiabatic rotations and I^+/I is the ratio of the moments of inertia for these rotations [3]. The value of k_{EJ} is relatively insensitive to fluctuations of $E_J - E_J^{\ddagger}$ about this mean, and the corresponding value of k_{EJ} obtained by making this replacement for $E_J - E_J^{\ddagger}$ is denoted by k_a; for the given value of E^+

$$k_a = \sum_g \frac{\alpha_g \sigma^*}{\sigma^+} \left[\sum_{E_n^+ \leqslant E^+} W^+(E_n^+)\right] \Big/ \{h F W^*(E^*)\}, \tag{17}$$

where

$$F = W^* \{E^* - l R T (I^+ - I)/2 I\} / W^*(E^*). \tag{18}$$

For the dissociation of ethane into methyl radicals and dissociation of N_2O_5 into NO_2 and NO_3 F has been calculated to be about 0.8 and 0.4, respectively, at $E^* = E_a$ [3].

Upon introducing (17) into (11) and integrating over J one finds

$$k_{\text{uni}} = \frac{kT}{hP} \sum_g \frac{\alpha_g P_1^{\ddagger} e^{-E_a/kT}}{\sigma_2^{\ddagger}} \int_{E^+ = 0}^{\infty} \frac{\sum\limits_{E_n^+ \leqslant E^+} W^+(E_n^+) e^{-\left(\frac{E^+}{RT}\right)}}{1 + k_a/\omega} \frac{dE^+}{kT}. \tag{19}$$

In (19) the properties of A^+ (e.g. α_g, E_a, P_1^+, W^+ and σ_2^+) may depend on the path g.

6. Concluding Remarks

The high pressure value of k_{uni}, given by (20), is obtained from (19) by setting $\omega = \infty$, interchanging order of summation and integration ($E^+ = E_n^+$ to ∞, and then $E_n^+ = 0$ to ∞), integrating and summing.

$$k_{\text{uni}, \infty} = \frac{kT}{h} \sum_g \frac{\alpha_g P^+}{P} \exp(-E_a/kT). \tag{20}$$

This expression can be used to discuss statistical factors in reaction rates. An alternative group theoretical description of the statistical effect has been given by other authors [12]. It should be noted that equation (20) is equivalent to that derived previously by EYRING [13].

If one wants to calculate k_{uni} as a function of pressure, the Arrhenius activation energy E_∞ must be known for the homogeneous reaction and the properties of the activated complex must be chosen in a reasonable manner. This choice is guided by the knowledge of the vibrational frequencies (and the properties of the internal rotations) for the reacting molecules and the product molecules and by the knowledge of the experimental value of $k_{uni,\,\infty}$ which must agree with the value calculated from equation (20).

Appendix

NOTATIONS:

A, A^*, A^+ = reactant, active molecule (an A with enough energy to react), and the activated complex.

E_0, E_0^+ = zero-point energy of A, A^+.

E, E^+ = energy of the active modes of A^* and of A^+, respectively.

E_J, E_J^+ = energy of adiabatic modes of A^* and of A^+, respectively.

E_a = energy of A^+ in its lowest vibrational, rotational and translational state minus that of A in its lowest state. (Hence, E_a also equals the activation energy at 0 °K).

E^* $= E_a + E^+$.

k_{EJ} = specific rate constant for molecules with energy E in their active modes and in quantum state J of their adiabatic modes (eq. 2).

k_{EJg} = specific rate constant as before, but only for a particular reaction path.

q, p = reaction coordinate and its conjugate momentum.

E_t^+ = internal translational energy of A^+; $E_t^+ = p^2/2\,m$.

E_n^+ = energy of the active vibrations and rotations of an A^+ in quantum state n.

g = label for a geometrically isomeric path.

α_g = number of optically isomeric reaction paths for each geometrically different path leading from the initial A molecule (or A isomer if there is more than one) through A^+ to the given product.

$\Omega^+(x)$ = number of states of active modes of A^+ (formed by a given path) in an energy interval $(x, x + dx)$.

$\Omega^*(y)$ = number of states of A^* per unit energy, when the energy of the active modes is y (energy density at y).

8*

$W^+(x), W^*(y) =$ number of such states of A^+ and A^* when symmetry numbers are ignored.

$W_v^*(y), W_r^*(y) =$ number of rotational and vibrational states per unit of energy.

σ^*, σ^+ $=$ symmetry numbers of A^* and A^+.

σ_2^+ $=$ symmetry number for the active rotations of A^+.

P, P^+ $=$ partition function of all the rotational and vibrational modes of A and A^+, respectively.

P_1, P_1^+ $=$ partition functions of the adiabatic modes of A and of A^+.

P_2, P_2^+ $=$ partition functions of the active modes of A and of A^+.
$$P = P_1 \cdot P_2, \; P^+ = P_1^+ P_2^+.$$

P_R $=$ rotational partition function for active rotations of A.

ω $=$ the specific collisional deactivation probability, which is for a strong collision mechanism equal to the probability of a collision per unit of time and hence approximately proportional to the total pressure.

References

1. MARCUS, R. A.: J. Chem. Phys. **20** 359 (1952).

2. WIEDER, G. M., and R. A. MARCUS: J. Chem. Phys. **37**, 1835 (1962).

3. MARCUS, R. A.: J. Chem. Phys. **43**, 2658 (1965).

4. BENSON, S. W.: The Foundations of Chemical Kinetics. New York: McGraw Hill, 1960.

5. RICE, O. K.: contribution in "Energy Transfer in Gases, 12th Solvay Congress. New York: Interscience Publishers 1962.

6. RABINOVITCH, B. S., and D. W. SETSER: Advances in Photochem. **3**, 1 (1964).

7. RABINOVITCH, B. S., R. F. KUBIN, and R. E. HARRINGTON: J. Chem. Phys. **38**, 405 (1963).

8. BUNKER, D. L.: Theory of Elementary Gas Reaction Rates. Braunschweig: Pergamon Press 1966.

9. MARCUS, R. A.: J. Chem. Phys. **43**, 1598 (1965).

10. RABINOVITCH, B. S., and R. W. DIESEN: J. Chem. Phys. **30**, 735 (1959); WHITTEN, G. Z., and B. S. RABINOVITCH: J. Chem. Phys. **38**, 2466 (1963); **41**, 1883 (1964); SCHLAG, E. W., and R. A. SANDSMARK: J. Chem. Phys. **37**, 168 (1962); THIELE, E: J. Chem. Phys. **39**, 3258 (1963); HAARHOFF, P. C.: Mol. Phys. **7**, 101 (1963); LIN, S. H., and H. EYRING: J. Chem. Phys. **43**, 2153 (1965).

11. MARCUS, R. A., and O. K. RICE: J. Phys. & Colloid Chem. **55**, 894 (1951).

12. SCHLAG, E. W.: J. Chem. Phys. **38**, 2480 (1963); SCHLAG, G. W., and G. L. HALLER: **42**, 584 (1965); cf. BISHOP, D. M., and K. J. LAIDLER: **42**, 1688 (1965).

13. GLASSTONE, S., K. J. LAIDLER, and H. EYRING: The Theory of Rate Processes. New York-London: McGraw Hill 1941.

R. A. Marcus
University of Illinois
Urbana, Illinois/USA

Zwischenmolekulare Energieübertragung bei Stößen unter besonderer Berücksichtigung von mehratomigen Molekülen in hochangeregten Schwingungszuständen

G. H. Kohlmaier

Mit 3 Abbildungen

1. Einführung

Chemische Reaktionen sind definitionsgemäß Vorgänge, bei denen die chemische Zusammensetzung durch Neuverknüpfung der die Moleküle aufbauenden Atome sich zeitlich ändert. Unter einem Elementarprozeß einer chemischen Reaktion versteht man einen Einzelschritt derselben, der von einem Ausgangszustand über einen Übergangszustand zu einem Endzustand nach den Gesetzen der Mechanik abläuft und der durch die Anfangsbedingungen der Ausgangskonfiguration vollkommen festgelegt ist. Die chemischen Elementarprozesse der unimolekularen, bimolekularen und trimolekularen Gasreaktionen unterscheiden sich formal durch ihren Übergangszustand, der aus ein, zwei oder drei Ausgangsmolekülen aufgebaut ist. Das Studium der molekularen Energieübertragung beim Stoß beschäftigt sich mit dem Austausch von Energie zwischen den verschiedenen Freiheitsgraden eines oder mehrerer Moleküle, wobei nach Voraussetzung die Zahl der Bindungen und die Art der Verknüpfungen von Anfangs- und Endzuständen identisch ist. Im Fall des Energieaustauschs innerhalb eines Moleküls spricht man von innermolekularer Energieübertragung, im Fall der Wechselwirkung von zwei oder mehr Molekülen von zwischenmolekularer Energieübertragung, wobei die Unterscheidung zwischen den beiden Fällen auch hier rein formal ist. Im Hinblick auf eine eintretende chemische Reaktion sind besonders jene Moleküle interessant, die schon eine hohe innere Energie besitzen, so daß deren Reaktionsvermögen durch den Schritt der Energieübertragung wesentlich beeinflußt werden kann. Beispiele hierfür sind zweiatomige Moleküle mit Energien nahe der Dissoziationsgrenze und mehratomige Moleküle mit einer inneren Gesamtenergie gleich oder größer der Aktivierungsenergie für einen unimolekularen Isomerisierungs- oder Zerfallsprozeß. Leider sind gerade in diesem Energiebereich nur wenige experimentelle Untersuchungen bekannt, so daß eine eindeutige theoretische Zuordnung dieser Resultate bis heute jedenfalls noch nicht gelungen ist. Sicher ist nur, daß eine direkte Übertragung der Theorie der Energieübertragung bei niedrigen Anregungsenergien nicht möglich ist.

Zum Verständnis der Problematik dieses Themenkreises werden zwei Extremfälle der Energieübertragung diskutiert, nämlich die für niedrige Energien geltenden Ultraschallexperimente und ihre theoretische Interpretation und die für hohe Anregungsenergien charakteristischen Experimente der chemischen Aktivierung und deren theoretische Deutung.

2. Phänomenologische Beschreibung der Energieübertragung [1, 2]

Die wichtigsten experimentellen Methoden zur Bestimmung von Stoß-Übergangswahrscheinlichkeiten sind die Ultraschall- [3] und Stoßwellentechnik [4], die Blitzlichtspektroskopie [5], die Fluoreszenzausbeutemessungen [6] und die Untersuchung spezieller chemisch aktivierter Systeme [7], bei denen eine Konkurrenz zwischen Desaktivierung und chemischer Reaktion auftritt.

Da über Stoßwellen [4] und über Blitzlichtphotolyse [5] hier in ausführlichen Einzeldarstellungen schon berichtet worden ist, soll es genügen, die Ultraschalltechnik und die Methode der chemischen Aktivierung zu diskutieren.

2.1. Ultraschalltechnik

Eine Schallquelle erzeugt mit einer bestimmten Frequenz ν_s Dilatationen und Kompressionen, die sich in einem gasförmigen Medium mit der Geschwindigkeit V ausbreiten, wobei die Geschwindigkeit von dem Molekulargewicht M und der Molwärme bei konstantem Volumen C abhängt:

$$V^2 = (R\,T/M)(1 + R/C) \tag{2.1}$$

hierbei ist

$$C = C_{\text{trans}} + C_{\text{rot}} + C_{\text{vib}} \tag{2.2}$$

Da die Dilatationen und Kompressionen adiabatisch verlaufen, erhält man eine periodische Abkühlung und Erwärmung, d. h. eine Änderung der Translationsenergie. Hat das System genügend lange Zeit (kleine Schall-Frequenz), so wird ein Gleichgewicht zwischen Translation einerseits und Rotation und Vibration andererseits eingestellt. Bei sehr hohen Schallfrequenzen ändert sich die Translationsenergie so schnell, daß die Geschwindigkeit zur Einstellung des Gleichgewichts Translation – Vibration zu klein wird; die effektive Wärmekapazität sinkt auf den Wert:

$$C_{\text{eff}} = C_{\text{trans}} + C_{\text{rot}} \qquad (2.3)$$

(Die Einstellung des Gleichgewichts Translation – Rotation findet im allgemeinen sehr schnell statt; eine Ausnahme etwa bildet H_2 mit seinen großen Rotationsquanten.) Trägt man V^2 gegen die Schallfrequenz ν_s auf, so erhält man eine Kurve, die zunächst horizontal verläuft, dann infolge der Änderung $C \to C_{\text{eff}}$ ansteigt und danach horizontal weiterläuft. Der Inversionspunkt, charakterisiert durch die Frequenz ν_{inv}, hängt mit der Relaxationszeit τ des Systems in folgender Weise zusammen:

$$\nu_{\text{inv}} = (1/\tau)\, C_{\textit{eff}}/C \ . \qquad (2.4)$$

2.1.1. Relaxationszeit für ein 2-Niveau-Schema

(Zweiatomige Moleküle, bei denen $h\nu \gg kT$, kommen dem Fall sehr nahe.)

Bezeichnet man mit N_0 die Zahl der Grundzustände und mit N_1 die Zahl der angeregten Zustände in einem System von $N = N_0 + N_1$ Gesamtzuständen, so gilt:

$$-dN_0/dt = dN_1/dt = k_{01}\, N_0 - k_{10}\, N_1 \ , \qquad (2.5)$$

wobei k_{01} und k_{10} die Geschwindigkeitskonstanten für Aktivierung und Desaktivierung sind.

Für das System im Gleichgewicht gilt:

$$-d\overline{N}_0/dt = d\overline{N}_1/dt = 0\,; \qquad k_{01}\,\overline{N}_0 = k_{10}\,\overline{N}_1 \qquad (2.6)$$

oder

$$k_{01}/k_{10} = \overline{N}_1/\overline{N}_0 = \exp\{-h\nu/kT\} \ . \qquad (2.7)$$

$\overline{N}_1$ und $\overline{N}_0$ bedeuten die Gleichgewichtskonzentrationen, deren Verhältnis durch den Boltzmann-Faktor gegeben ist. Wird das Gleichgewicht um den Betrag ΔN gestört (etwa bei Veränderung der Temperatur), so daß N_1 in Gleichung (2.5) gegeben ist durch

$$N_1 = \overline{N}_1 + \Delta N \ ,$$

so folgt aus dieser Gleichung:

$$dN_1/dt = d(\overline{N}_1 + \Delta N)/dt = k_{01}(\overline{N}_0 - \Delta N) - k_{10}(\overline{N}_1 + \Delta N) \qquad (2.8)$$

oder

$$-d\Delta N/dt = (k_{10} + k_{01})\,\Delta N = (1/\tau)\,\Delta N \ ; \qquad (2.9)$$

hierbei ist $1/\tau = k_{10} + k_{01}$.

Die Relaxationszeit ist also umgekehrt proportional zu der Summe der Geschwindigkeitskonstanten. Weiterhin ist die Stoß-Übergangswahrscheinlichkeit vom Grundzustand zum ersten angeregten Zustand definiert als

$$P_{01} = k_{01}/\omega \qquad (2.10)$$

120 G. H. Kohlmaier

oder

$$P_{01}\,\omega = k_{01}\ ,$$

wobei ω die Stoßfrequenz ist.

2.1.2. Relaxationszeit für ein n-Niveau-Schema

(Zweiatomige Moleküle, bei denen $h\nu \approx kt$, können als Beispiel dienen.)
Landau und Teller [8] haben gezeigt, daß die Stoß-Übergangswahrscheinlichkeiten des harmonischen Oszillators proportional dem Quadrat des Matrixelementes $\langle n|x|m\rangle$ sind. Dabei ist:

$$\langle n|x|m\rangle \equiv \int \chi_n\, x\, \chi_m\, d\tau\ ;$$

$\chi_n(x)$, $\chi_m(x)$ sind Eigenfunktionen des harmonischen Oszillators; x ist
die Auslenkung. Aus den Beziehungen der Matrixelemente folgt:

$$P_{n,m} = 0 \quad \text{außer für}\quad m = n \pm 1 \tag{2.11}$$

$$P_{01}:P_{12}:P_{23}\ldots = P_{10}:P_{21}:P_{32}\ldots = 1:2:3\ldots \tag{2.12}$$

oder

$$P_{n,n+1} = (n+1)\,P_{01}\ .$$

Mit dieser Beziehung können alle $P_{n,m}$ auf das 2-Niveau-Schema reduziert
werden. Dabei hat sich gezeigt, daß dieselbe Beziehung für τ gilt wie in
Gleichung (2.9), also

$$1/\tau = k_{10} + k_{01}\ .$$

2.1.3. Relaxationszeit τ für mehratomige Moleküle

Prinzipiell ist einerseits eine parallele und andererseits eine aufeinanderfolgende Anregung der Normalschwingungen denkbar. Im ersten Fall
ergibt die Rechnung [1], daß bei unabhängiger Anregung s verschiedener Normalschwingungen s verschiedene Relaxationszeiten gemessen
werden müßten. Für den zweiten Fall, in dem der Übergang Translation–Vibration zu einer bestimmten Normalschwingung stattfindet, von
der aus alle übrigen Schwingungsfreiheitsgrade gespeist werden, läßt sich
zeigen, daß, wenn $\tau_1 \gg \tau_{12}$ ist, wobei τ_1 den Anregungsprozeß und τ_{12}
den Verteilungsprozeß charakterisiert, nur eine Relaxationszeit gemessen
werden sollte. Tatsächlich wird experimentell der letztere Fall (von
wenigen Ausnahmen abgesehen) bestätigt. Dabei steht das gemessene τ zu
τ_1 in folgender Beziehung

$$\tau = \tau_1 \left(\sum_s C_{\text{vib},s} / C_{\text{vib},1} \right)\ . \tag{2.13}$$

Zusammenfassend kann man sagen, daß Ultraschallmessungen Auskunft über die Stoß-Übergangswahrscheinlichkeiten P_{01} (bzw. P_{10}) für zwei- und mehratomige Moleküle geben. Da nullter und erster angeregter Zustand die bei weitem wichtigsten Besetzungen sind, erhalten wir Informationen über Moleküle bei niedrigen Energien. Welche Normalschwingung bei mehratomigen Molekülen sich auf die Relaxationszeit τ_1 bezieht, ist aus dem Experiment nicht ableitbar, jedoch von der Theorie her erwartet man die niedrigste Frequenz.

2.2. Chemische Aktivierung [7]

Moleküle in Nichtgleichgewichtsverteilungen heißen chemisch aktiviert bezüglich des thermischen Gleichgewichts oder des Reaktionsvermögens, wenn ihre Anregung als Folge einer chemischen Reaktion auftritt.

Abhängig von dem Reaktionstyp, von der Aktivierungsenergie und von der Reaktionsenthalpie erhält man Moleküle in rotatorisch, vibratorisch und elektronisch angeregten Zuständen. Die thermische Relaxation der chemisch aktivierten Moleküle in einem Wärmebad kann immer dann bestimmt werden, wenn ein mit der Relaxation konkurrierender Prozeß existiert, dessen Geschwindigkeitskonstanten meßbar sind, oder wenn die Anregung direkt optisch verfolgt werden kann.

Als Beispiel für die chemische Aktivierung soll die bimolekulare Assoziationsreaktion mit anschließendem unimolekularem Zerfall oder Stabilisierung durch Stöße diskutiert werden, nämlich:

$$A + B \; \underset{k_C}{\overset{k_{AB}}{\rightleftharpoons}} \; C^* \begin{cases} \xrightarrow{k_E} & D \quad \text{(Zerfallsprodukte)} \\ \xrightarrow{s_E} & S \quad \text{(Stabilisierungsprodukte)} \end{cases} \qquad (2.14)$$

Das Potentialdiagramm für die Gesamtreaktion ist in Abb. 1 dargestellt. E_1 ist die Aktivierungsenergie für die Vereinigung der Moleküle A und B zu einem schwingungsangeregten Molekül C^*, dessen Mindestenergieinhalt $E_{min} = E_1 + \Delta H_1$ ist.

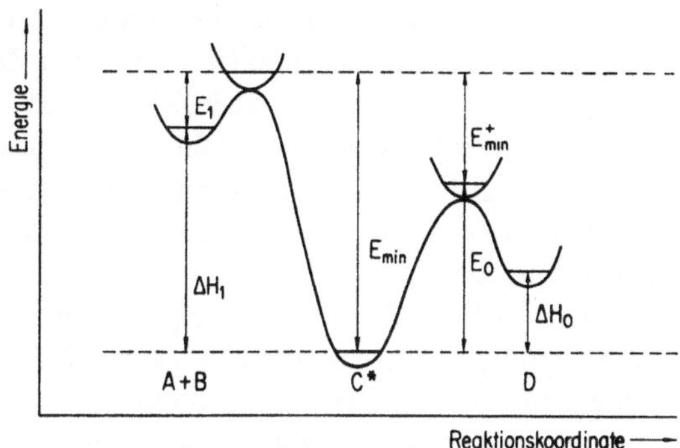

Abb. 1. Energiediagramm für die Reaktion $A + B \rightarrow C^* \rightarrow D$

E_0 ist die Aktivierungsenergie für den anschließenden unimolekularen Zerfallsprozeß und

$$E_{\min}^{+} = E_{\min} - E_0$$

die Mindestenergie für den entsprechenden aktivierten Komplex. k_E und s_E sind die spezifischen Geschwindigkeitskonstanten der Zerfalls- und Stabilisierungsreaktion. Nach der Theorie von KASSEL [9] oder nach der erweiterten Form, der Rice-Ramsperger-Kassel-Marcus-Theorie [10] (RRKM-Theorie), die an anderer Stelle ausführlich beschrieben worden ist, läßt sich k_E als Funktion der Energie berechnen. Es soll hier genügen, darauf hinzuweisen, daß in der Kassel-Formulierung, bei der das Molekül als System von s schwach gekoppelten Oszillatoren angesehen wird, k_E durch die Beziehung

$$k_E = \nu \left(\frac{E - E_0}{E} \right)^{s-1} \tag{2.15}$$

beschrieben wird, wobei E die innere Energie des Moleküls, und ν die Frequenz des Energieaustausches ist. Die Frequenz ν bleibt bei KASSEL offen, während sie bei MARCUS mit Hilfe der Theorie des Übergangszustands bestimmt wird. Die Verteilungsfunktion $f(E)$ der gebildeten Moleküle läßt sich ebenfalls mit der RRKM-Theorie berechnen.

Da im allgemeinen bei der chemischen Aktivierung (im Gegensatz zur thermischen Aktivierung) die Breite der Verteilungsfunktion klein gegenüber der Gesamtanregung ist, genügt es für die Zwecke dieser Illustration anzunehmen, daß die aktivierten Moleküle C^* in einer δ-Distribution bei E^* gebildet werden, wobei

$$E^* \geqslant E_{\min} .$$

Aus Abb. 1 geht hervor, daß C* in bezug auf den Zerfall chemisch aktiviert ist, solange seine Energie größer als E_0 ist. Gibt es nun aber einen zweiten Prozeß, nämlich den der Stoß-Desaktivierung, so können die Moleküle C^* so viel Energie verlieren, daß ihre Energie E kleiner wird als E_0; in diesem Falle sind Moleküle C^* stabilisiert ($C^* \rightarrow S$). In diesem Beispiel geht es also um die Konkurrenz zweier Prozesse, von denen die Geschwindigkeitskonstante für den einen Prozeß, nämlich k_E für den Zerfall bekannt ist. Die Geschwindigkeitskonstante für die Stabilisierung ist sicher proportional ω, der Stoßfrequenz des Moleküls C^*. Dabei ist aber nicht festgelegt, wieviel Energie pro Stoß (ΔE) übertragen wird. Je nach Größe von ΔE im Vergleich mit $E^* - E_0$ spricht man von einstufiger oder mehrstufiger Desaktivierung (siehe Abb. 2).

Die Zahl der Stufen im Bereich von $E^* - E_0$ bestimmt das Verhältnis der Produkte D und S und gestattet somit einen Vergleich mit dem Experiment. Bezeichnet man mit D und S die Wahrscheinlichkeit für den Zerfall und die Stabilisierung, so erhält man die in den beiden folgenden Abschnitten angegebenen Beziehungen.

2.2.1. Einstufige Desaktivierung

$$S = \frac{\omega}{k_1 + \omega} \; ; \quad D = \frac{k_1}{k_1 + \omega} \quad \text{wobei } k_1 = k_{E*} \; ; \qquad (2.16)$$

$$D + S = 1 \; ; \quad D/S = k_1/\omega \; ; \qquad (2.17)$$

$$\omega D/S = k_a = k_1 \; . \qquad (2.18)$$

Der Ausdruck $k_a = \omega\, D/S$ ist für den einstufigen Mechanismus über den ganzen Druckbereich konstant und wird deshalb als scheinbare Geschwindigkeitskonstante k_a definiert.

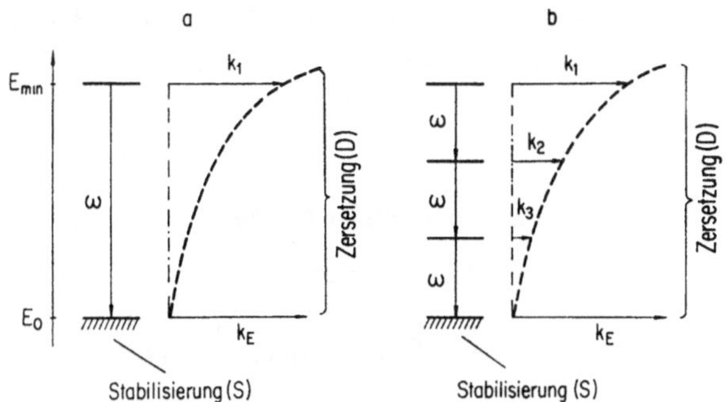

Abb. 2. a) Einstufige Desaktivierung, $T = 1$
b) Mehrstufige Desaktivierung, $T = 3$

E_0 — Aktivierungsenergie
E_{min} — Energieinhalt (abzüglich Nullpunktsenergie)
k_E (oder k_t) — energieabhängige Geschwindigkeitskonstante (1/sec)
ω — Stoßfrequenz (1/sec)

2.2.2. Mehrstufige Desaktivierung mit T Schritten

$$S_T = \prod_{t=1}^{T} \left\{ \frac{\omega}{\omega + k_t} \right\} \; ; \quad D_T = 1 - S_T \; ; \qquad (2.19)$$

Die Definition von k_t geht aus Abb. 2 hervor.

$$D_T/S_T = \prod_{t=1}^{T} (\omega + k_t)/\omega^T - 1 \; . \qquad (2.20)$$

Für $\omega \gg k_t$, d. h. bei hohem Druck, gilt:

$$(D_T/S_T)_\infty = \frac{1}{\omega} \sum_{t}^{T} k_t, \;\; \text{da} \prod_{t=1}^{T} (\omega + k_t) = \omega^T + \omega^{(T-1)} \sum_{t=1}^{T} k_t \; .$$

Für $\omega \ll k_t$ ist offensichtlich:

$$(D_T/S_T)_0 = \prod_{t=1}^{T} k_t/\omega^T \; . \qquad (2.21)$$

Bei kleinen Drucken geht $(D_T/S_T)_0$ um so schneller gegen unendlich, je größer die Zahl der Stufen T ist.

Bildet man analog zu (2.18) eine Geschwindigkeitskonstante $k_a = \omega\,(D_T/S_T)$, so sieht man, daß diese im Bereich kleiner Drucke nicht mehr druckunabhängig ist.

$$(k_a)_\infty = \omega\,(D_T/S_T)_\infty = \sum_{t=1}^{T} k_t\,,$$

$$(k_a)_0 = \omega\,(D_T/S_T)_0 = \prod_{t=1}^{T} k_t/\omega^{(T-1)}\,. \qquad (2.22)$$

Aus einem Vergleich der gemessenen D/S-Werte, bzw. der gemessenen k_a-Werte, mit den errechneten im Bereich niedrigen Drucks läßt sich die Zahl der Schritte im aktiven Bereich festlegen und somit die Durchschnittsenergie ΔE.

2.2.3. Experimentelles Beispiel

$$H + cis-C_4H_8 \;\rightleftharpoons\; C_4H_9{}^{*}$$

$$C_4H_9{}^{*} \xrightarrow{\;k_E\;} CH_3 + C_3H_6 \qquad (D) \qquad\qquad (2.23)$$

$$C_4H_9{}^{*} \xrightarrow{\;\omega\;} C_4H_9 \qquad\qquad (S)$$

Wasserstoffatome werden in einer Entladungsröhre erzeugt und entweichen durch einen Schlitz in einen Durchflußreaktor mit cis-Buten-Gas oder einem Gemisch von cis-Buten und inertem Gas $(1:50)$. Die erzeugten Butylradikale bilden je nach Druck und Temperatur und je nach Art des inerten Gases Produkte oder Folgeprodukte der Reaktionen D und S, deren Analyse mit Hilfe der Gaschromatographie geschieht. Die gemessenen Verhältnisse D/S als Funktion von ω werden mit der berechneten Kurvenschar verglichen und daraus Werte für ΔE ermittelt.

3. Mechanik der Stöße

3.1. Mechanische Grundlagen

Die Erhaltungssätze für den linearen Impuls $\boldsymbol{P}$ und den Drehimpuls $\boldsymbol{J}$ eines Systems von i Teilchen der Masse m_i und den Koordinaten $\boldsymbol{r}_i$ sind aus den Newtonschen Gesetzen der Bewegung herzuleiten. Aus der Definitionsgleichung des Massenzentrums (Masse M, Koordinate $\boldsymbol{R}$)

$$\sum_i m_i\,\boldsymbol{r}_i = \sum_i m_i\,\boldsymbol{R} = M\,\boldsymbol{R} \qquad\qquad (3.1)$$

folgt durch zweimalige zeitliche Ableitung

$$\sum_i m_i \dot{r}_i = \sum_i p_i \equiv P = M \dot{R} \tag{3.2}$$

und

$$\sum_i m_i \ddot{r}_i = \sum_i \dot{p}_i = \sum_i F_i = \dot{P} \ . \tag{3.3}$$

Gl. (3.3) folgt aus dem zweiten Newtonschen Gesetz:

$$F_i = \frac{dp_i}{dt} = m_i \ddot{r}_i \ , \tag{3.4}$$

wobei F_i bei konservativen Systemen sich als Gradient einer skalaren Potentialfunktion $V(r)$ darstellen läßt:

$$F_i = - \operatorname{grad}_i \ V(r_1, r_2, \ldots r_n) \tag{3.5}$$

mit

$$\oint F_i \cdot ds = 0 \tag{3.6}$$

Setzt sich $V(r_1, r_2, \ldots r_n)$ additiv aus Paarwechselwirkungen zusammen,

$$V(r_1, r_2 \ldots r_n) = \sum_{i<j} V_{ij}(|r_i - r_j|) \ , \tag{3.7}$$

so gilt für F_i:

$$F_i = \sum_{j \neq i} f_{ji} \ , \tag{3.8}$$

wobei man unter f_{ji} diejenige Kraft versteht, die das Teilchen j auf das Teilchen i ausübt. Nach dem dritten Newtonschen Gesetz von Aktion und Reaktion heben sich in einem System von n Teilchen die Kräfte immer gerade paarweise auf, d.h. $f_{ij} = - f_{ji}$, so daß in Gl. (3.3) $\dot{P}$ identisch Null und damit P konstant ist.

Der Drehimpuls eines abgeschlossenen Systems

$$J = \sum_i r_i \times p_i = \sum_i m_i (r_i \times \dot{r}_i) \tag{3.9}$$

ist für Zentralkräfte (f_{ij} und f_{ji} liegen auf der Verbindungslinie von $r_i - r_j$) ebenfalls konstant, denn

$$\dot{J} = \sum_i r_i \times m_i \ddot{r}_i = \sum_i \sum_{j \neq i} r_i \times f_{ji}$$
$$= \sum_{i>j} (r_i - r_j) \times f_{ji} = 0 \ . \tag{3.10}$$

Bei Einführung von Koordinaten relativ zum Schwerpunkt

$$\varrho_i = r_i - R \tag{3.11}$$

erhält man die den Gl. (3.1), (3.2) und (3.9) entsprechenden Beziehungen

$$\sum m_i\,\rho_i = 0 \tag{3.12}$$

$$\sum m_i\,\dot{\rho}_i = 0 \tag{3.13}$$

und

$$J = R \times P + \sum_i \rho_i \times m_i\,\dot{\rho}_i \; . \tag{3.14}$$

Der Gesamtdrehimpuls setzt sich also additiv aus dem Drehimpuls des Massenzentrums $P \times R$ und dem Drehimpuls der Teilchenbewegung relativ zum Schwerpunkt zusammen.

Für konservative Systeme ist die Gesamtenergie E ebenfalls eine Konstante der Bewegung,

$$E = T + V = T' + V' = E' \tag{3.15}$$

wobei die gestrichenen Größen für einen anderen Zeitpunkt gelten und die kinetische Energie T in kartesischen Koordinaten gegeben ist durch:

$$T = \frac{1}{2}\sum_i m_i\,\dot{r}_i \cdot \dot{r}_i = \frac{1}{2}\sum_i m_i\,(\dot{x}_i^2 + \dot{y}_i^2 + \dot{z}_i^2)\; . \tag{3.16}$$

Benutzt man nicht-kartesische, generalisierte Koordinaten q, so treten anstelle der Newtonschen Gleichungen die Lagrangeschen oder Hamiltonschen Gleichungen.

Die Lagrangesche Funktion $L\,(q, \dot{q}, t)$:

$$L = T - V \tag{3.17}$$

und die Hamiltonsche Funktion $H\,(q, p, t)$

$$H = \sum_j \dot{q}_j\,p_j - L \tag{3.18}$$

sind über die konjugierten oder kanonischen Impulse

$$p_j = \frac{\partial L}{\partial \dot{q}_j} \tag{3.19}$$

verknüpft.

Die Lagrangeschen Gleichungen für konservative Systeme

$$\frac{d}{dt}\left(\frac{\partial L}{\partial \dot{q}_j}\right) - \frac{\partial L}{\partial q_j} = 0 \tag{3.20}$$

bzw. deren kanonische Form

$$p_j = \frac{\partial L}{\partial \dot{q}_j}\,, \qquad \dot{p}_j = \frac{\partial L}{\partial q_j} \tag{3.21}$$

und die Hamiltonschen Gleichungen

$$\dot{q}_j = \frac{\partial H}{\partial p_j}\,, \qquad \dot{p}_j = -\frac{\partial H}{\partial q_j} \tag{3.22}$$

bleiben gegenüber einer Koordinatentransformation forminvariant.

Der Übergang von kartesischen Koordinaten x_i:

$$x_1 \to x_1 \qquad x_2 \to x_4 \qquad x_3 \to x_7$$
$$y_1 \to x_2 \qquad y_2 \to x_5 \qquad \text{etc.}$$
$$z_1 \to x_3 \qquad y_3 \to x_6$$

zu generalisierten Koordinaten q_i wird durch den metrischen Fundamentaltensor g beschrieben, dessen Komponenten g_{ij} folgende Form haben:

$$g_{ij} = \sum_l \frac{\partial x_l}{\partial q_i}\frac{\partial x_l}{\partial q_j} \tag{3.23}$$

Das Quadrat des Linienelements ds^2 ist durch

$$ds^2 = \sum_{i,j} g_{ij}\, dq_i\, dq_j \tag{3.24}$$

und die kinetische Energie T durch

$$2T = \sum_{i,j} m_i g_{ij}\, \dot{q}_i\, \dot{q}_j = \sum \bar{g}_{ij}\, \dot{q}_i\, \dot{q}_j \tag{3.25}$$

gegeben, wobei die Tensorkomponente durch

$$\bar{g}_{ij} = \sum_l \frac{\partial X_l}{\partial q_i}\frac{\partial X_l}{\partial q_j} \tag{3.26}$$

definiert ist mit der Zuordnung für X_i:

$$\sqrt{m_1}\, x_1 \to X_1 \qquad\qquad \sqrt{m_1}\, z_1 \to X_3$$
$$\sqrt{m_1}\, y_1 \to X_2 \qquad\qquad \sqrt{m_2}\, x_2 \to X_4 \qquad \text{etc.}$$

Die Transformation des Laplace-Operators wird von KEMBLE [11] und von WILSON, DECIUS und CROSS [12] beschrieben. Die Wurzel der Determinante von g, $|g|^{1/2}$, ist gleich der Funktionaldeterminante des x-Systems gegenüber dem q-System.

Bei Molekülen läßt sich die potentielle Energie V bis auf eine Konstante ebenfalls in quadratischer Form angeben:

$$2V = \sum_{i,j} k_{ij}\, q_i\, q_j\,, \tag{3.27}$$

solange die Annahme kleinerer Auslenkungen aus der Gleichgewichtslage gerechtfertigt ist. Der Bewegungslauf der q_i bei gegebenem T von Gl. (3.25) und V von Gl. (3.27) kann mit dem Lösungsansatz:

$$q_i = A_i \cos(\Lambda^{1/2} + \varepsilon) \tag{3.28}$$

aus den Lagrangeschen Gleichungen erhalten werden. Sind die Koeffizienten g_{ij} Konstanten, so führt das lineare Gleichungssystem der Amplituden A_i zu dem Säkularproblem:

$$\sum_{j=1}^{3N} k_{ij} - g_{ij}\Lambda = 0 \qquad \text{für} \quad i = 1, 2, \ldots 3N \,. \qquad (3.29)$$

Die Rotation eines Körpers wird durch Einführung eines körperfesten rotierenden Koordinatensystems beschrieben.

Mit der Einführung des Winkelgeschwindigkeitsvektors $\boldsymbol{\omega}$, der definitionsgemäß längs der augenblicklichen Drehachse liegt, läßt sich für einen starren Körper die zeitliche Änderung des Radiusvektors $\boldsymbol{r}_i$ durch die Beziehung:

$$\dot{\boldsymbol{r}}_i = \boldsymbol{\omega} \times \boldsymbol{r}_i \qquad\qquad (3.30)$$

ausdrücken. Für den Drehimpuls eines festen Körpers folgt aus Gl. (3.9) und (3.30)

$$\boldsymbol{J} = \sum_i m_i \left[\boldsymbol{r}_i \times (\boldsymbol{\omega} \times \boldsymbol{r}_i) \right]$$

$$\boldsymbol{J} = \sum_i m_i \left[\boldsymbol{\omega}\, r_i^2 - \boldsymbol{r}_i (\boldsymbol{r}_i \cdot \boldsymbol{\omega}) \right] \qquad (3.31)$$

oder in Komponentenschreibweise

$$J_x = \omega_x \sum_i m_i (r_i^2 - x_i^2) - \omega_y \sum_i m_i x_i y_i - \omega_z \sum_i m_i x_i z_i$$

$$J_x = I_{xx}\omega_x + I_{xy}\omega_y + I_{xz}\omega_z \qquad (3.32)$$

und J_y und J_z entsprechend.

Summarisch gilt also

$$\boldsymbol{J} = I\,\boldsymbol{\omega} \qquad\qquad (3.33)$$

Die Koeffizienten des Trägheitstensors I sind ebenso wie ω_x, ω_y und ω_z von der Richtung des gewählten Koordinatensystems abhängig. Meistens wird die Richtung so gewählt, daß die Trägheitsprodukte verschwinden; die dazu führende Transformation wird als Hauptachsentransformation bezeichnet. Die diagonalen Glieder sind die Hauptträgheitsmomente I_1, I_2 und I_3.

Die kinetische Energie eines starren rotierenden Körpers ist gegeben durch:

$$T = \frac{1}{2} \sum_i m_i (\boldsymbol{\omega} \times \boldsymbol{r}_i)(\boldsymbol{\omega} \times \boldsymbol{r}_i)$$
$$= \frac{1}{2} I\,\omega^2 \,, \qquad (3.34)$$

wobei das Trägheitsmoment um die Rotationsachse durch

$$I = \boldsymbol{n}\, I\, \boldsymbol{n} = \sum m_i (\boldsymbol{r}_i \times \boldsymbol{n})^2 \qquad (3.35)$$

(n ist der Einheitsvektor in Richtung $\boldsymbol{\omega}$) definiert ist. Ist I diagonal, so gilt:

$$T = \frac{1}{2} I_1 \omega_x^2 + \frac{1}{2} I_2 \omega_y^2 + \frac{1}{2} I_3 \omega_z^2 \tag{3.36}$$

Bei nicht starren Systemen kann für jede momentane Konfiguration a ein $\boldsymbol{\omega}_a$ festgelegt werden, so daß der Drehimpuls bezüglich des rotierenden Koordinatensystems verschwindet.

Diese von ECKART aufgestellte Beziehung

$$\sum m_i (r_i^a \times \dot{r}_i^a) = 0 \tag{3.37}$$

ermöglicht eine näherungsweise Separation von Rotation und Vibration. Wie WILSON, DECIUS und CROSS gezeigt haben, ist die kinetische Energie eines Moleküls damit gegeben durch:

$$2\,T = M\,\dot{R}^2 + I\,\boldsymbol{\omega}^{(a)\,2} + \sum_i m_i\,\dot{r}_i^2 + 2\,\boldsymbol{\omega}^a \sum m_i\,[(r_i - r_i^a) \times \dot{r}_i]\,, \tag{3.38}$$

wobei der letztere Term die Coriolis-Kopplung angibt. Die Gesamtenergie E läßt sich somit ebenfalls in Translations-, Rotations-, Vibrations- und Coriolis-Energie aufteilen.

3.2. Erhaltungssätze bei Stößen

Es soll zunächst gezeigt werden, daß der Drehimpuls zweier stoßender Moleküle A und B, $J(A,B)$, sich als vektorielle Summe der Einzeldrehimpulse S_A und S_B der Moleküle A und B bezüglich ihres Schwerpunkts und des Bahndrehimpulses L_{AB} darstellen läßt. Aus der Definition des Drehimpulses:

$$J(A,B) = \sum_i^{A,B} m_i\,r_i \times \dot{r}_i = \sum_i^{A} m_{ia}\,r_{ia} \times \dot{r}_{ia} + \sum_j^{B} m_{jb}\,r_{jb} \times \dot{r}_{jb} \tag{3.39}$$

(m_{ia} und r_{ia} bzw. m_{jb} und r_{jb} beziehen auf die Atome des Moleküls A bzw. B) folgt bei der Einführung von Relativkoordinaten zu den Schwerpunkten von A und B:

$$r_{ia} = R_A + \rho_{ia}; \qquad r_{jb} = R_B + \rho_{jb}; \tag{3.40}$$

$$J(A,B) = M_A R_A \times \dot{R}_A + M_B R_B \times \dot{R}_B + \sum_i^{A} m_{ia}\,\rho_{ia} \times \dot{\rho}_{ia} + \sum_j^{B} m_{jb}\,\rho_{jb} \times \dot{\rho}_{jb} \tag{3.41}$$

oder symbolisch

$$J(A,B) = L_{AB} + S_A + S_B \,. \tag{3.42}$$

Bezeichnet man mit gestrichenen Größen einen beliebig gewählten Anfangszustand eines nicht reaktiven Stoßablaufs (die Moleküle A und B

bleiben in ihrer Zusammensetzung erhalten) und mit doppelt gestrichenen Größen einen beliebigen Endzustand, so gilt der Drehimpulserhaltungssatz in der Form:

$$L'_{AB} + S'_A + S'_B = L''_{AB} + S''_A + S''_B \ . \tag{3.43}$$

Für den linearen Impuls des Systems (A, B), nämlich $P\,(A, B)$, der sich vektoriell aus den Einzelimpulsen P_A und P_B der Moleküle A und B zusammensetzt, gilt die Beziehung:

$$P'_A + P'_B = P''_A + P''_B \ . \tag{3.44}$$

Endlich erhält man für den Energieerhaltungssatz

$$E'_A + E'_B + E'_{AB} = E''_A + E''_B + E''_{AB} \ . \tag{3.45}$$

Hierbei ist E_{AB} die vom Abstand r der Moleküle A und B abhängige Wechselwirkung, die eine Funktion des Abstands r von A und B ist;

$$E_{AB} = V_{AB}(|r|) \ . \tag{3.46}$$

Bei einer Transformation vom Laboratoriumssystem der Moleküle A und B zum entsprechenden Schwerpunktssystem:

$$r = R_A - R_B \ ,$$
$$R = (M_A R_A + M_B R_B)/M \ , \quad M = M_A + M_B \ , \tag{3.47}$$

erhält man für die kinetische Energie $T\,(A, B)$ des Systems:

$$T\,(A,B) = \frac{1}{2} M_A \dot{R}_A \cdot \dot{R}_A + \frac{1}{2} M_B \dot{R}_B \cdot \dot{R}_B$$

$$= \frac{1}{2} \mu\, \dot{r} \cdot \dot{r} + \frac{1}{2} M \dot{R} \cdot \dot{R} \text{ mit } \mu = \frac{M_A M_B}{M_A + M_B} \ , \tag{3.48}$$

eine Form, bei welcher offensichtlich der Anteil der Schwerpunktsbewegung unabhängig von der Koordinate r der potentiellen Energie ist, die somit beim Stoßablauf nicht weiter berücksichtigt zu werden braucht.

Drückt man $^1/_2\,\dot{r} \cdot \dot{r}$ durch Polarkoordinaten aus, so erhält man nach Gl. (3.25)

$$\frac{1}{2} \mu\, \dot{r} \cdot \dot{r} = \frac{1}{2} \mu\, [\dot{r}^2 + r^2 \dot{\vartheta}^2 + r^2 \sin^2 \vartheta\, \dot{\varphi}^2] \tag{3.49}$$

oder, wenn man das Koordinatensystem so wählt, daß φ und $\dot{\varphi}$ Null sind:

$$\frac{1}{2} \mu\, \dot{r} \cdot \dot{r} = \frac{1}{2} \mu\, \dot{r}^2 + \frac{1}{2} \mu\, r^2 \dot{\vartheta}^2 \tag{3.50}$$

$$= \frac{1}{2} \mu\, \dot{r}^2 + \frac{1}{2} L_{AB}^2 / \mu\, r^2 \tag{3.51}$$

$$\text{mit } L_{AB} = \mu\, r^2 \dot{\vartheta} \tag{3.52}$$

Hierbei ist L_{AB} eine Konstante, solange keine Kopplung mit S_A und S_B besteht. Unter dieser Voraussetzung ist ihr Betrag ebenfalls durch

$$L_{AB} = \mu\, g\, b \tag{3.53}$$

gegeben, wobei g die anfängliche Relativgeschwindigkeit und b der Stoßparameter ist.

Mit der gemeinsamen Abkürzung E_i für die Rotationsenergie und Schwingungsenergie (und elektronische Anregungsenergie) lautet die Gl. (3.45) entsprechende Beziehung im Schwerpunktssystem:

$$\begin{aligned}
E'_{iA} + E'_{iB} + V'_{AB} + \frac{1}{2}\mu\, \dot{r}'^2 + \frac{1}{2} L'^2_{AB}/\mu\, r^2 \\
= E''_{iA} + E''_{iB} + V''_{AB} + \frac{1}{2}\mu\, \dot{r}''^2 + \frac{1}{2} L''^2_{AB}/\mu\, r^2 \ .
\end{aligned} \tag{3.54}$$

Für die Anfangszustände ($-$) und die Endzustände ($=$) gilt also

$$\bar{E}_{iA} + \bar{E}_{iB} + \frac{1}{2}\mu\, \dot{\bar{r}}^2 = \bar{\bar{E}}_{iA} + \bar{\bar{E}}_{iB} + \frac{1}{2}\mu\, \dot{\bar{\bar{r}}}^2 \ , \tag{3.55}$$

weil für $r \to \infty$: $V_{AB} \to 0$; $\frac{1}{2} L^2_{AB}/\mu\, r^2 \to 0$.

Für den Fall der Kopplung von L_{AB}, S_A, und S_B kann außer Schwingungsenergie auch Rotationsenergie in kinetische Energie übergehen.

3.3. Bewegungsablauf bei Stößen

Der Bewegungsablauf zweier stoßender Moleküle mit Rotations- und Schwingungsfreiheitsgraden ist bei Kenntnis der Potentialhyperflächen im Rahmen der Born-Oppenheimer-Näherung durch die Schrödinger-Gleichung der Kernbewegung festgelegt. In vielen Fällen läßt sich unter Vernachlässigung von Tunneleffekten und unter nachträglicher Einführung von Quantisierungsbedingungen (etwa bei Schwingungsenergieübertragung: ein Stoß ist nur erfolgreich, wenn die übertragene Energie ΔE mindestens $h\nu$ ist) die Schrödinger-Gleichung durch die klassischen Newtonschen oder Lagrangeschen Differentialgleichungen ersetzen, die bei mehr als zwei Teilchen, z. B. durch die Differenzenmethode gelöst werden können. Dabei werden die verschiedenen möglichen Anfangsbedingungen durch die Monte-Carlo-Methode [2] ausgewählt. Rein formal sind bei Kenntnis der Änderung der kinetischen Energie $\Delta T\,(AB)$ die Endgeschwindigkeiten zweier stoßender Moleküle durch Berücksichtigung des Energie- und Impulserhaltungssatzes festgelegt. $\Delta T(AB)$ hängt seinerseits mit der Änderung der inneren Energie E_{iA} oder E_{iB} der Stoßpartner A und B zusammen

$$\Delta T\,(AB) = \Delta E_{iA} + \Delta E_{iB} \ . \tag{3.56}$$

9*

Ist ΔE_{iA} oder ΔE_{iB} ungleich Null, so ist der Stoß definitionsgemäß unelastisch. Hierbei ist zu beachten, daß im Falle der exakten oder nahen Resonanzenergieübertragung $\Delta T(AB)$ Null oder sehr klein sein kann, weil

$$\Delta E_{iA} \simeq - \Delta E_{iB} \, . \tag{3.57}$$

Mit der im Abschnitt (3.2) eingeführten Definition von E_i, sind Rotations-, Schwingungs- und elektronische Anregungsübergänge unelastisch.

3.3.1. Beispiel: Eindimensionaler elastischer Stoß zwischen zwei Atomen

Nach dem Energie- und Impulserhaltungssatz erhält man als Endgeschwindigkeit $\bar{\bar{v}}_a$ und $\bar{\bar{v}}_b$ der Atome a und b:

$$\text{für} \quad m_a = m_b\colon \ \bar{\bar{v}}_a = \bar{v}_b; \ v_b = \bar{v}_a$$

$$\text{und für} \quad m_a \gg m_b\colon \ \bar{\bar{v}}_a \approx \bar{v}_a; \ \bar{\bar{v}}_b = 2\bar{v}_a - \bar{v}_b \, .$$

Dieses Resultat ist unabhängig von der Form des Potentials V_{ab}, da V_{ab} für den Anfangs- und Endzustand definitionsgemäß Null ist. Die Weg-Zeit-Kurven für verschiedene Potentiale unterscheiden sich aber wesentlich, indem nämlich beim Potential von harten Kugeln:

$$V_{ab} = 0 \quad \text{für} \quad x_b - x_a - r_{ab} > 0$$
$$V_{ab} = \infty \quad \text{für} \quad x_b - x_a - r_{ab} \gtrless 0 \tag{3.58}$$

die Geschwindigkeitsänderung in unendlich kurzer Zeit erfolgt, während beispielsweise beim exponentiellen Potential die Zeit des Energieaustausches endlich ist. Das exponentielle Potential wird beschrieben durch:

$$V_{ab} = V_0 \exp\left\{ - \alpha \, (x_b - x_a - x^0_{ab}) \right\} \, . \tag{3.59}$$

V_0 ist eine Konstante, α ist charakteristisch für die Steilheit des Potentials, x^0_{ab} ist der geringste Abstand während des Stoßes.

3.3.2. Beispiel: Eindimensionaler unelastischer Stoß zwischen einem Molekül und einem Atom

Ein Molekül $a\,b$ soll als klassischer Oszillator mit den Massen m_a und m_b dargestellt werden. Die Stoßrichtung des Moleküls $a\,b$ mit einem Atom c sei so festgelegt, daß $a\,b$ mit der b-Seite mit dem Atom c zusammenstößt.

Fall A: Potential harter Kugeln zwischen b und c

$$V_{bc} = 0 \quad \text{für} \quad x_c - x_b - r_{bc} > 0$$
$$\quad\quad = \infty \quad \text{für} \quad x_c - x_b - r_{bc} = 0 \ ;$$
$$V_{ac} = 0\colon \text{ keine Wechselwirkung angenommen;}$$
$$V_{ab} = {}^1\!/_2 \, k \, (x_b - x_a - r^0_{ab})^2\colon$$

k ist eine Kraftkonstante, und
r^0_{ab} der Gleichgewichtsabstand.

Für die Betrachtung des harten Potentials V_{bc} spielt jedoch die Form von V_{ab} keine Rolle, da beim momentanen Austausch zwischen b und c Atom b sich wie ein freies Atom verhält.

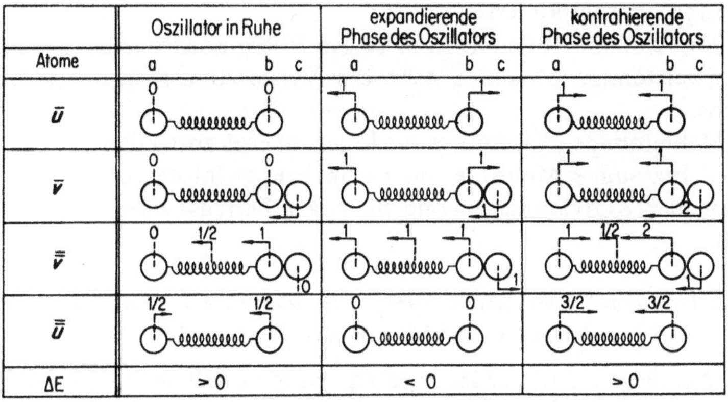

Atome	Oszillator in Ruhe			expandierende Phase des Oszillators			kontrahierende Phase des Oszillators		
	a	b	c	a	b	c	a	b	c
$\bar{u}$	0	0		1	1		1	1	
$\bar{v}$	0	0		1	1		1	1	2
$\bar{\bar{v}}$	0	1/2	1	1	1	1	1	1/2	2
$\bar{\bar{u}}$	1/2	1/2		0	0		3/2	3/2	
ΔE	> 0			< 0			> 0		

Abb. 3. Abhängigkeit des Energieaustausches von der Phase beim linearen Stoß eines klassischen Oszillators mit einem Atom Potential harter Kugeln; $\bar{v}$ und $\bar{\bar{v}}$ sind Absolut-, $\bar{u}$ und $\bar{\bar{u}}$ Relativgeschwindigkeiten bezüglich des Oszillatorschwerpunkts.

Der Stoßablauf ist für drei verschiedene Phasen des Oszillators in Abb. 3 dargestellt; für die Darstellung ist weiter angenommen, daß $m_a = m_b = m_c$. Aus Abb. 3 geht hervor, daß der Oszillator je nach Phase Energie aufnehmen oder abgeben wird ($\Delta E \gtrless 0$). Im Grenzfall sehr hoher relativer Geschwindigkeit und hartem Potential wird nur Energie aufgenommen.

Fall B: exponentielles Potential zwischen b und c,

$$V_{bc} = V_0 \exp \left\{ - \alpha \left(x_c - x_b - x_{bc}^0 \right) \right\}$$

Der Austausch zwischen b und c findet, wie schon gezeigt wurde, über eine endliche Zeit statt. Dies bedeutet aber, daß der Stoßprozeß über mehrere Perioden der Schwingung geht und somit über mehrere Phasen und deshalb im Mittel sich Energiegewinn und Energieverlust des Oszillators beinahe aufheben.

Es kann mathematisch gezeigt werden, daß dies in der Tat der Fall ist, und daß bei mittleren relativen Geschwindigkeiten und mittleren Schwingungsfrequenzen die Energie, die der Oszillator im Stoß aufnimmt oder abgibt, sehr gering ist. Es läßt sich schon aus diesem Beispiel abschätzen, daß die Energieübertragung besonders effektiv ist für ein steilabfallendes Potential, für eine hohe relative Geschwindigkeit v und für eine kleine Frequenz ν:

$$\Delta E \sim \frac{\alpha v}{\nu} . \tag{3.60}$$

4. Theorie der Energieübertragung bei Stößen

Im folgenden wird die Änderung der Schwingungsenergie (einschließlich der Energieänderung der inneren Rotation) bei den nachfolgenden vier Molekülgruppen diskutiert:

1. Zweiatomige Moleküle in niedrigen Energiezuständen.

2. Zweiatomige Moleküle nahe der Dissoziationsgrenze des Grundzustands.

3. Mehratomige Moleküle in niedrigen Energiezuständen

4. Mehratomige Moleküle mit einem Energieinhalt von der Größenordnung der Aktivierungsenergie für einen Zerfalls- oder Umlagerungsprozeß.

4.1. *Vibration-Translationsübergänge und Vibration-Vibrationsübergänge bei zweiatomigen Molekülen in niedrigen Energiezuständen*

Im Abschnitt 3.3 ist qualitativ gezeigt worden, daß die Aktivierung oder Desaktivierung eines zweiatomigen Moleküls, dargestellt als klassischer Oszillator, durch einen Stoß mit einem Atom von der Relativgeschwindigkeit und dem Wechselwirkungspotential der Stoßpartner und von der Frequenz der Amplitude des Oszillators abhängt. In der zeitabhängigen Störungsrechnung ist die Wahrscheinlichkeit eines Zustands k durch das Quadrat des Entwicklungskoeffizienten c_k gegeben:

$$c_k^2 = \frac{1}{\hbar^2}\left| \int dt \langle \psi_k | V(x,t) | \psi_j \rangle \exp\left\{-\hbar\pi i \nu t\right\}\right|^2 . \tag{4.1}$$

Hierbei sind ψ_j und ψ_k die Zustandsfunktionen des Ausgangs- und Endzustands, $V(x,t)$ ein zeitabhängiges Potential und $h\nu$ die Energiedifferenz der beiden betrachteten Zustände. Landau und Teller [8] haben mit der geschilderten Diracschen Methode die Übergangswahrscheinlichkeit p_{10} eines durch einen Stoß gestörten harmonischen Oszillators berechnet. Ihr Resultat lautet für einen Schwingungs-Translationsübergang:

$$p_{10} = c_0^2 = \frac{32\,\pi^4\,m^2\,\nu}{\hbar\,M\,\alpha^2}\exp\left\{-\frac{4\,\pi^2\,\nu}{\alpha\,v}\right\} \tag{4.2}$$

oder

$$p_{10} = \frac{64\,\pi^5\,m^2\,\Delta E}{\hbar^2\,M\,\alpha^2}\exp\left\{-\frac{8\,\pi^3\,\Delta E}{\hbar\,\alpha\,v}\right\} \tag{4.3}$$

mit

M = reduzierte Masse des Oszillators

m = reduzierte Masse des Systems Oszillator-Atom

ν = Frequenz des Oszillators

v = Relativgeschwindigkeit

α = Parameter des exponentiell angenommenen Potentials

ΔE = übertragene Schwingungsenergie;

ΔE ist gleich der Änderung der relativen Translationsenergie ΔT.

Bei einem Vergleich zweier Systeme mit verschiedenen v, α und ΔE ist der Exponentialanteil ausschlaggebend, solange, wie in der Herleitung von Gl. (4.2) und (4.3) angenommen, p_{10} klein ist. In Übereinstimmung mit den groben Abschätzungen von (3.60) ist p_{10} um so größer, je kleiner v, d. h. je kleiner ΔE und je größer α und v ist.

Aus p_{10} kann durch Mittelung über die Boltzmann-Besetzung die Übergangswahrscheinlichkeit bei einer bestimmten Temperatur, P_{10}, berechnet werden. Berücksichtigt man außer dem exponentiellen Abstoßungspotential noch die van-der-Waalssche Anziehung und die Änderung der kinetischen Energie beim Stoß, so erhält man nach SCHWARTZ, SLAWSKY und HERZFELD [13]:

$$P_{10} \sim \exp\left\{ -\frac{3}{2}\left[\frac{16\pi^4\, m\, v^2}{\alpha^2\, kT}\right]^{1/3} + \frac{\varepsilon}{kT} + \frac{hv}{2kT} \right\}, \qquad (4.4)$$

ε und σ sind die Parameter des Lennard-Jones-Potentials:

$$V_{L-J}(r) = 4\varepsilon\left[-\left(\frac{\sigma}{r}\right)^6 + \left(\frac{\sigma}{r}\right)^{12} \right].$$

Exakte oder nahe Resonanzübergänge zwischen zwei gleichen oder gleichartigen Molekülen sind wegen des kleinen Energieunterschieds des Anfangs- und Endzustands des gesamten Schwingungssystems besonders groß ($\Delta E = \Delta T \approx 0$). Sie sind von HERZFELD und Mitarbeitern ebenfalls mit Hilfe der Zenerschen [14] Methode berechnet worden.

Die Übergangswahrscheinlichkeiten P_{10} und P_{01} sind durch den Boltzmann-Faktor verknüpft.

$$P_{01} = P_{10} \exp\left\{ -hv/kT \right\} \qquad (4.5)$$

Im Abschnitt 2.1.2 wurde schon gezeigt, daß für Einquantenübergänge bei harmonischen Oszillatoren die Bezeichnung gilt:

$$P_{n,\,n+1} = n P_{10}$$
$$P_{n,\,m} = 0 \qquad \text{für} \quad m \neq n \pm 1$$

Zwei- und Mehrquantenübergänge sind entsprechend den in Gl. (4.1) auftretenden Matrixelementen immer dann erlaubt, wenn entweder Multipolübergänge bei harmonischen Oszillatoren oder Dipolübergänge (und höhere Übergänge) bei anharmonischen Oszillatoren berücksichtigt werden. SHULER und Mitarbeiter [15] haben gezeigt, daß solche Mehrquantenübergänge sehr wichtig sind für die Übereinstimmung von berechneten Dissoziationsraten zweiatomiger Moleküle mit den experimentellen Ergebnissen. Einzelheiten über die Berechnung von Übergangswahrscheinlichkeiten sowie Verfeinerungen der Modellvorstellungen sind in dem Buch "Molecular Energy Transfer in Gases" [1a] angegeben.

4.2. Zweiatomige Moleküle mit Energien in der Nähe der Dissoziationsgrenze

Die Theorie der höchstangeregten zweiatomigen Moleküle steht noch in ihrem Anfangsstadium [16, 17]. Die wenigen Experimente in diesem Bereich zeigen sehr hohe Übergangswahrscheinlichkeiten in der Nähe des Grenzwertes $P_{n,m} \approx 1$.

4.3. Erweiterung des Landau-Teller-Modells auf mehratomige Moleküle in niedrigen Energiezuständen

Die einfachste Modellvorstellung eines schwingenden Moleküls in niedrigen Energiezuständen ist die eines Systems von Normalschwingungen oder entkoppelten Oszillatoren. Die Aktivierung oder Desaktivierung des Moleküls ist dann von den Übergangswahrscheinlichkeiten der einzelnen Oszillatoren bestimmt, wobei nach Landau und Teller der Oszillator mit der niedrigsten Frequenz v die größte Übergangswahrscheinlichkeit hat.

Neben den reinen Schwingungs- und Translationsübergängen sind auch komplexe Übergänge dergestalt zu erwarten, daß die Desaktivierung eines Oszillators der Frequenz v_1 mit der gleichzeitigen Aktivierung eines Oszillators der Frequenz v_2 innerhalb desselben Moleküls gekoppelt ist, so daß insgesamt eine kleinere Energie ΔE in Translationsenergie ΔT überführt werden muß. Natürlich sind auch hier Resonanzübergänge zwischen zwei verschiedenen Molekülen zu erwarten. Im Abschnitt 2.1.3 wurde darauf hingewiesen, daß bei Ultraschalluntersuchungen von mehratomigen Molekülen nur eine oder in Ausnahmefällen zwei Relaxationszeiten gefunden werden, die in vielen Fällen der Desaktivierung der niedrigsten Normalschwingung und eventuell einer zweiten komplexen Desaktivierung zugeordnet werden können. Voraussetzung für ein solches Verhalten ist, daß die innermolekulare Energiewanderung zwischen den (durch den Stoß) gekoppelten Oszillatoren schnell ist im Vergleich zur Energieübertragung in relative Translationsenergie oder Schwingungsenergie des Stoßpartners. Die innermolekulare Energiewanderung in Molekülen niedriger Energiezustände ist außer für den Fall der exakten Resonanz immer an die Aufnahme oder Abgabe kleiner Energien an das Wärmebad gebunden, da die Energieniveaudichte des Gesamtschwingungszustands sehr gering ist.

4.4. Statistische Theorie für hochangeregte mehratomige Moleküle

Bei mehratomigen Molekülen mit großen inneren Energien spielen die anharmonischen Glieder der Potentialentwicklung eine wichtige Rolle und dürfen zumindest bei dynamischen Problemen, wie aus Monte-Carlo-Rechnungen für drei- und vieratomige Moleküle hervorgeht [2], nicht vernachlässigt werden. Im Gegensatz zur Slaterschen [18] Theorie der unimolekularen Reaktionen wird bei den Modellvorstellungen von KASSEL [9] ein schwingungsangeregtes Molekül als System schwach gekoppelter Oszillatoren dargestellt wird. Der Energieaustausch zwischen quantisierten Oszillatoren ist hier im Gegensatz zu den Molekülen bei niedrigen Energien wegen der hohen Entartung der Gesamtzustände des Moleküls auch ohne äußere Störung möglich. Die Energieverteilung in den Oszillatoren ist bei KASSEL allein durch die statistische Kombinatorik bestimmt, die freien Energiezufluß über das ganze Molekül voraussetzt. Die Lebensdauerverteilung energiereicher Moleküle, die eine unimolekulare Reaktion eingehen können, ist dann von exponentieller Form [2].

Es liegt nahe, die statistische Theorie der unimolekularen Reaktionen auf Stoßübergangswahrscheinlichkeiten zu übertragen. Dabei sind zwei Schritte zu unterscheiden, nämlich die Assoziation des stoßenden Moleküls mit dem Stoßpartner zu einem Stoßkomplex bestimmter Lebensdauer und die Dissoziation des Stoßkomplexes nach erfolgter statistischer Neuverteilung, wobei alle möglichen Endzustände als gleich wahrscheinlich angenommen werden. Sieht man zunächst von Translations- und Rotationsfreiheitsgraden ab, so wird die Wahrscheinlichkeitsverteilung der inneren Energie des Stoßkomplexes durch die Anfangsenergie des aktivierten Moleküls A und die Energieverteilung eines Wärmebades von Molekülen B bestimmt. Für Stoßzeiten, die lang im Vergleich zu den Perioden der Normalschwingung sind, kann eine völlige Neuverteilung der inneren Energie angenommen werden, für kurze Stoßzeiten eine nur unvollständige Wiederverteilung der Energie, deren Beschreibung mit der statistischen Methode unvollkommen ist. Die Wahrscheinlichkeit, daß nach Zerfall eines Stoßkomplexes (AB) mit der Gesamtenergie E die Energie E_2 im Molekül A mit n_A-Freiheitsgraden und die Energie $(E - E_2)$ im Molekül B mit n_B-Freiheitsgraden enthalten ist, ist von HOARE [19] berechnet worden.

$$P(E, E_2)\, dE_2 = \frac{\Gamma(n_A + n_B)(E - E_2)^{n_B - 1}\, E_2^{n_A - 1}}{\Gamma(n_A)\, \Gamma(n_B)\, E^{n_A + n_B - 1}}\, dE_2 \qquad (4.6)$$

Fragt man nach der größten Wahrscheinlichkeit von $P(E, E_2)$ bei festem E, so erhält man außer für den Grenzfall $n_A = n_B = 1$ die Beziehung:

$$E_{2\,\max} = \frac{(n_A - 1)}{(n_A - 1) + (n_B - 1)}\, E \qquad (4.7)$$

Für einen Stoß eines hochangeregten Moleküls A mit der Energie E_1 kann im Mittel E ungefähr E_1 gleichgesetzt werden, so daß

$$E_{2\,\max} \simeq \frac{n_A}{n_A + n_B}\, E_1 \qquad (4.8)$$

wobei angenommen ist, daß

$$n_A \gg 1 ; \quad n_B \gg 1 .$$

Daraus folgt die von Stevens, Neporent und Boudart [6] angegebene Beziehung für die übertragene Energie bei vollständiger Neuverteilung:

$$\Delta E_{\max} = E_1 - E_{2\,\max} = \frac{n_B}{n_A + n_B} E_1 \qquad (4.9)$$

Berücksichtigt man noch die Freiheitsgrade der Rotation und die Translation, die beim Begegnungskomplex verlorengehen, so erhält man die Beziehung

$$\Delta E'_{\max} = \left[\frac{n_B + 3 + n_r}{n_A + n_B + 3 + n_r} \right] E_1 , \qquad (4.10)$$

wenn n_r die Zahl der Rotationsfreiheitsgrade des Stoßpartners ist. Schließlich kann man eine nur unvollständige Neuverteilung, nämlich die mit den Übergangsschwingungen des Komplexes, durch:

$$\Delta E''_{\max} = \left[\frac{3 + n_r}{n_A + 3 + n_r} \right] E_1 \qquad (4.11)$$

beschreiben. Die zu den klassischen Überlegungen analogen Beziehungen für quantisierte Systeme sind ebenfalls von Hoare hergeleitet worden.

Es ist charakteristisch, daß die Übergangswahrscheinlichkeiten für das beschriebene Modell groß sein können, ohne daß wie bei der Landau-Teller-Theorie die übertragene Energie ΔE klein zu sein braucht. Die Energieübertragung ist in der statistischen Theorie maximal für lange Stoßzeiten und niedere Temperaturen, beides im Gegensatz zur Landau-Teller-Theorie. Die Theorie unterstützt das Modell der starken Stöße in dem Sinne, daß bei einem wirksamen Stoß eine erhebliche Reorganisation der Energie des hochangeregten Moleküls und des Stoßpartners stattfindet. Bei höchstangeregten Molekülen kann aber trotzdem eine stufenweise Desaktivierung mit einer durch Gleichung (4.9), (4.10) oder (4.11) bestimmten Schrittweite stattfinden, wie es im Abschnitt 2.2 beschrieben wurde.

5. Vergleich der experimentellen Resultate mit der Theorie

5.1. *Resultate für Moleküle bei niedrigen Energien: Ergebnisse von Ultraschallmessungen*

Die Fülle der experimentellen Daten in diesem Bereich ist sehr groß; eine Auswahl ist in Tabelle 1 wiedergegeben. Z_{10} in Tabelle 1 ist definiert durch:

$$Z_{10} = 1 / P_{10} .$$

Z_{10} gibt also die durchschnittliche Zahl der Stöße an, um Moleküle vom ersten angeregten Zustand zum Grundzustand zu desaktivieren. Z'_{10} bei mehratomigen Molekülen bezieht sich auf die niedrigste Frequenz ν' der

Moleküle. Aus Tabelle 1 kann der quantitative Zusammenhang von Z_{10} bzw. Z'_{10} mit ν bzw. ν' abgelesen werden; die Ergebnisse sind in Übereinstimmung mit der Landau-Teller-Theorie. Für zweiatomige Moleküle ist auch die Temperaturabhängigkeit von Z_{10} angegeben; mit steigender Temperatur wächst die relative Stoßgeschwindigkeit und hiermit auch die Wahrscheinlichkeit der Energieübertragung. Wie schon früher erwähnt wurde, ist in einigen Fällen von mehratomigen Molekülen eine zweite Relaxationszeit gemessen worden. Ein Beispiel hierfür ist SO_2, bei dem die niedrigste Frequenz 519 cm^{-1} der einen Relaxationszeit und die beiden anderen, nämlich 1151 und 1360 cm^{-1} der zweiten Relaxationszeit zugeordnet wurden. Eine Ausnahme in Tabelle 1 bildet H_2O, insofern als Z'_{10} viel zu klein ist im Vergleich zu der niedrigsten Frequenz. Weiter sind bei den Molekülen C_4H_{10}, C_5H_{12}, C_6H_{14} usw. die Relaxationszeiten τ so klein, daß sie nicht meßbar sind. In diesem Zusammenhang sollte noch erwähnt werden, daß die Reinheit der zu untersuchenden Gase eine große

Tabelle 1

Molekül	T° K	Z_{10}	ν (cm^{-1})	Molekül	T °K	Z'_{10}	ν' (cm^{-1})
N_2	556	2,7 (7) *	2331	CO_2	298	6 (4)	672
	761	9,6 (6)			393	2,5 (4)	
	1020	4,3 (6)			673	5,0 (3)	
	2153	1,5 (5)			1031	1,1 (2)	
	5545	2,2 (3)					
				CS_2	296	6 (3)	397
O_2	761	1,3 (5)	1554				
	1030	8,2 (4)		H_2O	486	2,7 (2)	1595
	2300	3,7 (3)					
	3112	1,0 (3)		SO_2 **	293	4,9 (2)	519
						3,5 (3)	1151, 1360
Cl_2	300	4,3 (4)	557				
	659	1,3 (3)		CH_4	303	1,5 (4)	1306
	1486	8,7 (1)			626	1,9 (3)	
				CH_3F	373	5,9 (3)	1048
Br_2	300	5,6 (3)	321	CH_3Cl	298	1,0 (3)	732
	529	1,2 (3)		$CHCl_2F$	300	5,1 (1)	274
I_2	385	1 (3)	213	C_3H_8	283	9 (0)	202
				C_4H_{10}, C_5H_{12}, C_6H_4	300	~1 (0)	

Z'_{10} bezieht sich auf ν', die niedrigste Frequenz des Moleküls.

   * 2,7 (7) $\equiv$ 2,7 $\times$ 10^7

** bei SO_2 gibt es 2 Relaxationszeiten.

Rolle spielt. Messungen an Gasgemischen haben ergeben, daß oft Fremdmoleküle einen wesentlich größeren Wirkungsquerschnitt haben als die Moleküle derselben Art. Einen großen Effekt haben manchmal Spuren von Wasser und höherer Kohlenwasserstoffe.

5.2. Resultate für mehratomige Moleküle bei hohen Energien: Ergebnisse der chemischen Aktivierung und der Fluoreszenzausbeutemessungen

Wie schon erläutert worden ist, bilden bei mehratomigen Molekülen in hochangeregten Schwingungszuständen die Energieniveaus des Gesamtsystems ein Quasikontinuum. Aussagen über den Energieaustausch werden deshalb formuliert als Energie ΔE, die im Durchschnitt pro Stoß übertragen wird.

In Tabelle 2 wird das System 3-Dimethylamino-6-aminophthalimid beschrieben, dessen erster angeregter elektronischer Zustand eine Schwingungsüberschußenergie E^* von 7,0 kcal/Mol bzw. 14,2 kcal/Mol je nach Benutzung des eingestrahlten Lichtes besitzen kann. Aus der Fluoreszenzausbeute dieses Moleküls in einem Überschuß von inerten Gasen ist ΔE bestimmt worden.

Tabelle 2. *Übertragene Energie ΔE als Funktion der Anregungsenergie E^**

	$E^* = 7,0$ kcal/Mol	$E^* = 14,2$ kcal/Mol
He	0,04 kcal/Mol	0,13 kcal/Mol
Ne	0,14	0,27
Ar	0,20	0,40
Kr	0,25	0,51
H_2	0,08	0,18
D_2	0,21	0,43
N_2	0,23	0,55
iso-C_5H_{12}	0,80	2,5

Mit 25 Bindungen und 72 Normalschwingungen pro Molekül ist die mittlere Anregungsenergie pro Bindung bzw. pro Normalschwingung für $E^* = 10$ kcal/Mol sehr klein. Der intramolekulare Energieaustausch ist dementsprechend relativ langsam; trotzdem werden schon bei diesem System relativ große mittlere Energien für entsprechend komplexe Stoßpartner gefunden. Die Anwendung der statistischen Vorstellung der Energieübertragung ist fragwürdig [20]; formal wird der unvollkommene Energieaustausch durch einen Adaptionsfaktor beschrieben. Im Vergleich dazu wird in Tabelle 3 das schon in Abschnitt 2 diskutierte Butylradikal-System aufgeführt. E^* ist 43 kcal/Mol; dabei sind nur 12 Bindungen bzw. 33 Normalschwingungen vorhanden, so daß die mittlere Energie pro Bindung bzw. pro Normalschwingung ungefähr 8mal so groß ist.

In Spalte 1 und 2 von Tabelle 3 sind die relativen Stoßquerschnitte β_h und β_{L-J} der Stoßpartner im Vergleich zu Buten aufgeführt. Die Indizes h und $L-J$ beziehen sich auf das zur Berechnung der Stoßfrequenz benutzte harte Kugelpotential und das Lennard-Jones-Potential. Die in den

Tabelle 3. *Übertragene Energie ΔE (kcal/Mol) und relativer Stoßquerschnitt β beim Butylradikalsystem (300 °K)*

Stoß-partner	β_h	β_{L-J}	$\Delta E_{\min h}$	$\Delta E_{\min L-J}$	ΔE	$\Delta E''_{\max}$
He	0,17	0,30	0,8	1,5	2–3	4,6
Ne	0,27	0,40	1,3	1,9	2–3	4,6
Ar	0,30	0,35	1,9	2,6	3–4	4,6
Kr	0,47	0,46	2,4	2,3	3–4	4,6
H_2	0,21	0,30	0,8	1,7	2–3	7,2
D_2	0,25	0,35	1,0	1,7	2–3	7,2
N_2	0,47	0,54	2,3	2,6	2–3	7,2
CO_2	0,75	0,83	4	4	–	8,3
CH_4	0,76	0,73	4	5	–	8,3
CD_3F	1,12	0,9	9	9	–	8,3
CH_3Cl	1,18	0,82	9	7	–	8,3
SF_6	1,06	1,00	9	9	–	8,3
C_4H_8	1,00	1,00	9	9	9	8,3

Spalten 3 und 4 angegebenen mittleren Energien $\Delta E_{\min h}$ und $\Delta E_{\min L-J}$ wurden aus dem Hochdruckbereich des Experiments unter der Annahme hergeleitet, daß jeder Stoß wirksam ist und daß die Stoßfrequenz durch die Potentiale (h) oder $(L-J)$ bestimmt ist.

ΔE folgt aus dem sehr schwierig meßbaren Niederdruckbereich und hängt theoretisch kaum von dem Wechselwirkungspotential ab; ΔE ist die mittlere pro Stoß übertragene Energie. Die erwähnten ΔE-Werte beziehen sich alle auf das Stufenmodell mit gleich großen Stufen und einer charakteristischen Schrittweite für jedes desaktivierende Molekül. Neben der δ-Distribution der Übergangswahrscheinlichkeiten wurden ebenfalls Gauss- und Exponentialverteilungen untersucht. Die mit dem statistischen Modell erhaltenen $\Delta E''_{\max}$-Werte sind von vergleichbarer Größenordnung. Sie liegen jedoch, wie zu erwarten und zu erhoffen war, noch über den gemessenen Werten.

Vergleichbare Resultate wurden bei chemisch aktivierten Cyclopropanen [7c] erhalten. Die Desaktivierung von Cyclopropanen bei $E^* \approx 100$ kcal/Mol ergab für ein- und zweiatomige Moleküle ΔE-Werte von ~5 kcal/Mol, bei mehratomigen solche von 20–30 kcal/Mol. Für das letztere System zeigte die Temperaturabhängigkeit an, daß die ΔE-Werte mit ansteigender Temperatur abfallen in Übereinstimmung mit dem statistischen Modell. Die Abhängigkeit von ΔE als Funktion der inneren Energie E^* wurde ebenfalls untersucht [7b], jedoch sind die bis jetzt vorhandenen Experimente nicht eindeutig.

Es wäre wünschenswert, in dem beschriebenen hochangeregten Energiebereich mehratomiger Moleküle mehr experimentelles Material zur Verfügung zu haben.

Literatur

1. a) Cottrell, T. L., and J. M. McCoubrey: Molecular Energy Transfer in Gases. London: Butterworths Scientific Publications, Ltd. 1961. (Übersicht mit besonderer Betonung der niedrig liegenden Energiezustände.)
   b) Transfert d'énergie dans les gaz (12. Solvay Konferenz, Institut International de Chimie, 1962), R. Stoops, Herausgeber. New York: Interscience 1964.
   c) Chemical Lasers (Supplement 2 der Applied Optics, 1965) J. N. Howard, Herausgeber, Op. Soc. of America, Wash. D. C.
2. Bunker, Don L.: Theory of Elementary Gas Reaction Rates, Int. Encyclopedia of Physical Chemistry and Chemical Physics. London: Pergamon Press 1966; Heydtmann, H.: Monte-Carlo-Rechnungen in der chemischen Kinetik, dieses Buch.
3. Herzfeld, K. P., and T. A. Litovitz: Absorption and Dispersion of Ultrasonic Waves. New York: Academic Press Inc. 1959.
4. Jost, W., u. H. Gg. Wagner: dieses Buch.
5. Thrush, B. A.: Flashphotolysis, dieses Buch.
6. a) Neporent, B. S., and S. O. Mirumyants: Opt. Spectry (USSR) 8, 336 (1960); 8, 414 (1960).
   b) Boudart, M., and J. T. Dubois: J. Chem. Phys. 23, 223 (1955).
   c) Stevens, B.: Molec. Phys. 3, 589 (1960).
7. a) Rabinovitch, B. S., and M. C. Flowers: Chemical Activation, Quart. Rev. 18, 122 (1964).
   b) Kohlmaier, G. H., and B. S. Rabinovitch: J. Chem. Phys. 38, 1692, 1709 (1963); J. Chem. Phys. 39, 490 (1963).
   c) Setser, D. W., J. W. Simons, and B. S. Rabinovitch: Bull. Soc. Chim. Belges 71, 662 (1962).
8. Landau, L., u. E. Teller: Phys. Z. Sowjetunion 10, 34 (1936).
9. Kassel, L. S.: The Kinetics of Homogeneous Gas Reactions. New York: Chemical Catalog Co. 1932.
10. Marcus, R. A.: Unimolecular Reaction Rate Theory, dieses Buch.
11. Kemble, E. C.: The Fundamental Principles of Quantum Mechanics. New York: Dover Publ. Inc. 1958.
12. Wilson, E. B., J. C. Decius, and P. C. Cross: Molecular Vibrations. New York: McGraw-Hill, 1955.
13. Schwartz, R. N., A. I. Slawsky, and K. F. Herzfeld: J. Chem. Phys. 20, 1591 (1952).
14. Zener, C.: Phys. Rev. 37, 556 (1931).
15. Shuler, K. E.: dieses Buch; ebenfalls Montroll, E. W. u. K. E. Shuler: Adv. Chem. Phys. 1, 361 (1958); Mies, F. H., u. K. E. Shuler: J. Chem. Phys. 37, 177 (1962).
16. Bauer, E., u. M. Salkoff: J. Chem. Phys. 33, 1202 (1960).
17. Brown, R. L., u. W. Klemperer: J. chem. Phys. 41, 3072 (1964).
18. Slater, N. B.: Theory of Unimolecular Reactions. Ithaca, N. Y.: Cornell University Press 1959.
19. Hoare, M.: Mol. Phys. 4, 465 (1961).
20. Serauskas, R. V., u. E. W. Schlag: J. Chem. Phys. 45, 3706 (1966).

G. H. Kohlmaier
Institut für physikalische Chemie
der Universität Frankfurt/Main

Monte-Carlo-Rechnungen in der chemischen Kinetik

H. Heydtmann

Mit 4 Abbildungen

1. Einleitung

Der theoretischen Berechnung von Geschwindigkeitskonstanten chemischer Reaktionen wurde in den dreißiger Jahren besondere Aufmerksamkeit geschenkt. In dem bekannten Buch von Glasstone, Laidler und Eyring, *The Theory of Rate Processes* [1], wurden die Ergebnisse dieser Entwicklung zusammengefaßt. Die Grundgedanken dieser Theorie der Reaktionsgeschwindigkeiten sind heute in jedem Lehrbuch der physikalischen Chemie zu finden: eine Potentialfläche[1] wird konstruiert, eine Gleichgewichtsverteilung für die molekularen oder atomaren Reaktionspartner wird auf dieser Potentialfläche bis in das Gebiet des sogenannten Aktivierungskomplexes oder Übergangszustandes angenommen. Aus der durchschnittlichen Zerfallsgeschwindigkeit dieser Komplexe folgt die makroskopische Geschwindigkeitskonstante.

Auf dieser Grundlage setzte eine neue Entwicklung mit den Arbeiten von Wall, Hiller und Mazur [2] ein, die eng verknüpft ist mit der Möglichkeit, elektronische Rechenanlagen zu benutzen. Die genannten Autoren konstruierten durch maschinelle, numerische Lösung der klassischen Bewegungsgleichungen eine große Anzahl von Wegzeitkurven (engl. trajectory) auf einer quantenmechanisch berechneten Potentialoberfläche. Aus Gründen der Einfachheit wurde für diese ersten Modellrechnungen die Reaktion von atomarem Wasserstoff mit molekularem Wasserstoff in verschiedenen Quantenzuständen der Rotation und der Schwingung und bei verschiedener Größe der Translationsenergie gewählt.

Untersucht man eine genügend große Anzahl von molekularen Systemen, und werden die Anfangsbedingungen — wie etwa relative räumliche Lage und relativer Geschwindigkeitsvektor für die Partner einer bimolekularen Reaktion, Schwingungszustände etc. — nach einem statistischen

---

[1] Es wird im Folgenden der Einfachheit halber gelegentlich von einer Fläche gesprochen, auch wenn von einer Hyperfläche die Rede ist.

Verfahren den Zufallsgesetzen entsprechend gewählt, spricht man von einer Monte-Carlo-Berechnung[2]. Monte-Carlo-Berechnungen simulieren also irgendein Reaktionssystem und erlauben dabei nicht nur eine Aussage über die Geschwindigkeitskonstante einer thermisch ausgelösten Reaktion sondern z. B. auch Aussagen über die Energieverteilung in den inneren Freiheitsgraden der Produkte, Ergebnisse, die mit modernen experimentellen Methoden heute ebenfalls nachgeprüft werden können. Ein solches Monte-Carlo-Verfahren wurde von WALL, HILLER und MAZUR zwar angestrebt, jedoch noch nicht verwirklicht. Es wurde auch bereits von diesen Autoren betont, daß dieses Verfahren keinen Ersatz für die rein quantenmechanische Lösung des Reaktionsproblems darstellt, sondern nur einen zur Zeit leichter gangbaren Weg der Näherung, der — wie die Zwischenzeit gezeigt hat — zu einem besseren Verständnis vieler Phänomene führt. Das Studium einzelner Wegzeitkurven, das in der neuesten Literatur wiederholt beschrieben ist, führt zwar auch zu interessanten Ergebnissen, soll jedoch in diesem Überblick nicht speziell behandelt werden.

Monte-Carlo-Methoden wurden auf drei- und neuerdings auch auf vieratomige Systeme vom unimolekularen sowie vom bimolekularen Reaktionstypus angewandt. Diese Typen sollen getrennt besprochen werden. Die Bedeutung der Monte-Carlo-Methode neben anderen, experimentellen und theoretischen Untersuchungen als Test für neue theoretische Vorstellungen über den Ablauf chemischer Reaktionen wurde kürzlich von BUNKER in dem Buch *Theory of Elementary Reaction Rates* in knapper Form herausgestellt [3].

2. Bimolekulare Reaktionen

Die Grundzüge des Berechnungsweges sollen anhand eines relativ einfachen Beispiels kurz erläutert werden [4]. Wir betrachten die Reaktion eines Atomes A von der Masse m_A mit einem zweiatomigen Molekül BC (Massen der Atome: m_B und m_C):

$$A + BC \longrightarrow AB + C. \tag{1}$$

Es sei vorausgesetzt, daß die drei Atome sich stets in derselben Ebene bewegen (zweidimensionale Näherung). Dieses System kann einschließlich seiner Lage bezüglich einer in der vorgegebenen Ebene fest gewählten Richtung (Vektor $\mathscr{A}$) zu jedem Zeitpunkt durch vier Koordina-

---

[2] Monte-Carlo-Rechnungen haben in der physikalischen Chemie einen weiten Anwendungsbereich. Häufig existieren physikalische Modellvorstellungen über das molekulare Geschehen, jedoch verbietet die ungeheure Kompliziertheit aller zu berücksichtigenden, statistisch bestimmten Effekte oft eine exakte Berechnung eines aus vielen Teilchen bestehenden Systems (z. B. einer Flüssigkeit). Einen Überblick über die Bedeutung der Monte-Carlo-Methode in der physikalischen Chemie gaben M. A. D. FLUENDY und E. B. SMITH [Quart. Rev. **16**, 241 (1962)].

ten beschrieben werden, beispielsweise durch die Abstände r_{AB} und r_{BC} und die Winkel α und φ (vgl. Abb. 1). Man interessiert sich nicht für die Bewe-

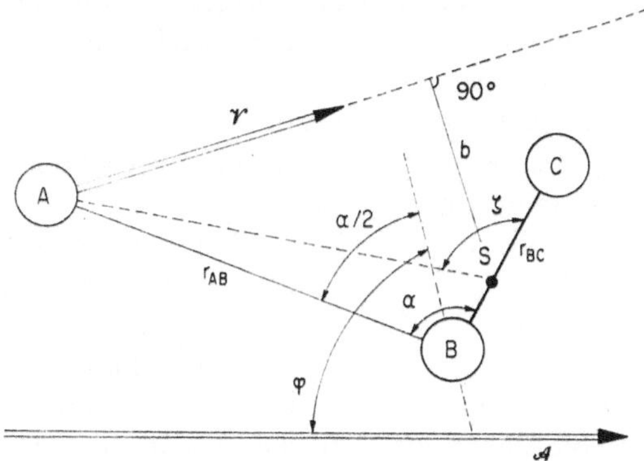

Abb. 1. Schema für eine Reaktion zwischen einem Atom und einem zweiatomigen Molekül in einer Ebene; S Schwerpunkt des Moleküls BC, $\mathscr{V}$ relativer Geschwindigkeitsvektor, $\mathscr{A}$ festgelegte Richtung in der Ebene $A-B-C$

gung des Schwerpunktes und geht deshalb bei der Aufstellung der Energie des Systems von einem kartesischen Koordinatensystem aus, in dem der Schwerpunkt keine Bewegung ausführt (Schwerpunktssystem). Betrachten wir die drei Atome unseres Modells als ein Quasimolekül, so geht daraus hervor, daß vier Freiheitsgrade — drei Schwingungen und eine Rotation — behandelt werden sollen.

Die Hamilton-Funktion für dieses System ist von der Form:

$$H(r_{AB}, r_{BC}, \alpha, p_1, p_2, p_\alpha, p_\varphi)$$
$$= T(r_{AB}, r_{BC}, \alpha, p_1, p_2, p_\alpha, p_\varphi) + V(r_{AB}, r_{BC}, \alpha). \tag{2}$$

Dabei bedeuten die p_i die zu den gewählten Koordinaten q_i konjugierten Impulse (Definition: $p_i = \partial T/\partial \dot{q}_i$). Die klassischen Bewegungsgleichungen in der Hamiltonschen Form lauten:

$$\frac{\partial H}{\partial q_i} = -\dot{p}_i \quad \text{und} \quad \frac{\partial H}{\partial p_i} = \dot{q}_i. \tag{3}$$

Die in den betrachteten vier Freiheitsgraden enthaltene Gesamtenergie H ist keine Funktion des Winkels φ. Aus den Bewegungsgleichungen folgt dann die zeitliche Konstanz von p_φ (Erhaltung des auf der Molekülebene senkrecht stehenden Drehimpulsvektors). Aus (3) erhalten wir ein System von sechs Differentialgleichungen erster Ordnung, die nach Wahl geeigneter Anfangsbedingungen beispielsweise nach dem Runge-Kutta-Verfahren durch schrittweises Voranschreiten um kleine Zeitintervalle δt gelöst werden können. Die Intervallgröße beträgt für typische Beispiele 10^{-15} sec, ist also klein gegenüber der Dauer jeder in dem Problem möglicherweise vorkommenden Normalschwingung.

Mit Hilfe von immer wieder neu gewählten Zufallszahlen R und S ($O \leq R \leq 1$; $S \equiv +1$ oder -1) werden die molekularen Hilfsgrößen ζ (Orientierung von BC zu A), $\dot{\zeta}$ (Rotationsgeschwindigkeit von BC), $\dot{r}_{BC}$ (Schwingungsgeschwindigkeit), V_R (Betrag des relativen Geschwindigkeitsvektors $\mathscr{V}$) und b (Stoßparameter) gewählt. Unter dem Stoßparameter versteht man die Länge des vom Schwerpunkt S des Moleküls BC auf den verlängerten, von A ausgehenden Geschwindigkeitsvektor gefällten Lotes; er gibt also die Bewegungsrichtung von $\mathscr{V}$ an.

Die Rolle der Zufallszahl soll für die Auswahl der Größen ζ und b näher erläutert werden [5]. Die Orientierung des Reaktionspartners BC läßt sich durch $\zeta = 2\pi R$ auf einfache Weise bestimmen.

Der Stoßparameter b soll innerhalb der Grenzen 0 und b_{max} liegen, wobei b_{max} so gewählt ist, daß für $b > b_{max}$ die Reaktionswahrscheinlichkeit in jedem Fall vernachlässigbar klein ist. Die Wahrscheinlichkeit, daß b in dem Intervall $b, b + db$ liegt, ist — wie aus der Geometrie des Stoßes leicht einzusehen — gegeben durch

$$W(b)\,db = d\,\frac{(\pi b^2)}{\pi b_{max}^2} = 2b\,db\,/\,b_{max}^2, \quad 0 \leq b \leq b_{max}\,. \tag{4}$$

Die Wahrscheinlichkeit, daß $0 \leq b \leq b'$, ist durch die kumulative Verteilungsfunktion gegeben:

$$KVF = \int\limits_0^{b'} W(b)\,db = (b'\,/\,b_{max})^2\,. \tag{5}$$

Diese Funktion kann nun gleich der Zufallszahl R gesetzt werden, und b ist mit Hilfe der Beziehung $b = b_{max}\,R^{1/2}$ zu wählen. Wenn die gewünschte variable Größe nicht als einfache Funktion von $KVF = R$ zu beschreiben ist, müssen fortgeschrittenere Methoden zur Bestimmung der Anfangswerte dieser Größe herangezogen werden. Solche Methoden werden in der zitierten Literatur ausführlicher beschrieben.

Die Verteilungen für ζ, $\dot{\zeta}$, $\dot{r}_{BC}$, V_R und b liefern zusammen mit festen, für alle Wegzeitkurven zu Beginn gleichen Werten r_{AB} und r_{BC} (Gleichgewichtswert r_{0BC} — es kann jedoch auch für r_{BC} eine Verteilung vorgenommen werden) die gewünschten Anfangsbedingungen. Zwischen 300 und 3000 Anfangsbedingungen wurden gewöhnlich für die Monte-Carlo-Behandlung eines Reaktionsproblems ausgewählt. Die Verteilungen lassen sich natürlich einzeln unterdrücken, und man kann auf diese Weise studieren, welche Verteilung das Endergebnis entscheidend beeinflußt.

Ein wichtiges Problem ist die Wahl der Potentialhyperfläche $V(r_{AB}, r_{BC}, \alpha)$ für die zu behandelnde Reaktion. Oft stellt der Potentialansatz ganz einfach eine zweckmäßige, gut zu handhabende Funktion dar, der es gestattet, die topographischen Einzelheiten der Hyperfläche wiederzugeben und — wenn erwünscht — durch Variation einzelner Parameter zu verändern. Als Beispiel soll eine Potentialfunktion, die für die Reaktion (1) verwandt werden kann, angegeben werden. Sie wurde von RAFF und KARPLUS [8] publiziert und ohne den vierten Term schon von BLAIS und BUNKER [4] benutzt:

$$
\begin{aligned}
V = D_{AB}&\left\{1 - \exp\left[-\beta_1\left(r_{AB} - r_{0AB}\right)\right]\right\}^2 + \\
&+ D_{BC}\left\{1 - \exp\left[-\beta_2\left(r_{BC} - r_{0BC}\right)\right]\right\}^2 + \\
&+ D_{BC}\left\{1 - \tanh\left(a \cdot r_{AB} + c\right)\right\} \cdot \exp\left[-\beta_2\left(r_{BC} - r_{0BC}\right)\right] + \\
&+ D_{AB}\left\{1 - \tanh\left(d \cdot r_{BC} + f\right)\right\} \cdot \exp\left[-\beta_1\left(r_{AB} - r_{0AB}\right)\right] + \\
&+ D_{AC}\exp\left[-\beta_3\left(r_{AC} - r_{0AC}\right)\right] .
\end{aligned}
\tag{6}
$$

Diese Funktion hat 13 Parameter und r_{AC} tritt als Variable anstelle von α auf. Die ersten beiden Terme stellen die beiden Potentialfunktionen vom Morse-Typ für die stabilen Moleküle AB und BC dar, und die in ihnen vorkommenden sechs Parameter lassen sich mit Hilfe experimenteller Daten angeben. Der dritte Term schwächt die Anziehung der Atome B und C ab für den Fall, daß sich das Atom A dem Molekül BC stark genähert hat. Ein analoger Term für die Anziehung von A und B liegt im folgenden Summanden vor. Der letzte, exponentielle Term beschreibt die Abstoßung zwischen A und C. Die potentielle Energie für den Fall unendlicher Entfernung der drei Teilchen voneinander beträgt $(D_{AB} + D_{BC})$ und gilt als Bezugsgröße.

Die Integration wird über eine große Folge von Zeitintervallen durchgeführt, bis r_{BC} einen kritischen Wert überschreitet (erfolgreicher Stoß) oder r_{AB} einen kritischen Wert, der natürlich größer als der Anfangswert gewählt werden muß, überschreitet (keine Reaktion) oder bis nach Durchlaufen von beispielsweise 2500 Zyklen die Begegnung als unentschieden angesehen werden kann. Letzterer Fall wurde — bei geeigneter Wahl aller Randbedingungen — nur selten beobachtet.

Monte-Carlo-Berechnungen wurden vor allem durchgeführt, um die experimentellen Aussagen von Reaktionen mit gekreuzten Molekularstrahlen zu interpretieren [6]. Hier ist an erster Stelle die Reaktion von Alkalimetallatomen mit Alkylhalogeniden, speziell die Reaktion von Kalium mit Methyljodid zu nennen [4—9].

$$
K + CH_3J \longrightarrow KJ + CH_3, \quad \Delta H_R = -22\,\text{kcal/mol} .
\tag{7}
$$

Die Methylgruppe wurde bei den Rechnungen als Atom behandelt. Anfängliche zweidimensionale Studien [4] erwiesen sich als unzulänglich und mußten durch dreidimensionale Studien [7, 8] ergänzt werden. Ausgewertet wurden insbesondere die Verteilungsfunktionen für die Schwingungsenergie E_{vib}, die Rotationsenergie E_{rot} und für die Gesamtenergie

10*

($E_{vib} + E_{rot}$) des KJ und die Winkelverteilung für die Produkte $n\,(\theta)$. Mit θ bezeichnet man den Winkel, den der Geschwindigkeitsvektor des CH_3-Radikals mit dem anfänglichen relativen Geschwindigkeitsvektor $\mathscr{V}$ im Schwerpunktssystem einschließt. Zur Interpretation der Laborergebnisse ist eine Umrechnung der Monte-Carlo-Ergebnisse für $n\,(\theta)$ aus dem Schwerpunktssystem in das Laborsystem erforderlich [6]. Für den Fall, daß in einem der beiden Strahlen eine wesentlich höhere Geschwindigkeit der Teilchen vorliegt als in dem anderen (kleine Masse der Teilchen oder höhere Temperatur), hat der Vektor $\mathscr{V}$ nahezu die Richtung dieses Strahles, und die Winkelverteilung $n\,(\theta)$ ist in Näherung der experimentell bestimmbaren Winkelverteilung im Laborsystem $n\,(\vartheta)$ gleichzusetzen. Es sei darauf hingewiesen, daß $n\,(\theta)$ von der relativen Geschwindigkeit der Teilchen

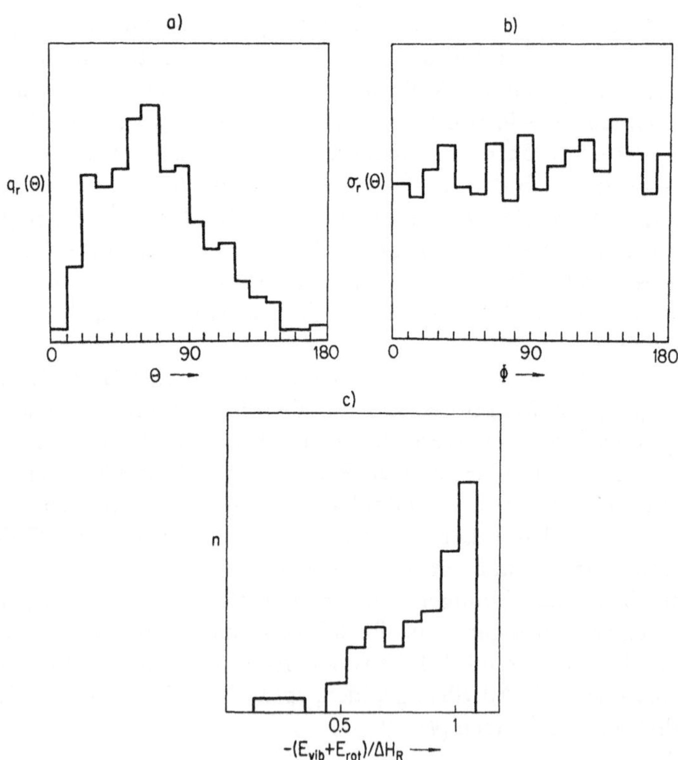

Abb. 2. Rechenergebnisse 3-dimensionaler Berechnungen für die Reaktion CH_3J + K mit 680 Wegzeitkurven [8]

a) Reaktionsquerschnitt in beliebigen Einheiten als Funktion von θ

b) Differentieller Reaktionsquerschnitt in beliebigen Einheiten als Funktion von Φ. Wie erwartet tritt keine Abhängigkeit von Φ auf, da keine bevorzugte Orientierung für CH_3J angenommen wurde

c) Innere Anregung des KJ-Moleküls, ausgedrückt in Einheiten von ΔH_R

wie von den Rotations- und Schwingungszuständen der Reaktionspartner abhängt. Falls eine Selektion bestimmter Zustände oder eine Vorzugsorientierung in den Molekularstrahlen vorliegt, ist dies bei der Berechnung von $n(\theta)$ zu berücksichtigen.

Die Reaktion (7) hat keine nennenswerte Aktivierungsenergie, und die Verteilungsfunktion für $E_{vib} + E_{rot}$ zeigt, daß im Durchschnitt etwa 90% der freiwerdenden Reaktionsenergie in den inneren Freiheitsgraden der Produkte angesammelt werden, daß also nur ein kleiner Bruchteil als relative kinetische Energie der Produkte auftritt (Abb. 2c). Die Verteilungsfunktion $n(\theta)$ und der dieser proportionale Reaktionsquerschnitt $q_r(\theta)$ haben bei einer dreidimensionalen Rechnung ein Maximum zwischen $\theta = 0°$ und 90° (Abb. 2a) [7—9].

$$q_r(\theta)\,\Delta\,\theta = \pi\,b^2_{max}\,n(\theta)\,/\,N, \tag{8}$$

wobei für N die Anzahl der untersuchten Wegzeitkurven und für $n(\theta)$ die Anzahl derjenigen Wegzeitkurven einzusetzen ist, die CH_3-Streuung in den Winkelbereich $\theta, \theta + \Delta\theta$ ergeben. Dieses Resultat entspricht einer Rückwärtsstreuung der CH_3-Radikale in die Richtung, aus der die CH_3J-Moleküle im Schwerpunktssystem kommen, und entsprechend einer Rückwärtsstreuung der KJ-Moleküle im Schwerpunktssystem in die Richtung, aus der der K-Strahl eintrifft (Rückstoßmechanismus).

Integriert man $q_r(\theta)$ über den ganzen Bereich von θ (0° bis 180°), erhält man den totalen Reaktionsquerschnitt $S_r = \pi\,b^2_{max}\,W_r(b_{max})$, wobei W_r die Reaktionswahrscheinlichkeit für ein gegebenes b_{max} bedeutet[3]. Der differentielle Reaktionsquerschnitt $\sigma_r(\theta)$ ist — unter der Voraussetzung, daß keine Vorzugsorientierung der Moleküle zueinander vorliegt — mit $q_r(\theta)$ über die Gleichung

$$\sigma_r(\theta) = q_r(\theta)\,(2\pi\sin\theta) \tag{9}$$

verknüpft und berücksichtigt die Streuung in ein Raumwinkelelement. Eine bevorzugte Orientierung der Moleküle bedingt eine Abhängigkeit des differentiellen Reaktionsquerschnittes vom azimutalen Winkel Φ, der von einer auf $\mathscr{V}$ senkrechten Richtung aus gezählt wird (vgl. Abb. 2b). $\sigma_r(\theta)$ sinkt in dem besprochenen Fall von einem maximalen Wert für $\theta = 0°$ für größere θ stark ab. Diese Aussagen sind in Einklang mit den Molekularstrahlergebnissen. Wie bereits aus der Anisotropie der Produktstreuung zu erwarten, ist der Verlauf der Wegzeitkurven meist wenig verwickelt, und es erscheint infolgedessen nicht gerechtfertigt, von einem molekülartigen Übergangskomplex zu reden.

Aus Berechnungen von RAFF [10] für die Reaktion

$$K + C_2H_5J \longrightarrow KJ + C_2H_5 \tag{10}$$

---

[3] W_r muß eingeführt werden, weil es neben der Streuung der Produkte auch eine physikalische Streuung durch nichtreaktive Stöße geben kann.

in einem 4-atomigen Modell ist zu entnehmen, daß die C—C-Schwingung ebenfalls Reaktionsenergie aufnimmt (durchschnittlich 14%). Für diesen Fall der Energieverteilung auf das Äthylradikal und das KJ treten längerlebige Komplexe auf. Zu den Untersuchungsergebnissen, deren experimentelle Nachprüfung zur Zeit noch nicht möglich ist, gehören Aussagen über die Verteilung des Gesamtdrehimpulses bezüglich der Rotation der Produktmoleküle und der relativen Bahnbewegung in Abhängigkeit von den Anfangsbedingungen.

Besonders interessant erscheint die Möglichkeit, durch Vergleich der Monte-Carlo-Berechnungen mit Molekularstrahlergebnissen etwas über die Gestalt der Potentialflächen zu erfahren. Karplus und Raff [8] bemerkten, daß der totale Reaktionsquerschnitt S_r für die Reaktion (7) mit dem Experiment wesentlich besser übereinstimmt, wenn die von Blais und Bunker [4] vorgeschlagene Potentialfunktion durch ein weiteres Glied ergänzt wird: die Anziehung des Alkaliatoms bei Annäherung an das Jodatom in einer relativ wenig gestreckten C—J-Bindung mußte abgeschwächt werden [Gleichung (6)]. Diese Potentialfläche, die auch bei der Berechnung der in Abb. 2 wiedergegebenen Resultate verwandt wurde, ist in Abb. 25 von Toennies [6] zu sehen. Eine eingehende Untersuchung des Einflusses der Potentialflächengestalt speziell für die Reaktion K + CH_3J [9] zeigte allerdings, daß eine größere Genauigkeit der Messungen erforderlich ist, um feinere Unterschiede zwischen den verschiedenen Potentialansätzen zu überprüfen.

Bunker und Blais [7, 11] benutzten für verschiedene hypothetische Reaktionen vom Typ (1), die ohne Aktivierungsenergie exotherm ablaufen sollen, im allgemeinen eine Potentialfläche, bei der die Energie im Anziehungsbereich der Reaktionspartner ($r_{BC} \simeq r_{oBC}$) frei wird. In diesem Falle wird — wie bereits für die Reaktion (7) angegeben — fast die ganze Reaktionsenthalpie, $-\Delta H_R$, in innere Energie der Produkte überführt. Dieser Potentialflächentyp wurde zuerst von Evans und Polanyi für die Reaktionen X + $Na_2 \rightarrow$ NaX + Na, Na + HgX $\rightarrow$ NaX + Hg vorgeschlagen[4], und später auf Grund neuerer experimenteller Ergebnisse von Simons diskutiert [12] (vgl. Abb. 3a). Bei einer Änderung der Potentialfunktion dergestalt, daß die Reaktionsenthalpie hauptsächlich bei Auseinanderbewegung der Produkte frei wird (vgl. Abb. 3b), erhielten Blais und Bunker z. B. für den speziellen Fall $m_A = m_B = m_C = 16$ a. E. und $\Delta H_R = -60$ kcal/mol eine deutliche Verschiebung der Maxima für die Verteilungsfunktionen von ($E_{vib} + E_{rot}$) und von E_{vib} zu kleineren Werten.

Eine solche Verteilungsfunktion für die Schwingungsenergie des Produktmoleküls wurde für die Reaktionen

$$H + X_2 \longrightarrow HX + X, \qquad X = Cl, Br$$
$$\text{für } X = Cl: \Delta H_R = -45\,\text{kcal/mol}, \; E_A = 2,5\,\text{kcal/mol} \qquad (11)$$

---

[4] X bedeutet ein Halogenatom.

durch Auswertung infraroter Chemilumineszenzmessungen festgestellt [*13*].
KUNTZ, NEMETH, POLANYI, ROSNER und YOUNG untersuchten ausgewählte
Wegzeitkurven für Reaktionen dieses Typs auf acht verschiedenen Potential-
flächen, für die in Prozent das Freiwerden der Energie im Anziehungs-,

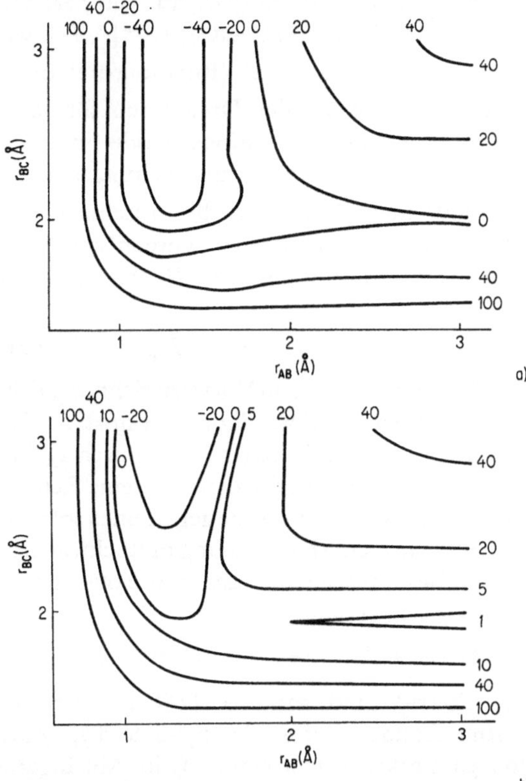

Abb. 3. Potentialflächen für zwei lineare exotherme Reaktionen $A + BC \rightarrow AB + C$ ohne Aktivierungsenergie. Die Reaktion beginnt unten rechts und endet oben links. Die Energien sind in kcal/mol angegeben. Ein starkes Anziehungs-potential (a) und ein starkes Abstoßungspotential (b) werden gezeigt [*14*]

Abstoßungs- und in einem „gemischten" Bereich jeweils angegeben wird
[*14*]. Von diesen Autoren wird ebenfalls für das Auftreten eines verhältnismäßig kleinen Bruchteils von $-\varDelta H_R$ als E_{vib} des Produktmoleküls in
erster Linie die Gestalt der Potentialfläche (Freiwerden eines großen Teiles
der Energie im Abstoßungsbereich, d. h. bei $r_{AB} \simeq r_{0\,AB}$) verantwortlich
gemacht. Bei Potentialflächen mit großem Abstoßungsanteil scheint außerdem eine kleine Masse von A die von AB durchschnittlich aufgenommene
Schwingungsenergie weiter zu verringern. Auch BLAIS und BUNKER [*11*]
machten für andere Potentialflächen letztgenannte Beobachtung und bezeichneten diese Erscheinung als „Energieanomalie".

Eine Studie der allgemeinen Reaktion $A + BC$ ohne Aktivierungsenergie beschäftigt sich mit dem Einfluß der Exothermizität, der Massen m_A, m_B, m_C und verschiedener Parameter einer allgemeinen Potentialfunktion auf die Ergebnisse [11].

Für den Fall, daß $m_B \ll m_A$ und $m_B \ll m_C$ für die Reaktion (1), wurde eine abnormale Winkelverteilung $n(\theta)$ vorausgesagt [7, 11], weil der durchschnittliche Gesamtdrehimpuls $\vec{\mathcal{J}} + \vec{\mathcal{L}}$ (Rotationsdrehimpuls $\vec{\mathcal{J}}$, Bahndrehimpuls $\vec{\mathcal{L}}$) so groß ist, daß $\vec{\mathcal{J}}'$, der Drehimpuls für das Molekül AB, bei Umwandlung der gesamten Reaktionsenergie in Rotationsenergie diesen Wert nicht erreicht. Der Drehimpulserhaltungssatz erzwingt einen entsprechenden Bahndrehimpuls $\vec{\mathcal{L}}'$ für die Auseinanderbewegung der Produkte. Die Monte-Carlo-Berechnungen wurden zur Deutung der experimentellen Untersuchungen von Grosser, Blythe und Bernstein [15] für die Reaktion

$$\text{K} + H Br \longrightarrow \text{KBr} + \text{H}; \quad \Delta H_R = -3,8 \text{ kcal} \tag{12}$$

herangezogen [5]: eine KBr-Streuung in Vorwärtsrichtung, d. h. in Richtung des ursprünglichen relativen Geschwindigkeitsvektors im Schwerpunktssystem (die in den Experimenten in erster Linie durch v_K bestimmt war, weil $v_K > v_{HBr}$) wurde festgestellt. Es gibt allerdings eine Reihe von Molekularstrahluntersuchungen dieser Reaktion, die gegen einen solchen Abstreifmechanismus sprechen und vielmehr eine Streuung in Rückwärtsrichtung wahrscheinlicher machen. Diese Arbeiten werden von Toennies [6] diskutiert.

Berechnungen für die Reaktionen

$$\text{H(D)} + \text{H}_2 \longrightarrow \text{H}_2(\text{HD}) + \text{H} \tag{13}$$

führten Karplus, Porter und Sharma [16] mit einem semiempirischen Potential vom London-Eyring-Polanyi-Sato-Typ durch. Sie berechneten den totalen Reaktionsquerschnitt S_r in Abhängigkeit von V_R und den Rotations- und Schwingungszuständen (J und v). Durch Integration von S_r über die Gleichgewichtsverteilungen von V_R, J und v bei einer gegebenen Temperatur wurden Geschwindigkeitskonstanten erhalten, die sowohl in ihrem Absolutwert wie auch in ihrer Temperaturabhängigkeit zwischen 300 und 1000 °K mit den experimentellen Daten gut übereinstimmen. Tunneleffekte wurden vernachlässigt. Bei einer Höhe der Potentialbarriere von 9,13 kcal/mol wurde die Arrheniussche Aktivierungsenergie zu 7,435 kcal/mol berechnet (experimenteller Wert 7,5 ± 1,0 kcal/ mol). Die Winkelabhängigkeit des differentiellen Reaktionsquerschnitts zeigte — in qualitativer Übereinstimmung mit Molekularstrahluntersuchungen an der Reaktion D + H_2 — eine starke Rückwärtsstreuung der Produktmoleküle im Schwerpunktssystem. In der letzten zitierten Arbeit der Autoren wurden die Berechnungen auf die Reaktionen heißer Tritiumatome mit H_2 und D_2 ausgedehnt.

3. Unimolekulare Reaktionen

In zwei Arbeiten behandelte BUNKER [*17, 18*] unimolekulare Zerfallsreaktionen dreiatomiger Moleküle mit Hilfe der Monte-Carlo-Methode.
Die Modelle wurden definiert durch die drei Atommassen, die Gleichgewichtswerte für die Abstände r_1 und r_2, den Gleichgewichtswert von α
und die drei Normalschwingungsfrequenzen. Für das Potential wurde
Separierbarkeit angenommen:

$$V(r_1, r_2, \alpha) = V_1(r_1) + V_2(r_2) + V_3(\alpha).$$

Während für die Deformation von α stets ein harmonisches Potential
gewählt wurde, wurden für V_1 und V_2 sowohl harmonische Potentiale als
auch Morsepotentiale getestet. Einige Modelle sind an den Molekülen N_2O
und O_3 orientiert, andere an instabilen Quasimolekülen wie HC_2 und C_3 etc.

Für unimolekulare Reaktionen ist es zweckmäßig, Eigenschaften reprä-
sentativer Gesamtheiten von Molekülen mit einer Energie in einem Intervall $E, E + dE$ in den inneren Freiheitsgraden der Schwingung und der
Rotation zu untersuchen. Wie in den Theorien unimolekularer Reaktionen
von SLATER [*19*] und von RICE, RAMSPERGER, KASSEL und MARCUS
(RRKM-Theorie [*20*]) wird angenommen, daß nach einem Stoß eine statistische Verteilung der Energie in den inneren Freiheitsgraden vorliegt,
daß sich das Molekül gewissermaßen nicht an seinen energetischen Zustand
vor dem Stoß „erinnert".

Die Geschwindigkeit der Dissoziation für eine Gesamtheit von N_0
Molekülen läßt sich in der von SLATER [*19*] eingeführten Terminologie wie
folgt ableiten: Der Anteil von Molekülen einer im thermischen Gleichgewicht vorliegenden Menge im Energieintervall $E, E + dE$ ist $G(E)$
$\exp(-E/kT)dE$, wobei $G(E)$ das relative Volumen des Phasenraumes zwischen den Hyperflächen E und $E + dE$ bedeutet. Es gilt die Normierungsbedingung

$$\int_0^\infty G(E) \exp(-E/kT) dE = 1. \tag{14}$$

Man definiert $n(\tau)d\tau$ als den Bruchteil der Moleküle, der bei Abwesen-
heit desaktivierender Stöße im Zeitintervall $\tau, \tau + d\tau$ zerfallen würde. Die
Verteilungsfunktion $n(\tau)$ ist abhängig von der Energie E. Die Zahl der
Moleküle mit innerer Energie im Intervall $E, E + dE$ und mit einer
Lebenserwartung, die im Intervall $\tau, \tau + d\tau$ liegt, ist dann gegeben durch

$$N_0 G(E) \exp(-E/kT) n(\tau) d\tau \, dE. \tag{15}$$

Unter Gleichgewichtsbedingungen werden aus diesem Zustand

$$\omega N_0 G(E) \exp(-E/kT) n(\tau) d\tau \, dE \tag{16}$$

Moleküle durch Stöße desaktiviert und gleichzeitig durch Stöße in diesen Zustand befördert. ω bedeutet die Stoßwahrscheinlichkeit pro Sekunde. Für die

Bildungsgeschwindigkeit gilt (16) auch für Nichtgleichgewichtszustände, wenn nur das Gleichgewicht bei Energien unterhalb der kritischen Energie E_0 (das ist das Minimum an Energie, das zum Zerfall des Moleküls benötigt wird) ungestört bleibt. Da die absolute Wahrscheinlichkeit für ein solches Molekül, die Zeit τ zu überdauern, d. h. zwischen $t = 0$ und $t = \tau$ keinen Stoß zu erleiden, $\exp(-\omega\tau)$ beträgt (der Stoß wird hier als ein Zufallsereignis mit gleich großer Wahrscheinlichkeit in jedem Zeitintervall aufgefaßt), ergibt sich für die Zerfallsgeschwindigkeit der Moleküle mit E, $E + dE$ in Abhängigkeit vom Gasdruck ($\omega \sim p$):

$$k(E)\,N_0\,G(E)\,\exp(-E/kT)dE$$
$$= \omega\,N_0\,G(E)\exp(-E/kT)\int_{\tau=0}^{\infty} n(\tau)\exp(-\omega\tau)\,d\tau\,dE\,. \tag{17}$$

Gleichung (17) definiert eine druckabhängige Geschwindigkeitskonstante $k(E)$.

Monte-Carlo-Berechnungen gestatten vor allem drei wichtige Aussagen:

1. Es läßt sich überprüfen, ob die Annahme der RRKM-Theorie zutrifft, daß alle Moleküle mit $E > E_0$ in den aktiven inneren Freiheitsgraden zerfallen.

2. Man kann Aussagen machen, in welchem Umfang die in allen unimolekularen Theorien enthaltene Hypothese von der Zufälligkeit der Lebensdauer aktivierter Moleküle gerechtfertigt ist. Hierzu ist folgendes zur Erklärung zu bemerken: Mit der Zerfallswahrscheinlichkeit pro Zeiteinheit $k^{\infty}(E) = \lim_{\omega\to\infty} k(E)$ — auch spezifische Geschwindigkeitskonstante genannt — ist für diese Hypothese die absolute Wahrscheinlichkeit für ein aktiviertes Molekül, in der Zeit von 0 bis τ nicht zu zerfallen, $\exp(-k^{\infty}(E)\tau)$ und daraus folgt

$$n(\tau)\,d\tau = k^{\infty}(E)\exp[-k^{\infty}(E)\tau]\,d\tau\,. \tag{18}$$

3. Die spezifischen Geschwindigkeitskonstanten $k^{\infty}(E)$ lassen sich auswerten und sowohl mit experimentellen wie auch theoretischen Werten, die man aus der Theorie von Slater oder aus der RRKM-Theorie (für den klassischen Grenzfall) erhalten kann, vergleichen.

Das Berechnungsverfahren verläuft ähnlich wie für die bimolekularen Prozesse. In einem Volumen des Phasenraumes, der begrenzt wird durch zwei Hyperflächen, auf denen E bzw. $E + dE$ konstant sind, wird eine gleichmäßige Verteilung von Punkten vorgenommen. Hierdurch gewinnt man $G(E)$ und zugleich die Ausgangsbedingungen, die in die klassischen Bewegungsgleichungen eingesetzt werden. Durch die jeweils gewählte Energie sind obere Grenzen für $p_1, p_2, p_a, p_\varphi, r_2$ und α gegeben. r_1, worunter wir den Abstand der in der Reaktion gelösten Bindung verstehen, kann bei der Wahl der Anfangsbedingungen durch den gleichen Wert nach oben begrenzt werden, der später als Kriterium für die erfolgte Dissoziation genommen wird. Das Integrationsverfahren liefert Zerfallszeiten, die in Histogramme eingetragen die Funktion $n(\tau)$ wiedergeben (Abb. 4).

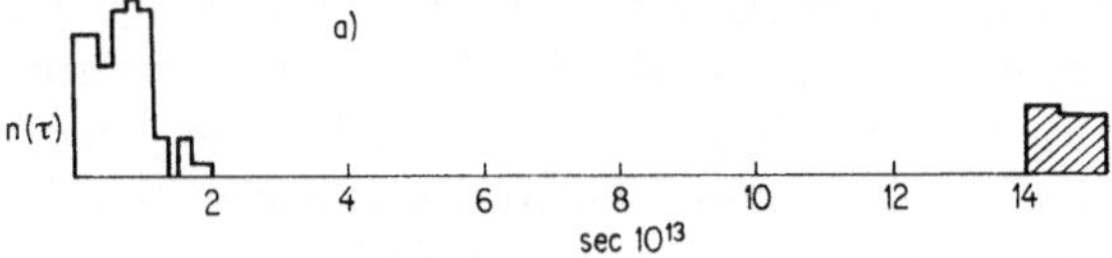

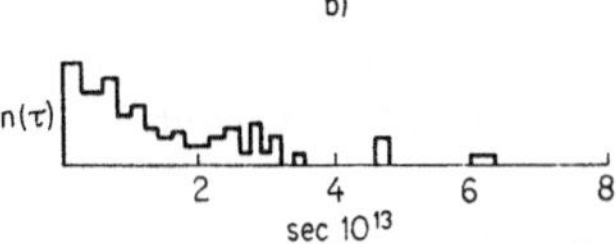

Abb. 4. Verteilungsfunktionen der natürlichen Lebenszeiten, berechnet für ein dreiatomiges Molekülmodell ohne Winkeldeformationsschwingungen [17]; $E_0 = 60$ kcal/mol, $E = 70$ kcal/mol; a) harmonische Bindungspotentiale; der schraffierte Block zeigt die Menge der nach $14 \cdot 10^{-13}$ sec noch unzersetzten Moleküle; b) anharmonische Bindungspotentiale; alle Moleküle sind nach $8 \cdot 10^{-13}$ sec zerfallen

Die Berechnungen ergaben für harmonische Potentiale $V_1(r_1)$ und $V_2(r_2)$, daß für innere Energien nur wenig oberhalb der kritischen Energie eine große Anzahl von Molekülen ($\lesssim 100\%$) sehr lange undissoziiert bleiben kann. Es ist daher notwendig, anharmonische Potentialfunktionen zu benutzen. Andererseits sind die Funktionen $n(\tau)$ für kleine τ bei anharmonischen und harmonischen Potentialen sehr ähnlich, d. h. der Anteil aktivierter Moleküle mit kurzer Lebenserwartung ist in beiden Fällen etwa gleich groß. Deshalb sind harmonische Potentiale bei hohen Drucken (etwa 100 at) als gute Näherung verwendbar, sie versagen jedoch mit Sicherheit bei niedrigen Drucken.

Der Grenzwert von $n(\tau)$ für $\tau \to 0$ ergibt k^∞ (E). Die mit der Monte-Carlo-Methode bestimmten Werte von $k^\infty(E)$ lassen sich, wie von der RRKM-Theorie gefordert, als lineare Funktion von $\{(E - E_0)/E\}^{n'}$ darstellen. Für n' erhält man innerhalb der Fehlergrenzen bei nicht zu hohen Temperaturen und für anharmonische Modelle einen Wert von 2 bei $s = 3$ (s Anzahl der Normalschwingungen). Es gilt also $n' = s - 1$, und daher können alle Schwingungen als aktiv im Sinne der RRKM-Theorie angesehen werden. Für die Rotationen findet man Inaktivität bei niedrigen, Aktivität bei hohen Temperaturen. Die spezifischen Geschwindigkeitskonstanten fallen bei Monte-Carlo-Berechnungen für anharmonische Modelle kleiner, die statistischen Gewichtsfaktoren $G(E)$ größer aus als nach der RRKM-Theorie, weil in letzterer die Anharmonizitätseffekte unberücksichtigt bleiben.

In einigen speziellen Fällen (z. B. bei sehr unterschiedlichen Werten für die drei Frequenzen) deutet die Form von $n(\tau)$ darauf hin, daß die

Hypothese der Zufälligkeit für das Eintreten des Zerfalles nicht immer gerechtfertigt ist. Bunker nimmt an, daß in diesen Fällen intramolekulare Relaxationswahrscheinlichkeiten für die Energieübertragung zwischen den Normalschwingungen kleiner als die Dissoziationswahrscheinlichkeiten sind. Er entwickelt eine Theorie, die eine Deutung der Berechnungsergebnisse erlaubt. Die intramolekulare Energierelaxation ist wahrscheinlich nicht nur in dreiatomigen, sondern auch in größeren Molekülen in 10^{-11} sec abgeschlossen. Da die spezifischen Geschwindigkeitskonstanten jedoch allgemein mit der Komplexität des Moleküls abnehmen, wird eine Lebenszeitverteilung, die nicht mehr den Zerfall als Zufallsereignis voraussetzt, nur für einige dreiatomige Moleküle erwartet.

Literatur

1. Glastone, S., K. J. Laidler, and H. Eyring: The Theory of Rate Processes, New York: McGraw-Hill Book Comp., Inc., 1941.

2. Wall, F. T., L. A. Hiller, and J. Mazur: J. Chem. Phys. **29**, 255 (1958); **35**, 1284 (1961).

3. Bunker, D. L.: Theory of Elementary Gas Reaction Rates. Oxford: Pergamon Press, 1966.

4. Blais, N. C., and D. L. Bunker: J. Chem. Phys. **37**, 2713 (1962).

5. Bunker, D. L.: Vorlesungsskriptum, Int. Sommerschule für Theoretische Chemie, Konstanz 1965.

6. Toennies, J. P.: Molecular Beam Studies of Chemical Reactions, dieses Buch.

7. Bunker, D. L., and N. C. Blais: J. Chem. Phys. **41**, 2377 (1964).

8. Karplus, M., and L. M. Raff: J. Chem. Phys. **41**, 1267 (1964).

9. Raff, L. M., and M. Karplus: J. Chem. Phys. **44**, 1212 (1966).

10. Raff, L. M.: J. Chem. Phys. **44**, 1202 (1966).

11. Blais, N. C., and D. L. Bunker: J. Chem. Phys. **39**, 315 (1963).

12. Evans, M. G., and M. Polanyi: Trans. Faraday Soc. **35**, 178 (1939); Simons, J.: Nature **186**, 551 (1960).

13. Airey, J. R., R. R. Getty, J. C. Polanyi, and D. R. Snelling: J. Chem. Phys. **41**, 3255 (1964).

14. Kuntz, P. J., E. M. Nemeth, J. C. Polanyi, S. D. Rosner, and C. E. Young: J. Chem. Phys. **44**, 1168 (1966).

15. Grosser, A. E., A. R. Blythe, and R. B. Bernstein: J. Chem. Phys. **42**, 1268 (1965).

16. Karplus, M., R. N. Porter, and R. Sharma: J. Chem. Phys. **40**, 2033 (1964); **43**, 3259 (1965); **45**, 3871 (1966).

17. Bunker, D. L.: J. Chem. Phys. **37**, 393 (1962).

18. — J. Chem. Phys. **40**, 1946 (1964).

19. Slater, N. B.: Theory of Unimolecular Reactions, London: Methuen & Co., Ltd., 1959.

20. Marcus, R. A.: J. Chem. Phys. **20**, 359 (1952); Wieder, G. M., and R. A. Marcus: J. Chem. Phys. **37**, 1835 (1962); Marcus, R. A.: Unimolecular Reaction Rate Theory, dieses Buch.

Horst Heydtmann
Institut für physikalische Chemie
der Universität Frankfurt/Main

Molecular Beam Studies of Chemical Reactions

J. P. Toennies

With 34 Figures

1. Introduction

1.1. General

The crossing of two molecular beams probably provides the most direct method for studying chemical reactions[1]. Although this was realized as early as 1926 it was as recent as 1955 that Taylor and Datz first succeeded in studying a chemical reaction in a crossed beam experiment[2]. In their experiment they detected KBr molecules from the reaction between a potassium and a HBr beam. Since then a large number of other reactions between alkali atoms and halogen containing molecules have been studied. For these systems direct information on the following properties of reactive collisions have been obtained.

1. Activation energies[3],
2. Energy dependence of the reaction probability and of the total reactive cross section,
3. Angular distribution of the products,
4. Translational, rotational and vibrational energies of the products.

In the years to come it is to be expected that these and other characteristic properties will be measured with improved resolution. For some systems it should even be possible to measure the populations of the various quantum states of the molecules formed in the reactions.

---

[1] Other less direct methods for studying chemical reactions using beams involves scattering from a gas and detecting the products directly (Bull and Moon, 1954) or indirectly by chemical analysis (Martin and Meyer, 1952). For some substances chemical analysis is more efficient than direct beam detection, and although less information is provided these methods are useful for measuring activation energies of reactions with small cross sections for which wall reactions can be neglected (Fristrom, 1966). They are not discussed further in this review.

[2] For a summary of the early history of chemical molecular beam experiments and other applications of molecular beams to problems of interest to chemists see Datz and Taylor (1959).

[3] Throughout this paper we use the term activation energy to mean the lowest energy at which the reaction can occur, neglecting tunneling. This is sometimes also called the threshold energy.

Unfortunately, however, because of limitations on source intensities and energies molecular beam techniques have been successfully applied only to reactions with low activation energies ($\lesssim 5$ kcal/mole), large cross sections (>10 Å^2) and ones which involve at least one alkali metal atom. Most chemical reactions do not, however, satisfy these conditions. Their investigation now appears to be possible by applying the recently developed nozzle beam source which provides higher beam intensities and energies than hitherto available. With this technique many reactions not satisfying the restrictive conditions above are now accessible to investigation.

The present review has three main aims: to show the capabilities of present day experimental techniques, to review the simple theoretical methods used in analyzing reactive scattering and, finally, to survey briefly the results of experiments on three representative reactions. In order to achieve these goals in a reasonable space many details have had to be left out. These may be found in the following recent reviews:

On experimental techniques: H. Pauly and J. P. Toennies in Vol. 8 of Methods of Experimental Physics (1968) and also the articles by J. B. Anderson, R. P. Andres and J. B. Fenn and by H. Pauly and J. P. Toennies in Vol. 1 of Advances in Atomic and Molecular Physics (1965).

On the theory of elastic scattering and its relation to chemical reactions: articles by R. B. Bernstein and by E. F. Greene, A. L. Moursund and J. Ross in "Molecular Beams", Vol. 10 of Advances in Chemical Physics and by H. Pauly and J. P. Toennies, ibid.

On reactive scattering: D. R. Herschbach in "Molecular Beams", Vol. 10 of Advances in Chemical Physics and D. R. Herschbach in Appl. Optics, Suppl. 2 (Chemical Lasers) 128 (1965).

References to the older original papers on topics discussed here have been omitted since they may be found in the reviews. In general only those papers are referred to which have either appeared after these reviews or which are discussed at some length in the text.

1.2. *Basic technique and definitions*

Fig. 1 shows a schematic diagram of the typical experimental layout used in a molecular beam scattering experiment. Molecules or atoms leaving the primary beam source move along straight line trajectories through the highly evacuated ($P \simeq 10^{-6}$ Torr) experimental chamber[4]. A

---

[4] At pressures of the order of 10^{-6} Torr the mean free path is of the order of 50 m. Since this is considerably greater than the dimensions of even the largest beam apparatus collisions of beam molecules with the residual gas are highly improbable. Furthermore, since the pressure within the beam is even smaller ($\sim 10^{-9}$ Torr) collisions among beam molecules can usually also be neglected.

secondary beam source with a large emission is used to produce an intense beam of a second reactant. For future reference we designate the reac-

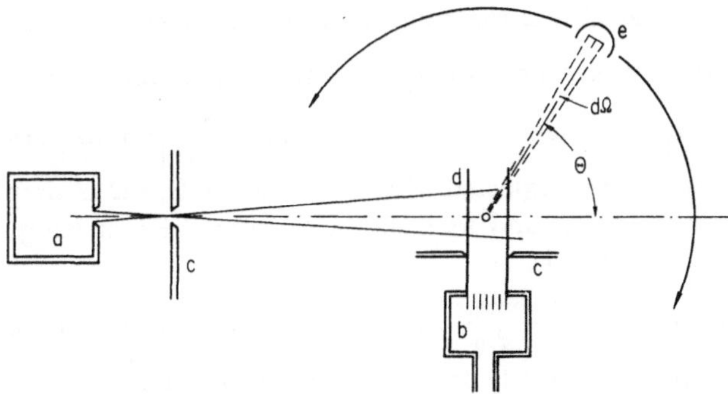

Fig. 1. Schematic diagram of a crossed beam apparatus for studying reactive collisions. *a* is the primary beam source; *b*, the secondary beam source; *c*, the collimating slits; *d*, the scattering region; and *e*, the detector. $d\Omega$ is the solid angle subtended by the detector at the angle Θ

tant beams by the subscripts 1 and 2 and the product beams by 3 and 4. Thus we write the reaction occurring at the region of intersection of the two beams as

$$1 + 2 \rightarrow 3 + 4$$

It is usually possible to arrange the conditions in such a way that several percent or more of the primary beam molecules are scattered. The number of product molecules scattered per second into a solid angle element $d\Omega$ is given by:

$$\dot{N} = n_1\, n_2\, g\, V \frac{d\sigma_{\text{react}}\,(g,\vartheta,\varphi)}{d\omega} \left| \frac{d\omega}{d\Omega} \right| d\Omega \tag{1}$$

where n is the particle density, g is the relative velocity defined by $g = |\vec{v_1} - \vec{v_2}|$ and V is the reaction volume. $\dfrac{d\sigma_{\text{react}}\,(g,\vartheta,\varphi)}{d\omega}$ has the dimensions of area per solid angle per colliding pair and is called the differential cross section. The probability of reaction is directly related to $\dfrac{d\sigma_{\text{react}}}{d\omega}$. It is a function of the relative velocity for practically all collision processes. The angle ϑ is defined with respect to the vector $\vec{g}$ before the collision, and φ is defined with respect to some laboratory axis perpendicular to $\vec{g}$. For beams of molecules which are not oriented (unpolarized) the observed scattering is an average over all orientations and is cylindrically symmetric with respect to $\vec{g}$. The coordinate system with an axis along the relative velocity is called the center of mass system. As we shall discuss later (section 3.5.) it is always possible to transform the angles from this system to the laboratory system in which the measurements are made. Thus the Jacobian $\left|\dfrac{d\omega}{d\Omega}\right|$ transforms the solid angle from the center of mass system into the laboratory system. For light primary beam molecules and heavy secondary beam molecules ($m_1 \ll m_2$) the factor $\left|\dfrac{d\omega}{d\Omega}\right|$ approaches 1 and the transformation can be neglected. Finally, for small V, $d\Omega$ is simply related to the detector area dF and its distance D from the scattering center by $d\Omega \simeq D^{-2} dF$ (see Fig. 1).

The dependence of $\dfrac{d\sigma_{\text{react}}}{d\omega}$ on the scattering angle and relative velocity is determined directly from measurements of $\dot{N}$ as a function of these parameters. This is the usual type of measurement. The absolute determination of $\dfrac{d\sigma_{\text{react}}}{d\omega}$ is a much more difficult task since it requires an absolute determination of both beam intensities and the volume of the interaction region. The intensity of the scattered beam can usually not be measured directly and estimates, good to within only a factor of two, have to be used. For some systems methods are available for circumventing these difficulties, and these are discussed in section 4.

So far we have neglected the internal energy states of the reactants and products. For taking these into account we introduce the collective quantum number [i] which stands for the rotational ($J. M.$) and vibrational (n) quantum numbers of the reactants and [f] which stands for the same quantum states and also for the relative translation energy of the products[5]. Thus for one reaction there are a large number of differential cross sections $\dfrac{d\sigma_{\text{react}}^{[i]\to[f]}}{d\omega}$ which presumably contain sufficient information to uniquely determine the energy hypersurface for the reactions (see section 3.1.). The differential cross section for an atom-molecule reaction involving a nonselected beam is therefore given by the following average.

$$\frac{d\sigma_{\text{react}}}{d\omega} = \sum_{J_i M_i} \frac{1}{2J_i + 1}\, F(J_i) \sum_{n_i} F(n_i) \sum_{J_f M_f n_f} \frac{d\sigma_{\text{react}}^{[i]\to[f]}}{d\omega} \tag{2}$$

where the F's are Boltzmann fractions and the sums are over the quantum states[6]. In recent experiments substantial progress has been made towards

---

[5] We also occasionally use i and f as an index to indicate initial and final conditions in general.

[6] In some cases the same reactants may form different sets of products (parallel reaction channels). All reactions leading to the observed products must then be included.

removing some of the sums in equation (2). Thus separate measurements have already been made of the vibrational and rotational energy distributions of the product molecule. Other experiments may permit the removal of the sum over initial vibrational levels by cooling the reactants sufficiently.

Another average quantity which can be measured in a beam experiment is the integral total cross section. As the name implies the integral cross section is related to the differential cross section by

$$\sigma_{tot}(\Theta_0, g) = \int\limits_{\vartheta_0}^{\pi} \int\limits_{0}^{2\pi} \frac{d\sigma_{tot}}{d\omega} \sin\vartheta \, d\vartheta \, d\varphi \tag{3}$$

where scattering events with deflection angles greater than the center of mass angle ϑ_0 (corresponding to Θ_0) contribute to the cross section.

The integral cross section generally does not contain as much information on the system under study as the differential cross section. It is, however, easier to measure since it is only necessary to observe the attenuation of the primary beam produced by the secondary beam or by a scattering gas in a scattering chamber. As in the measurement of the optical extinction coefficient the cross section is related to the attenuation by Beer's law

$$\frac{J}{J_0} = \exp\left(- \sigma_{tot} \, n \, l\right)$$

The apparatus is similar to the one shown in Fig. 1 with the detector placed at $\Theta = 0$.

There is some question as to the extent to which reactive collisions affect σ_{tot}. At least to a first approximation σ_{tot} is dependent only on the long range forces at distances of approach of the order of 10 to 20 Å. Thus no matter what occurs at closer distances these collisions are always included in σ_{tot} and the effect of reactive collisions which always appear to occur at smaller distances is negligibly small. Probably for this reason very few measurements of σ_{tot} on reactive collisions have been reported. Work is, however, in progress in several laboratories and it will be interesting to see the results.

2. The Essential Components
of a Molecular Beam Apparatus

Although the crossed beam method appears simple in concept the experiments prove to be extremely difficult. Difficulties arise because of the very small intensities of the product beams from which the entire information on the reaction has to be extracted. Even for the simple set-up in Fig. 1 product beam intensities for the alkali reactions are usually of the order of 10^{-12}—10^{-14} Amp. Signal intensities of this order of magnitude can still be measured with commercial apparatus without too much difficulty. Unfortunately, the simple apparatus shown in Fig. 1 does not always tell us very much about the detailed behavior of the reaction cross section. With such an apparatus it is possible to establish the approximate size of the reactive cross section and for certain reactions, as we shall see later, the approximate angular distribution of the products and their average energy. In order to measure activation energies unambiguously it is necessary to introduce a velocity selector into the beam with the highest velocity. For other systems it may be desirable or even necessary to velocity select the products in addition, in order to determine the angular distribution and the average energy of the products. This is approximately the present stage of technical development. The ultimate aim, of course, is to velocity select both beams, state select the molecular reactant beam and, furthermore, to state and velocity select the product beams. In this way it should be possible to measure the reactive cross section connecting reactant and product quantum states. The extent to which these goals can be achieved depends on the efficiency of each of the various components of the apparatus. In this section we discuss the beam sources for thermal and higher energies, the properties of velocity and state selectors and finally, the properties of beam detectors. Emphasis has been placed on the principles of the methods and the types of experiments which are potentially possible. Finally a comment on the relative importance of the various components: Since the product of the beam densities of both beams enters into equation 1 the sources are obviously the most important of all the components. This is sometimes overlooked and undue attention is placed on the detector. In judging the detector the signal to noise ratio is the most significant criterion. The question of signal to noise ratios is not discussed but may be found in the references on experimental techniques listed in the first part of the introduction.

2.1. *Effusive Beam Sources*

The commonly used method for producing a molecular beam is to allow the beam molecules to effuse through a small thin walled orifice of an otherwise closed container. Beams of condensable substances are usually produced by heating the container, whence the name oven. If operated at pressures at which the mean free path λ in the oven is greater than the orifice diameter d or slit width the velocity distribution is essentially Maxwellian. The curves for $M = 0$ in Figure 2 illustrate the velocity and angular distributions of such a source.

Typical intensities which can be achieved with such a source are of the order of $5 \cdot 10^{16} \left[\dfrac{\text{molecules}}{\text{sr} \cdot \text{sec}} \right]$. For a beam with a velocity of $5 \cdot 10^4$ cm/sec this corresponds to a particle density of $10^{10} \left[\dfrac{\text{molecules}}{\text{cm}^3} \right]$ at a distance of 10 cm from the source.

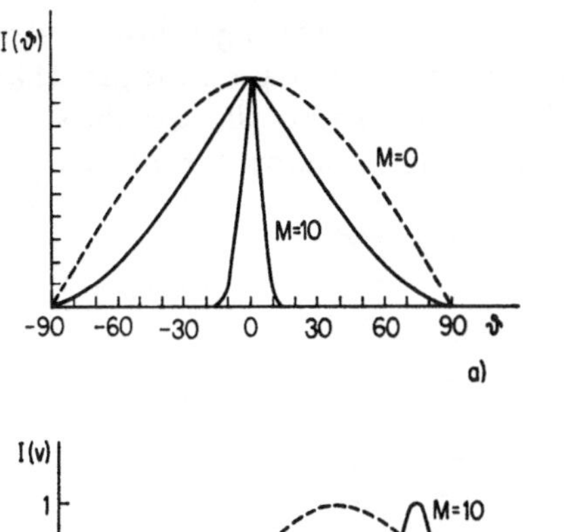

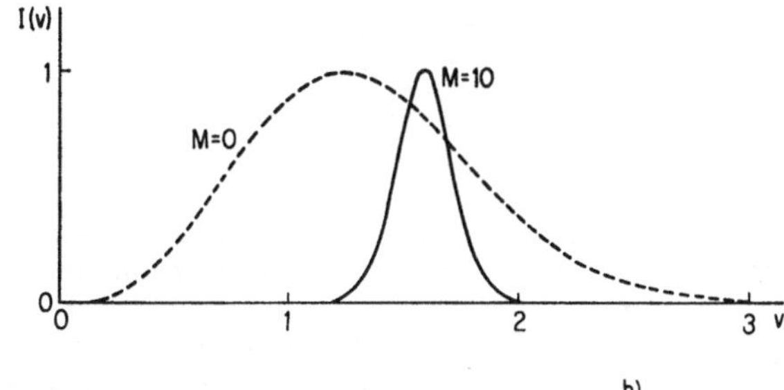

Fig. 2. Comparison of the properties of a molecular beam and a hydrodynamic beam source
a) Angular distributions for an oven (M = 0), nozzle (M = 10), and channel beam (KING and ZACHARIAS, 1956). M is the flow mach number
b) Velocity distributions for an oven and a nozzle beam (ANDERSON and FENN, 1965). The curves are normalized to the same maximum intensity

This source has the advantage that it can be adapted for use with any substance and that its behavior is easily predictable. For instance, even fairly pure hydrogen atom beams (85% depending on the pressure) have been produced with such a source by heating a tungsten oven to over 2500 °K.

Narrow primary beams are usually required, especially when a velocity selector with a high velocity resolution is used. The requirements on the secondary beam are different. It should be arranged to make V the interaction volume as large as possible. High beam directivity then becomes important to reduce the amount of excess gas which has to be pumped out of the system and to assure that the beam intersection angle is as uniform as possible. An oven satisfying these requirements is made up of many parallel channels each with a length l greater than the diameter d. When operated at pressures at which $\lambda \leq l$ the directivity shown in Fig. 2a can be achieved.

Recent work on nozzle beam sources show that by merely increasing the source pressure in a thin walled orifice oven a transition occurs from molecular to hydrodynamic flow. Of course, sufficient pumping speed must be available to handle the greatly increased gas flow. Several pumping stages are, therefore, usually provided to pump off the gas which does not pass the collimating slits. The first collimating slit after the source opening is generally called the skimmer. Fig. 3 shows the usual type of nozzle

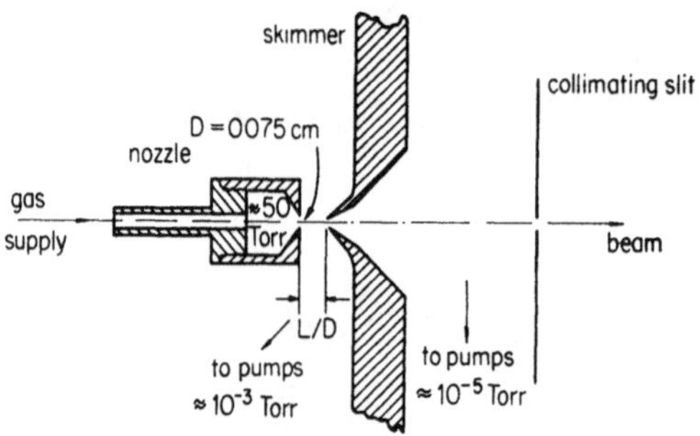

Fig. 3. Nozzle source arrangement (ANDERSON, ANDRES and FENN, 1965). L/D is the reduced distance between nozzle and skimmer

source. The important result of recent research on nozzle beams is that neither the design of the nozzle (a simple orifice is adequate) nor that of the skimmer are as critical as they had been thought to be at one time. The critical parameters are the source pressure and the nozzle-skimmer distance and these can be made to be easily varied.

The properties of such nozzle beams can be easily understood in terms of conservation of enthalpy. Since at the high densities the molecules in the flow are in constant collisional contact, the increase in kinetic energy is drawn from the translational motion and the internal degrees of freedom. The translational temperature and to a lesser extent the temperature of the internal degrees of freedom are reduced to about 1% of the source temperature resulting in greatly narrowed velocity and angular distributions (see Fig. 2). This, together with the higher gas densities, accounts for beam intensities increased by more than a factor of 100. For many experiments additional state and velocity selection may not be needed, resulting in a further intensity gain over the conventional source used together with selectors.

In the light of these advantages the infrequent application of nozzle beams in reactive scattering experiments is surprising. One reason is that they have only recently been fully understood. The other reason has less to do with the source than with the fact that an electron bombardment detector usually has to be used with substances not containing alkali metals. Electron bombardment detectors have a low efficiency and are particularly sensitive to the high background pressures expected from a nozzle source (see section 2.4.). This last technical difficulty can be avoided by providing additional pumping speed and new apparatuses are designed accordingly.

Intense beams of alkali metals have also been formed with nozzles. Under extreme conditions there is some evidence that these may contain a high concentration of dimers. As we shall see later detection is less of a problem and pumping is easily provided by cooled surfaces.

2.2. Beam Sources with Energies Greater than 10 kcal/mole

The beam sources discussed above are essentially limited to temperatures below 3000 °K. The energy range from 10 kcal/mole (0.5 eV) to 100 kcal/ mole (5 eV) is, however, of special interest for chemical scattering. Several methods have been tried and appear useful for producing beams with higher energies. These are:

1. Nozzle beams operating with gas mixtures: A nozzle using a heavy gas which is highly diluted by a light gas e.g. 99% He with Ar has been shown to have almost the same flow properties as the light gas. The energy increase is therefore proportional to the ratio of the mass of the heavy atom to that of the light atom and is often sufficient to provide the desired energies. The large amounts of the light gas is not a serious problem since the nozzle has the property that it concentrates the heavier species along the axis. Thus most of the lighter species is collected by the skimmer. Recently, argon beams with energies of 4 eV have been obtained (ANDERSON, ANDRES and FENN, 1966) by using a highly diluted mixture with He which had been heated to 2000 °K. Shock waves and arc discharges have also been used to increase the temperatures of the nozzle gas. Although both methods have been shown to work, further work is necessary before they can be applied to reactive experiments.

2. Ion neutralization: In this source ions formed by surface ionization (M. HOLLSTEIN and H. PAULY, 1966) or by electron bombardment are accelerated and subsequently neutralized. Space charge effects limit the available intensities at low energies $\lesssim 5$ eV and the method is best suited for higher energies. So far it has only been used in one instance to study a chemical reaction (UTTERBACK, 1966).

3. Sputtering: When ions of 5—100 keV energies strike a metal surface, atoms of the metal are ejected with energies in the range 0.5—20 eV. The method has been shown to work (POLITIEK, LOS and SCHIPPER, 1966) and appears to be promising as a source at intermediate energies, when the necessary velocity selection is feasible (see section 2.3.).

4. Rotating impeller. Mechanical acceleration of the beam has also been suggested for achieving higher beam energies. Maximum velocities are of the order of $6 \cdot 10^4$ cm/sec so that significant energy increases are only possible for heavy atoms or molecules.

Figure 4 shows a comparison of some of the beam sources discussed here. The beam intensity in molecules/sec. sr. is plotted as a function of the beam energy. From the figure it is evident that the simple oven and nozzle

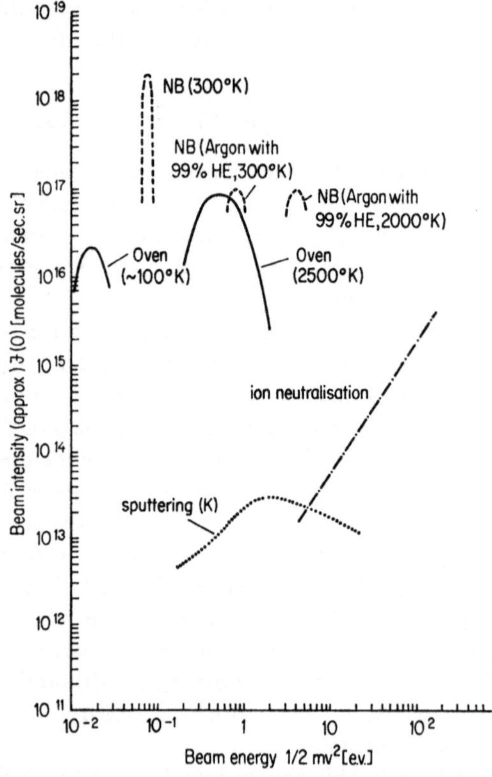

Fig. 4. Comparison of various continuous flow high energy beam sources. The intensity prior to velocity selection for four sources are shown. In general velocity selection is not necessary when working with the nozzle beam (NB) and ion neutralization sources. The data have been obtained from the following sources: nozzle beam, ANDERSON, ANDRES and FENN (1966); sputtering of K, POLITIEK, LOS and SCHIPPER (1966); ion neutralization, UTTERBACK and MILLER (1961); the data for the mixed gas nozzle beams and thermal molecular flow oven were estimated by the author

beams provide the highest intensities at energies up to 4 eV. Of course not all substances can be handled equally well by the various sources. For instance, a fast Li beam is not easily produced in a nozzle because of oven corrosion at the high temperatures required. In this case a sputtering or a neutralization source is probably easier to use.

2.3. Velocity Selectors

The most widely used device for velocity selecting molecular beams is the mechanical velocity selector similar to the one first used by FIZEAU. Fig. 5 shows the principle of a velocity selector.

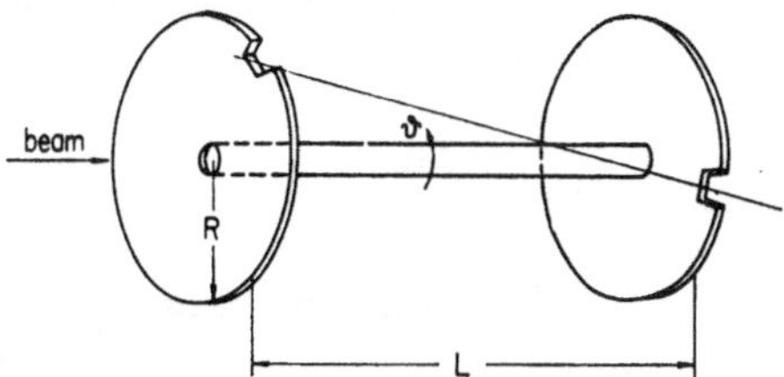

Fig. 5. Schematic diagram of a simple rotating disc velocity selector

Molecules moving parallel to the axis with velocity such that they arrive at the last disc when it has rotated by an amount d are transmitted. Because of the low transmission of a one slot selector modern selectors have many notches and in this case additional discs are necessary to prevent other velocities (higher harmonics) from being transmitted. The important properties of a many disc velocity selector are the maximum transmission T ($\approx 50\%$) at the selected velocity v_0, the velocity resolving power A which is given by $\dfrac{v_0}{\varDelta v_{1/2}}$ where $\varDelta v_{1/2}$ is the width of the transmission curve measured at half the maximum intensity, and finally the maximum velocity v_{max} that can be selected. These properties are simply related to the construction details of the velocity selectors. Their product is given by

$$T A v_{\mathrm{max}} = \zeta L f_{\mathrm{max}} \tag{4}$$

where ζ is the number of slits on the circumference of the rotor (the maximum ζ which can be achieved is of the order of 3000 for 20 cm diameter discs), L is the length of the rotor (for scattering experiments L should usually be as short as possible) and finally f_{max} is the maximum frequency of rotation. The essential limitation on v_{max} comes from f_{max} which is limited by the strength of the disc material and to a lesser extent by the ball bearings used. The present limit in the product $T \cdot A \cdot v_{\mathrm{max}}$ is of the order of 10^7 cm/ sec. A typical velocity selector has an overall length of 5 to 10 cm. With a resolving power of $v_0/\varDelta v_{1/2} = 20$ it reduces the beam density from a molecular flow source at the most probable velocity by factor of about $5 \cdot 10^{-2}$. Velocities as high as 9000 m/sec ($A = 16$) have been selected and one with a maximum velocity of over twice this value at the same resolving power is being constructed at Bonn.

2.4. State Selectors

Selection of individual rotational quantum states has been used in studies of inelastic, non-reactive collisions (TOENNIES, 1965). More simple selection techniques for estimating distributions over rotational energy states have also been recently applied to chemical scattering experiments. These are summarized in Table 1.

Table 1

States selected	Inhomogeneous field	Literature describing application
rotational state distribution	electric STERN-GERLACH	HERM and HERSCHBACH (1965)
rotational states of symmetric top molecules with negative $K. M.$*	electric six pole field	KRAMER and BERNSTEIN (1965)
$J. M$ states of polar molecules ($M < J$), especially $(J. M) = (1.0), (2.0), (3.0)$	electric four pole field	TOENNIES (1965)

* K is the projection of the total angular momentum (with quantum number J) along the molecular symmetry axis; M is the projection of the total angular momentum along the external field direction.

In the electric STERN-GERLACH field polar molecules are deflected by an amount approximately inversely proportional to J. The direction of deflection to higher or lower field strengths depends on whether $M \sim J$ or $M \sim 0$ respectively. The instrument is not suitable for picking out individual rotational states unless very narrow slits are used and the beam velocity selected. The observed deflection patterns (see Fig. 28) do however depend on the rotational energy distribution. They are narrow for highly excited beams and wide for one distributed largely among low lying rotational states.

The electric four pole and six pole fields select the molecule in the states indicated in Table 1. Molecules in these states which leave the source at a small angle ($\vartheta \sim 1°$) with respect to the center line are deflected back to the center line and are focussed (Fig. 6), whereas molecules in other states are defocussed. Thus some of the intensity which is lost by state selection is

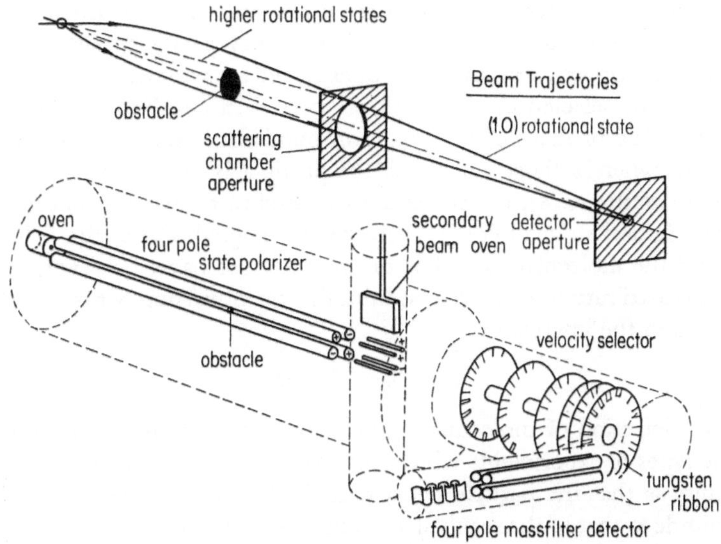

Fig. 6. Beam trajectories and apparatus used for doing experiments with state selected and oriented molecular beams

regained by the increased aperture. Typical intensities in Amp. (1 Amp. $\simeq 6 \cdot 10^{18}$ molecules/sec) for various arrangements but with the same slits at oven and detector are indicated below:

	0.1 m beam path	1 m beam path
without four pole field, molecules in all states	10^{-11}	10^{-13}
without four pole field, molecules in one low-lying rotational state	10^{-15}	10^{-17}
with four pole field and focussing action, molecules in one state	—	10^{-14}

Thus state selection typically reduces the intensities available for reactive scattering from 10^{-11} Amp. to about 10^{-14} Amp.

State selection of non-polar molecules is much more difficult and has not been used in scattering experiments. The selection of vibrational states, on the other hand, does not appear to be feasible at the present time. In this regard we recall that most molecules are in the ground vibrational state at room temperature and that vibrational state selection is not necessary in many cases. The identification of the vibrational state of a polar molecule is possible, at least in principle, with radiofrequency spectroscopy techniques.

Of particular interest to chemistry is the possibility of orienting molecules prior to a reactive collision. This is only partly possible for polar diatomic molecules[7], and is more easily achieved for symmetric top molecules. These can be selected in an electric six pole field (see Table 1). Molecules for which the product of the quantum numbers K and M is negative are focussed, whereas those for which the product is positive are defocussed. Stated in another way, molecules are focussed to an extent proportional to the expectation value of $\cos \vartheta$ where ϑ is the angle between the field direction and the molecular axis. The beam molecules have, however, a nearly thermal distribution over J states, since this quantum number is not directly involved in the interaction.

2.5. Beam Detectors

The detection of molecular beams provides the most severe limitation on the types of experiments which can be successfully performed. The basic difficulty is that the particle density in the beam is of the same order of magnitude as that of the residual gas which, in a typical apparatus, is present at a pressure of 10^{-7} Torr. Although a large number of detection schemes have been proposed and tried, the most sensitive methods involve conversion of the neutral beam atom or molecules into ions which can then be measured as a current or counted individually.

To discriminate against the residual gas two schemes are used. In the first method, the so-called Langmuir-Taylor detector, the atoms are permitted to strike a hot platinum or tungsten filament, both of which have a high work function. If the beam atoms striking the filament have an ionization potential lower than the work function they are ionized by tunneling of the electron from the atom into the surface. The ions evaporate subsequently because of the high filament temperature. If equilibrium on the surface is assumed, the ionization efficiency can be calculated (Langmuir-Saha equation) and approaches 100% for most of the alkalis and molecules containing them. Since residual gas molecules have much greater ionization potentials they are not ionized. This sensitive detector has made possible almost all the recent chemical scattering experiments.

Since the simple detector does not distinguish between products and reactants — one is and the other contains an alkali atom — a special technique was required to facilitate the study of chemical scattering. The problem was solved by TAYLOR and DATZ (1955) who discovered that whereas alkali-halide molecules and alkali atoms are detected with nearly equal efficiency on tungsten filaments, only the alkali atoms are detected on

---

[7] This can be seen from the uncertainty principle $\Delta \vartheta \cdot \Delta L \geqq \dfrac{\hbar}{2}$, where L is the angular momentum. For a state selected beam in field free space $\Delta L = 0$ and $\Delta \vartheta = \infty$. With increasing electric field $\Delta \vartheta \to 0$ but $\Delta L \to \infty$.

platinum filaments. Thus the difference in ion currents coming from both wires alternately exposed at the same scattering angle is a measure of the alkali-halide intensity. More recent work indicates that selective detection is a question of surface contamination and that one wire can be made to detect or not detect an alkali-halide depending on the wire's previous treatment (TOUW and TRISCHKA, 1963). Fig. 7 shows a two-filament surface ionization detector used by TAYLOR and DATZ for their study of the K + HBr reaction.

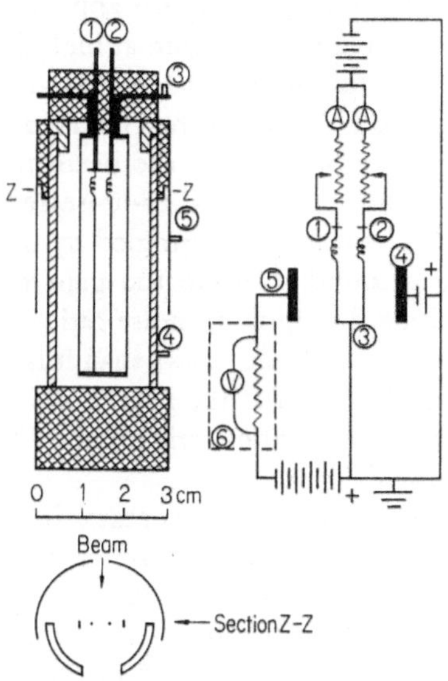

Fig. 7. Simple two wire Langmuir-Taylor surface ionization detector and circuit
(TAYLOR and DATZ, 1955)
(1) W filament 0.002 inch dia., (2) Pt filament 0.002 inch dia., (3) Common filament lead, (4) Guard ring, Ni, (5) Collector electrode, Ni, (6) Vibrating reed electrometer

Selective detection may also be achieved by passing the neutral molecular beam through an inhomogeneous magnetic (STERN-GERLACH) field prior to surface ionization. The method relies on the fact that alkali atoms have magnetic moments and are, therefore, deflected, whereas the molecules, which do not have magnetic moments, are not deflected. It is more reliable than selective ionization, which may give erroneous results because of filament contamination by the halogen reactants. A considerable loss in intensity has to be taken into account, however, since the product beam has to be narrowly collimated.

The detection of beams not containing an alkali is a much more difficult technical problem. In the most widely applicable method with the highest sensitivity the beam is ionized by electron bombardment; selection and discrimination are achieved by subsequent mass selection of the ion beam. Typically only one in a thousand beam atoms are ionized. Furthermore because of the relatively large residual gas density the background current from ionization of the residual gas is a serious problem. Thus the pumping speed available in the apparatus is of extreme importance for keeping the residual gas pressure as small as possible.

A further improvement in signal-to-noise ratio can be achieved with the modulated beams method. The method relies on the fact that the frequency spectrum of the noise in a typical apparatus is strongly peaked at low frequencies. By modulating the beam at high frequencies and amplifying the detector signal with a narrow band-pass amplifier a large part of the noise is filtered out. The a.c. signal is then rectified at the same frequency but with a phase shift corresponding to the beam transit time. An order of magnitude improvement in the signal-to-noise ratio is possible in this way. The remaining noise is usually determined by the statistical (random) fluctuations of the background. Despite the gain made possible by the modulated beam technique the signal-to-noise ratio with a universal beam detector is many orders of magnitude worse than it is with a surface ionization detector. Chopping the beam at the source has one other advantage. Since particles with different velocities arrive with different phases a rough velocity selection is possible.

As is often practised in high energy scattering experiments a considerable improvement in the signal to noise ratio is always possible by increasing the over-all measuring time. Generally the gain is proportional to the square root of the measuring time. Such measurements are easily possible with counting techniques. Since a direct current signal can be converted to a frequency, long time measurements are possible even if a simple electrometer amplifier is used. Despite this averaging techniques have only rarely been used in chemical scattering experiments.

2.6. Some Typical Values Needed for Estimating Product Intensities

Below are summarized some typical values of quantities which are useful for estimating available product intensities. The values are only rough estimates and can be exceeded by careful design of the apparatus:

$$n_1 \simeq 10^{10} \left[\frac{\text{molecules}}{\text{cm}^3} \right] \quad \text{(for a molecular flow source)},$$

$$n_2 \simeq 10^{11} \left[\frac{\text{molecules}}{\text{cm}^3} \right] \quad \text{(for a molecular flow source)},$$

$$g \simeq 5 \cdot 10^4 \ \text{cm/sec},$$

$$V \simeq 4 \cdot 10^{-2} \ \text{cm}^3,$$

$$d\Omega \simeq 2 \cdot 10^{-4} - 2 \cdot 10^{-5} \ \text{sr}.$$

Attenuation factors for a thermal beam
    Langmuir-Taylor detector $0.1-1$
    electron bombardment detector[8] $10^{-2}-10^{-3}$
    velocity selector at most probable velocity $5 \cdot 10^{-2}$
    rotational state selector including focussing[9] $10^{-3}-10^{-4}$

For a typical reaction with $\dfrac{d\sigma_{\text{react}}}{d\omega} = \dfrac{5\,\text{Å}^2}{sr.}$, and if an alkali halide product

is formed in an experiment without state but with velocity selection the numbers above predict a signal intensity of 10^4 mol./sec or 10^{-15} Amp. Signals down to 10^{-17} Amp. can be detected with commercially available vibrating reed electrometers.

3. Elastic Scattering and Chemical Reactions

3.1. The Intermolecular Potential

In most present day theories of chemical reactions the assumption is made that there is an intermolecular potential field, which is a function only of the distances between the nuclei of the atoms involved in the reaction. Thus for the reaction

$$A + BC \;\rightarrow\; AB + C$$

the potential is given by $V(R_{AB}, R_{BC}, R_{AC})$. To calculate the scattering and the cross sections the nuclei are then treated classically as particles which are acted on by the forces corresponding to the potential. One of the most important early results coming from simple classical considerations of this type (EVANS and POLANYI, 1939) is repeated here since we have occasion to refer to it later on. In exothermic reactions the kinetic energy gained in the attractive force field prior to reaction is converted into vibrational excitation of the products. Conversely if repulsion of the products is the force responsible for the energy release then only a small amount of vibrational excitation of the products is expected.

Since the available experimental techniques only permit the investigation of reactions with low activation energies, encounters in which the distance of closest approach is of the order of $4-6$ Å may still have trajectories which cross the potential barrier leading to reaction. This distance is about 3 to 5 times larger than the internuclear distance of the molecule so that to a first approximation we need only consider the distance R_{A-BC} between

---

[8] An electron bombardment detector measures beam density whereas the Langmuir-Taylor detector measures the flux. Thus the detection efficiency of the former is inversely proportional to the beam velocity.

[9] Most of the intensity loss (10^{-2}) comes from the greater beam path length needed.

the atom A and the center of mass of the molecule BC. Furthermore by neglecting the anisotropy of the potential we can reduce the problem to one with a central potential, which can be treated exactly both classically and quantum mechanically. We call this model the central field impact theory. We first discuss elastic scattering between two atoms and later we take account of the effect of a reaction.

The interaction energy between two atoms with several electrons cannot be calculated accurately, so we rely on potential models for which the parameters are obtained from experiment. The most widely used potential is the Lennard-Jones $(n, 6)$ potential:

$$V(R) = \frac{6\,\varepsilon}{n-6}\left[\left(\frac{R_m}{R}\right)^n - \frac{n}{6}\left(\frac{R_m}{R}\right)^6\right] \tag{5}$$

The meaning of the potential parameter n can be seen from Fig. 8. From experiments it appears that the value of n lies between 7 and 12. ε is the depth of the potential minimum and R_m is its position.

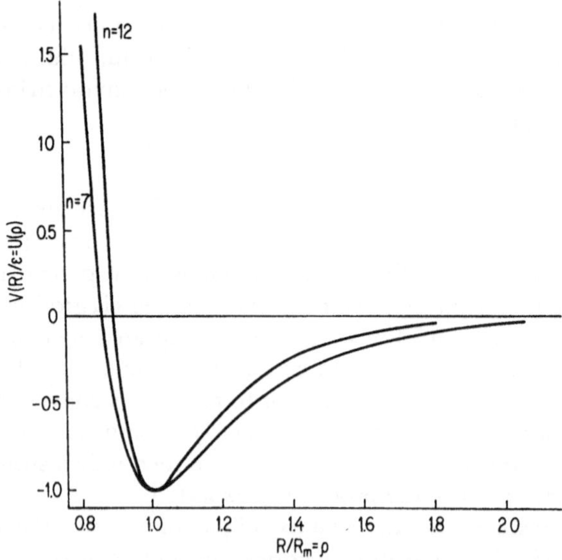

Fig. 8. The Lennard-Jones (n, 6) intermolecular potential between two atoms for $n = 7$ and 12. The potential parameters R_m and ε are defined as the position and depth of the potential minimum, respectively

3.2. Classical Theory of Elastic Scattering

By transforming to the center of mass system (see section 3.5) the two particle problem is reduced to a one body problem. The corresponding mass is given by the reduced mass $\mu = \dfrac{m_1\,m_2}{m_1 + m_2}$ and its velocity is the relative velocity $\vec{g} = \vec{v}_1 - \vec{v}_2$, which is measured with respect to the center of mass. The angle of deflection, as a function of the impact parameter b, is obtained directly from the equations of conservation of energy and angular momentum (KENNARD, 1938):

$$\vartheta = \pi - 2\beta \int_{\varrho_{min}}^{\infty} \frac{d\varrho}{\varrho^2 \sqrt{1 - \dfrac{U(\varrho)}{K} - \dfrac{\beta^2}{\varrho^2}}} \tag{6}$$

where the following reduced parameters have been used:

$$\beta = \frac{b}{R_m}, \quad \varrho = \frac{R}{R_m}, \quad U(\varrho) = \frac{V(\varrho)}{\varepsilon}, \quad K = \frac{{}^{1}/_{2}\mu g^2}{\varepsilon}$$

and where ϱ_{min} is the distance of closest approach. It is given by the smallest value of ϱ for which the denominator is zero. The deflection angle as a function of impact parameter for several energies is shown in Fig. 9.

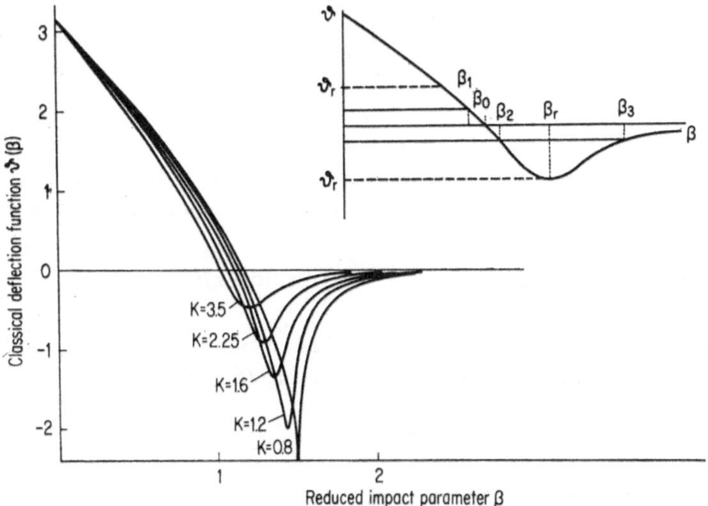

Fig. 9. Calculated classical deflection function $\vartheta(\beta)$ for various reduced energie-K and for a Lennard-Jones (10,6) potential. The diagram in the upper rights hand corner gives the definition of various reduced impact parameters

Fig. 10 shows some typical trajectories which have been reconstructed from the classical deflection function for a Lennard-Jones (10.6) potential for an incident energy of $K = 3.5$.

Equation (6) may be used to calculate the differential scattering cross section $\dfrac{d\sigma(\vartheta)}{d\omega}$. Conservation of particles requires that

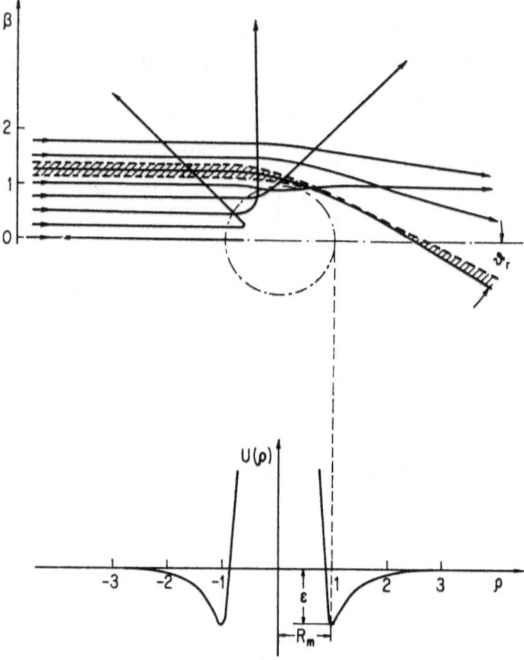

Fig. 10. Classical trajectories in a potential of the Lennard-Jones type (estimated from the classical deflection function) for several different impact parameters

$$2\pi\, b\,(\vartheta)\, db = \frac{d\sigma\,(\vartheta)}{d\omega}\, 2\pi\, \sin\vartheta\, d\vartheta \tag{7}$$

for cylindrically symmetric scattering. The meaning and derivation of equation (7) are apparent from Fig. 11. Equations (6) and (7), which together

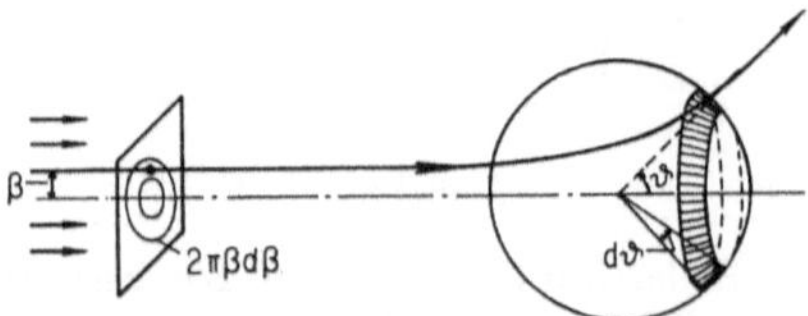

Fig. 11. Diagram illustrating calculation of classical differential cross sections

take account of conservation of mass, momentum and energy, are then sufficient to calculate the differential cross section. For a potential with one or more extrema we get in reduced notation

$$\frac{d\sigma\,(\vartheta)}{d\omega} = \frac{R_m^2}{\sin\vartheta}\sum_j \left| \beta_j \frac{d\beta_j}{d\vartheta} \right| \tag{8}$$

where the sum extends over all impact parameters — on both sides of the center of the potential field — which lead to a scattering at a given angle ϑ. An example is shown in the insert of Fig. 9 where the impact parameters β_1, β_2 and β_3 all lead to scattering at the same angle.

Fig. 12 shows the classical differential cross section for several different energies. The scattering near the sharp maximum at intermediate angles is called rainbow scattering and the angle at which it occurs is the rainbow angle, ϑ_r. It may be attributed to the discontinuity in $\dfrac{d\beta}{d\vartheta}$ at β_r (see insert of Fig. 9). The shaded range of impact parameters in Fig. 10 illustrates the origin of the rainbow, which results from the folding over of the deflection function at ϑ_r, leading to a concentration, or funneling, of particles into a narrow angular region near ϑ_r. Finally we note that the classical rainbow scattering angle is strongly dependent on K or for a given energy on ε. As we shall see later an exact quantum mechanical calculation gives essentially the same dependence on ε.

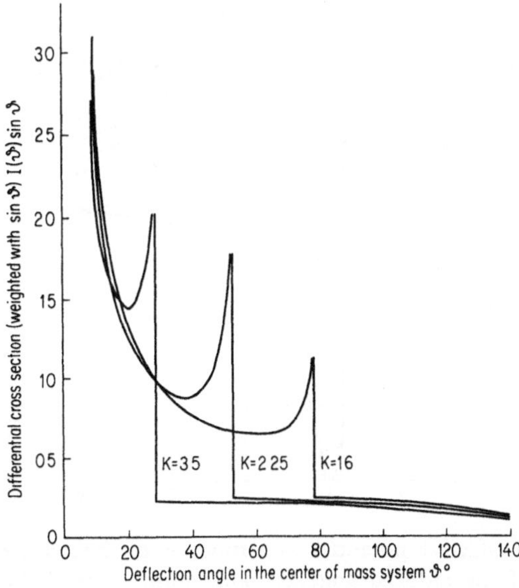

Fig. 12. Calculated differential scattering cross section (weighted with sin ϑ) as a function of the angle of deflection in the center of mass system for different reduced energies K, calculated classically for a Lennard-Jones (10,6) potential

The small angle scattering from collisions with large impact parameters contributes the largest part of the differential cross section at small angles. A second contribution comes from impact parameters near $\beta = 1$ which are also weighted heavily since $1/\sin\vartheta$ in equation (8) goes to infinity as ϑ goes to 0. This singularity is called glory scattering.

At energies below about $K = 0.8$ particles with a narrow range of impact parameters may be trapped by the attractive potential field and orbit or spiral about the scattering center. Orbiting can best be described in terms of the effective potential. From conservation of angular momentum and energy an equation for the radial motion during scattering is obtained:

$$\frac{1}{2}\mu\,\dot{R}^2 = E - \left[V(R) + \frac{1}{2}\mu\frac{b^2 g^2}{R^2}\right] \tag{9}$$

The entire term in brackets is called the effective potential. The second term inside the brackets takes account of the rotational energy around the center of mass and is referred to as the centrifugal potential. Fig. 13 shows the effective potential for a Lennard-Jones potential and

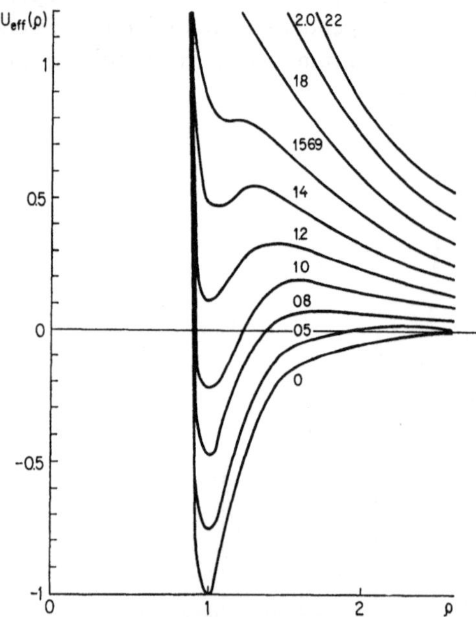

Fig. 13. Reduced effective potential as a function of the reduced distance ϱ, for various values of the reduced angular momentum $= \sqrt{K}\cdot\beta$. For $\sqrt{K}\cdot\beta \geq 1.569$ the effective potential is monotonic, the point of inflection is characterized by $K = 0.8$. For greater values of the energy the orbiting phenomenon gives way to rainbow scattering (FORD and WHEELER, 1959)

for various values of the initial angular momentum. Particles with an initial kinetic energy nearly equal to that of the maximum of the effective potential are slowed down to a small or vanishing radial velocity in the vicinity of the peak. Conservation of angular momentum requires that they continue to circle the potential, however. Thus they may undergo several revolutions before their radial velocity is large enough to remove them quickly from the potential. The scattering angle is large therefore and for $K < 0.8$ approaches ∞.

3.3. Quantum Mechanical Theory of Elastic Scattering

So far we have neglected quantum mechanical effects. Essentially, two types of interference are possible if we take the wave nature of the particles into account. The easiest to understand is the interference between the various terms in the differential cross section leading to scattering at a certain angle. This is the essence of the semi-classical approximation (FORD and WHEELER, 1959). The interference of waves scattered on both sides of the rainbow impact parameter (β_2 and β_3 in Fig. 9) leads to widely spaced oscillations in the differential cross section. The interference of waves scattered on opposite sides of the potential e.g. β_1, with β_2 and β_3 leads to closely spaced oscillations. Fig. 14 shows the results of quantum

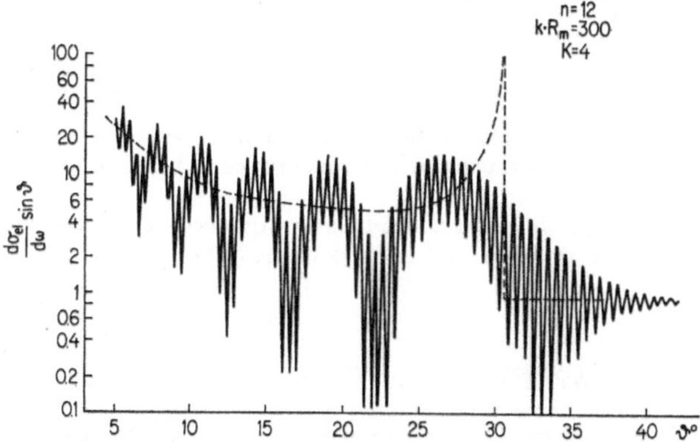

Fig. 14. Quantum mechanically calculated differential scattering cross sections (weighted with $\sin \vartheta$). The oscillations with large spacing are called rainbows. The dotted curve shows the classical differential cross section for the same Lennard-Jones (12,6) potential (HUNDHAUSEN and PAULY, 1965)

mechanical calculations of elastic cross sections. In most experiments the angular resolving power is generally not great enough to resolve the small oscillations, and an average over these is observed. A comparison with classical differential cross sections (the dotted line in Fig. 14) shows that these agree with the averaged quantum mechanical cross sections at all angles except in the vicinity of the rainbow angle and at very small angles (not seen in Fig. 14). Calculations show that the smoothed quantum mechanical rainbow peak $\vartheta_{\max}$ always appears at angles smaller than ϑ_r, and that ϑ_r is situated on the large angle side of $\vartheta_{\max}$ at a position where the intensity has

12*

dropped to 44% of its peak value. Thus ϑ_r, and from this ε, can be determined from the measurements. It can also be shown that $\vartheta_R - \vartheta_{\max} \sim \dfrac{\varepsilon}{R_m^2}$ and, provided the rainbow peak is well resolved, R_m can be measured as well. Other better methods for determining R_m from the oscillations of the differential cross section are also available. Table 2 lists some recently determined potential parameters for unreactive and reactive systems.

Table 2. *Potential parameters for unreactive and reactive systems from recent rainbow scattering measurements (1 erg/molecule = 1.439 · 10¹³ kcal/mole)*

System	$\varepsilon \left[\dfrac{\text{kcal}}{\text{mole}}\right]$	R_m [Å]
K + Xe[a]	0.26	5.8
K + Hg[a]	1.25	4.8
K + HBr[a]	0.56	4.4
Na + HBr[a]	0.45	(5.6)
K + CH_3I[b]	0.51 ± 0.06	—
K + CCl_4[b]	0.69 ± 0.06	5.5 ± 1.8
K + $SiCl_4$[b]	0.65 ± 0.03	2.8 ± 1.4
K + SF_6[b]	0.31 ± 0.03	6.3 ± 0.8

[a] The values were obtained from the review article by BERNSTEIN and MUCKERMAN (1967) and were evaluated using a Lennard-Jones (12,6) potential; the errors are smaller than 10%.

[b] From AIREY et al., (1967b) evaluated using BUCKINGHAM's exp-six ($\alpha = 12$) potential, which is very similar to a Lennard-Jones (12,6) potential.

3.4. *The Central Field Impact Theory of Chemical Reactions*

For describing a chemical reaction we assume that the Lennard-Jones potential holds for relative kinetic energies (along the line of centers) less than the activation energy. For greater energies the reactive cross section is assumed to be given by

$$\frac{\sigma_{\text{react}}}{R_m^2} = 2\pi \int\limits_0^{\beta_{\text{act}}} P(\beta, K)\beta \, d\beta \ , \tag{10}$$

where $P(\beta, K)$ is the reaction probability and β_{act} is the impact parameter corresponding to a potential energy at the point of closest approach ($\varrho_{\min}$) which is equal to the reduced activation energy E_{act}.

If $P(\beta, K) = 1$ for all impact parameters less than β_{act} and is zero elsewhere then

$$\frac{\sigma_{\text{react}}}{R_m^2} = \pi \beta_{\text{act}}^2 \ . \tag{11}$$

We can write this result in terms of energies if we recall that $\varrho_{\min}$ is defined by (see equation 6)

$$\beta_{\mathrm{act}} = \varrho_{\min} \sqrt{1 - U(\varrho_{\min})/K} \ .$$ (12)

With $U(\varrho_{\min}) = E_{\mathrm{act}}$ equation (11) becomes

$$\frac{\sigma_{\mathrm{react}}}{R_m^2} = \pi \, \varrho_{\mathrm{act}}^2 \left(1 - \frac{E_{\mathrm{act}}}{K}\right) \ .$$ (13)

For positive activation energies ϱ_{act}^2 is fairly insensitive to the kinetic energy K and E_{act}. This can be seen from the fact that the Lennard-Jones repulsive potential approaches a hard sphere potential, which is no longer dependent on the energy. The hard sphere dependence of the reactive cross section on the energy is also a good approximation to the real behavior for which $P(\beta, K) < 1$ and agrees well with the experimental data (see Fig. 18).

$P(\beta, K)$ is a quantity of fundamental importance for describing a chemical reaction. BECK, GREENE and ROSS (1962) have suggested a method for determining $P(\beta, K)$ from measurements of elastic scattering. The method involves the assumption, which is correct in the limit of strictly classical scattering, that if reaction occurs the observed elastic scattering is reduced correspondingly. Equation (8) is then modified to read

$$\frac{d\sigma_{\mathrm{obs}}(\vartheta)}{d\omega} = \frac{R_m^2}{\sin\vartheta} \sum_j \beta_j \left[1 - P(\beta_j, K)\right] \frac{d\beta_j}{d\vartheta}$$ (14)

At angles larger than the rainbow angle only one impact parameter appears in the sum and $P(\beta, K)$ may be evaluated from the measured cross section

$$P(\beta, K) = \frac{\beta \dfrac{d\beta}{d\vartheta} - \dfrac{\sin\vartheta}{R_m^2} \dfrac{d\sigma_{\mathrm{obs}}(\vartheta)}{d\omega}}{\beta \dfrac{d\beta}{d\vartheta}} = \frac{\dfrac{d\sigma_{\mathrm{el}}}{d\omega} - \dfrac{d\sigma_{\mathrm{obs}}}{d\omega}}{\dfrac{d\sigma_{\mathrm{el}}}{d\omega}}$$ (15)

$\dfrac{d\sigma_{\mathrm{el}}}{d\omega}$ is calculated from potential parameters, which have either been estimated or deduced from small angle scattering where reaction does not occur.

The application of equation (15) in general implies a number of assumptions the importance of which is difficult to assess at the present time. Probably the most serious assumption is that the potential at small distances where a reaction can occur is the same as that between two atoms. As we shall see in the next section most of the reactions studied by molecular beams lead to highly vibrationally excited products so that according to the rule proposed by EVANS and POLANYI the reactants are expected to attract each other strongly prior to reaction. Other evidence for a deviation

182 J. P. TOENNIES

of the potential from the atom-atom behavior is provided by the very interesting quantum mechanical calculations of the intermolecular potential for He—H_2 by KRAUSS and MIES (1965). Their results show that even for this simple non-reactive system a slight reduction in the repulsive potential results from a contraction in the molecular bond distance at atom-molecule distances of the order of twice the bond distance. For reactive systems the effect will be larger and will very likely lead to a significant reduction of the repulsive potential, which, of course, must take place in order to make a reaction possible. Unfortunately, although the techniques are available, the highly successful trajectory calculations have as yet not been used to calculate the elastic differential cross sections at wide angles. At any rate we would expect that the pure elastic scattering will be different from that calculated on the basis of an atom-atom potential with parameters estimated from the small angle scattering.

Quantum effects must also be taken into account. Using an optical model and semi-classical arguments ROSENFELD and Ross (1966) have shown that equation (15) holds for reactive $P(\beta, K)$ curves having the following property:

$$\frac{dP(\beta, K)}{d\beta} \ll 2k(\pi - \vartheta)[1 - P(\beta, K)], \quad E = \frac{k^2 \hbar^2}{2\mu} . \qquad (16)$$

Recent quantum mechanical calculations (HUNDHAUSEN and PAULY, 1965) of elastic differential cross sections give an indication of what happens

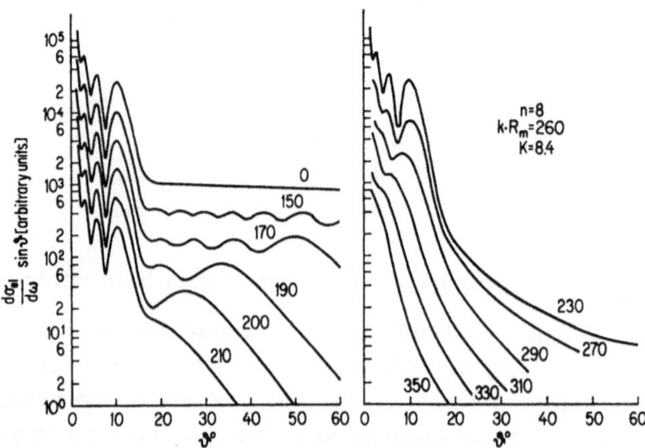

Fig. 15. Quantum mechanically calculated differential scattering cross sections for a Lennard-Jones (8,6) potential with complete absorption for all partial waves (impact parameters) smaller than the value of l_i given on each curve. The curves have been shifted vertically for the sake of clarity (HUNDHAUSEN and PAULY, 1965)

when the condition expressed by equation (16) is not fulfilled. Fig. 15 shows the results for the simple step probability function used previously, i.e.:

$$P(l, K) = 1, \quad l < l_i; \quad P(l, K) = 0, \quad l > l_i \, ,$$

where l_i is the orbital angular momentum quantum number which is related to the classical angular momentum $\mu \cdot b \cdot g \simeq l\hbar$. The curves, which have been calculated for increasing values of l_i, show that there is no sharp borderline between total absorption and complete elastic scattering. The gradual transition and the oscillations in the classically forbidden region are the result of wave mechanical effects. Finally we note that with increasing l_i the break in the elastic scattering shifts to successively smaller values of ϑ until at large l_i the rainbow maximum begins to disappear.

Rotational and vibrational excitation are neglected in equation 15. ROSENFELD and ROSS (1966) show, however, that if the probabilities for excitation are small it is still possible to use equation (15) in analysing the data. Recent calculations of rotational excitation which have been carried out for the system He—N_2 show that the cross sections for rotational excitation are an order of magnitude smaller than the elastic cross sections at large angles (ERLEWEIN, VON SEGGERN and TOENNIES, 1967). For reactive partners the rotational inelastic cross sections will be considerably larger depending on their multipole moments and the spacing of the rotational levels. Because of the bond length adjustment occuring with near head-on collisions, vibrational excitation will also be more probable than in non-reactive collisions. If inelastic scattering is probable and the deviations from the atom-atom potential are serious then equation (15) has to be replaced by

$$P(\beta, K) = \frac{\dfrac{d\sigma'_{el}}{d\omega} + \dfrac{d\sigma'_{inel}}{d\omega} - \dfrac{d\sigma_{obs}}{d\omega}}{\dfrac{d\sigma'_{el}}{d\omega} + \dfrac{d\sigma'_{inel}}{d\omega}} \, , \tag{17}$$

where $\dfrac{d\sigma'_{el}}{d\omega}$ and $\dfrac{d\sigma'_{inel}}{d\omega}$ are the differential cross sections which hold for the reactive potential. Since these differential cross sections are not known it is difficult, at the present time, to assess the extent to which equation (15) is valid, especially in cases in which inelastic and reactive collisions are highly probable. Under circumstances in which $P(\beta, K)$ approaches 1 its evaluation no longer depends on specific assumptions concerning the differential cross sections which hold when reactive scattering does not occur. This, fortunately, appears to be the case in all reactions to which the method has so far been applied and justifies the use of equation (15).

3.5. The Transformation of Cross Sections from the Center of Mass Coordinate System to the Laboratory Coordinate System

We come now to the problem of relating measured laboratory angular distributions to those in the center of mass system. These relationships are somewhat more involved than in high energy collisions, where the velocity of the scattering target (secondary beam) can be neglected. They are important, however, for an understanding of chemical scattering experiments, and we discuss briefly here the transformation equations and some of the complications which arise in interpreting the experimental data.

From classical mechanics we recall that if the potential between two particles 1 and 2 is only a function of their relative distance $\vec{r_1} - \vec{r_2}$ the Hamiltonian and the equation of motion can be separated by transforming to a new frame of reference. The equations describing the transformation are

$$\vec{v_1} = \vec{V} + \vec{v}_{1\,cm}, \quad \vec{v_2} = \vec{V} + \vec{v}_{2\,cm} \tag{18}$$

where $\vec{v_1}$ and $\vec{v_2}$ are the laboratory velocities and $\vec{V}$ is the velocity of the new frame of reference, called the center of mass system. $\vec{v}_{1\,cm}$ and $\vec{v}_{2\,cm}$ are the velocities in this moving frame. The Hamiltonian is transformed accordingly from

$$\mathcal{H} = \frac{\vec{p_1}^2}{2\,m_1} + \frac{\vec{p_2}^2}{2\,m_2} + V(|\vec{r_1} - \vec{r_2}|) \tag{19}$$

laboratory system

$$\text{to} \qquad \mathcal{H} = \frac{\vec{\mathscr{P}}^2}{2\,\mathscr{M}} + \frac{\vec{\pi}^2}{2\,\mu} + V(|\vec{R}|), \tag{20}$$

center of mass system

where the mass, position and momentum of the center of mass are given by $\mathscr{M} = m_1 + m_2$, $\vec{\mathscr{R}} = \dfrac{m_1 \vec{r_1} + m_2 \vec{r_2}}{m_1 + m_2}$ and $\vec{\mathscr{P}} = \vec{p_1} + \vec{p_2}$. Furthermore, in the center of mass system the mass, position and momentum are given by $\mu = \dfrac{m_1 m_2}{m_1 + m_2}$, $\vec{R} = \vec{r_1} - \vec{r_2}$ and $\pi = \mu \, \dot{\vec{R}} = \dfrac{m_2 \vec{p_1} - m_1 \vec{p_2}}{m_1 + m_2}$. The first term on the right in equation (20) is the energy of the center of mass and the second term is the energy of the relative motion. The last is the only term affected by the potential so that $\dot{\vec{\mathscr{P}}} = 0$ and consequently

$$\dot{\vec{\pi}} = -\operatorname{grad} V(|\vec{R}|) \tag{21}$$

Thus the kinematic problem has been formally reduced to the one body problem, which has already been discussed in the previous section.

Differentiating $\vec{R}$ with respect to time and using $\dot{\vec{\mathscr{P}}} = 0$ in the center of mass system we have that in this system

$$m_1 \vec{v}_{1\,cm} = -\, m_2 \vec{v}_{2\,cm} \tag{22}$$

where $\vec{v}_{1\,cm} - \vec{v}_{2\,cm} = \vec{g}$. $\vec{v}_{1\,cm}$ and $\vec{v}_{2\,cm}$ are therefore oppositely directed in the center of mass system and are parallel or antiparallel to $\vec{g}$.

An observer in the center of mass system sees the two particles approaching the system's origin, collide at the origin and then move away from each other in opposite directions. Due to the potential of interaction the vector $\vec{g}$ is rotated by an angle equal to the scattering angle.

The vector equations (18) above are sufficient for transforming velocities and scattering angles from one system to the other. Since equation (22) holds after the collision as well, it and equation (18) enable us to reconstruct $\vec{v}_3$ and $\vec{v}_4$ after a collision for a given scattering angle. Fig. 16 shows such a construction for an elastic collision. The solid lines in Figure 16

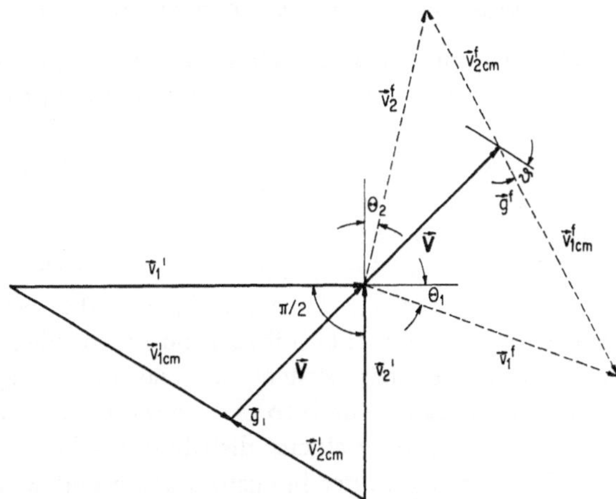

Fig. 16. Velocity diagram for crossed beam kinematics. The example shown is for elastic scattering. The solid lines represent velocity vectors before collision; the dotted lines represent the velocity vectors after collision

show the vectors before collision whereas the dotted lines show the vectors after collision. This type of diagram shows clearly the sequence of events in the laboratory and center of mass systems. For instance at time $t = -\,\tau$ the two molecules are at the origins of the vectors at the left hand part of the diagram; at $t = 0$ they collide and at $t = +\,\tau$ they are located at the tips of the vectors at the right.

In an inelastic collision g is usually increased or decreased. Conservation of energy requires then that we have

$$\frac{1}{2}\,\mu\,g_f^2 - \frac{1}{2}\,\mu\,g_i^2 = \Delta E \tag{23}$$

where ΔE is the change in internal energy and f and i refer to final and initial relative velocities, respectively.

For a chemical reaction where the masses change during collision we easily obtain the following result for the change in velocities from equation (22) and (23):

$$\frac{v_{4\,cm}}{v_{2\,cm}} = \sqrt{\frac{m_2\,m_3}{m_1\,m_4}\left(1 + \frac{\Delta E}{\frac{1}{2}\,\mu_i\,g_i^2}\right)} \tag{24a}$$

$$\frac{v_{3\,cm}}{v_{1\,cm}} = \sqrt{\frac{m_1\,m_4}{m_2\,m_3}\left(1 + \frac{\Delta E}{\frac{1}{2}\,\mu_i\,g_i^2}\right)} \tag{24b}$$

Thus the vector diagram of Fig. 16 has to be modified depending on the size of $\frac{\Delta E}{^1/_2\mu_i g_i^2}$ and the mass changes accompanying the reaction. Fig. 22b shows a vector diagram in which circles corresponding to different ΔE have been drawn to show the locus of energetically possible center of mass velocity vectors.

On the basis of the equation above we expect in general that the heavier of the products moves in the direction of the center of mass to an extent depending on the ratio $m_3/m_4\,(m_3 > m_4)$. The angular distribution of $\vec{v_3}$ is, therefore, similar to that of the distribution in $\vec{V}$ which depends only on the distributions of $\vec{v_1}$ and $\vec{v_2}$. Thus for $m_3 \gg m_4$ the distribution in $\vec{v_3}$ does not contain easily retrievable information on the distributions of the products in the center of mass system, whereas that of $\vec{v_4}$ does. For $m_3 \sim m_4$ both distributions can be used to deduce the reaction mechanism.

So far we have neglected the velocity distribution of the primary and secondary beams. The small product intensities associated with the small cross sections usually permits velocity selection of only one of the beams[10]. Thus there is a distribution in $\vec{g}$ and $\vec{V}$ resulting from that of $\vec{v_2}$ and this has to be taken into account in an analysis of the data[11]. For a given distribution in $\vec{v_2}$ the narrowest distributions in $\vec{g}$ are obtained when the angle between $\vec{v_1}$ and $\vec{v_2}$ is somewhat less than 90° and such that $\vec{g}$ is perpendicular to $\vec{v_2}$. It can also be shown (GREENE and LAU, 1964) that the

---

[10] The case that both are not selected is treated by DATZ, HERSCHBACH and TAYLOR (1961).

[11] The distribution in the angle between the two beams can, depending on the relative masses, lead to more blurring than the distribution in the magnitude of v_2.

narrowest velocity distributions after scattering are obtained if the product intensities are measured in a plane perpendicular to $\vec{v_2}$. This is especially important when measuring elastic and inelastic differential cross sections.

Fig. 17 shows the vector identities (equation (22)) in three dimensional space for an exothermic (ΔE positive), endothermic, and an elastic process with $m_1 = m_3$ and $m_2 = m_4$. With the aid of Fig. 17 we can visualize

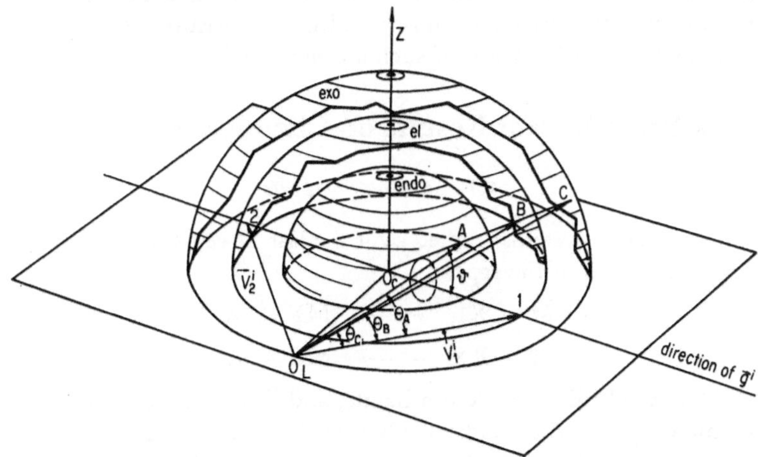

Fig. 17. Velocity diagram for crossed beam kinematics. The vectors correspond to velocities. O_L is the origin of the laboratory system. The laboratory velocities v_1 and v_2 define the plane to which Z is perpendicular. The relative velocity before collision is equal to the vector 1,2. O_c is the origin in the center of mass system. The spheres are the locus of the tops of all possible velocity vectors of the molecule 3 in the center of mass system after collision. Thus, the vectors $O_c A$, $O_c B$ and $O_c C$ correspond to the center of mass velocities for an endothermic, elastic ($\Delta E_0^0 = 0$) and an exothermic collision, respectively all with scattering angles ϑ and φ (φ is usually measured with respect to Z). The vectors $O_L A$, $O_L B$ and $O_L C$ are the corresponding observable laboratory system velocities. Note that to one angle in the center of mass system there are different angles Θ and Φ (Φ is not shown) in the laboratory system depending on the collision process. Also to one set of laboratory angles there may be up to 6 different angles in the center of mass system

how $d\Omega$ is related to $d\omega$ and that their ratio is strongly dependent on the angles and relative velocities. The actual expression for $\left| \dfrac{d\omega}{d\Omega} \right|$ is quite complicated. It is discussed in the papers of RUSSEK (1960), HELBING (1967) and MORSE and BERNSTEIN (1962) and is not presented here.

From Fig. 17 it is possible to construct situations in which the transformation from the laboratory system to the center of mass system is not unique. Consider the laboratory angle in the general direction $O_L O_C$ in Fig. 17. For an elastic process two center of mass velocity vectors lead to the same laboratory angle, and these can only be disentangled by measuring the velocity distribution of the products in the laboratory system. For this reason it is also customary to carry out the transformation from the center of mass system to the laboratory system, in which case there is always a unique transformation.

In the actual interpretation of measured data these transformations have to be carried out using numerical procedures until a best fit between assumed center of mass distributions and those measured is obtained (see section 4.3.2, for a discussion of such a calculation).

4. Some Recent Results on Chemical Reactions

4.1. Summary of Data

Of the reactive systems studied over the last 12 years the overwhelming majority have been of the types:

$$M + X_2 \longrightarrow MX + X$$

$$M + RX \longrightarrow MX + R$$

where M is an alkali atom, X is a halide, and R is either a hydrogen atom or a simple hydrocarbon radical. One of the reasons for choosing these reactions has to do with the high efficiency of the Langmuir-Taylor detector for detecting alkali atoms and molecules containing them. In addition the reactions with the halogens have low activation energies and large reaction cross sections. Finally considerable information on these reactions was already available from observations made with the much simpler *sodium flame* technique (M. Polanyi, 1933) over thirty years ago.

Table 3 lists the reactions which have been studied using molecular beams. The table contains information on the change in the dissociation energies accompanying the reaction (column 2) and the method used in studying the reaction (column 3). PA refers to product analysis or the direct measurement of product distributions. ESA refers to elastic scattering analysis, or in other words, the measurement of the angular distribution of the unreacted alkali atom. We discuss these methods more fully further on in this section. v_1 implies that the measurement was carried out with a velocity selected primary beam. The next two columns summarize briefly the results of the beam experiments on cross sections and activation energies. The last column characterizes the observed angular distributions of the products. In the *rebound* reactions the observed distributions indicate that in the center of mass system the incoming particle M reverses direction after the collision.

$$A \rightarrow \quad \leftarrow BC \qquad \text{before collision}$$
$$AB \leftarrow \quad \rightarrow C \qquad \text{after collision}$$

In the *stripping* reactions the opposite situation appears to hold most frequently

$$A \rightarrow \quad \leftarrow BC \qquad \text{before collision}$$
$$C \leftarrow \quad \rightarrow AB \qquad \text{after collision}$$

Cases *intermediate* to these have also been observed.

The classification according to rebound and stripping mechanism is suggested by the history of the experiments and, for reasons presented below, may lead to confusion. It is now known that the observed mechanism is directly related to the size or range of the reactive cross section. Thus when the reactive cross section is small so that reaction can only occur at small impact parameters the products must rebound. When it is large, on the other hand, most of the products appear in the forward direction at small angles corresponding to large impact parameters, but, of course, some will also rebound. In general, we expect the production of rebounding products of a stripping reaction to have a larger differential cross section than in a pure rebound type reaction.

The list of reactions in Table 3 is not complete. Additional reactions are tabulated in the review by HERSCHBACH (1966).

Instead of attempting a survey of all the reactions we have chosen to concentrate on the following three reactions: $K + HBr$, $K + CH_3I(Br)$, and $K + Br_2$, which have been investigated in considerably more detail than the others. The experimental techniques used and the observed properties of the reactive collisions in the three systems are representative of most of the other reactions listed in Table 3.

4.2. Rebound Reactions

4.2.1 The Reaction $K + HBr \rightarrow H + KBr$[12]

The HBr reaction is the first to have been studied by scattering in crossed molecular beams. It is slightly exothermic (3.8 kcal/mole) and from earlier sodium flame studies it was known to have a low activation energy. The KBr molecule is much heavier than the H atom and since the earlier flame experiments indicated that much of the reaction energy is absorbed by internal degrees of freedom, the product moves only slowly and its motion in the laboratory system was expected to be quite close to that of the center of mass. This makes detection easier but makes a determination of the product angular distribution in the center of mass system difficult.

In their pioneering experiment TAYLOR and DATZ (1955) crossed a well collimated potassium beam $(T \sim 700\,°K)$ with a short path length

---

[12] The classification of this reaction as a rebound reaction is actually not quite certain at the present time as discussed in the text.

Table 3. *Summary of experimental results on chemical reactions from crossed beam experiments*

Reaction	ΔE_0^0 [kcal/mole][a]	Method	activation energy [kcal/mole]	σ_{react} [Å^2] at $E_{trans(i)}$ [kcal/mole][b]	product angular distribution	most recent literature
$K + HBr \rightarrow H + KBr$	3.8	PA	3.0	—	—	1
	3.8	PA v_1	<0.4	31 at 2	—	2
	3.8	ESA v_1	—	35 ±9 at 2.6—5.8	—	3
	3.8	PA v_1	—	—	stripping	4
$K + DBr \rightarrow D + KBr$	2.5	ESA v_1	—	26 ±8 at 2.6—5.8	—	3
$K + TBr \rightarrow T + KBr$	—	PA	—	—	rebound	5
	Tritium was detected					
$Cs + HBr \rightarrow H + CsBr$	—	PA	—	140	—	6
$K + HCl \rightarrow H + KCl$	−0.7 ±1.2	ESA v_1	0.55 ±0.1	3.9 at 2	—	7
$K + HI \rightarrow H + KI$	+5.9 ±1.2	ESA, PA v_1	0.2 ±0.1	28 at 2	—	7
$M + RI \rightarrow R + MI$	17—40	PA	—	—	rebound	8
$M = Na, K, Rb, Cs$						
$R = CH_3, C_2H_5, \ldots C_5H_{11},$						
$CH_2CH:CH_2$						
$K + CH_3Br \rightarrow CH_3 + KBr$	+23	ESA, v_1	—	23 at 2	—	9
$M + CH_3I \rightarrow CH_3 + MI$	+22	PA	—	orientation dependence was measured	rebound	10
$M = Rb$ (ref. 10a)						
$M = K$ (ref. 10b)						
$K + CH_3I, CCl_4,$	22, 31,	ESA v_1	—	50 (4); 50 (4.9);	—	11
$SiCl_4, SnCl_4, SF_6$	14, 25,			6 (3.4); 100 (3.9);		
$\rightarrow KX + \cdots$	44 resp.			60 (2) resp.		

Table 3 (Continuation)

Reaction						
$M + X_2 \rightarrow MX + X$ $M = K, Rb, Cs$ $X = Cl, Br, I$	≈ 40	PA	—	$\approx 200, 100$	stripping	*12, 13*
$K + Br_2 \rightarrow KBr + Br$	44.8	PA	—	product energy distri- butions were measured		*14, 15, 16*
$M + ICl, IBr \rightarrow ?$	?	PA	—	100	stripping	*17, 13*
$K + SCl_2, PCl_3, PBr_3$ $CH_2I_2, SnCl_4, CBr_4$ and $SF_6 \rightarrow KX + \ldots$	40, 22, 27 30, 24 40, 50	PA	—	—	stripping	*12* (p. 339)
$K + CBr_4 \rightarrow ?$	40	ESA v_1		236 at 2		*9*
$K + CHCl_3, CCl_4 \rightarrow$ $KX + \ldots$	32 32	PA	—	—	intermediate between rebound and stripping	*12* (p. 339)
$H + D_2 \rightarrow H + HD$	-1.09	PA	—	$\dfrac{d\sigma}{d\omega} \approx 10^{-2}\text{Å}^2/\text{sr}$		*18*
$D + H_2 \rightarrow D + HD$	$+0.83$	PA	—	—		*19*
Molecular beam experiments, but not with crossed beams:						
$Cs + CCl_4 \rightarrow CsCl + CCl_3$		essentially PA v.	—	—		*20*
$N_2 + CO \rightarrow NO^+ + CN^-$		essentially PA v.	—	$2 \cdot 10^{-2}$ at 300		*21*
$Na + CH_3I \rightarrow CH_3 + NaI$		essentially PA v.	—	1.3		*22*

[a] ΔE_0^0 positive corresponds to an endothermic reaction. This convention is opposite to the one adopted in Pauly and Toennies (1965).
[b] $E_{trans(i)} = \frac{1}{2}\mu g^2$.

Literature to Table 3

1. Taylor and Datz (1955) and Datz, Herschbach and Taylor (1961).
2. Beck, Greene and Ross (1962).
3. Airey et al. (1967a).
4. Grosser, Blythe and Bernstein (1965).
5. Kinsey and Martin (1966).
6. Herschbach (1966).
7. Ackermann et al. (1964).
8. Herschbach (1962).
9. Ackermann et al. (1963) and Greene, Moursund and Ross (1966).
10a. Beuhler, Bernstein and Kramer (1966), *10b.* Brooks and Jones (1966).
11. Airey et al. (1967b).
12. Herschbach (1966).
13. Datz and Minturn (1964), Minturn, Datz and Becker (1966).
14. Grosser and Bernstein (1965).
15. Birely and Herschbach (1966).
16. Herm and Herschbach (1965).
17. Wilson et al. (1964).
18. Fite and Brackmann (1963).
19. Datz and Taylor (1963).
20. Bull and Moon (1954).
21. Utterback (1966).
22. Fristrom (1966).

(1 cm long) intense HBr beam ($T \sim 460\,°K$). Both ovens had two chambers so that the beam temperature could be varied without changing the source pressure. Velocity selectors were not used. The KBr angular distribution was measured in the plane of the two beams with the two wire detector. The distributions showed a narrow ($\Delta\,\Theta_{1/2} \sim 30°$) maximum in the approximate direction of the center of mass. Because of the peaking the maximum product intensities were high and of the order of 10^{-12} Amp. From the shape of the distributions it was possible to estimate the activation energy.

Thus only collisions with sufficient relative kinetic energy[13] $E_{\mathrm{rel}} \sim \frac{1}{2}\mu$ $(v_1^2 + v_2^2) > E_{\mathrm{act}}$ produce a reaction. Since $v_1 > v_2$ the center of mass of the product distribution is shifted in the direction of the primary beam depending on the activation energy (Datz, Herschbach and Taylor, 1961). Their result of 2.5—3.0 kcal/mole is considerably larger than the upper limit of 0.4 kcal/mole determined by Greene, Roberts and Ross (1960). In their experiments the latter authors used a velocity selector in the primary beam and measured the product intensity as a function of the primary beam velocity. In this way they were able to cover the range 0.3 $< E_{\mathrm{trans}(i)} < 5\,\dfrac{\mathrm{kcal}}{\mathrm{mole}}$. The results are shown in Fig. 18. A comparison

---

[13] neglecting the angular spread of the secondary beam.

of results analyzed with different assumed angular distributions suggests
that possibly the angular spread of the HBr beam, which was not accounted

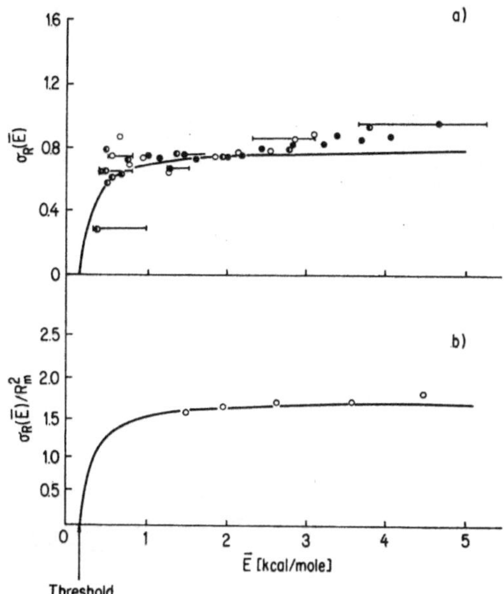

Fig. 18. Energy dependence of the reaction cross section for
K + HBr → KBr + H (BECK, GREENE and ROSS, 1962)
a) From measurements of KBr intensity and for an assumed angular distribution
of the secondary beam; • T_{HBr} = 292 °K, ○ T_{HBr} = 175 °K, ◉ T_{HBr} = 176 °K.
The ordinate is in relative units. The horizontal bars represent approximately
the range of energy which contributes to the cross section at the given $\bar{E}_{trans}$. The
chemical reaction cross section for the model of hard spheres is shown by the
solid curve (see equation (13) in the text)
b) From measurements of the elastic scattering (BECK, GREENE and ROSS, 1962)

for in the DATZ, HERSCHBACH and TAYLOR analysis, may explain the dis-
crepancy between their activation energy and that of GREENE, ROBERTS
and ROSS. With the apparatus used in the previous experiments BECK,
GREENE and ROSS (1962) measured the elastic differential cross section for
K scattered on HBr. Fig. 19 shows the observed distributions. Since
the general shape of the cross sections at small angles (<40°) agreed essentially with those obtained with K on Kr, these authors assumed that only
elastic scattering occured at the small angles. This is an important result
since it suggested that the central field impact model could be used to
explain these reactions (see sections 3.4. and 4.2.3.). Thus the elastic rainbow
scattering at small angles could be used to estimate the potential parameters ε and R_m. The small dipole moment of the HBr molecule (μ =
0.79 D.) and especially the large spacing of the rotational levels (for

$J = 1 \rightarrow J = 2$, $\Delta E_{rot} \sim 10^{-1}$ kcal/mole) suggest that the amount of inelastic scattering is not large, at least not at small angles.

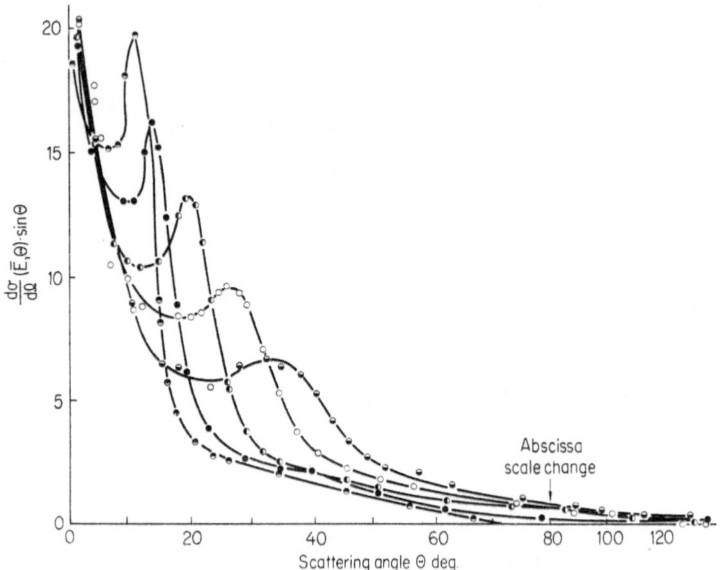

Fig. 19. The measured differential elastic scattering cross section, in arbitrary units, for the system K + HBr at five different relative energies. $E_{trans(i)}$ in kcal/mole: ⊙: 1.49, ○: 1.96, ◐: 2.64, ●: 3.61, ◑: 4.49. The decrease in the cross section beyond 60° is attributed to the removal of K by reactive scattering (BECK, GREENE and ROSS, 1962)

With these potential parameters the large angle elastic differential cross sections were estimated, and from the difference between these and the observed cross sections the reaction probability was finally calculated from equation (15). Provided we take $\vartheta > \vartheta_r$ the observed scattering angles can be assigned to impact parameters of the atom before collision. In so doing the assumption is implicitly made that the interaction potential of the non-reacting atoms is the same as that in elastic scattering. Fig. 20 shows $P(\beta, K)$ versus β for the HBr reaction obtained in this way. The interpretation of this result is discussed at the end of this section.

These experiments may also be interpreted in terms of a total reactive cross section

$$\frac{\sigma_{react}(K)}{R_m^2} = 2\pi \int_0^{\beta_{act}} P(\beta, K)\beta\, d\beta \tag{25}$$

where for K + HBr, β_{act} corresponds to a distance of closest approach of 2.8 Å. The result for this reaction was found to be $\sigma_{react} = 35 \pm 9$ Å² in the energy range 2.6 to 5.8 kcal/mole. This method of determining

σ_{react} has the advantage over integrating $\dfrac{d\,\sigma_{\text{react}}}{d\omega}$ in that it is absolute, requiring only knowledge of R_m which in turn can be obtained from measurements of the rainbow angular position (see section 3.3.). The

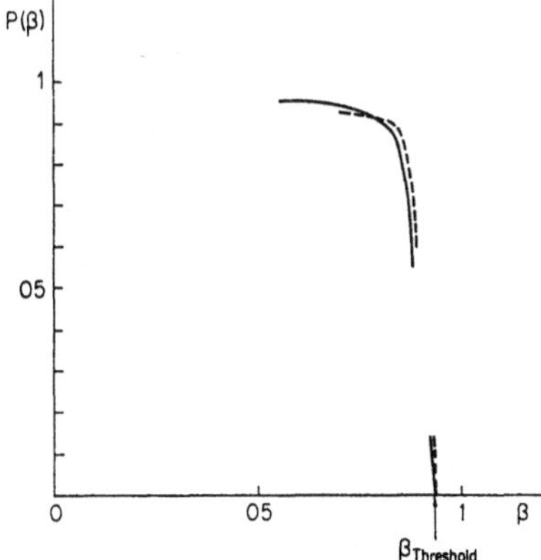

Fig. 20. The probability of reaction versus the reduced impact parameter at the distance of closest approach for the reactants K + HBr. Estimated from data presented in AIREY et al. (1967a). The solid and dotted lines are for $E_{\text{trans}\,(i)} =$ 2.64 and 4.9 kcal/mole, respectively

energy dependence obtained from equation (25) is in surprisingly good agreement with that deduced from the product intensity (see Fig. 18).

The depletion only of large angle scattering (small impact parameters) by reaction strongly suggests that the alkali atom picks up the halide atom and rebounds in the course of the reaction. Recently GROSSER, BLYTHE and BERNSTEIN (1965) have measured the velocity distribution of the KBr formed in the reaction. Fig. 21b) shows the apparatus and Fig. 21a) shows a diagram with a typical measured velocity distribution. Of course most of the laboratory velocity is that of the center of mass and this has to be subtracted in order to obtain the KBr velocity in the center of mass system. When this is done the translational energy of the KBr appears to be small. From conservation of energy

$$E_{\text{int\,(KBr)}} - E_{\text{int\,(HBr)}} = E_{\text{trans}\,(i)} - E_{\text{trans}\,(f)} + \Delta E_0^0, \qquad (26)$$

the initial conditions, and the measured result for $E_{\text{trans}\,(f)}$, $E_{\text{int\,(KBr)}}$ was estimated to be 4.6 ± 0.7 kcal/mole indicating that most of the reaction energy goes into internal excitation. These authors further conclude that the KBr is scattered into the forward direction, with respect to the center of mass velocity of the K atom before collision, (stripping reaction mechanism) in disagreement with the elastic scattering results discussed earlier. The angular spread of the HBr beam was large ($\approx 30°$) in these early experiments and as in the case of the experiments of BECK, GREENE,

13*

and Ross, may lead to considerable smearing of the center of mass vectors and consequently considerable loss of information on the angular distribution on the center of mass system[14]. To avoid this difficulty, Kinsey and Martin (1966) investigated the angular distribution of radioactive

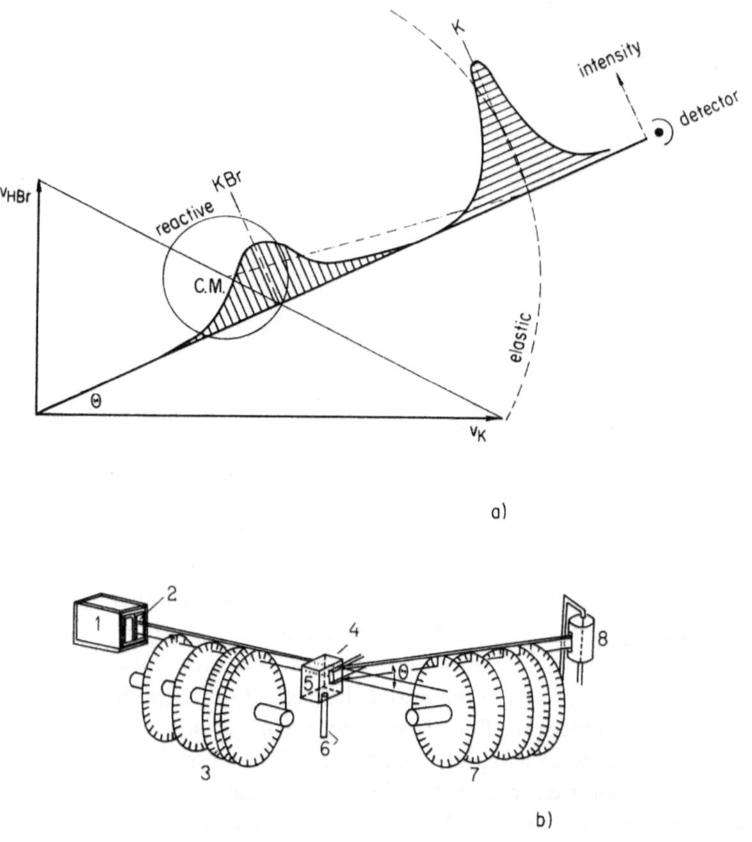

Fig. 21. a) Velocity vector diagram for the crossed beam reaction K + HBr → KBr + H. A typical velocity distribution for reactively scattered KBr and elastically scattered K is shown (Grosser, Blythe, and Bernstein, 1965)
b) Schematic diagram of the apparatus used in the measurements of the velocity distribution of reaction products. The numbers refer to the following parts: 1, primary beam oven; 2, oven slit; 3, primary beam velocity selector; 4, secondary beam channels; 5, secondary beam oven; 6, secondary oven gas inlet; 7, velocity selector for the scattered beam; 8, Langmuir-Taylor detector. Θ is the laboratory scattering angle

---

[14] Recently the experiments were repeated with reduced angular spread with similar results. The new results also provide some evidence for "backward" as well as "forward" scattering. (C. Riley, K. Gillen and R. Bernstein, private communication, 1967).

tritium formed in the reactions K + TBr and Cs + TBr. The tritium atoms produced in the reaction were first selectively absorbed on MoO_3 surfaces distributed at various scattering angles, and from the beta decay activity the intensities were estimated. For both reactions the distributions show that the reaction is of the rebound type. This result is also supported by results on the $CH_3I(Br)$ reaction.

4.2.2. The Reaction $K + CH_3I(Br) \rightarrow CH_3 + KI(Br)$

These reactions are somewhat more exothermic ($\Delta E_0^0 = 22\ (23)$ kcal/ mole) than the HBr reaction, but the mechanism is expected to be much the same as for HBr. As we shall see this is also borne out by the experiments. The increased mass of the CH_3 radical compared to hydrogen means, however, that the KI angular distribution in the laboratory provides more information on the reaction mechanism. The reactions with CH_3I and with successively heavier aliphatic organic radicals were first studied by HERSCHBACH, KWEI and NORRIS in 1961.

Fig. 22a) shows that the KI product distribution is strongly peaked at 83° from the direction of the potassium beam. This result can be easily

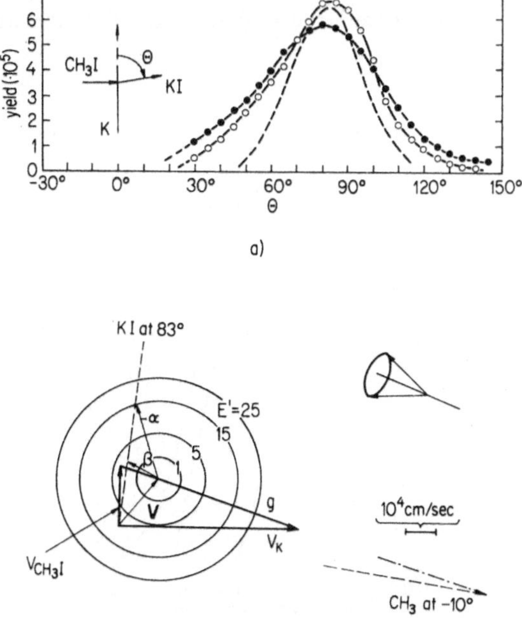

Fig. 22. a) Observed (points) and calculated (dashed line) angular distributions of KI formed in the reaction K + CH_3I (HERSCHBACH, 1962)
b) Velocity diagram showing two possible center of mass velocities α and β which can explain the observed angular distributions (HERSCHBACH, 1962)

198 J. P. TOENNIES

interpreted with the help of the velocity diagram shown in Fig. 22b).
From the diagram it is evident that two different velocity vectors α and β
both lead to the same laboratory scattering angle. Since the scattering of
the unpolarized molecules is symmetric with respect to $\vec{g}$, the possibility
indicated by α should lead to considerable out of plane scattering. This was
not observed experimentally, and thus the scattering angle and also the
average velocity of the products was unambiguously identified. In summary, then, the single observation of the angular location of the product
maximum enables us to draw three firm conclusions: 1. The product
peaking indicates that the reaction occurs in a time which is short compared
with the rotational period of the complex. 2. The direction of peaking
implies that the mechanism is of the rebound type. 3. The low velocity
of the KI means that most of the reaction energy (3/4) is converted into
internal excitation of the reactants. These properties are similar to those
found for the HBr reaction. Unfortunately, the results of similar experiments
for CH_3Br have not as yet been reported.

The elastic scattering of K from CH_3Br (ACKERMANN et al., 1963) and
from CH_3I (AIREY et al., 1967b) provides information on the energy
dependence of the reactive cross section. As in the case of HBr there is a
distinct drop off in intensity at large angles. Two differences are observable,

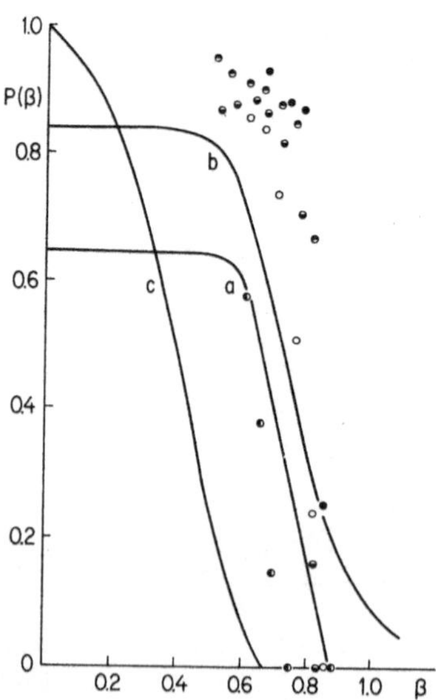

Fig. 23. The probability of reaction versus the reduced impact parameter at the
distance of closest approach for the reactants $K + CH_3I$ for five different relative
energies $E_{trans(i)}$ in kcal/mole: $\circ$: 1.42, $\circ$: 2.99, $\circ$: 4.77, $\circ$: 5.55, $\bullet$: 6.73
(AIREY et al. 1967b). The calculated curves a, b and c are for three different
potential energy surfaces (RAFF and KARPLUS, 1966)

however. First of all the wide angle scattering decreases with increasing energy and, secondly, the rainbow scattering although recognizable, is less distinct than in the case of HBr. The energy parameter ε was estimated from the rainbow position and from it the wide angle scattering cross sections could be calculated[15]. Probabilities of reaction were obtained as before using equation (15). The results for $K + CH_3I$ are shown in Fig. 23 and also show an increase in reactivity with increasing energy. Finally a total reactive cross section was determined by integrating over all impact parameters[16]. These were also found to have the same general energy dependence as for HBr (see Fig. 18). Another value of the total reactive cross section was reported by HERSCHBACH (1966) who integrated the measured differential reactive cross section, transformed into the center of mass coordinate system, over all angles. The table presented below shows that the two values are in reasonable agreement:

reaction	method	$\sigma_{\text{react}}\,[\text{Å}^2]$	$E_{\text{trans}\,(i)}\left[\dfrac{\text{kcal}}{\text{mole}}\right]$
$K + CH_3Br$	ESA	23[a]	2.0
$K + CH_3I$	PA	30[b]	6.3
$K + CH_3I$	ESA	50[c]	4.2

[a] GREENE, MOURSUND and ROSS (1966)   [b] HERSCHBACH (1966)
[c] AIREY et al. (1967 b).

Preliminary measurements on the reactive scattering of oriented CH_3I on an alkali atom (M) have recently been reported for $M = Rb$ by BUEHLER, BERNSTEIN and KRAMER (1966) and for $M = K$ by BROOKS and JONES (1966). The beam of oriented CH_3I was produced by state selection in an electric six pole field, which has been discussed in section 2.4.

Fig. 24 is a schematic diagram of the scattering region, showing the relative orientation of the electric field and of the atom beam, which comes from above. After leaving the six pole field the CH_3I molecules enter from behind the plane of the page and proceed to the right. An orienting electric field is provided in the region of scattering by an arrangement of four parallel rods. With adjacent rods at the same polarity a two pole field is formed in which the selected molecules are oriented. By switching the electrical polarity of the rods the orientation can be easily rotated by 180° without changing the geometry of the scattering region. Thus it is possible to compare cross sections for two relative orientations of the molecular axis with respect to the direction of approach of the atom. The two possible

---

[15] $P\,(\beta, K)$ is not very sensitive to the choice of ε since the wide angle elastic differential cross section is also nearly independent of ε (see Fig. 12).

[16] Since R_m could not be obtained from the rainbow scattering it had to be estimated.

relative orientations of the electric field and the relative velocity are shown in Fig. 24. The incomplete polarization of the primary beam and the angular and velocity distributions of $\vec{v_2}$ have to be taken into account in a fina

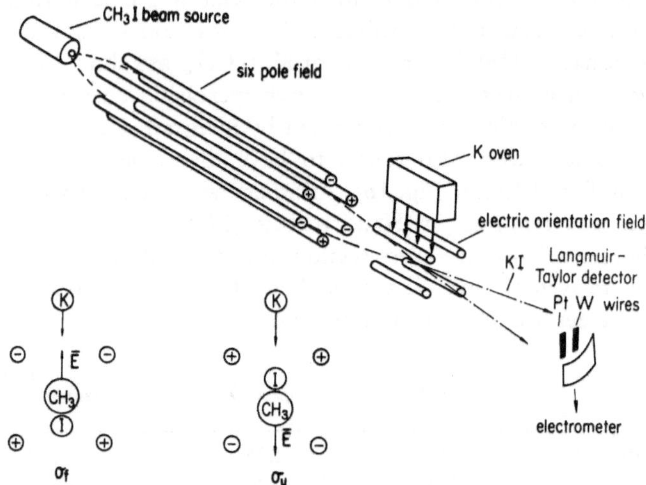

Fig. 24. Schematic drawing of the apparatus used for measuring the orientation dependence of the reactive cross section for $K + CH_3I$. The CH_3I molecules transmitted by the six pole field are polarized and then lined up in the orientation field where they are struck by K atoms. The reactive cross sections for the two relative orientations shown in the bottom left can be measured

interpretation of the results. The arrangement used is similar to one used earlier (BENNEWITZ et al. 1964) to measure the anisotropy in the potential of diatomic molecules in collisions with atoms.

The RI formed in the collision was detected by a differential Langmuir-Taylor detector placed at almost 90° to the Rb beam, where the maximum product intensity is to be expected. The preliminary results suggest that the reactive cross sections for the oriented reactions

$$Rb \rightarrow\leftarrow ICH_3, \; \sigma_f; \quad Rb \rightarrow\leftarrow CH_3I, \; \sigma_u$$

are in the ratio $\sigma_f/\sigma_u \geq 1.5$ as is to be expected if the reaction occurs in a time short compared with the rotational period of the complex. This experiment demonstrates nicely the type of detailed information which can be obtained by the molecular beam method.

4.2.3. Discussion

The experimental findings on these two systems are briefly:

1. The reaction cross sections are small and depend on the relative kinetic energy as $\sigma_{\text{react}} \sim \left(1 - \dfrac{E_{\text{act}}}{E}\right)$ where E_{act} is of the order of 0.5 kcal/mole.

2. The reaction probability increases (CH_3I) or is constant (HBr) with the relative kinetic energy.

3. The angular distribution of the products is strongly peaked in the backward direction (rebound reaction).

4. Most of the energy released goes into internal excitation of the products.

Result 3 suggests that the products "remember" the direction of approach of the reactants. From this it can be concluded that the collision complex has a lifetime which is short compared to its rotational period ($< 10^{-12}$ sec).

The majority of these results can be explained in terms of the central field impact model discussed in section 3.4. For this model to hold it must be assumed that, at energies above threshold, reaction occurs with a high probability at small impact parameters. Steric factors have the effect of reducing the total reaction probability, which is of course an average over all orientations. Another assumption which is not essential to the model but which enables the results from product analysis experiments to be linked with those from elastic scattering measurements is that the reaction occurs immediately upon impact. The central field model says nothing about the excitation of the products. These can either be deduced from the rules set up by EVANS and POLANYI (Section 3.1) or from exact classical calculations.

HBr and all reactions producing H atoms are a special case. For these reactions strict limits on the product excitation and the reaction probability as a function of relative energy are imposed by conservation of energy and angular momentum (HERSCHBACH, 1962, and BECK, GREENE and Ross, 1962). Conservation of energy requires that

$$E_{\text{trans}(i)} + E_{\text{rot(HBr)}} + E_{\text{vib(HBr)}} + \Delta E_0^0 = E_{\text{trans}(f)} + E_{\text{rot(KBr)}} + E_{\text{vib(KBr)}} \quad (27)$$

where (i) and (f) refer to initial and final conditions. At the low source temperature of the HBr beam it is not vibrationally excited and therefore we have $E_{\text{vib (HBr)}} = 0$. Conservation of angular momentum requires moreover that

$$\vec{P}_i \times \vec{b_i} + I_{\text{HBr}}\,\vec{\omega}_{\text{HBr}} = \vec{P}_f \times \vec{b_f} + I_{\text{KBr}}\,\vec{\omega}_{\text{KBr}} \quad (28)$$

with $\vec{P} = \mu \cdot \vec{g}$, I is the molecular moment of inertia and $\vec{\omega}$ is the angular velocity of rotation of the molecule. In the case of almost all reactions with the halogens we find $\vec{P}_i \times \vec{b_i} \geqslant 5\,I\,\omega_{\text{HX}}$ so that we can neglect the internal angular momentum of the reactant molecule. If hydrogen atoms are produced the reduced mass of the products is approximately 1, and since to a first approximation $b_f \approx b_i$, the orbital angular momentum of the products is negligible compared to the angular momentum of the KBr. Thus (28) reduces to

$$\vec{P}_i \times \vec{b_i} \simeq I_{\text{KBr}}\,\vec{\omega}_{\text{KBr}}$$

Substituting into (27) and eliminating b_i^2 by means of equation (12) we obtain

$$E_{\text{trans}(i)} \cong E_{\text{trans}(f)} + E_{\text{vib}(\text{KBr})} +$$

$$+ \frac{R_{\min}^2}{2 R_{\text{KBr}}^2} [E_{\text{trans}(i)} - V(R_{\min})] - \Delta E_0^0 \tag{29}$$

where $R_{\min}$ and $V(R_{\min})$ are the distance of closest approach and the potential energy at the reaction threshold. R_{KBr} is the internuclear distance of the highly vibrationally excited KBr molecule. The third term on the right can be interpreted as the energy converted into rotational excitation of the products. Equation (29) predicts, therefore, that with increasing initial kinetic energy the rotational energy increases proportionally and less energy becomes available for vibrational excitation. Thus in this case the partitioning of energy among the internal degrees of freedom is not affected by the nature of the potential energy hypersurface.

The restrictions imposed by the conservation equations can now be used to explain the differences in the energy dependence of the reaction probability in the two systems HBr and $CH_3I(Br)$. The recent data shown in Fig. 20 indicate that the HBr reaction probability is nearly constant with energy, and, in any case, does not show as great an increase in reaction probability as observed with $CH_3I(Br)$. An explanation is provided by considerations (GREENE, MOURSUND and ROSS, 1966) based on a statistical theory of chemical reaction similar to that of LIGHT (1964). In these theories it is assumed that the probability of formation of any given product in a "strong coupling" collision is proportional to the phase space available to the product. Thus in the HBr reaction the vibrational part of the product phase space is increasingly restricted by the conservation equations at higher energies and the reaction probability decreases accordingly. Of course, the very short reaction time makes the assumption of strong coupling questionable, but there is some evidence that some of the conclusions of the phase space theory concerning the products may nevertheless still apply. For instance, PECHUKAS, LIGHT and RANKIN (1966) have shown that the theory gives results for vibrational excitation cross sections which are in reasonable agreement with the more rigorous quantum mechanical model calculations. Their calculations on K + HBr lead to conclusions which are essentially in agreement with those presented here.

Additional evidence for the statistical model is provided by similar studies of the K + HCl and HI reaction probabilities. These increase sharply once the interaction energy $V(R_{\min})$ is greater than the threshold for vibrational excitation of the products. Also of interest in this connection is the direct observation of high rotational excitation of the KBr formed in the reaction Cs + HBr (HERSCHBACH, 1966 p. 356) which is also predicted by the statistical model (PECHUKAS, LIGHT and RANKIN, 1966).

We now turn our attention to the $K + CH_3I(Br)$ reaction. Its understanding is more complicated for two reasons: For one, the orbital angular momentum of the product cannot be neglected as with HBr. Denoting the orbital angular momentum by $\vec{L}$ and the rotational angular momentum by $\vec{J}$ (L and J being the absolute values), conservation of angular momentum now requires

$$L_i^2 \simeq J_{KBr}^2 + L_f^2 - 2 J_{KBr} L_f \cos\alpha \tag{30}$$

where α is the angle between $\vec{J}_{KBr}$[17] and $\vec{L}_f$. The impact parameter associated with $\vec{L}_f$ can be estimated from an impact theory (HERSCHBACH, 1966, p. 363), but $\cos\alpha$ or $\vec{J}_{KBr}$ can only be obtained from a calculation which takes account of the nature of the potential energy surface. Such calculations do indicate that $\cos\alpha = 1$ (parallel vectors), but it would appear unlikely that $\cos\alpha = -1$ (antiparallel) if the reaction time is short. Since there are now no obvious restrictions on the internal energy of the product molecule as in the HBr reaction, the $CH_3I(Br)$ reaction can proceed in accordance with the simple statistical theory which predicts an increase in reaction probability with increasing energy. The other important difference comes about because of the highly anisotropic structure of the symmetric top molecule (e.g. for CH_3Br $l_{C\text{-}Br} = 1.9\text{Å}$, $l_{C\text{-}H} = 1.1\text{ Å}$ and $> HCH = 111°$). The interaction potential is therefore also expected to be highly anisotropic. Furthermore because of its symmetric top structure and the close spacing of the rotational levels, the inelastic cross section for rotational excitation can be expected to be considerably larger (TOENNIES, 1966) than with HBr. This is a possible explanation for the observed reduction in the elastic cross section down to angles in the vicinity of the rainbow angle (AIREY et al., 1967b). The reduction cannot be explained in terms of a larger reaction cross section. The results 1, 2, and 3 can nevertheless be explained in terms of the simple central field impact theory and the statistical theory discussed above.

Stimulated by the large amount of experimental data available for this reaction extensive classical calculations of collision trajectories have been carried out to simulate the reactive collision (BUNKER, 1964; HEYDTMANN, 1968). To initiate the trajectory calculations it is necessary to assume a potential energy surface which, as indicated by property 4 and the rule of EVANS and POLANYI, shows a strong attraction between reactants. Early calculations yielded integral reaction cross sections which were more than an order of magnitude larger than the experimental values (KARPLUS and RAFF, 1964). The potential model was then modified to reduce the range of the attraction between the reactants and in this way the integral cross sections were made to agree with the measured values. In the most

---

[17] The angular momentum of the metylbromide molecule has been neglected.

recent work of this type (RAFF and KARPLUS, 1966) the rotational and vibrational degrees of freedom of the reactants[18], and a large number of different velocities and directions of approach in three dimensions have been considered. Altogether four potentials, all with a strong attraction of the reactants at a distance of approach of about 4 Å, but differing in shape details were used in the calculations. An example is given in Fig. 25.

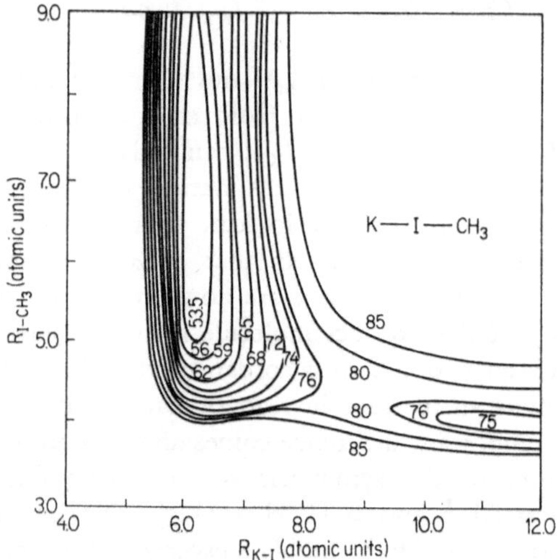

Fig. 25. Potential energy hypersurface for linear K—I—CH$_3$. Energy contours are in kilocalories (RAFF and KARPLUS, 1966)

The results for the four potentials predict: a) reaction cross sections in the range $13-36$ Å^2, b) backward peaking of the product in the center of mass system, c) short reaction times, and d) efficient conversion of the heat of reaction into internal degrees of freedom. These theoretical conclusions are in essential agreement with the experimental results. The calculated probability of reaction curves for three of the potential models are shown in Fig. 23. These are also in rough agreement with the experiments. The agreement is even better if we take account of the fact that inelastic processes lead to a reduction in the measured values.

An important result of these recent calculations is that most of the reaction properties including some not mentioned above are fairly insenfitive to details in the potential model. An exception is the absolute value of the reactive cross section, which depends directly on the distance or range at which the strong attraction occurs. Unfortunately as we have already pointed out, the reactive cross section is difficult to measure

---

[18] CH$_3$ is, however, still regarded as a point particle and rotation about the symmetry axis is neglected.

precisely. Other properties which appear to depend on the size and shape of the potential energy surface are the ratio of the internal product angular momentum $\vec{J}$ to the orbital angular momentum and also the polarization of angular momentum of the product. The ratio of the angular momenta cannot be measured directly, but the distribution among individual rotational quantum numbers $P(J)$ can be determined, at least in principle, using an electric four pole field as a rotational state selector. HERSCHBACH (1961) has also suggested the use of a Stern-Gerlach field to measure the polarization of the product angular momentum.

4.3. Stripping Reactions

4.3.1. The Reaction $K + Br_2 \rightarrow Br + KBr$

Although reactions of this type were known from the early sodium flame experiments to have very large cross sections and small activation energies, molecular beam techniques have only recently been used successfully to study these reactions. Early attempts at studying this reaction were frustated by a poisoning of the hot wire detector by the halogen gas. With the use of the TOUW and TRISCHKA technique (see section 2.5.) this is usually no longer a problem.

The first experiments on this reaction were reported almost simultaneously by the groups at Oak Ridge (DATZ and MINTURN, 1964) and Harvard (WILSON et al. 1964). Fig. 26a) shows the measured angular distribution of KBr in the laboratory system. Fig. 26b) shows the associated velocity diagram with circles corresponding to different center of mass translational energies of the products. These experimental results have been transformed into the center of mass system by the methods discussed in section 3.5. (HERSCHBACH 1966). A fixed velocity of both reactant beams is assumed and with this approximation the angular distribution and the average translational energy of the products in the center of mass system is estimated. The center of mass system angular distributions are strongly peaked in the forward direction and have a width at half intensity of 60° (BIRELY and HERSCHBACH, 1966). The center of mass curves can be integrated over all angles (assuming cylindrical symmetry about $\vec{g}$) and from the densities in the two beams the total reaction cross section was estimated. Similar estimates have been made by MINTURN, DATZ and BECKER. Both results are summarized below:

reaction	σ_{react} [Å²]	$E_{trans(i)} \left[\dfrac{kcal}{mole} \right]$
$K + Br_2$	210[a], 100[b]	3.6[a]
$Cs + Br_2$	100[c]	0.5—2.0[c]

[a] HERSCHBACH (1966).   [b] MINTURN, DATZ and BECKER (1966).
[c] DATZ and MINTURN (1964).

206 J. P. Toennies

In similar experiments on the system Cs + Br_2 with a velocity selected Cs beam, Datz and Minturn (1964) observed a shift in the product distribution to smaller angles at higher relative kinetic energies. The observed shift can be explained if the products have low translational energy

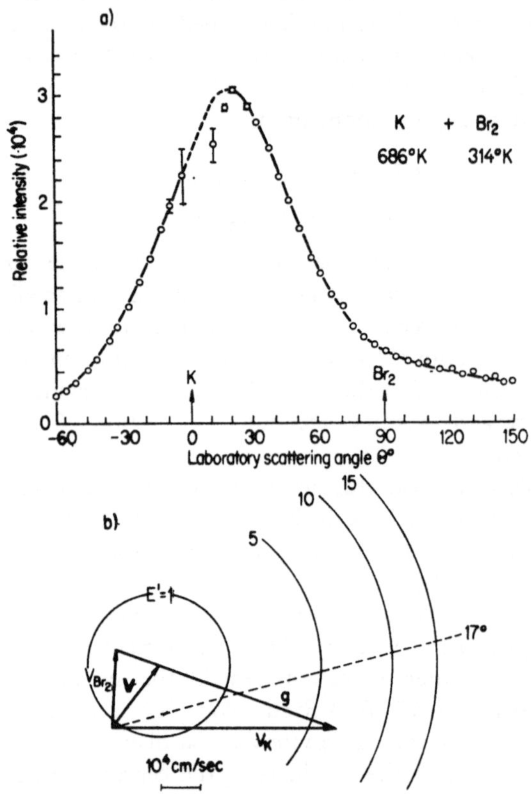

Fig. 26. Measured laboratory angular distribution of KBr from the K + Br_2 reaction (measured in the plane of the incident beams). The lower panel b) gives the kinematic diagram corresponding to the most probable velocities in the reactant beams; the circles indicate the length of the recoil velocity for KBr with various values of E_{trans} (f) (Herschbach, 1966)

and are scattered only in the forward direction (Herschbach, 1966, p. 333). The large cross sections and the strong peaking in the forward direction found in these experiments imply not only a stripping reaction but also a short lived intermediate state.

Additional evidence for the reaction occurring at large impact parameters comes from measurements of elastic scattering (Datz, Minturn and Becker, 1966, also see Herschbach, 1966). Fig. 27 shows differential elastic cross sections measured for $K + Br_2$ ($K = 6.5$) and for $K + Xe$ ($K = 5$). Since the reduced energies are nearly the same, the rainbow if present should appear at about the same angle for the two systems. The

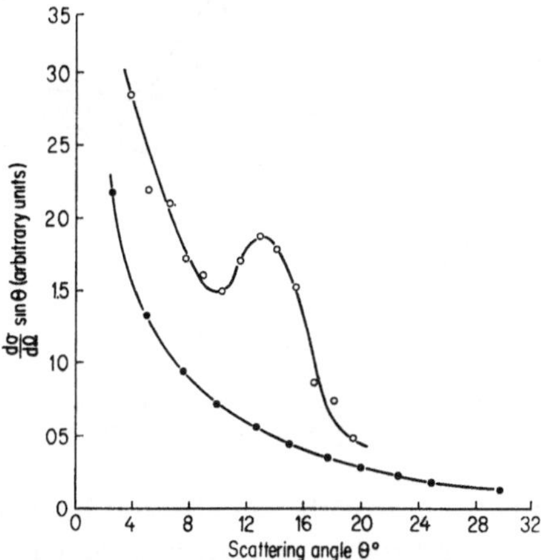

Fig. 27. Measured elastic differential scattering cross section in arbitrary units for non-reactive scattering of $K + Xe$ and reactive scattering $K + Br_2$. Since the K-value for both systems is about the same the Br_2 rainbow should appear at about the same angle. Its absence is evidence for reaction at large impact parameters (MINTURN, DATZ, and BECKER, 1966)

● : $K + Br_2$ for $K = 6.5$ ○ : $K + Xe$ for $K = 5$

data however show no rainbow in the scattering of K from Br_2. The disappearance of the rainbow cannot be explained by rotational excitation since the inelastic cross sections are not expected to be large for the non-polar Br_2 molecule. It appears reasonable to assume that the vanishing of the rainbow is due to reactive scattering. Thus the estimated impact parameter for rainbow scattering leads to a lower limit of 80 Å² for the reaction cross section.

The large product intensities available with this reaction have made it possible to measure the velocity distribution of the product and even its rotational energy distribution. In one of the first investigations of this type (GROSSER and BERNSTEIN, 1965) the apparatus shown in Fig. 21b was used to measure the velocity distribution of the KBr beam. Fig. 28 shows a kinematic diagram with a typical velocity distribution. The distribution is relatively broad with a maximum at low velocities of about 500 m/sec. The velocities and energies at the tails of the distribution are of the order of 200 m/sec (0.57 kcal/mole) and 1200 m/sec (21 kcal/mole).

Similar experiments have also been carried out by BIRELY and HERSCH-BACH (1966), but without a velocity selector in the primary beam. Their results are in good agreement with those from experiments with a velocity selected primary beam, as is to be expected for the broad distributions of final velocities.

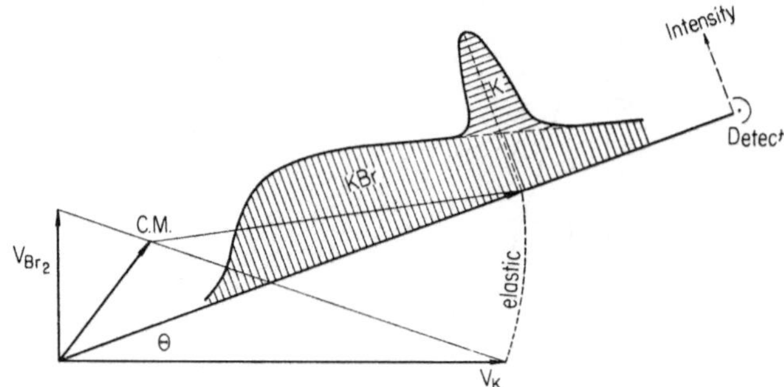

Fig. 28. Schematic velocity vector diagram for in-plane scattering of K + Br$_2$ (Grosser and Bernstein, 1965)

If the assumption is made that the angular and internal energy distributions are uncorrelated, the product velocity distributions can be analyzed to obtain information on the internal excitation of the product molecule. Conservation of energy requires

$$E_{int\,(KBr)} = E_{trans\,(i)} + E_{int\,(Br_2)} + \Delta E_0^0 - E_{trans\,(f)} \qquad (31)$$

In a first approximation the first two terms on the right can be neglected. They equal 0.5 and 1.2 kcal/mole respectively in a typical case. Thus the value of $E_{int\,(KBr)}$ is given for each value of $E_{trans\,(f)}$. The transformation from center of mass to laboratory velocities was carried out by Bernstein's group (Warnock, Bernstein and Grosser, 1967) using a computer simulation technique. They introduce a parameter for the fraction of final energy in the internal degrees of freedom defined by

$$F_{int} = \frac{E_{int\,(KBr)}}{\Delta E_0^0 + \overline{E}_{rot\,(Br_2)} + \overline{E}_{trans\,(i)}} \qquad (32)$$

Thus for a given v_{Br_2} (since the potassium beam was velocity selected v_K was fixed), scattering angle ϑ, and F_{int} a point in the laboratory phase space with coordinates v_{KBr}, Θ was calculated using the transformation equations (24). Each point received a probability weighting factor proportional to the product $P(v_{Br_2}) \cdot P(\vartheta) \cdot P(F_{int})$, where $P(v_{Br_2})$ is the Maxwell velocity distribution. This procedure was repeated for all possible values of v_{Br_2}, ϑ, and F_{int}. The laboratory phase space was then divided up into small segments and the sum of the weighting factors of all points in the segments were added up to give a distribution in laboratory space. From this the laboratory velocity distribution could be easily calculated. The process was repeated until a best fit between $P(\vartheta)$ and $P(F_{int})$ and measured distributions was obtained. The results are shown with the error range in Fig. 29. BIRELY and HERSCHBACH (1966) carried out the same transformation using essentially a graphical technique.

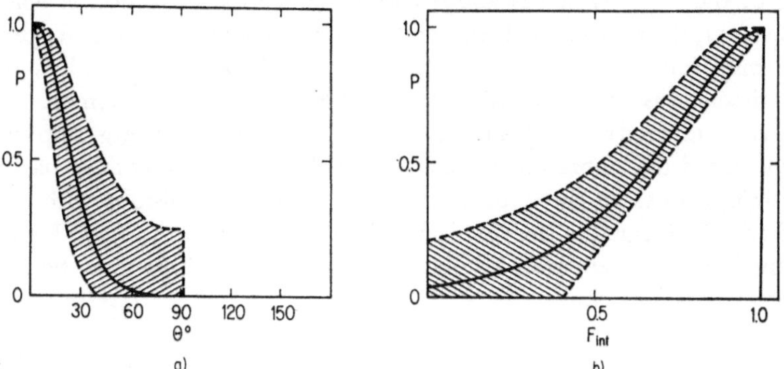

Fig. 29. Probability distribution curves obtained from measurements of the KBr velocity distribution for the reaction K + Br$_2$. a) Probability of scattering for various angles (differential reactive cross section). b) The probability of scattering for various values of F_{int} (see text) (WARNOCK, BERNSTEIN and GROSSER, 1967)

Of particular interest for the theoretical interpretation of the experiments, however, is a knowledge of the partitioning of energy among the vibrational and rotational degrees of freedom. Using an electric Stern-Gerlach field of the type described in section 2.4., HERM and HERSCHBACH (1965) were able to make some preliminary measurements of the rotational energy distribution. Fig. 30 shows a comparison of deflection distributions

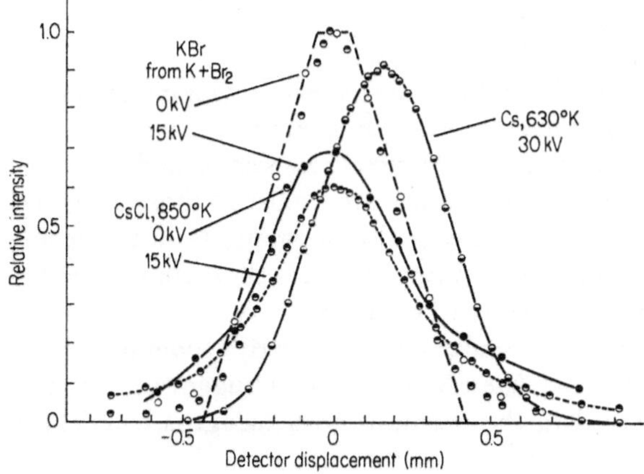

Fig. 30. Measured deflection patterns for different beams in a Stern-Gerlach electric field (HERM and HERSCHBACH, 1965). By comparing the deflection distribution of KBr formed in the reaction with that of CsCl from a thermal oven the rotational energy distribution of the KBr can be estimated

for a CsCl ($\mu = 10.5$ D) beam from a thermal oven and for a KBr ($\mu = 10.4$ D)[19] product beam. From the similarity in the deflection pattern it appears that the KBr has a rotational temperature roughly equivalent to that of the CsCl, which is about 850 °K. Since the deflection is velocity dependent, the translational temperature must be known for a more precise determination. A velocity distribution which had been measured in another experiment was used to obtain a value of $E_{\mathrm{rot(KBr)}} \sim 2.2$ kcal/mole which is small compared to ΔE_0^0. The remainder of the reaction energy apparently is converted into vibrational excitation of the KBr[20]. Recently it has been possible to observe the optical excitation of Na by collisions with highly excited KBr formed in the reaction of K with Br_2 (MOULTON and HERSCH-BACH, 1966). The energies involved indicate that the KBr must be highly excited vibrationally.

4.3.2. Discussion

The experiments indicate that the $K + Br_2$ reaction has the following properties:

1. The reaction cross section is unusually large (100–250 Å²) and is independent of the mass of the alkali atom.

2. The angular distribution is strongly peaked in the forward direction with width of 60°.

3. The product velocity distribution is quite broad with a maximum of 500 m/sec at $\Theta = 10°$ which shifts to smaller velocities at larger laboratory angles.

4. The product velocity distribution is insensitive to the initial velocity of the potassium atom.

5. The mean rotational excitation is small.

6. From 3. and 5. it follows finally that the vibrational energy distribution is strongly peaked near the reaction energy.

With the exception of the larger cross section and predominant forward scattering the reaction properties are similar to those of the HBr and CH_3I reactions. The strong peaking also suggests, as in the rebound reaction, that the reaction occurs in a time which is short compared to the rotational period of a possible activated complex.

In attempting to understand the early sodium flame experiments POLANYI (1932) and later MAGEE (1940) suggested that the large cross sections observed in the reactions with halogens could be explained by

---

[19] If the molecules are vibrationally excited as other experiments mentioned below suggest, then the molecules will have considerably larger dipole moments.

[20] So far there is no evidence for the formation of excited halogen atoms, which could also absorb a part of the reaction energy.

assuming an intermediate step involving the transfer of an electron and the formation of an ion pair:

$$K + Br_2 \;\rightarrow\; K^+ + Br_2^- \;\rightarrow\; K^+Br^- + Br$$

The mechanism was called harpooning by POLANYI. The idea has been recently taken up and used for calculating some of the properties discussed above (MINTURN, DATZ and BECKER, 1966; and HERSCHBACH, 1966).

That such a process can occur is well known from the spectroscopically measured potential energy curves for the alkali halides (see Fig. 31). The attractive portion can be explained by a large contribution of the ionic

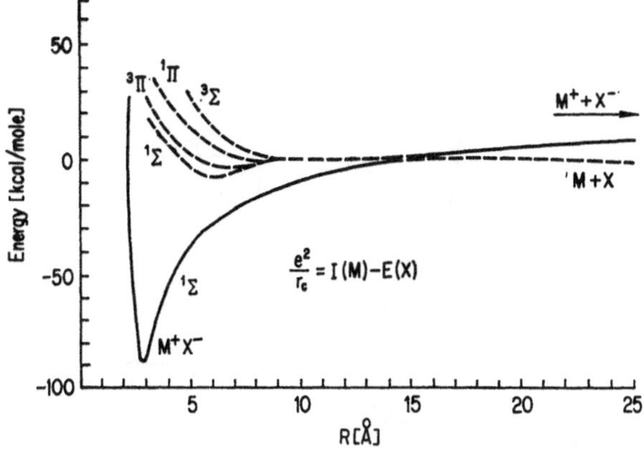

Fig. 31. Potential energy curve for the interaction of K and Br. The dashed curves correspond to the covalent states and the solid curve to the ionic state. Only states with the same symmetry and angular momentum (e.g. $^1\Sigma$) do not cross as shown at R_c. Atoms forming these states are expected to produce molecules in the ionic configuration (MAGEE, 1940)

configuration K^+Br^- for R near the equilibrium distance, whereas the magnitude of the dissociation energies indicate that the molecules dissociate to neutral atoms. It is well known that configuration interaction prevents the curves from crossing (non-crossing rule) for two configurations having the same symmetry. Thus in the adiabatic approach of two neutral atoms these become ionized at a distance R_c given by

$$\frac{e^2}{R_c} = I_{(K)} - \mathscr{E}_{(Br)}$$

where $I_{(K)}$ is the ionization potential of K (= 4.3 eV) and $\mathscr{E}_{(Br)}$ is the electron affinity of the bromine atom (= 3.36 eV). In this case $R_c = 15$ Å. Of course in fast collisions transitions from one curve to the other can occur and their probability can be calculated in a first approximation from the Landau-Zener formula.

In the case of $K + Br_2$ information on the potential energy curves is not available but a calculation of R_c suggests that a similar situation should hold here as well (MAGEE, 1940). Since we are now dealing with a three body problem involving a ^2S and $^1\Sigma$ molecule angular momentum and symmetry restrictions are less stringent. Except for a few special arrangements (e.g. all three atoms in one line) the covalent and ionic potential surfaces can cross and radiationless transitions from one to the other can occur with a high probability. There is however some question concerning the correct value of the electron affinity to be used in the case of a halogen molecule. Fig. 32 shows the experimentally measured potential curve for a neutral

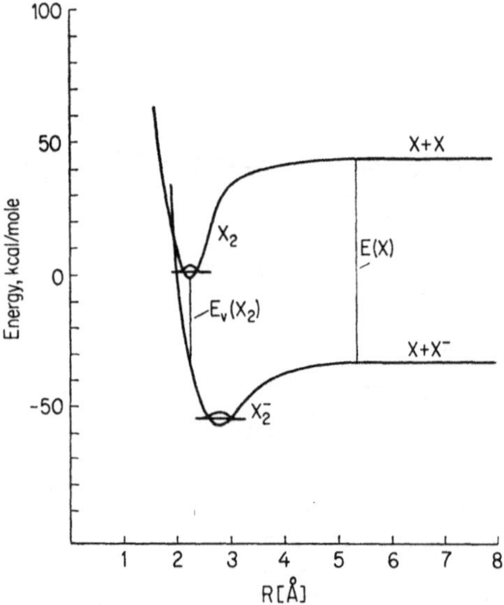

Fig. 32. Potential energy curves for Br_2 and Br_2^- showing two possible values of the electron affinity

Br_2 molecule and the postulated curve for an ionized Br_2^- molecule as well as the definitions of the various electron affinities. A number of arguments based on the short times involved can, however, be presented in favor of the vertical electron affinity[21]

$$\frac{e^2}{R_c} = I_{(K)} - \mathscr{E}_{v(Br_2)}$$

where $\mathscr{E}_{v(Br_2)} = 1.1 \pm 0.5$ eV. This formula yields $R_c = 4.5 \pm 0.6$ Å[22].

---

[21] This choice is also favored by consideration of the Franck-Condon factor.

[22] DATZ, MINTURN, BECKER (p. 24) list R_c and calculated cross sections for vertical and adiabatic electron affinities.

From the value of R_c a crude estimate of the reaction cross section can be made:

$$\sigma_{react} \lesssim \pi R_c^2 = 64\,\text{Å}^2$$

Furthermore, we can now postulate a potential energy hypersurface. Fig. 33 shows two sections through the postulated hypersurface. Fig. 33a

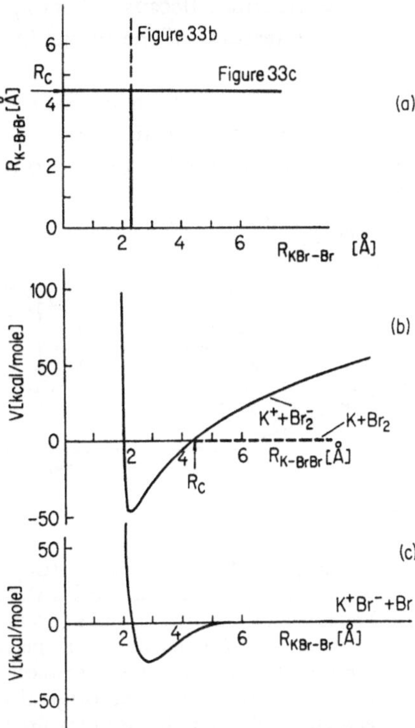

Fig. 33. Two sections of the K + Br$_2$ potential energy hypersurface. Part a) shows the locations of the sections in the usual representation of the hypersurface for a collinear collision. The potential coordinate is perpendicular to the plane of the page. Part b) shows *postulated* potential energy curves for K + Br$_2$ for the ionic configuration. Part c) shows the potential energy curve for the Br$_2^-$ ion. In the spectator stripping model this is assumed to be the potential curve for the KBr—Br coordinate in the reaction

shows the location of the two sections. Fig. 33b shows the postulated potential for the K-Br$_2$ coordinate (R_{KBr-Br} is fixed) and Fig. 33c shows the potential for the KBr-Br coordinate (R_{K-Br_2} is fixed). The latter is the same as the bottom curve in Fig. 32.

The partitioning of angular and linear momentum and the reaction energy among the various degrees of freedom after the formation of the two ionized atoms cannot be simply calculated from the conservation equation, and additional assumptions are required. The well defined distance R_c and the sudden large increase in attractive potential between two of the atoms suggests however that a simple model may be successful in this case. The spectator stripping model which has been worked out in considerable detail by Minturn, Datz and Taylor (1966) and has been discussed critically by Herschbach (1966), is quite successful in explaining many of the features of the reaction. Because of its appealing conceptual simplicity we present here the features of the model in a slightly modified version.

Fig. 34 is a velocity vector diagram showing the sequence of events in the center of mass system during the reaction according to the spectator stripping model. The distances are roughly drawn to scale. In the typical

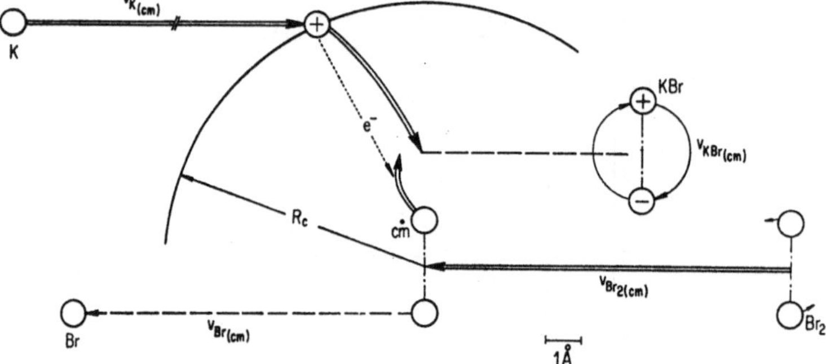

Fig. 34. Schematic velocity diagram for a spectator stripping reaction model. The sequence of events is as follows. A K atom (from the left) and a bromine molecule (from the right) approach the center of mass (point c.m.). At a distance R_c between the centers of mass electron transfer takes place. The K^+ and Br^- ions attract each other forming a highly excited K^+Br^- molecule. The remaining Br atom acts as a spectator and continues in the original direction of the Br_2 molecule with the same velocity. All velocities are measured with respect to the center of mass. The velocities chosen for the example discussed in the text are

$$\left. \begin{array}{l} v_{K\ cm}\ \ = 640\ \text{m/sec} \\ v_{Br_2\ cm} = 160\ \text{m/sec} \end{array} \right\} \text{before collision}$$

$$\left. \begin{array}{l} v_{KBr\ cm} = 107\ \text{m/sec} \\ v_{Br\ cm}\ \ = 160\ \text{m/sec} \end{array} \right\} \text{after collision}$$

example chosen, a potassium atom enters from the upper left with a velocity $v_{cm} = 640$ m/sec. A slowly rotating $(\bar{\tau}_{rot} \sim 10^{-11}$ sec) bromine molecule with a velocity $v_{cm} = 160$ m/sec enters from the bottom right. When the distance between the two is equal to R_c an electron jumps from the atom to the molecule. As a result of the electron transfer the Br_2 ion is in a highly excited, barely stable state (see Fig. 32). The strong electric field originating at the K^+ ion is sufficient to provide whatever remaining energy is necessary for complete dissociation and this occurs in a time less than 10^{-14} sec. Due to the strong coulomb forces the ions K^+ and Br_2^- then approach each other quickly and arrive to within 3 Å in 10^{-13} sec. The Br atom which is left behind after dissociation is not subject to strong forces and is therefore regarded as a "spectator" to the interaction of the ions. The momentum of the KBr molecule after collision is given simply by conservation of momentum in the center of mass system

$$m_{Br}\, \vec{v}_{Br\ cm} = -\, m_{KBr}\, \vec{v}_{KBr\ cm}$$

($\dot{\vec{\mathscr{P}}} = 0$; equation (20)) and $\vec{v}_{KBr}$ equals 107 m/sec for the example used here. The internal (vibrational and rotational) momentum of the Br_2 molecule also makes a contribution to $\vec{v}_{KBr}$ but this is negligible. We can therefore estimate the orbital angular momentum of the products from the velocities and impact parameters. The result is $L_f = 150\, ℏ$. Substracting this from the initial angular momentum yields $300\, ℏ$ for the rotational angular momentum of the KBr molecule[23] corresponding to half of the reaction energy or 21 kcal/mole in rotational excitation. Conservation of energy requires that the rest of the available energy, 24 kcal/mole, be used for vibrational excitation. Thus if all interactions between the departing products are neglected, the spectator-stripping model predicts property 2 above, while properties 1 and 4 are direct consequences of the electron transfer mechanism. The predicted velocity of the KBr is, on the other hand, considerably less than that observed, property 3, while the rotational angular momentum is too large, property 5. We note, however, that an increase in the KBr velocity by a factor of 5, to bring it in agreement with the observation, is sufficient to absorb most of the rotational angular momentum. This might be explained by the Br^- ion imparting some of its momentum to the Br atom (corresponding to a slight shift of the X_2^- curve to the right in Fig. 32) just after reaction resulting in rotational deexcitation of the KBr molecule.

In summary the simple spectator-stripping model explains the forward peaking of the products and presumably describes the sequence of events during the reactive collision as far as this is possible with classical mechanics. On the other hand it does not provide a satisfactory explanation of the internal energy distribution of the product nor does it fully explain the large cross sections. Apparently more elaborate classical calculations will be necessary to provide a more satisfactory explanation of all the properties of this reaction[24].

The success of the electron transfer model in explaining the $K + Br_2$ reaction raises the question as to the applicability of this model to the reac-

---

[23] assuming a planar collision and $\cos \alpha = -1$ (equation (30)).

[24] Such calculations are underway. See footnote on page 1212 of RAFF and KARPLUS (1966).

tions of K with HBr and CH_3I (see HERSCHBACH, 1966). In agreement with the observed smaller cross sections the R_c calculated from the estimated electronegativities of HBr and CH_3I are considerably smaller than for $K + Br_2$. On the other hand the deformation of the molecular wave functions, which is expected at the close distances of approach, could however lead to large errors in the estimates leading to the R_c. A partial answer to the above question is provided by the potential hypersurface for $K + CH_3I$ shown in Fig. 25. We recall that this potential was essentially the result of a comparison between calculated reaction properties for various potential models and the molecular beam results. The sudden steep drop in the potential is entirely consistent with electron transfer at about 4 Å (≈ 8 atomic units) and, surprisingly, this distance is in satisfactory agreement with the estimated R_c for this system. There is therefore considerable evidence for the conclusion that electron transfer occurs in most of the molecular beam reactions studied so far. Indeed the low ionization potential which makes the alkali beams easy to detect favors their participation in an electron transfer reaction.

What is the situation with regards to systems in which alkali atoms are not involved? Since most atoms and molecules have considerably greater ionization potentials it appears unlikely that the same mechanism and reaction properties hold for most of the other chemical reactions. Essentially because of the detection problem their exploration requires considerably more sophisticated apparatus. In principle, however, there are no basic experimental difficulties so that there is reason to believe that results will be forthcoming in the next few years.

Acknowledgements

The author gratefully acknowledges the many valuable suggestions and comments made by Prof. E. F. GREENE and Dr. A. L. MOURSUND and Prof. R. B. BERNSTEIN and his research group.

References

ACKERMAN, M., E. F. GREENE, A. L. MOURSUND, and J. Ross: Proc. 9th. Intern. Symp. Combustion, Cornell Univ., 1962; New York: Academic Press 1963.

— — — — J. Chem. Phys. **41**, 1183 (1964).

AIREY, J. R., E. F. GREENE, K. KODERA, G. P. RECK, and J. Ross: J. Chem. Phys. **46**, 3295 (1967a).

— —, G. P. RECK, and J. Ross: J. Chem. Phys. **46**, 3287 (1967b).

ANDERSON, J. B., R. P. ANDRES, and J. B. FENN: in Adv. Atomic and Mol. Phys., Vol. 1., New York: Academic Press 1965.

— — — Proc. 28th. Propulsion and Energetics Panel on Sel. Topics in Aerothermochem. (AGARD), Oslo, in press (1966).

—, and J. B. FENN: The Physics of Fluids **8**, 780 (1965).

BECK, D., E. F. GREENE, and J. ROSS: J. Chem. Phys. **37**, 2895 (1962).
BENNEWITZ, H. G., K. H. KRAMER, W. PAUL, and J. P. TOENNIES: Z. Physik **177**, 84 (1964).
BERNSTEIN, R. B.: Adv. Chem. Phys. **10**, 75 (1966).
—, and J. T. MUCKERMAN in Adv. Chem. Phys. **12**, in preparation (1967).
BIRELY, J. H., and D. R. HERSCHBACH: J. Chem. Phys. **44**, 1690 (1966).
BROOKS, P. R., and E. M. JONES: J. Chem. Phys. **45**, 3449 (1966).
BUNKER, D. L.: Scientific American, July, 100 (1964).
BEUHLER, R. J., JR., R. B. BERNSTEIN, and K. H. KRAMER: J. Am. chem. Soc. **88**, 5331 (1966).
BULL, T. H., and P. B. MOON: Discussions Faraday Soc. **17**, 54 (1954).
DATZ, S., D. R. HERSCHBACH, and E. H. TAYLOR: J. Chem. Phys. **35**, 1549 (1961).
—, and R. E. MINTURN: J. Chem. Phys. **41**, 1153 (1964).
—, and E. H. TAYLOR: In "Recent Research in Molecular Beams" (Estermann I., ed.) p. 157. New York: Academic Press 1959.
— — J. Chem. Phys. **39**, 1896 (1963).
ERLEWEIN, W., M. von SEGGERN, and J. P. TOENNIES: Z. Physik, in press (1967).
EVANS, M. G., and M. POLANYI: Trans. Faraday Soc. **35**, 178 (1939).
FITE, W. L., and R. T. BRACKMANN: Proc. 3rd Intern. Conf. Phys. Electron. and At. Collisions, London, p. 955. Amsterdam: North-Holland Publ. 1963.
FORD, K. W., and J. A. WHEELER: Ann. Phys. **7**, 259 and 287 (1959).
FRISTROM, R. M.: Private communication (1966).
GREENE, E. F., A. L. MOURSUND, and J. ROSS: Adv. Chem. Phys. **10**, 135 (1966).
—, R. W. ROBERTS, and J. ROSS: J. Chem. Phys. **32**, 940 (1960).
—, E. F., and M. H. LAU: Private communication (1964).
GROSSER, A. E., and R. B. BERNSTEIN: J. Chem. Phys. **93**, 1140 (1965).
—, A. E., A. R. BLYTHE, and R. B. BERNSTEIN: J. Chem. Phys. **42**, 1268 (1965).
HELBING, R.: J. Chem. Phys., in press (1967).
HERM, R. R., and D. R. HERSCHBACH: J. Chem. Phys. **43**, 2139 (1965).
HERSCHBACH, D. R.: Vortex **22**, 348 (1961).
— Discussions Faraday Soc. **33**, 149 (1962).
— Appl. Opt. Supp. **2**, 128 (1965).
— Adv. Chem. Phys. **10**, 319 (1966).
—, G. H. KWEI, and J. A. NORRIS: J. Chem. Phys. **34**, 1842 (1961).
HEYDTMANN, H.: Monte-Carlo-Rechnungen in der chemischen Kinetik, dieses Buch.
HOLLSTEIN, M., and H. PAULY: Z. Physik **196**, 353 (1966).
HUNDHAUSEN, E., and H. PAULY: Z. Physik **187**, 305 (1965).
KARPLUS, M., and L. M. RAFF: J. Chem. Phys. **41**, 1267 (1964).
KENNARD, E. M.: Kinetic Theory of Gases. New York: McGraw-Hill 1938.
KING, J. G., and J. R. ZACHARIAS: Adv. Electron. Phys. **8**, 1—85 (1956).
KINSEY, J. C., and L. R. MARTIN: Private Communication (1966).
KRAMER, K. H., and R. B. BERNSTEIN: J. Chem. Phys. **42**, 767 (1965).
KRAUSS, M., and F. H. MIES: J. Chem. Phys. **42**, 2703 (1965).
LIGHT, J. C.: J. Chem. Phys. **40**, 3221 (1964).
MAGEE, J. C.: J. Chem. Phys. **8**, 687 (1940).
MARTIN, H., and H. J. MEYER: Z. Elektrochemie **56**, 740 (1952); Naturwiss. **39**, 85 (1952).
MINTURN, R. E., S. DATZ, and R. L. BECKER: J. Chem. Phys. **44**, 1149 (1966).
MORSE, F. A., and R. B. BERNSTEIN: J. Chem. Phys. **37**, 2019 (1962).
MOULTON, M. G., and D. R. HERSCHBACH: J. Chem. Phys. **44**, 3010 (1966).
PAULY, H., and J. P. TOENNIES: in Adv. Atomic and Mol. Phys., Vol. 1., New York: Academic Press 1965.

— — in Methods of Experimental Physics, Vol. **8**, in press (1968).

Pechukas, P., J. C. Light, and C. Rankin: J. Chem. Phys. **44**, 794 (1966).

Polanyi, M.: Atomic Reactions. London: Williams and Norgate, Ltd. 1932.

Politiek, J., J. Los, and J. J. M. Schipper: Proc. 28th. Propulsion and Energetics Panel on Sel. Topics in Aerothermochem. (AGARD), Oslo, in press 1966.

Raff, L. M., and M. Karplus: J. Chem. Phys. **44**, 1212 (1966).

Rosenfeld, J. L. J., and J. Ross: J. Chem. Phys. **44**, 188 (1966).

Russek, A.: Phys. Rev. **120**, 1536 (1960).

Taylor, E. H., and S. Datz: J. Chem. Phys. **23**, 1711 (1955).

Toennies, J. P.: Z. Physik **182**, 257 (1965).

— Z. Physik **193**, 76 (1966).

Touw, T. R., and J. W. Trischka: J. Appl. Phys. **34**, 3635 (1963).

Utterback, N. G.: J. Chem. Phys. **44**, 2540 (1966).

—, and G. H. Miller: Rev. Sci. Instrum. **32**, 1101 (1961).

Warnock, T. T., R. B. Bernstein, and A. E. Grosser: J. Chem. Phys. (1967) in press.

Wilson, K. R., G. H. Kwei, J. A. Norris, R. R. Herm, J. H. Birely, and D. R. Herschbach: J. Chem. Phys. **41**, 1154 (1964).

J. P. Toennies
Physikalisches Institut
der Universität Bonn

Einige Beispiele für die Anwendung von Stoßwellen zur Untersuchung chemischer Reaktionen

W. Jost und H. Gg. Wagner

Mit 2 Abbildungen

1. Beschreibung von Stoßwellen

Die Anwendung von Stoßwellen hat für die Untersuchung chemischer Reaktionen einen Bereich von Versuchsbedingungen eröffnet, der mit anderen Methoden nicht oder nur schwer zugänglich ist. Im Stoßwellenrohr lassen sich in Gasen leicht Temperaturen von über 10000 °K erreichen, hinter Stoßwellen in Flüssigkeiten oder festen Körpern kommt man zu Drucken von einigen Millionen Atmosphären. Die Versuchstechnik ist, soweit es die Erzeugung von Stoßwellen betrifft, verhältnismäßig einfach. Für die eigentliche Messung ist im allgemeinen ein etwas höherer Aufwand erforderlich als bei Messungen mit klassischen Methoden, wie man aus der für die Messung zur Verfügung stehenden Zeit, die zwischen 10^{-3} und 10^{-6} Sekunden liegt, erkennt. Ein für die Untersuchung von Gasreaktionen entscheidender Vorteil von Stoßwellenversuchen ist die Möglichkeit, Wandreaktionen während der chemischen Umsetzung praktisch völlig auszuschalten. Die Begrenzung der Meßzeit auf Werte um bzw. unter 10^{-3} Sekunden stellt keinen wirklichen Nachteil dar, denn von hier zu längeren Reaktionszeiten kann man sich dann der Methode der adiabatischen Kompression bedienen. Die erste zusammenfassende Darstellung über die Anwendung von Stoßwellen als allgemeines Hilfsmittel zur Untersuchung chemischer Reaktionen war das 1958 erschienene Buch „Chemische Reaktionen in Stoßwellen" [1] von E. F. Green und J. P. Toennies. Daran schlossen sich mehrere Monographien und Referate an. In diesen Darstellungen sind neben allgemeinen Grundlagen auch zahlreiche neue Anwendungsbeispiele und -möglichkeiten für Stoßwellen besprochen worden. Da hier nur einige spezielle Gesichtspunkte behandelt werden, sei wegen Einzelheiten auf diese Zusammenfassungen verwiesen.

Die Entstehung einer Stoßwelle kann man sich an dem von R. Becker stammenden Modell veranschaulichen. Ein Rohr mit beweglichem Kolben sei mit Gas gefüllt. Schiebt man den Kolben etwas in das Gas hinein, dann entsteht eine Verdichtungswelle, die vom Kolben weg mit Schallgeschwindigkeit in das Gas hineinläuft und das Gas etwas erhitzt. Wird kurz danach der Kolben auf etwas höhere Geschwindigkeit gebracht, so läuft eine zweite Verdichtungswelle mit der Schallgeschwindigkeit des schon etwas erwärmten Gases, also etwas schneller als die erste Verdichtungswelle, hinter dieser ersten Welle her. Eine Folge solcher Beschleunigungen der Kolbens ergibt eine Reihe immer schneller werdender Verdichtungswellen, die einander einholen und schließlich eine plötzliche Änderung der Zustandsgrößen des Gases hervorrufen; eine Stoßwelle ist entstanden. Für eine mathematisch saubere Abteilung des Vorgangs der Entstehung einer Stoßwelle sei verwiesen auf Courant-Friedrichs „Supersonic Flow and Shock Waves" [2]. Es sei nur darauf hingewiesen, daß man aus der, die Ausbreitung von Schallwellen beschreibenden linearen, partiellen Differentialgleichung die Entstehung einer Stoßwelle nicht ableiten kann, denn es handelt sich bei einer Stoßwelle um einen „nicht-linearen" Vorgang. Die charakteristischen Eigenschaften einer Stoßwelle sind die folgenden:

1. Sie läuft mit einer Geschwindigkeit in das Frischgas, die größer ist als die Schallgeschwindigkeit und auch größer als die Kolbengeschwindigkeit.

2. Vom komprimierten Gas hinter der Stoßwelle aus gesehen läuft die Stoßwelle mit einer Geschwindigkeit, die kleiner ist als die Schallgeschwindigkeit im komprimierten Gas.

3. Die Kompression des Gases in der Stoßfront verläuft nicht isentrop, sondern unter Entropievermehrung; der Vorgang ist aber adiabatisch.

4. Experimente und Berechnungen zeigen, daß die Dicke der Stoßfront, d. h. die Ausdehnung des Bereiches, in welchem das Gas vor der Stoßwelle durch die Stoßfront komprimiert und auf den Zustand hinter der Stoßfront gebracht wird, für ein einatomiges Gas nur wenige freie Weglängen beträgt. Die Kompression erfolgt also außerordentlich schnell.

5. Für eine Stoßwelle, die sich mit konstanter Geschwindigkeit ausbreitet, kann man durch Anwendung der Erhaltungssätze für Masse, Impuls und Energie die Zustandsgrößen des Gases vor und hinter der Stoßfront miteinander in Verbindung bringen. Dazu betrachtet man die Stoßwelle in einem mit der Stoßfront mitbewegten Koordinatensystem; die Stoßfront stehe am Ort $x = 0$. Das Frischgas mit dem Druck P_1, der Dichte ϱ_1, der Temperatur T_1 und der Enthalpie H_1 pro Gramm ströme mit der Geschwindigkeit v_1 in die bei $x = 0$ stehende Stoßfront ein. Die entsprechenden Größen für das aus der Stoßfront austretende Gas haben den Index 2. Dann gilt

$$\text{für die Massenerhaltung: 1.} \quad \varrho_1 v_1 = \varrho_2 v_2$$

$$\text{für die Impulserhaltung: 2.} \quad (\varrho_1 v_1) v_1 + P_1 = (\varrho_2 v_2) v_2 + P_2$$

$$\text{für die Enthalpieerhaltung: 3.} \quad \frac{1}{2} v_1^2 + H_1 = H_2 + \frac{1}{2} v_2^2$$

Mit Hilfe von 1. und 2. kann man aus 3. v_1 und v_2 eliminieren und erhält

4. $\quad \Delta H = H_2 - H_1 = \dfrac{1}{2}(P_2 - P_1)\left(\dfrac{1}{\varrho_1} + \dfrac{1}{\varrho_2}\right)$ (Hugoniot-Kurve)

Aus 1. und 2. folgt

5. $\quad (P_2 - P_1)\left(\dfrac{1}{\varrho_1} - \dfrac{1}{\varrho_2}\right)^{-1} = (\varrho_2 v_2)^2 = (\varrho_1 v_1)^2$ (Rayleigh-Gerade)

In Abb. 1 ist eine Hugoniot-Kurve und eine Rayleigh-Gerade dargestellt.

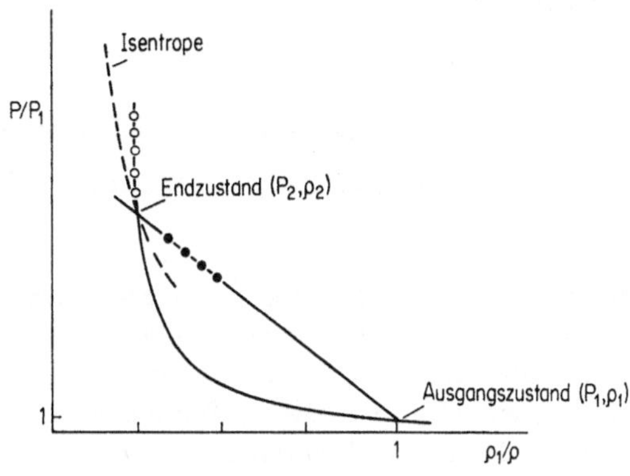

Abb. 1. Hugoniot-Kurve und Rayleigh-Gerade
$- - -$ Isentrope
$-\circ-\circ$ Hugoniot-Kurve
$-\bullet-\bullet$ Rayleigh-Gerade

Wenn man z. B. die Geschwindigkeit $v_1 = -U$ (U = Geschwindigkeit der Stoßwelle im ruhenden Koordinatensystem) gemessen hat, liegt die Neigung der Rayleigh-Geraden fest, und man kann leicht P_2, T_2 und ϱ_2 für den Zustand hinter der Stoßwelle berechnen. Trifft die Stoßwelle senkrecht auf eine starre Wand, dann wird sie von dort reflektiert. Die Reflexion erfolgt so, daß das Gas hinter der zurücklaufenden Stoßwelle in Ruhe ist. Unter den oben angegebenen Voraussetzungen erhält man, wenn die Größen hinter der reflektierten Welle mit dem Index 5 versehen werden:

$$\frac{P_5 - P_1}{P_2 - P_1} = 1 + \frac{1 + \dfrac{\varkappa - 1}{\varkappa + 1}}{\dfrac{P_1}{P_2} + \dfrac{\varkappa - 1}{\varkappa + 1}}$$

Solange die einfallende Stoßwelle schwach ist ($P_2 \sim P_1$), ist der Ausdruck auf der linken Seite ~ 2 wie für die Reflexion normaler Schallwellen. Wird $P_2 \gg P_1$, dann geht der Ausdruck gegen $\left(2 + \dfrac{\varkappa + 1}{\varkappa - 1}\right)$, d. h., er wird 6 für $\varkappa = 1{,}67$; 8 für $\varkappa = 1{,}4$; 13 für $\varkappa = 1{,}2$ usw.). Der Druckanstieg in der

reflektierten Welle kann also den in der einfallenden beträchtlich über-
steigen. Die Temperatur hinter der reflektierten Welle ist grob gesehen
doppelt so hoch wie die hinter der einfallenden. Zur Berechnung der
Zustandsdaten hinter einer Stoßwelle aus der gemessenen Stoßwellengeschwindigkeit geht man von den Gln. 4., 5. und der Gasgleichung aus,
entnimmt die Enthalpiewerte aus Tabellen und bestimmt den Zusammenhang zwischen der Stoßwellengeschwindigkeit und den Daten der einfallenden und reflektierten Welle durch Iteration. Das macht etwas Arbeit,
jedoch keinerlei Schwierigkeiten.

2. Erzeugung von Stoßwellen und Messungen im Stoßwellenrohr zur Untersuchung chemischer Reaktionen

Für die Erzeugung von Stoßwellen zur Untersuchung von Gasreaktionen bedient man sich meist des sog. Stoßwellenrohres, eines Rohres
von z. B. 6 m Länge und 10 cm Durchmesser. Die Rolle des schiebenden
Kolbens übernimmt dabei ein leichtes Gas, das in einem Teil des Rohres
unter erhöhtem Druck eingefüllt ist. Dieser Teil des Rohres ist gegen den
anderen Teil durch eine Membran abgetrennt. In Abb. 2 sind verschiedene
Phasen der Ausbreitung einer Stoßwelle dargestellt.

Wenn die Membran geplatzt ist, läuft in den Niederdruckteil eine Stoßwelle, während sich das Gas im Hochdruckteil entspannt. Die Stoßwelle
trifft schließlich auf die Endplatte und wird dort reflektiert. Aus dem
Diagramm kann man die zeitliche Reihenfolge der Vorgänge entnehmen.
Wenn die Stoßwelle die Teilchen erfaßt, strömen sie mit der Geschwindigkeit der Kontaktfläche hinter der Stoßwelle her. Als Kontaktfläche bezeichnet man die Zone, in der das Gas aus dem Hoch- und dem Niederdruckteil
zusammenstoßen. In Wirklichkeit handelt es sich dabei um eine ausgedehnte
Zone, denn beim Platzen der Membran erfolgt an dieser Stelle Vermischung
der beiden Gase. Durch die reflektierte Welle werden die Teilchen wieder
zur Ruhe gebracht.

Dieses Diagramm ist sehr wichtig für die Bestimmung der Dimensionen
des Stoßwellenrohres. Die Beobachtung der chemischen Reaktion in einer
Stoßwelle erfolgt an einem festen Ort x_i des Rohres. Für Messungen in der
einfallenden Welle ist das verfügbare Zeitintervall von t_{i0} bis t_{i1} in der
reflektierten Welle von t_{i1} bis t_{i2}. Während in der reflektierten Welle das
Gas in Ruhe ist, eine zeitliche Änderung im Gas also direkt registriert
wird, sind in der einfallenden Welle Laboratoriumszeit t und wirkliche
Reaktionszeit τ verschieden, denn im Laufe der Zeit kommen immer andere
Volumenelemente an die Meßstelle bei x_i; solange die Teilchenspuren
(Stromlinien) gerade sind, gilt $\tau = \dfrac{\varrho_2}{\varrho_1} \cdot t$.

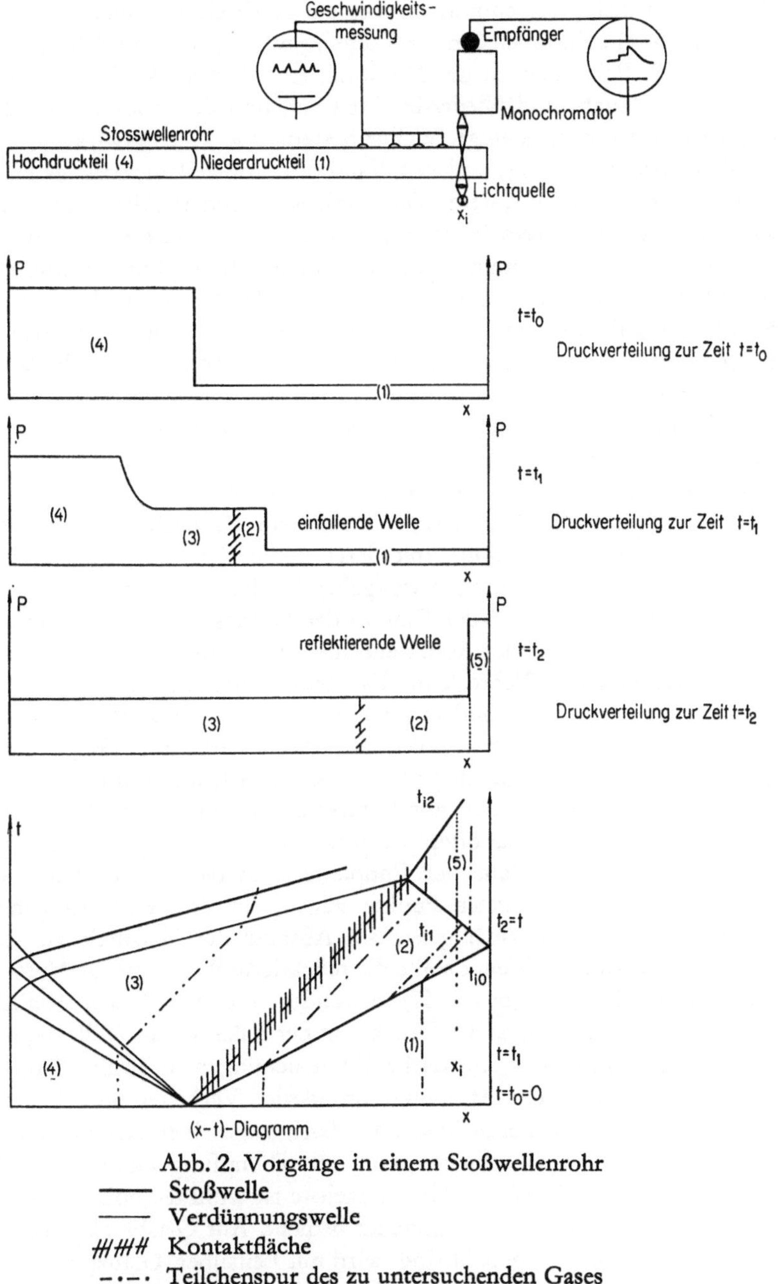

Abb. 2. Vorgänge in einem Stoßwellenrohr

——— Stoßwelle

——— Verdünnungswelle

///// Kontaktfläche

–·–· Teilchenspur des zu untersuchenden Gases

–··– Teilchenspur des Treibgases

(1) Frischgas; (2) Gas hinter der einfallenden Welle; (3) expandiertes Treibgas; (4) Gas im Hochdruckteil (nicht expandiert); (5) Gas hinter der reflektierten Stoßwelle

Wir haben bisher angenommen, daß die Ausbreitung einer Stoßwelle in einem Rohr ein eindimensionaler Vorgang wäre. Das ist in Wirklichkeit nicht der Fall. Beim Platzen der Membran tritt dort starke Vermischung der Gase von Hoch- und Niederdruckteil auf, und die Stoßwelle wird in der Anfangsphase beschleunigt. Weiter besteht eine Wechselwirkung zwischen dem strömenden Gas und der Wand des Stoßwellenrohres; es bildet sich eine Grenzschicht aus, deren Dicke mit wachsendem Abstand von der Stoßfront zunimmt. Dieses in der Grenzschicht verzögerte Gas wird von der Kontaktzone aufgenommen und vermindert die verfügbare Meßzeit. Hinzu kommt, daß in einigem Abstand hinter der Stoßfront das Gas bei entsprechenden Bedingungen turbulent wird. In hinreichender Entfernung von der Membran nimmt die Geschwindigkeit der Stoßwelle ab, die Welle wird gedämpft.

Der Einfluß der Grenzschicht wird kleiner, wenn der Rohrdurchmesser hinreichend groß gewählt wird und die Rohrwand gleichzeitig sehr glatt gemacht werden kann. Bei sehr niedrigen Drucken ($< 0,1$ Torr) wird der Einfluß der Grenzschicht so stark, daß man zur Beobachtung besondere Vorrichtungen anbringen muß, mit deren Hilfe sich der Grenzschichteinfluß unter diesen Bedingungen weitgehend reduzieren läßt.

Sehr viel kritischer wird der Einfluß der Grenzschicht für reflektierte Stoßwellen, da hier nahe der Rohrwand die reflektierte Welle in praktisch ruhendes Gas höherer Dichte läuft. Es können komplizierte Strömungsvorgänge auftreten, die besonders bei $\varkappa$-Werten, die merklich unter 1,67 liegen, so starke Störungen in der reflektierten Welle hervorrufen können, daß sie zu kinetischen Messungen nicht mehr geeignet ist. Insbesondere bei der Untersuchung von Reaktionen mit Explosionscharakter, z. B. bei Selbstzündungsvorgängen hat das öfters zu völlig falschen Interpretationen geführt. In unmittelbarer Nähe der Endplatte wird das Gas ebenfalls abgekühlt. Die hier angedeuteten Punkte zeigen, daß es für Messungen in reflektierten Wellen zweckmäßig ist, den Abstand der Meßstelle von der Endplatte zu verändern. Weiter sollte die reflektierte Welle nur für Messungen in Gasen mit hohem $\varkappa$ (nahe 1,67) verwendet werden, und nach Möglichkeit sollte auch die Konstanz des Druckes in der reflektierten Welle geprüft werden. Es gibt zahlreiche kinetische Untersuchungen in Stoßwellen, bei denen die Abweichung der Strömung vom idealen Verhalten nicht berücksichtigt wurde und die daher nicht als zuverlässig angesehen werden können.

Zur Messung von Geschwindigkeit und Dämpfung einer Stoßwelle gibt es verschiedene Verfahren. Das einfachste ist wohl dasjenige, das sich der Oberflächenwiderstandsthermometer bedient. Auf Glasblöcke, die gut in das Stoßwellenrohr eingepaßt sind, wird mit Leitsilber, Glanzplatin u. a. ein feiner Strich gezeichnet und eingebrannt. Man kann dann die Schicht auf den gewünschten Widerstand abschmirgeln. Kommt die Stoßwelle an den Meßstreifen, dann ändert sich der Widerstand. Das resultierende Signal wird verstärkt, differenziert und kann dann entweder auf einem Oszillographen registriert werden oder zum Ein- bzw. Ausschalten eines elektronischen Zählers verwendet werden. Man kann, wenn man die Signalform genau geprüft hat, die Geschwindigkeit der Stoßwelle so auf 0,1 Prozent genau bestimmen und daraus dann die Daten der Gase hinter der Stoßwelle berechnen.

Für die Messung des Konzentrationsverlaufes der einzelnen Teilchenarten werden hauptsächlich optische Methoden herangezogen, und zwar Emissions- oder Absorptionsmessungen. Man kann dabei drei verschiedene Anordnungen wählen, die hier für Absorptionsmessungen beschrieben werden sollen und von denen meist die erste verwendet wird.

1. Ein schmales Lichtbündel, das genau senkrecht zur Rohrachse verläuft und Licht des gewünschten Wellenlängenintervalls enthält, durchdringt das Rohr. Seine Intensität wird mit einem geeigneten Empfänger als Funktion der Zeit gemessen, und aus der Absorption wird die zeitliche Änderung der Konzentration bestimmt. Die Zeitauflösung wird bestimmt durch die Bündelbreite bzw. Ansprechzeit des Empfängers.

2. Ein schmales Bündel „weißen“ Lichtes durchdringt das Rohr wie bei 1. Das Bündel wird spektral zerlegt, und in kurz aufeinanderfolgenden Zeitabständen wird die Intensität über einen größeren Wellenlängenbereich auf einem Film registriert. Aus den aufeinanderfolgenden Aufnahmen kann man die Absorption als Funktion der Zeit bestimmen. Die Zeitauflösung wird bestimmt durch die Bündelbreite bzw. durch die Zeit, die für aufeinanderfolgende Registrierungen gebraucht wird.

3. Ein breites Bündel von Licht eines bestimmten Wellenlängenintervalls durchdringt das Stoßrohr quer zur Rohrachse. Das Rohrinnere wird auf einen Film abgebildet. Mit einer einzigen, möglichst kurzzeitigen Belichtung bekommt man die Änderung der Absorption längs des Stoßwellenrohres und damit als Funktion der Zeit. Die Zeitauflösung wird bestimmt durch die Belichtungsdauer.

Die Übertragung dieser Verfahren auf Emissionsmessungen braucht nicht diskutiert zu werden.

Während man bei den unter 2 und 3 genannten Methoden im wesentlichen auf die Empfindlichkeit des Photomaterials und dadurch auf den sichtbaren Spektralbereich und das nahe UV angewiesen ist, steht für Methode 1 der ganze Spektralbereich von etwa 1000 Å bis 30 μ zur Verfügung. Nach Methode 3 besteht dann, wenn bei den gewünschten Wellenlängen Laser verfügbar sind, durch Kombination mit einer Kerr-Zelle prinzipiell die Möglichkeit, zu Zeitauflösungen von etwa 10^{-8} Sekunden zu kommen. Da der große Dichtesprung in der Stoßfront jedoch Ablenkung des Lichtes bewirkt (Schliereneffekt), kann man diese hohe Zeitauflösung, wenigstens nahe der Stoßfront nicht ausnützen. Auch dahinter kann der Schliereneffekt noch stören. Für die beiden anderen Methoden wird die Zeitauflösung durch die Breite des Lichtbündels, durch die Zeitauflösung der Registriereinrichtung oder durch die Anzeigeempfindlichkeit des Empfängers bzw. die Lichtstärke der Lichtquelle bestimmt. Man kann die Breite des Lichtbündels nicht wesentlich unter 1 mm wählen, da sonst Beugungserscheinungen störend wirken. Bei einer Stoßwellengeschwindigkeit von $1\,\dfrac{km}{sec} = 1\,\dfrac{mm}{\mu sec}$ erhält man damit also eine Zeitauflösung von etwa $1\,\mu sec$, wobei auch hier beim Passieren der Stoßfront ein Schliereneffekt auftritt. Mit diesem Schliereneffekt kann man die Ankunft der Stoßfront an der Meßstelle sehr genau festlegen; gleichzeitig erkennt man, ob die Stoßfront eben ist und senkrecht zur Rohrachse steht. Diese Zeitauflösung legt auch die obere Grenze der meßbaren Reaktionsgeschwindigkeiten fest, denn die Reaktionszeit soll ein Mehrfaches der Zeitauflösung t_z betragen. Bei einem Geschwindigkeitsausdruck $\dot{n} = -k \cdot n^m$ muß also $-\dfrac{1}{k}\displaystyle\int_{n_1}^{n_2}\dfrac{dn}{n^m} = \tau > t_z$ sein, wobei n_1 und n_2 soweit auseinander liegen müssen, daß man ihren Unterschied gut messen kann. Zu langen Reaktionszeiten hin begrenzen Abweichungen der Strömung vom idealen Verhalten die Meßzeit.

Viele interessierende Reaktionen laufen nicht thermoneutral ab. Die durch die Reaktion freigesetzte oder verbrauchte Wärme beeinflußt die Ausbreitung der Stoßwelle; sie kann zu instationären Vorgängen führen; auf jeden Fall ändern sich aber die Zustandsgrößen während der Reaktion. Es ist grundsätzlich möglich, diese Einflüsse rechnerisch zu erfassen und zu korrigieren. Diese Korrekturen sind aber nur zuverlässig, solange sie nicht zu groß sind, und es empfiehlt sich, die Versuchsbedingungen so zu wählen, daß man sie ganz vernachlässigen kann. Das erfordert im allgemeinen kleine Konzentrationen der Reaktionspartner und damit hohe Nachweisempfindlichkeit bei gleichzeitig hoher Zeitauflösung. Im sichtbaren Spektralbereich und im UV sind Sekundärelektronenvervielfacher, für Messungen von Absorption und Emission im IR sind die neu entwickelten Indium-Arsenid, Indium-Antimonid und die mit Gold, Quecksilber oder Kupfer dotierten Germanium-Infrarotempfänger sehr gut geeignet. Mit Kupfer-dotierten Germanium-Empfängern kann man bei guter Empfindlichkeit Messungen bis $30\,\mu$ ausführen. Alle diese Empfänger sind mit Zeitauflösungen von besser als $1\,\mu sec$ zu bekommen. Einen Eindruck der Empfindlichkeit eines Indium-Antimonid-Empfängers geben folgende Daten.

In einem Stoßwellenrohr von 8 cm Durchmesser kann man bei 2000 °K, einer Gesamtdichte von 10^{-5} Mol·cm^{-3}, einer Bündelbreite von 3 mm bei $4{,}4\,\mu$ in einem Wellenlängenbereich von $0{,}1\,\mu$ den Zerfall von 0,02 Prozent N_2O ($2 \cdot 10^{-5}$ Mol·cm^{-3}) gut verfolgen. Sehr wichtig für Stoßwellenversuche

ist die Auswahl eines geeigneten Trägergases. Zweckmäßig ist ein einatomiges Gas hohen Molekulargewichtes. Bei zweiatomigen Trägergasen
(und auch Reaktionspartnern) sind bei hohen Temperaturen die Schwin-
gungsrelaxationszeiten oft so, daß sie mit dem Ablauf der zu untersuchenden
Reaktion, für die ja ein Zeitbereich durch experimentelle Bedingungen
vorgegeben ist, zusammenfallen. Das ist für Messungen von Schwingungs-
relaxationsvorgängen und für die Untersuchung chemischer Reaktionen
in relaxierenden Gasen, also in einer Umgebung, die nicht im thermodynamischen Gleichgewicht ist, sehr günstig; bei anderen Untersuchungen
wirkt es jedoch störend. Bei mehratomigen Trägergasen sollte man wegen
der oben genannten Einflüsse nur in der einfallenden Welle messen.

Es wurden hier einige Gesichtspunkte zur Anwendung von Stoßwellen
für die Untersuchung chemischer Reaktionen gegeben. Daneben gibt es
noch einige andere Möglichkeiten für die Anwendung von Stoßwellen zur
Untersuchung von Reaktionen, z. B. das chemische Stoßwellenrohr oder
die seit langer Zeit bekannten Gasdetonationen. Auch die Anwendung
schnell registrierender Massenspektrometer zur Konzentrationsmessung in
Stoßwellen hat zu interessanten Ergebnissen geführt. Wegen Einzelheiten
sei auf die unter [1] genannte Literatur verwiesen.

3. Beispiele für Untersuchungen an unimolekularen Reaktionen

Die optimalen Bedingungen zur Messung einer bestimmten Reaktion
hängen natürlich von dem entsprechenden Problem ab. Wenn man sie
gefunden hat, kann man bei Untersuchungen in Stoßwellen aber in den
meisten Fällen eine erstaunlich hohe Meßgenauigkeit erzielen. Im folgenden
sollen einige Beispiele dafür gegeben werden. Wir wollen den in Stoßwellen
gemessenen unimolekularen Zerfall einiger kleiner Moleküle besprechen
und die Ergebnisse mit Werten vergleichen, die aus statischen Messungen
erhalten wurden.

3.1. Theoretische Beschreibung unimolekularer Reaktionen

Für die Beschreibung unimolekularer Reaktionen schließen wir uns an
die Bezeichnungsweise von SLATER [3] an, die hier kurz wiedergegeben
werden soll. Es ist c die Konzentration der reagierenden Moleküle, $c \cdot f_r$
ist der Bruchteil der Moleküle im Zustand r mit der Energie ε_r im Gleichgewicht, k_r die Zerfallswahrscheinlichkeit aus dem Zustand r pro Zeiteinheit, k_r^{-1} ist also die Lebensdauer dieses Zustandes bezüglich des Zerfalls.
Wenn eine Reaktion, z. B. ein Zerfall, stattfindet, dann wird bei niedrigen
Dichten die Besetzung des Zustandes r nicht mehr durch den Gleichgewichtswert $c \cdot f_r$, sondern durch $c \cdot g_r$ gegeben sein, und pro Zeiteinheit

15*

zerfallen $c \cdot g_r \cdot k_r$ Moleküle aus dem Zustand r. Gleichzeitig werden Moleküle durch Stoß aus dem Zustand r heraus befördert und zwar $c \cdot g_r \cdot \omega_r$ pro Sekunde. Damit verschwinden aus dem Zustand r pro Sekunde $c \cdot g_r \cdot k_r + c \cdot g_r \cdot \omega_r$ Teilchen. Die Zahl der Moleküle, die pro Zeiteinheit durch Stoß in den Zustand r gebracht werden, ist unter Anwendung des Prinzips der mikroskopischen Reversibilität gegeben durch $c \cdot f_r \cdot \omega_r$. Im stationären Zustand gilt damit $c \cdot f_r \cdot \omega_r = c \cdot k_r \cdot g_r + c_r \cdot \omega_r$ und es wird

$$g_r = \frac{f_r \cdot \omega_r}{k_r + \omega_r}$$

Die Reaktionsgeschwindigkeit ist gegeben durch

$$\frac{1}{[c]} \cdot \frac{dc}{dt} = - \sum_r \cdot k_r \cdot \frac{f_r \cdot \omega_r}{k_r + \omega_r} = - k$$

Man kann zwei Grenzfälle unterscheiden:

1. Wenn $\omega_r \gg k_r$, wird $k \approx \sum_r \cdot k_r f_r = k_H$

2. Wenn $\omega_r \ll k_r$, wird $k = \sum_r \omega_r \cdot f_r = \omega \cdot \Sigma f_r = k_N \cdot [M]$

Im Fall 2. is k proportional ω und ω ist proportional zur Gesamtdichte $[M]$. Damit gilt $\frac{1}{c} \cdot \frac{dc}{dt} = -k_N \cdot [M]$; die Reaktion erfolgt nach einem Geschwindigkeitsgesetz 2. Ordnung. Der Dichtebereich, in dem das gilt, wird als Niederdruckbereich der unimolekularen Reaktion bezeichnet im Gegensatz zu dem ersten Fall, bei dem die Reaktion einem Geschwindigkeitsgesetz 1. Ordnung folgt. Diesen Bereich nennt man den Hochdruckbereich der unimolekularen Reaktion. Zwischen beiden gibt es natürlich ein Übergangsgebiet.

Bei dieser Art der Beschreibung unimolekularer Reaktionen wird, wie in allen „klassischen" Theorien, angenommen, daß die Besetzung der Zustände unterhalb der Energieschwelle E_0 eine Gleichgewichtsbesetzung ist. Für eine richtige Interpretation der Geschwindigkeitskonstanten muß man diese Annahme fallenlassen und, wie das am Beispiel des Zerfalls zweiatomiger Moleküle besonders deutlich wird, die Verarmung in der Besetzung der Zustände unterhalb E_0 in Rechnung stellen. Ohne auf die Berechnung im einzelnen einzugehen, sollen hier einige Ergebnisse zusammengestellt werden (s. a. [4]).

Wenn man den Ausdruck für die Geschwindigkeitskonstante im Niederdruckbereich des unimolekularen Zerfalls in der folgenden Form schreibt:

6. $$k^0_{\text{uni}} = \omega(T) \cdot \frac{\varrho_{\text{vib}}(E_0)}{Q_{\text{vib}}} \cdot A(T) \cdot \exp - \left(\frac{E_0}{RT}\right)$$

wobei $Q_{\text{vib}} = \prod_{j=1}^{s} \left(1 - e^{-\frac{\varepsilon_i}{RT}}\right)^{-1}$ die Schwingungszustandssummen des Moleküls sind und $\varrho_{vib}(E_0)$ die Dichte der Zustände um E_0, die sich in guter Näherung darstellen läßt als $\varrho_{vib}(E_0) = \dfrac{(E_0 + E_z)^{s-1}}{(s-1)!\, \Pi\, \varepsilon_j}$ mit $E_z = $ Nullpunktsenergie der Schwingungen und $s = $ Zahl der Oszillatoren, und wobei $\omega(T)$ die Stoßzahl ist, dann erhält man für $A(T)$ die folgenden Ausdrücke:

Bei gutem intramolekularem Energieaustausch und exponentieller Abhängigkeit der Übergangswahrscheinlichkeit von der im Stoß übertragenen Energie (bestimmt durch den Parameter α, der eine Art mittlerer pro desaktivierendem Stoß übertragene Energie darstellt) ergibt sich, wenn A_{anh} und A_{rot} die Korrekturen für Anharmonizität und Rotation sind

6a) im Fall starker Stöße ($\alpha \gg RT$):

$$A(T) = A_{anh} \cdot A_{rot} \cdot RT$$

6b) im Fall schwacher Stöße ($\alpha \ll RT$):

$$A(T) = A_{anh} \cdot A_{rot} \cdot \frac{\alpha^2}{RT}$$

6c) für ein Stufenmodell mit der mittleren (geometrisch) Stufenbreite $\bar{\varepsilon}$:

$$A(T) = A_{anh} \cdot A_{rot} \cdot \frac{\alpha}{RT}\, \bar{\varepsilon}$$

Man erhält also eine zusätzliche Temperaturabhängigkeit der Geschwindigkeitskonstanten gegenüber den „klassischen" Theorien. Das gilt auch, wenn man das Slatersche Modell streng harmonischer Oszillatoren ohne Kopplung zugrunde legt.

3.2. Ergebnisse für den Zerfall vier- und mehratomiger Moleküle

Über unimolekulare Reaktionen größerer Moleküle wurden in den vergangenen Jahren von B. S. RABINOVITCH [5] und seinen Mitarbeitern sorgfältige Untersuchungen angestellt, die sowohl den Hochdruck- als auch den Niederdruckbereich der unimolekularen Reaktion dieser Teilchen erfaßten und an denen man das allgemeine Verhalten einer solchen Reaktion gut erkennen kann. Demgegenüber sind die verfügbaren Daten über unimolekulare Reaktionen kleinerer Moleküle, die aus drei oder vier Atomen aufgebaut sind, relativ unvollständig. Dabei sollten gerade die Reaktionen dieser Moleküle, von denen im allgemeinen mehr Daten, die man zur theoretischen Behandlung heranziehen kann, bekannt sind, leichter zu interpretieren sein.

Von vieratomigen Molekülen wurden NO_2Cl [6], FO_2Cl [7] und F_2O_2 [8] im Übergang vom Niederdruck- zum Hochdruckbereich untersucht; beim NO_2Cl wurde auch der Niederdruckbereich praktisch erreicht. In diesen Fällen ist es möglich, auf die Reaktionsgeschwindigkeit im Hochdruckbereich zu extrapolieren. Die unimolekularen Reaktionen von C_2N_2 [9] und H_2O_2 [10] wurden nur im Niederdruckbereich untersucht, wobei im Falle des C_2N_2 die Interpretation dadurch erschwert wird, daß für die Bindungs-Dissoziationsenergie zwei Werte möglich sind. Für NH_3 [11] und ND_3 ist der Niederdruckbereich gut untersucht; ein Übergang in den Hochdruckbereich zeichnet sich bei 20—30 atm und 2000 °K deutlich ab. Die Interpretation der Messungen ist aber unsicher, da Folgereaktionen im ganzen Bereich nicht sicher ausgeschlossen werden können. Einige der für den unimolekularen Zerfall von 4-atomigen Molekülen erhaltenen Daten sind in Tab. 1 zusammengestellt. Dort bedeutet k_N die Geschwindigkeits-konstante im Niederdruckbereich, k_∞ die im Hochdruckbereich.

Die scheinbare Aktivierungsenergie für eine unimolekulare Reaktion im Niederdruckbereich wird u. a. bestimmt durch 1. die Dissoziationsenergie der entsprechenden Bindung, 2. die sich aus den Zustandssummen ergebende Temperaturabhängigkeit, zu der im wesentlichen diejenigen inneren Frei-heitsgrade beitragen, bei denen $h\nu < kT$, 3. die Temperaturabhängigkeit der Stoßzahl, 4. die Stoßwirksamkeit und 5. eine Verarmung höherer Energieniveaus unterhalb E_0. Betrachtet man die in Tab. 1 zusammen-gestellten Ergebnisse unter diesem Gesichtspunkt, dann bieten sich für die Interpretation der Resultate Schwierigkeiten, von denen man wahrschein-lich annehmen muß, daß sie zum Teil auf Fehler in der experimentellen Bestimmung der Geschwindigkeitskonstanten liegen. Allerdings dürfte es in einigen Fällen schwierig sein, diese Fehler zu beheben, und es ist zu erwarten, daß sich der Zerfall von dreiatomigen Molekülen genauer unter-suchen läßt.

3.3. Ergebnisse für den Zerfall dreiatomiger Moleküle

Als erstes Beispiel für den unimolekularen Zerfall eines dreiatomigen Moleküls wurde die Dissoziation von F_2O im statischen System gefunden [12]. Es ergab sich eine Geschwindigkeitskonstante im Niederdruckbereich von $10^{17,4} \cdot \exp -\dfrac{40600}{RT}$ [$cm^3 \cdot mol^{-1} \cdot sec^{-1}$]. O. K. Rice [13] hat versucht, diesen Wert der Geschwindigkeitskonstanten zu interpretieren, wobei er neben dem Einfluß der Rotation des Moleküls den Übergang einer Schwin-gung in eine Rotation bei hochangeregten Molekülen, engeres Zusammen-rücken der Schwingungsniveaus zu höheren Energien u. a. berücksichtigte und eine Stoßwirksamkeit von 1 annahm. Damit ließ sich die gemessene Geschwindigkeitskonstante deuten. Allerdings muß man berücksichtigen, daß die Bindungs-Dissoziationsenergie für F_2O nicht genau bekannt ist. Aus massenspektrometrischen Messungen ergab sich ein Wert, der höher ist als die gefundene scheinbare Aktivierungsenergie für die Reaktion im Niederdruckbereich. Weiter fällt auf, daß der praeexponentielle Faktor für

Tabelle 1

Gemessene Geschwindigkeiten unimolekularer Reaktionen vieratomiger Moleküle

	$\Delta H^0 \left[\dfrac{\text{kcal}}{\text{Mol}}\right]$	Temperaturbereich	Geschwindigkeitskonstante	Literatur
F_2O_2	17,3	$210-250\,^\circ$K	$k_\infty \approx 10^{13,08} \cdot \exp - \dfrac{45\,000}{RT}\ [\text{sec}^{-1}]$	[8]
FO_2Cl	51	$570-620\,^\circ$K	$k_\infty \approx 10^{12,77} \cdot \exp - \dfrac{17\,300}{RT}\ [\text{sec}^{-1}]$	[7]
NO_2Cl	29,5	$450-520\,^\circ$K	$k_N \approx 10^{16,76} \cdot \exp - \dfrac{27\,500}{RT}\left[\dfrac{\text{cm}^3}{\text{Mol}\cdot\text{sec}}\right]$ in NO_2Cl und in Ar, O_2	[6]
NO_2Cl	29,5	$870-1270\,^\circ$K	$k_N \approx 10^{14,85} \cdot \exp - \dfrac{26\,000}{RT}\left[\dfrac{\text{cm}^3}{\text{Mol}\cdot\text{sec}}\right]$	—
C_2N_2	125,8	$1700-2500\,^\circ$K	$k_N \approx 10^{17,23} \cdot \exp - \dfrac{100\,400}{RT}\left[\dfrac{\text{cm}^3}{\text{Mol}\cdot\text{sec}}\right]$ in Ar	[9]
H_2O_2	50,3	$700-950\,^\circ$K	$k_N \approx 10^{17,33} \cdot \exp - \dfrac{46\,300}{RT}\left[\dfrac{\text{cm}^3}{\text{Mol}\cdot\text{sec}}\right]$ in N_2	[10]
NH_3	104	$2000-3000\,^\circ$K	$k_N \approx 10^{15,64} \cdot \exp - \dfrac{79\,500}{RT}\left[\dfrac{\text{cm}^3}{\text{Mol}\cdot\text{sec}}\right]$ in Ar	[11]

Tabelle 2

Gemessene Geschwindigkeiten des unimolekularen Zerfalls dreiatomiger Moleküle

Moleküldaten			Versuchsbedingungen			Ergebnisse	
Molekül	$\Delta H \left[\frac{kcal}{Mol}\right]$	Θ in °K [31]	Temperatur in 1000 °K	$\log k_1$	$E \left[\frac{kcal}{Mol}\right]$		Bemerkungen
O_3	24,3	1030, 1500, 2460	0,69−0,91	14,77	23,15		[14] Dazu Werte bei 360 °K
NO_2	71,6	930, 1900, 2340	1,4−2,3	16,48	65,4		[15] Folgereaktionen schwer zu eliminieren
BrCN	90,5	493, 825, 3170	2,6−4,2	14,08	90		[16] indirekte Bestimmung
ClCN	97		2,0−2,8	16,53	91,5		[17] indirekte Bestimmung
NOCl	37,2	476, 850, 2540	0,88−1,35 0,57−0,68	14,63 16,3	31 35,8		[18] statisch, in N_2
F_2O		663, 1190, 1335	0,52−0,54	17,4	39		[12] statisch, kein Trägergas
SO_2	131	750, 1655, 1960	4,5−7,5	14,4	110		[19] in Argon
H_2O	117,6	2300, 5250, 5400	2,8−7	14,7	105		[20] in Argon
N_2O	38,6	860, 1870, 3280	0,9−1,05 1,2−2,5	15,65 14,7	59 58		[21a] statisch, kein Trägergas [21b] in Argon, Hochdruckbereich gemessen
CO_2	125,4	960, 1995, 3380	2,8−4,4	14,7	99		[22] in Argon, Hochdruckbereich gemessen [30]
COS	71	755, 1235, 2970	1,8−3,0	14,2	61		[23] in Argon
CS_2	92	570, 945, 2200	1,8−3,7	15,56	80,3		[24] in Ar, Hochdruckbereich gemessen [30]

diese Reaktion, verglichen mit dem der Zerfallsreaktionen anderer Moleküle, z. B. H_2O, die allerdings meist bei höheren Temperaturen untersucht wurden, ziemlich groß ist.

In Tab. 2 sind die gemessenen Geschwindigkeitskonstanten für den Zerfall dreiatomiger Moleküle zusammengestellt. Es sind angegeben: die Reaktionsenthalpie für $T = 0\,°K$, die charakteristischen Temperaturen Θ der Oszillatoren, der Temperaturbereich, in dem die Messungen ausgeführt wurden und die Geschwindigkeitskonstanten für die Reaktion im Niederdruckbereich, also $[\dot{A}] = -k \cdot [A] \cdot [M]$, wobei $[A]$ die Konzentration des betrachteten Moleküls, $[M]$ die Gesamtkonzentration ist. k wurde in der Form $k = k_1 \cdot \exp - \dfrac{E}{RT}$ dargestellt mit k_1 in $[cm^3 \cdot mol^{-1} \cdot sec^{-1}]$.

Von den aufgeführten Werten sind die für BrCN und ClCN indirekt bestimmt worden. Bei SO_2 und besonders bei NO_2 ist es schwierig, den unimolekularen Zerfall von gleichzeitig ablaufenden anderen Reaktionen abzutrennen.

Im folgenden soll auf die Reaktion einiger der angeführten Moleküle etwas genauer eingegangen werden.

3.3.1. Der Zerfall von CO_2

Wegen der hohen Reaktionsenthalpie für die Reaktion $CO_2 \rightarrow CO + O$ erwartet man, daß der CO_2-Zerfall erst bei Temperaturen von über 2500 °K so schnell wird, daß man ihn gut verfolgen kann. Für die Untersuchung dieser Reaktion hat man daher praktisch nur noch die Stoßwellentechnik verfügbar, die gleichzeitig garantiert, daß man den Einfluß von Wänden auf die Reaktion ausschalten kann und einen großen Temperaturbereich überstreichen kann. Der gewünschte, weitgehend isotherme, Reaktionsablauf und die für die Interpretation notwendige, möglichst vollständige, Ausschließung von Folgereaktionen macht eine hinreichend starke Verdünnung des CO_2 in einem geeigneten Trägergas, z. B. Ar, notwendig. Das kommt auch der aus der Aerodynamik stammenden Forderung nach möglichst hohen $\varkappa$-Werten entgegen. Man kann nun die CO_2-Konzentration um so niedriger wählen, je empfindlicher mehrere oder wenigstens eines der in der Reaktion auftretenden Teilchen nachgewiesen werden können. Sowohl CO_2 als auch CO und O sowie das als Folgeprodukt möglicherweise auftretende O_2 haben starke Absorptionsbanden im Schumann-UV. Abgesehen davon, daß sich dort die Banden teilweise überlappen, wird die Nachweisempfindlichkeit, wenn gleichzeitig hohe Zeitauflösung verlangt wird, wegen der relativ geringen Intensität der Lichtquellen in diesem Wellenlängenbereich verhältnismäßig schlecht. Es ist bekannt, daß bei hohen Temperaturen die Absorption des CO_2 zu längeren Wellenlängen hin zunimmt, jedoch bleibt auch hier der Absorptionskoeffizient relativ klein. Die Absorption des CO_2 im Ultraroten (4,25 μ) ist stark, die von CO ist davon deutlich getrennt, jedoch wesentlich schwächer. Als Lichtquelle für diesen Wellenlängenbereich würde sich ein Kohlebogen eignen. Da es sich jedoch zeigte, daß Versuchstemperaturen über 4000 °K erforderlich waren, mußte, wenigstens für die hohen Temperaturen, eine andere Nachweismethode gefunden werden. Versuche haben nun ergeben, daß die Infrarotemission des CO_2 ein geeignetes Maß für Messung der CO_2-Konzentration als Funktion der Zeit ist [22]. Bei Konzentrationen von CO_2, die um und unter 1 % lagen (Rohrdurchmesser 8 cm, Niederdruckbereich) konnte die Selbstabsorption vernachlässigt werden; wegen der hohen Temperaturen, bei denen die Reaktion stattfindet, ist die Emission des CO_2 soweit von der bei Zimmertemperatur beobachteten Absorption entfernt, daß diese nicht stört. Es zeigte sich weiter, daß die Emission des CO_2 bei Temperaturen zwischen 2500 und 4500 °K nur sehr wenig von der Temperatur abhängt. Da diese Nachweismethode so empfindlich ist, daß man auch noch bei Konzentrationen weit unter 0,1 % den Reaktionsablauf gut verfolgen kann, ist es prinzipiell möglich, eventuelle Folgereaktionen soweit auszuschalten, daß man den CO_2-Zerfall am Anfang der Reaktion messen kann. Nun ist beim CO_2 die Situation auch insofern günstig, als man die Reaktionsgeschwindigkeit der möglichen Folgeschritte $O + O + M \rightarrow O_2 + M$; $O + CO_2 \rightarrow CO + O_2$ und $O + CO + (M) \rightarrow CO_2 + (M)$ wenigstens größenordnungsmäßig kennt. Die zuletzt genannte Reaktion wird in den späteren Phasen des CO_2-Zerfalls wichtig, da die Umsetzung zu einem Gleichgewicht strebt; am Anfang sind jedoch beide Reaktionspartner in so geringen Konzentrationen vorhanden, daß man dort diesen Schritt vernachlässigen kann. Auch die beiden anderen Reaktionen haben sich bei den gewählten Versuchsbedingungen nicht wesentlich bemerkbar gemacht. Wichtig ist, daß H_2 bzw. H_2O aus dem Gemisch ferngehalten wird.

Zur Prüfung, ob der CO_2-Zerfall tatsächlich als unimolekulare Reaktion im Unterdruckbereich abläuft, wurden sowohl die CO_2-Konzentrationen als auch die Gesamtkonzentrationen, die hier im wesentlichen durch die Argonkonzentration bestimmt waren, jeweils unabhängig voneinander um einen Faktor 10 variiert. Weiter wurde geprüft, ob die CO_2-Konzentration bei jedem einzelnen Versuch nach einem Geschwindigkeitsgesetz 1. Ordnung abnahm, zumindest in einem größeren Zeitintervall hinter der Stoßwelle. Diese Messungen ergaben [22], daß der CO_2-Zerfall unter den angewandten Bedingungen einem Zeitgesetz der Form $[\dot{C}O_2] = -k_N \cdot [CO_2] \cdot [Ar]$ folgte, wobei zwischen 2800 und 4400 °K für k gilt: $k = 10^{14,7} \cdot \exp - \left(\dfrac{99}{RT} \right)$ $[cm^3 \cdot mol^{-1} \cdot sec^{-1}]$ ($[R] = [kcal \cdot mol^{-1} \cdot grad^{-1}]$).

Dieser Wert wurde inzwischen bei Dichten um 10^{-5} $Mol \cdot cm^{-3}$ wiederholt nachgeprüft und bestätigt. Nach der Kasselschen Theorie sollte man bei einer Reaktionsenthalpie von 125,4 kcal/Mol und unter Berücksichtigung der Tatsache, daß die charakteristischen Temperaturen der Oszillatoren unterhalb der Meßtemperatur liegen, mit Ausnahme der eines Oszillators, die im Bereich der Meßtemperatur liegt, aus $E = D - (n-1) \cdot RT$, wobei D die Bindungsdissoziationsenergie, n die Zahl der Oszillatoren bedeuten und die mittlere Temperatur $T \approx 3600\,°K$ beträgt, eine scheinbare Aktivierungsenergie E_a von 110 kcal/Mol erhalten. Dieser Wert liegt deutlich außerhalb der Fehlergrenzen des experimentell bestimmten Wertes von E_a.

Berechnet man nach der vorne angegebenen Beziehung für „schwache" Stöße aus der gemessenen scheinbaren Aktivierungsenergie die Höhe der Energieschwelle E_0 für die Reaktion, dann folgt aus $E_a = E_0 + R \left(\dfrac{\partial \ln Q}{\partial \dfrac{1}{T}} + \dfrac{T}{2} \right)$

ein Wert von $E_0 \approx 117$ kcal/Mol. Dieser Wert liegt ebenfalls unter dem Wert für die Bindungsdissoziationsenergie von 125 kcal/Mol. Nun ist bekannt, daß für die Dissoziation von CO_2 aus dem Grundzustand heraus 171 kcal/Mol erforderlich sind. Die Dissoziation von CO_2 erfolgt also nicht aus dem Grundzustand heraus, sondern durch Übergang auf eine andere Potentialfläche mit anderer Multiplizität. Wenn nun der Zerfall aus diesem Zustand schnell erfolgt, verglichen mit dem Übergang von einem Potential auf das andere, dann ist für die Reaktion im Niederdruckbereich die effektive Besetzung der Niveaus im Grundzustand maßgebend für die Reaktionsgeschwindigkeit, und der Wert von $E_0 \approx 117$ kcal/Mol entspräche der Energieschwelle für diesen Übergang. Für den Faktor $A(T)$ (vgl. [4]) ergibt sich ein Wert von 0,2 kcal/Mol, d. h., die mittlere im desaktivierenden Stoß übertragene Energie ist etwa 1,2 kcal/Mol, während RT etwa 6 kcal/Mol ist. Den Einfluß der Rotation braucht man bei dieser Art von Reaktion nicht zu berücksichtigen. Inzwischen wurde für CO_2 ebenso wie für CS_2 und N_2O auch der Hochdruckbereich des unimolekularen Zerfalls erreicht.

3.3.2. Der Zerfall von H_2O

Der Zerfall von H_2O wurde von Bauer und Mitarbeitern [25] untersucht durch Messung der Absorption des gebildeten OH. Dabei zeigte sich, daß eine zusammengesetzte Reaktion gemessen wurde, die keinen Schluß auf den Reaktionsschritt $H_2O \rightarrow OH + H$ zuläßt. Das steht in Übereinstimmung mit älteren Untersuchungen von uns, die auf gleiche Weise ausgeführt wurden. Allerdings deuteten diese Versuche darauf hin, daß eine ähnliche Situation vorliegt wie bei SO_2 und NO_2, und daß eventuell bei höheren Temperaturen und bei kleineren Konzentrationen von Wasser im Trägergas hier die Einleitungsreaktion deutlicher hervortritt. Es wurde daher der thermische Zerfall von Wasser in einfallenden und reflektierten Stoßwellen bei Temperaturen zwischen 2700 und 6000 °K gemessen [20]. Die Wasserkonzentrationen in Argon lagen zwischen 0,02 und 0,2 %, die Gesamtdichte zwischen $2 \cdot 10^{-6}$ und $6 \cdot 10^{-5}$ Mol/cm³. Die Registrierung der Wasserkonzentration in Abhängigkeit von der Zeit erfolgte mit einem, mit flüssigem N_2 gekühlten Indium-Antimonid-Infrarotempfänger bei 2,8 μ (Bandbreite 0,1 μ), aber auch bei anderen Wellenlängen im Infraroten.

Im ersten Teil des Zerfalls konnten die Infrarot-Signale durch ein Geschwindigkeitsgesetz erster Ordnung dargestellt werden. Die erhaltenen Geschwindigkeitskonstanten waren unabhängig von der H_2O-Konzentration und proportional zur Ar-Konzentration. Bei Temperaturen oberhalb 3500 °K gingen die Infrarot-Signale praktisch auf Null zurück; für niedrigere Temperaturen strebten sie einem endlichen Wert zu, der mit steigender Wasser-Konzentration anstieg. Aus der Auftragung der gemessenen Geschwindigkeitskonstanten gegen die reziproke Temperatur kann man erkennen, daß unter den bei diesen Versuchen angewandten Bedingungen unterhalb 4500 °K der doppelte Wert der Geschwindigkeit der Reaktion $H_2O \rightarrow OH + H$, oberhalb der einfache Wert gemessen wird. Der Übergangsbereich kann durch Variation der Ausgangskonzentration von H_2O verschoben werden. Ähnliche Beobachtungen ergaben sich auch bei anderen Molekülen. Sie sind darauf zurückzuführen, daß Folgereaktionen ablaufen, bei Wasser dürfte es sich um $H + H_2O \rightarrow OH + H_2$ handeln, die niedrige Aktivierungsenergien haben und die bei geringen Konzentrationen und bei hohen Temperaturen nicht ins Gewicht fallen. Die Geschwindigkeitskonstante für den unimolekularen Zerfall von Wasser im Niederdruckbereich ergab sich zu

$$k^0_{uni} = [Ar] \cdot 10^{14,7} \cdot \exp -\frac{105}{RT} \, [cm^3 \cdot mol^{-1} \cdot sec^{-1}] \; [R] = [kcal \cdot mol^{-1} \cdot grad^{-1}]).$$

Die experimentell bestimmte scheinbare Aktivierungsenergie von 105 kcal/Mol für den Zerfall von Wasser steht einer Bindungsdissoziationsenergie von 117,6 kcal/Mol gegenüber. Bei der Interpretation werden wegen der Lage der charakteristischen Temperaturen der einzelnen Schwingungen Quanteneffekte wichtig. Man bekommt für $A(T)$ bei 5000 °K einen Wert von 1,7 kcal/Mol, und der Wert von α aus $A(T) \sim T^\alpha$ ist praktisch gleich Null. Für die mittlere, im desaktivierenden Stoß übertragene Energie ergibt sich daraus etwa 4 kcal/Mol, die mit $RT \sim 10$ kcal/Mol zu vergleichen sind.

3.3.3. Der Zerfall von SO_2

Der Zerfall von SO_2 wurde bei Temperaturen zwischen 3000 und 7000 °K in Argon als Trägergas untersucht. Der Verlauf der SO_2- und SO-Konzentration wurde durch Messung der Absorption im langwelligen UV als

Funktion der Zeit bestimmt. Wie schon aus Messungen von GAYDON [26] und Mitarbeitern bekannt ist, erfolgt bei Konzentrationen von über 1 Prozent und Temperaturen unterhalb 4500 °K die Abnahme der SO_2-Konzentration in einer zusammengesetzten Reaktion, deren Mechanismus im einzelnen noch nicht genau bekannt ist. Bemerkenswert ist jedoch, daß die Besetzung des Zustandes $^3B_1$ von SO_2, wie LEVITT und SHEEN [27] fanden, sehr schnell einen stationären Zustand erreicht.

Vermindert man die Konzentration von SO_2 im Argon unter 0,3% und wertet den Verlauf der SO_2-Konzentration nach einem Zeitgesetz 1. Ordnung aus, dann sieht man, daß mit fallender Konzentration die Temperaturabhängigkeit dieser Geschwindigkeitskonstanten größer wird. Bei noch höheren Temperaturen fallen die gemessenen Geschwindigkeitskonstanten für verschiedene SO_2-Konzentrationen im $(\lg k - \frac{1}{T})$-Diagramm auf eine Kurve. Dieser Übergang erfolgt bei um so niedrigeren Temperaturen, je niedriger die SO_2-Konzentration ist. In diesem Gebiet folgt die Reaktion sowohl bezüglich des zeitlichen Verlaufes als auch bezüglich der SO_2-Konzentration einem Geschwindigkeitsgesetz $[\dot{SO_2}] = -k \cdot [SO_2] = -k_N \cdot [Ar] \cdot [SO_2]$. Dabei ist also k unabhängig von $[SO_2]$ und bei Gesamtkonzentrationen zwischen 10^{-5} und $7 \cdot 10^{-5}$ $[Mol \cdot cm^{-3}]$ der Argonkonzentration streng proportional. Für k_N ergibt sich $k_N = 10^{14,4} \cdot \exp -\frac{110}{RT}$ $[cm^3 \cdot mol^{-1} \cdot sec^{-1}]$, $([R] = [kcal \cdot mol^{-1} \cdot grad^{-1}])$, als Geschwindigkeitskonstante für den unimolekularen Zerfall von SO_2 nach SO_2 $(^1A_1) \rightarrow SO$ $(^3\Sigma) + O$ (^3P) im Niederdruckbereich [19]. Es ist unwahrscheinlich, wie man aus den Messungen bei verschiedenen SO_2-Konzentrationen schließen kann, daß die Reaktion $O + SO_2 \rightarrow SO + O_2$ eine Rolle spielt.

Ebenso wie SO_2 dissoziieren auch O_3, NO_2, $BrCN$, $ClCN$, F_2O und H_2O vom Grundzustand aus. Während aber für SO_2 die charakteristischen Temperaturen aller Oszillatoren niedriger als die Versuchstemperatur liegen, sind bei O_3 und F_2O die charakteristischen Temperaturen aller Oszillatoren höher; bei NO_2 liegt nur eine, bei $BrCN$ und $NOCl$ liegen zwei niedriger als die Versuchstemperaturen. Für H_2O liegen zwei charakteristische Temperaturen höher als die Versuchstemperaturen. Von den hier beschriebenen Molekülen sollte sich daher SO_2 noch am ehesten durch die klassische Form der Kasselschen Theorie darstellen lassen. Vernachlässigt man die Temperaturabhängigkeit der Stoßausbeute, dann erhält man nach $E = D - (n-1) RT$ einen Wert von $n = 3$ für die Zahl der Kasselschen Oszillatoren. Eine realistischere Auswertung [4] ergibt für $A(T)$ bei 4500 °K einen Wert von 0,6 mit einer geringen positiven Temperaturabhängigkeit, und die „mittlere" im desaktivierenden Stoß übertragene Energie beträgt etwas mehr als 2 kcal/Mol, die mit $RT \sim 9$ kcal/Mol zu vergleichen sind.

3.3.4. Der Zerfall von N_2O

Von allen dreiatomigen Molekülen ist der Zerfall des N_2O bisher am besten untersucht [21]. Wie beim CO_2, COS und CS_2 erfolgt die Reaktion durch Übertritt auf einen Elektronenzustand anderer Multiplizität. Beim N_2O erfolgt dieser Übertritt auf den anderen Elektronenzustand in einem Bereich der Energie, in dem man die Schwingungen noch einigermaßen als harmonisch ansehen kann. Hinzu kommt, daß der angeregte Zustand abstoßend ist.

Es hat sich gezeigt, daß die Absorption von N_2O im UV mit steigender Temperatur stark anwächst, so daß man den N_2O-Zerfall bei Konzentrationen weit unter 1% im Trägergas verfolgen kann. Weiter ist die Infrarotemission von N_2O sehr gut geeignet zur Messung des zeitlichen Verlaufes der N_2O-Konzentration. Deshalb konnte diese Reaktion noch bei Konzentrationen von 0,02% N_2O in Argon gemessen werden, so daß auch hier die Voraussetzungen für sehr „saubere" Stoßwellen gegeben sind. Der Zerfall verläuft im ganzen untersuchten Bereich nach 1. Ordnung in N_2O. Bei den höheren N_2O-Konzentrationen um 1% und darüber erfordert die freiwerdende Reaktionswärme eine Korrektur der Geschwindigkeitskonstanten. Außer bei sehr niedrigen Konzentrationen von N_2O schließt sich an die Reaktion (I)

$$\text{I.} \quad N_2O \to N_2 + O$$

der Schritt

$$\text{II.} \quad O + N_2O \to N_2 + O_2 \text{ bzw. } 2\,NO$$

an. Wie man anhand der Messung des NO-Konzentrationsverlaufes und der Signale über die Abnahme der N_2O-Konzentration schließen kann, ist dabei die O-Atomkonzentration quasi-stationär, so daß man hier den doppelten Wert der Geschwindigkeitskonstanten von Reaktion I mißt. Beim Übergang zu kleinen N_2O-Konzentrationen wird der Einfluß von Reaktion II geringer, bis man schließlich bei sehr kleinen Werten der N_2O-Konzentration direkt den Wert von k_I mißt.

Für den unimolekularen Zerfall von N_2O im Niederdruckbereich, in dem N_2O nach $[\dot{N_2O}] = -k_1 \cdot [NO_2] \cdot [Ar]$ zerfällt, ergab sich:

$$k_1 = 10^{14,7} \cdot \exp - \frac{58}{RT} \ [cm^3 \cdot mol^{-1} \cdot sec^{-1}], \quad ([R] = [kcal \cdot mol^{-1} \cdot grad^{-1}]).$$

Messungen mit der Methode der adiabatischen Kompression gestatteten es, den Temperaturbereich, in dem diese Reaktion untersucht ist, bis herunter zu 1300 °K auszudehnen [28]. Dabei ergab sich sehr gute Übereinstimmung mit den Werten, die mit Stoßwellen erhalten wurden. Dies läßt sich als Bestätigung der Zuverlässigkeit der mit Stoßwellen erhaltenen Ergebnisse ansehen.

Wird bei der Untersuchung des N_2O-Zerfalls die Gesamtdichte erhöht, dann findet man eine Abnahme der Abhängigkeit der Geschwindigkeitskonstante von der Gesamtdichte, und etwa bei $[Ar] \sim 10^{-3}$ Mol/cm³ wird die Geschwindigkeitskonstante unabhängig von der Ar-Konzentration. Man erhält ein Zerfallsgesetz $[\dot{N_2O}] = -k_H \cdot [N_2O]$ mit

$$k_H = 10^{11,2} \exp - \frac{60}{RT} \text{ [sec}^{-1}\text{]}, \quad ([R] = [\text{kcal} \cdot \text{mol}^{-1} \cdot \text{grad}^{-1}]).$$

Das entspricht dem Geschwindigkeitsgesetz einer unimolekularen Reaktion im Hochdruckbereich. Wie beim CO_2 erfolgt die Dissoziation des N_2O über eine „spinverbotene" Reaktion durch Übertritt auf ein Potential anderer Multiplizität. Dieses Potential ist beim N_2O, wie aus spektroskopischen Untersuchungen bekannt, abstoßend. Die minimale Energie E_0, bei der sich die Potentiale durchdringen, ist jedoch nicht bekannt und muß aus kinetischen Messungen bestimmt werden. Aus der gemessenen scheinbaren Aktivierungsenergie ergibt sich unter Anwendung der eingangs angegebenen Formeln ein Wert $E_0 = 63$ kcal/Mol $\pm$ 1/2 RT. In diesem Energiebereich kann man die Schwingungen des N_2O noch in hinreichender Näherung als harmonisch annehmen. Man bekommt dann für die Dichte der Schwingungszustände nahe E_0 den Wert $\varrho_{\text{vib}}(E_0) \approx 820/(\text{kcal/Mol})$. Damit folgt für $A(T) \sim 0,1$ kcal/Mol bei 2000 °K, und die „im Mittel im desaktivierenden Stoß übertragene Schwingungsenergie" ist 0,7 kcal/Mol bei einem RT von 4 kcal/Mol.

Den Einfluß der Rotation kann man bei diesen „spinverbotenen" Reaktionen vernachlässigen. Damit werden die in $A(T)$ enthaltenen Anteile $A_{\text{rot}} \cdot A_{\text{anh}}$ nur wenig von eins verschieden sein.

Die Tatsache, daß der Hochdruckbereich für die Dissoziation schon bei relativ niedrigen Drucken von wenigen hundert Atmosphären beobachtet werden kann, ist aufs engste verknüpft mit der „spinverbotenen" Reaktion und der relativ geringen Übergangswahrscheinlichkeit von einem Potential auf das andere. Das ist nicht nur ein Vorteil für die Messung der Reaktionsgeschwindigkeit, sondern ebenfalls für die Interpretation. Bei normalen Reaktionen, z. B. auch bei der Dissoziation von H_2O und SO_2 ist die funktionale Abhängigkeit der Zerfallswahrscheinlichkeit von der Energie nicht bekannt. In den klassischen Theorien wird dafür eine plausible Funktion angenommen, die man aber anhand der experimentellen Ergebnisse nicht prüfen kann. Im Fall der spinverbotenen Reaktionen ist der funktionelle Zusammenhang zwischen Übergangswahrscheinlichkeit und Energie jedoch aus den Arbeiten von LANDAU [29] bekannt, und man braucht aus den gemessenen Reaktionsgeschwindigkeitsdaten nur einen Faktor zu bestimmen, der das Matrix-Element der Spin-Bahn-Wechselwirkung sowie Ableitungen der potentiellen Energie der beiden beteiligten Zustände nach dem Abstand an der Stelle der Überschneidung beider Potentiale enthält. Die Größe des Matrixelementes kann damit um so genauer bestimmt werden, je besser der Verlauf beider Potentialflächen bekannt ist. Für das N_2O erhält man mit bisher bekannten Daten des Potentialverlaufes für das Matrixelement einen Wert in der gleichen Größenordnung wie für die Spin-Bahn-Kopplung im O-Atom.

Es fällt weiter auf, daß die gemessene Aktivierungsenergie für den Hochdruckbereich verschieden ist von der aus Messungen im Niederdruckbereich bestimmten Energieschwelle E_0. Das wird aber ebenfalls verständlich, wenn man sich überlegt, wie die einzelnen Schwingungen beim Übergang auf das andere Potential mitwirken. Im vorliegenden Fall ergibt sich für den Unterschied der gemessenen Aktivierungsenergie von dem Wert E_0 ein Betrag von etwa 3 kcal/Mol, so daß man für E_0 jetzt aus Nieder- und Hochdruckbereich einen Wert von $E_0 = 63 \pm 1$ kcal/Mol erhält.

Wenn man für N_2O zu wesentlich höheren Temperaturen oder Drucken übergeht, dann ist zu erwarten, daß ein Teil der Dissoziation aus dem Potential des Grundzustandes heraus erfolgt ($\Delta H_0^0 = 83,7$ kcal/Mol), daß man bei den hier angewendeten Drucken also nicht mehr einen reinen Hochdruckbereich der unimolekularen Reaktion hat, sondern der spinverbotenen Reaktion überlagert sich eine unimolekulare Reaktion im Niederdruckbereich, eine Reaktion zweiter Ordnung, die erst bei noch höheren Drucken in den Hochdruckbereich übergeht.

Wir haben hier einige Beispiele angeführt, die zeigen sollten, daß Stoßwellen die Untersuchung von Reaktionen ermöglichen, die mit klassischen Methoden nicht zugänglich sind. Der große verfügbare Bereich der Versuchsbedingungen und bei die vernünftiger Anwendung von Stoßwellen erreichbare hohe Genauigkeit der einzelnen Messung erlauben es, kinetische Daten für Gasreaktionen mit relativ geringem Fehler anzugeben, so daß in geeigneten Fällen ein Vergleich mit theoretischen Behandlungen dieser Reaktionen durchaus sinnvoll erscheint. Dabei ist die Methode selbstverständlich nicht auf die hier besprochenen unimolekularen Reaktionen beschränkt.

Literatur

1. Greene, E. F., u. J. P. Toennies: Chemische Reaktionen in Stoßwellen. Steinkopff Darmstadt: 1959; Chemical Reactions in Shock Waves. London: Edward Arnold, Publishers, Ltd., 1964; Gaydon, A. G., and I. R. Hurle: The Shock Tube in High-Temperature Chemical Physics. London: Chapman & Hall 1963; Bradley, J. N.: Shock Waves in Chemistry and Physics. New York: John Wiley & Sons 1962; ACARDograph No. 41: Fundamental Data obtained from Shock-Tube Experiments. London: Pergamon Press 1961.
2. Courant, R., and K. O. Friedrichs: Supersonic Flow and Shock Waves. New York: Interscience Publ. Inc., 1948.

3. SLATER, N. B.: Theory of Unimolecular Reactions. Ithaca, New York: Cornell Univ. Press 1959.

4. TROE, J., and H. GG. WAGNER: 28th Propulsion and Energetics Meeting der AGARD, Oslo, 1966. J. TROE, H. GG. WAGNER: Ber. Bunsenges. physik. Chem. **71**, 937 (1967). Dort ist eine ausführlichere Diskussion neuer Ergebnisse über unimolekulare Reaktionen gegeben.

5. SCHNEIDER, F. W., and B. S. RABINOVITCH: J. Amer. chem. Soc. **84**, 4215 (1962); **85**, 2365 (1963); RABINOVITCH, B. S., P. W. GILDERSON, and F. W. SCHNEIDER: J. Amer. chem. Soc. **87**, 158 (1965); RABINOVITCH, B. S., and R. W. DIESEN: J. chem. Physics **30**, 735 (1959); **32**, 1245 (1960); RABINOVITCH, B. S., D. W. SETSER, and F. W. SCHNEIDER: Can. J. Chem. **39**, 2609 (1961); RABINOVITCH, B. S., R. F. KUBIN, and R. E. HARRINGTON: J. chem. Physics **3**, 405 (1963); PLACZEK, D. W., B. S. RABINOVITCH, G. Z. WHITTEN, and E. TSCHUIKOW-ROUX: J. chem. Physics **43**, 4071 (1965); WHITTEN, G. Z., and B. S. RABINOVITCH: J. chem. Physics **38**, 2466 (1963); THIELE, E.: J. chem. Physics **39**, 3258 (1963).

6. CORDES, H. F., and H. S. JOHNSTON: J. Amer. chem. Soc. **76**, 4264 (1954).

7. HERAS, M. J., P. J. AYMONINO, u. H. J. SCHUMACHER: Z. physik. Chem. [Frankfurt/M.] **22**, 161 (1959).

8. SCHUMACHER, H. J., u. P. FRISCH: Z. physik. Chem. B **37**, 1 und 18 (1937).

9. TSANG, W., S. H. BAUER, and M. COWPERTHWAITE: J. chem. Physics **36**, 1768 (1962).

10. HOARE, D. E., J. B. PROTHEROE, and A. D. WALSH: Trans. Faraday Soc. **55**, 548 (1959).

11. MICHEL, K. W., and H. GG. WAGNER. X. Symp. on Combustion, 353: Pittsburgh 1965.

12. KOBLITZ, W., u. H. J. SCHUMACHER, Z. physik. Chem. B **25**, 283 (1934).

13. RICE, O. K.: Monatsh. Chem. **90**, 330 (1959).

14. GLISSMANN, A., u. H. J. SCHUMACHER: Z. physik. Chem. B **21**, 323 (1933); BENSON, S. W., and A. E. AXWORTHY: J. chem. Physics **26**, 1718 (1957); JONES, W. M., and N. DAVIDSON: J. Amer. chem. Soc. **84**, 2868 (1962).

15. HUFFMAN, R. E., and N. DAVIDSON: J. Amer. chem. Soc. **81**, 2311 (1959), vgl. a.: LEVITT, B. P.: J. chem. Physics **42**, 1038 (1965).

16. PATTERSON, W. L., and E. F. GREENE: J. chem. Physics **36**, 1146 (1962).

17. SCHOFIELD, D., W. TSANG and S. H. BAUER: J. chem. Physics **42**, 2132 (1965).

18. DEKLAN, P., and H. B. PALMER: VIII. Sympos. on Combustion, 139. Baltimore: Williams & Wilkinson 1962; ASHMORE, P. G., u. M. G. BURNETT: Trans. Faraday Soc. **58**, 1801 (1962).

19. OLSCHEWSKI, H. A., J. TROE u. H. GG. WAGNER: Z. physik. Chem. [Frankfurt/M.] **44**, 173 (1965).

20. OLSCHEWSKI, H. A., J. TROE u. H. GG. WAGNER: Z. physik. Chem. [Frankfurt/M.] **47**, 383 (1965); Z. physik. Chem. [Frankfurt/M.], im Druck.

21. a) JOHNSTON, H. S.: J. chem. Physics **19**, 663 (1951).
b) OLSCHEWSKI, H. A., J. TROE u. H. GG. WAGNER: Ber. Bunsenges. physik. Chem. **70**, 450 (1966); s. a.: FISHBURNE, E. S., and R. EDSE: J. chem. Physics **41**, 1297 (1964).

22. MICHEL, K. W., H. A. OLSCHEWSKI, H. RICHTERING u. H. GG. WAGNER: Z. physik. Chem. [Frankfurt/M.] **39**, 129 [1963]; **44**, 160 (1965); s. a.: LOSEV, S. A., N. A. GENERALOV u. V. A. MAKSIMENKO: Doklad. Akad. Nauk. SSSR **150**, 839 (1963); BRABBS, T. A., F. E. BELLES, and S. A. ZLATARICH: J. chem. Physics **38**, 1939 (1963); DAVIES, W. O.: J. chem. Physics **41**, 1840 (1964); **43**, 2809 (1965); FISHBURNE, E. S., K. R. BILWAKESH, and R. EDSE: J. chem. Physics **45**, 160 (1966).

23. Schecker, H. Gg.: Dissertation, Göttingen 1966.
24. Olschewski, H. A., J. Troe u. H. Gg. Wagner: Z. physik. Chem. [Frankfurt/M.] **45**, 329 (1965).
25. Bauer, S. H., G. L. Schott, and R. E. Duff: J. chem. Physics **28**, 1089 (1958).
26. Gaydon, A. G., G. H. Kimbell, and H. B. Palmer: Proc. Roy. Soc. [London] A **279**, 131 (1964).
27. Levitt, B. P., and D. B. Sheen: J. chem. Physics **41**, 584 (1964).
28. Martinengo, A., J. Troe u. H. Gg. Wagner: Z. physik. Chem. [Frankfurt/M.] **51**, 104 (1966).
29. Landau, L. D., and E. M. Lifshitz: Quantum Mechanics. London-Paris: Pergamon Press 1958.
30. Olschewski, H. A., J. Troe u. H. Gg. Wagner: Ber. Bunsenges. physik. Chem. **70**, 1060 (1966).
31. G. Herzberg: Molecular Spectra and Molecular Structure, New York: van Norstand, 1945.

W. Jost
Universität Göttingen
H. Gg. Wagner
Ruhr-Universität Bochum

Atom Reactions in Discharge-Flow Systems

B. A. THRUSH

1. Experimental Methods

The use of electric discharges as sources of free atoms originated in the 1920's with the work of BONHOEFFER [1] and of HARTECK [2] in Germany and of WOOD [3] in England. Interest in these systems died out during the 1940's but has revived tremendously during the last ten years. This resurgence of activity has arisen with the development of more specific methods for measuring the concentrations of free atoms which have made accurate and detailed kinetic studies possible.

In these systems, free atoms are usually produced by passing gases at about 1—5 mm Hg pressure through an electrodeless discharge, which may operate at microwave or radio frequencies. Such a discharge will dissociate hydrogen, nitrogen, oxygen, chlorine or bromine into atoms to the extent of about 1—2%. Large concentrations of the undissociated parent molecules can be avoided by passing a few per cent of this gas in an argon or helium carrier through the discharge. The reactions of the discharge products are normally studied in a cylindrical flow line down which they are pumped with a velocity typically of 1—5 metres per second. Such a flow line might have a diameter of 2—3 cm and be equipped with a series of jets through which known flows of reactants could be added, rate constants being determined by the change of atom concentration downstream from the point of addition. In the absence of added reactant, atoms recombine only in three body collisions in the gas-phase or at a small fraction of collisions with the walls. With clean glass this fraction (γ) is only about 10^{-5} for nitrogen or oxygen atoms; similar values for hydrogen or halogen atoms can be obtained by coating the surface with Teflon or an oxyacid respectively.

In the early studies of discharge flow systems only non-specific methods were available for measuring the concentrations of free atoms. These were calorimetric detectors which measured their heat of recombination on a metal surface and the Wrede gauge. The latter measures the pressure difference established across a small orifice which separates the flow system from a catalyst for atom recombination. This pressure difference arises from the different mass rates of effusion of atoms and molecules and can be directly converted into atom concentrations.

16*

In recent years, mass spectrometric and chemiluminescence methods have been developed; electron spin resonance is also a promising technique [4]. All these methods rely to some extent on rapid stoichiometric titration reactions for determining atom concentration. In this discussion, I will concentrate on methods based on chemiluminescence, since these need nothing more elaborate than a photomultiplier cell which provides a simple linear method of measuring low light intensities with a very wide range of sensitivity.

Chemiluminescence methods are based on the measurement of the intensities of emission by electronically excited molecules formed in the combination reactions of atoms. These combination reactions require a third body and need not therefore be so rapid as to contribute significantly to the overall rate of loss of atoms. In most of these processes, excited molecules are formed by three body recombinations and are removed either by radiation or collisional quenching. The intensity of the associated light emission is therefore proportional to the product of the concentrations of the active species. Its variation with pressure, temperature and nature of the carrier gas used, will depend partly on the relative importance of radiation and quenching. The emissions which can be used include the yellow nitrogen afterglow [5, 6] from $N + N$, the blue nitric oxide afterglow [6] from $N + O$, the green air afterglow [7] from $O + NO$, red HNO emission [8] from $H + NO$, and the red glows associated with chlorine atom recombination [9] and with bromine atom recombination [10].

In practice, the use of nitric oxide addition to measure hydrogen or oxygen atom concentrations is very convenient, since any nitric oxide removed by the combination reactions which accompany the chemiluminescence:

$$H + NO + M = HNO + M \quad \text{or} \quad O + NO + M = NO_2 + M$$

is rapidly regenerated in the secondary reactions [7, 11]

$$H + HNO = H_2 + NO \quad \text{or} \quad O + NO_2 = O_2 + NO \ .$$

This considerably simplifies a point-by-point determination of atom concentration in a flow tube, since nitric oxide is conserved, and the measured emission intensity is directly proportional to the atom concentration.

These chemiluminescence methods may be calibrated by using a calorimetric method to measure absolute atom concentrations. Another method is to use gas-phase titration.

The titration of nitrogen atoms with nitric oxide was first described in 1957 by Kistiakowsky and Volpi [12] and by Kaufman and Kelso [13]. They showed that these species underwent a rapid stoichiometric reaction

$$N + NO = N_2 + O$$

which has since been found to occur at one collision in four between these species [14]. As increasing amounts of nitric oxide are added to active nitrogen, the yellow nitrogen afterglow is progressively replaced by the blue glow from N + O. When the stoichiometric amount of nitric oxide is being added, no glow is observed downstream since oxygen atom recombination gives no visible emission (if it did, the sky would glow brightly at night). Addition of excess nitric oxide produces the green air afterglow from the combination of O and NO. It should be noted that an alternative method for measuring nitrogen atom concentration, based on the limiting HCN yield when ethylene is added at elevated temperatures, has recently been shown to be inaccurate.

KAUFMAN [7] has shown that oxygen atoms can be titrated with nitrogen dioxide according to the stoichiometric equation

$$O + NO_2 = NO + O_2 \ .$$

In this case, the air afterglow from O + NO reaches its maximum intensity for half the stoichiometric addition of nitrogen dioxide and is extinguished at the end-point. This titration reaction occurs at one collision in two hundred between reactants [15] and it must be allowed enough time for completion.

The afterglows discussed above all arise from ground state atoms. Metastable excited atoms do not appear to have significant concentrations in any discharge-flow experiments. Excited molecular species can be important in discharged oxygen, where the low-lying $^1\Sigma_g^+$ state (energy 37 kcal/mole) and $^1\Delta_g$ state (energy 22 kcal/mole) of O_2 can constitute 0.1% and 10% of the gases [19, 20]. These species can be studied using their infra-red emission, also two $^1\Delta_g$ O_2 molecules give a dimole emission band at 6340 Å. O_2 ($^1\Delta_g$) is best obtained by passing the products of a discharge through oxygen over a mercuric oxide surface.

Powerful discharges through nitrogen give vibrationally excited ground state N_2 molecules. In general, there is no real evidence for high concentrations of electronically excited nitrogen molecules. H_2 has no low-lying metastable states.

Apart from OH [16, 17] the study of free radical reactions in flow systems by measurements on their ultra-violet absorption spectra is in its infancy. This is however a promising technique for several species. The free radical SO, which is a comparatively stable species, can be studied by the chemiluminescent emission which it gives with oxygen atoms (the sulphur dioxide afterglow) and with ozone [18]. It can be seen that discharge flow systems are complementary to flash photolysis as a technique, being largely concerned with the rapid reaction of atoms, while only free radicals are normally studied by flash photolysis.

2. Applications

The use of discharge-flow systems to determine the rate constants of the elementary steps in the hydrogen-oxygen reaction provides an excellent example of these methods. From direct studies it was deduced that the reaction between hydrogen and oxygen at elevated temperatures proceeds by a branched chain mechanism,

$$H + O_2 = OH + O -\ 17 \text{ kcal/mole} \tag{1}$$
$$O + H_2 = OH + H -\ \ \ 2 \text{ kcal/mole} \tag{2}$$
$$OH + H_2 = H_2O + H + 15 \text{ kcal/mole} \quad \text{twice} \tag{3}$$

$$\overline{H + 3\,H_2 + O_2 = 2\,H_2O + 3\,H} \ \ .$$

As the pressure is increased this reaction shows three explosion limits at temperatures between 400 and 550° C. At the first explosion limit the rate of production of hydrogen atoms in the chain reaction is balanced by their removal at the walls. For pressures above this, explosion occurs until the second limit is reached, when the reaction

$$H + O_2 + M = HO_2 + M + 47 \text{ kcal/mole} \tag{4}$$

can remove hydrogen atoms rapidly enough. The third explosion limit at high pressure has been less fully investigated.

The strongly endothermic reaction (1) controls the rate of chain branching; its rate constant can be determined from the first explosion limit where it competes with wall diffusion. Once k_1 has been found, k_4 can be obtained at the second explosion limit. The rate of one of the chain propagation steps, (2) or (3) can be found from the detailed kinetics, but this reaction cannot be positively identified.

Reaction (2) provides the best illustration of the use of chemiluminescence methods [21]. To study this reaction in solution, it is necessary to work in the absence of molecular oxygen, since the occurrence of reaction (4) followed by the subsequent reactions of HO_2 would complicate the interpretation. Oxygen atoms in a molecular nitrogen carrier can be obtained by titrating active nitrogen with nitric oxide as described in the previous section. If a slight excess of nitric oxide is added, the resultant air afterglow can be used to monitor the oxygen atom concentration. Addition of molecular hydrogen then produces an additional decay of the oxygen atoms which is first order both in O and in H_2. Before these decays can be converted into rate constants for reaction (2), the overall stoichiometry of the reaction has to be established. Hydrogen atoms will not undergo further reaction at the times involved, but the hydroxyl radical can react in two possible ways

$$O + OH = O_2 + H + 17 \text{ kcal/mole} \tag{-1}$$
or
$$OH + H_2 = H_2O + H + 15 \text{ kcal/mole} \ . \tag{3}$$

If the former occurs the stoichiometry will be $2\,O + H_2 = O_2 + 2\,H$ whereas the latter gives $O + 2\,H_2 = H_2O + 2\,H$. In fact, measurement of the concentrations of hydrogen atoms produced shows that the former stoichiometry is correct and that the rate of removal of oxygen atoms is twice that of reaction (2).

The rate expression obtained [21] for reaction (2) was $k_2 = 1.2 \times 10^{13} \exp\,(-9400/R\,T)\,\mathrm{cm^3\,mole^{-1}sec^{-1}}$. This is in good agreement with BALDWIN's value [22] for the rate of the propagating step in the hydrogen-oxygen reaction and with determinations in flames and from the effect of hydrogen on the carbon monoxide-oxygen reaction [23].

Reaction (1) cannot be studied directly, since its equilibrium lies too far to the left, but its rate constant can be deduced from measurements on the reverse process (−1, above) and the calculated equilibrium constant. Hydroxyl radicals are most readily obtained from the reaction of hydrogen atoms with nitrogen dioxide

$$H + NO_2 = NO + OH \ .$$

In the absence of molecular hydrogen (that is, when hydrogen atoms are obtained from a discharge through a trace of hydrogen in an argon carrier) the dominant reactions removing OH are

$$OH + OH = H_2O + O \qquad\qquad (5)$$
$$O + OH = H + O_2 \ . \qquad\qquad (-1)$$

Reaction (−1) is faster than (5), and measurement of the decay of the small steady oxygen atom concentration which is established between them gives a value [23] of $k_{-1} = 3 \times 10^{13}\,\mathrm{cm^3\,mole^{-1}\,sec^{-1}}$ between 200° K and 300 °K. This value is in good agreement with determinations of k_1 from explosion limit data when combined with its calculated equilibrium constant.

Rate constants for reaction (3) are obtained using measurements of the OH concentration by electronic absorption spectroscopy. Unfortunately there is very poor agreement between the results of workers who used the $H + NO_2$ reaction as the source of OH radicals [16] and those who used a discharge through water vapour [17]. Further work on this problem is clearly needed.

The homogeneous chain termination reaction (4) can be studied

$$H + O_2 + M = HO_2 + M \qquad\qquad (4)$$

by using HNO emission to follow the reaction between molecular oxygen and hydrogen atoms in an argon carrier [24]. As with reaction (2), the overall stoichiometry is a problem, since the subsequent reactions of HO_2 in these systems are:

$$H + HO_2 = OH + OH$$
$$= H_2O + O$$
$$= H_2 + O_2$$

followed by reactions (5) and (-1). Measurements of the small steady-state oxygen atom concentration present in this reaction and product analysis have resolved this satisfactorily. The rate constant obtained at room temperature, $k_4 = 1.3 \times 10^{16}$ cm^6 mole^{-2} sec^{-1} for $M = $ Ar agrees very well with explosion limit data if the overall temperature dependence is about T^{-2}, and such a dependence was found at and below room temperature [24].

3. Chemiluminescent Reactions

The formation of vibrationally excited molecules in atom transfer reactions has been studied over the last ten years by flash photolysis and by the infra-red chemiluminescence technique of J. C. Polanyi. Monte Carlo calculations of the relevant types of potential surface have been made by Bunker and others to find how the energy distribution in the products is affected by the surface chosen.

The chemiluminescent reactions described above are all recombination processes yielding electronically excited molecules. In almost every case the emitting state is not one which correlates with ground state dissociation products, but the emission comes from levels just below this limit. From the temperature dependence of the emission, it is clear that ground state reactants must approach each other on an attractive potential curve, and then cross from this into the state from which radiation occurs. Both the kinetics of the emission and the fact that it comes from levels below the dissociation limit show that stabilisation by a third body is necessary, and there is evidence that this stabilises the molecule into the state which correlates with normal atoms.

One interesting point is that the measurement of the absolute intensity of the chemiluminescence combined with a knowledge of the fluorescence efficiency of the emitting state, yields a value of the absolute rate of recombination via the emitting state. This quantity is surprisingly large in many cases. For the nitrogen afterglow, it is about half the total rate of recombination [25]. Recent work has shown that the main emitting state ($B\,^3\Pi_g$) is populated from the metastable triplet state ($A\,^3\Sigma_u^+$). On purely statistical grounds, one would expect three-quarters of the recombination to occur into the A state and one quarter into the ground state ($^1\Sigma_g^+$), but the proportion is probably lower than this because the A state is shallower than the ground state. This work shows clearly how readily crossing can occur between the A and B states.

In the air afterglow, almost all the recombination appears to occur via the excited state of NO_2 responsible for light emission [26, 27]. This excited state probably becomes degenerate with the ground state of NO_2 in a linear configuration, where both would correspond to a $^2\Pi$ state. As a result, free crossing between the states occurs, both in fluorescence and in chemiluminescence experiments. This is, of course, the phenomenon of internal conversion familiar in the photochemistry of larger molecules. This occurs in both directions (into and out of the excited electronic state) for a triatomic molecule, whereas for a more complex species, the density of levels at high energies of the ground state is so great that the crossing to the ground state is effectively irreversible.

The formation of electronically excited molecules in transfer reactions is comparatively rare [28]. It can readily be seen that most transfer reactions should have higher activation energies to yield electronically excited products than to yield products in their ground states. This arises because the potential curves for excited states of the product will generally lie above that of the ground state at all the internuclear distances concerned. CN is an exception to this [29], at internuclear distances greater than 1.5 Å the $A\,^2\Pi$ state lies below the $X\,^2\Sigma^+$ state. This is because promotion of an electron from the $\pi_u 2p$ to the $\sigma_g 2p$ orbital increases bonding at large distances and decreases it at small distances. It is not surprising that the strong CN emission in the reaction between active nitrogen and halogenated hydrocarbons can be attributed to the process

$$XC + N = CN^*(A^2\Pi) + X$$

where

$$X = H,\ Cl\ or\ Br.$$

Spin conservation has yet to be shown to be important in governing the formation of electronically excited molecules in transfer reactions in the gas phase. The other factor which operates is correlation of potential surfaces. This is illustrated by our recent work in low pressure flow systems on the reactions

$$NO + O_3 = NO_2 + O_2 +\ \ 48\ kcal/mole$$

$$SO + O_3 =\ \ SO_2 + O_2 + 106\ kcal/mole\ .$$

These yield NO_2 and SO_2 respectively both electronically excited and in their ground states. There was no evidence of the formation of electronically excited oxygen. In the $NO + O_3$ reaction, the formation of ground state and electronically excited NO_2 have activation energies of 2.4 and 4.1 kcal/mole respectively [30], but both processes apparently have similar pre-exponential factors $(5 \times 10^{11}\ cm^3\ mole^{-1}\ sec^{-1})$ [31]. For the $SO + O_3$ reaction, the activation energy for formation of ground state SO_2 was 2.1 kcal/mole, while that for the excited singlet state was 4.2 kcal/mole [32]. This state has an excitation energy of 85 kcal/mole, and is formed with up to 16 kcal/mole of vibrational energy. In this system, the frequency factor for the reaction yielding ground state SO_2 is 1.5×10^{12} cm^3 mole^{-1} sec^{-1}; for the excited singlet state it is 10^{11} cm^3 mole^{-1} sec^{-1} and it is even smaller for the population of the $^3B_1$ state of SO_2.

The essential difference between these two reactions is that the degeneracy associated with orbital angular momentum in the ground state of NO ensures that the reactants $NO\,(^2\Pi) + O_3\,(^1A)$ give two doublet potential surfaces. One of these correlates with ground state products $NO_2\,(^2A_1) + O_2\,(^3\Sigma_g^-)$, the other with excited $NO_2\,(^2B_1) + O_2\,(^3\Sigma_g^-)$. In the other reaction, $SO\,(^3\Sigma^-) + O_3\,(^1A)$ give only one triplet surface which must correlate with ground state products. Formation of electronically excited SO_2 can only occur by crossing to another potential surface, which is believed to occur at the top of the potential barrier.

The techniques described here can be applied to a wide range of gas reactions, and publications in this field are making a noticeable contribution to the current "explosion" in the chemical literature. For this reason, only 'key' references to original work are given in the text. There are a few recent review articles [17, 28, 33–36].

References

1. BONHOEFFER, K. F.: Z. phys. Chem. 113, 199, 492 (1924); 116, 391 (1925).
2. HARTECK, P., and V. KOPSCH: Naturwiss. 1929, 727.
3. WOOD, R. W.: Proc. Roy. Soc., A, 102, 1 (1922).
4. WESTENBERG, A. A., and N. DE HAAS: J. Chem. Phys. 42, 1785 (1965).
5. BERKOWITZ, J., W. A. CHUPKA, and G. B. KISTIAKOWSKY: J. Chem. Phys. 25, 457 (1956).
6. YOUNG, R. A., and R. L. SHARPLESS: J. Chem. Phys. 39, 1071 (1963).
7. KAUFMAN, F.: Proc. Roy. Soc. A, 247, 123 (1958).
8. CLYNE, M. A. A., and B. A. THRUSH: Disc. Faraday Soc. 33, 139 (1962).
9. BADER, L. W., and E. A. OGRYZLO: J. Chem. Phys. 41, 2926 (1964).
10. GIBBS, D. B., and E. A. OGRYZLO: Canad. J. Chem. 43, 1905 (1965).
11. CLYNE, M. A. A., and B. A. THRUSH: Trans. Faraday Soc. 57, 1305 (1961).
12. KISTIAKOWSKY, G. B., and G. G. VOLPI: J. Chem. Phys. 27, 1141 (1957).
13. KAUFMAN, F., and J. R. KELSO: J. Chem. Phys. 27, 1209 (1957).
14. CLYNE, M. A. A., and B. A. THRUSH: Proc. Roy. Soc. A, 261, 259 (1961).
15. KLEIN, F. S., and J. T. HERRON, J. Chem. Phys. 41, 1285 (1964).
16. KAUFMAN, F., and F. P. DEL GRECO: Disc. Faraday Soc. 33, 128 (1962).
17. AVRAMENKO, L. I., and R. V. KOLESNIKOVA: Adv. in Photochem. 2, 25 (1964).
18. HALSTEAD, C. J., and B. A. THRUSH: Proc. Roy. Soc. A, 295, 380 (1966).
19. CLYNE, M. A. A., B. A. THRUSH, and R. P. WAYNE: Nature 199, 1057 (1963).
20. BADER, L. W., and E. A. OGRYZLO: Disc. Faraday Soc. 37, 46 (1964).
21. CLYNE, M. A. A., and B. A. THRUSH: Proc. Roy. Soc. A, 275, 544 (1963).
22. BALDWIN, R. R.: Trans. Faraday Soc. 52, 1344 (1956).
23. AZATYAN, V. V., V. V. VOEVODSKII, and A. B. NALBANDYAN: Kinetika i Kataliz 2, 340 (1961).
24. CLYNE, M. A. A., and B. A. THRUSH: Proc. Roy. Soc. A, 275, 559 (1963).

25. CAMPBELL, I. M., and B. A. THRUSH: Chem. Comm. **1965**, 250.
26. CLYNE, M. A. A., and B. A. THRUSH: Proc. Roy. Soc. A, **269**, 404 (1962).
27. HARTLEY, D. B., and B. A. THRUSH: Disc. Faraday Soc. **37**, 220 (1964).
28. THRUSH, B. A.: Chem. in Britain **2**, 287 (1966).
29. SETSER, D. W., and B. A. THRUSH: Proc. Roy. Soc. A, **288**, 256 (1965).
30. CLYNE, M. A. A., B. A. THRUSH, and R. P. WAYNE: Trans. Faraday Soc. **60**, 359 (1964).
31. CLOUGH, P. N., and B. A. THRUSH: Chem. Commun. **1966**, 783.
32. HALSTEAD, C. J., and B. A. THRUSH: Photochem. and Photobiol. **4**, 1007 (1965).
33. KAUFMAN, F.: Prog. in Reaction Kinetics **1**, 3 (1961).
34. JENNINGS, K. R.: Quart. Revs. **12**, 116 (1958).
35. THRUSH, B. A.: Prog. in Reaction Kinetics **3**, 63 (1965).
36. CAMPBELL, I. M., and B. A. THRUSH: Ann. Reps. Chem. Soc. **62**, 17 (1965).

B. A. Thrush
University of Cambridge
Cambridge/Great Britain

Flash Photolysis

B. A. Thrush

1. Introduction

If the rate constants of elementary chemical reactions are expressed in the simple Arrhenius form, $A \cdot \exp(-E/RT)$, the value of A is frequently as high as $10^9 - 10^{11}$ l mole^{-1}sec^{-1} both for bimolecular reactions in the gas phase and for diffusion controlled processes in solution. Conventional methods of studying reaction kinetics are therefore limited to processes which have comparatively large activation energies, where it is rarely possible to determine A accurately enough to justify the use of any expression other than the simple Arrhenius one. Other limitations in conventional kinetics are lack of information on the nature of intermediates in multistep reactions and on the energy distribution in the reaction products.

Various techniques for studying rapid elementary processes have been developed over the last twenty years. Of these, only molecular beams [1] are not limited by time resolution, since the observed quantities are probabilities of scattering. The relaxation methods of EIGEN [2] cover an exceptionally wide range of reaction times, but are limited to systems in which the reaction studied participates in an equilibrium or steady state which is affected by some applied perturbation e.g. temperature change, pressure or electric field. Such studies have provided much information about rapid reactions in solution, but cannot be applied to systems which are far from equilibrium.

Shock tubes [3] provide a very convenient method of studying many high-temperature gas reactions, particularly under conditions which are far from equilibrium. Difficulties are often encountered in separating the chemical processes from relaxation phenomena, and the problem of the inter-relation of sensitivity and time resolution are generally greater than in flash photolysis because the optical path across the shock tube is smaller.

The technique of flash photolysis was developed by NORRISH and PORTER [4, 5] some fifteen years ago. Since then the technique has been used in a considerable number of laboratories to study the kinetics of reactions in solution and in the gas phase, and to obtain the absorption spectra of free radicals and of other transient species. The present aim is to show how a wide range of problems can be tackled with comparatively simple apparatus.

2. The Apparatus

The basic apparatus for flash photolysis consists of a cylindrical quartz reaction vessel, anything from 25 cm to 2 m long, with one or more quartz flash discharge lamps of equal length placed alongside it. The whole is surrounded by a cylindrical reflector coated with magnesium oxide to concentrate the light on the reaction vessel. Large electrodes at the ends of the lamp are connected by short heavy leads to a condenser bank. For a typical inert-gas filled flash tube 50 cm long and about 15 mm i.d., a 40 μF 10 kV condenser could be used giving a flash duration of about 20 μ sec with careful layout. Thus 2000 J of energy is dissipated in about 20 μ sec, and some 10% of this is converted to photochemically active light between 2000 Å and 5000 Å. This corresponds to nearly 10^{-3} Einsteins, so it would be possible for a substance with a reasonably intense absorption spectrum to absorb between 10^{-4} and 10^{-5} Einsteins of photochemically active light in 20 μ sec, the volume of the vessel being about 200 ccs. For a substance with an f-value of 0.1 at a concentration of 10^{-4} moles 1^{-1}, this would correspond to excitation of about 30% of the molecules in one photolytic flash, depending on the form of its absorption spectrum. Although aromatic molecules frequently have absorptions as strong as this, many other molecules of photochemical interest, for instance aldehydes and ketones, have much weaker absorptions. More intense flashes are then necessary to achieve large decompositions. Several very ingeneous designs have been described by CLAESSON [6], but these require considerable workshop facilities. One metre long flash tubes capable of dissipating 10000 J in 10 μ sec have been constructed in Ottawa for the photolysis of gases [7]. These consist of quartz tubes about 25 mm diameter containing air at low pressures. Arrangements to minimize the circuit inductance include return leads consisting of two bars placed close to opposite ends of a diameter of the flash tube.

The simplest method of studying radical reactions by flash photolysis is product analysis and comparison with the products of steady photolysis. For this aldehydes and ketones provide a useful method of distinguishing the products of radical-molecule reactions which predominate in steady photolysis from the products of radical-radical reactions which are mainly formed at the higher radical concentrations produced in flash photolysis [8, 9]. One reason that this method has not been used more extensively is that the nature and energy of the primary products in aldehyde and ketone photolysis vary with the wavelength of the existing radiation. Flash sources give essentially continuous emission over a wide range of frequencies, and it is difficult to obtain really high intensities of light which is monochromatic enough to compare with the results of steady photolysis where line sources are used.

In the vast majority of studies using flash photolysis, a spectroscopic method is used to follow the reaction. Observations can be made in two ways, either by following absorption at one wavelength using a continuous source (xenon or mercury arc, quartz-iodine lamp), monochromator, photomultiplier cell and oscilloscope, or by recording the absorption spectrum in known time after photolysis with a second flash and a photographic spectrograph. Both methods have adequate time resolution in comparison with the photolytic flash. Combination of the two methods in a rapid scanning system, leads to an awkward compromise between scanning time and spectral resolution. This method is only suitable for studies in the infra-red where photographic methods are not available; progress is now being made in this field [10].

In the photographic method, a photocell receiving light from the photolytic flash triggers an electric delay which fires a spark gap or hydrogen thyratron in series with the spectroscopic flash tube. This is commonly a capillary discharge through an inert gas having an energy of 100—200 J ($2-4\ \mu$F at 10 kV). For vacuum ultra-violet spectroscopy, the low pressure ($10-30\ \mu$) wide bore flash sources developed by Garton [11] are particularly suitable.

The choice of spectrograph depends on the nature of the problem. For reactions in solution medium quartz or glass instruments provide adequate resolution and high optical speed. For work in the gas phase, the high resolution needed to analyse free radical spectra and to detect small quantities of discrete absorbers is best obtained with a grating instrument. Ebert mountings of plane gratings are now favoured because of the high speeds obtainable in high orders since they are stigmatic instruments. In these studies, the detection sensitivity is frequently increased by the use of multiple transversal cells with the monitoring light beam. With such systems, the experiment may have to be repeated many times to achieve adequate plate density. For kinetic studies in gases, when a discrete spectrum has to be observed at many different times after the photolytic flash and in systems of differing composition, a compromise between spectroscopic resolution and recording speed has to be made. For this purpose large Littrow prism spectrographs prove very useful.

From the viewpoint of recording information, the flash photographic method of following a reaction is very impressive, since the absorption at 10^4 resolved wavelengths can be recorded in ca. $5\ \mu$ sec. Unfortunately, most of this information is redundant in a kinetic study, where the time dependence of the absorption at a few selected wavelengths will provide the kinetic data needed. Nevertheless, preliminary experiments using photographic methods are highly desirable and in fact almost essential to ensure that the absorptions measured in a photoelectric study correspond to a particular absorption spectrum. With this reservation photoelectric methods offer real advantages for kinetic studies since the problems of calibration and of reproducibility which occur in photographic studies are avoided. The chief difficulty in photoelectric measurements arises from scattered light and electrical and magnetic interference from the photolytic flash. These can be overcome with care. Photoelectric measurements need not be confined to relatively broad absorptions which obey the Beer-Lambert Law, but can also be applied to complex discrete spectra such as those of CN [12].

Without specially designed apparatus it is difficult to study reactions with half-lives less than 10^{-5} sec. Thus it is possible to study first order reactions with rate constants up to 10^5 sec^{-1}; bimolecular rate constants up to the gas phase collision number (10^{11} l $mole^{-1}$ sec^{-1}) can be measured providing a reactant or product has a molar extinction coefficient of not less than 1000. For bimolecular processes in solutions and for third order reactions the extinction coefficients needed are an order of magnitude less.

3. Spectroscopic Studies

Purely spectroscopic observations provide useful information in the study of elementary processes in reaction kinetics. Apart from the identification of intermediates in many photochemical processes, e.g. CH_3 and CHO from CH_3CHO [13], ClO from ClO_2 [14], CH_2 from CH_2N_2 [13], N_3 from HN_3 [15], the spectra themselves can yield useful information about the stability of these intermediates. The determination of the structures of free radicals by rotational analysis of their spectra, mainly in Ottawa [13, 16] has strikingly confirmed WALSH's predictions [17] as to the configurations of triatomic species in their ground and excited states. In addition, the interaction between the electronic orbital angular momentum and the angular momentum associated with bending vibrations of a linear polyatomic molecule, which was predicted by RENNER [18] has been detected in triatomic radical spectra.

The dissociation energies of several radicals (notably ClO [19]) have been found by flash photolysis, but unfortunately little information has been obtained about bond energies in polyatomic species. Flash photolysis has almost settled the vexed question as to whether the ground state of CH_2 is a linear triplet or a bent singlet. It is clear that these states have closely similar energies, the former being almost certainly the ground state [13, 20]. In contrast to the short-lived and highly reactive methylene, CF_2 which apparently has a bent singlet ground state is remarkably unreactive.

The ionisation potentials of three polyatomic radicals observed by flash photolysis, CH_2, CH_3 [13] and C_7H_7 radicals [21] have been determined from Rydberg series. It is to be hoped that more transient species will be studied in this way to provide an independent check on mass spectrometric studies.

Many aromatic radicals have been identified by their spectra in solution, and in some cases these spectra have also been observed in the gas phase. Where a detailed analysis is not possible, the carrier is normally identified by which members of a group of related compounds yield its spectrum [22]. This method can be dangerous since the cyclic C_5H_5 radical [23] is obtained from both aromatic species [24] and linear conjugated hydrocarbons [25].

4. The Recombination of Atoms

One of the simplest reactions which can be studied by flash photolysis is the three body recombination of iodine atoms

$$I + I + M = I_2 + M$$

This can be followed quite simply by photoelectric measurements of the reappearance of the strong absorption spectrum of molecular iodine following flash photolysis. Unfortunately, several complications arise which illustrate the potential difficulties and how these may be overcome in practice.

The absorption spectrum of atomic iodine lies in the vacuum ultra-violet, which is a difficult region for quantitative absorption spectroscopy. Even if the atomic spectrum lay in a readily accessible region, problems of calibration would arise since the absorption coefficients for atomic and free radical spectra are generally unknown. These absorption coefficients are normally found by material balance, which in this case is the assumption that the concentration of iodine atoms is given directly by the difference between the initial and transient absorptions by I_2. This assumption produces the problem that the iodine atom has a low-lying metastable $^2P_{1/2}$ state in addition to its $^2P_{3/2}$ ground state, and these two states probably have different rates of recombination. Light absorption by I_2 in the banded region can only give two $^2P_{3/2}$ ground state atoms by predissociation, but light absorption in the continuum yields one $^2P_{3/2}$ and one $^2P_{1/2}$ iodine atom, which may persist. Fortunately, recent studies by vacuum ultra-violet spectroscopy [26] show that the $^2P_{1/2}$ atom is rapidly quenched to the ground state, and that this is not a real difficulty. However, the general problem of identifying all the photolysis products, so that mechanisms can be identified uniquely and absolute rate constants determined, is a serious one in more complex systems. Fortunately the decay of most transient species is first order with respect to that species, and its absolute concentration is not needed to determine absolute rate constants.

One more problem remains, that of temperature rise which is confined to gaseous systems. Most of the light energy absorbed from the flash appears as heat, either directly as kinetic energy of photolysis products, or indirectly by degradation of the absorbed energy or heat released in subsequent reactions. Even with large excesses of inert gas, this temperature rise will be appreciable.

Since heat is lost to the walls, a radial temperature gradient is established in the vessel, and as the gas pressure is constant throughout the vessel, the gas density rises at the walls and falls at the centre through which the monitoring beam passes. The reduction in the number of absorbing molecules in the light path increases the apparent concentration of iodine atoms. This effect is negligible in solution, but in gases it accounts for the very high apparent decomposition found in some early flash photolysis experiments, where the temperature rise was sufficient to drive a significant fraction of the parent molecules from the centre to the vessel walls.

What has been said above should not be taken to imply that flash photolysis is a difficult technique or that its accuracy is limited. In fact, the data on iodine atom recombination [27, 28, 29] obtained from flash photolysis cover a wide range of third bodies, and yield more accurate temperature coefficients and absolute rate constants than have yet been obtained for any other recombination process. It is the only such system for which it has been established that the relative efficiencies for different third bodies change with temperature. If the third order rate constant is expressed in the form $A \exp (+E/RT)$, differences in E rather than in A determine the relative efficiencies [29]. It is not yet established that such a relationship will apply to other recombinations where charge transfer interactions are less important.

Bromine atom recombination has also been studied by flash photolysis. Chlorine atoms are less easy to work with, molecular chlorine being a much weaker absorber, so that it is less readily decomposed and the recombination is harder to monitor. With these systems excited $^3\Pi_{o^+,u}$ molecules of the Cl_2 or Br_2 are formed, probably in the recombination [30].

5. Radical Reactions in Gases

Our knowledge of the chemistry of the ClO radical comes largely from flash photolysis. It is an important intermediate in the photochemistry of chlorine oxides and chlorine-oxygen systems [31]. There are still a number of unsolved problems connected with its reactions, but these arise because of the difficulty of measuring chlorine atom concentration in flash photolysis.

The reactions of a number of other radicals have been studied under essentially isothermal conditions, these include CN, CS and SO, but much work remains to be done in this field.

Quite a lot of information about the reactions of free radicals in combustion systems has been obtained from reactions initiated by flash photolysis. A sensitizer such as nitrogen dioxide [32] or an alkyl nitrite [33] is added to a mixture of fuel and oxygen. In this way, explosive combustion can be initiated homogeneously throughout a reaction vessel and the absorption spectra of the free radicals involved can be obtained with a very much longer optical path than would be practicable in a flame. Unfortunately there is a limit to the homogeneity which can be achieved [34], but the method has proved very useful for investigating the mechanisms of carbon formation [32] and of the action of anti-knock additives [33].

Apart from fluorescence studies which provide information about vibrational energy transfer in excited electronic states, most methods of studying vibrational relaxation only yield information about the lowest vibrational levels.

Flash photolysis was first used to study the formation and quenching of vibrationally excited oxygen molecules in the photolysis of NO_2, ClO_2 and O_3, where these species are formed in the secondary reaction [14]

$$OXO + h\nu = O + XO$$
$$O + OXO = O_2^* + XO \ .$$

In these reactions the vibrational excitation appears primarily in the newly formed bond. Vibrationally excited nitric oxide is produced in the straight photolysis of nitrosyl halides [35],

$$NOX + h\nu = NO^* + X$$

although this process appears to be an exception. Another method of populating higher vibrational levels of the ground state relies on the fluorescence (and collisional quenching) of a molecule such as nitric oxide where the excited state is somewhat displaced from the ground state [36]. This is also possible with free radicals such as CN.

6. Reactions in Solution

Although flash photolysis has been used to study iodine atom recombination in solution, few other processes involving small molecules have been investigated in the liquid state. Solution studies have been largely of aromatic molecules, for which the gas phase studies have been mainly spectroscopic rather than kinetic.

The energy of electronic excitation of an aromatic molecule can be lost in several ways, by radiation as fluorescence, by quenching, by internal conversion to high vibrational levels of the ground state from which decomposition can occur [37]. Fluorescence, its quenching and internal conversion occur with almost all aromatic molecules. Decomposition to yield an aromatic radical occurs mainly with substituted monocyclic hydrocarbons. This can be identified by the appearance of the same transient spectrum from a range of related molecules; for instance, benzyl from benzyl derivatives but not from say benzal chloride [22]. It is also accompanied by a net decomposition of the parent molecule.

By contrast, the formation of a triplet state which is characteristic of polycyclic aromatic hydrocarbons, is a reversible process [38]. The behaviour of aromatic triplet states have been studied much more thoroughly than those of aromatic radicals, although little effort has been made to investigate heterocyclics. In fluid solvents, the triplet states of aromatic molecules decay by two mechanisms: a second order diffusion limited triplet-triplet annihilation process which sometimes yields an excited singlet molecule which emits delayed fluorescence and a first-order process which occurs even in highly purified solvents [39, 40]. This latter process is not fully understood and may be quenching by some exceptionally persistent trace impurity, since its rate appears to be governed by the viscosity of the solvent and hence the diffusion time. Such interesting phenomena as triplet-triplet energy transfer, and the reaction or quenching of triplet states with various additives can be studied. It should be noted that the absorption spectrum of a triplet state is a transition to a higher triplet level and provides no information about the energy of the lowest triplet state, a quantity which is found from the energy of the phosphorescence spectrum.

No attempt has been made to give either a comprehensive review or more than selected references in the subjects discussed. Further details of the applications of flash photolysis can be found in a number of reviews [41—45].

References

1. Toennies, J. P.: this book.
2. Eigen, M., L. De Maeyer, J. Heidberg, and G. Kohlmaier: this book.
3. Greene, E. F., and J. P. Toennies: Chemische Reaktionen in Stoßwellen, Darmstadt: Steinkopff, 1959.
4. Norrish, R. G. W., and G. Porter: Nature 164, 658 (1949).
5. Porter, G.: Proc. Roy. Soc., A, 200, 284 (1950).
6. Claesson, S.: Svensk Naturvetenskap 13, 252 (1960).
7. Details can be obtained from Division of Pure Physics, National Research Council, Ottawa.
8. Khan, M. A., G. Porter, and R. G. W. Norrish: Proc. Roy. Soc., A, 219, 312 (1953).
9. Shilman, A., and R. A. Marcus: J. Chem. Phys. 39, 996 (1963).

10. Herr, K. C., and G. C. Pimentel: 7th Symposium on Free Radicals, Padua, (1965).
11. Garton, W. R. S.: J. Sci. Instr. **36**, 11 (1959).
12. Paul, D. E., and F. W. Dalby: J. Chem. Phys. **37**, 592 (1962).
13. Herzberg, G.: Proc. Roy. Soc., A, **262**, 291 (1961).
14. Lipscomb, F. J., R. G. W. Norrish, and B. A. Thrush: Proc. Roy. Soc., A, **233**, 455 (1956).
15. Thrush, B. A.: Proc. Roy. Soc., A, **235**, 143 (1956).
16. Ramsay, D. A.: Determination of Organic Structures by Physical Methods. Academic Press, New York and London, **2**, 246 (1962).
17. Walsh, A. D.: J. Chem. Soc., **1953**, 2260 *et seq.*
18. Renner, R.: Z. Physik **92**, 172 (1934).
19. Porter, G.: Disc. Faraday Soc. **9**, 60 (1950).
20. Herzberg, G.: Disc. Faraday Soc. **35**, 7 (1963).
21. Thrush, B. A., and J. J. Zwolenik: Disc. Faraday Soc. **35**, 196 (1963).
22. Porter, G., and F. J. Wright: Trans. Faraday Soc. **51**, 1469 (1955).
23. Thrush, B. A.: Nature **155**, 178 (1955).
24. Porter, G., and B. Ward: Proc. Chem. Soc. **1964**, 288.
25. Currie, C. L., and D. A. Ramsay, unpublished results.
26. Donovan, R. J., and D. Husain, Nature **206**, 171 (1965).
27. Christie, M. I., A. J. Harrison, R. G. W. Norrish, and G. Porter: Proc. Roy. Soc., A, **231**, 446 (1955).
28. Russell, K. E., and J. Simons: Proc. Roy. Soc., A, **217**, 271 (1953).
29. Porter, G., and J. A. Smith: Proc. Roy. Soc., A, **261**, 28 (1961).
30. Briggs, G. A., and R. G. W. Norrish: Proc. Roy. Soc., A, **276**, 51 (1963).
31. Edgecombe, F. H. C., R. G. W. Norrish, and B. A. Thrush: Spec. Publ. Chem. Soc. **9**, 121 (1958).
32. Norrish, R. G. W., G. Porter, and B. A. Thrush: Proc. Roy. Soc., A, **216**, 165 (1953); **227**, 423 (1955).
33. Erhard, K. H. L., and R. G. W. Norrish: Proc. Roy. Soc., A, **234**, 178 (1956); **259**, 297 (1960).
34. Thrush, B. A.: Proc. Roy. Soc., A, **233**, 147 (1955).
35. Basco, N., and R. G. W. Norrish: Proc. Roy. Soc., A, **268**, 291 (1962).
36. —, A. B. Callear, and R. G. W. Norrish: Proc. Roy. Soc., A, **260**, 459 (1961).
37. Bowen, E. J.: Chemical Aspects of Light, London: Oxford University Press 1940.
38. Porter, G., and M. W. Windsor: Disc. Faraday Soc. **17**, 178 (1954).
39. Jackson, G., and R. Livingston: J. Chem. Phys. **35**, 2182 (1961).
40. Hilpern, J. W., G. Porter, and L. J. Stief: Proc. Roy. Soc., A, **277**, 437 (1964).
41. Norrish, R. G. W., and B. A. Thrush: Quart. Revs. Chem. Soc. **10**, 149 (1956).
42. Porter, G.: Technique of Organic Chemistry, Vol. III, part. II, 1055. New York: Interscience (1963).
43. Norrish, R. G. W.: Chem. in Britain **1**, 289 (1965).
44. Porter, G.: R. G. W. Norrish Commemorative Volume, Cambridge University Press 1966.
45. Thrush, B. A.: R. G. W. Norrish Commemorative Volume, Cambridge University Press 1966.

B. A. Thrush
University of Cambridge
Cambridge/Great Britain

Molekülstruktur und chemische Reaktivität

F. Becker

Mit 2 Abbildungen

1. Reaktionstypen und Reaktionsmechanismen in der organischen Chemie

Voraussetzung für eine Diskussion der Zusammenhänge zwischen Molekülstruktur und chemischer Reaktivität ist eine Kenntnis der Reaktionsmechanismen, d. h. der Folge von Elementarprozessen, durch welche die stabilen Ausgangsprodukte in die stabilen Endprodukte gelangen. Es ist das besondere Verdienst der englischen Schule unter INGOLD, durch eine große Zahl reaktionskinetischer Arbeiten zur Aufklärung vieler Reaktionsmechanismen in der organischen Chemie beigetragen und durch die Schaffung einer Systematik eine Übersicht in die Vielfalt der Reaktionsmöglichkeiten gebracht zu haben. Tatsächlich lassen sich zahlreiche grundlegende Reaktionstypen, die bei den verschiedenartigsten Verbindungsklassen wiederkehren, im Rahmen einer solchen Systematik behandeln und verstehen, auch wenn sich das angewandte Schema in manchen Fällen nur als ein stark vereinfachender Grenzfall erweist.

1.1. *Klassifizierung organischer Reagenzien und Reaktionen*

Im allgemeinen vollzieht sich bei den betrachteten Reaktionen die Schließung oder die Trennung einer oder mehrerer kovalenter Bindungen. Da eine solche Bindung durch ein Elektronenpaar vermittelt wird, sind zwei Grenzfälle zu unterscheiden:

a) *Homolytische Reaktionen* $\qquad A - B \rightleftharpoons A\cdot + \cdot B$

Bei der Trennung der Bindung verbleibt bei jedem der Bruchstücke ein ungepaartes Valenzelektron, d. h. es bilden sich Radikale; entsprechend wird bei der Schließung der Bindung von jedem Partner eines der Bindungselektronen geliefert.

b) *Heterolytische Reaktionen* $\qquad A - B \rightleftharpoons A\colon + B$ oder $A + \colon B$

Bei der Trennung verbleibt das bindende Elektronenpaar bei einem der Bruchstücke bzw. es wird bei der Schließung der Bindung das Elektronenpaar von einem der Partner geliefert.

Bei der Untersuchung der Zusammenhänge zwischen Struktur und Reaktivität kann man im allgemeinen eine Unterscheidung der Reaktionspartner nach „Substrat" und „Reagens" treffen, wobei z. B. die Struktur des Substrates innerhalb einer Reaktionsserie systematisch variiert wird, während das Reagens unverändert bleibt. Bei heterolytischen Reaktionen kommt es darauf an, ob das bindende Elektronenpaar vom Reagens oder vom Substrat geliefert wird. Je nachdem unterscheidet man

a) *Elektrophile Reaktionen,* bei denen das Substrat ein einsames Elektronenpaar besitzt und als Elektronendonator wirkt, während das Reagens Elektronenacceptoreigenschaft besitzt.

b) *Nukleophile Reaktionen,* bei denen das Substrat Elektronenacceptor und das Reagens Elektronendonator ist.

Die Bezeichnung der Reaktion richtet sich also nach der Eigenschaft des Reagens, d. h., man spricht von *nukleophilen Reagenzien,* wenn diese Elektronendonatoreigenschaft haben und das zur Bindung erforderliche Elektronenpaar liefern, von *elektrophilen Reagenzien,* wenn diese Elektronenacceptoreigenschaften haben und eine Elektronenlücke besitzen. Im Sinne der Lewisschen Säure-Basen-Theorie sind nukleophile Reagenzien Basen, elektrophile Reagenzien Säuren. So verhält sich z. B. im Falle der Adduktbildung aus Bortrimethyl und Trimethylamin $B(CH_3)_3$ elektrophil oder als Lewis-Säure, $N(CH_3)_3$ dagegen nukleophil oder als Lewis-Base:

$$\begin{matrix} & CH_3 & & CH_3 & & & CH_3 & CH_3 \\ & | & & | & & & | & | \\ H_3C{-}B & & + & |N{-}CH_3 & \longrightarrow & H_3C{-}B & {-}N{-}CH_3\,. \\ & | & & | & & & | & | \\ & CH_3 & & CH_3 & & & CH_3 & CH_3 \end{matrix}$$

Betrachtet man die Reaktionen vom Standpunkt der Trennung und Schließung kovalenter Bindungen, so ergibt sich hieraus eine Einteilung in eine Anzahl grundlegender Reaktionstypen, von denen die wichtigsten aufgezählt werden sollen:

1. *Substitutionsreaktionen.* Hierunter sind Prozesse zu verstehen, bei denen eine bestimmte kovalente Bindung getrennt und dafür eine neue geschlossen wird, so daß ein neuer Ligand an die Stelle des ursprünglich vorhandenen tritt. Es ist in diesem Falle klar, daß das angreifende Reagens den neuen Liganden, das Substrat den austretenden Liganden liefert. Entsprechend der oben gegebenen Klassifizierung unterscheidet man

a) *Nukleophile Substitutionen* (Symbol S_N), bei denen das angreifende Reagens nukleophilen Charakter hat und das bindende Elektronenpaar liefert. Die vorhandene kovalente Bindung im Substrat wird heterolytisch getrennt, wobei das Elektronenpaar bei der austretenden Gruppe verbleibt:

Substrat:	Reagens (= Base):	Produkt:	austretende Gruppe:
$(CH_3)_3C{-}Br$	$+ \ |\overline{O}H^{\ominus}$	$\longrightarrow \ (CH_3)_3C{-}OH$	$+ \ |\overline{Br}|^{\ominus}$
$(CH_3)_2CH{-}J$	$+ \ |N(CH_3)_3$	$\longrightarrow \ (CH_3)_2CH{-}\overset{\oplus}{N}(CH_3)_3$	$+ \ |\overline{\underline{J}}|^{\ominus}\,.$

b) *Elektrophile Substitutionen* (Symbol S_E), bei denen das angreifende Reagens elektrophilen Charakter hat und das bindende Elektronenpaar vom Substrat geliefert wird. Die vorhandene kovalente Bindung wird heterolytisch getrennt, wobei das Elektronenpaar beim Substrat verbleibt:

$$\text{Substrat:} \qquad \text{Reagens} \atop (= \text{Säure}): \qquad \text{Produkt:} \qquad \text{austretende} \atop \text{Gruppe:}$$

$$C_6H_5-H + SO_3 \longrightarrow C_6H_5-\overset{\ominus}{S}O_3 + H^{\oplus}$$

$$C_6H_5-H + HO^{\ominus}\overset{\oplus}{N}O_2 \longrightarrow C_6H_5-NO_2 + H-OH .$$

Im allgemeinen kann die Entscheidung über den nukleophilen oder elektrophilen Charakter einer Substitutionsreaktion ohne nähere Kenntnis des Mechanismus auf Grund der Eigenschaften von Reagens und austretender Gruppe getroffen werden.

c) *Radikalische Substitutionen*, bei der das Reagens radikalischen Charakter hat und ein ungepaartes Elektron besitzt. Die vorhandene kovalente Bindung wird homolytisch getrennt, wobei ein ungepaartes Elektron beim Substrat verbleibt. Radikalische Substitutionen vollziehen sich meistens in mehreren Schritten:

$$\text{Substrat:} \qquad \text{Reagens:} \qquad \text{Produkt:} \qquad \text{austretende} \atop \text{Gruppe:}$$

$$(CH_3)_3C-H + \cdot Cl \longrightarrow (CH_3)_3C\cdot + H-Cl$$

$$(CH_3)_3C\cdot + Cl-Cl \longrightarrow (CH_3)_3C-Cl + \cdot Cl .$$

Den radikalischen Charakter erkennt man an der Art der Initiierung des Prozesses (z. B. photochemische Radikalbildung) und an dem Auftreten von Radikalen als Zwischenstufen (Kettenreaktionen).

2. *Additionsreaktionen.* Hierunter sind Prozesse zu verstehen, bei denen sich ein Reagens kovalent an eine Mehrfachbindung anlagert, hauptsächlich handelt es sich um C=C-, C≡C- oder C=O-Bindungen, wobei zwei Atome je einen zusätzlichen Liganden erhalten. Eine Klassifizierung nach den oben genannten Gesichtspunkten hat gewisse Annahmen über den Reaktionsmechanismus, etwa über die Reihenfolge des Eintretens der Liganden, zur Voraussetzung.

a) *Nukleophile Addition.* Hierbei tritt ein nukleophiles Reagens mit seinem einsamen Elektronenpaar in die Elektronenlücke, die bei der heterolytischen Trennung der vorhandenen π-Bindung an dem als Reaktionszentrum angesehenen C-Atom entsteht:

$$\text{Substrat:} \qquad\qquad \text{Reagens:} \qquad\qquad \text{Produkt:}$$

$$CH_3-CH=O + \; |NH_3 \longrightarrow CH_3-CH{\overset{\textstyle OH}{\underset{\textstyle NH_2}{\big\langle}}}$$

$$CH_3-CH=O + H-\overline{O}-H \longrightarrow CH_3-CH{\overset{\textstyle OH}{\underset{\textstyle OH}{\big\langle}}} .$$

Der nukleophile Charakter dieser Prozesse ergibt sich einerseits aus der Eigenschaft der Reagenzien NH_3 und H_2O als Elektronendonatoren. Andererseits kann auf Grund der Polarität der $C=O$-Bindung angenommen werden, daß bei der heterolytischen Trennung der π-Bindung das Elektronenpaar beim O-Atom verbleibt und das C-Atom als Elektronenacceptor wirkt.

b) *Elektrophile Addition.* Hierbei besteht der erste Schritt in der Anlagerung eines elektrophilen Reagens an das Atom, welches bei der heterolytischen Trennung der π-Bindung das Elektronenpaar übernimmt:

$$\text{Substrat:} \qquad \text{Reagens:} \qquad \text{Produkt:}$$

$$\text{>C=C<} \quad + \quad H_3O^{\oplus} \longrightarrow \quad \underset{\underset{\text{H}}{|}}{-\overset{|}{C}}-\underset{\underset{\text{OH}}{|}}{\overset{|}{C}}- \; + \; H^{\oplus}$$

$$CH_3-CH=CH_2 \; + \; Cl^{\ominus}H^{\oplus} \longrightarrow \qquad CH_3-\underset{\underset{\text{Cl}}{|}}{CH}-\underset{\underset{\text{H}}{|}}{CH_2} \; .$$

In beiden Fällen wird angenommen, daß zunächst die elektrophile Addition eines Protons erfolgt.

c) *Radikalische Addition.* In diesem Fall wird die π-Bindung homolytisch getrennt; die Anlagerung des radikalischen Reagens erfolgt in zwei Schritten:

$$\text{Substrat:} \quad \text{Reagens:} \quad \text{Produkt:}$$

$$\text{>C=C<} \; + \; \cdot Br \longrightarrow \underset{\underset{\text{Br}}{|}}{-\overset{|}{C}}-\overset{\cdot}{\underset{|}{C}}-$$

$$\underset{\underset{\text{Br}}{|}}{-\overset{|}{C}}-\overset{\cdot}{\underset{|}{C}}- \; + \; Br-Br \longrightarrow \qquad \underset{\underset{\text{Br}}{|}}{-\overset{|}{C}}-\overset{\overset{\text{Br}}{|}}{\underset{|}{C}}- \; + \; \cdot Br \; .$$

3. *Abspaltungsreaktionen.* Man versteht hierunter Prozesse, bei denen Liganden austreten, ohne daß die entstandene Lücke von einem neuen Liganden besetzt wird. Bei solchen Reaktionen entstehen entweder Radikale, Ionen oder ungesättigte Verbindungen. Besonders wichtig sind die 1,2- oder β-Eliminationen, bei denen zwei benachbarte Liganden unter Bildung einer Doppel- oder Dreifachbindung abgespalten werden. Die Abspaltung erfolgt häufig unter der Einwirkung eines Reagens, das mit einer der austretenden Gruppen eine kovalente Bindung eingeht:

$$\text{Substrat:} \qquad \text{Reagens:} \qquad \text{Olefin:} \qquad \overset{\text{austretende}}{\text{Gruppen:}}$$

$$H-CH_2-\underset{\underset{\text{Br}}{|}}{CH}-CH_3 \; + \; {}^{\ominus}|\underline{O}-C_2H_5 \longrightarrow H_2C=CH-CH_3 \; + \; H\underline{O}-C_2H_5 \; + \; |\overline{Br}|^{\ominus}$$

$$H-CH_2-CH_2-\overset{\oplus}{N}(CH_3)_3 \; + \; {}^{\ominus}|\underline{O}H \longrightarrow H_2C=CH_2 \qquad + \; |N(CH_3)_3 + H-OH \; .$$

Da in diesen Fällen eine C=C-Bindung entsteht, spricht man von *olefin-bildenden Abspaltungsreaktionen*. Es gibt ferner carbonylbildende Abspaltungen, wie z. B.

$$H_3C-CH-O + {}^{\ominus}|\overline{O}H \longrightarrow H_3C-CH=O + N\overset{\ominus}{O}_2 + H-\overline{O}H \, ,$$
$$| \quad |$$
$$H \quad NO_2$$

bei denen eine C=O-Bindung entsteht oder nitrilbildende Abspaltungen, durch die eine C≡N-Gruppe erzeugt wird.

Bei den angegebenen olefinbildenden Eliminationen hat das Reagens nukleophilen Charakter; durch den Angriff von $HO^\ominus$ oder $H_5C_2O^\ominus$ wird dem Substrat ein Proton entzogen, wobei das bindende Elektronenpaar am Substrat verbleibt. Dementsprechend sind diese Prozesse als *nukleophile Abspaltungsreaktionen* zu bezeichnen. Wird die Abspaltung durch die Einwirkung eines Radikals ausgelöst und erfolgt die Abtrennung des Liganden homolytisch, so liegt eine *radikalische Abspaltungsreaktion* vor:

$$H_3C-CH-CH-CH_3 + \cdot J \longrightarrow H_3C-\overset{\bullet}{C}H-CH-CH_3 + J-J$$
$$| \quad | \qquad\qquad\qquad\qquad |$$
$$J \quad J \qquad\qquad\qquad\qquad J$$

$$H_3C-\overset{\bullet}{C}H-CH-CH_3 \longrightarrow H_3C-CH=CH-CH_3 + J\cdot .$$
$$|$$
$$J$$

4. *Umlagerungsreaktionen.* Von den zahlreichen Umlagerungsreaktionen sollen hier nur *isomere Umlagerungen* betrachtet werden. Bei diesen Prozessen erfolgt zunächst eine Trennung kovalenter Bindungen, dann eine Umordnung der entstandenen Molekülbruchstücke und schließlich eine Rekombination zu einem Molekül mit veränderter Struktur. Ein Sonderfall ist die Vertauschung des Ortes zweier Substituenten an einem größeren Molekül. Auch hier kann eine Klassifizierung nach dem Modus der Trennung der Bindung des zuerst abgelösten Liganden erfolgen. Wird diese Bindung heterolytisch in der Weise gelöst, daß das bindende Elektronenpaar bei dem Substituenten verbleibt, der seinen Platz wechselt, so verhält sich dieser bei der Schließung der neuen Bindung nukleophil; demnach ist der Prozeß als eine *nukleophile Umlagerung* zu bezeichnen. Ein Beispiel hierfür ist die Umlagerung des Phenylhydroxylamins in p-Aminophenol, die sich unter der katalytischen Wirkung von $H^\oplus$-Ionen vollzieht:

$$C_6H_5-NH-OH + H^\oplus \rightleftharpoons C_6H_5-NH-O^\oplus H_2 \quad (\text{schnell})$$
$$C_6H_5-NH-O^\oplus H_2 \longrightarrow [C_6H_5=NH]^\oplus + H_2O \longrightarrow$$
$$HO-C_6H_4-NH_2 + H^\oplus .$$

Erfolgt dagegen die Abtrennung des Liganden unter Verbleib des bindenden Elektronenpaars beim Substrat, so wird die neue Bindung im Sinne einer elektrophilen Reaktion geschlossen; dementsprechend hat man es mit einer *elektrophilen Umlagerung* zu tun. Als Beispiel ist die Stevens-Umlagerung zu nennen, bei welcher der Benzylrest mit einem Elektronensextett wandert, während das Elektronenpaar am N-Atom verbleibt:

$$C_6H_5-CO-CH_2-\overset{|}{N}^{\oplus}(CH_3)_2OH^{\ominus} \quad\longrightarrow\quad C_6H_5-CO-\overset{|}{C}H-\underline{N}(CH_3)_2 + H_2O\,.$$
$$\qquad CH_2C_6H_5 \qquad\qquad\qquad\qquad CH_2C_6H_5$$

Schließlich gibt es auch *radikalische Umlagerungen*, bei denen ein zunächst durch Homolyse, etwa auf pyrolytischem Wege, gebildetes Radikal eine Strukturänderung erleidet, bevor die Rekombination erfolgt:

$$C_6H_5-C(CH_3)_2-CH_3 \quad\longrightarrow\quad C_6H_5-C(CH_3)_2-CH_2\cdot + H\cdot \quad\longrightarrow$$
$$(CH_3)_2\overset{\cdot}{C}-CH_2-C_6H_5 + H\cdot\,.$$

Erhebliches Interesse beanspruchen Umlagerungsreaktionen an ungesättigten Verbindungen, bei denen das mesomeriefähige System durch die Abtrennung eines Substituenten vergrößert wird. Unter diesen Bedingungen kann der Substituent an ein anderes Atom des mesomeren Systems wandern und dort gebunden werden, ohne daß vorher ein zweiter Ligand abgetrennt werden muß. Auch bei solchen „tautomeren Umlagerungen" ergeben sich drei Möglichkeiten:

a) *Anionotropie*. Der Substituent wandert als Anion, d. h. er behält bei der heterolytischen Trennung der Bindung das Elektronenpaar. Ein Beispiel ist die in den meisten Fällen anionotrop verlaufende Allylumlagerung:

$$H_3C-\overset{|}{C}H-CH=CH_2 \quad\longrightarrow\quad [H_3C-CH\text{⋯}CH\text{⋯}CH_2]^{\oplus} \quad\longrightarrow\quad H_3C-CH=CH-\overset{|}{C}H_2\,.$$
$$\;Cl \qquad\qquad\qquad\qquad Cl^{\ominus} \qquad\qquad\qquad\qquad\qquad Cl$$

b) *Kationotropie*. In diesem Fall wandert der Substituent als Kation, d. h. das bindende Elektronenpaar bleibt beim Molekülrumpf. Am wichtigsten sind solche Fälle, in denen das wandernde Kation ein Proton ist. Als Beispiel für eine solche *Prototropie* sei die Umwandlung des Nitromethans aus der Nitro-Form in die aci-Form angeführt:

$$CH_3-\overset{\oplus}{N}\overset{\overline{O}|^{\ominus}}{\diagdown_{O}} \longrightarrow \left[\overset{\ominus}{|}CH_2-\overset{\oplus}{N}\overset{\overline{O}|^{\ominus}}{\diagdown_{O}} \longleftrightarrow CH_2=\overset{\oplus}{N}\overset{\overline{O}|^{\ominus}}{\diagdown_{\underline{O}|^{\ominus}}} \right] H^{\oplus} \longrightarrow CH_2=\overset{\oplus}{N}\overset{\overline{O}|^{\ominus}}{\diagdown_{OH}}\,.$$

c) *Radikalische Umlagerung*. Wird der Prozeß durch Radikale katalysiert, so ist ein radikalischer Mechanismus wahrscheinlich, z. B.

$$CH_2=CH-CHBr-CH_3 \quad\longrightarrow\quad [CH_2\text{⋯}CH\text{⋯}CH-CH_3]\cdot\,Br\cdot$$
$$\longrightarrow\quad Br-CH_2-CH=CH-CH_3\,.$$

1.2. Beispiele für Reaktionsmechanismen

1. Nukleophile Substitution am gesättigten C-Atom

Bei einer nukleophilen Substitution am gesättigten C-Atom wird der Substituent X heterolytisch unter Mitnahme des Elektronenpaares gelöst, während das in Form einer Lewis-Base angreifende Reagens Y mit seinem einsamen Elektronenpaar an das Reaktionszentrum herantritt:

$$Y| + -\overset{|}{\underset{|}{C}}-X \longrightarrow Y^{\oplus}-\overset{|}{\underset{|}{C}}- + |X^{\ominus} \; .$$

Stellt das Reagens Y gleichzeitig das Lösungsmittel dar, so spricht man von einer *solvolytischen Reaktion*. Da bei einer solchen Substitution stets eine vorhandene Bindung gelöst und eine neue Bindung geschlossen werden muß, kommt es bei der Diskussion der möglichen Reaktionsmechanismen darauf an, ob diese beiden Vorgänge gleichzeitig oder nacheinander ablaufen.

a) *Bimolekularer Mechanismus* (S_N2). Es wird hierbei angenommen, daß die Trennung der alten und die Schließung der neuen Bindung synchron verlaufen und sich der Prozeß somit in einem einzigen Schritt vollzieht. Der Übergangszustand enthält den alten und den neuen Substituenten zu beiden Seiten des Reaktionszentrums in symmetrischer Anordnung:

$$Y| + R-X \longrightarrow \left[\overset{\delta+}{Y} \ldots R \ldots \overset{\delta-}{X}\right] \longrightarrow Y^{\oplus} - R + |X^{\ominus} \; .$$

Das C-Atom des Reaktionzentrums besitzt also im Übergangszustand 5 Liganden, die in Form einer trigonalen Bipyramide angeordnet sind: $X \ldots C \ldots Y$ liegen auf einer Geraden; die Bindungen sind länger als in stabilen Molekülen und stark polar; die drei übrigen Liganden sind trigonal auf einer Ebene senkrecht hierzu angeordnet: $\overset{\delta+}{Y} \ldots \overset{|}{C} \ldots \overset{\delta-}{X}$. Die treibende Kraft ist der Energiegewinn bei der Schließung der neuen Bindung R−Y; daher hängt die Reaktionsgeschwindigkeit sehr stark von der Natur des Reagens Y und von dessen Solvatation ab. Der angegebene bimolekulare Mechanismus verläuft unter Umkehrung der Konfiguration am Reaktionszentrum streng stereospezifisch.

b) *Unimolekularer Mechanismus* (S_N1). In diesem Fall wird angenommen, daß zuerst die heterolytische Trennung der Bindung R−X erfolgt, wobei sich ein Carboniumion $R^{\oplus}$ bildet. Im zweiten, schnellen Reaktionsschritt wird die R−Y-Bindung geschlossen:

$$(1) \qquad R-X \longrightarrow \left[\overset{\delta+}{R} \ldots \overset{\delta-}{X}\right] \longrightarrow R^{\oplus} + |X^{\ominus} \quad \text{langsam}$$

$$(2) \qquad R^{\oplus} + |Y \qquad\qquad \longrightarrow R-Y^{\oplus} \qquad \text{schnell} \; .$$

Der geschwindigkeitsbestimmende Schritt ist also die unimolekulare Heterolyse der Bindung R−X; als treibende Kraft muß dabei der Gewinn an Solvatationsenergie der entstehenden Ionen angesehen werden. Das Carboniumion $R^{\oplus}$ reagiert als instabiles Zwischenprodukt sehr schnell mit dem Reagens Y weiter. Da der Übergangszustand des geschwindigkeitsbestimmenden Schrittes das Reagens Y nicht enthält, muß die Reaktionsgeschwindigkeit von dessen chemischer Natur weitgehend unabhängig sein, dagegen spielt die Solvatation von $R^{\oplus}$ und $X^{\ominus}$ eine ausschlaggebende Rolle. Das Carboniumion enthält nur 3 Liganden in ebener Anordnung; da die Wahrscheinlichkeit für die Annäherung von Y von beiden Seiten dieselbe ist, tritt bei optisch aktiven Substraten eine vollständige Racemisierung ein.

Nukleophile Substitutionen an einem primären C-Atom verlaufen bevorzugt nach einem S_N2-Mechanismus, solche an einem tertiären C-Atom bevorzugt nach einem S_N1-Mechanismus, während sekundäre Alkylderivate eine Zwischenstellung einnehmen:

$$S_N2: \; HO|^{\ominus} + CH_3-Br \longrightarrow \left[HO \ldots \underset{H \diagup \; \diagdown H}{\overset{\overset{\displaystyle H}{|}}{C}} \ldots Br \right]^{\ominus} \longrightarrow HO-CH_3 + |Br|^{\ominus}.$$

$$S_N1: \; (CH_3)_3C-Cl \longrightarrow [(CH_3)_3C^{\oplus} + Cl^{\ominus}] \overset{H_2O}{\longrightarrow} (CH_3)_3C-OH + HCl \;.$$

Haben bei einer S_N2-Reaktion beide Reaktanten kleine und kontrollierbare Konzentrationen, so ergibt sich ein kinetisches Zeitgesetz 2. Ordnung. Liegt ein Partner in großem Überschuß vor, wie dies z. B. bei einer solvolytischen Reaktion der Fall ist, oder wird dieser in konstanter Konzentration durch ein vorgelagertes Gleichgewicht geliefert, so ergibt sich auch bei einem S_N2-Mechanismus ein Zeitgesetz 1. Ordnung. Die Entscheidung über den tatsächlich vorliegenden Mechanismus kann in solchen Fällen auf Grund von Untersuchungen über die Strukturabhängigkeit der Reaktionsgeschwindigkeit getroffen werden. Als Beispiel seien die Geschwindigkeitskonstanten für die oben formulierte Hydrolyse von Alkylbromiden (bei 55° in 80%igem wäßrigem Äthanol) angeführt:

		R = Me	Et	iPr	tBu	
(1) R–Br + H$_2$O,	1. Ordnung	0,35	0,14	0,24	1010	(S_N1)
	anfangs neutral $10^5\,k_1$ (sec^{-1})					
(2) R–Br + OEt$^{\ominus}$,	2. Ordnung $10^5\,k_2$ (l·Mol^{-1}·sec^{-1})	2140	170	4,7	–	(S_N2)

Bei einem bimolekularen Mechanismus erschweren Methylgruppen am Reaktionszentrum die Annäherung des nukleophilen Partners und setzen die Reaktionsgeschwindigkeit herab. Diese sterischen Effekte spielen bei einem unimolekularen Mechanismus keine Rolle; dagegen wird das sich bildende Carboniumion durch die auf induktivem Wege als Elektronendonatoren wirksamen Methylgruppen stabilisiert. Beide Mechanismen zeigen also eine gegenläufige Abhängigkeit von der Zahl der α-Methylgruppen, die sich darin auswirkt, daß beim Übergang von $(CH_3)_2CHBr$ zu $(CH_3)_3CBr$ ein Wechsel vom bimolekularen zum unimolekularen Reaktionsmechanismus erfolgt.

Eines der wichtigsten Kriterien für den S_N2-Mechanismus ist die hierbei zu erwartende Konfigurationsumkehr. Zum Beweis wurde eine Lösung von optisch aktivem D-2-Jodoctan mit radioaktiv indiziertem Jodid versetzt:

$$J^{*\ominus} + \underset{\underset{H}{|}}{\overset{\overset{C_6H_{13}}{|}}{C}}{-}J \longrightarrow \left[J^{*}\ldots \underset{\underset{H}{|}}{\overset{\overset{C_6H_{13}}{|}}{C}} \ldots J \right]^{\ominus} \longrightarrow J^{*}{-}\underset{H}{\overset{\overset{C_6H_{13}}{|}}{C}}{\diagdown}{CH_3} + J^{\ominus} .$$

Jeder bimolekulare Austausch des gebundenen mit dem in der Lösung vorhandenen markierten Jod führt zur Bildung eines Moleküls L-2-Jodoctan, so daß die Racemisierungsgeschwindigkeit genau doppelt so groß sein sollte wie die Austauschgeschwindigkeit. Tatsächlich wurde das Verhältnis der Racemisierungsgeschwindigkeit zur Austauschgeschwindigkeit zu 1,93 ± 0,16 bestimmt, was mit dem angenommenen Mechanismus innerhalb der Fehlergrenzen übereinstimmt.

Wenn beim S_N1-Mechanismus als Zwischenstufe ein freies Carbonium-Ion mit ebener Anordnung der Valenzen auftritt, so muß dies zu einer völligen Racemisierung optisch aktiver Verbindungen führen. Dies ist auch bei manchen S_N1-Reaktionen der Fall. So beobachtet man z. B. bei der Hydrolyse von optisch aktivem α-Chloräthylbenzol, $C_6H_5{-}CHCl{-}CH_3$, in 80%igem wäßrigem Aceton eine Racemisierung zu etwa 97%. In anderen Fällen findet man neben dem racemischen Gemisch zum Teil ein Produkt mit Konfigurationsumkehr, z. B. tritt bei der Solvolyse des tertiären Alkylchlorids $(CH_3)_2CH{-}(CH_2)_3{-}C^{*}(CH_3)(C_2H_5)Cl$ in wäßrigem Äthanol zu 70—80% Racemisierung ein. Daraus geht hervor, daß sich trotz des sonst eindeutigen S_N1-Charakters der Reaktion kein völlig freies Carboniumion bildet, sondern die austretende Gruppe ihre Orientierung gegenüber dem Substrat eine gewisse Zeit lang beibehält:

$$S + {-}\overset{|}{\underset{|}{C}}{-}X \longrightarrow \left[S\ldots \overset{|\oplus}{\underset{\diagup \diagdown}{C}} \ldots X^{\ominus} \right] \begin{cases} \nearrow & \overset{\oplus}{S}{-}\overset{\overset{|}{}}{\underset{|}{C}}{-} + X^{\ominus} \quad \text{Umkehrung} \\ \searrow & \left[S\ldots \overset{|}{\underset{\diagup \diagdown}{C}} \ldots S \right]^{\oplus} + X^{\ominus} \, \substack{\text{Racemi-} \\ \text{sierung.}} \end{cases}$$

Es bildet sich im ersten Reaktionsschritt ein solvatisiertes Ionenpaar (S = Solvens). Erfolgt ein Austausch von $X^{\ominus}$ gegen S, bevor die neue Bindung S—C geschlossen ist, so tritt Racemisierung ein. Wird dagegen die austretende Gruppe erst dann frei, wenn S an C gebunden ist, so erhält man ein Produkt mit Konfigurationsumkehr.

2. Olefinbildende Abspaltungsreaktionen

Diese, auch als 1,2- oder β-Eliminierungen bezeichneten Prozesse verlaufen nach dem Schema:

$$H{-}\overset{\overset{|\beta}{}}{\underset{|}{C}}{-}\overset{\overset{|\alpha}{}}{\underset{|}{C}}{-}X \longrightarrow \underset{\diagup}{\overset{\diagdown}{C}}{=}\overset{\diagup}{\underset{\diagdown}{C}} + H^{\oplus} + |X^{\ominus} ,$$

d. h. es werden zwei Liganden unter Ausbildung einer Doppelbindung abgelöst. Auch hier kommt es bei der Diskussion der möglichen Reaktionsmechanismen darauf an, ob die beiden Substituenten gleichzeitig oder nacheinander abgetrennt werden.

a) *Bimolekularer Mechanismus* (E2). Das angreifende nukleophile Reagens B bewirkt die Ablösung eines Protons am β-C-Atom, während gleichzeitig die elektronenanziehende Gruppe X unter Mitnahme des Bindungselektronenpaars austritt:

$$B| + H-C^{\beta}R_2-C^{\alpha}R_2-X \longrightarrow B^{\oplus}H + CR_2{=}CR_2 + |X^{\ominus}.$$

Der Prozeß läuft somit in einem einzigen bimolekularen Schritt ab, dessen Übergangszustand sowohl die Base B wie die austretende Gruppe X enthält.

b) *Unimolekularer Mechanismus* (E1). Hierbei wird angenommen, daß die Reaktion in zwei Schritten verläuft, wobei im ersten, geschwindigkeitsbestimmenden Schritt die Ionisierung des Substrats in ein Carbonium-Kation und ein Anion $X^{\ominus}$ erfolgt:

$$(1) \quad H-C^{\beta}R_2-C^{\alpha}R_2-X \longrightarrow H-CR_2-C^{\oplus}R_2 + |X^{\ominus} \quad \text{geschwindigkeits-bestimmend}$$

$$(2) \quad H-CR_2-C^{\oplus}R_2 \longrightarrow H^{\oplus}+CR_2{=}CR_2 \quad \text{schnell}.$$

Der Übergangszustand für den Schritt (1) muß als stark polar angenommen werden, wobei die Solvatisierung der entstehenden Ionen stabilisierend wirkt.

Der wesentliche Unterschied zwischen den beiden Reaktionsmechanismen besteht darin, daß bei E2 die Trennung der $H-C^{\beta}$-Bindung geschwindigkeitsbestimmend ist, bei E1 dagegen die Trennung der $C^{\alpha}-X$-Bindung. Der Nachweis eines solchen Unterschiedes im reaktiven Verhalten konnte durch Untersuchung des kinetischen Deuterium-Isotopeneffektes erbracht werden. Während eine Deuterium-Substitution an C^{α} die Geschwindigkeit der bimolekularen E2-Elimination nur geringfügig beeinflußt, verlangsamen Deuteriumatome an C^{β} die Reaktion um den Faktor 4—8, hauptsächlich, weil die Trennungsenergie einer C—D-Bindung wegen der kleineren Nullpunktsenergie größer ist als diejenige einer C—H-Bindung. Bei E1-Eliminationen beträgt der Effekt von β-Deuteriumatomen nur $k_H/k_D = 1,3-2,3$, da es sich hier nur um einen indirekten Einfluß auf die Trennung der $C^{\alpha}-X$-Bindung handelt.

c) *Stereochemie der Abspaltungsreaktionen.* Die energetisch günstigste Anordnung für den Übergangszustand einer E2-Elimination in Lösung enthält die angreifende Base B, das abzuspaltende Proton, die beiden C-Atome der entstehenden Doppelbindung und die austretende Gruppe X in einer ebenen trans-Konfiguration. Bei Aufsicht auf die $C^{\alpha}-C^{\beta}$-Bindung ergibt sich unter Verwendung der Newmanschen Projektionsformeln („staggered"-Konformationen) folgendes Bild:

Der Nachweis für die Bevorzugung der trans-Elimination konnte bei diastereomeren Verbindungen mit zwei benachbarten asymmetrischen C-Atomen erbracht werden. So entsteht z. B. bei der E2-Elimination von HBr aus dem inaktiven („meso-artigen") Stilbendibromid, $C_6H_5-CHBr-CHBr-C_6H_5$, ausschließlich das cis-Olefin, während das racemische („D,L-artige") Stilbendibromid nur trans-Olefin liefert. Beides ist nur zu verstehen, wenn die austretenden Gruppen im Übergangszustand trans-ständig sind: (⃝ austretende Gruppen)

Bei einer cis-Elimination ist die umgekehrte Zuordnung der Reaktionsprodukte zu erwarten, im Widerspruch zum experimentellen Ergebnis:

Bei pyrolytischen, in der Gasphase verlaufenden Eliminationen konnte dagegen in ganz analoger Weise der Nachweis für eine cis-Konformation der austretenden Gruppen erbracht werden. Die erythro-Form des Deuterium-markierten Esters $C_6H_5-CHD-CH(OCOCH_3)-C_6H_5$ spaltet CH_3COOH ab, wobei das Deuterium größtenteils in dem entstehenden Olefin verbleibt; die threo-Form dagegen liefert ein weitgehend D-freies Olefin, während CH_3COOD abgespalten wird. Beides läßt sich zwanglos erklären, wenn man annimmt, daß die austretenden Gruppen im Übergangszustand cis-ständig sind; im Falle einer trans-Elimination ergäbe sich die umgekehrte Zuordnung der Reaktionsprodukte. Wegen des großen Raumbedarfs der Phenylgruppen können Konformationen, in denen diese cis-ständig sind, als unwahrscheinlich außer Betracht bleiben.

"erythro"   →   trans-Olefin
D-frei
trans-Elimination

"erythro"   →   trans-Olefin
D-haltig
cis-Elimination

"threo"   →   trans-Olefin
D-haltig
trans-Elimination

"threo"   →   trans-Olefin
D-frei
cis-Elimination

Mit großer Wahrscheinlichkeit vollzieht sich die pyrolytische cis-Elimination intramolekular unter Ausbildung einer Brückenstruktur zwischen den austretenden Liganden:

Während also die E2-Elimination (analog der S_N2-Substitution) und die intramolekulare cis-Elimination streng stereospezifisch verlaufen, trifft dies für die nach E1 verlaufenden Reaktionen nicht zu. Das im ersten Schritt entstehende Carboniumion enthält die Liganden am C^α-Atom in einer trigonal-ebenen Anordnung, wodurch eine dort vorhandene Asymmetrie verlorengeht:

Selbst wenn die Ablösung der Gruppe X unter dem Einfluß des Lösungsmittels noch bevorzugt aus einer trans-Konformation erfolgt, ergibt sich ein gemischtes cis-trans-Olefin, sofern während der Lebensdauer des

Carboniumions eine innere Rotation um die entstehende C=C-Bindung möglich ist. E1-Eliminationen werden vorzugsweise bei tertiären Halogeniden und Alkoholen und in stark polaren Lösungsmitteln beobachtet, d. h. unter Bedingungen, die auch die unimolekulare Substitution S_N1 begünstigen. Durch Anwendung einer genügend hohen Konzentration der Base B kann aber auch bei tertiären Alkylderivaten ein bimolekularer Ablauf der Abspaltungsreaktion nach E2 erzwungen werden.

2. Empirische Struktur-Reaktivitäts-Beziehungen

Die zahlreichen experimentellen Untersuchungen über systematische Zusammenhänge zwischen Molekülstruktur und chemischer Reaktivität stellen das empirische Zahlenmaterial dar, auf das sich der Theoretiker bei der Prüfung der Richtigkeit seiner Berechnungen stützen muß. Ein solcher Vergleich zwischen Theorie und Experiment ist besonders dann nicht frei von Problematik, wenn es sich um die Ergebnisse von Modell- und Näherungsrechnungen handelt. Ebenso wichtig wie das Problem, welche strukturellen Effekte mit einem bestimmten Modell erfaßt werden können, ist eine kritische Analyse der experimentellen Aussagemöglichkeiten. Die folgenden Ausführungen sollen sich mit der Frage der Aufstellung empirischer Zusammenhänge zwischen Struktur und Reaktivität befassen, insbesondere mit den Möglichkeiten einer Klassifizierung struktureller Effekte.

Weitaus die Mehrzahl aller Experimentalarbeiten auf diesem Gebiet wurden an organischen Reaktionen durchgeführt, da man hier am besten eine systematische Variation der Struktur eines Reaktionspartners durchführen kann, während alle übrigen Versuchsbedingungen, wie der zweite Reaktionspartner, der Reaktionsmechanismus, das Reaktionsmilieu und die Reaktionsbedingungen, konstant gehalten werden. Es ist also typisch für derartige Untersuchungen, daß man die Geschwindigkeitskonstante k_0 einer Standardreaktion bzw. die Gleichgewichtskonstante K_0 eines Standardgleichgewichtes:

$$R-X_0 + Y \rightleftharpoons P_0^{\ddagger} \longrightarrow \text{Produkte}$$

$$\text{bzw. } R-X_0 + Y \rightleftharpoons P_0 \tag{1}$$

mit den entsprechenden Größen k bzw. K vergleicht, die man erhält, wenn der Substituent X_0 durch einen anderen Substituenten X ersetzt wird:

$$R-X + Y \rightleftharpoons P^{\ddagger} \longrightarrow \text{Produkte}$$

$$\text{bzw. } R-X + Y \rightleftharpoons P\,. \tag{2}$$

2.1. *Thermodynamische Beziehungen*

Zwischen den Geschwindigkeitskonstanten k_0 und k der Reaktionen (1) und (2) und der Änderung der freien Aktivierungsenthalpie $\delta \Delta G^{\ddagger}$ besteht die Beziehung

$$\delta \Delta G^{\ddagger} = \Delta G^{\ddagger} - \Delta G_0^{\ddagger} = - RT \ln k/k_0 \, , \tag{3}$$

wobei sich der Index $_0$ auf die Standardreaktion bezieht. Entsprechend lautet der Zusammenhang zwischen den Gleichgewichtskonstanten K und K_0 und der freien Reaktionsenthalpie

$$\delta \Delta G^0 = \Delta G^0 - \Delta G_0^0 = - RT \ln K/K_0 \, . \tag{4}$$

Nach der Theorie der absoluten Reaktionsgeschwindigkeit gilt für die Geschwindigkeitskonstante k der Reaktion (2):

$$k = \varkappa \, \frac{kT}{h} \, \frac{Q_P^{\ddagger}}{Q_{RX} \cdot Q_Y} \, e^{-\Delta E_0^{\ddagger}/RT} \, , \tag{5}$$

wobei $\varkappa$ der (als konstant angenommene) Transmissionskoeffizient, $Q_P^{\ddagger}$, Q_{RX} und Q_Y die Verteilungsfunktionen von Übergangszustand und Ausgangsprodukten und $\Delta E_0^{\ddagger}$ die Aktivierungsenergie bei 0 °K sind. Somit erhält man für den Quotienten der Geschwindigkeitskonstanten von Vergleichs- und Standardreaktion:

$$\delta \Delta G^{\ddagger} = - RT \ln k/k_0 = \delta \Delta E_0^{\ddagger} - RT \ln \Pi Q^{\ddagger}$$

$$\text{mit} \quad \Pi Q^{\ddagger} = \frac{Q_P^{\ddagger} \cdot Q_{RX_0}}{Q_{P_0}^{\ddagger} \cdot Q_{RX}} \, . \tag{6}$$

Für die Änderungen der Aktivierungsenthalpie, $\delta \Delta H^{\ddagger}$, und der Aktivierungsentropie, $\delta \Delta S^{\ddagger}$, gelten die Ausdrücke:

$$\delta \Delta H^{\ddagger} = \delta \Delta E_0^{\ddagger} + RT^2 \, \frac{d}{dT} \ln \Pi Q^{\ddagger} \tag{7}$$

$$\delta \Delta S^{\ddagger} = R \ln \Pi Q^{\ddagger} + RT \, \frac{d}{dT} \ln \Pi Q^{\ddagger} \, . \tag{8}$$

Die Größe $\delta \Delta E_0^{\ddagger}$ enthält nur Änderungen der potentiellen (Elektronen)-Energie, während das Glied $RT \ln \Pi Q^{\ddagger}$ auch Anteile an kinetischer Energie umfaßt, die temperaturabhängig sind. Bei der Bildung von $\delta \Delta S^{\ddagger}$ fällt die Größe $\delta \Delta E_0^{\ddagger}$ heraus; das gleiche gilt für Faktoren in $\Pi Q^{\ddagger}$, die reine Exponentialfunktionen sind.

Hat man es mit chemischen Gleichgewichten zu tun, so gelten die analogen Beziehungen:

$$\delta \Delta G^0 = - RT \ln K/K_0 = \delta \Delta E_0 - RT \ln \Pi Q \, ,$$

$$\delta \Delta H^0 = \delta \Delta E_0 + RT^2 \, \frac{d}{dT} \ln \Pi Q \, ,$$

$$\delta \Delta S^0 = R \ln \Pi Q + RT \, \frac{d}{dT} \ln \Pi Q \, , \qquad\qquad \text{mit } \Pi Q = \frac{Q_P \cdot Q_{RX_0}}{Q_{P_0} \cdot Q_{RX}} \, , \tag{9}$$

wobei $\delta \Delta E_0$ die Änderung der potentiellen Reaktionsenergie bei 0 °K darstellt.

Das primäre Ergebnis der experimentellen Untersuchungen über den Struktureinfluß auf Geschwindigkeits- bzw. Gleichgewichtskonstanten stellt also die Änderung der freien Enthalpie dar, während die quantentheoretischen Rechnungen sich gewöhnlich auf die Größe $\delta \Delta E_0$, d. h. auf die Änderung der Elektronenenergie beziehen. Jeder Vergleich zwischen Theorie und Experiment bedarf also einer zusätzlichen Diskussion der strukturellen Einflüsse auf $R T \ln \Pi Q$.

2.2. Klassifizierung der strukturellen Effekte

Bei den zur Aufstellung von Struktur-Reaktivitäts-Beziehungen ausgewählten Reaktionsserien wird ein bestimmtes Detail in der Molekülstruktur des Reaktionspartners RX systematisch variiert, während die funktionelle Gruppe, die das Reaktionszentrum darstellt, unverändert bleibt. Man untersucht also den Einfluß einer Strukturänderung auf das Reaktionszentrum, genauer gesagt, die Änderung dieses Struktureinflusses bei der Bildung des Übergangszustandes aus den Ausgangsprodukten bzw. — wenn es sich um ein Gleichgewicht handelt — die Unterschiede der strukturellen Einflüsse in End- und Ausgangszustand. Dabei bedient man sich im allgemeinen folgender Klassifizierung:

a) *Polare oder induktive Effekte:* Sie werden durch Änderungen der Polarität bei der Bildung des Übergangszustandes bzw. des Endzustandes aus den Ausgangsprodukten hervorgerufen und haben ihre Ursache in der unterschiedlichen Elektronegativität der Atome bzw. in dem mehr oder weniger ausgeprägten polaren Charakter der zwischen ihnen bestehenden Bindungen. Induktive Effekte werden entweder über σ-Bindungen (innere Polarisation) oder durch den Raum (Feldeffekte) übertragen und können mit Hilfe elektrostatischer Modelle beschrieben werden.

b) *Mesomere Effekte oder Resonanzeffekte:* Hierbei handelt es sich um eine Delokalisierung von Valenzelektronen infolge einer quantenmechanischen Wechselwirkung mit denjenigen benachbarter Bindungen oder mit einsamen Elektronenpaaren. Ihre Übertragung erfolgt hauptsächlich durch π-Bindungen oder durch Atome mit Elektronenzuständen, die eine Wechselwirkung mit π-Bindungen erlauben. In die Größen $-R T \ln k/k_0$ bzw. $-R T \ln K/K_0$ gehen nur die Änderungen der mesomeren Effekte zwischen Übergangs- bzw. Endzustand und Ausgangsprodukten ein. Da mesomere Effekte von Änderungen der Polarität stark beeinflußt werden, überlagern sie sich den induktiven Effekten in bestimmter Weise.

c) *Sterische Effekte:* Hierunter versteht man meistens abstoßende, gegebenenfalls auch anziehende Kräfte zwischen Atomen und Atomgruppen der Moleküle, die nicht durch kovalente Bindungen verknüpft

18*

sind. Abstoßende Kräfte treten auf, wenn sich die Wirkungssphären bestimmter Atome des Moleküls überlappen, so daß es zu einer Änderung von Atomabständen oder einer Deformation von Valenzwinkeln kommt. Sterische Effekte beeinflussen sowohl die potentielle wie auch die kinetische Energie der Moleküle, letztere durch Änderung von Schwingungsfrequenzen und durch Einschränkung der gehemmten inneren Rotation. Sterische Effekte hängen in erheblichem Maße vom Reaktionsmechanismus und vom Raumbedarf des Reaktionspartners im Übergangszustand ab.

Die meisten experimentellen Untersuchungen über Struktur-Reaktivitätsbeziehungen sind in Lösungsmitteln durchgeführt worden. Damit überlagern sich den angeführten strukturellen Effekten auch Lösungsmitteleffekte. Auch diese können in ähnlicher Weise klassifiziert werden:

a) *Polare oder induktive Lösungsmitteleffekte:* Es handelt sich hier um Einflüsse auf die Solvatation der Moleküle, die durch Änderungen der Polarität des Reaktionszentrums bei der Bildung des Übergangszustandes bzw. des Endproduktes aus den Ausgangsprodukten bedingt werden. Hierzu gehören die Orientierung der permanenten Dipole polarer Lösungsmittel im Felde von Ionen oder Dipolen, die Induktion von Dipolmomenten in den Lösungsmittelmolekülen und die Elektrostriktion des Lösungsmittels. Diese Effekte betreffen sowohl die potentielle wie auch die kinetische Energie des Lösungsmittels (Einschränkung der Beweglichkeit der Dipole) und treten daher in ΔG, ΔH und ΔS in Erscheinung.

b) *Mesomere und spezifische Lösungsmitteleffekte:* Hierzu ist eine Bindung bestimmter Lösungsmittelmoleküle im Ausgangs-, Übergangs- oder Endzustand an das Reaktionszentrum über einsame Elektronenpaare, über H-Brücken oder in Form eines Elektronendonator-Acceptorkomplexes erforderlich.

c) *Sterische Lösungsmitteleffekte:* Derartige Effekte machen sich bemerkbar, wenn infolge des Raumbedarfs des Reaktionspartners im Übergangszustand oder im Endzustand die Solvatation eingeschränkt wird (sterische Hinderung der Solvatation); sie sind stark von der Größe der Lösungsmittelmoleküle abhängig.

Die genannten Lösungsmitteleffekte beeinflussen die Zusammenhänge zwischen der Struktur der Reaktionspartner und ihrer Reaktivität nur dann nicht, wenn sie bei Standard- und Vergleichsreaktion identisch sind. Es ist klar, daß innerhalb einer Reaktionsserie eine Variation des Lösungsmittels nicht erfolgen darf und daß gleichartige Reaktionsserien in verschiedenen Lösungsmitteln zu unterschiedlichen Resultaten für $-RT \ln k/k_0$ bzw. $-RT \ln K/K_0$ führen müssen. Zweifellos sind zahlreiche in der Literatur beschriebene Reaktionsserien so ausgeführt, daß die Lösungsmitteleffekte konstant bleiben, jedoch bedarf dieser Umstand in jedem Fall einer besonderen Diskussion.

2.3. Lineare ΔG-Beziehungen

a) *Aromatische Reaktivität:* Nach Gl. (6) u. (9) sind die Logarithmen der relativen Geschwindigkeits- bzw. Gleichgewichtskonstanten den Änderungen der freien Enthalpie, $\delta \Delta G^{\ne}$ bzw. $\delta \Delta G^0$, proportional. Es ist deshalb richtig, zunächst nach einem Zusammenhang zwischen Strukturparametern und ΔG zu suchen. Eine erste Beziehung dieser Art wurde 1935 von L. P. HAMMETT in der Form

$$\frac{-\delta \Delta G^{\ne}}{2{,}303\,RT} = \log k/k_0 = \varrho\,\sigma$$

$$\frac{-\delta \Delta G^0}{2{,}303\,RT} = \log K/K_0 = \varrho\,\sigma \tag{10}$$

aufgestellt. Dabei ist σ eine „polare Substituentenkonstante", die unabhängig von der jeweiligen Reaktion ein Maß für den induktiven Substituenteneinfluß darstellen soll. ϱ ist eine „Reaktionskonstante", die als Maß für die Empfindlichkeit des betreffenden Prozesses gegenüber Polaritätsänderungen der Substituenten angesehen werden kann. HAMMETT beschränkte den Anwendungsbereich dieser Beziehung auf aromatische Reaktionen und Gleichgewichte, bei denen das Reaktionszentrum unmittelbar mit einem Benzolkern verknüpft ist, und berücksichtigte lediglich die induktiven Einflüsse von Substituenten in meta- und para-Stellung zum Reaktionszentrum. Auf diese Weise wird angesichts der Starrheit des Benzolgerüstes ein ausreichender Abstand von Substituenten zum Reaktionszentrum gewährleistet, so daß sterische Effekte in den meisten Fällen ausgeschlossen sind. Als Standardreaktion dient diejenige des unsubstituierten Benzolderivats, d. h. es wird für den Substituenten $R_0 = H$ die Substituentenkonstante $\sigma = 1$ gesetzt. Sind mehrere Substituenten oder Benzolringe am Reaktionszentrum vorhanden, so verhalten sich die induktiven Effekte in guter Näherung additiv, d. h. es gilt

$$\log k/k_0 = \varrho\,\Sigma\,\sigma \quad \text{bzw.} \quad \log k/k_0 = n\,\varrho\,\sigma\,. \tag{11}$$

Zahlenwerte für die Substituentenkonstanten σ nach Gl. (10) sind in Tabelle 1 zusammengestellt. Substituenten mit Elektronendonatoreigenschaft (relativ zu H) haben negative, solche mit Elektronenacceptoreigenschaft positive σ-Werte. In Abb. 1 (S. 280) sind für eine Reaktionsserie bei zwei verschiedenen Temperaturen die Hammett-Geraden gezeichnet; man erkennt, daß die Abhängigkeit der Geschwindigkeitskonstanten von Substitutionen in meta- und para-Stellung am Benzolkern durch Gl. (10) gut wiedergegeben wird. Tabelle 2 enthält einige Beispiele für Reaktionskonstanten ϱ nach Gl. (10); sie sind positiv, wenn im Verlauf der Reaktion an dem (mit dem aromatischen Ring verknüpften) Reaktionszentrum eine negative Ladung entsteht bzw. eine positive Ladung verschwindet. Negative ϱ-Werte haben Reaktionen, in deren Verlauf das Reaktionszentrum positiver geladen wird.

In einer umfangreichen Untersuchung hat H. H. Jaffé (1953) an Hand von 42 000 Geschwindigkeits- und Gleichgewichtskonstanten aus der Literatur an 3180 Reaktionen die Gültigkeit der Hammett-Gleichung geprüft und im Mittel eine Übereinstimmung zwischen experimentellen und berechneten $\log k/k_0$-Werten von $\pm 15\%$ gefunden. Offenbar lassen sich also die induktiven Substituenteneffekte bei aromatischen Reaktionen recht gut in Form einer einzigen Substituentenkonstanten σ erfassen, wenn man den Einfluß weiterer struktureller Einflüsse ausschaltet.

Tabelle 1. *Beispiele für Substituentenkonstanten nach der Hammett-Gleichung und ihren Varianten*

Substituent	σ_{para}	σ_{meta}	σ^+_{para}	σ^-_{para}
$NHCH_3$	$-0,84$			
$N(CH_3)_2$	$-0,83$		$-1,7$	
NH_2	$-0,66$	$-0,16$	$-1,3$	
OH	$-0,37$	$+0,12$	$-0,92$	
OC_6H_5	$-0,32$	$+0,25$	$-0,5$	
OCH_3	$-0,27$	$+0,12$	$-0,78$	
OC_2H_5	$-0,24$	$+0,10$		
$C(CH_3)_3$	$-0,20$	$-0,10$	$-0,26$	
CH_3	$-0,17$	$-0,07$	$-0,31$	
C_2H_5	$-0,15$	$-0,07$	$-0,30$	
$CH(CH_3)_2$	$-0,15$		$-0,28$	
C_6H_5	$-0,01$	$+0,06$	$-0,18$	
H	$0,00$	$0,00$	$0,00$	
F	$+0,06$	$+0,34$	$-0,07$	
Cl	$+0,22$	$+0,37$	$+0,11$	
Br	$+0,23$	$+0,39$	$+0,15$	
J	$+0,27$	$+0,35$	$+0,14$	
$COCH_3$	$+0,50$	$+0,38$		$+0,87$
$COOC_2H_5$	$+0,45$	$+0,37$	$+0,48$	$+0,68$
CF_3	$+0,54$	$+0,43$	$+0,61$	
CN	$+0,66$	$+0,56$	$+0,66$	$+1,00$
SO_2CH_3	$+0,72$	$+0,60$		$+1,05$
NO_2	$+0,78$	$+0,71$	$+0,79$	$+1,27$
$N^+(CH_3)_3$	$+0,82$	$+0,88$		

σ_{para}, σ_{meta} Hammettsche Substituentenkonstanten, basierend auf der Ionisierung substituierter Benzoesäuren; nach einer Zusammenstellung von D. H. McDaniel u. H. C. Brown: J. Org. Chemistry **23**, 420 (1958).

σ^+_{para} Elektrophile Substituentenkonstanten nach H. C. Brown u. Y. Okamoto: J. Amer. Chem. Soc. **80**, 4979 (1958); hauptsächlich basierend auf der Solvolyse von substituierten 2-Phenyl-2-propylchloriden.

σ^-_{para} Nukleophile Substituentenkonstanten nach H. H. Jaffé: Chem. Reviews **53**, 191 (1953).

Mesomere Substituenteneffekte machen sich vor allem bei p-ständigen Substituenten bemerkbar. Da in log k/k_0 nur solche mesomeren Effekte eingehen, die sich während der Bildung des Übergangszustandes bzw. des Endzustandes ändern, spielen die mesomeren Wechselwirkungen der Substituenten mit dem ungestörten Benzolring keine Rolle. Nun hängen die mesomeren Wechselwirkungen stark von der Ladungsverteilung im Benzolring ab und laufen daher in gewissem Umfang den induktiven Effekten parallel. Es ist daher nicht verwunderlich, daß sich der Einfluß auch mesomer wirksamer Substituenten in vielen Fällen mit Hilfe der σ-Konstanten erfassen läßt. Dieser, auch als „polarer Resonazeffekt" bezeichnete Substituenteneinfluß gibt sich durch einen größeren Unterschied zwischen den σ-Werten für para- und meta-Substitution zu erkennen.

Handelt es sich jedoch um Reaktionen, in deren Ablauf sich eine wesentliche Änderung der Mesomeriefähigkeit des Reaktionszentrums mit dem Benzolring vollzieht (z. B. durch Übergang eines C-Atoms von der sp^2- in die sp^3-Hybridisierung) ohne entsprechend große Änderung der Polarität, so müssen mesomer wirksame Substituenten in para-Stellung Abweichungen von der Hammett-Beziehung aufweisen. Man kann dies durch eine Erweiterung der Hammett-Gleichung in der Form

$$\log k/k_0 = \varrho\,\sigma + \varphi \qquad (12)$$

berücksichtigen, wobei $\varrho\,\sigma$ der induktive und φ der mesomere Anteil an $\delta\varDelta G^{\neq}/2{,}303\,RT$ ist. Letzterer läßt sich in der Weise bestimmen, daß man

Tabelle 2

Beispiele für Reaktionskonstanten nach der Hammett-Gleichung
[nach einer Zusammenstellung von H. H. Jaffé, Chem. Reviews 53, 191 (1953)]

Gleichgewicht bzw. Reaktion	Lösungsmittel	Temp.	ϱ	$-\log k_0$
N.N-Dimethylaniline + CH_3J	$C_6H_5NO_2$	24,8°	−3,069	6,802
subst. Aniline + Benzoylchlorid	C_6H_6	25°	−2,781	2,888
säurekat. Hydrolyse v. Benzamiden	60% EtOH	52,4°	−0,483	5,606
säurekat. Hydrolyse v. Arylacetaten	60% Aceton	25°	−0,198	4,744
Ionisierung von Benzoesäuren	H_2O	25°	+1,000	4,203
subst. Benzoylchloride + Anilin	C_6H_6	25°	+1,219	2,980
basenkat. Hydrolyse v. Benzamiden	60% EtOH	52,8°	+1,364	5,190
Äthanolyse von Benzoylchloriden	60% EtOH 40% Et_2O	25°	+1,499	2,969
Hydrolyse von Benzoylchloriden	95% Aceton	25°	+1,782	4,200
Ionisierung von Phenolen	H_2O	25°	+2,113	9,847
Verseifg. v. Methylbenzoaten	60% Aceton	25°	+2,229	2,075
Verseifg. v. Äthylbenzoaten	75% MeOH	25°	+2,193	3,239
Verseifg. v. L-Menthylbenzoaten	MeOH	30°	+2,628	4,257
Ionisierung von $ArNH_3^{\oplus}$	H_2O	25°	+2,767	4,557

$\log k_0$ = Ordinatenabschnitt der Hammett-Geraden, d. h. der Wert von $\log k$ für $\sigma = 0$.

$\log k/k_0 = f(\sigma)$ für „normale", d. h. nicht mesomeriefähige Substituenten aus der betreffenden Reaktionsserie aufträgt und den vertikalen Abstand der Meßwerte mesomerer Substituenten von der erhaltenen Hammett-Geraden ermittelt.

Eine genauere Betrachtung läßt erkennen, daß diese Versuche, experimentell eine Trennung der strukturellen Einflüsse in induktive und mesomere Effekte vorzunehmen, unvollständig bleiben müssen. So rührt z. B. die Differenz der σ-Konstanten desselben Substituenten in meta- und para-Stellung nicht ausschließlich von dessen mesomerer Wirksamkeit her. Erstens ist nämlich der induktive Effekt in m-Stellung wegen des geringe-

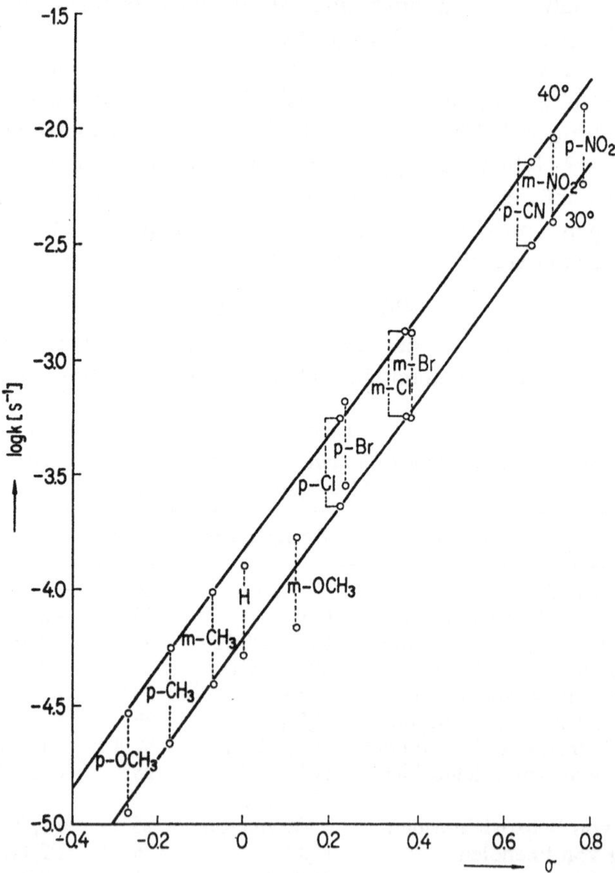

Abb. 1. Beispiel für die Anwendung der Hammettschen Gleichung. Auftragung von log k [s^{-1}] als Funktion der Substituentenkonstanten σ (vgl. Tabelle 1) im Falle der Solvolyse von meta- und para-substituierten L-Menthylbenzoaten in Methanol bei 30 °C und 40 °C. [Nach R. W. TAFT, M. S. NEWMAN u. F. H. VERHOEK: J. Amer. Chem. Soc. 72, 4511 (1950).] Steigung der Geraden: $\varrho_{30°} = +2{,}587$; $\varrho_{40°} = +2{,}552$

ren Abstandes etwas größer als in para-Stellung und zweitens beeinflussen auch meta-Substituenten die Mesomerie des Reaktionszentrums mit dem Benzolkern, weil sie auf induktivem Wege die Ladungsverteilung im aromatischen Ring ändern, die ihrerseits in die mesomere Wechselwirkung eingeht.

In jüngerer Zeit sind eine Reihe von Varianten der Hammett-Gleichung vorgeschlagen worden, die bei bestimmten Reaktionstypen eine verbesserte Korrelation zwischen Substituentenkonstanten und relativen Geschwindigkeitskonstanten ergeben. So haben H. C. BROWN und Y. OKAMOTO (1953) für elektrophile aromatische Substitutionen eine Beziehung der Form

$$\log k/k_0 = \varrho\,\sigma^+ \tag{13}$$

angegeben, die immer dann gelten soll, wenn im Übergangszustand der Reaktion am Reaktionszentrum — das mit dem Benzolkern direkt verbunden ist — eine positive Ladung entsteht. Während für meta-Substitutionen $\sigma^+ \sim \sigma$ ist, ergeben sich für para-Substitutionen stärker negative σ^+-Konstanten als nach der Hammettschen Skala. Eine analoge Bedeutung hat die 1953 von H. H. JAFFÉ aufgestellte Beziehung

$$\log k/k_0 = \varrho\,\sigma^- \, , \tag{14}$$

die für nukleophile aromatische Substitutionen gelten soll, bei denen am Reaktionszentrum eine negative Ladung entsteht. Es charakterisieren offenbar die Konstanten σ^+ mesomere Elektronendonatoren und die Konstanten σ^- mesomere Elektronenacceptoren in para-Stellung am Benzolring in besserer Weise als die Hammettschen σ-Konstanten.

b) *Aliphatische Reaktivität:* HAMMETT hatte bereits ausdrücklich darauf hingewiesen, daß die von ihm aufgestellte Beziehung sich weder auf ortho-Substituenten am Benzolring noch auf aliphatische Reaktionen anwenden läßt, bei denen Reaktionszentrum und Substituent nicht durch einen dazwischenliegenden Benzolring getrennt sind. Selbst in m- und p-Stellung zeigen aliphatische Substituenten Abweichungen von der linearen ΔG-Beziehung, wenn sie eine Kettenlänge von mehr als zwei Gliedern besitzen. Offensichtlich stören in diesen Fällen sterische Substituenteneffekte, die nicht durch eine lineare ΔG-Beziehung erfaßt werden können. Daß es nicht so sehr auf den aromatischen Benzolring ankommt, konnten J. D. ROBERTS und W. T. MORELAND (1953) an Hand von Reaktionen 4-substituierter Bicyclo-[2.2.2]-octan-1-carbonsäuren (I) zeigen.

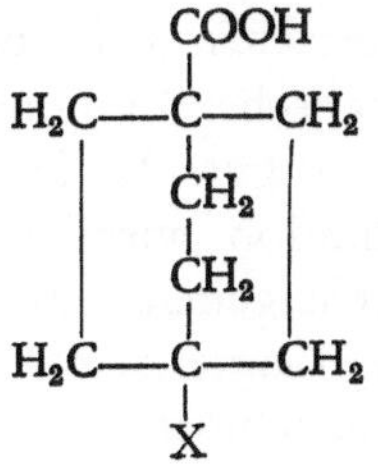

$$
\begin{array}{c}
\text{COOH} \\
| \\
H_2C - C - CH_2 \\
|\quad\ |\quad\ | \\
\quad\ CH_2 \\
|\quad\ |\quad\ | \\
\quad\ CH_2 \\
|\quad\ |\quad\ | \\
H_2C - C - CH_2 \\
| \\
X
\end{array}
\qquad (I)
$$

Hier lassen sich die Substituenteneffekte sehr gut durch eine lineare Beziehung der Form

$$\log k/k_0 = \varrho' \, \sigma' \tag{15}$$

wiedergeben, wobei die σ'-Konstanten in 4-Stellung den Hammettschen σ-Konstanten in meta-Stellung sehr ähnlich sind. Es kommt also darauf an, daß durch das starre Ringgerüst eine genügende räumliche Trennung von Reaktionszentrum und Substituent erfolgt, damit diese sich sterisch nicht beeinflussen können.

Um auch bei aliphatischen Reaktionen polare Substituenteneffekte in Form von linearen ΔG-Beziehungen erfassen zu können, schlug R. W. Taft (1953) einen Weg zur näherungsweisen Eliminierung sterischer Effekte vor. Er verglich die Substituenteneffekte auf die säurekatalysierte Esterhydrolyse und auf die basische Esterverseifung, deren Übergangszustände sich nur durch zwei zusätzliche Protonen bei der ersteren Reaktion unterscheiden. Dabei nahm er an, daß 1. eine Trennung der Substituenteneffekte in einen polaren, mesomeren und sterischen Anteil möglich sei, 2. die sterischen und mesomeren Substituenteneffekte bei der sauren Esterhydrolyse und bei der alkalischen Verseifung gleich sind und sich daher bei der Differenzbildung herausheben und 3. die polaren Effekte bei der sauren Hydrolyse viel schwächer sind als bei der alkalischen Verseifung. Für die letztere Annahme spricht die Tatsache, daß die polare Reaktionskonstante ϱ im Falle der sauren Hydrolyse der Benzoesäureester einen nahe Null gelegenen, bei der alkalischen Verseifung dagegen einen stark positiven Wert besitzt. Definiert man daher nach Taft die polaren Substituentenkonstanten σ^* in der Weise:

$$\sigma^* = \frac{1}{2{,}48} \left[\log (k/k_0)_{\text{bas}} - \log (k/k_0)_{\text{sauer}} \right], \tag{16}$$

wobei der Faktor 1/2,48 lediglich der Anpassung der so erhaltenen Skala an die Hammett-Skala dient, so erhält man bei einer Reihe aliphatischer Reaktionen eine Korrelation

$$\log k/k_0 = \varrho^* \, \sigma^*, \tag{17}$$

die der Hammettschen völlig analog ist. Die Kennzeichnung der Konstanten ϱ^* und σ^* durch einen Stern dient lediglich als Hinweis für ihre andersartige Definition: Sie beziehen sich auf die Hydrolyse bzw. Verseifung von Alkylestern aliphatischer Carbonsäuren, R'COOR, wobei für R = CH_3 die Festsetzung $\sigma^* = 0$ gilt. Als Stütze für die Auffassung, daß man mit der Taftschen Definition wirklich nur polare Substituenteneffekte erfaßt, kann die Tatsache gelten, daß $\sigma^*_X \approx {}^1/_3 \, \sigma^*_{XCH_2}$ ist, d. h., daß die polaren Effekte in der zu erwartenden Weise mit zunehmendem Abstand vom Reaktionszentrum abnehmen.

Tabelle 3. *Beispiele für polare Substituentenkonstanten σ* nach* TAFT *(vgl.* R. W. TAFT *in* M. S. NEWMAN, *Steric Effects in Organic Chemistry, S. 619. New York: J. Wiley 1956).*

Substituent	σ*	Substituent	σ*
CCl_3	+2,65	H	+0,49
$COOCH_3$	+2,00	$CH_2C_6H_5$	+0.22
$COCH_3$	+1,65	CH_3	0,00
CH_2Cl	+1,05	C_2H_5	−0,10
CH_2Br	+1,00	$CH(CH_3)_2$	−0,19
CH_2J	+0,85	$C(CH_3)_3$	−0,30
C_6H_5	+0,60	$CH_2CH(CH_3)_2$	−0,13
CH_2OH	+0,56	Cyclohexyl	−0,15

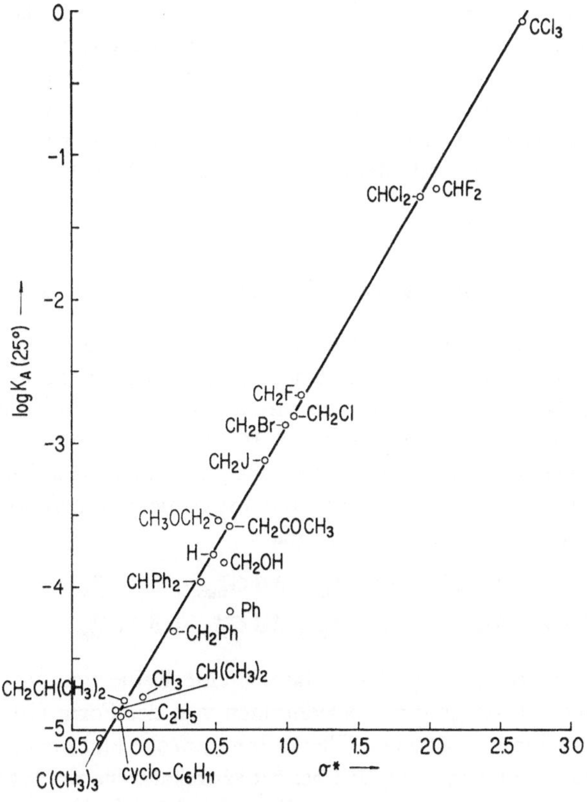

Abb. 2. Beispiel für die Anwendung der Taftschen Gleichung. Auftragung von $\log K_A$ (= $-pK_A$) für aliphatische Carbonsäuren in Wasser bei 25 °C als Funktion der polaren Substituentenkonstanten σ* (vgl. Tab. 3). Steigung der Geraden: $\varrho^* = +1{,}71$; $\log K_A^0 = -4{,}60$. (pK_A-Werte nach einer Zusammenstellung von H. C. BROWN, D. H. MCDANIEL u. O. HÄFLIGER in E. A. BRAUDE u. F. C. NACHOD: Determination of Organic Structures by Physical Methods, Bd. 1, S. 567−662. New York: Academic Press, 1955)

284   F. Becker

In Tabelle 3 sind einige Zahlenwerte für polare Substituentenkonstanten σ^* nach Gl. (17) zusammengestellt. Abb. 2 gibt die Taftsche Gerade für den Fall der Dissoziation aliphatischer Carbonsäuren in Wasser bei 25° wieder.

Spätere Untersuchungen haben ergeben, daß zwischen den aliphatischen und den aromatischen polaren Substituentenkonstanten ein Zusammenhang von der Form

$$\sigma_{I(X)} = 0{,}45\ \sigma^*_{(XCH_2)} \tag{18}$$

besteht, wobei $\sigma_{I(X)}$ der polare Anteil der Substituentenkonstanten des Substituenten X am aromatischen Kern und $\sigma^*_{(XCH_2)}$ die Taftsche Substituentenkonstante für den Substituenten XCH_2 an einem aliphatischen Reaktionszentrum ist. Eine weitgehend übereinstimmende Skala der Substituenteneffekte erhält man auch, wenn man die chemischen Verschiebungen der NMR-Signale in der Reihe der meta-substituierten Fluorbenzole durch σ_I-Konstanten beschreibt. Es ändert sich also nichts an der Form der linearen ΔG-Beziehung, wenn sich diese — wie im Falle der NMR-Spektren — nur auf Messungen an stabilen Molekülen stützt.

2.4. Die Trennung von induktiven, mesomeren und sterischen Effekten

In den vorangegangenen Abschnitten wurde bereits wiederholt die Frage der Aufteilung der experimentell gefundenen Substituenteneinflüsse auf ΔG in einen induktiven, mesomeren und sterischen Anteil angeschnitten, da die Aufstellung linearer ΔG-Beziehungen von allgemeinerer Anwendbarkeit offenbar von einer solchen Möglichkeit abhängig ist. Es wird also angenommen, daß sich die experimentell gefundenen strukturellen Effekte in das additive Schema:

$$\delta\Delta G^0 = \delta\Delta G^0_{ind} + \delta\Delta G^0_{mes} + \delta\Delta G^0_{ster} \tag{19}$$

$$\text{bzw.} \quad \delta\Delta G^+ = \delta\Delta G^+_{ind} + \delta\Delta G^+_{mes} + \delta\Delta G^+_{ster} \tag{20}$$

einfügen. Insbesondere von Taft stammt eine Reihe von Versuchen, die für induktive Effekte gültigen Beziehungen von der Form (10), (13), (14), (15) und (17) durch zusätzliche Glieder für sterische und mesomere Effekte zu erweitern. So nimmt er an, daß bei der säurekatalysierten Hydrolyse von aliphatischen Carbonsäureestern — die sich nicht durch eine Gleichung vom Hammett-Typ beschreiben läßt — induktive und mesomere Effekte praktisch keine Rolle spielen. Er setzt daher den gemessenen Substituenteneinfluß auf k einem sterischen Effekt gleich und definiert eine sterische Substituentenkonstante E_s gemäß

$$\log (k/k_0)_{Hydrolyse} = E_s\ . \tag{21}$$

Die Taftschen Konstanten E_s erweisen sich bei anderen, allerdings ähnlichen Reaktionsserien als brauchbar, wenn man die Gleichung in der Form

$$\log k/k_0 = E_s \cdot \Theta \tag{22}$$

benützt, wobei Θ eine „sterische Reaktionskonstante" ist, die ein Maß für die Empfindlichkeit der betreffenden Reaktion gegenüber sterischen Substituenteneffekten darstellt. Treten daneben polare Effekte auf, so hat man Gl. (22) durch Gl. (17) zu ergänzen:

$$\log k/k_0 = \varrho^* \sigma^* + E_s \Theta, \tag{23}$$

wodurch sich ein Ausdruck von erweitertem Anwendungsbereich ergibt. So läßt sich Gl. (23) z. B. sehr gut auf die basische Verseifung der L-Menthylester aliphatischer Carbonsäuren in Methanol anwenden (PAVELICH u. TAFT, 1957). Bemerkenswerterweise eignen sich die Taftschen E_s-Konstanten auch zur Beschreibung der Strukturabhängigkeit der Bildungswärmen von Additionsverbindungen zwischen Bortrimethyl und aliphatischen Aminen. Einige Zahlenwerte für sterische Substituentenkonstanten E_s nach Gl. (22) bzw. (23) sind in Tabelle 4 zusammengestellt.

Tabelle 4. *Beispiele für sterische Substituentenkonstanten (aliphatische Reihe) E_s nach* TAFT *(vgl.* R. W. TAFT *in* M. S. NEWMAN, *Steric Effects in Organic Chemistry, S. 598. New York: J. Wiley 1956)*

Substituent	E_s	Substituent	E_s
H	$+1,24$	$CH_2CH(CH_3)_2$	$-0,93$
CH_3	$0,00$	CF_3	$-1,16$
C_2H_5	$-0,07$	$C(CH_3)_3$	$-1,54$
CH_2Br	$-0,27$	$CH_2C(CH_3)_3$	$-1,74$
CH_2J	$-0,37$	$CH(C_6H_5)_2$	$-1,76$
$CH_2C_6H_5$	$-0,38$	CCl_3	$-2,06$
$CH(CH_3)_2$	$-0,47$	CBr_3	$-2,43$
Cyclohexyl	$-0,79$	$C(C_2H_5)_3$	$-3,8$

KREEVOY und TAFT (1955) haben ferner versucht, den Anteil der induktiven und der mesomeren Effekte bei geeigneten Reaktionsserien formelmäßig zu erfassen. Als Beispiel wählten sie den Substituenteneinfluß auf die Hydrolyse von Diäthylacetalen und -ketalen, bei der das Carbonyl-C-Atom vom tetraedrischen in den trigonalen Bindungszustand übergeht. Dabei macht sich eine Hyperkonjugation zwischen der entstehenden $C=O$-Bindung und den α-ständigen $C-H$-Bindungen bemerkbar. Nimmt man an, daß die sterischen Effekte im Ausgangs- und Übergangszustand gleich sind, so gelingt es, induktive und mesomere Effekte mittels der Gleichung

$$\log k/k_0 = \varrho^* \Sigma \sigma^* + (n - 6) h$$

zu beschreiben. Die Summation erstreckt sich über die durch Gl. (16) definierten polaren Substituentenkonstanten aller mit dem Reaktionszentrum direkt verknüpften Liganden. h ist ein Hyperkonjugationsparameter, der die mesomere Wechselwirkung einer α-ständigen C—H-Bindung berücksichtigt; der Faktor $(n-6)$ rührt von der Verwendung von Acetonal als Bezugssubstanz her, das die Maximalzahl von 6 C—H-Bindungen in α-Stellung aufweist. Die gute Korrelation einer größeren Zahl von experimentellen Daten mit der obigen Gleichung spricht zwar dafür, daß es einen mit der Zahl der α-ständigen C—H-Bindungen parallel laufenden strukturellen Einfluß gibt, jedoch ist dessen Interpretation problematisch. Der nach Gl. (16) definierte Parameter σ^* hat für die CH_3-Gruppe den Wert Null, obwohl diese bei der Esterhydrolyse ebenfalls einer sich bildenden C=O-Bindung benachbart ist, wenn auch der Grad der Ungesättigtheit im Übergangszustand geringer angenommen wird als bei der Hydrolyse der Acetale und Ketale. Jedenfalls muß damit gerechnet werden, daß ein Teil des Einflusses der α-C—H-Bindungen bereits in den polaren Konstanten σ^* enthalten ist.

Neuere Versuche von Taft, bei aromatischen Reaktionen eine Trennung von induktiven und mesomeren Effekten vorzunehmen, gehen von folgenden Annahmen aus: 1. Substituenten in m- und p-Stellung am Benzolkern üben nur induktive und mesomere Effekte aus, die man in der Form

$$\log k_{\text{meta}}/k_0 = I^m + R^m \; ; \; \log k_{\text{para}}/k_0 = I^p + R^p \tag{24}$$

ansetzen kann. 2. Die induktiven Effekte I^m und I^p in meta- und para-Stellung sind gleich. 3. Das Verhältnis der mesomeren Effekte R^m und R^p in meta- und para-Stellung ist konstant: $R^m/R^p = \alpha$. Damit erhält man für den induktiven Effekt $I = I^m = I^p$ alleine:

$$I = \frac{1}{1-\alpha} \left(\log \frac{k_m}{k_0} - \alpha \log \frac{k_p}{k_0} \right) = \sigma_I \varrho_I . \tag{25}$$

Berechnet man I mittels der Parameter σ_I und ϱ_I aus mesomeriefreien Reaktionen, so ergibt sich für die mesomeren Effekte:

$$R^m = \log k_m/k_0 - \sigma_I \varrho_I \; ; \; R^p = \log k_p/k_0 - \sigma_I/\varrho_I \; . \tag{26}$$

Bei „normalen" Reaktionen hat $\alpha = R^m/R^p$ einen Wert von 1/3; bei elektrophilen Reaktionen erhält man kleinere Werte bis zu 1/10. Es resultiert also auch in meta-Stellung noch ein gewisser mesomerer Effekt, der indirekt, durch die Beeinflussung der Mesomerie des Reaktionszentrums mit dem aromatischen Kern, zustande kommt.

Die Rechtfertigung für alle angeführten Formulierungen wird aus der Tatsache hergeleitet, daß die ermittelten Strukturparameter bei einer mehr oder weniger größeren Anzahl von Reaktionsserien eine gute Korrelation mit den gemessenen $\log k/k_0$-Werten ergeben. Die empirischen Struktur-Reaktivitäts-Beziehungen haben also zweifellos einen praktischen Wert, jedoch dürfen aus ihnen theoretische Folgerungen nur mit großer Vorsicht gezogen werden.

Die Tatsache, daß man in den empirischen Gleichungen eine Aufteilung des gesamten Struktureinflusses in Einzeleffekte vornehmen kann, beweist noch nicht, daß man es mit wirklich unabhängigen Größen zu tun hat. Streng genommen, ist es wohl auf experimentellem Wege überhaupt nicht möglich, einen einzelnen (d. h. von einem bestimmten theoretischen Modell erfaßbaren) Struktureinfluß zu isolieren. Da aber der Theoretiker auf den Vergleich mit dem Experiment nicht verzichten kann, ist eine kritische Analyse der Aussagemöglichkeiten gezielter Experimente von großer Wichtigkeit. Unter diesem Gesichtspunkt sollen die linearen ΔG-Beziehungen zum Schluß noch einmal zusammenfassend betrachtet werden, wobei wir annehmen wollen, daß es mit den Begriffen des induktiven, mesomeren und sterischen Effektes möglich sei, die Zusammenhänge zwischen Struktur und Reaktivität einigermaßen vollständig zu erfassen.

Am klarsten sind die Verhältnisse offenbar bei rein induktiven Effekten, was auch aus der sehr großen Zahl von Reaktionen hervorgeht, die der einfachen Hammett-Gleichung (10) gehorchen. Die polare Substituentenkonstante σ gibt an, in welchem Maße der betreffende Substituent auf induktivem Wege als Elektronendonator (negative σ-Werte) bzw. als Elektronenacceptor (positive σ-Werte) wirken kann. In welchem Umfang dieser Effekt wirksam wird, hängt nur von der Änderung der Polarität des Reaktionszentrums während des betreffenden Prozesses ab. Ein Maß hierfür stellt die polare Reaktionskonstante ϱ dar, die positiv ist, wenn das Reaktionszentrum bei der Reaktion stärker negativ wird, und negativ, wenn es eine positive Ladung annimmt. Die Konstante σ läßt sich auch ohne Bezugnahme auf eine chemische Reaktion, z. B. an Hand der chemischen Verschiebungen der NMR-Signale bestimmen, wenn man Moleküle vergleicht, die alle dieselbe stark polare Bindung (z. B. C—F) enthalten. Bei Prozessen in Lösung kann man damit rechnen, daß die polaren Solvatationseffekte des Substituenten seinen polaren Einflüssen auf das Reaktionszentrum proportional und damit in dem für das betreffende Lösungsmittel gültigen ϱ-Wert enthalten sind.

Mesomeriefähige Substituenten am Benzolkern können hierdurch zu Elektronendonatoren oder -acceptoren werden. Diese mesomere Wechselwirkung ist eine Funktion der Ladungsverteilung im Benzolkern und daher ebenfalls dem induktiven Einfluß einer am Reaktionszentrum vorhandenen oder entstehenden Ladung unterworfen. Da aber die mesomere Wechselwirkung in anderer Weise von der Ladung abhängt als die induktive, läßt sie sich nicht ohne weiteres in den polaren σ-Konstanten erfassen. Dies geht z. B. aus der Tatsache hervor, daß man gezwungen war, für mesomeriefähige Substituenten in p-Stellung zum Reaktionszentrum besondere σ-Konstanten einzuführen, wenn es sich um Reaktionen mit starker Ladungsänderung am Reaktionszentrum, also um nukleophile oder elektrophile Reaktionen, handelt. Andererseits kann sich die Mesomeriefähigkeit

des Reaktionszentrums während der Reaktion erheblich ändern, ohne daß
damit eine starke Polaritätsänderung verbunden ist. Auch dann geht der
mesomere Substituenteneinfluß nicht mit dem induktiven parallel. Man
kann also nicht erwarten, daß eine Gleichung der Form (13) oder (14)
induktive und mesomere Effekte gleichzeitig allgemeingültig erfaßt; hierzu
wäre wenigstens ein Ansatz notwendig, der den induktiven und mesomeren
Anteil unabhängig in Form von Produkten $\sigma_I \varrho_I$ und $\sigma_R \varrho_R$ enthält. Auch
dann bleibt, wie schon auf S. 280 ausgeführt, die Frage offen, wie weit sich
beide Effekte experimentell überhaupt trennen lassen. Grundsätzlich besteht
an einer Klärung dieser Frage erhebliches Interesse, da z. B. über die Abhängigkeit der Mesomerie von einer vorgegebenen Ladungsverteilung im
Benzolring noch recht wenig bekannt ist.

Sterische Substituenteneffekte lassen sich durch einen hinreichend großen
Abstand vom Reaktionszentrum weitgehend ausschalten. Diese Bedingung
läßt sich am besten durch einen dazwischen geschobenen Benzolring erfüllen,
da dieser induktive und mesomere Effekte weiterzuleiten vermag. Bei
Substituenten, die eine innere Beweglichkeit besitzen, ist eine unabhängige
Betrachtung sterischer Effekte an die Voraussetzung geknüpft, daß sie
keine orientierungsabhängige Mesomeriefähigkeit aufweisen und keine
Ladung oder stark polare Bindung enthalten, deren Abstand vom Reaktionszentrum konformationsabhängig ist. Auf jeden Fall hängen sterische
Effekte in anderer Weise vom Reaktionsmechanismus ab als induktive oder
mesomere. So spielen z. B. der Raumbedarf des Reaktionspartners im
Übergangszustand oder die Möglichkeit einer Änderung der Valenzwinkel
zwischen den Liganden — z. B. durch Übergang eines C-Atoms aus dem
trigonalen in den tetraedrischen Zustand — eine ausschlaggebende Rolle.
Ferner muß die Frage erörtert werden, in welcher Weise das Lösungsmittel
in den Reaktionsmechanismus eingreift und ob sterische Wechselwirkungen
mit den Substituenten von einer Änderung des Solvatationszustandes des
Reaktionszentrums während des Reaktionsablaufes herrühren können.
Aus diesen Überlegungen geht hervor, daß der Versuch, sterische Substituenteneffekte in Form einer „sterischen Substituentenkonstanten E_s"
[vgl. Gl. (23)] zu erfassen, notwendigerweise auch die Einführung einer
„sterischen Reaktionskonstanten Θ" erfordert, da ein bestimmter Reaktionsablauf in durchaus verschiedener Weise auf induktive, mesomere und sterische Einflüsse anspricht. Die vollständige Gleichung lautet also:

$$\log k/k_0 = \sigma_I \varrho_I + \sigma_R \varrho_R + E_s \Theta \ .$$

Wenn man sich zusammenfassend Rechenschaft darüber ablegen will,
welche Bedeutung den hier erörterten empirischen Struktur-Reaktivitäts-
Beziehungen beizumessen ist, so liegt diese wohl in erster Linie in der
systematischen Ordnung, die sie in das sonst nur schwer zu durchschauende
Tatsachenmaterial gebracht haben. Sie gaben außerdem die Anregung zur

Durchführung zahlreicher systematischer Versuchsserien und haben dazu beigetragen, daß diese in einer gezielteren und im Hinblick auf eine Interpretation besser zu überblickenden Weise vorgenommen wurden als früher. Das zusammengetragene experimentelle Material bildet die Grundlage für theoretische Untersuchungen, wie z. B. die Behandlung des Problems der Übertragung induktiver Effekte als reine Feldeffekte oder auf dem Wege über die Polarisation dazwischen liegender Bindungen, die Anwendung halbempirischer quantenmechanischer Näherungsmethoden zur Berechnung von Mesomerieenergien oder die quantitative Behandlung sterischer Effekte mit Hilfe statistisch-thermodynamischer Modelle.

Literatur

Zusammenfassende Literatur zu Teil 1.

1. INGOLD, C. K.: Structure and Mechanism in Organic Chemistry. Ithaca: Cornell University Press 1953.
2. GOULD, E. S.: Mechanismus und Struktur in der organischen Chemie. Weinheim: Verlag Chemie 1962 (Übersetzt von G. Koch).
3. BUNTON, C. A.: Nucleophilic Substitution at a Saturated Carbon Atom. Amsterdam: Elsevier Publishing Company 1963.
4. BANTHORPE, D. V.: Elimination Reactions. Amsterdam: Elsevier Publishing Company 1963.

Zusammenfassende Literatur zu Teil 2.

1. HAMMETT, L. P.: Physical Organic Chemistry. New York: McGraw-Hill. 1940.
2. TAFT, R. W. JR.: Separation of Polar, Steric, and Resonance Effects in Reactivity. In M. S. NEWMAN, Steric Effects in Organic Chemistry, pp. 556—675. New York: J. Wiley 1956.
3. LEFFLER, J. E., and E. GRUNWALD, Rates and Equilibria of Organic Reactions. New York: J. Wiley 1963.
4. RITCHIE, C. D., and W. F. SAGER: An Examination of Structure-Reactivity Relationships. In COHEN-STREITWIESER-TAFT, Progress in Physical Organic Chemistry, Bd. 2, S. 232—400. New York: Interscience 1964.
5. JAFFÉ, H. H.: Chem. Reviews **53**, 191 (1953).
6. EHRENSON, S.: Theoretical Interpretation of the Hammett and Derivative Structure-Reactivity Relationships. In COHEN-STREITWIESER-TAFT, Progress in Physical Organic Chemistry, Bd. 2., S. 195—251. New York: Interscience 1964.
7. PALM, V.: Russ. Chem. Reviews (Engl. Transl.) **9**, 471 (1961).

F. Becker
Institut für physikalische Chemie
der Universität Saarbrücken

Abschätzung relativer freier Aktivierungs-enthalpien mittels der HMO-Methode

Oskar E. Polansky und Peter Schuster

Mit 10 Abbildungen

1. Einleitung

Seit längerer Zeit stehen im wesentlichen zwei verschiedene Methoden zur Verfügung, um die chemische Reaktivität konjugierter organischer Verbindungen aufgrund der Ergebnisse von Hückel-Molekülorbital-(HMO)-Rechnungen zu diskutieren: die Näherung der isolierten Moleküle und die Näherung der Lokalisierungsenergien [9].

Bei der *Näherung der isolierten Moleküle* wird die durch die Annäherung des Reaktanten R an das Substratmolekül S bewirkte Energieänderung δE des Substratmoleküls näherungsweise in Form einer Taylorreihe entwickelt.

$$\delta E = \sum_r \frac{\partial E}{\partial \alpha_r} \overline{\delta \alpha_r} + \sum_{r<s} \frac{\partial E}{\partial \beta_{rs}} \overline{\delta \beta_{rs}} + \sum_{r<t} \frac{\partial^2 E}{\partial \alpha_t \partial \alpha_r} \overline{\delta \alpha_r \delta \alpha_t} +$$
$$+ \frac{1}{2} \sum_r \sum_{t<u} \frac{\partial^2 E}{\partial \beta_{tu} \partial \alpha_r} \overline{\delta \alpha_r \delta \beta_{tu}} + \frac{1}{2} \sum_{r<s} \sum_{t<u} \frac{\partial^2 E}{\partial \beta_{tu} \partial \beta_{rs}} \overline{\delta \beta_{rs} \delta \beta_{tu}} + \cdots \tag{1}$$

Hierin numerieren r, s, t und u verschiedene Zentren des π-Elektronensystems des Substratmoleküls S; $\delta \alpha_r$ und $\delta \beta_{rs}$ stellen die Änderung der Coulomb-Integrale α_r bzw. der Resonanzintegrale β_{rs} dar, welche durch die Annäherung von R verursacht werden; die Summierungen erstrecken sich über alle Zentren des π-Elektronensystems; die ersten partiellen Differentialquotienten der Gesamt-π-Elektronenenergie E sind als π-Elektrondichten q_r bzw. π-Bindungsordnungen p_{rs}, die zweiten partiellen Differentialquotienten als Polarisierbarkeiten $\Pi_{r,t}$, $\Pi_{r,tu}$ und $\Pi_{rs,tu}$ bekannt. Da üblicherweise nur wenige Parameter $\delta \alpha_r$ und $\delta \beta_{rs}$ des Substratmoleküls S als durch den Angriff des Reaktanten R geändert betrachtet werden, die meisten $\delta \alpha_r$ und $\delta \beta_{rs}$ also Null sind, reduziert sich die Zahl der Summanden in der vorstehenden Gleichung erheblich.

In die Taylor-Entwicklung gehen die für die π-Elektronenstruktur des Substratmoleküls S charakteristischen Größen ein. Ferner schmiegt sie sich dem Energieprofil des Reaktionsablaufes umso besser an, je kleiner die absoluten Werte von $\delta\alpha_r$ und $\delta\beta_{rs}$ sind. Sie beschreibt daher einen Zustand im Energie-Reaktionskoordinaten-Diagramm, welcher zwischen dem Ausgangs- und dem Übergangszustand und zwar eher dem ersteren als dem letzteren benachbart liegt (vgl. Abb. 1). Außerdem bleiben durch sie alle Energieänderungen des Reaktanten R unberücksichtigt.

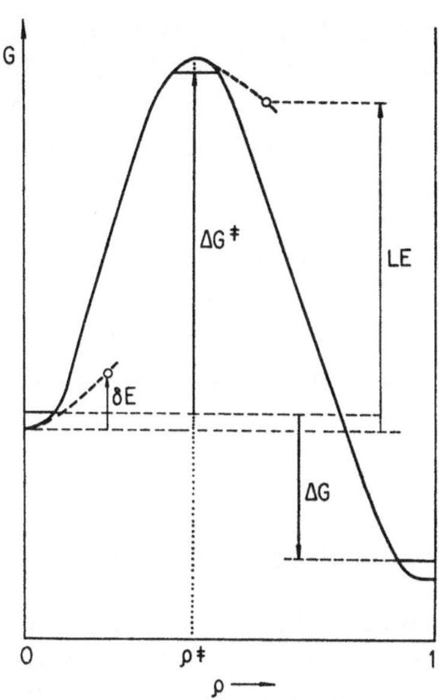

Abb. 1. Energieprofil einer elementaren Reaktion (schematisch). $\Delta G^{\ddagger}$ = Freie Aktivierungsenthalpie, ΔG = Freie Reaktionsenthalpie, $\varrho^{\ddagger}$ = normierte Reaktionskoordinate des aktivierten Komplexes (Übergangszustandes), ϱ = normierte Reaktionskoordinate; δE = Energieänderung nach der Taylor-Entwicklung (Methode der isolierten Moleküle), LE = Lokalisierungsenergie

In der *Näherung der Lokalisierungsenergien* diskutiert man die Reaktivität aufgrund des Unterschiedes der π-Elektronenenergien des Substratmoleküls S und einer aus S und dem Reaktanten R gebildeten intermediären Spezies, welche durch die Knüpfung einer perfekten σ-Bindung zwischen dem r-ten Zentrum von S und dem attackierenden Zentrum des Reagens R entstanden gedacht wird. Als Grundlage der Diskussion wird hier also ein Zustand benutzt, welcher am ehesten einem dem Übergangszustand folgenden σ-Komplex als diesem selbst entspricht und im Energieprofil der Reaktion knapp hinter dem Übergangszustand liegt (vgl. Abb. 1). Auch hier bleiben etwaige Änderungen der Energie des Reaktanten R unberücksichtigt.

In beiden Näherungen wird die Energieänderung δE bzw. die Lokalisierungsenergie LE als ein ungefähres Maß der Freien Aktivierungsenthalpie ΔG^+ betrachtet; gemäß dem aus der Theorie der absoluten Reaktionsgeschwindigkeiten (Transition State Theory) folgenden Ausdruck für die Reaktionsgeschwindigkeitskonstante k_j, in welchem k die Boltzmann-

$$k_j = \frac{kT}{h}\, e^{-\Delta G^+/RT} \qquad (2)$$

Konstante, h das Plancksche Wirkungsquantum, R die Gaskonstante und T die absolute Temperatur bedeuten, wird diejenige Reaktion j als kinetisch bevorzugt angesehen, welche der kleinsten Freien Aktivierungsenthalpie ΔG^+ bedarf. Die diesen beiden Näherungen anhaftenden Unsicherheiten und methodischen Fehler wurden vor einiger Zeit von R. D. Brown [9] ausführlich diskutiert.

Im Gegensatz zu den beiden oben erwähnten Methoden zielt die im folgenden beschriebene Näherung darauf ab, die relativen Größen der Freien Aktivierungsenthalpie selbst abzuschätzen. Zugleich wird versucht, die Voraussetzungen der Anwendbarkeit und die Grenzen dieser Näherungsmethode zu umreißen.

2. Reaktionstheoretische Grundlagen

Die Theorie des aktivierten Komplexes ist auf elementare Reaktionen, welche aus der reversiblen Bildung des aktivierten Komplexes T^+ aus den Ausgangsstoffen A und seinem Zerfall zu den Reaktionsprodukten P bestehen,

$$A \rightleftharpoons [T^+] \rightarrow P \; . \qquad (3)$$

direkt anwendbar. Zwei Gruppen von Eigenschaften charakterisieren eine bestimmte elementare Reaktion:

1. der Typus der Ausgangs- und Endstoffe, sowie der Mechanismus ihres Ablaufes — diese allgemeine Charakteristik wird üblicherweise als *Reaktionsweise* bezeichnet und faßt die Art und relative Lage der Reaktionszentren und die Vorgänge, welche während des Ablaufes der Reaktion an diesen stattfinden, zusammen;

2. die Zahl, Anordnung und Natur der chemischen Gruppen der Ausgangsstoffe, welche sich dem Mechanismus entsprechend im aktivierten Komplex und in den Endprodukten in gesetzmäßiger Weise wiederfinden und welche auf den Ablauf der Reaktion nur graduellen, aber keinen prinzipiellen Einfluß haben — in der Folge wird diese zweite Gruppe von Eigenschaften als *Substitution der Reaktionszentren* bezeichnet.

Eine *individuelle Reaktion* ist somit durch die ihr zugrunde liegende Reaktionsweise und die Substitution der Reaktionszentren eindeutig bestimmt. Bezeichnet der Index j eine bestimmte Reaktionsweise, der Index μ eine bestimmte Substitution der Reaktionszentren, so schreibt sich eine individuelle elementare Reaktion gemäß Gl. (3) wie folgt:

$$A_{j,\mu} \rightleftharpoons [T_{j,\mu}^{\neq}] \rightarrow P_{j,\mu} \; . \tag{3a}$$

Als Illustrationsbeispiel für eine Reaktionsweise diene die Bildung von Δ^1-Pyrazolinen durch 1,3-Cycloaddition von Diazoalkanen an Olefine [2, 3]:

$$\left.\begin{array}{c} R_4 \diagdown \; {}^{\beta} \; {}^{\alpha} \diagup R_5 \\ \mathrm{C}=\mathrm{C} \\ R_3 \diagup \qquad \diagdown R_6 \\ + \\ R_2 \diagdown \; {}^{\gamma} \; {}^{\delta} \; {}^{\varepsilon} \\ \mathrm{C} \cdots \mathrm{N} = \!\!= \mathrm{N}| \\ R_1 \diagup \end{array}\right\} \rightleftharpoons [T_{p,\mu}^{\neq}] \longrightarrow \begin{array}{c} R_4 \quad R_5 \\ R_3-\mathrm{C}-\!\!-\!\!-\mathrm{C}-R_6 \\ R_2-\mathrm{C} \qquad \mathrm{N}| \\ R_1 \qquad \mathrm{N} \\ \underline{} \end{array} \; . \tag{4}$$

Eine individuelle Reaktion μ dieser Reaktionsweise p ist z. B. die Umsetzung von Diphenyldiazomethan mit trans-Zimtsäureäthylester ($R_1 = R_2 = R_3 = C_6H_5$; $R_4 = R_6 = H$; $R_5 = COOC_2H_5$).

Der Mechanismus beschreibt die Änderungen der gegenseitigen Lage der Reaktanten und der Atome und Atomgruppen in den Reaktanten während des Ablaufs der Reaktion. Der Reaktionsmechanismus läßt sich so als Gesamtheit der Geometrien aller derjenigen Zustände auffassen, welche bei der Reaktion durchlaufen werden. Er beinhaltet daher auch die Geometrie des Übergangszustandes. Da hierbei aus der Vielfalt der möglichen, einander ähnlichen räumlichen Anordnungen diejenigen ausgewählt werden, welche ihre typischen Merkmale am reinsten widerspiegeln, beschreibt der Mechanismus die Geometrie der bei der Reaktion durchlaufenen Zustände, darunter auch die des Übergangszustandes in vereinfachender und idealisierender Weise. Durch die Selektion der typischen Merkmale wählt er gleichzeitig aus der Zahl der physikalischen Meßgrößen, welche sich während der Reaktion ändern, diejenigen aus, welche für den Fortschritt der Reaktion charakteristisch und in entsprechender Verknüpfung als Reaktionskoordinaten geeignet sind. Für die Cycloaddition (4), bei welcher die C_β–C_γ- und die C_α–N_ε-Bindung *gleichzeitig* geknüpft werden, folgt auf diesem Wege ein Übergangszustand, dessen Geometrie etwa der in Abb. 2 dargestellten entspricht. Als normierte[1] Reaktionskoordinate ϱ bietet sich

$$\varrho = \frac{r_{\beta\gamma}^E + r_{\varepsilon\alpha}^E}{r_{\beta\gamma} + r_{\varepsilon\alpha}} = \frac{(r_{\beta\gamma}^E + r_{\varepsilon\alpha}^E)\sin\vartheta}{(r_{\gamma\delta} + r_{\delta\varepsilon})\cos 2\vartheta - r_{\alpha\beta}} \tag{5}$$

---

[1] Als normiert soll die Reaktionskoordinate ϱ dann bezeichnet werden, wenn sie für das Ausgangssystem den Wert $\varrho = 0$ und für das Endsystem $\varrho = 1$ besitzt und im Bereich $0 \leq \varrho \leq 1$ stetig ist.

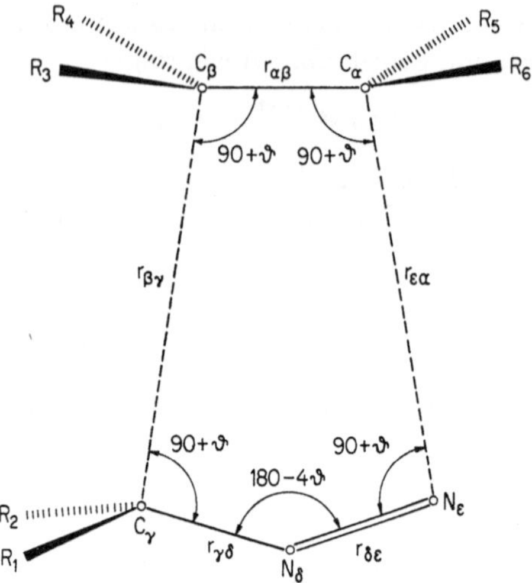

Abb. 2. Übergangszustand der Cycloaddition (4). Idealisierung der Geometrie durch Annahme exakter Planarität und gleich großer Valenzwinkel (90 + ϑ) an den Reaktionszentren (C_α, C_β, C_γ und N_ε)

an, worin $r_{\mu\nu}$ und ϑ die in Abb. 2 angegebene Bedeutung besitzen, $r_{\beta\gamma}^E$ und $r_{\varepsilon\alpha}^E$ die Längen dieser Bindungen im Endprodukt (Pyrazolin) darstellen.

Die Freie Enthalpie G aller, bei der Reaktion durchlaufenen Zustände ist im Prinzip berechenbar und als eine Funktion der Reaktionskoordinate (ϱ) darstellbar. In der Regel durchläuft sie ein bei einem bestimmten Wert der Reaktionskoordinate ($\varrho = \varrho^+$) liegendes Maximum; der diesem Maximum entsprechende Zustand wird als Übergangszustand hervorgehoben (Abb. 1). Da für jede elementare Reaktion, deren Mechanismus hinreichend bekannt ist, in dieser oder ähnlicher Weise plausible Vorstellungen über die Geometrie des Übergangszustandes und der anderen, bei der Reaktion durchlaufenden Zustände gewonnen werden können, sollte im Prinzip für jede elementare Reaktion das Energieprofil (Abb. 1) berechenbar sein.

Der Unterschied der Freien Enthalpien des aktivierten Komplexes im Übergangszustand und der des Ausgangssystems ist als Freie Aktivierungsenthalpie ΔG^+ definiert:

$$\Delta G^+ \equiv G_{\text{akt. Kompl.}} - G_{\text{Ausgangsstoffe}} \; . \tag{6}$$

Bekanntlich stellt die Freie Aktivierungsenthalpie eine Summe aus einem rein enthalpischen und einem entropischen Glied dar

$$\Delta G^+ = \Delta H^+ - T\Delta S^+ \; , \tag{7}$$

von welchem zunächst die Aktivierungsenthalpie ΔH^+ näher betrachtet werden soll.

Die Aktivierungsenthalpie $\Delta H^{\neq}$ setzt sich aus dem Energieunterschied zwischen dem aktivierten Komplex und den Ausgangsstoffen, der Aktivierungsenergie $\Delta U^{\neq}$ also, und der Volumsarbeit $p\Delta V^{\neq}$ zusammen:

$$\Delta H^{\neq} = \Delta U^{\neq} + p\Delta V^{\neq} \; . \tag{8}$$

Für Reaktionen in kondensierter Phase (z. B. Lösungen) ist $\Delta V^{\neq}$ in der Regel klein und daher $\Delta H^{\neq} \approx \Delta U^{\neq}$ zumeist eine gute Näherung.

Die Aktivierungsenergie $\Delta U^{\neq}$ läßt sich als eine Summe von Beiträgen auffassen, welche den Energieänderungen der σ-, π-, einsamen ($lp =$ „lone pair"), usw. Elektronen und der Änderung der Kernabstoßungsenergie entsprechen. Entsprechend Gl. (8) tragen alle diese Energieänderungen sowohl der Art als auch dem Betrage nach in gleicher Weise zur Aktivierungsenthalpie $\Delta H^{\neq}$ als zur Aktivierungsenergie $\Delta U^{\neq}$ bei, so daß Gl. (8) detailliert wie folgt angesetzt werden kann:

$$\Delta H^{\neq} = \Delta H_{\sigma}^{\neq} + \Delta H_{\pi}^{\neq} + \cdots + \Delta H_{lp}^{\neq} + \Delta H_{K}^{\neq} + p\Delta V^{\neq} \; . \tag{9}$$

Von allen diesen Beiträgen ist $\Delta H_{\pi}^{\neq}$ näherungsweise mit Hilfe der HMO-Methode erfaßbar. Aus diesem Grunde ziehen wir alle anderen Beiträge zu dem Restglied $\Delta H_{R}^{\neq}$ zusammen, so daß für die Aktivierungsenthalpie

$$\Delta H^{\neq} = \Delta H_{\pi}^{\neq} + \Delta H_{R}^{\neq} \tag{10}$$

folgt, worin $\Delta H_{R}^{\neq}$ definiert ist zu:

$$\Delta H_{R}^{\neq} = \Delta H_{\sigma}^{\neq} + \cdots + \Delta H_{lp}^{\neq} + \Delta H_{K}^{\neq} + p\Delta V^{\neq} \; . \tag{11}$$

Der Beitrag der σ-Elektronen $\Delta H_{\sigma}^{\neq}$ geht auf Energieänderungen im System der σ-Bindungen zurück: bindende Beiträge liefern stets die σ-Bindungen, welche die Reaktanten zum aktivierten Komplex verknüpfen; weitere Beiträge verschiedenen Vorzeichens folgen aus den Umarrangements (Änderungen der Bindungslängen, der Hybridisierungszustände usw.) in den Reaktanten. Da der Übergangszustand in der Regel mit großen Bindungslängen für die zwischen den Reaktanten neu geknüpften σ-Bindungen korrespondiert und die in den Reaktanten stattgehabten Änderungen zumeist noch sehr klein sind, ist auch der gesamte Beitrag der σ-Elektronen $\Delta H_{\sigma}^{\neq}$ dem absoluten Werte nach klein und besitzt je nach der Reaktionsweise bindenden oder gegenbindenden Charakter.

Der Beitrag der einsamen Elektronenpaare $\Delta H_{lp}^{\neq}$ folgt aus den Hybridisierungsänderungen, soweit diese die in den Reaktanten vorhandenen einsamen Elektronenpaare miterfassen bzw. Bindungselektronen der Reaktanten in einsamen Elektronenpaaren des aktivierten Komplexes lokalisieren oder umgekehrt einsame Elektronenpaare der Reaktanten in Bindungselektronen umwandeln. Im Falle der Cycloaddition (4) treten beide Typen der Änderungen auf: Ohne Zweifel ändert das einsame Elektronenpaar am N_ε im Laufe der Reaktion seinen Hybridisierungszustand; ferner werden aber am N_δ zwei Elektronen lokalisiert, welche dem dreizentrigen π-MO des Diazoalkans entstammen.

Der Beitrag $\Delta H_K^{\pm}$ entspricht der Änderung der Kernabstoßungsenergie und ist für multimolekulare Reaktionen stets positiv (d. h. gegenbindend). Zerfällt der aktivierte Komplex in zwei oder mehrere Reaktionsprodukte, so erreicht der Beitrag der Kernabstoßungsenergie ΔH_K ein Maximum, das — wie leicht einzusehen ist — nicht mit $\Delta H_K^{\pm}$ identisch zu sein braucht. Stabilisiert sich aber — wie z. B. im Falle von Cycloadditionen — der aktivierte Komplex zu einem einzigen Reaktionsprodukt, steigt ΔH_K stetig an[2].

Es ist leicht einzusehen, daß alle diese Enthalpieänderungen und auch $\Delta V(\varrho)$ stetige, eindeutige und differenzierbare Funktionen der Reaktionskoordinate ϱ sein müssen, was durch Formulierungen wie z. B.

$$\Delta H_\sigma = \Delta H_\sigma(\varrho) \tag{12a}$$

bzw.

$$\Delta H_R^{\pm} = \Delta H_R(\varrho^{\pm}) \tag{12b}$$

usw.

ausgedrückt sein soll. Für alle individuelle Reaktionen $[j, \mu]$ einer bestimmten Reaktionsweise $[j]$, deren aktivierte Komplexe $[T_{j,\mu}^{\pm}]_{j\,=\,\text{const.};\,\mu\,=\,\text{var.}}$ immer wieder bei dem gleichen Wert der Reaktionskoordinate liegen, für die also $\varrho^{\pm}$ keine Funktion der Substitution $[\mu]$, sondern nur eine der Reaktionsweise $[j]$ ist

$$\varrho_j^{\pm} = [\varrho(j)]^{\pm} \neq \varrho(j,\mu) \tag{13}$$

läßt sich nun zeigen, daß die Größe aller Beiträge zum Restglied der Aktivierungsenthalpie $\Delta H_R^{\pm}$ und damit auch die Größe von $\Delta H_R^{\pm}$ selbst, nur von der Reaktionsweise $[j]$, nicht aber von der Substitution $[\mu]$ abhängen, für sie also unter der Voraussetzung der Gl. (13) gilt:

$$(\Delta H_R^{\pm})_j = \Delta H_R(\varrho_j^{\pm}) \equiv [\Delta H_R(\varrho, j)]^{\pm} \neq \Delta H_R(\varrho, j, \mu) \ . \tag{14}$$

Liegen nämlich die aktivierten Komplexe $T_{j,\mu}^{\pm}$ der betrachteten individuellen Reaktionen $[j, \mu]$ stets bei der gleichen Reaktionskoordinate $\varrho_j^{\pm}$, so besitzen sie bezüglich ihrer Reaktionszentren kongruente Geometrie. Es treten daher in ihnen allen qualitativ und quantitativ dieselben Bin-

---

[2] Beschreibt man die Energieänderungen der σ-Bindungen mit der Bindungslänge durch Morse-Potentiale oder ähnlichen Potentialfunktionen, die ein Minimum durchlaufen, ist zu beachten, daß diese Potentialfunktionen, die Abstoßungsenergie der aneinander gebundenen Kerne mit berücksichtigen; in diesem Falle setzt sich ΔH_K nur aus den Energien der Abstoßung nicht unmittelbar aneinander gebundener Kerne zusammen, und $\Delta H_K^{\pm}$ ist dann stets verhältnismäßig klein.

dungs-, Hybridisierungs- und andere Änderungen auf, so daß die einzelnen enthalpischen Beiträge zu $\Delta H_R^{\ddagger}$ stets die gleichen sein müssen. Falls die betrachteten Reaktionen in der Gasphase stattfinden und die Beschreibung des Gases als Idealgas zulässig ist, kann das Aktivierungsvolumen $\Delta V^{\ddagger}$ einfach zu

$$(\Delta V^{\ddagger})_j = -(n-1)\frac{RT}{p} \tag{15}$$

berechnet werden, worin n die Molarität der Reaktionsweise [j] darstellt. Gl. (15) zeigt, daß $\Delta V^{\ddagger}$ für Reaktionen in der Gasphase tatsächlich von der Substitution [μ] unabhängig ist. Die Abhängigkeit des Aktivierungsvolumens $\Delta V^{\ddagger}$ von der Substitution kann dann bei Realgasen unter den in Frage kommenden Zustandsbedingungen nur vernachlässigbar klein sein.

Finden die betrachteten Reaktionen in kondensierter Phase (z. B. in Lösungen) statt, entspricht $\Delta V^{\ddagger}$ dem teilweisen oder vollständigen Abbau der Solvathüllen der Reaktanten an den Reaktionszentren, welcher der Bildung des aktivierten Komplexes notwendigerweise vorausgehen muß und dessen Ausmaß mit den Längen der zwischen den Reaktanten neu geknüpften Bindungen korrespondiert. Da diese letzteren aber in den zur Diskussion stehenden aktivierten Komplexen von gleicher Länge sind, folgt auch für Reaktionen in kondensierter Phase (Lösungen), daß unter den Voraussetzungen der Gl. (13) $\Delta V_K^{\ddagger}$ von der Substitution [μ] der Reaktionszentren unabhängig ist. Unter den Voraussetzungen der Gl. (13) ist somit $\Delta H_R^{\ddagger}$ in Gl. (10) eine nur von der Reaktionsweise [j] abhängige Konstante $(\Delta H_R^{\ddagger})_j$, so daß $(\Delta H^{\ddagger})_{j,\,\mu}$ nur durch den Beitrag der π-Elektronen $(\Delta H_\pi^{\ddagger})_{j,\,\mu}$ variiert:

$$(\Delta H^{\ddagger})_{j,\,\mu} = (\Delta H_\pi^{\ddagger})_{j,\,\mu} + (\Delta H_R^{\ddagger})_j \; . \tag{16}$$

Weniger schlüssig als die Aussagen über die Aktivierungsenthalpie sind die, welche über die Aktivierungsentropie $\Delta S^{\ddagger}$ erhalten werden. Spaltet man die Aktivierungsentropie $\Delta S^{\ddagger}$ in die Beiträge der Translation, der Rotation und der Vibration auf

$$\Delta S^{\ddagger} = \Delta S_{tr}^{\ddagger} + \Delta S_{rot}^{\ddagger} + \Delta S_{vib}^{\ddagger} \; , \tag{17}$$

so entspricht jedem dieser Beiträge ein Quotient aus der entsprechenden Zustandssumme des aktivierten Komplexes (z. B.: $\sigma_{tr}^{\ddagger}$) und dem Produkt der entsprechenden Zustandssummen der Ausgangsstoffe A_l (z. B.: $\Pi\sigma_{tr}^{A_l} = \sigma_{tr}^{A_1} \cdot \sigma_{tr}^{A_2} \cdots \sigma_{tr}^{A_n}$, worin n die Molarität der betrachteten Reaktionsweise darstellt).

Für den Beitrag der Translation $\Delta S_{tr}^{\ddagger}$ zur Aktivierungsentropie folgt dieser Quotient zu

$$\frac{\sigma_{tr}^{\ddagger}}{\Pi\sigma_{tr}^{A_l}} = \left[\frac{V}{h^3}(2\pi k\,T)^{3/2}\right]^{1-n}\left\{\frac{\Sigma\,m_{A_l}}{\Pi\,m_{A_l}}\right\}^{3/2} , \tag{18}$$

worin m_{A_i} die Masse des Ausgangsstoffes A_i, die Summe $\Sigma\, m_{A_i}$ die des aktivierten Komplexes bedeuten. Selbstverständlich hängen die Massen m_{A_i} stark von der Substitution $[\mu]$ ab, doch variiert der Quotient $(\Sigma\, m_{A_i})/(\Pi\, m_{A_i})$ nur wenig mit der Substitution, so daß der Beitrag $\Delta S_{\mathrm{tr}}^{\ddagger}$ gerade noch als von der Substitution $[\mu]$ unabhängig angesehen werden kann; es sei besonders betont, daß diese Aussage nicht an die Voraussetzung der Gl. (13) geknüpft ist.

Da für den zumeist interessierenden Temperaturbereich um $T = 300\ {}^{\circ}\mathrm{K}$ die Beziehung

$$T \gg \frac{h^2}{8\pi^2 k I_{\mathrm{q}}} \ , \tag{19}$$

in welcher I_{q} das Trägheitsmoment um die Haupt-Trägheitsachse $q = x, y, z$ darstellt, stets erfüllt ist, kann der Berechnung der Zustandssummen das Modell des klassischen Rotators unterlegt werden. Unter dieser Voraussetzung folgt

$$\frac{\sigma_{\mathrm{rot}}^{\ddagger}}{\Pi\, \sigma_{\mathrm{rot}}^{A_i}} = \left(\frac{h^2}{8\pi^2 k T}\right)^{3n-3} \frac{I_x^{\ddagger} I_y^{\ddagger} I_z^{\ddagger}}{\Pi\{I_{x_i}^{A_i} I_{y_i}^{A_i} I_{z_i}^{A_i}\}} \ . \tag{20}$$

Hierin stellen $I_x^{\ddagger}$, $I_{x_i}^{A_i}$ usw. die Trägheitsmomente des aktivierten Komplexes bzw. die des Reaktanten A_i dar. Der aus den Trägheitsmomenten gebildete Faktor in Gl. (20) variiert stark mit der Substitution $[\mu]$, und zwar um so stärker, je kleiner der Wert der normierten Reaktionskoordinate $\varrho^{\ddagger}$ des Übergangszustandes ist.

Der Vibrationsbeitrag $\Delta S_{\mathrm{vib}}^{\ddagger}$ zur Aktivierungsentropie läßt sich am besten auf der Grundlage der charakteristischen Gruppenfrequenzen diskutieren. Der Berechnung der Zustandssummen wird das Modell des harmonischen Oszillators mit

$$\sigma_{\mathrm{vib}} = \frac{1}{1 - e^{-h\nu/kT}} \tag{21}$$

und

$$\nu = \frac{1}{2\pi} \sqrt{\frac{f}{\mu}} \tag{22}$$

worin f die Federkraftkonstante und μ die reduzierte Masse des Schwingers darstellen, zugrunde gelegt.

Die Schwingungen des aktivierten Komplexes lassen sich zu drei Gruppen zusammenfassen:

1. *Schwingungen, an welchen nur die Atomkerne der Substituenten teilnehmen:* Sie finden bei den charakteristischen Gruppenfrequenzen statt. Da bei Gültigkeit von Gl. (13) in der Regel keine Änderungen in den Substituenten stattfinden, sind die Federkraftkonstanten dieser Schwingungen im aktivierten Komplex und in den Reaktanten gleich groß. Da die an diesen Vibrationen beteiligten reduzierten Massen nahezu gleich bleiben (Gruppen-

frequenzen!), sind auch die Frequenzen dieser Schwingungen und die ihnen entsprechenden Zustandssummen gemäß Gl. (21) im aktivierten Komplex und in den betreffenden Ausgangsstoffen gleich, so daß der Quotient dieser Zustandssummen etwa den Wert Eins besitzt. Die in dieser Gruppe zusammengefaßten Schwingungen leisten somit keinen Beitrag zur Aktivierungsentropie.

2. *Schwingungen, an denen Atomkerne der Substituenten und Reaktionszentren beteiligt sind, alle schwingenden Zentren aber im Bereich eines der Reaktanten liegen:* Durch die bindungslockernden oder -festigenden Änderungen während der Bildung des aktivierten Komplexes werden die Federkraftkonstanten, nicht aber die reduzierten Massen geändert, so daß diese Schwingungen des aktivierten Komplexes bei anderen Frequenzen liegen als die entsprechenden Oszillationen der Reaktanten. Gilt aber Gl. (13) als erfüllt, liegen die aktivierten Komplexe der betrachteten individuellen Reaktionen also bei der von der Substitution $[\mu]$ unabhängigen Reaktionskoordinate $\varrho_j^{\ddagger}$, so finden alle hier erwähnten Änderungen der Federkraftkonstanten und der Schwingungsfrequenzen in allen zur Diskussion stehenden individuellen aktivierten Komplexen in qualitativ und quantitativ gleicher Weise statt; der Quotient der diesen Schwingungen entsprechenden Zustandssummen besitzt somit einen für die Reaktionsweise [j] charakteristischen, von der Substitution $[\mu]$ unabhängigen Wert. Die in dieser Gruppe zusammengefaßten Schwingungen leisten einen für die Reaktionsweise [j] charakteristischen Beitrag zur Aktivierungsentropie.

3. *Schwingungen, an denen Zentren verschiedener Reaktanten beteiligt sind:* Diese Schwingungen treten nur im aktivierten Komplex auf und sind für diesen charakteristisch. Sie bauen sich aus den 6 Translations- und Rotationsfreiheitsgraden je Reaktant auf, soweit diese nicht zu den 6 Translations- und Rotationsfreiheitsgraden des aktivierten Komplexes und zum Freiheitsgrad des Durchgangs entlang der Reaktionskoordinate kombinieren; wenn n die Molarität der betrachteten Reaktionsweise darstellt, beträgt ihre Anzahl $(6n-7)$ und ist somit von der Anzahl der Kerne im aktivierten Komplex unabhängig. Gilt Gl. (13), so sind die Federkraftkonstanten dieser Schwingungen von der Substitution $[\mu]$ unabhängig. Da aber die bei diesen Vibrationen schwingenden Massen — es handelt sich hier in der Regel um die Massen der einzelnen Reaktanten! — von der Substitution $[\mu]$ abhängen, sind auch die Frequenzen dieser $(6n-7)$ Schwingungen[3] und die ihnen entsprechenden Zustandssummen substitutionsabhängig.

---

[3] Hier und in der Folge sind stets nicht-linear gebaute Reaktanten angenommen. Bei Übertragung der Betrachtung auf linear gebaute Reaktanten ergeben sich andere Zahlen, z. B. $(5n-7)$ anstatt von $(6n-7)$, im Grundsätzlichen ändern sich jedoch nichts. Nur in sehr wenigen elementaren Reaktionen ist ein linear gebauter aktivierter Komplex denkbar; für einen solchen ergäbe sich hier $(5n-6)$.

Diese Verhältnisse lassen sich wie folgt zusammenfassen: Der Vibrationsbeitrag zur Aktivierungsentropie baut sich aus drei Gruppen-Beiträgen auf

$$\Delta S_{\text{vib}}^{\mp} = (\Delta S_{\text{vib}}^{\mp})_1 + (\Delta S_{\text{vib}}^{\mp})_2 + (\Delta S_{\text{vib}}^{\mp})_3 \ , \tag{23}$$

von welchen unter den durch Gl. (13) ausgedrückten Verhältnissen der erste gleich Null, der zweite nur eine Funktion der Reaktionsweise [j], der dritte aber eine Funktion der Reaktionsweise [j] und der Substitution [μ] ist:

$$\varrho^{\mp} = [\varrho\,(j)]^{\mp} \begin{cases} (\Delta S_{\text{vib}}^{\mp})_1 = 0 & \text{(24a)} \\[2ex] (\Delta S_{\text{vib}}^{\mp})_2 = [\Delta S_{\text{vib}}^{\mp}\,(j)]_2 & \text{(24b)} \\[2ex] (\Delta S_{\text{vib}}^{\mp})_3 = [\Delta S_{\text{vib}}^{\mp}\,(j,\mu)]_3 \ . & \text{(24c)} \end{cases}$$

Von allen Beiträgen zur Aktivierungsentropie sind also nur $(\Delta S_{\text{vib}}^{\mp})_1$ und $(\Delta S_{\text{vib}}^{\mp})_2$ von der Substitution [μ] unabhängig. Es ist daher noch die Frage zu diskutieren, wie weit die Änderungen der drei anderen Beiträge $\Delta S_{\text{tr}}^{\mp}$, $\Delta S_{\text{rot}}^{\mp}$ und $(\Delta S_{\text{vib}}^{\mp})_3$ mit der Substitution sich gegenseitig kompensieren können. Der Einfluß der Substitution auf diese drei Entropiebeiträge erfolgt auf Grund der Massen der Substituenten und ihrer geometrischen Anordnung in diesen. Der Einfluß der geometrischen Anordnung der Massenpunkte in den Substituenten kann nicht generell diskutiert werden, scheint aber von geringerer Bedeutung zu sein als der der Massen selbst. Vergrößert man die Massen der Substituenten, so verkleinert sich gemäß Gl. (18) der Translationsbeitrag $\Delta S_{\text{tr}}^{\mp}$, während der Rotationsbeitrag $\Delta S_{\text{rot}}^{\mp}$ entsprechend Gl. (20) und der Vibrationsbeitrag $(\Delta S_{\text{vib}}^{\mp})_3$ gemäß Gl. (22) und (21) zunehmen. Ein Teil dieser gegenläufigen Substitutionsabhängigkeiten der Entropiebeiträge kompensieren sich gegenseitig, wodurch die Abhängigkeit der Aktivierungsentropie $\Delta S^{\mp}$ von der Substitution [μ] relativ gering bleibt, zumal in $\Delta S^{\mp}$ ja auch substitutionsunabhängige Beiträge eingehen. Wie detaillierte Rechnungen für einige Reaktionsbeispiele [31] zeigen, variiert $\Delta S^{\mp}$ bei Gültigkeit von Gl. (13) nur mäßig mit der Substitution [μ] innerhalb eines relativ engen, für die Reaktionsweise [j] charakteristischen Bereiches.

Innerhalb gewisser Grenzen kann somit bei Bestehen der Gl. (13) auch die Aktivierungsentropie $\Delta S^{\mp}$ näherungsweise als eine nur von der Reaktionsweise [j] abhängige Größe angesehen werden:

$$(\Delta S^{\mp})_{j,\,\mu} = [\Delta S\,(\varrho, j)]^{\mp} \doteq (\Delta S^{\mp})_j \ . \tag{25}$$

Diese Beziehung ist, wie aus der obigen detaillierten Betrachtung folgt, um so besser erfüllt, je geringer die Variation der Massen mit der Substitution ist. Dies wird z. B. für individuelle Reaktionen [j, μ] in einem Lösungsmittel zutreffen, welches sowohl den aktivierten Komplex als auch die Ausgangsstoffe gut solvatisiert, da in diesem Falle die Massen der

Solvathüllen mit zu den effektiven Massen der Reaktanten und des akti-
vierten Komplexes gerechnet werden müssen — eine Massenänderung der
Reaktanten infolge geänderter Substitution also einen geringeren relativen
Wert und daher auch geringeren Einfluß auf ΔS^+ besitzt.

Die Vereinigung der Gl. (7), (16) und (25) führt zu

$$(\Delta G^+)_{j,\mu} = (\Delta H_{\pi}^{\neq})_{j,\mu} + (\Delta H_{R}^{\neq})_j - T(\Delta S^+)_j \ . \tag{26}$$

Da die beiden letzten Glieder der Gl. (26) bei Bestehen der Gl. (13) näherungsweise nur von der Reaktionsweise [j] abhängen, können sie in Analogie
zu ΔH_R zu einer (temperaturabhängigen) Restgröße der Freien Aktivierungsenthalpie zusammengefaßt werden:

$$(\Delta G_{R}^{\neq})_j \equiv (\Delta H_{R}^{\neq})_j - T(\Delta S^+)_j \ . \tag{27}$$

Gl. (26) schreibt sich dann schließlich

$$(\Delta G^+)_{j,\mu} = (\Delta H_{\pi}^{\neq})_{j,\mu} + (\Delta G_{R}^{\neq})_j \ . \tag{28}$$

In ihr stellen $(\Delta H_{\pi}^{+})_{j,\mu}$ den Unterschied der π-Elektronen-Energien des
aktivierten Komplexes und der Ausgangsstoffe in der individuellen Reak-
tion $[j,\mu]$, $(\Delta G_{R}^{\neq})_j$ eine für eine bestimmte Reaktionsweise [j] charakteristische, bei Bestehen der Gl. (13) von der Substitution $[\mu]$ näherungsweise
unabhängige Größe dar, welche bei einer bestimmten Temperatur T eindeutig empirisch bestimmbar ist. In Worten ausgedrückt besagt Gl. (28),
daß bei Bestehen von Gl. (13) die Freie Aktivierungsenthalpie $(\Delta G^+)_{j,\mu}$
einer individuellen Reaktion $[j,\mu]$ nur über ihren, durch HMO-Rechnungen
abschätzbaren π-Elektronen-Beitrag $(\Delta H_{\pi}^{+})_{j,\mu}$ von der Substitution $[\mu]$ abhängt.

Der Einfluß etwaiger Lösungsmittel auf das Aktivierungsvolumen
ΔV^+ und die Aktivierungsentropie ΔS^+ wurde bereits oben kurz diskutiert. Vergleicht man mittels Gl. (28) einige individuelle Reaktionen
$[j,\mu]$ einer einzigen Reaktionsweise [j] miteinander, so werden die Beiträge
der Solvatisierung zu ΔV^+, ΔS^+ und ΔH^+ von Lösungsmittel zu Lösungsmittel verschieden, für ein bestimmtes Lösungsmittel aber im allgemeinen
von der Substitution $[\mu]$ unabhängig sein. Zusammenfassend wird sich
$(\Delta G_{R}^{\neq})_j$ daher bei einem derartigen Vergleich nur dann als invariant erweisen, wenn folgende Voraussetzungen erfüllt sind:

1. Alle mittels Gl. (28) verglichenen individuellen Reaktionen der
Reaktionsweise [j] müssen bei derselben Temperatur T ablaufen, damit
der in Gl. (27) ausgedrückten Temperaturabhängigkeit von $(\Delta G_{R}^{\neq})_j$ Rechnung getragen wird.

2. Alle bei diesen Reaktionen durchlaufenen Übergangszustände müssen
bei demselben Wert ϱ_j^+ der Reaktionskoordinate liegen, damit die Voraussetzung der Gl. (13) erfüllt ist.

3. Damit Gl. (13) bestehen kann, dürfen in keiner der verglichenen Reaktionen die Substituenten einen sterischen Effekt auf die Bildung des aktivierten Komplexes ausüben.

4. Werden in Lösungen ablaufende Reaktionen miteinander verglichen, so müssen alle Reaktionen im gleichen Lösungsmittel ablaufen gelassen werden.

Diese vier Voraussetzungen für eine Invarianz von $(\Delta G_R^{\ddagger})_j$ grenzen die Anwendbarkeit von Gl. (28) einerseits in gewisser Weise ein, andererseits sind sie in überraschend vielen praktischen Fällen erfüllt bzw. erfüllbar. Sie erlauben stets eine rasche Orientierung, ob Gl. (28) und die auf ihr beruhenden, weiter unten beschriebenen Verfahren in einem zur Diskussion stehenden konkreten Fall angewandt werden dürfen oder nicht.

3. Konkurrenzreaktionen

Die Theorie des aktivierten Komplexes verknüpft die Reaktionsgeschwindigkeitskonstante $(k)_{j,\mu}$ einer individuellen Reaktion mit deren Freien Aktivierungsenthalpie $(\Delta G^{\ddagger})_{j,\mu}$ bekanntlich durch

$$- R T \ln (k)_{j,\mu} = - R T \ln \left(\frac{kT}{b} \right) + (\Delta G^{\ddagger})_{j,\mu} , \tag{29}$$

woraus nach Eintragen von Gl. (28) folgt

$$- R T \ln (k)_{j,\mu} = - R T \ln \left(\frac{kT}{b} \right) + (\Delta H_{\pi}^{\ddagger})_{j,\mu} + (\Delta G_R^{\ddagger})_j . \tag{30}$$

In der Folge soll die Anwendung von Gl. (30) auf zwei typische Fälle von Konkurrenzreaktionen besprochen werden: den Fall der Kreuzungsversuche (competition reaction) und den Fall der konkurrierenden Reaktionsweisen.

Der Fall der Kreuzungsversuche (competition reaction) liegt vor, wenn zwei individuelle Ausgangssysteme $A_{j,\mu}$ und $A_{j,\lambda}$, welche unter Umständen einen Reaktanten gemeinsam haben, in einer Phase vorliegen und gleichzeitig zu den entsprechenden Endprodukten $P_{j,\mu}$ bzw. $P_{j,\lambda}$ nach der gleichen Reaktionsweise [j] reagieren:

$$\left\{ \begin{array}{lll} A_{j,\mu} & \rightleftharpoons \; [T_{j,\mu}^{\ddagger}] \; \longrightarrow & P_{j,\mu} \\ A_{j,\lambda} & \rightleftharpoons \; [T_{j,\lambda}^{\ddagger}] \; \longrightarrow & P_{j,\lambda} \end{array} \right\} . \tag{31}$$

Wendet man die Gln. (29) und (30) auf jede dieser individuellen Reaktionen an und bildet man die Differenz, so erhält man

$$- R T \ln \frac{(k)_{j,\mu}}{(k)_{j,\lambda}} = (\Delta G^{\ddagger})_{j,\mu} - (\Delta G^{\ddagger})_{j,\lambda}$$
$$= [\Delta H_{\pi}^{\ddagger})_{j,\mu} - (\Delta H_{\pi}^{\ddagger})_{j,\lambda}] . \tag{32}$$

In diesem Falle ist also die relative Freie Aktivierungsenthalpie $[(\Delta G^{\neq})_{j,\mu} - (\Delta G^{\neq})_{j,\lambda}]$ der Differenz der π-Elektronenbeiträge zu den Aktivierungsenthalpien gleich. In diesem Falle von Konkurrenzreaktionen wird diejenige Reaktion schneller, für die $(\Delta H_{\pi}^{\neq})$ kleiner ist.

Von größerer praktischer Bedeutung als die eben besprochenen „competition reactions" ist der Fall der konkurrierenden Reaktionsweisen. In der Regel sind für jedes Ausgangssystem vom Typus seiner Komponenten her mehrere Reaktionsweisen möglich, je nach der Art der Substitution $[\mu]$ ist zumeist eine von diesen besonders bevorzugt. So reagieren Diazoalkane mit Olefinen nicht nur nach der Cycloaddition (4), bei bestimmten Substitutionen bilden sich aus diesen Ausgangsstoffen auch Cyclopropane, Dihydrofurane, Alkylierungsprodukte u. a. m. Soll ein bestimmtes materielles Ausgangssystem nach zwei Reaktionsweisen $[j]$ bzw. $[l]$ reagieren können, müssen wir es bezüglich dieser Reaktionsweisen einmal mit $A_{j,\mu}$ und ein andermal mit $A_{l,\mu}$ bezeichnen. Da $A_{j,\mu}$ und $A_{l,\mu}$ ein und dasselbe Ausgangssystem bezeichnen, muß die Identität $A_{j,\mu} \equiv A_{l,\mu}$ bestehen. Der Fall der konkurrierenden Reaktionsweisen läßt sich analog zu Gl. (31) darstellen durch

$$\left\{ \begin{array}{l} A_{j,\mu} \; \rightleftharpoons \; [T_{j,\mu}^{\neq}] \; \longrightarrow \; P_{j,\mu} \\ \| \\ A_{l,\mu} \; \rightleftharpoons \; [T_{l,\mu}^{\neq}] \; \longrightarrow \; P_{l,\mu} \end{array} \right\} . \tag{33}$$

Die Anwendung der Gln. (29) und (30) auf jede dieser Reaktionen und Differenzbildung führt hier zu

$$-RT \ln \frac{(k)_{j,\mu}}{(k)_{l,\mu}} = (\Delta G^{\neq})_{j,\mu} - (\Delta G^{\neq})_{l,\mu}$$
$$= [(\Delta H_{\pi}^{\neq})_{j,\mu} - (\Delta H_{\pi}^{\neq})_{l,\mu}] + [(\Delta G_{R}^{\neq})_{j} - (\Delta G_{R}^{\neq})_{l}] . \tag{34}$$

Hierin ist

$$[(\Delta G_{R}^{\neq})_{j} - (\Delta G_{R}^{\neq})_{l}] \equiv G_{j,l}^{\neq} \tag{35}$$

eine von der Temperatur abhängige Konstante. Sie ist für den Wettbewerb der konkurrierenden Reaktionsweisen charakteristisch und spannt zwei Enthalpieniveaus auf, von welchen aus die Freien Aktivierungsenthalpien ausgemessen werden (Abb. 3).

Folgen die konkurrierenden Reaktionsweisen $[j]$ und $[l]$ dem gleichen Geschwindigkeitsgesetz, so stehen die Konzentrationen der gebildeten Produkte jederzeit in demselben Verhältnis wie die Reaktionsgeschwindigkeitskonstanten

$$\frac{[P_{j,\mu}]}{[P_{l,\mu}]} = \frac{(k)_{j,\mu}}{(k)_{l,\mu}} . \tag{36}$$

Die Produktverteilung ist dann gemäß Gln. (34) und (35) direkt mit den relativen Freien Aktivierungsenthalpien dieser konkurrierenden Reaktionsweisen verknüpft.

$$\frac{[\mathrm{P}_{j,\mu}]}{[\mathrm{P}_{1,\mu}]} = \frac{(k)_{j,\mu}}{(k)_{1,\mu}} = \exp\left\{ -\frac{[(\varDelta H_\pi^{\pm})_{j,\mu} - (\varDelta H_\pi^{\pm})_{1,\mu}] + G_{j,1}^{\pm}}{RT} \right\}. \quad (37)$$

Auf diese Beziehung gründet sich eine rechnerische Vorhersage des Ablaufes konkurrierender Reaktionsweisen gleichen Geschwindigkeits-

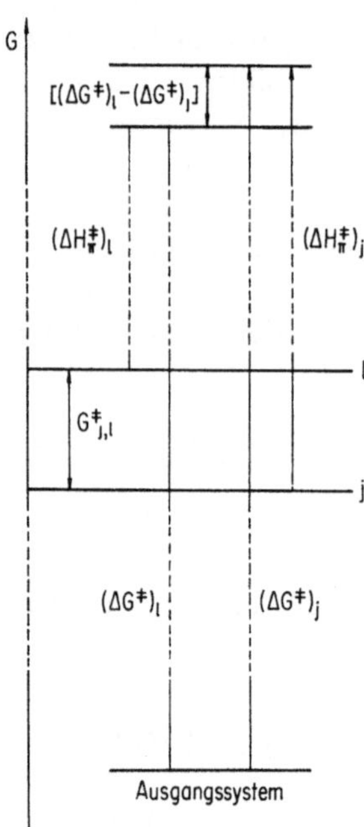

Abb. 3. Enthalpiediagramm für konkurrierende Reaktionsweisen [j] bzw. [l]

gesetzes. Voraussetzung ihrer Anwendung ist die Kenntnis der Konkurrenzkonstante $G_{j,1}^{\pm}$. Falls ein individuelles System $A_{j,\nu} = A_{1,\nu}$ gefunden werden kann, aus welchem sich die Produkte $P_{j,\nu}$ und $P_{1,\nu}$ in gleicher Konzentration

$$[\mathrm{P}_{j,\nu}] = [\mathrm{P}_{1,\nu}] \quad (38)$$

bilden, für das also der Quotient der Gl. (36) den Wert Eins annimmt, kann $G_{j,1}^{\pm}$ einfach bestimmt werden. Gl. (34) nimmt in diesem Fall die Form

$$0 = [(\varDelta H_\pi^{\pm})_{j,\nu} - (\varDelta H_\pi^{\pm})_{1,\nu}] + G_{j,1}^{\pm} \quad (39\,\mathrm{a})$$

an, woraus folgt

$$G_{j,1}^{\pm} = (\varDelta H_\pi^{\pm})_{1,\nu} - (\varDelta H_\pi^{\pm})_{j,\nu}. \quad (39\,\mathrm{b})$$

Kann kein derartiges Ausgangssystem gefunden werden, läßt sich der Wert von $G_{j,l}^{\ddagger}$ durch den Vergleich einiger Rechendaten mit experimentellen Befunden ziemlich gut eingrenzen; ein Beispiel dafür ist weiter unten (siehe Tab. 1) gegeben.

Stehen mehrere Reaktionsweisen in Konkurrenz miteinander, kann für jedes Paar eine Konkurrenzkonstante $G_{j,l}^{\ddagger}$ definiert werden, die aber nicht unabhängig voneinander sind. Wie leicht einzusehen ist, gilt

$$\sum G_{j,l}^{\ddagger} = 0 \tag{40}$$

bei vollständiger cyclischer Vertauschung aller j und l.

4. Abschätzung der Beiträge $\Delta H_{\pi}^{\ddagger}$ mittels der HMO-Methode

Wie bereits wiederholt erwähnt wurde, sind die Beiträge der π-Elektronen zur Aktivierungsenthalpie mittels der HMO-Methode näherungsweise berechenbar. Als Beispiel derartiger Rechnungen soll hier $\Delta H_{\pi}^{\ddagger}$ für die Cycloaddition (4) abgeschätzt werden. Als Realbeispiel für eine derartige Reaktion wählen wir die Umsetzung von Diazomethan [$R_1 = R_2 = H$] und Benzylidenmeldrumsäure [$R_3 = H$, $R_4 = C_6H_5$, $R_5 + R_6 = (-COO)_2C(CH_3)_2$]

$$CH_2N_2 \;+\; \underset{H}{\overset{C_6H_5}{}} C{=}C \underset{\underset{O}{\overset{\|}{C}}-O}{\overset{\overset{O}{\overset{\|}{C}}-O}{}} C \underset{CH_3}{\overset{CH_3}{}} \quad \rightleftharpoons \quad [T_{p,\mu}^{\ddagger}] \;\longrightarrow\; \cdots \tag{41}$$

Der größeren Allgemeinheit wegen diskutieren wir die von uns benützte Arbeitstechnik für beliebige Substituenten $R_1, R_2 \ldots R_6$, die Zahlenwerte beziehen sich aber hier stets auf das Realbeispiel (41).

Wie bereits weiter oben ausgeführt, stellt der π-Elektronenbeitrag $(\Delta H_{\pi}^{\ddagger})_{p,\mu}$ dieser Reaktion die Differenz der π-Elektronen-Gesamtenergien des aktivierten Komplexes $T_{p,\mu}^{\ddagger}$ einerseits und der Ausgangsstoffe A_{μ} andererseits dar. Werden diese beiden Größen durch $E_{\pi}(T_{p,\mu}^{\ddagger})$ bzw. $E_{\pi}(A_{\mu})$ bezeichnet, so folgt

$$(\Delta H_{\pi}^{\ddagger})_{p,\mu} = N_L \cdot \{E_{\pi}(T_{p,\mu}^{\ddagger}) - E_{\pi}(A_{\mu})\} \;.$$

$E_{\pi}(A_{\mu})$ stellt darin die Summe der Gesamt-π-Elektronen-Energien aller an der Umsetzung beteiligten Reaktanten dar, im Realbeispiel (41) also die Summe der π-Elektronenenergien von Diazomethan und Benzylidenmeldrumsäure. Da die π-Elektronen-Gesamtenergien molekulare Größen sind, $(\Delta H_{\pi}^{\ddagger})_{p,\mu}$ aber definitionsgemäß auf 1 Mol bezogen ist, tritt in der vorstehenden Beziehung die Lohschmidtsche Zahl N_L als Proportionalitätsfaktor auf.

Wir wenden uns vorerst der Berechnung von E_π (A_μ) zu. Das Diazo-alkan enthält zwei orthogonale π-Elektronensysteme: das eine (π_x) er-

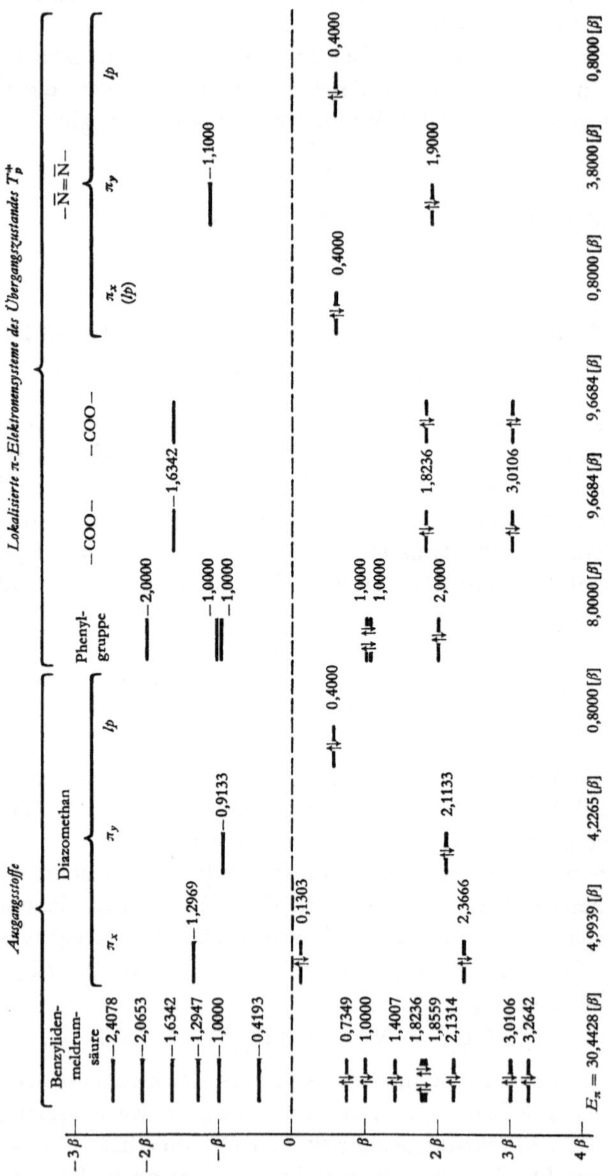

Abb. 4. π-Elektronenniveaus der Ausgangsstoffe und des π-Elektronenmodells des Übergangszustandes T_p^+ bei der Cycloaddition von Diazomethan an Benzyliden-meldrumsäure (die gesamten π-Elektronenenergien E_π sind mit mehr Dezimalen berechnet worden, als in der Abbildung angegeben sind)

streckt sich über die drei Zentren der CN_2-Gruppierung und steht mit R_1 und R_2 in Konjugation, während das zweite, dazu orthogonale System (π_y) sich nur über die beiden N-Atome erstreckt [29]. In Abb. 4a sind die durch eine entsprechende HMO-Rechnung erhaltenen Energieniveaus dieser beiden π-Elektronensysteme gesondert aufgeführt sowie das eine doppelt besetzte (lp)-Niveau des einsamen Elektronenpaares am N_ε. Wir haben dieser Rechnung die üblichen Parameter [29] zugrunde gelegt.

Für die Berechnung der π-Elektronenniveaus der Benzylidenmeldrumsäure mittels HMO-Methode wurde die ideale Koplanarität der $C_6H_5CH{=}C(COO-)_2$-Gruppierung angenommen und der übliche Parametersatz [28] verwendet. Auch sie sind in Abb. 4a angegeben.

Für $E_\pi(A_\mu)$ ergibt sich somit für unser Realbeispiel:

1. π-Elektronen-Gesamt-Energie der
   Benzylidenmeldrumsäure   30,4428 [β]
2. π_x-Elektronen-Energie des Diazomethans   4,9939 [β]
3. π_y-Elektronen-Energie des Diazomethans   4,2265 [β]
4. Energie des einsamen Elektronenpaares am N_ε des
   Diazomethans   0,8000 [β]

somit: $$E_\pi(A_\mu) = \underline{\underline{40,4632\ [\beta]}}$$

Im folgenden schreiten wir nun zur Berechnung der π-Elektronenenergie des Übergangszustandes $E_\pi(T^{\neq}_{p,\mu})$. Vorerst benötigen wir aber konkrete Modellvorstellungen bezüglich der Struktur des aktivierten Komplexes. Im allgemeinen werden diese Modelle immer nur eine grobe Näherung der tatsächlichen Verhältnisse darstellen, da wir nur durch indirekte Schlüsse Aussagen über die Geometrie von Übergangszuständen machen können. Wie oben ausführlich dargelegt, ist für eine sinnvolle Berechnung von $E_\pi(T^{\neq}_{p,\mu})$ im Rahmen der HMO-Methode jedoch erforderlich, daß die Änderung von $\varrho^{\neq}$ mit den Substituenten R_1 bis R_6 vernachlässigbar klein ist.

Bei der 1,3-Cycloaddition werden im Übergangszustand zwei neue σ-Bindungen geknüpft: die eine ist die $C_\beta{-}C_\gamma$-Bindung, die andere die $C_\alpha{-}N_\varepsilon$-Bindung. Am mittleren N-Atom des Diazoalkans (N_δ) bildet sich ein einsames Elektronenpaar (lp) aus. Zur Beschreibung des Übergangszustandes verwenden wir ein Modell (vgl. Abb. 7), das eine vollständige Unterbrechung des π-Elektronensystems des Olefins an den Zentren C_α- und C_β-Atom und eine vollständige Unterbrechung des π_x-Systems des Diazoalkans am C_γ- und N_ε-Atom vorsieht. Sechs Elektronen der ursprünglichen zwei ausgedehnten π-Elektronensysteme von Olefin und Diazoalkan werden in den beiden σ-Bindungen und in dem einsamen Elektronenpaar am N_δ-Atom untergebracht. Die restlichen π-Elektronen verteilen sich auf die isolierten π-Elektronensysteme in den Resten R_1 bis R_6.

20*

In dem gewählten Realbeispiel finden sich nur drei von H verschiedene Substituenten, nämlich $R_3 = C_6H_5$- bzw. $(R_5 + R_6) = (-COO)_2C(CH_3)_2$. Dementsprechend ergeben sich auch nur drei isolierte π-Elektronensysteme, und zwar eine Phenyl- und zwei Carbalkoxygruppen.

Das π_y-Elektronensystem des Diazoalkans wird von der Reaktion nicht direkt berührt; durch die Ausbildung des einsamen Elektronenpaares am N_δ-Atom und die σ-$(C_\alpha-N_\varepsilon)$-Bindung treten — wie bereits oben erwähnt — Umhybridisierungen auf, welche durch die entsprechenden Änderungen der Zahlenwerte der Coulombparameter [29] der N-Atome berücksichtigt werden.

Die Berechnung von $E_\pi(T^+_{p,\mu})$ ergibt dementsprechend (vgl. Abb. 4)

1. π-Elektronen-Gesamtenergie des Phenylrestes (R_3)   8,0000 $[\beta]$
2. π-Elektronen-Gesamtenergie der beiden Carbalkoxygruppen   19,3368 $[\beta]$
3. Energie des einsamen Elektronenpaares (lp) am N_δ,
   das aus dem π_x-System des Diazomethans stammt   0,8000 $[\beta]$
4. π-Elektronenenergie der N=N-Doppelbindung,
   welche aus dem π_y-System des Diazomethans hervorgeht   3,8000 $[\beta]$
5. Energie des einsamen Elektronenpaares (lp) am N_ε,
   welches auch schon im Diazomethan vorhanden war   0,8000 $[\beta]$

Somit $E_\pi\,(T^+_{p,\mu}) =$ 32,7368 $[\beta]$

Durch Differenzbildung erhalten wir aus

$$
\begin{aligned}
E_\pi\,(T^+_{p,\mu}) &\quad 32{,}7368\ [\beta] \\
E_\pi\,(A_\mu) &\quad 40{,}4632\ [\beta] \\
\hline
&\quad -7{,}7264\ [\beta]
\end{aligned}
$$

schließlich den π-Elektronenbeitrag zur Aktivierungsenthalpie:

$$(\Delta H^+_\pi)_{p,\mu} = -\,N_L \cdot 7{,}7264\ [\beta].$$

Bei allen späteren zahlenmäßigen Angaben wird der Faktor $-N_L$ kürzehalber weggelassen.

5. Anwendungsbeispiel: Reaktionen von Diazoalkanen mit Olefinen

Neben der bereits wiederholt als Illustrationsbeispiel angezogenen Cycloaddition (4) reagieren Diazoalkane mit Olefinen auch noch in anderer Weise: Je nach der Substitution bilden sich Pyrazoline (P) oder Cyclopropane (C) oder Dihydrofurane (F) oder andere Verbindungen in jeweils guter Ausbeute. Einige experimentelle Beobachtungen und einige theoretische Überlegungen, auf welche weiter unten noch einzugehen sein wird, haben

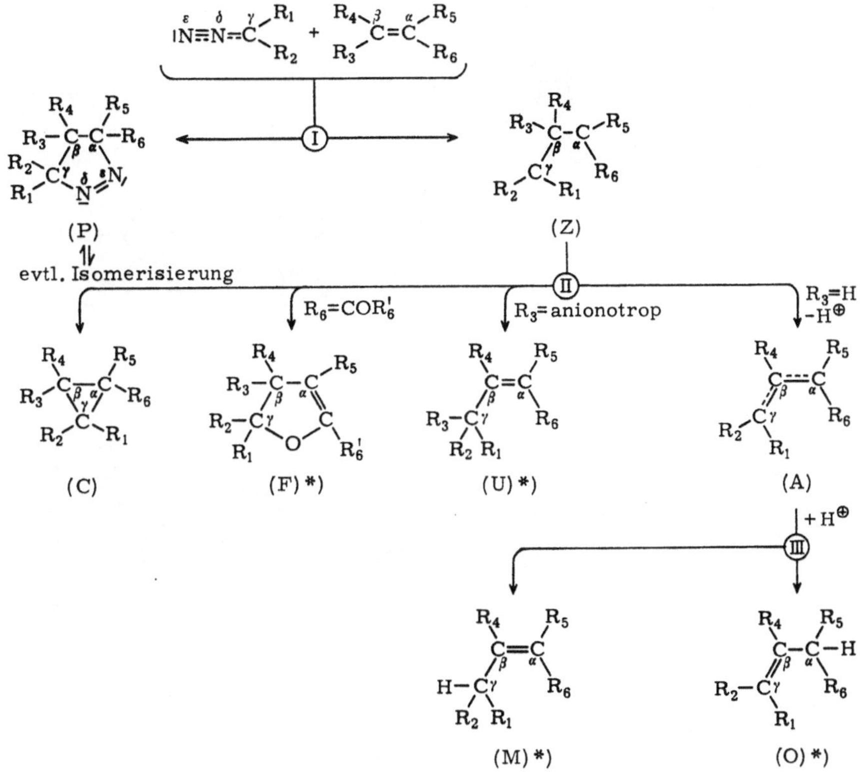

Abb. 5. Reaktionsschema (Umsetzung von Diazoalkanen mit Olefinen)

*) Umsatz mit einem oder mehreren weiteren Molekülen Diazoalkan möglich. (A) trägt eine negative Ladung

zu dem in Abb. 5 dargestellten Reaktionsschema geführt [12, 36], welches die alternative Bildung eines der darin erwähnten Produkte aufgrund der kinetischen Bevorzugung einer der miteinander konkurrierenden Reaktionsweisen beinhaltet. Obgleich die Bildung der verschiedenen Reaktionsprodukte eine unterschiedliche Anzahl elementarer Folgereaktionen erfordert, finden die Konkurrenzen stets zwischen elementaren Reaktionen statt. Das erwähnte Reaktionsschema und das in den vorhergehenden Abschnitten beschriebene Verfahren zur Abschätzung relativer freier Aktivierungsenthalpien mittels der HMO-Methode wurden einer theoretischen Untersuchung zugrunde gelegt, welche sich auf die 150 individuellen Ausgangssysteme erstreckte, die aus den in Abb. 6 dargestellten 30 Olefinen und den folgenden 5 Diazoalkanen aufgebaut werden können: Diazomethan, Diazoessigester, Phenyldiazomethan, Diphenyldiazomethan und Diazofluoren. Die mit Phenyl- und Diphenyldiazomethan erzielten Ergebnisse

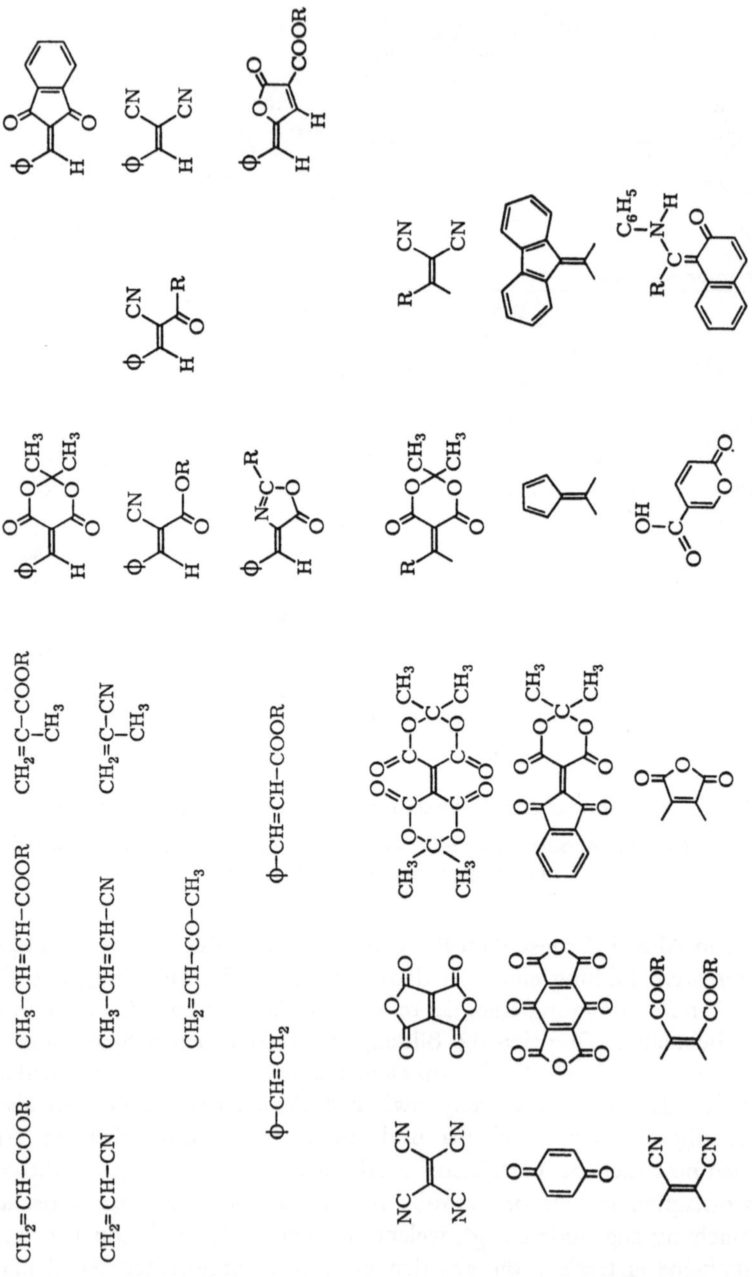

Abb. 6. Olefinische Systeme (s. Text!)

waren unbefriedigend — die mutmaßlichen Gründe hierfür werden weiter unten diskutiert. In der Folge werden daher nur diejenigen Ergebnisse kurz berichtet, die an den verbleibenden 90 Ausgangssystemen erhalten wurden.

Die Diskussion der Konkurrenz I des Reaktionsschemas (Abb. 5) wird außerdem benützt, um die Eingrenzung des Wertes der durch Gl. (35) definierten Konkurrenzkonstanten $G_{j,1}^{\ddagger}$ zu zeigen, wenn diese nicht mit Hilfe der Gln. (39a, b) ermittelt werden kann.

5.1. Konkurrenz I

Konkurrenz I des Reaktionsschemas entscheidet über die alternative Bildung von Δ^1-Pyrazolin (P) oder eines stickstoffreien Zwischenkörpers (Z)[4]. Die zu den Pyrazolinen (P) führende Reaktionsweise [p] wurde wiederholt als Illustrationsbeispiel herangezogen; das π-Elektronenmodell des bei ihm durchlaufenen Übergangszustandes $T_p^{\ddagger}$ wurde bereits oben diskutiert und ist in Abb. 7 zusammen mit den π-Elektronenmodellen der anderen Übergangszustände des Reaktionsschemas dargestellt. In $T_p^{\ddagger}$ sind die π-Elektronen in den sechs Substituenten R_1 bis R_6 und in der N=N-Doppelbindung ohne gegenseitiger Konjugation lokalisiert. Für $(\Delta H_\pi^{\ddagger})_p$ als Differenz der π-Elektronen-Energien (E) von $T_p^{\ddagger}$ und des Ausgangssystems ergibt sich somit

$$(\Delta H_\pi^{\ddagger})_p = E\,(T_p^{\ddagger}) - E\,(\text{Ausgangssystem})$$

$$= \left[\sum_1^6 i\, E(R_i) + E(N{=}N) + E(lp{:}N) \right] \tag{42}$$

$$- [E(R_1 R_2 CN_2) + E(R_3 R_4 C{=}CR_5 R_6)]\ .$$

Der Mechanismus der mit [p] konkurrierenden, zum zwitterionischen Zwischenkörper Z führenden Reaktionsweise [z] ist durch einen nucleophilen Angriff des C_γ am C_β ohne simultane Knüpfung der N_ε–C_α-Bindung charakterisiert. Mit dem Fortschritt der Reaktion finden Umhybridisierungen am C_β und C_γ statt. Im Olefin wird dadurch die Konjugation von R_3 mit R_4 und die von R_3 und R_4 mit C_α und dessen Substituenten (R_5 und R_6), im Diazoalkan die von R_1 mit R_2 und die von R_1 und R_2 mit der N_2-Gruppierung sukzessive verringert. Mit diesen Konjugationsbehinderungen geht die zunehmende Lokalisierung eines π-Elektronenpaares in der N_2-Gruppe einher, so daß die Bindungsordnung der N–N-Bindung dem Wert 3 zustrebt. Die C_γ–N_δ-Bindung wird hingegen in ihrer Ordnung erniedrigt; sie wird dadurch und durch die Umhybridisierung am C_γ geschwächt und verlängert. Die Linearität der C_γ–N_δ-Gruppierung bleibt dabei erhalten. Eine Annäherung von N_ε an C_α infolge einer geeigneten Knickschwingung

---

[4] „stickstofffrei" im Zusammenhang mit Z bezieht sich hier und in der Folge immer auf den Stickstoff der CN_2-Gruppierung, nicht aber auf N-Atome, welche in den Substituenten R_i enthalten sein mögen.

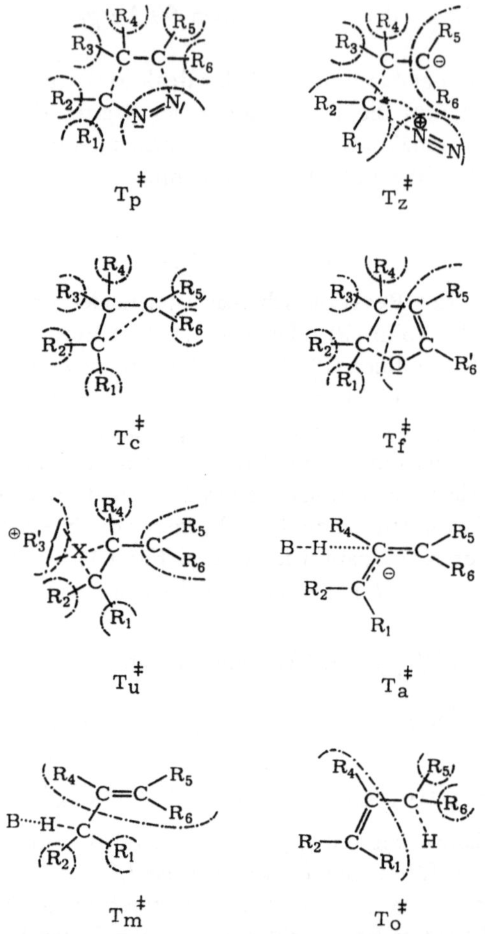

Abb. 7. π-Elektronenmodelle der Übergangszustände. (In $T_u^{\ddagger}$ ist eine negative Ladung in der Gruppe CR_5R_6 lokalisiert.)

der C—N—N-Bindungen führt zu keiner Bindungsknüpfung und ist daher für den Fortschritt der Reaktion ohne Bedeutung. Da eine permanente Abwinkelung der CN_2-Gruppierung infolge einer Änderung des Hybridisierungszustandes am N_δ die Energie der Partikel erhöht, ist sie statistisch benachteiligt und es verbleibt somit N_δ im linearen Hybridisierungszustand. N_δ ist aber stark elektronegativ, seine Elektronegativität wird durch die formale Ladung von $+1$ weiter erhöht, so daß die aus den oben erwähnten Gründen bereits geschwächte C_γ—N_δ-Bindung stark zum N_δ polarisiert wird. Hierdurch wird die Eliminierung von N_2 vorbereitet und es ist zu erwarten, daß mit dem Fortschritt des nucleophilen Angriffes des C_γ am C_β die Eliminierung von N_2 einhergeht.

Auf die Reaktionsweise [z] kann auch auf anderem Wege geschlossen werden: Will man der Vielfalt der Substitutionsmöglichkeiten eines Olefins nur einigermaßen Rechnung tragen, muß man der Reaktionsweise [p] eine alternative Reaktionsweise gegenüberstellen, welche aber keine Cyclo-addition sein kann. Von allen zweizentrigen Reaktionsweisen erweist sich bei Berücksichtigung der Elektronenstruktur der Reaktanten die Reaktionsweise [z] als die wahrscheinlichste.

Für den Übergangszustand $T_z^{\ddagger}$ sind vier π-Elektronen (paarweise zueinander orthogonal) in der N_2-Gruppe, die anderen in $[R_1 CR_2]^{\oplus}$, R_3, R_4 und in der $[R_5-C_\alpha-R_6]^{\ominus}$-Gruppierung lokalisiert.

$(\Delta H_\pi^{\ddagger})_z$ folgt somit zu

$$(\Delta H_\pi^{\ddagger})_z = [E(R_1 CR_2^{\oplus}) + E(R_3) + E(R_4) + E(R_5 CR_6^{\ominus}) + E(-N \equiv \overline{N})]$$
$$- [E(R_1 R_2 CN_2) + E(R_3 R_4 C = CR_5 R_6)] \ . \tag{43}$$

Der erste Schritt des oben angegebenen Mechanismus der Reaktionsweise [z] kann als elektrophiler Angriff des Olefins am C aufgefaßt werden. So betrachtet reagiert das Olefin mit seinem C_β als Lewis-Säure. Dies steht im Einklang mit der Beobachtung, daß Olefine mit ausgeprägtem Lewis-säurencharakter bei der Reaktion mit Diazoalkanen keine Pyrazoline bilden [35]. Wie an einer größeren Anzahl von Beispielen gezeigt werden konnte, besitzt ein Olefin $R_3 R_4 C_\beta = C_\alpha R_5 R_6$ an seinem C_β-Atom dann ausgeprägteren Lewis-Säurecharakter, wenn R_5 und R_6 starke π-Elektronenacceptoren und R_3 und R_4 schwache π-Elektronendonatoren sind [30]. In derartig substituierten Äthylenen besitzt C_β eine ausgeprägte positive π-Elektronen-Nettoladung von etwa $q_\beta \doteq +0,2$; die negative Ladung ist über die Zentren von R_5 und R_6 verteilt, auf C_α entfällt etwa $q_\alpha \doteq -0,05$ [28, 30]. Mit einer solchen Ladungsverteilung provoziert zwar das Olefin einen nucleophilen Angriff an seinem C_β-Atom, ist aber nicht imstande, gleichzeitig mit seinem C_α an einem anderen Zentrum (etwa dem N_ϵ) nucleophil anzugreifen. Ein derartig substituiertes Olefin neigt daher eher zur Reaktionsweise [z] als zur Reaktionsweise [p].[5]

Durch die Gln. (42) und (43) sind $(\Delta H_\pi^{\ddagger})_p$ bzw. $(\Delta H_\pi^{\ddagger})_z$ gegeben. Man sieht sofort, daß sich bei der Differenzbildung die π-Elektronenenergien der Ausgangsstoffe $[E(R_1 R_2 CN_2) + E(R_3 R_4 C = CR_5 R_6)]$ wegheben und daher

$$(\Delta H_\pi^{\ddagger})_p - (\Delta H_\pi^{\ddagger})_z = E(T_p^{\ddagger}) - E(T_z^{\ddagger}) \tag{44}$$

---

[5] Anmerkung bei der Fahnenkorrektur: Inzwischen wurden bei genügend tiefen Temperaturen (−15° bis −60°C) von einer Reihe organischer Lewissäuren Pyrazoline durch Diazoalkan-Einwirkung erhalten; die so dargestellten Pyrazoline zersetzen sich spontan beim Erwärmen auf Raumtemperatur. [J. Bus, H. Steinberg und Th. J. de Boer, Mh. Chem. 98, 1155 (1967); H. Kisch (unveröffentlicht)].

wird. Berücksichtigt man die in Gln. (42) bzw. (43) angegebenen Summen für $E\,(\mathrm{T_p^{\neq}})$ und $E\,(\mathrm{T_z^{\neq}})$, so heben sich in Gl. (44) auch $E\,(\mathrm{R_3})$ und $E\,(\mathrm{R_4})$ weg, während $[E\,(\mathrm{-N{=}N-}) - E\,(\mathrm{-N{=}N})] = \varDelta E\,(\mathrm{N_2})$ eine konstante Größe ist. Berücksichtigt man ferner, daß die Lokalisierungsenergien

$$L^+ = L^+(\mathrm{R_1}\,C R_2^{\oplus}) = [E\,(\mathrm{R_1}) + E\,(\mathrm{R_2}) - E\,(\mathrm{R_1}\,C R_2^{\oplus})]$$

und

$$L^- = L^-(\mathrm{R_5}\underline{C}R_6^{\ominus}) = [E\,(\mathrm{R_5}) + E\,(\mathrm{R_6}) - E\,(\mathrm{R_5}\underline{C}R_6^{\ominus})]$$

in der Differenz (44) vorkommen, läßt sich hier die relative Aktivierungsenthalpie gemäß Gln. (34), (35) und (44) darstellen durch:

$$(\varDelta G^{\neq})_\mathrm{p} - (\varDelta G^{\neq})_\mathrm{z} = L^+ + L^- + \varDelta E\,(\mathrm{N_2}) + G_{\mathrm{p,z}}^{\neq}\,; \qquad (45)$$

ist sie positiv, wird Z, ist sie negativ wird P bevorzugt gebildet. Eine von L^+ und L^- aufgespannte Ebene wird durch die Gerade

$$(\varDelta G^{\neq})_\mathrm{p} - (\varDelta G^{\neq})_\mathrm{z} = 0 = L^+ + L^- + \varDelta E\,(\mathrm{N_2}) + G_{\mathrm{p,z}}^{\neq}$$

in zwei Felder geteilt, welche der bevorzugten Bildung von R bzw. Z entsprechen (Abb. 8). Die Entscheidung der Konkurrenz I zwischen der Bildung des Pyrazolins P oder des Zwischenkörpers Z wird also durch die Substituenten $\mathrm{R_1}$ und $\mathrm{R_2}$ des Diazoalkans einerseits, und $\mathrm{R_5}$ und $\mathrm{R_6}$ des Olefins andererseits bestimmt, da diese die Größe von L^+ bzw. L^- bedingen.

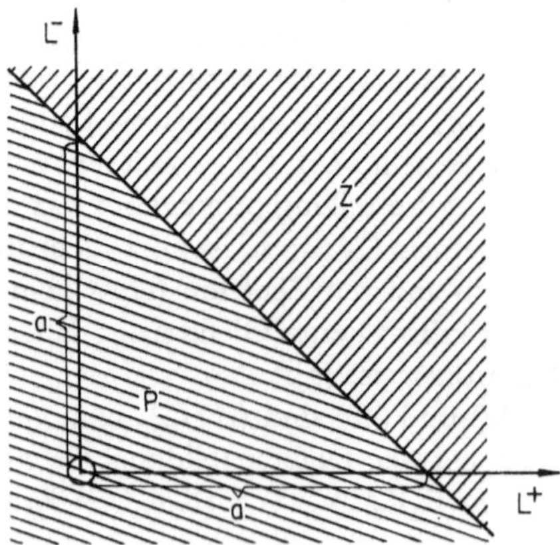

Abb. 8. Zur Konkurrenz der Bildung von Pyrazolinen oder Zwischenkörper (Z)

Von den Summanden in Gl. (37) oder Gl. (45), welche die Konkurrenz I beschreiben, ist $G_{\mathrm{p,z}}^{\neq}$ noch unbestimmt. Da keine der experimentell untersuchten Reaktionen die für die Anwendung von Gl. (39) notwendigen Voraussetzungen erfüllt, wurde $[(\varDelta H_\pi^{\neq})_\mathrm{p} - (\varDelta H_\pi^{\neq})_\mathrm{z}]$ für die in Tab. 1 zu-

Tabelle 1. Zur Ermittlung von $G_{p,z}^{\pm}$

Diazoalkan	Olefin	$(\Delta H_\pi^{\pm})_p - (\Delta H_\pi^{\pm})_z$ in $[\beta]$	gebildetes Produkt	Lit.
Diazofluoren	Acrylsäureester	4,783	Z	18
Diazoessigester	Benzyliden-meldrumsäure	4,454	Z	40
Diazomethan	Benzyliden-meldrumsäure	4,166	Z	39
Diazomethan	Methacrylnitril	3,937	P	15
Diazoessigester	Acrylsäureester	3,910	P	10
Diazomethan	Acrylsäureester	3,623	P	27

sammengefaßten sechs Ausgangssysteme nach Gl. (44) berechnet, nach der Größe geordnet und mit den experimentellen Resultaten verglichen. Dabei wurde entsprechend dem Reaktionsschema die Bildung eines stickstofffreien Reaktionsproduktes als eine Entscheidung der Konkurrenz I zugunsten des Zwischenkörpers Z aufgefaßt. Aus dieser Gegenüberstellung experimenteller und theoretischer Daten folgt für $G_{p,z}$ die Eingrenzung

$$- 4{,}166\ [\beta] < G_{p,z}^{\pm} < -3{,}937\ [\beta] \ . \tag{46}$$

Nimmt man $G_{p,z}^{\pm}$ etwa in der Mitte dieses Intervalles an, erhält man:

$$G_{p,z}^{\pm} \doteq - 4{,}047 \pm 0{,}115\ [\beta] \ . \tag{47}$$

Mit Hilfe dieses Wertes kann nun die Entscheidung der Konkurrenz I gemäß Gln. (37) oder (45) berechnet und damit vorerst vorhergesagt werden, ob bei der betrachteten individuellen Reaktion Pyrazoline oder stickstoffreie Produkte zu erwarten sind. Die so erhaltenen Rechendaten sind in Tabelle 2 den experimentellen Befunden gegenübergestellt. Positive Werte für $[(\Delta G_\pi^{\pm})_p - (\Delta G_\pi^{\pm})_z]$ zeigen eine Bevorzugung von Z, negative die von P an. Infolge Gl. (45) sind die in Tab. 2 wiedergegebenen Rechendaten voneinander abhängig.

Für ein bestimmtes Olefin ist der für Diazoessigester errechnete Wert von $[(\Delta G_\pi^{\pm})_p - (\Delta G_\pi^{\pm})_z]$ um $L^+[\text{CHCOOEt}^{\oplus}] = +0{,}2877\ [\beta]$, der für Diazofluoren um $L^+[(o\text{-}C_6H_4)_2C^{\oplus}] = +1{,}1603\ [\beta]$ größer als der für Diazomethan erhaltene. Der Anstieg von $[(\Delta G^{\pm})_p - (\Delta G^{\pm})_z]$ von oben nach unten geht auf den Anstieg von $L^-(\text{R}_5\underline{\text{C}}\text{R}_6^{\ominus})$ in dieser Reihung zurück. Er hängt nur von R_5 und R_6 ab; daher sind auch die relativen freien Aktivierungsenthalpien für Olefine, welche die gleichen Substituenten R_5 und R_6 enthalten (wie z. B. alle α,β-ungesättigten Nitrile, Ester, Malodinitrile, Meldrumsäuren usw.), untereinander gleich. Auf die oben diskutierte Bedeutung von R_5 und R_6 für den Lewissäurencharakter des Olefins und dessen Zusammenhang mit der alternativen Bildung von P oder Z sei hier rückverwiesen.

Tabelle 2. *Relative Aktivierungsenthalpien für die Konkurrenz I beim Umsatz von Diazoalkanen mit Olefinen (Abb. 5)*

Olefin	Diazomethan $(\Delta G^+)_p - (\Delta G^+)_z$ (ber.)	exp.: Prod.	Lit.	Diazoessigester $(\Delta G^+)_p - (\Delta G^+)_z$ (ber.)	exp.: Prod.	Lit.	Diazofluoren $(\Delta G^+)_p - (\Delta G^+)_z$ (ber.)	exp.: Prod.	Lit.
Acrylnitril	−0,7890	P	19	−0,5013	n. b.	—	+0,3713	Z	8
Crotonnitril	−0,7890	n. b.	—	−0,5013	n. b.	—	+0,3713	n. b.	—
Maleinsäuredinitril	−0,7890	n. b.	—	−0,5013	n. b.	—	+0,3713	Z	8
Acrylsäureester	−0,4243	P	13	−0,1366	P	12	+0,7360	Z	8
Crotonsäureester	−0,4243	P	18	−0,1366	n. b.	—	+0,7360	n. b.	—
Zimtsäureester	−0,4243	P	17	−0,1366	P	20	+0,7360	Z	21
Maleinsäureester	−0,4243	P	24	−0,1366	P	17	+0,7360	Z	8
Maleinsäureanhydrid	−0,4199	n. b.	—	−0,1322	n. b.	—	+0,7404	P*	25
Styrol	−0,3794	P	22	−0,0917	n. b.	—	+0,7809	n. b.	—
p-Benzochinon	−0,3748	P	23	−0,0871	n. b.	—	+0,7855	P*	8
Methacrylsäureester	−0,2827	P	17	+0,0050	n. b.	—	+0,8776	Z	8
Methyl-vinylketon	−0,2625	n. b.	—	+0,0252	n. b.	—	+0,8978	Z	8
Methacrylnitril	−0,1096	P	11	+0,1781	n. b.	—	+1,0507	n. b.	—
2-Benzylidenindandion-1,3	+0,1025	Z	26	+0,3902	Z	27	+1,2628	Z	28
Methylenmeldrumsäure**	+0,1191	Z	9	+0,4068	Z	9	+1,2794	n. b.	—
Benzylidenmeldrumsäure	+0,1191	Z	10	+0,4068	Z	9	+1,2794	n. b.	—

Tabelle 2. Fortsetzung

Ph. CH=C(O—C=O)(CH=C-COOR)	+0,1219	Z	32	+0,4096	n. b.	—	+1,2822	n. b.	—
(CH₃)₂C(O-CO)C=C(COO)C(CH₃)₂ ...	+0,1355	Z	37	+0,4232	n.b.	—	+1,2958	n. b.	—
Benzylidencyanessigester	+0,1394	Z	29	+0,4271	n. b.	—	+1,2997	n. b.	—
Methylenmalodinitril **	+0,1657	Z	6	+0,4534	n. b.	—	+1,3260	n. b.	—
Benzylidenmalodinitril	+0,1657	Z	30	+0,4534	n. b.	—	+1,3260	n. b.	—
Tetracyanoäthylen ***	+0,1657	Z	36	+0,4534	n. b.	—	+1,3260	n. b.	—
Benzylidencyanaceton	+0,1787	n. b.	—	+0,4664	n. b.	—	+1,3390	n. b.	—
Benzylidenazlacton der Hippursäure	+0,2937	Z	31	+0,5814	n. b.	—	+1,4540	n. b.	—
α-Pyron-5-carbonsäureester	+0,3634	Z	33	+0,6511	n. b.	—	+1,5237	n. b.	—
HC=NPh-OH ⇌ HC-NHPh-O (Naphthalin)	+0,3731	Z	34	+0,6608	n. b.	—	+1,5334	n. b.	—
9-Methylfluoren	+0,4225	Z	35	+0,7102	n. b.	—	+1,5828	n. b.	—
Fulven	+0,9000	n. b.	—	+1,1877	n. b.	—	+2,0603	n. b.	—

* Keine Übereinstimmung zwischen Rechnung und Experiment.

** Die für die Methylenverbindung ausgeführten Rechnungen sind hier mit den an Alkylidenderivaten erhobenen experimentellen Befunden verglichen.

*** Z kann sich hier aus strukturellen Gründen nur zu C stabilisieren.

Die Übereinstimmung mit den verfügbaren experimentellen Befunden ist, wie Tab. 2 ausweist, sehr gut: nur bei zwei der 38 Reaktionen, für welche experimentelle Befunde vorliegen, und zwar bei der Umsetzung von Diazofluoren mit p-Benzochinon bzw. Maleinsäureanhydrid, treten Abweichungen auf.

Bei den Pyrazolinen wurde kein Unterschied gemacht, ob sie die Struktur der Δ^1- oder die der Δ^2-Verbindung haben, da die letzteren durch Isomerisierung der ersteren entstehen können, die Konkurrenz I sich aber nur auf die primär gebildeten Produkte bezieht.

Vergleicht man die Zahlenwerte für die Reaktionen verschiedener Diazoalkane mit ein und demselben Olefin, folgt aus Tab. 2, daß im allgemeinen Diazomethan schneller reagieren sollte als Diazofluoren und dieses schneller als Diazoessigester, was mit der chemischen Erfahrung übereinstimmt.

5.2. Konkurrenz II

Der Zwischenkörper Z kann sich durch einfachen Ringschluß zu einem Cyclopropanderivat (C) oder — wenn $R_6 = -CO-R_6'$ ist — zu einem Dihydrofuranderivat (F) stabilisieren. Die π-Elektronenmodelle der dabei durchlaufenden Übergangszustände $T_c^{\ddagger}$ bzw. $T_f^{\ddagger}$ sind in Abb. 7 dargestellt. In $T_c^{\ddagger}$ sind die Substituenten R_1 bis R_6 ohne gegenseitige Konjugation; daher folgt $(\Delta H_\pi^{\ddagger})_c$ zu

$$(\Delta H_\pi^{\ddagger})_c = \sum_1^6 E(R_i) - E(Z) . \tag{48}$$

In $T_f^{\ddagger}$ sind die π-Elektronensysteme der Substituenten R_1 bis R_4 isoliert, R_5, R_6 und C_α bilden ein konjugiertes $[R_5C{=}CR_6'O-]$-System aus; daher gilt

$$(\Delta H_\pi^{\ddagger})_f = \sum_1^4 E(R_i) + E(R_5\overset{|}{C}{=}CR_6'\overset{|}{O}) - E(Z) . \tag{49}$$

Mit diesen beiden im Wettbewerb stehenden Reaktionsweisen [c] und [f] des Zwischenkörpers Z konkurrieren zwei weitere Reaktionsweisen [u] und [a].

Die Reaktionsweise [u] stellt eine Umlagerungsreaktion dar, welche als Anionotropie des Substituenten R_3 vom C_β zum C_γ aufzufassen ist. Geläufige Beispiele für diese Reaktionsweise liefern die auf der Ringerweiterung von Aromaten durch Diazoalkane basierenden Cycloheptatriensynthesen. Diese Reaktionsweise haben F. Wessely und Mitarbeiter ferner bei einigen olefinischen Ausgangssystemen beobachtet [12, 35, 36, 39]. Die Umsetzung von Diazomethan mit Benzylidenmeldrumsäure führt zu einem Cyclopropanderivat des Typs C und einem weiteren, das durch

Reaktion eines zweiten Moleküls Diazomethan mit dem Umlagerungs-
produkt U gebildet wird. Durch Markierung des C_β der Benzylidenmeldrumsäure mit C^{14} und Abbau des Benzylcyclopropanderivates (UC)
konnte die Phenylwanderung vom C_β zum C_γ sichergestellt werden [1]:

$$CH_2N_2 \;+\; Ph\overset{*}{C}H{=}C\big(\underset{COO}{\overset{COO}{}}\big)C\big(\underset{CH_3}{\overset{CH_3}{}}\big) \longrightarrow \tag{50}$$

$$Ph{-}\overset{*}{C}H{-}C\big(\underset{\underset{CH_2}{COO}}{COO}\big)C\big(\underset{CH_3}{\overset{CH_3}{}}\big) \;+\; PhCH_2\overset{*}{C}H{-}C\big(\underset{\underset{CH_2}{COO}}{COO}\big)C\big(\underset{CH_3}{\overset{CH_3}{}}\big)\;.$$

Das π-Elektronenmodell des bei der Umlagerung durchlaufenen Übergangszustandes T_u ist in Abb. 7 dargestellt: Die Substituenten R_1R_2 und
R_4 sind isoliert, R_5 mit R_6 über C_α konjugiert. Das Atom X, mit dem in Z
der Substituent R_3 an C_β gebunden war, steht mit den anderen Zentren
von R_3, die zu R_3' zusammengefaßt sind, nicht mehr in Konjugation und
ist gleichzeitig an C_β und C_γ gebunden; dadurch ist die in Z im Bereich
von R_1CR_2 lokalisierte positive Ladung in den Rest R_3' verschoben. Für
$(\Delta H_\pi^{\ddagger})_u$ folgt dementsprechend

$$(\Delta H_\pi^{\ddagger})_u = E(R_1) + E(R_2) + E(R_3'^{\oplus}) + E(R_4) + E(R_5CR_6^{\ominus}) - E(Z)\;. \tag{51}$$

Obgleich bei Zimmertemperatur die Reaktion (50) etwa äquimolare Men-
gen von C und UC liefert, $G_{u,c}^{\ddagger}$ also nach Gl. (39) ziemlich scharf bestimmbar sein sollte, wird mit diesem Wert keine gute Übereinstimmung der
experimentellen und rechnerischen Daten erzielt. Auch der Versuch,
$G_{u,c}^{\ddagger}$ in der oben für $G_{p,z}^{\ddagger}$ geschilderten Weise zu bestimmen, ändert nichts
an der Situation: Bei der Vorhersage der Reaktionsweise [u] versagt das
hier beschriebene Verfahren. Bei den in Tab. 3 zusammengefaßten Rechendaten ist aus diesem Grunde die Reaktionsweise [u] nicht berücksichtigt.

Reaktionsweise [a], die letzte der Konkurrenz II, ist als Deprotonisierung von Z am C_β unter Ausbildung des vollständig konjugierten Anions
A charakterisiert. Als Protonenacceptor wirkt hierbei das Lösungsmittel
oder das zumeist mindestens spurenweise vorhandene Wasser mit. Struk-
turell ist die Deprotonisierung [a] natürlich an das Vorhandensein eines
H-Atoms am C_β (d. h. $R_3 = H$) geknüpft. Der präparative Nachweis
dieses Reaktionsweges wurde durch den Umsatz von Benzylidenmalodi-
nitril mit einer zuerst sorgfältig getrockneten, mit D_2O angefeuchteten
ätherischen Lösung von Diazomethan erbracht [12]:

$$\text{Ph}\underset{H}{\overset{}{C}}{=}C\big(\underset{CN}{\overset{CN}{}}\big) \;+\; CH_2N_2 \;\xrightarrow{(D_2O\ \text{anwesend})}\; \text{Ph}\underset{DH_2C}{\overset{}{C}}{=}C\big(\underset{CN}{\overset{CN}{}}\big) \tag{52}$$

Das bei der Reaktion gebildete α-Phenyläthyliden-malodinitril enthielt etwa 60 bis 70% des Deuteriums, welches theoretisch für einen Reaktionsablauf über A zu erwarten ist.

Es ist fraglich, ob die Reaktionsweise [a] kinetisch kontrolliert wird. In diesem speziellen Falle stört diese Unsicherheit aber nicht, da das π-Elektronenmodell des Übergangszustandes $T_a^{\ddagger}$ (Abb. 7) in erster Näherung mit der π-Elektronenstruktur des Anions A identisch ist, die π-Elektronenenergien von $T_a^{\ddagger}$ und A also gleich sind:

$$E(T_a^{\ddagger}) = E(A) \ . \tag{53}$$

Es sei bemerkt, daß Gl. (53) nur durch die von uns gewählte Vorgangsweise bei der Konstruktion der π-Elektronenmodelle der Übergangszustände verursacht wird. In Wirklichkeit werden sich $T_a^{\ddagger}$ und A unterscheiden, und zwar hauptsächlich bezüglich ihrer Planarität. Infolge der Existenz der Gl. (53) liefert aber im Rahmen des hier beschriebenen Näherungsverfahrens die Annahme der thermodynamischen Kontrolle der Reaktionsweise [a] die gleichen Ergebnisse wie die einer kinetischen Kontrolle; für beide Annahmen folgt die kritische Größe — hier als $(\Delta H_\pi^{\ddagger})_a$ ausgedrückt — zu

$$(\Delta H_\pi^{\ddagger})_a = E(A) - E(Z) \ . \tag{54}$$

Da die Konzentration der Base B als praktisch konstant angesehen werden kann, ist die Reaktionsweise [a] quasi-monomolekular, so daß für die Reaktionsweisen [c], [f] und [a] die formal gleichen Geschwindigkeitsgesetze folgen. Die Behandlung der Konkurrenz II nach diesem Verfahren ist somit gerechtfertigt. Die in ihr auftretenden relativen Freien Aktivierungsenthalpien ergeben sich aus den Gln. (48), (49) und (54) formelmäßig zu:

$$(\Delta G^+)_c - (\Delta G^+)_f = \left[E(R_5) + E(R_6)\right] - E(R_5C{=}CR_6'O-) + G_{c,f}^{+} \tag{55}$$

$$(\Delta G^+)_f - (\Delta G^+)_a = \left[\sum_1^4 E(R_i) + E(R_5C{=}CR_6'O-)\right] - E(A) + G_{f,a}^{+} \tag{56}$$

$$(\Delta G^+)_a - (\Delta G^+)_c = E(A) - \sum_1^6 E(R_i) + G_{a,c}^{+} \ . \tag{57}$$

Wie die Addition der letzten drei Gleichungen zeigt, ist entsprechend Gl. (40) die Summe $G_{c,f}^{+} + G_{f,a}^{+} + G_{a,c}^{+} = 0$. In analoger Weise wie bei $G_{p,z}^{+}$ wurde der Zahlenwert dieser drei Konstanten bestimmt (Tabelle 3). Kürzt man die in den Gl. (55) bis (57) auftretenden Summen mit L_1 bzw. L_2 ab

$$L_1 = E(R_5) + E(R_6) - E(R_5C{=}CR_6'O-) \tag{58a}$$

$$L_2 = \sum_1^6 E(R_i) - E(A) \tag{58b}$$

Tabelle 3. *Ermittlung von $G_{c,f}^{+}$ $G_{f,a}^{+}$ und $G_{a,c}^{+}$ (Energien in Einheiten von [ß])*

Reaktion von Diazomethan mit	$(\Delta H_{\pi}^{+})_c - (\Delta H_{\pi}^{+})_f$	$(\Delta H_{\pi}^{+})_f - (\Delta H_{\pi}^{+})_a$	$(\Delta H_{\pi}^{+})_a - (\Delta H_{\pi}^{+})_c$	Gebildetes Produkt	Lit.
Benzyliden-malodinitril	*	*	−4,2141	A	7
Benzyliden-meldrumsäure	+0,9803	+3,1766	−4,1569	C	*39*
2-Benzyliden-indandion-1,3	+0,9931	+3,1517	−4,1448	F	*22*
Mittelwert von $G_{j,1}^{+}$	−0,992	−3,168	+4,160		

* entfällt hier, da infolge $R_6 \neq -COR_6'$ die Bildung von F unmöglich ist.

und führt diese Abkürzungen in die Gln. (55) bis (57) ein, folgen diese zu:

$$(\Delta G^{+})_c - (\Delta G^{+})_f = L_1 + G_{c,f}^{+} \qquad (59\,a)$$

$$(\Delta G^{+})_f - (\Delta G^{+})_a = (L_2 - L_1) + G_{f,a}^{+} \qquad (59\,b)$$

$$(\Delta G^{+})_a - (\Delta G^{+})_c = - L_2 + G_{a,c}^{+} \; . \qquad (59\,c)$$

Diese drei Gleichungen stellen in einer durch L_1 und L_2 aufgespannten Ebene Geraden dar, die sich in einem Punkt P ($L_1 = -G_{c,f}^{+}$, $L_2 = +G_{a,c}^{+}$) schneiden und die Ebene in drei Felder teilen, welche der Bevorzugung von C bzw. F bzw. A entsprechen (Abb. 9).

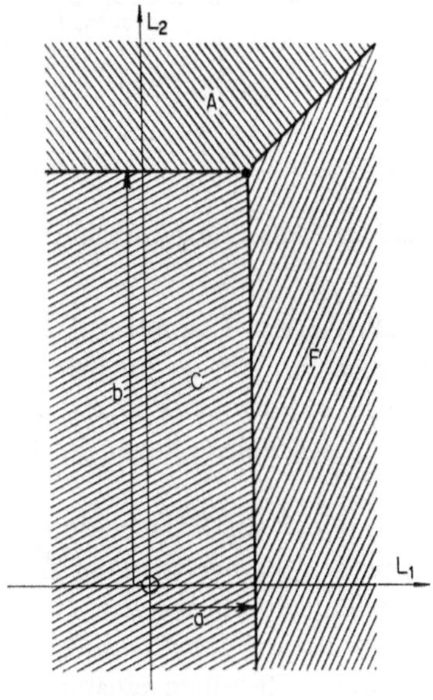

Abb. 9. Zur Konkurrenz bei der Stabilisierung des Zwischenkörpers Z (s. Text)!

Durch die drei so bestimmten Konkurrenzkonstanten $G^{\pm}_{j,l}$ (j, l = c, f, a) wird die relative Lage der Niveaus der Restgrößen der Freien Aktivierungsenthalpien $(\Delta G^{\pm}_{R})_j$ (j = c, f, a) festgelegt. Die relativen Lagen der Niveaus $(\Delta G^{+})_j$ der Freien Aktivierungsenthalpien der aktivierten Komplexe $T^{\pm}_j$ werden entsprechend Gl. (28) gefunden, indem die Beiträge der π-Elektronen zur Aktivierungsenthalpie $(\Delta H^{\pm}_{\pi})_j$ von $(\Delta G^{\pm}_{R})_j$ aus vorzeichenrichtig aufgetragen werden (Abb. 10). Für eine übersichtliche Darstellung ist es

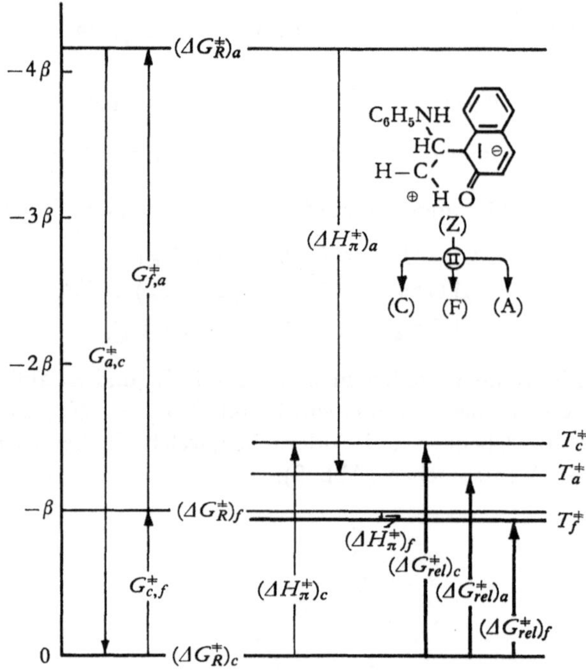

Abb. 10. Ermittlung der relativen Aktivierungsenthalpien im Falle der Konkurrenz II

günstig, die Lage der aktivierten Komplexe $T^{\pm}_j$ vom tiefsten der drei $(\Delta G^{\pm}_{R})_l$-Niveaus aus zu messen. Die so erhaltenen Zahlenwerte sind als $(\Delta G^{\pm}_{rel})_j$ bezeichnet. In der Konkurrenz II korrespondiert das tiefste Niveau mit $(\Delta G^{\pm}_{R})_c$; die aktivierten Komplexe liegen in der Regel 1 bis 2 β-Einheiten darüber.

Es ist natürlich nur sinnvoll, die Entscheidung der Konkurrenz II für diejenigen Fälle zu untersuchen, bei welchen ein Zwischenkörper Z durchlaufen wird. Da die meisten experimentellen Befunde für Diazomethan vorliegen, sind in Tabelle 4 nur die für dieses Diazoalkan erhaltenen rechnerischen und experimentellen Resultate einander gegenübergestellt; die Reihenfolge der Olefine ist die gleiche wie in Tab. 2. Verbindungen, welche

Tabelle 4. *Relative Freie Aktivierungsenthalpien für Konkurrenz II bei der Umsetzung von Olefinen mit Diazomethan (Energien in* $[|\beta|]$*)*

Olefin des Ausgangssystems	$(\Delta G^{\pm}_{rel})_c$	$(\Delta G^{\pm}_{rel})_a$	$(\Delta G^{\pm}_{rel})_f$	exp.: Prod.	Lit.
2-Benzylidenindandion-(1,3)	1,2025	1,2177	1,2014	F	22
Methylenmeldrumsäure***	1,2191	1,5990	1,2308	C	35, 36
Benzylidenmeldrumsäure	1,2191	1,2222	1,2308	C	39
Ph–CH=C(O–C=O)(CH=C–COOR)	1,2219	1,1669	1,2319	A	2
Benzylidencyanessigester	1,2394	1,2157	*	A	32
Benzylidenmalodinitril	1,2657	1,2116	*	A	7
Benzylidencyanaceton	1,2787	1,2234	1,2583	n. b.	—
Benzylidenazlacton der Hippursäure	1,3937	1,2355	1,0457	C**	5
α-Pyron-5-carbonsäureester	1,4552	1,3621	*	A	14
Naphthol ⇌ Naphthochinon-anil	1,4731	1,2751	0,9499	F	34
9-Methylfluoren	1,5225	1,6453	*	C	41
Fulven	2,0000	1,7147	*	n. b.	—

* Da hier $R_6 \neq -COR_6$, Bildung von F nicht möglich.

** Keine Übereinstimmung zwischen Rechnung und Experiment.

*** Rechendaten an Methylenmeldrumsäure, experimentelle Befunde an Alkylidenmeldrumsäuren erhoben.

aus strukturellen Gründen weder F noch A bilden können (Di-isopropylidenacylal der Äthylentetracarbonsäure, *sek*-Alkylidenmalodinitrile, Tetracyanoäthylen), sind in Tab. 4 nicht aufgenommen, da sich der aus ihnen gebildete Zwischenkörper Z nur zu C stabilisieren kann. Die Rechenresultate stimmen mit den experimentellen Befunden gut überein; eine einzige Abweichung tritt auf: aus Diazomethan und dem Benzylidenazlacton der Hippursäure bildet sich anstelle des nach der Rechnung zu erwartenden Di-hydrofurans ein Cyclopropanderivat.

5.3. Konkurrenz III

Das Allylanion A stabilisiert sich durch Protonenaufnahme. Sowohl das C_α als das C_γ kommen als Positionen der Protonenaufnahme in Frage und sind im Reaktionsschema (Abb. 5) berücksichtigt. Alle in Tab. 4 angeführten Systeme, bei welchen die Entscheidung der Konkurrenz II zugunsten von A fällt, nehmen das Proton am C_γ auf. Mangels geeigneten Erfahrungsmaterials kann also $G^{\pm}_{m,o}$ nicht bestimmt werden.

21*

Die Bevorzugung von C_γ bei der Protonisierung von A ist leicht verständlich. A bildet sich nur, wenn die Konkurrenz I zugunsten von Z entschieden wurde. Diese wird hauptsächlich von dem mit der Konjugation von R_5 und R_6 über C_α erzielten Energiegewinn bestimmt; in Tab. 2 sind die Olefine nach dieser Größe geordnet. Der mit der Konjugation $R_1-C_\gamma-R_2$ erzielte Energiegewinn markiert auf der Skala dieser Reihung nur einen Punkt, von dem aus auf der einen Seite diejenigen Olefine liegen, bei welchen die Bildung von P, auf der anderen Seite diejenigen Olefine liegen, bei welchen die Bildung von Z kinetisch bevorzugt ist. Im Diazomethan sind $R_1 = R_2 = H$, die Konjugation $R_1-C_\gamma-R_2$ ist mit keinem Energiegewinn verbunden. Bei der Konkurrenz I (P oder Z) kann also nur die Konjugationsenergie ($R_5-C_\alpha-R_6$) zum Zuge kommen. Die Protonenakzeptanz am C_α würde daher stets energetisch ungünstiger als die am C_γ sein. Es ist also im Falle der Umsetzung von Olefinen mit Diazomethan nur M, nicht aber O zu erwarten. Die größte Konjugationsenergie ($R_1-C_\gamma-R_2$) besitzt von allen üblicherweise eingesetzten Diazoalkanen Diazofluoren; bei dessen Umsatz mit Olefinen wird aber die Konkurrenz II in der Regel zuungunsten von A entschieden, so daß über die Konkurrenz III nicht befunden werden kann. Das Olefin O erscheint somit als ein Reaktionsprodukt, das zwar theoretisch in Betracht zu ziehen ist, praktisch aber kaum eine reelle Chance zur Bildung besitzt.

5.4. Reaktionen mit weiteren Molekülen Diazoalkan

Von den bisher beschriebenen Reaktionsprodukten enthalten die Dihydrofurane F, die Umlagerungsprodukte U und die Alkylierungsprodukte M und O eine Doppelbindung und können somit als die Olefinkomponente eines neuen Ausgangssystems betrachtet werden.

Aus den Umlagerungsprodukten sollten sich keine Pyrazoline bilden können, da die Substituenten R_3 und R_4 an der Entscheidung der Konkurrenz I nicht beteiligt sind, das Olefin U aber in R_5 und R_6 mit dem ursprünglichen übereinstimmt. Für das neue Ausgangssystem sind also R_1,R_2,R_5 und R_6 dieselben, und daher ist bezüglich der Konkurrenz I die gleiche Situation wie für das ursprüngliche Ausgangssystem. Gleiches trifft für die Alkylierungsprodukte M, nicht unbedingt aber für F und O zu.

Ob die Reaktion von U, F, M und O mit einem weiteren Molekül Diazoalkan in merklichem Umfang eintritt, ist eine Frage der absoluten Größe der entsprechenden Reaktionsgeschwindigkeitskonstanten, liegt somit außerhalb der hier angestellten Betrachtungen und kann mit Hilfe des hier beschriebenen Verfahrens nicht diskutiert werden.

Selbstverständlich können alle im Verlaufe der Reaktion gebildeten Produkte mit einem weiteren Molekül Diazoalkan in der hier beschriebenen Weise reagieren, wenn sie reaktive Doppelbindungen in den Substituenten enthalten. In derartigen Fällen taucht die Frage nach der reaktiveren Doppelbindung auf, welche als erste reagieren sollte. Im Prinzip ist das hier beschriebene Verfahren auch für die Diskussion solcher Probleme geeignet; seine praktische Anwendung kann gelegentlich daran scheitern, daß zu wenig Erfahrungsmaterial vorliegt, um die betreffende Konkurrenzkonstante $G_{j,1}^{\ddagger}$ zu bestimmen.

Ein charakteristisches Beispiel für die Vielfalt derartiger Reaktionsfolgen bietet die Umsetzung von Diazomethan mit 1,3,5-Tri-nitrobenzol [8]:

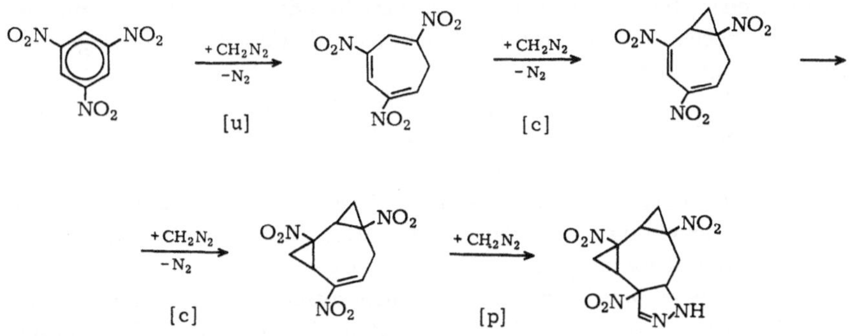

Einer Ringerweiterung [u] folgen zwei Cyclopropanbildungen [c]; die verbleibende Doppelbindung reagiert schließlich nach [p] zu einem Pyrazolin.

6. Diskussion

Das hier beschriebene Verfahren hat viel Ähnlichkeit mit der Methode der Lokalisierungsenergien. Zum Unterschied von dieser sind aber die Lokalisierungsenergien selbst nicht entscheidend, sondern stellen nur Rechengrößen in einer auch andere Größen enthaltenden Summe dar. Daß Lokalisierungsenergien hier überhaupt auftreten, ist nur eine Folge der bei der Konstruktion der π-Elektronenmodelle der Übergangszustände eingeschlagenen Vorgangsweise. Sie verschwinden sofort aus den Ausdrücken für die relativen Freien Aktivierungsenthalpien, sobald eine andere (eventuell verfeinerte) Methode der Berechnung von $E\,(T_j^{\ddagger})$ zugrunde gelegt wird. Das von uns gewählte π-Elektronenmodell der Übergangszustände bietet den Vorteil des relativ geringen Aufwandes für die Berechnung von $E\,(T_j^{\ddagger})$.

Die erzielten Ergebnisse sind überraschend gut. Von den in den Tabellen 2 und 4 angestellten 48 Vergleichen mit experimentellen Befunden sind alle bis auf 3 richtig, d. h. die Treffsicherheit der Methode liegt über 93%.

Wie oben erwähnt, sind die Rechenergebnisse für Phenyldiazomethan und Diphenyldiazomethan als Diazokomponente nicht so verläßlich. Vermutlich hängt dies damit zusammen, daß sich die Neigungswinkel zwischen den Ebenen der Phenylgruppen und der Ebene der CN_2-Gruppierung während der Reaktion ändern; jedenfalls ließ sich an einigen ausgewählten Beispielen zeigen, daß durch Vergrößerung des Twistwinkels der Phenylgruppen Übereinstimmung mit den experimentellen Befunden erzielt werden kann.

Die Konkurrenz zwischen der Umlagerungsreaktion [u] und den Reaktionsweisen [c], [f] und [a] wird nicht beherrscht. Dies kann daran liegen, daß

1. die mechanistische Interpretation von [u] als anionotrope Umlagerung von R_3 unzutreffend ist;

2. [u] einem anderen Geschwindigkeitsgesetz als [c], [f] und [a] folgt;

3. die Substituenten einen sterischen Effekt auf $T_u^{\ddagger}$ ausüben und somit die Bedingung $\varrho_k^{\ddagger} = [\varrho(k)]^{\ddagger}$ nicht erfüllt ist; und schließlich

4. die als Rechenverfahren benutzte HMO-Methode auf die Konjugationsunterbrechung im (üblicherweise aromatischen) Substituenten R_3 in fehlerhafter Weise reagiert.

Welche dieser Möglichkeiten tatsächlich zutreffen und für die Unstimmigkeiten verantwortlich sind, kann zur Zeit nicht gesagt werden. Da die relativen Wanderungstendenzen von R_3 in befriedigender Weise wiedergegeben werden, beziehen sich die erwähnten Unstimmigkeiten nur auf die Konkurrenz zwischen [u] und [c] bzw. [f] bzw. [a].

Die von uns gewählten π-Elektronenmodelle der Übergangszustände sind den π-Elektronenstrukturen der Reaktionsprodukte sehr ähnlich. Von diesem Umstand her betrachtet, ist die exzellente Übereinstimmung der Rechendaten mit den experimentellen Befunden besonders überraschend. Die Güte der Übereinstimmung kann verursacht sein durch:

1. eine gute lineare Korrelation zwischen den Aktivierungsenthalpien und den Reaktionsenthalpien bei allen hier untersuchten Reaktionsweisen. Derartige Verhältnisse sind zu erwarten, wenn $\varrho_j^{\ddagger}$ hohe, d. h. nahe bei Eins liegende Werte besitzt. Für das konjugierte Anion A wurden sie oben als wahrscheinlich diskutiert. Für die 1,3-Cycloaddition [p], für welche kinetische Daten [20] in ausreichender Menge vorliegen, wurde eine solche lineare Korrelation gefunden [36].

2. eine geringe Konjugationsenergie der im π-Elektronenmodell des Übergangszustandes als isoliert bzw. vollständig konjugiert betrachteten π-Elektronensysteme. Dadurch würde die Vernachlässigung des Bruchteils an Konjugationenergie, welche durch die Annäherung der tatsächlichen π-Elektronenstruktur des Übergangszustandes durch deren π-Elektronenmodell verursacht wird, nicht allzusehr ins Gewicht fallen; ein Teil dieses methodischen Fehlers würde überdies in die Konkurrenzkonstante $G_{j,1}^{\ddagger}$ überwälzt und durch deren halbempirische Bestimmung kompensiert werden.

3. ein Wechselspiel der beiden vorgenannten Möglichkeiten.

Bevor nicht umfangreiches Erfahrungsmaterial über die Anwendung des hier beschriebenen Verfahrens vorliegt, kann auch über diese Frage nicht weiter befunden werden. Die Güte der Übereinstimmung zwischen den Rechendaten und den experimentellen Befunden darf aber sicher als eine Bestätigung der Gl. (28) angesehen werden. Nun konnte zwar die Unabhängigkeit der Restgröße $\Delta H_R^{\ddagger}$ von der Substitution unter den durch die Gl. (13) symbolisierten Bedingungen recht plausibel gemacht werden, die entsprechenden Überlegungen für die Konstanz von $\Delta S^{\ddagger}$ sind aber weit weniger schlüssig. Es hieße jedoch die Bedeutung der erzielten guten Übereinstimmung überschätzen, wollte man aus ihr auf die Substitutionsunabhängigkeit des Wertes von $\Delta S^{\ddagger}$ schließen.

Eine andere Erklärungsmöglichkeit bietet überdies die sogenannte „isokinetische Beziehung" [21], nach welcher die durch die Substitution R bedingten Änderungen der Aktivierungsenthalpie und der Aktivierungsentropie einander proportional sind:

$$\delta_R(\Delta H^{\ddagger}) = \varkappa \cdot \delta_R(\Delta S^{\ddagger}) \ . \tag{60}$$

Darin stellt δ_R einen Differenzoperator, z. B. $\delta_R(\Delta H^{\ddagger}) = \Delta H^{\ddagger}(RX) - \Delta H^{\ddagger}(R_0X)$ und $\varkappa$ eine stets positive, substitutions- und temperaturunabhängige Konstante[6] von der Dimension einer absoluten Temperatur dar. Da $\delta_R(\Delta G^{\ddagger}) = \delta_R(\Delta H^{\ddagger} - T \cdot \Delta S^{\ddagger}) = \delta_R(\Delta H^{\ddagger}) - T \cdot \delta_R(\Delta S^{\ddagger})$ gilt und somit aus Gl. (60)

$$\delta_R(\Delta H^{\ddagger}) = \left(1 - \frac{T}{\varkappa}\right) \cdot \delta_R(\Delta H^{\ddagger}) \tag{61}$$

folgt, ist bei Gültigkeit der isokinetischen Beziehung ebenfalls eine strenge Proportionalität zwischen der Freien Aktivierungsenthalpie und der Aktivierungsenthalpie zu erwarten. Da ferner unter den in Gl. (13) ausgedrückten Voraussetzungen $\delta_R(\Delta H_R^{\ddagger}) = 0$ ist und somit gemäß Gl. (16)

$$\delta_R(\Delta H^{\ddagger}) = \delta_R(\Delta H_\pi^{\ddagger}) \tag{62}$$

gilt, findet man schließlich für die relative Freie Aktivierungsenthalpie $\delta_R(\Delta G^{\ddagger})$ den Ausdruck

$$\delta_R(\Delta G^{\ddagger}) = \left(1 - \frac{T}{\varkappa}\right) \cdot \delta_R(\Delta H_\pi^{\ddagger}) \tag{63}$$

als alternative theoretische Basis des oben beschriebenen Rechenverfahrens, von welcher aus die gute Übereinstimmung zwischen berechneten und experimentellen Daten ebenfalls gut verstanden werden kann. Gl. (63) beinhaltet dann allerdings implicit eine Zerlegung der Aktivierungsentropie in zwei Beiträge gemäß

$$\Delta S^{\ddagger} = \Delta S_{(\pi)}^{\ddagger} + \Delta S_R^{\ddagger} \tag{64}$$

---

[6] Um einer Verwechslung mit dem Resonanzintegral β vorzubeugen, wird hier der Proportionalitätsfaktor durch $\varkappa$ und nicht wie in [21] durch β dargestellt.

deren Änderungen mit der Substitution entsprechend Gl. (60) den Änderungen von ΔH_π^+ bzw. ΔH_R^+ mit der Substitution proportional sein müßten

$$\delta_R(\Delta S_\pi^+) = \varkappa^{-1} \cdot \delta_R(\Delta H_\pi^+) ; \tag{60a}$$

$$\delta_R(\Delta S_R^+) = \varkappa^{-1} \cdot \delta_R(\Delta H_R^+) = 0 . \tag{60b}$$

Da aber — wie bereits oben vermerkt — unter den durch Gl. (13) ausgedrückten Voraussetzungen $\delta_R(\Delta H_R^+) = 0$ ist, folgt aus Gl. (60b) auch für $\delta_R(\Delta S_R^+) = 0$. Somit läßt sich aber ΔS_R^+ als die Summe der in den Gln. (24a) und (24b) angegebenen Gruppen von Beiträgen zu ΔS_{vib}^+ auffassen:

$$\Delta S_R^+ = (\Delta S_{vib}^+)_1 + (\Delta S_{vib}^+)_2 . \tag{65}$$

Die Restgröße ΔG_R^+ ist dann abweichend von Gl. (27) zu definieren durch

$$\Delta G_R^+ = \Delta H_R^+ - T \cdot \Delta S_R^+ . \tag{66}$$

Sie besitzt mit

$$\delta_R(\Delta G_R^+) = 0 \tag{67}$$

die für Gl. (63) als notwendig vorauszusetzenden Eigenschaften.

Welche der beiden Alternativen — Substitutionsunabhängigkeit von ΔS^+ unter den Voraussetzungen der Gl. (13) bzw. Gültigkeit der isokinetischen Beziehung — im untersuchten Realbeispiel tatsächlich die physikalische Basis für die Anwendung des beschriebenen Rechenverfahrens darstellt, ist Gegenstand noch laufender Untersuchungen. Jedenfalls geben die im Abschnitt 2 angestellten Überlegungen zur Substitutionsabhängigkeit von ΔH^+ und ΔS^+ einen neuen und interessanten Aspekt für eine molekular-theoretische Diskussion der isokinetischen Beziehung [21].

Abschließend sei noch darauf hingewiesen, daß das Reaktionsschema der Abb. 5 nur die wahrscheinlichsten Reaktionsweisen eines aus einem Diazoalkan und einem Olefin bestehenden Systems wiedergibt; Reaktionen, die in den als Substituenten R_1 bis R_6 aufgefaßten Molekülteilen stattfinden, sind nicht berücksichtigt. Als Beispiel für eine derartige Reaktion sei die aromatische Substitution des Diazo-cyclopentadiens beim Umsatz mit Benzylidenmeldrumsäure angeführt, bei welcher das folgende substituierte Diazo-cyclopentadien

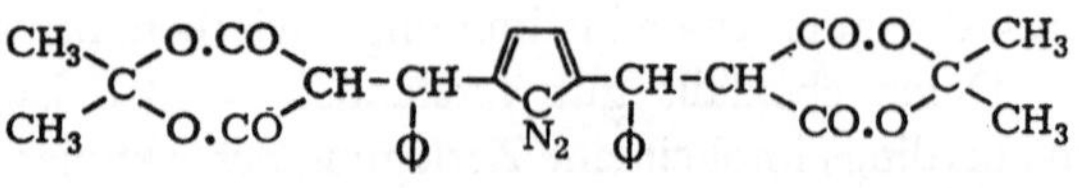

$$\begin{array}{c}
CH_3\diagdown\qquad\diagup O.CO\diagdown\qquad\qquad\qquad\qquad\diagup CO.O\diagdown\qquad\diagup CH_3\\
\qquad C\qquad\qquad\qquad CH-CH-\!\!\!\langle\;\rangle\!\!\!-CH-CH\qquad\qquad C\\
CH_3\diagup\qquad\diagdown O.CO\diagup\quad\;\;|\qquad\;|\quad\;|\qquad\;\diagdown CO.O\diagup\qquad\diagdown CH_3\\
\qquad\qquad\qquad\qquad\qquad\phi\quad N_2\quad\phi
\end{array}$$

gebildet wird (Ausbeute 16%) [13]. Fragte man mit Hilfe des oben beschriebenen Verfahrens der relativen Freien Aktivierungsenthalpien auf Grund des in Abb. 5 wiedergegebenen Reaktionsschemas nach den bei dieser Umsetzung zu erwartenden Produkten, könnte man keine richtige Vorhersage erhalten; eine derartige Fehlprognose wäre aber nicht dem genannten Rechenverfahren, sondern nur der Unvollständigkeit des benützten Reaktionsschemas (Abb. 5) anzulasten und sollte durch dessen Ergänzung behebbar sein.

Literatur

1. ADAMETZ, G., G. BILLEK, A. EITEL, O. E. POLANSKY, O. SAIKO, J. SWOBODA u. F. WESSELY: Mh. Chem. **94**, 334 (1963).

2. ALGUERO, M., J. BOSCH, J. CASTAÑER, J. CASTELLÁ, J. CASTELS, R. MESTRES, J. PASCUAL u. F. SERRATOSA: Tetrahedron **18**, 1381 (1962).

3. AUWERS, K. v., u. E. CAUER: Ann. **470**, 284 (1929).

4. —, u. F. KÖNIG: Ann. **496**, 27 (1932).

5. BALTAZZI, E.: Quart. Rev. **9**, 162 (1955).

6. BALTZLY, R., N. B. MEHTA, P. B. RUSSEL, R. E. BROOKS, E. M. GRIVSKY u. A. M. STEINBERG: J. Org. Chem. **26**, 3669 (1961).

7. BASTUS, J.: Tetrahedr. Lett. **15**, 955 (1963).

8. BOER, TH. J. DE, J. C. VAN VELZEN: Rec. trav. chim. **78**, 947 (1959).

9. BROWN, R. D.: Quart. Rev. **6**, 63 (1952).

10. BUCHNER, E., u. A. PAPENDIECK: Ann. **273**, 232 (1892).

11. —, u. H. DESSAUER: Ber. **26**, 259 (1893).

12. EITEL, A.: Diss. Univ. Wien, 1964.

13. —, u. F. WESSELY: Mh. Chem. **95**, 1382 (1964).

14. FRIED, J., u. R. ELDERFIELD: J. Org. Chem. **6**, 577 (1941).

15. GOTHIS, D., u. J. CLOKE: J. Amer. Chem. Soc. **56**, 2710 (1934).

16. GURVICH, A., u. S. TERENTEW: Sbornik Statei Obsh. Khim. Akad. Nauk SSSR **1**, 409 (1953); zitiert: Chem. Abstr. **49**, 1047 (1955).

17. HEDGE, J., C. CRUSE u. H. SNYDER: J. Org. Chem. **26**, 992 (1961).

18. HORNER, L., u. E. LINGAU: Ann. **591**, 21 (1954).

19. HUISGEN, R.: Angew. **67**, 439 (1955).

20. —, H. STANGL, H. STURM u. H. WAGENHOFER: Angew. **73**, 170 (1961).

21. LEFFLER, J. E.: J. Org. Chem. **20**, 1202 (1955).

22. MUSTAFA, A., and M. KAMAL-HILMY: J. Chem. Soc. [London] **1952**, 1434.

23. —, u. A. H. E. HARHASH: J. Amer. Chem. Soc. **78**, 1649 (1956).

24. OLIVERI-MANDALA, E.: Gazz. chim. Ital. **40** (I), 117 (1910).

25. PECHMANN, H. v., u. E. SEEL: Ber. **32**, 2295 (1899).

26. —, u. E. BURKHARD: Ber. **33**, 3590 (1900).

27. — —: Ber. **33**, 3595 (1900).

28. POLANSKY, O. E., u. P. SCHUSTER: Mh. Chem. **95**, 281 (1964).

29. — —: Mh. Chem. **96**, 396 (1965).

30. — —: Mh. Chem., im Druck.

31. —: (in Vorbereitung).

32. POPP, F., u. A. CATALA: J. Org. Chem. **26**, 2738 (1961).

33. RADULESCU, D.: Bull. Soc. Stiinte Cluj. **3**, 129 (1926); zitiert: Chem. Zentrbl. **1927 I**, 1453.

34. SCHÖNBERG, A., A. MUSTAFA u. M. HILMY: J. Chem. Soc. [London], **1947**, 1045.

35. SCHUSTER, P., O. E. POLANSKY u. F. WESSELY: Mh. Chem. **95**, 53 (1964).
36. —: Diss. Univ. Wien, 1966.
37. SCRIBNER, R., G. SAUSEN u. W. PRICHARD: J. Org. Chem. **25**, 1440 (1960).
38. STAUDINGER, H., u. A. GAULE: Ber. **49**, 1958 (1916).
39. SWOBODA, G. (geb. ADAMETZ), A. EITEL, J. SWOBODA u. F. WESSELY: Mh. Chem. **95**, 1355 (1964).
40. WESSELY, F., u. A. EITEL: Mh. Chem. **95**, 1577 (1964).
41. WIELAND, H., u. O. PROBST: Ann. **530**, 274 (1937).

Oskar E. Polansky
Peter Schuster

Institut für Theoretische
Chemie der Universität Wien
und Max-Planck-Institut
für Kohleforschung,
Mülheim (Ruhr)

Reaktionen in Lösungen unter erhöhten statischen Drucken

H. Heydtmann

Mit 5 Abbildungen

1. Einleitung

Die theoretische Behandlung von chemischen Reaktionen in Lösungen kann nach der Stoßtheorie oder nach der Theorie des Übergangszustandes erfolgen. Da der Druck eine Zustandsvariable der klassischen Thermodynamik ist, liegt es nahe, den Einfluß hoher Drucke auf Lösungsreaktionen nach der Theorie des Übergangszustandes zu behandeln. In dieser Theorie nimmt man an, daß ein Gleichgewicht zwischen den Ausgangsstoffen und dem Übergangszustand existiert, also z. B. für eine bimolekulare Reaktion

$$A + B \rightleftharpoons X^{\pm} \rightarrow \text{Produkte,} \tag{1}$$

und das Gleichgewicht kann mit den Gleichungen der Thermodynamik beschrieben werden.

Für ein chemisches Gleichgewicht in einer Lösung, $A + B \rightleftharpoons C + D$, wird die Gleichgewichtskonstante in Molenbrüchen durch folgenden Ausdruck wiedergegeben:

$$K_x = \frac{x_C \gamma_C \cdot x_D \gamma_D}{x_A \gamma_A \cdot x_B \gamma_B} = \prod_i (x_i \gamma_i)^{\nu_i} \, , \tag{2}$$

$x_i \gamma_i$ sind die Gleichgewichtsaktivitäten und K_x ist von der Temperatur, vom Druck und auch vom Lösungsmittel abhängig; als hypothetischer Standardzustand wird der reine gelöste Stoff mit den Eigenschaften einer ideal verdünnten Lösung gewählt. Mit den chemischen Standardpotentialen μ_{iS} wird K_x ausgedrückt durch

$$\ln K_x = -(RT)^{-1} \sum \nu_i \mu_{iS}, \quad \text{da ja}$$
$$\mu_i = \mu_{iS} + RT \ln x_i \gamma_i \quad \text{und} \quad \sum \nu_i \mu_i = 0 \, . \tag{3}$$

Nun ist weiter

$$\left(\frac{\partial \mu_{iS}}{\partial P} \right)_{T, LM} = \overline{V}_i \tag{4}$$

das partielle Molvolumen der i-ten Komponente; also ergibt sich

$$RT \left(\frac{\partial \ln K_x}{\partial P} \right)_{T, LM} = (\overline{V}_A + \overline{V}_B) - (\overline{V}_C + \overline{V}_D) = -\sum \nu_i \overline{V}_i . \tag{5}$$

Benutzt man statt der Molenbrüche x_i die Konzentrationen c_i, so gilt bei großen Verdünnungen $c_i = x_i/V_{LM}$ und

$$RT \left(\frac{\partial \ln K_c}{\partial P} \right)_{T,\,LM} = - \sum \nu_i \, \overline{V}_i + \left(\sum \nu_i \right) RT \cdot \beta \qquad (6)$$

mit der isothermen Kompressibilität des reinen Lösungsmittels $\beta = - \partial \ln V_{LM}/\partial P$. Die Gleichungen (5) und (6) sagen aus, was nach dem Prinzip von LE CHATELIER qualitativ zu erwarten ist, daß nämlich K mit dem Druck dann zunimmt (abnimmt), wenn das Volumen der Lösung zu Beginn der Reaktion größer (kleiner) ist als nach Abschluß der Reaktion.

Die erste Anwendung dieser Beziehung auf die Kinetik stammt von VAN'T HOFF (1898) [1], jedoch geben erst EVANS und POLANYI (1935) [2] die Gleichung in ihrer heute noch benutzten Form an. Die Theorie des Übergangszustandes gibt für die Geschwindigkeitskonstante der Reaktion (1) folgenden Ausdruck

$$k_x = - \frac{d\,x_A}{dt} \cdot \frac{1}{x_A \cdot x_B} = \varkappa \, \frac{kT}{h} \, K_{\rm x}^{\ast} \, \frac{\gamma_A \gamma_B}{\gamma_X} \, , \qquad (7)$$

$K_x^{\ast}$ ist eine quasi-Gleichgewichtskonstante für das Gleichgewicht zwischen den stabilen Molekülen der Reaktanten und dem hypothetischen Übergangszustand[1]. Für genügend verdünnte Lösungen ist $\gamma_i = 1$ und $\partial \ln \gamma_i/\partial P = 0$. Außerdem wird angenommen $\partial \ln \varkappa/\partial P = 0$. Die Differentiation nach dem Druck ergibt

$$\left(\frac{\partial \ln k_x}{\partial P} \right)_{T,\,LM} = - \varDelta \, \overline{V}^{\ast}/RT = \left(\frac{\partial \ln k_c}{\partial P} \right)_{T,\,LM} + (n-1)\beta \, , \qquad (8)$$

n ist die Molekularität der Reaktion, $\varDelta \overline{V}^{\ast}$ wird das Aktivierungsvolumen genannt. $\varDelta \overline{V}^{\ast} = \overline{V}_X - (\overline{V}_A + \overline{V}_B)$ bei einer bimolekularen Reaktion. Es ist zu beachten, daß eine solche, einfache Interpretation von $-RT \times \times (\partial \ln k_x/\partial P)_{T,\,LM}$ nur unter der oft unbewiesenen Voraussetzung möglich ist, daß die Geschwindigkeit der Umsetzung gleich der Geschwindigkeit eines Elementarprozesses ist. Bei Benutzung eines Konzentrationsmaßes zur Beschreibung von k bedeutet der Summand $(n-1)\beta$ meist nur eine geringfügige Korrektur. Durch Bestimmung der Anstiege von Tangenten an eine Kurve im $\ln k_x - P$-Diagramm kann man die $\varDelta \overline{V}^{\ast}$ bei verschiedenen Drucken erhalten. Zwei vor kurzem erschienene Arbeiten beschäftigen sich mit der analytischen Gestalt der Funktionen $\ln k_x (P)$ und zeigen, daß die Differentiation dieser Funktion nach dem Druck ein besseres Verfahren zur Bestimmung der Aktivierungsvolumina ist als die graphische Methode [3, 4].

---

[1] In der Gleichung für k_x nach der Theorie des Übergangszustandes treten keine Aktivitäten, sondern nur Molenbrüche auf; eine thermodynamische Gleichgewichtskonstante enthält jedoch die Aktivitäten und deshalb ist eine Multiplikation mit Aktivitätskoeffizienten in (7) notwendig.

Eine Änderung der Reaktionsgeschwindigkeit um etwa den Faktor zwei ist erst bei Drucken von 1000 at zu erwarten, wie man leicht aus Gleichung (8) errechnen kann mit der vernünftigen Annahme, daß $\Delta \overline{V}^+$ von der Größenordnung des Molvolumens einer chemischen Verbindung mit wenigen Atomen sein wird (20 bis 50 cm³ mol⁻¹). Der übliche Arbeitsbereich liegt zwischen 1 und 10000 at. Dem Experimentator, der an Reaktionen in Lösungen interessiert ist, sind außer durch den erforderlichen experimentellen Aufwand Grenzen durch die Kristallisation des Lösungsmittels gesetzt. So kristallisiert Benzol z. B. bei 33,4 °C unter einem Druck von 1000 at. Andererseits sind auch Untersuchungen an Lösungen bei Temperaturen möglich, die oberhalb des Siedepunktes unter Normalbedingungen liegen. Abb. 1 zeigt vier Beispiele für den Verlauf von $\log k_P/k_1$ in Abhängigkeit vom Druck.

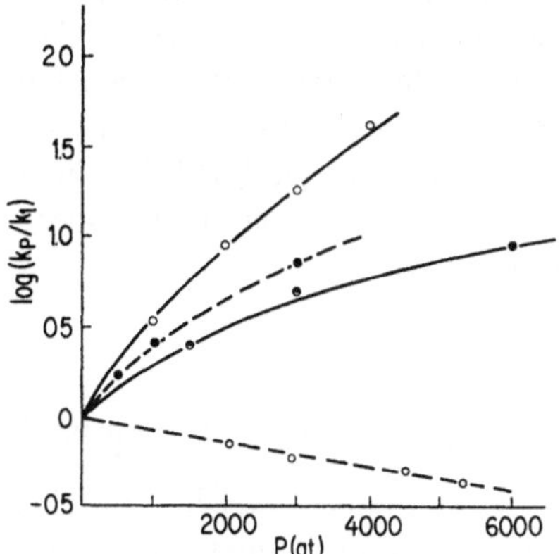

Abb. 1. Druckabhängigkeit für die Geschwindigkeit von vier Reaktionen

o — o — o   Dimerisierung von Cyclopentadien bei 20 °C
RAISTRICK, SAPIRO, NEWITT, 1939 [*11*]

● — — ● — — ●   Äthyljodid + Pyridin → Äthylpyridiniumjodid (in Aceton)
GONIKBERG, ZHULIN, 1958 [*12*]

◑ — ◑ — ◑   Solvolyse von tert.-Butylchlorid in Äthanol/Wasser (20 Volumenprozente Wasser), DAVID, HAMANN, 1954 [*13*]

o — — o — — o   Radikalzerfall von Di-tert.-Butylperoxyd in Toluol
WALLING, METZGER, 1959 [*14*]

Wenn die partiellen Molvolumina der Reaktionspartner A und B sowie deren Druckabhängigkeit aus Dichtemessungen bekannt sind, erhält man das partielle Molvolumen und die partielle molare Kompressibilität des

Übergangszustandes[2]. Im allgemeinen wird die Kompressibilität des Übergangszustandes nicht genau die gleiche sein wie die Kompressibilität der Ausgangsstoffe; dann ist $\Delta \overline{V}^{+}$ eine Funktion des Druckes. In den meisten zur Zeit veröffentlichten Arbeiten wird $\Delta \overline{V}^{+}$ nur für 1 at angegeben. Die Bedeutung und die Interpretation der Druckabhängigkeit des Aktivierungsvolumens wird in einigen neueren Arbeiten diskutiert [3, 5, 6].

Hamann (1958) [7] hat bei einigen zunächst druckbeschleunigten Reaktionen bei sehr hohen Drucken ($> 10\,000$ at) eine Verlangsamung gefunden. Es handelte sich dabei um die Bildung von Äthern beispielsweise in dem Lösungsmittel Isopropanol:

$$C_2H_5Br + RO^- \longrightarrow C_2H_5OR + Br^-$$

Die Viskosität des Isopropylalkohols steigt mit dem Druck so stark an, daß schließlich die Diffusion der Reaktionspartner zueinander zum geschwindigkeitsbestimmenden Schritt wird. Die Viskosität des Mediums spielt vor allem bei schnellen Reaktionen, z. B. bei der Fluoreszenzlöschung, eine entscheidende Rolle. Bei solchen Prozessen bewirkt eine Erhöhung des Druckes eine Erhöhung der Viskosität, und dies hat eine verminderte Zahl von Begegnungen zur Folge [8, 9, 10]. Wir wollen uns im folgenden mit den Fällen beschäftigen, in denen Diffusionserscheinungen die Geschwindigkeit nicht beeinflussen.

2. Deutung des Aktivierungsvolumens

Es soll nun die Frage behandelt werden, ob eine qualitative oder gar quantitative Deutung der Aktivierungsvolumina chemischer Reaktionen heute bereits möglich ist. Es ist üblich — seit der Arbeit von Evans und Polanyi [2] —, das Aktivierungsvolumen als eine Summe von zwei Größen anzusehen

$$\Delta \overline{V}^{+} = \Delta \overline{V}_1^{+} + \Delta \overline{V}_2^{+} \,. \tag{9}$$

Unter ΔV_1^{+} versteht man den Anteil, der von den Reaktionspartnern herrührt, und unter ΔV_2^{+} den Anteil, der auf die Veränderungen im Lösungsmittel beim Übergang von den gelösten Ausgangsmolekülen zum Übergangszustand zurückzuführen ist. Gleichung (9) hat den Charakter einer Arbeitshypothese. Wie die folgende Diskussion zeigt, ist eine strenge Trennung in einen Reaktantenanteil und einen Lösungsmittelanteil nicht durchführbar.

---

[2] Es ist zu betonen, daß eine Kompressibilität des Übergangszustandes nur in Analogie zu derjenigen gelöster Moleküle gemeint ist, und daß insbesondere die Bewegung im Freiheitsgrad der inneren Translation als druckunabhängig angesehen wird.

2.1. Der Einfluß der Änderung von Abständen zwischen im Anfangs- oder Endzustand gebundenen Atomen

In $\Delta \overline{V}_1^{\ddagger}$ ist zunächst der Volumenanteil enthalten, der auf die Knüpfung oder Lösung von Bindungen zurückgeht; er läßt sich nur schwer und mit völlig unzureichender Genauigkeit theoretisch ermitteln. HAMANN [15] hat eine Zusammenstellung veröffentlicht, in der solche theoretischen $\Delta \overline{V}_1^{\ddagger}$-Werte angegeben sind. Es wurden berechnete Potentialflächen für Reaktionen zwischen einfachen Molekülen zur Berechnung der $\Delta \overline{V}_1^{\ddagger}$-Werte herangezogen. Aus den Atomabständen bei der Konfiguration des Sattelpunktes ist zu entnehmen, daß bei der homolytischen Spaltung kovalenter Bindungen eine Dehnung dl von etwa 35% gegenüber der ursprünglichen Länge zu erwarten ist. Nimmt man an, daß eine solche Dehnung einer Bindung $A-B$ das Molekülvolumen um ein zylindrisches Stück vergrößert, so kann man für diesen Beitrag näherungsweise schreiben

$$\Delta V_i = \frac{\pi\,(r_A^2 + r_B^2)}{2}\,dl \qquad (10)$$

(r_A und r_B sind die van der Waalsschen Radien der Atome A bzw. B).

Ebenso läßt sich $\Delta \overline{V}_i$ für die Knüpfung einer neuen kovalenten Bindung berechnen. Es ergibt sich in diesem Fall ein negatives Aktivierungsvolumen von allgemein größerem Absolutbetrag. Beispiel: $N_2O \rightarrow N_2 + O$, $\Delta \overline{V}_i^{\ddagger} = 2{,}2$ cm³/mol; $N_2 + O \rightarrow N_2O$, $\Delta \overline{V}_i^{\ddagger} = -4{,}8$ cm³/mol. Solche Berechnungsergebnisse sind mit den Unzulänglichkeiten der Berechnung von Potentialflächen und den Unsicherheiten bei der Größenangabe von atomaren Wirkungsradien behaftet. Bei Reaktionen, die ohne sterische Hinderung ablaufen und die elektrisch neutrale Anfangs- und Endprodukte haben, wird oft angenommen, daß $\Delta \overline{V}^{\ddagger}$ auf die beschriebene Weise erklärt werden kann. Tabelle 1 zeigt hierfür einige Beispiele. Eine erhebliche Änderung des Volumens der Solvathülle ist jedoch immer dann zu erwarten, wenn Ladungen auftreten oder verschwinden, also starke elektrostatische Kräfte wirksam werden oder entfallen.

2.2. Der Einfluß sterischer Behinderung in der Umgebung des Reaktionszentrums

In $\Delta \overline{V}_1^{\ddagger}$ wird noch ein weiterer Anteil diskutiert. Die Beobachtung, daß sterisch behindernde Gruppen in Nachbarschaft der reagierenden Atome von Reaktionspartnern bimolekularer Reaktionen eine besonders große Druckbeschleunigung mit sich bringen, wurde vor allem von GONIKBERG und seinen Mitarbeitern hervorgehoben [18, 19, 20].

Es handelt sich bei den Beispielen der Tabelle 2 um Menschutkin-Reaktionen, bei denen aus einem Amin und einem Alkyljodid ein quartäres

Tabelle 1. *Aktivierungsvolumina einiger Reaktionen, die ohne starke Veränderung der Ladungsverteilung ablaufen*

Reaktion	LM	T (°C)	$\Delta \overline{V}^{\ddagger}$ (cm³/mol) für 1 at
Dissoziationen			
1. $t\text{-Bu}-O-O-t\text{-Bu} \rightarrow 2\,t\text{-Bu}-O\cdot$	Toluol	120	$+\ 5{,}4 \pm 0{,}6$
2. $t\text{-Bu}-O-O-t\text{-Bu} \rightarrow 2\,t\text{-Bu}-O\cdot$	Benzol	120	$+\ 12{,}6 \pm 1{,}3$
Walling und Metzger, 1959 [14][a]			
3. $C_6H_5-\overset{\overset{\displaystyle O}{\|}}{C}-N_3 \longrightarrow C_6H_5-\overset{\overset{\displaystyle O}{\|}}{C}-N: +\ N_2$	Ligroin	64	$+\ 5\ \ \pm 1$
4. $C_6H_5-\overset{\overset{\displaystyle O}{\|}}{C}-N_3 \longrightarrow C_6H_5-\overset{\overset{\displaystyle O}{\|}}{C}-N: +\ N_2$	Äthanol/ Wasser	50	$+\ 2\ \ \pm 1$
Brower, 1961 [16][b]			
Dimerisierung			
5. (Cyclopentadien-Dimerisierung)	Cyclopentadien	20	$-31{,}0$
Raistrick, Sapiro, Newitt, 1939 [11][c]			
Austauschreaktion			
6. $n\text{-Bu}-SH\ +\ (C_6H_5)_2N{-}N\cdot{-}C_6H_2(NO_2)_3 \longrightarrow$ $n\text{-Bu}-S\cdot +\ \text{Diphenylpikrylhydrazin}$	Toluol	35−75	$-14{,}5$
Ewald, 1959 [17][d]			

[a] Bei der Zersetzung des Di-t-Butylperoxyds ist das Aktivierungsvolumen zwar wie zu erwarten positiv, doch überrascht der Lösungsmitteleffekt. Die Autoren kommen zu dem Schluß, daß nur in Toluol die Volumenveränderung der Elementarreaktion ROOR → 2RO· gemessen wird. Dies ergibt sich aus einer Berücksichtigung der Einzelschritte der Reaktion: Bildung der Radikale im *LM*-Käfig gefolgt von entweder Rekombination, Diffusion oder Reaktion mit dem Lösungsmittel. Die Reaktion mit Toluol ist viel schneller als mit Benzol.

[b] Bei der Zersetzung von Benzazid wird im geschwindigkeitsbestimmenden Schritt eine Bindung gelöst; die Umlagerung zum Isocyanid wird als Folgeschritt diskutiert.

[c] Aus dem Wert des Aktivierungsvolumens für die Dimerisierung von Cyclopentadien wurde geschlossen, daß zwei Bindungen zur gleichen Zeit geknüpft werden.

[d] Bei der Reaktion des Diphenylpikrylhydrazylradikals mit Butylmercaptan wird eine Bindung geknüpft und eine andere Bindung gelöst. In Übereinstimmung mit der oben geschilderten Vorstellung hat die neu geknüpfte Bindung den entscheidenden Einfluß, und das Aktivierungsvolumen wird negativ.

Tabelle 2. *Aktivierungsvolumina bei 1 at für einige Menschutkin-Reaktionen mit und ohne sterische Hinderung nach* GONIKBERG *und* ÉL'YANOV, *1961* [*19*]

Reaktionspartner (*LM* ist in jedem Fall Nitrobenzol)	$\Delta \overline{V}^{\ddagger}_{\text{exp.}}$	$(\Delta \overline{V}^{\ddagger} - \Delta \overline{V}^{\ddagger}_G)_{\text{exp.}}$	$(\Delta \overline{V}^{\ddagger} - \Delta \overline{V}^{\ddagger}_G)_{\text{ber.}}$
		cm³ pro Mol	
Pyridin + Äthyljodid* $\Delta \overline{V}^{\ddagger}_G = -23{,}8$		0	0
4-Methylpyridin + Äthyljodid	$-24{,}3$	$-0{,}5$	0
2,6-Dimethylpyridin + Äthyljodid	$-27{,}2$	$-3{,}4$	$-5{,}7$
2,6-Dimethylpyridin + Isopropyljodid	$-30{,}5$	$-6{,}7$	$-11{,}5$

* Modellrechnung ergab $\Delta \overline{V}^{\ddagger}_1 = -14$ cm³/mol; $\Delta \overline{V}^{\ddagger}_{\text{exp.}}$ dieser Reaktion ist als Bezugsgröße $\Delta \overline{V}^{\ddagger}_G$ definiert.

Ammoniumsalz gebildet wird. Der Übergangszustand hat ein starkes Dipolmoment. Deshalb ist hier der Lösungsmittelanteil $\Delta \overline{V}^{\ddagger}_2$ sicher nicht vernachlässigbar, doch ist er speziell im Lösungsmittel Nitrobenzol klein im Vergleich zu anderen Lösungsmitteln (vgl. Tab. 3). Es wird angenommen, daß $\Delta \overline{V}^{\ddagger}_2$ bei allen Reaktionen der Tab. 2 gleich groß ist und bei der Differenzbildung (Spalte 3) deswegen fortfällt. Die Werte der letzten Spalte sind mit Hilfe starrer Molekülmodelle berechnet worden (Annahme bestimmter Atomabstände, Winkel, atomarer Wirkungsradien; Modell des Übergangszustandes wie am Fuß der Tabelle 2 skizziert). Die mit sterischer Hinderung zunehmende Volumenabnahme kommt nach der Auffassung von GONIKBERG und seinen Mitarbeitern durch eine Verminderung des Abstandes zwischen nicht miteinander verbundenen Atomen — also formal durch Überlappung der van der Waalsschen Wirkungsradien solcher Atome — zustande. HARRIS und WEALE [*21*] haben diese Interpretation kritisiert und erklärten die erhöhte Beschleunigung durch den Druck bei zunehmender sterischer Hinderung mit einem Abschirmeffekt, ausgeübt durch die in Nachbarschaft des Reaktionszentrums eingeführten Substituenten. Ein solcher Abschirmeffekt, der nur bei starker Elektrostriktion des Lösungsmittels durch die Reaktionspartner bzw. den Übergangszustand wirksam werden kann (s. unten), sollte aber gerade zu einer Verlangsamung dieser Reaktionen führen. Auf diese Weise lassen sich außerdem die Beobachtungen bei Radikalreaktionen nicht erklären, für die qualitativ das gleiche Verhalten — nämlich eine starke Beschleunigung sterisch behinderter Reaktionen durch den Druck — festgestellt werden konnte [22].

2.3. Der Einfluß der Solvatation

Wir wollen uns jetzt dem Lösungsmittelanteil $\Delta \overline{V}_2$ zuwenden. Es ist vorgeschlagen worden, auch in diesem Anteil wieder zwei Einflüsse zu unterscheiden. Man kann sich vorstellen, daß bei der Bildung von einem Mol Übergangszustand aus den Reaktanten sich das freie Volumen der Lösung ändert. Man versteht unter dem freien Volumen die Differenz zwischen dem makroskopischen Volumen und dem Volumen starrer Molekülmodelle mit hartem Abstoßungspotential. Auf die mögliche Existenz eines solchen Packungseffektes haben Gonikberg und Él'Yanov [23] hingewiesen. Ein spezieller Packungseffekt ist von Le Noble [24] durch Messung der Druckabhängigkeit des Keto-Enol-Gleichgewichts von Acetessigsäureäthylester nachgewiesen worden. In der ringförmigen Enol-Form dieser Verbindung ist offenbar das Ringinnere für Lösungsmittelmoleküle (als Lösungsmittel diente bei den Messungen die Keto-Form) schwer zugänglich. Die Gleichgewichtskonstante K (keto/enol) wächst mit dem Druck. Die Volumenänderung beträgt 1,5 cm³ mol⁻¹ bei 1 at.

Die elektrostriktive Wirkung geladener Teilchen auf die umgebende Lösung war bereits vor der Jahrhundertwende bekannt. Kohlrausch und Hallwachs (1894) [25] stellten fest, daß $ZnSO_4$ — obwohl es im festen Zustand ein Volumen von 45 cm³/mol einnimmt — in verdünnter wäßriger Lösung ein partielles Molvolumen von -11 cm³/mol hat. Man kann die elektrostriktive Wirkung eines kugelförmigen Teilchens mit dem Radius r und der Ladung $z\,e$ auf ein homogenes Medium, das nur durch seine Dielektrizitätskonstante D beschrieben wird, mit Hilfe der Bornschen Beziehung für die molare freie Enthalpie der Solvatation solcher Teilchen angeben [26]

$$\Delta \overline{G}_{\text{solv}} = -\frac{N z^2 e^2}{2r}(1 - 1/D) \qquad N = \text{Loschmidtsche Zahl} \qquad (11)$$

$$\frac{\partial \Delta \overline{G}_{\text{solv}}}{\partial P} = \Delta \overline{V}_{\text{solv}} = -\frac{N z^2 e^2}{2r D^2}\frac{\partial D}{\partial P} + \frac{N z^2 e^2 (1 - 1/D)}{2 \cdot r^2}\frac{\partial r}{\partial P} \qquad (12)$$

Mit Hilfe dieser Kontinuumstheorie ist es z. B. möglich, die partiellen Molvolumina gelöster Salze bei unendlicher Verdünnung in vielen Fällen befriedigend zu beschreiben [27, 28]. Es stellte sich heraus, daß der Kompressibilitätsterm in Gleichung (12) vernachlässigt werden kann.

Durch zahlreiche Arbeiten ist belegt, daß die Dissoziationsgleichgewichte schwacher Säuren und Basen durch hohe Drucke in Richtung auf die geladenen Teilchen verschoben werden. Auch die spezifische Leitfähigkeit des Wassers steigt mit dem Druck [29]. In Stoßwellenexperimenten wurde bei 770 °C und 300 000 at eine Leitfähigkeit von 10 Ohm^{-1} cm^{-1} gemessen [30].

Ein interessantes Beispiel für den Elektrostriktionseffekt in der Reaktionskinetik ist die neutrale Hydrolyse substituierter Benzoylchloride (vgl. Abb. 2) [4]. Diese Reaktionen werden im Bereich niedriger Wasser-

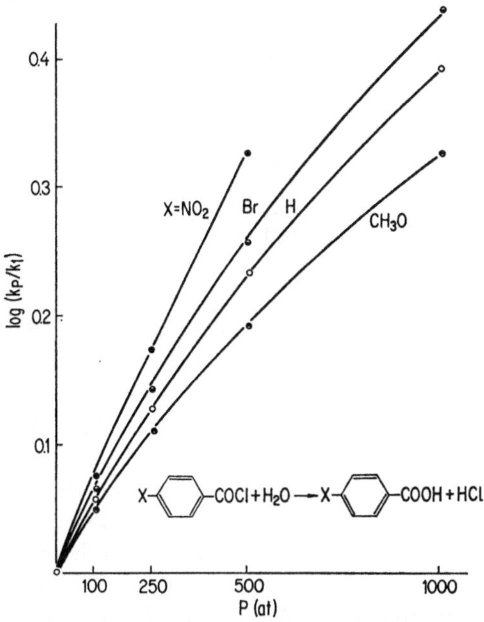

Abb. 2. Druckabhängigkeit der Hydrolysegeschwindigkeit von p-substituierten Benzoylchloriden in Tetrahydrofuran mit 2,2% Wasser [4]

konzentrationen in Lösungsmitteln mit niedriger DK für S_N2-Reaktionen gehalten [31]. Eine Druckbeschleunigung wird wegen der Bildung von Cl$^-$ und H$_3$O$^+$ erwartet. Aus der hohen Druckbeschleunigung, die man für den „elektronenanziehenden" Substituenten p-NO$_2$ gemessen hat ($\Delta \overline{V}^{\pm}_{P=1} = -43,0$ cm^3 mol^{-1}), und der demgegenüber geringeren Druckbeschleunigung, die man für den „elektronenliefernden" Substituenten p-OCH$_3$ ermittelt hat ($\Delta \overline{V}^{\pm}_{P=1} = -27,5$ cm^3 mol^{-1}) kann man auf einen unterschiedlichen Elektrostriktionseffekt, ausgeübt von einem polarisierten Übergangszustand, schließen. Man erkennt, daß die Substituenten nicht nur die Elektronendichte am Carbonylkohlenstoff der stabilen Verbindung bestimmen, sondern daß sie auch einen starken Einfluß auf das Dipolmoment des Übergangszustandes besitzen.

22*

Zu den Reaktionen, die unter Bildung von Ionen ablaufen, gehören auch die Menschutkin-Reaktionen. Ihre Geschwindigkeiten und Aktivierungsvolumina sind stark vom Lösungsmittel abhängig (Tabelle 3). Das Aktivierungsvolumen besitzt stets negative Werte; dies ist zu erwarten,

Tabelle 3. *Aktivierungsvolumina für die Reaktion von ω-Bromacetophenon mit α-Picolin bei 30 °C und Normaldruck* [34]

Lösungsmittel	DK	$10^3 \cdot k \cdot$ (l/mol · min)	$-\Delta \overline{V}^{\ddagger}$ (ml/mol)
Toluol	2,362	0,57	30,2
Brombenzol	5,289	2,7	26,3
Chlorbenzol	5,519	2,4	26,7
Methylenchlorid	8,654	5,0	27,8
Cyclohexanon	15,43	6,1	26,2
Aceton	20,23	6,3	30,2
Benzonitril	24,91	12	24,3
Propionitril	27,91	9,9	26,0
Nitrobenzol	33,90	14	21,1

denn sowohl $\Delta V_1^{\ddagger}$ als auch $\Delta V_2^{\ddagger}$ sollten negativ sein. Über die Größe der beiden Anteile besteht keine endgültige Klarheit. Menschutkin-Reaktionen haben außerdem immer eine negative Aktivierungsentropie, deren Absolutwert im allgemeinen mit steigendem Druck abnimmt. Dies wird so gedeutet, daß der Ordnungsgrad der Solvathüllen von Ausgangsstoffen und Übergangszustand sich immer mehr einander angleichen. Die Aktivierungsenergie steigt nur geringfügig mit dem Druck und vermag die Zunahme von $\Delta S^{\ddagger}$ nicht zu kompensieren. Beispielsweise besitzen die genannten Aktivierungsgrößen für die Reaktion von Äthyljodid mit Pyridin in Aceton die folgenden Werte [18, 23]: $\Delta S^{\ddagger} = -34{,}4$ cal·mol^{-1}· grad^{-1} und $\Delta V^{\ddagger} = -34$ cm^3·mol^{-1} für 1 at; $\Delta S^{\ddagger} = -28{,}4$ cal·mol^{-1}· grad^{-1} und $\Delta V^{\ddagger} = -10$ cm^3·mol^{-1} für 3000 at.

Gonikberg und Él'Yanov [23] zeigten als erste, daß in der Regel eine lineare Beziehung zwischen dem Molvolumen des gewählten Lösungsmittels und $\Delta V^{\ddagger}$ bei verschiedenen Drucken besteht; dies macht den großen Einfluß von $\Delta V_2^{\ddagger}$ ebenfalls deutlich. Brower [32] hat ganz allgemein festgestellt, daß eine deutliche Abhängigkeit des Aktivierungsvolumens vom Lösungsmittel zu beobachten ist, wenn geladene Teilchen bei der Reaktion verschwinden oder neu gebildet werden. Aus seinen Überlegungen und auch aus den Untersuchungen von Stewart und Weale, die die Druckabhängigkeit zweier Rückreaktionen vom Menschutkin-Typ gemessen haben [33], geht hervor, daß der Übergangszustand der Menschutkin-Reaktionen eine hohe Polarität besitzt, daß er jedoch noch nicht eine vollständige Dissoziation zum Ionenpaar darstellt.

Angeregt durch die Erfolge, die HARTMANN, BRAUER, KELM und RINCK [37] mit der Anwendung einer Kontinuumstheorie für das Lösungsmittel bei der Interpretation von Aktivierungsvolumina hatten, wurde versucht, eine quantitative Aussage über die Polarität des Übergangszustandes für die Reaktion von ω-Bromacetophenon mit α-Picolin zu erhalten [34]. KIRKWOOD [35] hat einen der Bornschen Gleichung entsprechenden Ausdruck angegeben, der die Änderung der partiellen molaren freien Enthalpie für eine polare Substanz beim Übergang von einer verdünnten Lösung in einem Medium der DK $= 1$ zu einem solchen mit der DK $= D$ beschreibt. Für den Spezialfall, daß sich zwei punktförmige entgegengesetzte elektrische Ladungen in einer Kugel mit dem Radius r und der DK $= 1$ im Innern aufhalten, lautet die Gleichung:

$$\Delta \overline{G} = \overline{G} - \overline{G}_0 = - \frac{N\mu^2}{r^3} \cdot \frac{D-1}{2D+1} \; ; \tag{13}$$

dabei bedeutet μ das Dipolmoment. Durch Einsetzen in die Gleichung für die Geschwindigkeitskonstante nach der Theorie des Übergangszustandes erhält man [36]

$$\ln k_x = \ln k_{xo} + \frac{1}{kT} \left(\mu_X^2/r_X^3 - \mu_A^2/r_A^3 - \mu_B^2/r_B^3 \right) \frac{D-1}{2D+1} + \frac{\Sigma v_i \Phi_i}{RT} \, , \tag{14}$$

wobei mit dem letzten Term die sogenannten spezifischen Wechselwirkungen berücksichtigt worden sind. Durch Differentiation nach dem Druck ergibt sich bei Vernachlässigung einer möglichen Druckabhängigkeit von r:

$$\Delta \overline{V}^\ddagger = \Delta \overline{V}_1^\ddagger - N \left(\mu_X^2/r_X^3 - \mu_A^2/r_A^3 - \mu_B^2/r_B^3 \right) q_P + \left(\frac{\partial \Sigma v_i \Phi_i}{\partial P} \right)_{T, LM} ,$$

$$\text{mit} \quad q_P = \frac{\partial}{\partial P} \left(\frac{D-1}{2D+1} \right) = \frac{3}{(2D-1)^2} \left(\frac{\partial D}{\partial P} \right)_{T, LM} . \tag{15}$$

Wenn in einer Reihe von Lösungsmitteln für eine bestimmte Reaktion der letzte Term einen konstanten Wert besitzt, erwartet man in der Auftragung von $\Delta \overline{V}^\ddagger$ gegen q_P eine Gerade mit dem Anstieg $-N\Sigma v_i \mu_i^2/r_i^3$. Dies ist für die Umsetzung von ω-Bromacetophenon mit α-Picolin in den Lösungsmitteln Toluol, Methylenchlorid, Chlorbenzol, Brombenzol und Nitrobenzol der Fall. Außerhalb der Geraden liegen die Meßpunkte für die Ketone Aceton und Cyclohexanon sowie für Benzonitril und Propionitril als Lösungsmittel. Es wurden außerdem die partiellen Molvolumina in den verschiedenen Lösungsmitteln für das Amin und die Halogenkomponente bestimmt und daraus das partielle Molvolumen des Übergangszustandes $\overline{V}_X = \Delta \overline{V}^\ddagger + \overline{V}_A + \overline{V}_B$ berechnet, für das man schreiben muß

$$\overline{V}_X = \overline{V}_{X, 1} - N \mu_X^2/r_X^3 q_P + \left(\frac{\partial \Phi_X}{\partial P} \right)_{T, LM} . \tag{16}$$

In der Auftragung von $\overline{V}_i$ gegen q_P (Abb. 3 und 4) sieht man für ω-Bromacetophenon und für den Übergangszustand der besprochenen Reaktion deutlich den linearen Zusammenhang für die genannten fünf

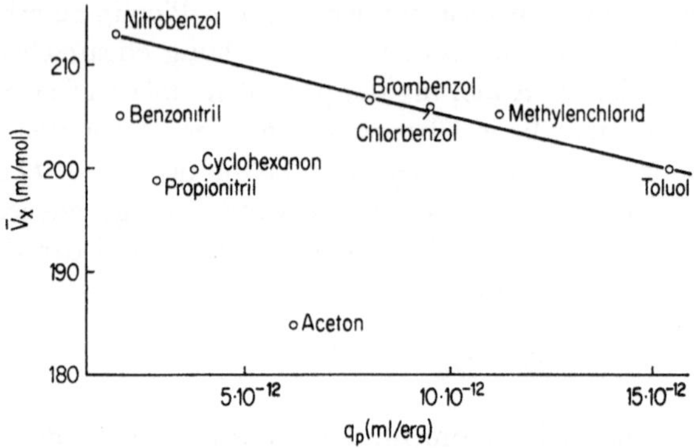

Abb. 3. Partielles Molvolumen des Übergangskomplexes der Reaktion von ω-Bromacetophenon mit α-Picolin aufgetragen gegen q_P

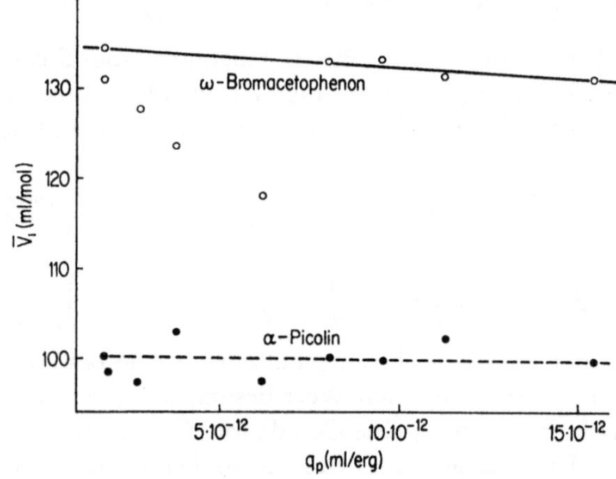

Abb. 4. Partielle Molvolumina aufgetragen gegen q_P; die Lösungsmittel sind die gleichen wie in Abb. 3 angegeben

Lösungsmittel. Mit Hilfe von Stuart-Brieglebschen Modellen für die Moleküle wurden Radien berechnet, und mit diesen Werten konnten aus den Anstiegen der Geraden nach Gleichung (16) ein Dipolmoment von $3,2 \pm 0,4$ Debye für ω-Bromacetophenon und ein solches von $7,8 \pm 0,5$ Debye für den Übergangszustand ermittelt werden. Für α-Picolin, das ein niedriges Dipolmoment besitzt, ist eine solche Aussage nicht möglich. Das in Benzol gemessene Dipolmoment von ω-Bromacetophenon beträgt

3,14 Debye in guter Übereinstimmung mit der von uns durchgeführten Bestimmung. Für die Übergangszustände einiger anderer Menschutkin-Reaktionen wurden von HARTMANN und Mitarbeitern nach einem ähnlichen Verfahren Dipolmomente von 7,7 bis 8,1 Debye ermittelt [37]. Da es nicht sinnvoll ist, für die Reaktionen in Ketonen und Nitrilen andere Übergangszustände mit sehr viel höheren Dipolmomenten anzunehmen, führten wir die in Abb. 3 und 4 sichtbaren Abweichungen auf spezifische Lösungsmitteleffekte zurück, die offenbar bei der Halogenkomponenten von derselben Art sind wie beim Komplex. Unter spezifischen Lösungsmitteleffekten sind Assoziationen, Wechselwirkungen kurzer Reichweite (durch Dispersionskräfte) und lokale Störungen der Lösungsmittelstruktur zu verstehen.

Gleichung (15) kann in der Form

$$\Delta \overline{V}^{\ddagger} = \Delta \overline{V}_1^{\ddagger} + \Delta \overline{V}_2^{\ddagger} + \Delta \overline{V}_3^{\ddagger} \tag{17}$$

geschrieben werden und stellt eine Erweiterung der Gleichung (9) dar. Die beiden letzten Terme beschreiben die Wechselwirkungen mit dem Lösungsmittel in der eben erläuterten Weise. Daraus folgt, daß nunmehr $\Delta \overline{V}_1^{\ddagger}$, das keine Mediumeffekte enthält, eine Differenz von Eigenvolumina darstellt und als praktisch inkompressibel angenommen werden muß, da die Kompressibilität der Anteile $\overline{V}_{i,1}$ nur noch von der Veränderung der innermolekularen Abstände herrühren kann. Abb. 5 zeigt, daß bei der Anwendung von Gleichung (15) auf Ergebnisse bei hohen Drucken die Geraden den gleichen Anstieg behalten wie bei $P = 1$ at.

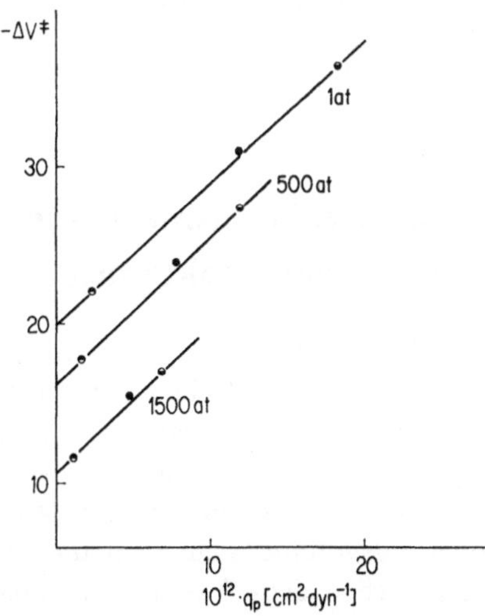

Abb. 5. Aktivierungsvolumina für die Reaktion von Pyridin mit Methyljodid bei drei Drucken aufgetragen gegen den *LM*-Parameter q_P [6]; ● Nitrobenzol, ● Chlorbenzol, ○ Benzol

Whalley [*38, 39, 40*] berücksichtigt bei der Differentiation von $\Delta \overline{G}$ nach dem Druck (Gl. (12)) eine Druckabhängigkeit von r. Man erhält in dem Ausdruck für das partielle Molvolumen eines Dipolmodells dann einen zusätzlichen Term $N \cdot q \, (\mu^2/r^3)(\partial \ln r/\partial P)$ [*6*]. Es ist nicht vorauszusehen, ob ein solcher Term zweckmäßig ist. Da in dem gewählten Modell der Radius r in besonderem Maße den Charakter einer Arbeitshypothese besitzt, ist jede weitere Annahme — wie z. B. diejenige einer Kompressibilität von r — eine Erweiterung des Modells, die der Nachprüfung bedarf. Auch die oben genannten Annahmen zur Bestimmung von r stellen eine solche Erweiterung dar. Die hier kurz zusammengefaßten Ergebnisse machen deutlich, daß ein solcher Kompressibilitätsterm unzweckmäßig ist. Das bedeutet nicht, daß die Wechselwirkung zwischen dem gelösten Ion oder Dipolmolekül und den unmittelbar benachbarten Lösungsmittelmolekülen keinerlei Auswirkung auf das partielle Molvolumen hat — eben diese Wechselwirkung bringt $\Delta \overline{V}_3^{\ddagger}$ zum Ausdruck — nur ist eine Proportionalität dieses Terms zu $q = (D-1)/(2D + 1)$ offenbar nicht vorhanden. Die starke Druckabhängigkeit von $\Delta \overline{V}_3^{\ddagger} = \partial \, \Sigma \, v_i \, \Phi_i/\partial P$ ist aus der Zunahme von $\Delta \overline{V}^{\ddagger}$ mit dem Druck für $q_P = 0$ zu sehen (Abb. 5).

Zusammenfassend kann gesagt werden, daß von einem quantitativen Verständnis der Aktivierungsvolumina heute noch nicht gesprochen werden kann. Von besonderem Interesse sind die in den letzten Jahren zunehmenden Bemühungen, partielle Molvolumina gelöster Salze und Moleküle zu interpretieren. Von diesen Untersuchungen werden auch die Erklärungsversuche für die Druckabhängigkeit von Geschwindigkeitskonstanten profitieren.

3. Interpretation von Reaktionsmechanismen mit Hilfe der Druckabhängigkeit von Reaktionsgeschwindigkeiten

Das halbquantitative Verständnis der Aktivierungsvolumina hat zu Versuchen geführt, mit dieser neu entdeckten Größe Reaktionsmechanismen von Lösungsreaktionen besser zu verstehen. In einem Übersichtsartikel von Whalley [*39*] sind diese Bemühungen — soweit es sich um organische Reaktionen handelt — zusammengefaßt. Whalley sieht in der Bestimmung des Aktivierungsvolumens ein allen anderen Methoden überlegenes Verfahren, um zwischen dem A1- und A2-Mechanismus für die säurekatalysierten Solvolysen zu unterscheiden. Im ersten Fall zerfällt das protonierte Substrat unimolekular

$$RH^+ \; \dashrightarrow \; \text{Produkte} \, ,$$

und ein $\Delta \overline{V}^{\ddagger} \geq 0$ cm³ mol⁻¹ wird erwartet; im zweiten Fall ist ein Lösungs-
mittelmolekül an der Reaktion beteiligt

$$\text{LM} + \text{RH}^+ \longrightarrow \text{Produkte} ,$$

und ein $\Delta \overline{V}^{\ddagger} \leq -5$ cm³ mol⁻¹ wird vorausgesagt. Allerdings wird darauf
hingewiesen, daß in jedem einzelnen Fall noch eine Reihe zusätzlicher Informationen berücksichtigt werden muß, bevor man den Reaktionsmechanismus als einigermaßen gesichert ansehen kann.

Aus der anorganischen Chemie soll eine Untersuchung von CANDLIN
und HALPERN [41] erwähnt werden. Die Autoren hoffen über die Druckabhängigkeit der Reduktion von Co(III)-Komplexen mit Fe(II) ein Kriterium für die Unterscheidung zwischen innersphärischen und außen-
sphärischen Übergangszuständen zu erhalten.

Schließlich scheint die Größe $\Delta \overline{V}^{\ddagger}$ noch eine weitere Bedeutung für
die Reaktionskinetik zu gewinnen. Sie kann nämlich dazu benutzt werden,
Aktivierungsgrößen bei konstantem Volumen, die der direkten Messung
schwer zugänglich sind, zu bestimmen [39]. Aus der Aktivierungsenthalpie
$\Delta \overline{H}_P^{\ddagger}$ und der Aktivierungsentropie bei konstantem Druck $\Delta \overline{S}_P^{\ddagger}$, die in der
üblichen Weise aus den Experimenten bei Normaldruck erhalten werden,
kann man mit den Gleichungen

$$\Delta \overline{U}_V^{\ddagger} = \Delta \overline{H}_P^{\ddagger} - T\alpha\Delta \overline{V}^{\ddagger}/\beta , \tag{18}$$

$$T\Delta \overline{S}_V^{\ddagger} = \Delta \overline{S}_P^{\ddagger} - T\alpha\Delta \overline{V}^{\ddagger}/\beta$$

die innere Energie der Aktivierung und die Aktivierungsentropie bei
konstantem Volumen für die betreffende Reaktion berechnen, wenn
neben dem Aktivierungsvolumen noch der thermische isobare Aus-
dehnungskoeffizient α und die isotherme Kompressibilität der Lösung β
bekannt sind.

Es hat sich gezeigt, daß $\Delta \overline{U}_V^{\ddagger}$ und $T\Delta \overline{S}_V^{\ddagger}$ im Fall der säurekatalysierten
Hydrolyse von Methylacetat in Aceton-Wasser-Gemischen mit der Vari-
ation der Lösungsmittelzusammensetzung einen ganz anderen Verlauf
nehmen als $\Delta \overline{H}_P^{\ddagger}$ und $T\Delta \overline{S}_P^{\ddagger}$ [42]. Es ist daher in Frage gestellt, ob eine
mechanistische Interpretation von Minima bei den Größen $\Delta \overline{H}_P^{\ddagger}$ und
$\Delta \overline{S}_P^{\ddagger}$ mit wechselnder Lösungsmittelzusammensetzung sinnvoll ist. Zweifellos ist eine weitere Aufklärung der Strukturen der in der Kinetik besonders
häufig verwendeten wäßrigen und alkoholischen Lösungsmittelgemische
und ihre Beeinflußbarkeit durch Temperatur, Druck und gelöste Stoffe
zum besseren Verständnis der in diesen Lösungsmitteln ablaufenden
kinetischen Vorgänge erforderlich.

Literatur[3]

1. VAN'T HOFF, J. H.: Vorlesungen über theoretische und physikalische Chemie, 1. Heft, 238, Braunschweig 1898.
2. EVANS, M. G., and M. POLANYI: Trans. Faraday Soc. **31**, 875 (1935).
3. GOLIKIN, H. S., W. G. LAIDLAW u. J. B. HYNE: Can. J. Chem. **44**, 2193 (1966).
4. HEYDTMANN, H., u. H. STIEGER: Ber. Bunsenges. **70**, 1095 (1966).
5. BENSON, S. W., u. J. A. BERSON: J. Am. Chem. Soc. **84**, 152 (1962); **86**, 259 (1964); WILLIAMS, G.: Trans. Faraday Soc. **60**, 1548 (1964).
6. HEYDTMANN, H., H. D. BRAUER u. H. KELM: Z. physik. Chem. (Frankfurt) **54**, 237 (1967).
7. HAMANN, S. D.: Trans. Faraday Soc. **54**, 507 (1958).
8. EWALD, A. H.: J. Phys. Chem. **67**, 1727 (1963).
9. HAWORTH, D. W., u. W. S. METCALF: J. Chem. Soc. London **1965**, 4678.
10. LEIBER, C. O., D. REHM u. A. WELLER: Ber. Bunsenges. **70**, 1086 (1966).
11. RAISTRICK, B., R. H. SAPIRO u. D. M. NEWITT: J. Chem. Soc. London **1939**, 1761.
12. GONIKBERG, M. G., u. V. M. ZHULIN: Austr. J. Chem. **11**, 285 (1958).
13. DAVID, H. G., u. S. D. HAMANN: Trans. Faraday Soc. **50**, 1188, (1954).
14. WALLING, C., and G. METZGER: J. Am. Chem. Soc. **81**, 5365 (1959).
15. HAMANN, S. D.: Chemical Kinetics, in: High Pressure Physics and Chemistry, R. S. BRADLEY ed., London: Acad. Press 1963.
16. BROWER, K. R.: J. Am. Chem. Soc. **83**, 4370 (1961).
17. EWALD, A. H.: Trans. Faraday Soc. **55**, 792 (1959).
18. GONIKBERG, M. G.: Chemical Equilibria and Reaction Rates at High Pressures, Israel Progr. Scient. Transl., Jerusalem 1963.
19. GONIKBERG, M. G., u. B. S. ÉL'YANOV: Dokl. Akad. Nauk SSSR **138**, 1103 (1961).
20. —, V. M. ZHULIN u. B. S. ÉL'YANOV: Symposium Olympia, London 1962, Soc. Chem. Ind. 1963, 212.
21. HARRIS, A. P., u. K. E. WEALE: J. Chem. Soc. London **1961**, 146.
22. GONIKBERG, M. G., N. I. PROKHOROVA u. E. F. LITVIN: Izv. Akad. Nauk SSSR, Chem. Ser. **1962**, 1495.
23. —, u. B. S. ÉL'YANOV: Izv. Akad. Nauk SSSR, Chem. Ser. **1960**, 629.
24. LE NOBLE, W. J.: J. Am. Chem. Soc. **82**, 5253 (1960).
25. KOHLRAUSCH, F., u. W. HALLWACHS: Ann. Phys. Chem. **53**, 1 u. 14 (1894).
26. BORN, M.: Z. Physik **1**, 45 (1920).
27. MUKERJEE, P.: J. Phys. Chem. **65**, 740 (1961).
28. PADOVA, J.: J. Chem. Phys. **39**, 1552 (1963).
29. HOLZAPFEL, W., u. E. U. FRANCK: Ber. Bunsenges. **70**, 1105 (1966).
30. HAMANN, S. D.: Chemical Equilibria in Condensed Systems, in: High Pressure Physics and Chemistry, R. S. BRADLEY ed., London: Acad. Press 1963.
31. HUDSON, R. F.: Ber. Bunsenges. **68**, 215 (1964).
32. BROWER, K. R.: J. Am. Chem. Soc. **85**, 1401 (1963).

---

[3] Ein Verzeichnis der wichtigsten Monographien, die zu dem Thema „Materie unter Hohen Drucken" erschienen sind, gibt FRANCK, E. U.: Ber. Bunsenges. **70**, 944 (1966); über Kinetik unter hohen Drucken gibt es zwei neue Übersichtsreferate: E. WHALLEY, Adv. Physical Chem. **18**, 205 (1967), und W. J. Le NOBLE, Progr. Physical Org. Chem. **5**, 207 (1967); besonders hingewiesen sei auf das erste Buch, das sich ausschließlich diesem Thema widmet: K. E. WEALE, Chemical Reactions at High Pressures, London: SPON, 1967.

33. STEWART, J. M., u. K. E. WEALE: J. Chem. Soc. London **1965**, 2854.
34. HEYDTMANN, H., A. P. SCHMIDT u. H. HARTMANN: Ber. Bunsenges. **70**, 444 (1966).
35. KIRKWOOD, J. G.: J. Chem. Phys. **2**, 351 (1934).
36. LAIDLER, K. J., u. H. EYRING: Ann. New York Acad. Sci. **39**, 303 (1940); GLASSTONE, S., K. J. LAIDLER, u. H. EYRING: The Theory of Rate Processes, Chap. VIII. New York: McGraw-Hill Book C., 1941.
37. HARTMANN, H., H. KELM u. G. RINCK: Z. physik. Chem. (Frankfurt) **44**, 335 (1965); HARTMANN, H., H. D. BRAUER u. G. RINCK: Z. physik. Chem. (Frankfurt), im Druck; HARTMANN, H., H. D. BRAUER, H. KELM, u. G. RINCK: Z. physik. Chem. (Frankfurt), im Druck.
38. WHALLEY, E.: J. Chem. Phys. **38**, 1400 (1963).
39. WHALLEY, E.: Adv. Phys. Org. Chem. **2**, 93 (1964).
40. WHALLEY, E.: Ber. Bunsenges. **70**, 958 (1966).
41. CANDLIN, J. P. u. J. HALPERN: Inorg. Chem. **4**, 1086 (1965).
42. BALIGA, B. T., R. J. WITHEY, D. POULTON u. E. WHALLEY: Trans. Faraday Soc. **61**, 517 (1965).

Horst Heydtmann
Institut für physikalische Chemie
der Universität Frankfurt/Main

Electron Transfer Reactions

R. A. Marcus

With 2 Figures

1. Introduction

Electron transfer processes of chemical interest are of various types: certain homogeneous redox reactions, electron exchange reactions, electrode processes, solvated electron reactions, certain chemiluminescent processes, intramolecular electron transfers, and electron transfers between ions in solution and semiconductor electrodes. The distinction between redox reactions of the atom transfer type and those of the electron transfer type should be noted.

This contribution will discuss theoretical and experimental results in this field and will be based on several recent papers of the author (Ann. Rev. Phys. Chem. **15**, 155 (1964), J. Phys. Chem. **67**, 853, 2889 (1963), J. Chem. Phys. **43**, 679 (1965)).

In the present paper, we outline the assumptions, aspects of the derivation, and deductions of a theory of electrochemical and homogeneous reactions of the redox type. The theory has been developed with increase in generality in several recent papers [1]. The last of these [1e] encompasses the others and is the one summarized here. Because of the techniques devised [2] it became possible to present the theory of homogeneous reactions and of electrochemical ones in a single formulation, thereby emphasizing their similarity. We shall follow this procedure here. A variety of applications of the theory has already been described [3].

2. Potential Energy Surfaces and Mechanism of Electron Transfer

We consider an electron transfer between two "reacting" species. These species consist of two dissolved ions (molecules) in the homogeneous case and one such ion and electrode in the electrochemical one. Some insight into the mechanism of electron transfer is provided by examination of the behavior of the entire system on its potential energy surface [1c, 1d]. The surface for the system of reactants and surrounding medium is a function of all the translational, rotational, and vibrational coordinates in this system. Similarly, the surface for a system consisting of the products and the medium is a function of these coordinates also.

In the absence of electronic coupling between the orbitals of the two reacting species, the two potential energy surfaces intersect. It is then not possible for the system to go from one surface to the other, i.e., to undergo electron transfer (Fig. 1). A suitable coupling (Fig. 2) removes the de-

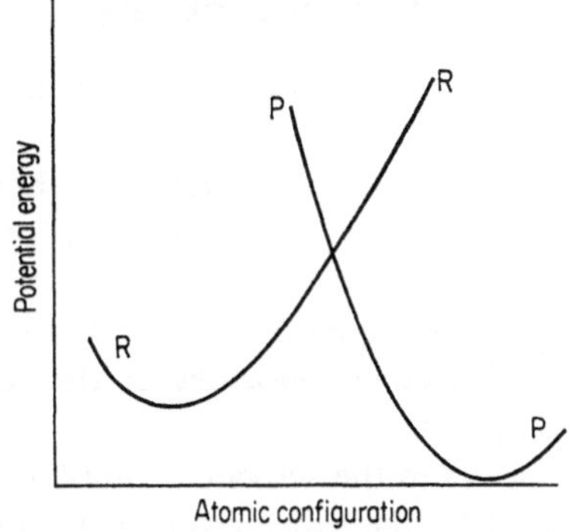

Fig. 1. R-reactants, P-products, no coupling between the states ψ_R and ψ_P

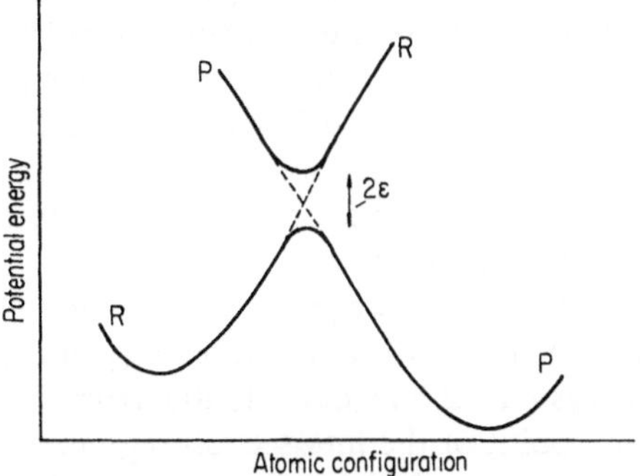

Fig. 2. R-reactants, P-products, coupling between the states ψ_R and ψ_P, ε-exchange energy

generacy at this intersection, in the usual quantum mechanical manner [4], and the system may now go from one original potential energy surface to the other by a fluctuation which permits it to pass through the "intersection" region of coordinate space. If the coupling is strong enough the product is thereby formed adiabatically from the reactant; otherwise the formation occurs nonadiabatically. For an adiabatic process in reaction kinetics there is continuous quantum mechanical equilibrium between electrons and nuclei and no abrupt electronic rearrangement. Such processes are primarily of interest here.

In summary, electron transfer can occur if there is suitable electronic coupling between the reacting species and if there is a suitable fluctuation of coordinates (e.g. bond distances, orientation of solvent molecules, position of ions in atmosphere, including those in the double layer) leading from values appropriate to the reactants to ones appropriate to the products.

3. Assumptions

To calculate the rate of the reaction, it is convenient to introduce several assumptions:

1. The rotational, translational and *relevant* vibrational coordinates of both reacting species and of the molecules and ions in the medium are treated classically.

2. The probability of the system being in the intersection region and having any specified molecular velocities is that computed from equilibrium statistical mechanics.

3. Each reacting species in the activated complex has an inner coordination shell [5] not shared by the other reactant. Consequently, any "bridging" of the reactants is of the outer sphere type [6].

4. The potential energy of the entire system is the sum of *intra* and *inter* particle terms [2a], the second term being a quadratic function of the permanent charge distributions on the particles. (This particle description is much more general than the usual one of polar media, which treats solvent molecules as having only induced and permanent *dipoles* and which was used in reference [1d]. The entire electrode and the entire medium can each be treated as a single particle [2a].

5. The splitting of the doubly degenerate energy level at the intersection surface is sufficiently small that the potential energy of the lowest electronic state of the entire system differs but slightly from that at the original intersection, for each point of the intersection region.

6. Reaction occurs primarily by the system passing over the barrier rather than by a nuclear tunnelling motion through it.

7. The reaction coordinate *in the region near the intersection surface* does not involve the rupture of a chemical bond.

Assumption 1 is reasonable under the usual experimental conditions, since the highest pertinent molecular frequency is usually a metal-ligand one in the inner coordination shell of an ionic reactant [7], and available data indicate this frequency to be typically $300-500$ cm^{-1}. Assumption 2 is the standard one in activated complex theory. Assumption 3 could be removed by extending the theory. When assumption 5 is not fulfilled, a condition which could conceivably happen sometimes when inner type bridging occurs (i.e. when assumption 3 fails), the theory could again be extended by calculating independently the splitting of the two energy levels, as a function of coordinates of the intersection surface. Assumption 6 will undoubtedly fail at sufficiently low temperatures. Assumption 7 can be removed by calculating the adiabatic potential energy surface for the case of bond ruptures. In the current literature of chemical kinetics such calculations are presently of a semi-empirical nature, and we wish to avoid computations based on adjustable parameters at present.

Supplementing these assumptions, it is convenient to introduce several approximations which convert the equations for the reaction rate into a rather simple functional form, a form particularly useful for theoretical predictions of correlations among the experimental data:

(I) The partial dielectric (or, more precisely, electric) saturation approximation [2b, c] for the entire medium outside the inner coordination shells is valid.

(II) The effective potential energy function for coordinates in the inner coordination shell of each reactant is a quadratic function of the coordinate displacements.

(III) In any given species, the force constant of any bond s may differ when that species is a reactant, k_s^r, as compared with when it is a product, k_s^p. These two force constants are expressed in terms of symmetric and anti-symmetric functions of them, namely of $2\,k_s^r\,k_s^p/(k_s^r + k_s^p)$ and of $(k_s^r - k_s^p)/(k_s^r + k_s^p)$, respectively, and it is shown in ref. [1e] that the second of these can be neglected. This approximation simplifies the equations of reference [1e] considerably.

Using these assumptions and approximations, and using the method of equivalent equilibrium distribution described in reference [1d] and further refined in reference [1e], one eventually obtains Eq. (3) for the rate constant [8a] of the electrochemical reaction (1) or of the homogeneous reaction (2) where 1 and 2 denote different soluble redox couples.

$$Ox + ne \longrightarrow Red \tag{1}$$

$$Ox_1 + Red_2 \longrightarrow Red_1 + Ox_2 \tag{2}$$

$$k_r = Z\gamma\varrho \exp\left(-\Delta F^*/R_g T\right) \tag{3}$$

where
electrochemical:

$$\Delta F^* = \frac{w^r + w^p}{2} + \frac{\lambda_{el}}{4} - \frac{nF(E - E_0')}{2} + \frac{[-nF(E - E^{0'}) + w^p - w^r]^2}{4\,\lambda_{el}} \tag{4}$$

homogeneous:

$$\Delta F^* = \frac{w^r + w^p}{2} + \frac{\lambda}{4} + \frac{\Delta F^{0'}}{2N_a} + \frac{(\Delta F^{0'} + w^p - w^r)^2}{4\lambda} \tag{5}$$

$R_g = k N_a$, where k is Boltzmann's constant and N_a is Avogadro's number.

w^r = work required to bring reactants together until they are at the most probable of the separation distances contributing effectively to reaction.

w^p = corresponding quantity for the products, for the same separation distance.

λ, λ_{el} = reorganization factors, described in reference [3] and, with more generality, in reference [1e]. They are independent of E, E_0' and $\Delta F^{0'}$. λ as well as λ_{el} is a sum of the contributions from the inner coordination shell (λ_i) and of the contributions from the medium outside this coordination shell (λ_0). λ_i is dependent on the changes in bond distances and bond angles and on the force constants of all the vibrational coordinates of each reacting species in its reacting state and in its product state. λ_0 is given for a dielectric continuum treatment (D_{op} – square of the refractive index, D_s – static dielectric constant) by

$$\lambda_0 = \left(\frac{1}{2\,a_1} + \frac{1}{2\,a_2} - \frac{1}{r} \right) \left(\frac{1}{D_{op}} - \frac{1}{D_s} \right) (n\,e)^2 \; ;$$

a_1 and a_2 are the radii of the spherical particles undergoing reaction including the coordination shell and $r = a_1 + a_2$ is taken as the mean distance between the centers of the reactants in the activated complex. In the electrochemical case $a_1 = a_2$, and r is twice the distance from the center of the reacting particle to the electrode surface. A formal statistical mechanical expression for λ_0 is given in reference [1e].

n = number of electrons transferred,

F = Faraday,

E= potential of half-cell (American convention, sign bivariant),

E_0' = standard potential of half-cell,

$\Delta F^{0'}$ = the apparent standard free energy of reaction [9], equal to nF times the difference of standard potentials of the two redox systems, 1 and 2.

Z = collision number in solution (10^{11} l mole^{-1} sec^{-1}) or, in the electrochemical case, the collision frequency with the electrode (10^4 cm sec^{-1}).

γ = an averaged probability of remaining on the lowest potential energy surface per passage across the intersection region [1d] ($\gamma = 1$ for an adiabatic reaction).

$\varrho = \langle \Delta r \rangle / \langle \Delta q \rangle$, where $\langle \Delta r \rangle$ is the average range of separation distances contributing effectively to reaction, and $\langle \Delta q \rangle$ is the average range of distances along reaction coordinate effectively occupied by the equivalent equilibrium distribution. Explicit formal expressions for ϱ are given in reference [1e]. Typically, it is probably of the order of magnitude of unity; an exception would occur if a low frequency orientation dielectric relaxation coordinate were the major component of the reaction coordinate, thereby making ϱ less than one. We shall take $\varrho\gamma$ to be unity. Further work on its evaluation is in progress. In all cases, a weaker assumption suffices to predict the correlations below.

Sometimes reactions (1) and (2) are preceded or followed by other reactions, so that then neither Ox nor Red denote the stable species present, merely those participating in the electron transfers (1) and (2). Then, k_r is the rate per unit concentration of these species rather than of the stable ones, and w^r, w^p, $\Delta F^{0'}$, nFE_0', λ and λ_{el} refer to the free energy terms involving such species and not to the stable ones.

4. Deductions

The characteristics of λ most important for predicting correlations are:

(I) In a homogeneous reaction λ is essentially an additive property [10] of the two redox couples, 1 and 2.

$$\lambda = \lambda_1 + \lambda_2 \tag{6}$$

(II) Let λ_{ex} be the value of λ for the homogeneous isotopic exchange reaction:

$$Ox_j^* + Red_j = Ox_j + Red_j^* \tag{7}$$

where the starred species denote isotopic labelling. If this reaction corresponds to an electrode reaction (1) then

$$\lambda_{el} = \frac{1}{2}\lambda_{ex} , \tag{8}$$

when the most probable distance of the reactant from the electrode in the pre-electrode layer equals twice that between the reactants in the homogeneous reaction.

As noted in reference [3], a very common condition experimentally is that the last term in Eqs. (4) and (5) can be neglected. From the resulting equation and the above properties of λ and λ_{el} one would then predict:

1. When the effect of E on the work terms can be ignored the electrochemical transfer coefficient should be 1/2. The slope of the electrochemical plot of $(R_g T/nF)\ln k_r$ vs. electrode potential is called the electrochemical transfer coefficient.

2. The k_{el} deduced from the electrochemical exchange current at standard conditions should bear a simple relation to rate constant of the homogeneous isotopic exchange reaction, k_{ex} for the case that the work terms are small in both experiments.

$$k_{el}/Z = \sqrt{k_{ex}/Z_{ex}} \tag{9}$$

3. In the comparison of the redox reaction of a series of reagents with a given chemical agent and with an electrode at a given E, the ratio of rate constants should be the same for each member of the series, regardless of whether or not the reaction is irreversible.

4. The "chemical transfer coefficient" [3] — defined as the slope of the ΔF^* vs. ΔF^0 plot in a series of homogeneous redox-reactions — should be 1/2, a result which provides some insight into the electrochemical coefficient when the work terms cannot be ignored.

5. The bimolecular rate constant k_{12} of the cross-reaction of two redox systems, 1 and 2,

$$Ox_1 + Red_2 = Red_1 + Ox_2, \tag{10}$$

is related to the rate constants of the isotopic exchanges in the two systems, k_{11} and k_{22}, and to the equilibrium constant K_{12}, as in (11), when the work terms in $\Delta F_{12}^* - (\Delta F_{11}^* + \Delta F_{22}^*)/2$ cancel or can be ignored.

$$k_{12} = \sqrt{k_{11} k_{22} K_{12}} \tag{11}$$

When these work terms cannot be ignored, the appropriately corrected form of (11) is given in Eq. (12) of reference [3]. A more elaborate equation based on Eq. (5) itself, rather than on the approximate form of (5), is given in Eq. (13) of reference [3].

Any numerical calculation of the rate constant k_r itself requires a knowledge of equilibrium bond distances and force constants in the coordination shells of the reactant and the product, as well as of certain properties of the medium, for these contribute to λ. Some illuminating calculations based on uncertain values of these constants have been given [6a, 11] and provide some insight into the problem of *a priori* calculation of reaction rates. In the present paper, we are particularly interested in correlations, however, for they are independent of the present uncertainties of *a priori* estimates of k_r's.

Eqs. (3) to (5) apply when each reacting species has a given inner coordination shell. However, sometimes a species may be present in several forms which differ in their inner coordination shell. Again, fluctuations in the number of strongly adsorbed ions on the electrode correspond to fluctuations in the "inner coordination shell" of the electrode [5]. Subject to conditions summarized in footnote 9 of reference [3], the above five deductions still apply when these fluctuations in composition of the inner shells occur.

Comparison of the five deductions with the experimental data has been given [3]. The agreement with the data may be considered encouraging. A considerable quantity of data supporting deduction 5 has since been obtained [12].

References

1. a) MARCUS, R. A.: J. Chem. Phys. **24**, 966 (1956), Part I;
   b) O.N.R. Technical Report No. 12. Project NR 051-337 (1957);
   c) Trans. Symposium Electrode Processes, May 1959, p. 239. E. YEAGER, ed., New York: John Wiley and Sons, 1961.
   d) Discussions Faraday Soc. **29**, 21 (1960), Part IV of reference [1a].

e) J. Chem. Phys., **43**, 679 (1965), Part VI of reference [*1 a*].
In a) the contribution of the medium (outside the inner coordination shells of the reacting species) to the free energy of activation of homogeneous electron transfers was computed, using nonequilibrium polarization theory and treating the medium as a continuum. In b), a similar calculation was made for electrochemical reactions. The problem was then formulated for electrode reactions in c) and for homogeneous ones in d) in terms of the behavior on many-dimensional potential energy surfaces. A detailed treatment, including the contributions of the inner coordination shells and of the medium outside to the free energy of activation, was given for homogeneous reactions in d), which automatically included a). A unified treatment for both homogeneous and electrochemical reactions, encompassing and generalizing a) to d), is given in e).

2. a) MARCUS, R. A.: J. Chem. Phys. **38**, 1335 (1963);
   b) **38**, 1858 (1963);
   c) Part III of b), **39**, 1734 (1963);
   d) Part II of a), **39**, 460 (1963). Reference [*2c*] is the statistical mechanical counterpart of [*2b*].

3. MARCUS, R. A.: J. Phys. Chem. **67**, 853 (1963), Part V of ref. [*1 a*] η_a there should be replaced by $E - E^{0'}$.

4. e.g. GLASSTONE, S., K. J. LAIDLER, and H. EYRING: The Theory of Rate Processes, New York-London: McGraw-Hill 1941.

5. When one of the "reacting" species participating in an electron transfer is an electrode, any strongly adsorbed solvent molecules and ions can be considered as the "inner coordination shell" of this "reactant".

6. a) Discussions of inner sphere and outer sphere reactions are given in several recent reviews: SUTIN, N.: Ann. Rev. Nuclear Sci. **12**, 285 (1962);
   b) HALPERN, J.: Quart. Rev. (London), **15**, 207 (1961);
   c) TAUBE, H.: Advances Inorg. Chem. Radiochem. **1**, 1 (1959).

7. When an equilibrium bond length in a reacting species differs in the reactant and product forms of that species, the corresponding stretching or compressing of the bond will contribute to the reaction coordinate. For recent data on force constants and frequencies, see SHIMANOUCHI, T., and I. NAKAGAWA: Spectrochim. Acta **18**, 89, 101 (1962); International Symposium on Molecular Structure and Spectroscopy (September 1962, Tokyo). KAKIUTI, Y., S. KIDA, and T. V. QUAGLIANO: Spectrochim. Acta **19**, 201 (1963).

8. a) When no "dynamic ψ-effect" [*8b*] occurs, it is convenient to define an electrochemical rate constant k_r as the rate per unit concentration of soluble reactant just outside the electrical double layer region [*8c*]. When the dynamic ψ-effect occurs, a rate constant k_r, defined as the rate per unit concentration of soluble reactant in some defined pre-electrode layer, is more useful since it can be incorporated into the theory of the dynamic ψ-effect. k_r is given by Eq. (3) and k'_r equals $k_r \exp(w^r/kT)$. Analogous remarks apply to the rate constants of homogeneous reactions, as noted elsewhere [*8d*]; the one defined by Eq. (3) is the rate for unit concentrations of both reactants in the bulk of the solution;
   b) See R. PARSONS, in Advances in Electrochemistry and Electrochemical Engineering, ed. P. DELAHAY, (New York: Interscience 1961), pp. 52—54, for references to the work of LEVICH and of SPARNAAY on the dynamic ψ-effect; *cf* GIERST, L.: in Trans. Symposium Electrode Processes, ed. E. YEAGER, (New York: John Wiley and Sons 1961), p. 109.
   c) Occasionally, it is asserted in the literature that reaction rate should rigrously beexpressed in terms of the activity of the reacting ion. This state-

ment is not only incorrect but also presupposes a knowledge of the single ion activity coefficient. Rather, the rate should be expressed operationally in terms of concentrations, and all *pertinent* salt effects should be incorporated into the theory of the proportionality constant, i.e. of the rate constant. It is readily verified that in careful transition state theory formulations of the latter constant no single ion activity coefficients appear.

d) MARCUS, R. A.: Discussions Faraday Soc. **29**, 129 (1960).

9. For brevity, $\Delta F^{0\prime}$ was denoted by $\Delta F0$ in reference [*1a*—*1d*].

10. This additivity is described in detail in reference [*1e*], but is evident from the less general equations in reference [*3*]. The presence of one term there, λ_r, depending on the most probable separation distance r of the two homogeneous reactants make λ not quite additive, because of differences in size of the reactants. It is easily shown that this deviation from additivity is small, using the dielectric continuum expression for λ_r. It is then assumed that the same small deviation from additivity will obtain in the statistical mechanical evaluation.

11. HUSH, N.: Trans. Faraday Soc. **57**, 557 (1961).

12. See work of N. SUTIN et al., cited in Ann. Rev. Phys. Chem. **15**, 155 (1964), and subsequent work of SUTIN. Exceptions are cited in the Ann. Rev. article.

R. A. Marcus
University of Illinois
Urbana, Illinois/USA

Chemische Elementarprozesse und kernmagnetische Resonanz in Flüssigkeiten

Joachim Heidberg

Mit 5 Abbildungen

1. Einleitung

Mit Hilfe der kernmagnetischen Resonanz lassen sich die Geschwindigkeiten gewisser intra- und intermolekularer Umlagerungen von Atomen, Atomgruppen, Ionen und Elektronen messen [1]. Es gelang durch ihre Anwendung, die Aktivierungsparameter der für die Eiweißchemie bedeutenden Rotationsumwandlungen von Amiden, der Inversionen organischer Ringsysteme, schneller Ligandenaustauschvorgänge in Komplexen sowie rascher Elektronen- und Protonenübertragungen zu bestimmen [2]. Diese keineswegs vollständige Übersicht soll andeuten, daß die Anwendung der kernmagnetischen Resonanz-Methode die Erschließung eines bisher schwer zugänglichen Bereiches der chemischen Kinetik möglich macht.

Es besteht ein Zusammenhang zwischen der Form kernmagnetischer Resonanzlinien und der Geschwindigkeit der genannten chemischen Vorgänge. An dieser Stelle sollen die quantitativen Beziehungen zwischen der Linienform und der Reaktionsgeschwindigkeit diskutiert und einige Beispiele angegeben werden. Die Grenzen der Anwendbarkeit werden aufgezeigt. Auf die konventionelle Methode zur Bestimmung von Reaktionsgeschwindigkeiten über die Messung von Linienintensitäten wird hier nicht näher eingegangen.

2. Die Umlagerung von Spins in adiabatischen Theorien

Im Rahmen der adiabatischen Theorien werden Umlagerungen von Spins zwischen Lagen oder Umgebungen behandelt, deren Abschirmkonstanten verschieden sind. Die lokalen magnetischen Felder sind parallel, unterscheiden sich aber in ihrer Größe durch die chemischen Verschiebungen. Bei der Umlagerung eines Spins aus einer Lage A in eine Lage B werden keine Übergänge zwischen den Zeeman-Termen der Spins induziert;

man nennt deshalb eine Theorie, die sich auf die Annahme paralleler Felder
gründet, adiabatisch. Die Wirkung der Umlagerung im Spektrum rührt
allein von der willkürlichen Modulation der Larmorfrequenzen der Spins
her.

Verursachen skalare Spin-Spin-Kopplungen die Unterschiede der lokalen Felder, so verlieren die adiabatischen Theorien ihre strenge Gültigkeit.
In diesen Fällen ist eine quantenstatistische Behandlung des Problems erforderlich. Die Bedingungen, unter denen die quantenstatistische Theorie
zur adiabatischen sich vereinfacht und letztere bei Anwesenheit von Spin-
Spin-Kopplung anwendbar ist, werden in Kapitel 7.1 behandelt. Im folgenden geben wir einen kurzen Überblick über diejenige adiabatische Theorie,
die der Existenz von Spin-Umlagerungen einfach durch die direkte Einführung von Dämpfungstermen Rechnung trägt. Dies geschieht nicht nur
um ihrer historischen Bedeutung willen, sondern auch weil wesentliche
Züge der adiabatischen Theorie in der expliziten quantenstatistischen Behandlung wiederkehren.

2.1. Blochsche Gleichungen

Die klassischen phänomenologischen Blochschen Gleichungen [3] stellen den Ausgangspunkt für die Ableitung der Relation zwischen Linienform und Reaktionsgeschwindigkeit dar. Die Blochschen Gleichungen
beschreiben die Zeitabhängigkeit der makroskopischen Kernmagnetisierung
$\mathbf{M}(t)$ in einem zeitabhängigen äußeren Magnetfeld $\mathbf{H}(t)$,

$$\mathrm{d}\,\mathbf{M}/\mathrm{d}\,t = \gamma\,\mathbf{M} \times \mathbf{H} + \partial\,\mathbf{M}/\partial\,t \ \text{(Relaxation)}, \tag{2.1}$$

wobei γ das gyromagnetische Verhältnis der betrachteten Kerne und
$\mathbf{M} \times \mathbf{H}$ das Vektorprodukt von $\mathbf{M}$ und $\mathbf{H}$ ist. Bei Resonanzexperimenten
hat $\mathbf{H}(t)$ die Form

$$\mathbf{H}(t) = \mathbf{k}\,H_0 + \mathbf{H}_1(t)\,, \tag{2.2}$$

wo H_0 ein starkes, konstantes und $\mathbf{H}_1(t)$ ein schwaches, in der Ebene
senkrecht zur k-Richtung rotierendes Magnetfeld bedeuten. Der erste
Term in Gl. (2.1) stellt die aus der klassischen Elektrodynamik folgende
Beschreibung der Präzession eines Systems mit dem magnetischen Moment
$\mathbf{M}$ und einem dazu parallelen Drehimpuls $\mathbf{J}$ in einem Magnetfeld $\mathbf{H}$ dar.
Der zweite Term in Gl. (2.1) beschreibt die Kinetik eines Prozesses zur
Einstellung des thermischen Gleichgewichts. Er wurde eingeführt, um der
Erfahrung Rechnung zu tragen, daß das System in einem konstanten Ma-
gnetfeld vermöge der Dämpfung durch die Umgebung allmählich in einer
zum äußeren Feld parallelen Stellung zur Ruhe kommt, unabhängig von
der Ausgangsrichtung von $\mathbf{M}$ zur Zeit $t = 0$. Zur Beschreibung des zeit-

lichen Ablaufs der Relaxationsprozesse macht man versuchsweise einen Ansatz 1. Ordnung:

$$\frac{\partial M_z}{\partial t} \text{ (Relaxation)} = \frac{M_0 - M_z}{T_1} \, , \tag{2.3a}$$

$$\frac{\partial M_x}{\partial t} \text{ (Relaxation)} = - \frac{M_x}{T_2} \, , \tag{2.3b}$$

$$\frac{\partial M_y}{\partial t} \text{ (Relaxation)} = - \frac{M_y}{T_2} \, . \tag{2.3c}$$

In Abwesenheit von $\mathbf{H}_1$ wird die z-Komponente von $\mathbf{M}$ sich dem Gleichgewichtswert M_0 im Feld H_0 nähern, während die x- und y-Komponenten von $\mathbf{M}$ verschwinden. T_1 bezeichnet man als die longitudinale, T_2 als die transversale Relaxationszeit. Für den longitudinalen und den transversalen Relaxationsprozeß werden verschiedene Relaxationszeiten angesetzt, da sich nur beim ersten Prozeß die Energie des Systems im Magnetfeld ändert.

Nimmt man nach (2.1) an, daß sich die Bewegung der Kerne als Folge der Relaxation der Bewegung im Magnetfeld $\mathbf{H}(t)$ überlagert, so lautet die Blochsche Gleichung für die Zeitabhängigkeit der Magnetisierung $\mathbf{M}$ der Kerne X in der Lage oder Umgebung A

$$d\,\mathbf{M}^{(A)}/d\,t = \gamma\,\mathbf{M}^{(A)} \times \mathbf{H}^{(A)} - \mathbf{i}\,\frac{M_x^{(A)}}{T_2^{(A)}} - \mathbf{j}\,\frac{M_y^{(A)}}{T_2^{(A)}} +$$
$$+ \mathbf{k}\,\frac{M_0^{(A)} - M_z^{(A)}}{T_1^{(A)}} \, . \tag{2.4a}$$

Die Vektoren $\mathbf{i}$, $\mathbf{j}$ und $\mathbf{k}$ sind die Einheitsvektoren in der x-, y- bzw. z-Richtung. Die Relaxationszeit und das am Kernort wirkende Magnetfeld werden von der Umgebung des Kernes beeinflußt. Analog erhält man für die Kerne X in der Umgebung B

$$d\,\mathbf{M}^{(B)}/d\,t = \gamma\,\mathbf{M}^{(B)} \times \mathbf{H}^{(B)} - \mathbf{i}\,\frac{M_x^{(B)}}{T_2^{(B)}} - \mathbf{j}\,\frac{M_y^{(B)}}{T_2^{(B)}} +$$
$$+ \mathbf{k}\,\frac{M_0^{(B)} - M_z^{(B)}}{T_1^{(B)}} \, . \tag{2.4b}$$

In einer typischen experimentellen Anordnung wird die Magnetisierung beobachtet, indem man die elektromotorische Kraft mißt, die in einer fixierten, z.B. längs der x-Achse angebrachten Spule induziert wird. Diese elektromotorische Kraft läßt sich bei Kenntnis von

$$M_x^{(A)} + M_x^{(B)} \equiv M_x = u\cos\omega\,t - v\sin\omega\,t \tag{2.5}$$

berechnen. $M_x(t, \omega)$ ergibt sich aus Gl. (2.4a) bzw. (2.4b), die vereinfacht werden können, da sich nach einer gewissen Einschwingzeit ein stationärer

Zustand derart einstellt, daß sich die Magnetisierung **M** mit dem rotierenden Feld H_1 mit der Kreisfrequenz $(= -\omega_z)$ bewegt. u und v sind die (x)- bzw. $(-y)$-Komponenten der Magnetisierung im mit dem Feld H_1 rotierenden Koordinatensystem. Im Absorptionsspektrum wird (v) in Abhängigkeit von ω dargestellt. Ein aus den Kernen X in den Umgebungen A und B bestehendes System besitzt ein Spektrum mit einem Signal bei $\omega = \gamma H_0^{(A)} = \omega_A$ und einem Signal bei $\omega = \gamma H_0^{(B)} = \omega_B$. Für die Form der Absorptionslinie von A ergibt sich aus (2.4a) bei vernachlässigbarer Sättigung,

$$ H_1 \ll \frac{1}{\gamma \sqrt{T_1^{(A)} T_2^{(A)}}} \; , \tag{2.6} $$

der Ausdruck einer Lorentzkurve

$$ v_A = -\gamma H_1 M_0^{(A)} \frac{T_2^{(A)}}{1 + (T_2^{(A)})^2 (\omega_A - \omega)^2} \tag{2.7} $$

mit der Halbwertsbreite $2/T_2^{(A)}$ in rad/sec bzw. $(\pi T_2^{(A)})^{-1}$ in Hz; für v_B folgt eine analoge Beziehung. Die Komponente der Gesamtmagnetisierung längs der $(-y)$-Achse im rotierenden Koordinatensystem, v, ist durch $v = v_A + v_B$ gegeben. Die zugehörigen Dispersionslinien werden durch die Komponente

$$ u_A = \gamma H_1 M_0^{(A)} \frac{(T_2^{(A)})^2 (\omega_A - \omega)}{1 + (T_2^{(A)})^2 (\omega_A - \omega)^2} \tag{2.8} $$

und die Komponente u_B beschrieben, deren Bestimmungsgleichung man formal aus (2.8) erhält, indem man A durch B ersetzt. Es gilt $u_A + u_B = u$. Vgl. insbesondere [9c].

2.2. *Blochsche Gleichungen mit Berücksichtigung chemischer Umlagerungen*

Die einfachste Form einer chemischen Umlagerung ist wahrscheinlich der Austausch der Kerne X zwischen ihren Umgebungen A und B:

$$ X \text{ (bei } A) \longrightarrow X \text{ (bei } B) $$

$$ X \text{ (bei } B) \longrightarrow X \text{ (bei } A). $$

Die mittlere Verweilzeit oder Lebensdauer von X bei A sei τ_A und bei B τ_B. Da das totale magnetische Moment **M** gleich der Summe der Einzelmomente **µ** der Kerne,

$$ \mathbf{M} = \sum_k \boldsymbol{\mu}_k N_k \; , \tag{2.9} $$

ist, wo N_k die Anzahl der Kerne mit dem magnetischen Moment μ_k bezeichnet, muß sich bei einem unimolekularen Austauschvorgang die totale Magnetisierung nach

$$\left(\frac{\partial \mathbf{M}^{(A)}}{\partial t}\right)_{\text{Austausch}} = \sum_r \mu_r^{(A)} \frac{d N_r^{(A)}}{d t} - \sum_s \mu_s^{(B)} \frac{d N_s^{(B)}}{d t},$$

$$= -\frac{\mathbf{M}^{(A)}}{\tau_A} + \frac{\mathbf{M}^{(B)}}{\tau_B} = -\left(\frac{\partial \mathbf{M}^{(B)}}{\partial t}\right)_{\text{Austausch}} \tag{2.10}$$

ändern. In diesem Ansatz ist die Annahme enthalten, daß die Lebensdauern von X in jeder anderen Umgebung als A und B, z. B. im Übergangszustand, so kurz sind und die Magnetfelder in diesen Zuständen derart beschaffen sind, daß sich die Magnetisierung beim Übergang der Kerne X von A nach B und umgekehrt nicht ändert. Die Präzession der Kerne während des Sprungs wird vernachlässigt, die Präzessionsperiode als groß gegen die Lebensdauer von X im Übergangszustand angesehen. Setzen wir wieder, in Anlehnung an den ursprünglichen Blochschen Ansatz, voraus, daß die Bewegungen der Kerne sich einfach überlagern, so lauten die erweiterten Blochschen Gleichungen (HAHN und MAXWELL [1c], McCONNELL [4])

$$\frac{d \mathbf{M}^{(A)}}{d t} = \gamma \mathbf{M}^{(A)} \times \mathbf{H}^{(A)} + \frac{\partial \mathbf{M}^{(A)}}{\partial t} \text{ (Relaxation)}$$

$$-\frac{\mathbf{M}^{(A)}}{\tau_A} + \frac{\mathbf{M}^{(B)}}{\tau_B}, \tag{2.11 a}$$

$$\frac{d \mathbf{M}^{(B)}}{d t} = \gamma \mathbf{M}^{(B)} \times \mathbf{H}^{(B)} + \frac{\partial \mathbf{M}^{(B)}}{\partial t} \text{ (Relaxation)}$$

$$-\frac{\mathbf{M}^{(B)}}{\tau_B} + \frac{\mathbf{M}^{(A)}}{\tau_A}. \tag{2.11 b}$$

Die erweiterten Blochschen Gleichungen können nun den Versuchsbedingungen angepaßt werden. Selbst für den Fall des stationären Zustandes bei langsamem Durchgang (slow passage) und vernachlässigbarer Sättigung ergibt sich für die transversale Komponente G der Magnetisierung im rotierenden Koordinatensystem, deren Imaginärteil im Absorptionsspektrum als Funktion von ω aufgezeichnet wird, der von GUTOWSKY und SAIKA [1b] erstmals angegebene, komplizierte Ausdruck

$$G = u + i v = -i \omega_1 M_0 \frac{(\tau_A + \tau_B) + \tau_A \tau_B (\alpha_A p_B + \alpha_B p_A)}{(1 + \alpha_A \tau_A)(1 + \alpha_B \tau_B) - 1} \tag{2.12}$$

mit

$$\alpha_A = (T_2^{(A)})^{-1} - i\,(\omega_A - \omega)\,,$$

$$\alpha_B = (T_2^{(B)})^{-1} - i\,(\omega_B - \omega)$$

und

$$\omega_1 = \gamma\, H_1\,.$$

p_A und p_B, die relativen Häufigkeiten der Komponenten A bzw. B, sind im Zustand des chemischen Gleichgewichts mit den mittleren Lebensdauern τ_A und τ_B durch die Gleichungen

$$p_A = \frac{\tau_A}{\tau_A + \tau_B}\,, \quad p_B = \frac{\tau_B}{\tau_A + \tau_B}$$

verknüpft. In Gl. (2.12) bezeichnet M_0 die Gleichgewichtsmagnetisierung der Kerne X in den Lagen A und B; M_0 setzt sich nach

$$M_0 = M_0^{(A)} + M_0^{(B)}$$

aus den Gleichgewichtsmagnetisierungen $M_0^{(A)}$ und $M_0^{(B)}$ der Komponenten A und B zusammen. Die Differenz

$$\omega_A - \omega_B = 2\,\delta$$

ist die chemische Verschiebung der Resonanz von A bezüglich der Resonanz von B.

Mittlere Lebensdauern τ_i lassen sich auch ohne Analyse der vollständigen Linienformgleichung (2.12) mit Hilfe der Ausdrücke für den langsamen und schnellen Austausch näherungsweise bestimmen. Die Bedingung für den langsamen Austausch ist

$$|2\,\delta| \gg \frac{1}{\tau_A} + \frac{1}{\tau_B} + \frac{1}{T_2^{(i)}} \quad i = A, B\,, \tag{2.13}$$

für den schnellen Austausch

$$|2\,\delta| \ll \frac{1}{\tau_A} + \frac{1}{\tau_B} \left(\text{für } |2\,\delta| > \frac{1}{T_2^{(A)}}\,,\ \frac{1}{T_2^{(B)}}\right). \tag{2.14}$$

Führt man die Bedingung (2.13) in Gl. (2.12) ein, so erhält man für v bei langsamem Austausch die Summe von zwei Gliedern der Form von Gl. (2.7), je ein Glied für die Linie A und die Linie B, in welchen an Stelle der transversalen Relaxationszeit $T_2^{(i)}$ in Abwesenheit von chemischem Austausch die nach

$$1 / T_2'^{(i)} = 1 / T_2^{(i)} + 1 / \tau_i \tag{2.15}$$

definierte effektive Relaxationszeit $T_2'^{(i)}$ auftritt. Langsamer chemischer Austausch bedingt eine Verbreiterung der ursprünglichen Linien A und B, da die Halbwertsbreite der Linien proportional $(T_2'^{(i)})^{-1}$ ist. Gl. (2.15) erlaubt auch bei Unkenntnis der chemischen Verschiebung $2\,\delta$ die näherungsweise Ermittlung von τ_i. Bei raschem Austausch sind die beiden ursprünglichen Linien zu einer Linie koalesziert, die ebenfalls durch einen Ausdruck der Form der Gl. (2.7) beschrieben wird. An Stelle von $T_2^{(i)}$ erscheint nun die durch (MEIBOOM, LUZ und GILL [5])

$$\frac{1}{T_2'} = \frac{p_A}{T_2^{(A)}} + \frac{p_B}{T_2^{(B)}} + p_A^2 p_B^2 \, 4 \, \delta^2 \, (\tau_A + \tau_B) \tag{2.16}$$

gegebene effektive Relaxationszeit T_2'. Im Gegensatz zum langsamen Austausch verursacht der rasche Austausch eine Verschärfung der durch Koaleszenz der ursprünglichen Linien entstandenen Resonanzlinie, bis für $\tau_i \to 0$ $(T_2')^{-1} \to p_A/T_2^{(A)} + p_B/T_2^{(B)}$. Die Gleichungen (2.15) und (2.16) machen die Grenzen der Anwendbarkeit der Methode sichtbar. Nach der Methode des langsamen Durchgangs (slow passage) können Reaktionsgeschwindigkeiten u. a. nur gemessen werden, wenn die Lebensdauer τ_i vergleichbar der reziproken chemischen Verschiebung $(2\,\delta)^{-1}$ der koaleszierenden Linien ist, d. h. wenn

$$2\,\tau_i\,\delta \approx 1 \; ,$$

wobei $2\,\delta$ etwa zwischen 5 und 100 sec^{-1} liegt. Der Bereich der erfaßbaren Lebensdauern erstreckt sich über drei bis vier Größenordnungen. Ferner müssen δ und $T_2^{(i)}$ bekannt sein. Zur Ermittlung dieser Parameter wird im allgemeinen der chemische Austausch durch Temperaturerniedrigung „eingefroren". Es ist nicht bekannt, mit welcher Genauigkeit die Annahme, daß δ und T_2 temperaturunabhängig sind, erfüllt ist. Die Lebensdauern von Kern-Konfigurationen, die an Umlagerungen höher als 1. Ordnung teilnehmen, lassen sich vielfach durch geeignete Variation der Konzentrationen der Reaktionspartner in den gewünschten Bereich bringen. Auf diese Weise können Geschwindigkeitskonstanten von Reaktionen in Lösung gemessen werden, die diffusionskontrolliert, d. h. mit der maximal möglichen Geschwindigkeit, ablaufen.

Bei der Auswertung der Spektren zur Ermittlung der mittleren Lebensdauern bestimmter Konfigurationen zieht man nicht nur die Gleichungen für die Grenzfälle, sondern auch die vollständigen Gleichungen heran und löst sie numerisch für verschiedene Werte von τ_i, $T_2^{(i)}$ etc. Die berechneten Spektren werden mit den beobachteten verglichen und bei Übereinstimmung τ_i unmittelbar abgelesen.

Verallgemeinerungen im Rahmen der adiabatischen Theorie für Systeme mit N inäquivalenten Lagen, zwischen denen Umlagerungen möglich sind, wurden von P. W. ANDERSON [6], KUBO [7] sowie von PIETTE und W. A. ANDERSON [8] vorgeschlagen. Die von PIETTE und ANDERSON angegebenen Gleichungen, die auf der Annahme beruhen, daß die Übergangswahrscheinlichkeiten $(\tau_{ij})^{-1}$ proportional den relativen Besetzungszahlen p_i der Lagen i sind, sich mithin auf Systeme im Gleichgewicht beziehen, für die überdies das Prinzip der mikroskopischen Reversibilität gelten muß, werden gewöhnlich zur Analyse des N Lagen Problems verwandt.

3. Quantenstatistische Beschreibung eines Ensembles gekoppelter Kerne mit dem Spin 1/2. Die Boltzmanngleichung

3.1. Reine Zustände

Ein beliebiger Spinzustand eines Teilchens mit dem Spin $I = {}^1/_2$ wird durch die Linearkombination

$$c_1 \, \alpha \, (\sigma_1) + c_2 \, \beta \, (\sigma_1)$$

dargestellt [vgl. Anhang Gl. (A. 6)]. Die einfachste orthonormierte Basis zur Beschreibung des Spinzustands eines Systems von q Teilchen, die alle den Spin $I = {}^1/_2$ besitzen, bildet der Satz der 2^q Produktfunktionen, von denen die n-te Funktion die Gestalt [9]

$$\psi_n = \alpha \, (\sigma_1) \, \alpha \, (\sigma_2) \, \ldots \, \beta \, (\sigma_q) \tag{3.1}$$

haben möge. Alle Produktfunktionen ψ_n sind Eigenfunktionen des Operators F_z, der Summe der Operatoren der Spinkomponenten in z-Richtung, $I_z^{(s)}$,

$$F_z = \sum_{s=1}^{q} I_z^{(s)} \, . \tag{3.2}$$

Die Eigenzustände u zu den Eigenwerten E des Spinhamiltonoperators $\mathscr{H}$ müssen der Gleichung

$$\mathscr{H} \, u = E \, u \tag{3.3}$$

genügen. Für u machen wir den Ansatz

$$u = \sum_{n=1}^{2^q} a_n \, \psi_n \, . \tag{3.4}$$

Führt man (3.4) in (3.3) ein und bildet die skalaren Produkte

$$(\psi_m \, , \, \mathscr{H} \, u) = \sum_n a_n \, \mathscr{H}_{mn} = E \, a_m \tag{3.5}$$

mit

$$\mathscr{H}_{mn} = (\psi_m, \, \mathscr{H} \, \psi_n) \, ,$$

so erhält man ein lineares Gleichungssystem, das nur dann nicht verschwindende Lösungen besitzt, wenn die aus den Koeffizienten der unbekannten Größen a gebildete Determinante Null ist[1]:

$$|\mathscr{H}_{mn} - \delta_{mn} E| = 0 \, . \tag{3.6}$$

---

[1] $\delta_{mn} = $ für $\begin{cases} 1 & m = n \\ 0 & m \neq n \end{cases}$

Die Säkulargleichung (3.6) ist eine algebraische Gleichung 2^q Grades für E. Die Wurzeln von (3.6), die Eigenwerte E_l der Energie, ergeben, nacheinander in (3.5) eingesetzt, die Überlagerungskoeffizienten $a_{l\,n}$ für die betreffenden Eigenzustände u_l.

Die möglichen Zustände φ des Mehrspinsystems lassen sich durch die Linearkombination

$$\varphi = \sum_l b_l\, u_l \tag{3.7}$$

beschreiben. Es erweist sich als zweckmäßig, neben der Energiedarstellung u_l auch die I_z-Darstellung ψ_n zu benutzen, um φ auszudrücken:

$$\varphi = \sum_n c_n\, \psi_n \;. \tag{3.8}$$

In (3.7) und (3.8) sind die Koeffizienten b_l bzw. c_n einschließlich ihrer relativen Phasen festgelegt, wenn sich das Mehrspinsystem in einem definierten oder — in der Notation der Quantenstatistik — reinen Zustand φ befindet [vgl. Anhang Gl. (A. 21)]. Dann sind auch die zeitlichen Veränderungen des Systems so vollständig und exakt vorgegeben wie im Rahmen der Quantenmechanik möglich. Überdies ist es bei Vorliegen eines reinen Zustands immer möglich, einen hermitischen Operator zu finden, der einer physikalischen Größe entspricht, für welche der betreffende Zustand Eigenzustand ist[2]. Wenn diese Bedingung nicht erfüllt ist, nennt man den Zustand gemischt.

3.2. Gemischte Zustände. Die Kernspindichtematrix

Experimente zur kernmagnetischen Resonanz führen wir im allgemeinen an makroskopischen (thermodynamischen) Systemen aus. Ein makroskopisches System enthält so viele Teilchen, daß eine vollständige Spezifizierung ihrer quantenmechanischen Zustände in den meisten Fällen unmöglich ist. Mit einem vorgegebenen thermodynamischen Zustand eines makroskopischen Systems ist eine große Anzahl quantenmechanischer Zustände der Teilchen, die das System bilden, vereinbar. Man steht mithin vor der Aufgabe, das Verhalten eines Systems zu beschreiben, über das weniger Information als maximal möglich vorliegt. Dies gelingt, indem man statistische Methoden benutzt und sog. repräsentative Ensemble einführt. Ein repräsentatives Ensemble ist eine Gesamtheit isolierter Exemplare eines Systems, dessen bekannte Parameter für alle Exemplare den gleichen Wert besitzen, während die übrigen Parameter Werte haben, die sich innerhalb

---

[2] Ein reiner Zustand wird durch eine Dichtematrix ρ_r mit einem Eigenwert gleich 1 und allen anderen gleich 0 dargestellt ($\rho_r^2 = \rho_r$). Vgl. Kap. 3.2.

eines geeignet gewählten Bereichs für verschiedene Exemplare unterscheiden. Hier interessiert das Ensemble eines Systems, dessen Exemplare
den gleichen Hamiltonoperator haben und sich nur durch ihren quantenmechanischen Zustand φ unterscheiden. Da sich die Beschreibung eines
quantenmechanischen Zustands durch φ vom Standpunkt der klassischen
Theorie bereits auf ein repräsentatives Ensemble bezieht, liegen für ein
quantenmechanisches System im allgemeinen zwei Gründe für eine statistische Behandlung vor: Unvollständige Information und der Wahr-
scheinlichkeitscharakter der Quantenmechanik. Die statistische Behandlung führt man mit Hilfe eines Operators aus, den man zweckmäßig durch
seine Matrixelemente in einer geeigneten Darstellung definiert. Die Matrix
nennt man die Dichtematrix. Sie wurde 1927 durch v. Neumann in einer
Reihe von Arbeiten in den Göttinger Nachrichten eingeführt [10].

Wir betrachten das Ensemble eines Systems, das aus q Kernspins
bestehe, die untereinander und mit ihrer molekularen Umgebung gekoppelt
sein mögen. Die molekulare Umgebung fassen wir als ein Wärmereservoir
auf, das sich stets im thermischen Gleichgewicht befinde[3].

Der Zustand des k-ten Exemplars des Mehrspinsystems oder — kürzer — des k-ten Mehrspinsystems sei durch die normierte Wellenfunktion
$\varphi^{(k)}$ ($k = 1, 2, \ldots, N$) beschrieben. Der Erwartungswert $\langle A \rangle_k$ einer
physikalischen Größe des k-ten Systems ist durch

$$\langle A \rangle_k = (\varphi^{(k)}, A\,\varphi^{(k)}) \tag{3.9}$$

gegeben. Der Ensemble-Mittelwert berechnet sich nach

$$\overline{\langle A \rangle} = \frac{1}{N} \sum_{k=1}^{N} \langle A \rangle_k = \frac{1}{N} \sum_{k=1}^{N} (\varphi^{(k)}, A\,\varphi^{(k)}) \,. \tag{3.10}$$

Substituieren wir $\varphi^{(k)}$ durch die Entwicklung nach dem vollständigen
Orthogonalsystem der ψ_n (3.8),

$$\varphi^{(k)} = \sum_n c_n^{(k)} \psi_n \,,$$

so ergibt sich

$$\overline{\langle A \rangle} = \frac{1}{N} \sum_{k=1}^{N} \sum_{m,\,n} c_m^{(k)*} c_n^{(k)} A_{mn} \,, \tag{3.11}$$

wobei

$$A_{mn} = (\psi_m, A\,\psi_n) \,.$$

---

[3] Wir führen die Dichtematrix auf dem „statistischen" Wege ein, da dieser für
die uns interessierende Betrachtung von Transport- und Gleichgewichtsproblemen
in Spinsystemen der geeignete ist. Bezüglich des „quantenmechanischen" Weges
und des sog. „operational approach", ersterer zur Behandlung von Vielkörper-
problemen, letzterer von Polarisationsproblemen geeignet, verweisen wir auf Ref.
[10c] [10d].

Die Dichtematrix ρ definieren wir durch ihre Matrixelemente, die hier ebenso wie die A_{mn} in der ψ_n-Darstellung gegeben sind,

$$\varrho_{mn} = \frac{1}{N} \sum_{k=1}^{N} c_m^{(k)} c_n^{(k)*} \tag{3.12a}$$

oder

$$\varrho_{mn} = \sum_i p^{(i)} c_m^{(i)} c_n^{(i)*} \tag{3.12b}$$

mit

$$p^{(i)} = \frac{N^{(i)}}{\sum\limits_i N^{(i)}} \qquad \text{und} \qquad \sum_i p^{(i)} = 1 \; .$$

$p^{(i)}$ ist die relative Häufigkeit des i-ten Zustands, $\varphi^{(i)}$, oder sein statistisches Gewicht, $N^{(i)}$ die Anzahl der Systeme, die sich im Zustand $\varphi^{(i)}$ befinden. Mit der Definition (3.12) nimmt (3.11) die Gestalt

$$\overline{\langle A \rangle} = \sum_{m,n} A_{mn}\, \varrho_{nm} = \mathrm{Sp}\,(\rho\, A) \tag{3.13}$$

an; die Spur Sp ist die Summe der Diagonalelemente einer Matrix.

Wir haben ρ in Gleichung (3.12) in der ψ_n-Darstellung definiert. Mit Hilfe der Gleichung

$$\rho' = U^{-1}\, \rho\, U \tag{3.14}$$

findet man die Dichtematrix in jeder anderen Darstellung g, die mit ersterer durch die unitäre Transformation U nach $g_n = U\,\psi_n = \sum\limits_m U_{mn}\,\psi_m$ verknüpft ist. Es läßt sich leicht zeigen, daß der Ensemble-Mittelwert $\langle A \rangle$ invariant gegen eine solche Transformation ist.

Die Elemente der Dichtematrix unterliegen einer Reihe einschränkender Bedingungen. Aus der Definition (3.12) folgt, daß ρ hermitisch,

$$\varrho_{mn} = \varrho_{nm}^{*} \; , \tag{3.15a}$$

und normiert,

$$\mathrm{Sp}\,\rho = 1 \; , \tag{3.15b}$$

ist. Nach (3.15a) sind die Diagonalelemente ϱ_{nn} reell; sie geben die Wahrscheinlichkeit für die Realisierung des Zustands ψ_n in dem Ensemble an. Da die Wahrscheinlichkeit keine negative Größe sein kann, muß ϱ_{nn} positiv definit sein, so daß im Hinblick auf (3.15b)

$$0 \le \varrho_{nn} \le 1 \; . \tag{3.16}$$

Der Wert sowohl der Diagonal- als auch der Nichtdiagonalelemente wird durch die Ungleichung

$$1 = (\mathrm{Sp}\,\rho)^2 \ge \mathrm{Sp}\,\rho^2 = \sum_{m,n} |\varrho_{mn}|^2 \tag{3.17}$$

begrenzt, die man unter Benutzung von (3.15a) erhält.

Die zeitliche Veränderung der Dichtematrix wird durch die Schrödinger-Gleichung

$$\mathscr{H}\,\varphi^{(i)} = i\,\hbar\,\frac{\mathrm{d}}{\mathrm{d}\,t}\,\varphi^{(i)} \qquad (3.18)$$

bestimmt. (3.18) lautet in der ψ_n-Darstellung

$$\sum_l H_{nl}\,c_l^{(i)} = i\,\hbar\,\frac{\mathrm{d}}{\mathrm{d}\,t}\,c_n^{(i)}\;. \qquad (3.19)$$

Wenn sich die statistischen Gewichte $p^{(i)}$ mit der Zeit nicht ändern, folgt aus (3.12) und (3.19) bei Beachtung der Hermitizität von $\mathscr{H}$

$$i\,\hbar\,\frac{\mathrm{d}}{\mathrm{d}\,t}\,\varrho_{mn} = i\hbar\sum_i p^{(i)}\left\{c_n^{(i)*}\,\frac{\mathrm{d}\,c_m^{(i)}}{\mathrm{d}\,t} + c_m^{(i)}\,\frac{\mathrm{d}\,c_n^{(i)*}}{\mathrm{d}\,t}\right\} \qquad (3.20\,\mathrm{a})$$

$$= [\mathscr{H},\varrho]_{-mn}\,, \qquad (3.20\,\mathrm{b})$$

wo $[\mathscr{H},\varrho]_- = \mathscr{H}\varrho - \varrho\mathscr{H}$ den Kommutator von $\mathscr{H}$ und ϱ bezeichnet.

Bisher hatten wir die Kopplung des Kernspinsystems an seine molekulare Umgebung, die als Wärmereservoir wirkt, nicht zu berücksichtigen. Wir betrachten nun den wichtigen Fall, daß das Kernspinsystem mit seiner Umgebung ungehindert Energie austauscht und sich mit ihr im thermischen Gleichgewicht befindet. Der Hamiltonoperator $\mathscr{H}$ für das stationäre System im thermischen Gleichgewicht ist zeitunabhängig. Dann postuliert man für die Dichtematrix[4]

$$\varrho_0 = \frac{1}{Z}\,e^{-\mathscr{H}/(kT)}\;, \qquad (3.21)$$

wo T die absolute Temperatur des Systems, k die Boltzmannkonstante und Z die durch die Normierungsbedingung (3.15b) festgelegte Zustandssumme

$$Z = \mathrm{Sp}\;e^{-\mathscr{H}/(kT)} \qquad (3.22)$$

bedeuten[5].

Wir betrachten als einfaches Beispiel ein Ensemble von AB-Systemen in einem statischen Magnetfeld H_0, das in die negative z-Richtung weist. Als AB-System bezeichnen wir ein System aus zwei gekoppelten gleich-

---

[4] Durch (3.21) wird das kanonische Ensemble definiert.

[5] Die Dichtematrix ist nach (3.21) in der Energiedarstellung diagonal. Die Beziehung

$$\varrho_{0\,ll'}^{(E)} = \sum_i p^{(i)}\,b_{l'}^{(i)*}\,b_l^{(i)} = \sum_i p^{(i)}\,r_{l'}^{(i)}\,r_l^{(i)}\,e^{i(\gamma_l^{(i)} - \gamma_{l'}^{(i)})} = \varrho_{0\,ll'}^{(E)}\,\delta_{ll'}$$

wird am einfachsten erfüllt, wenn man annimmt, daß die Phasen $\gamma_l^{(i)}$, für jedes l, vollkommen willkürlich über die Exemplare des Ensembles verteilt sind. Man nennt die Annahme das Postulat von der willkürlichen Verteilung der Phasen.

artigen Kernen mit dem Spin $I = {}^1/_2$. Für dieses System lautet der Hamilton-operator $\mathscr{H}_0$

$$\mathscr{H}_0 = -\gamma \hbar H_z \sum_s I_z^{(s)} + \hbar \sum_s \delta^{(s)} I_z^{(s)} + \hbar J (\mathbf{I}^{(A)}, \mathbf{I}^{(B)}) \qquad (3.23\,\text{a})$$

und in Kreisfrequenzeinheiten rad / sec, d. h. mit $\hbar = 1$

$$\mathscr{H}_0 = \omega_0 (I_z^{(A)} + I_z^{(B)}) + \delta (I_z^{(A)} - I_z^{(B)}) + J (\mathbf{I}^{(A)}, \mathbf{I}^{(B)}) \ , \qquad (3.23\,\text{b})$$

wo

$$\omega_0 \equiv \gamma H_0 + (\delta^{(A)} + \delta^{(B)}) \, {}^1/_2 \approx \gamma H_0 \ , \qquad \delta^{(A)} - \delta^{(B)} = 2\delta$$

und γ das gyromagnetische Verhältnis der betreffenden Kerne, h das Plancksche Wirkungsquantum ($\hbar = h/2\pi$), $\pm\,\delta$ die Parameter der chemischen Verschiebung der Kerne A bzw. B im Feld H_0 und J die Spinkopplungskonstante für die als isotrop angenommene Wechselwirkung zwischen den Spins bedeuten; $\mathbf{I}^{(s)}$ ist der Spinvektoroperator des Kerns s.

Die Eigenwerte des Hamiltonoperators (3.23 b) ergeben sich zu

$$
\begin{aligned}
E_1 &= \omega_0 + \frac{1}{4} J, \\[2mm]
E_2 &= A_0 - \frac{1}{4} J, \\[2mm]
E_3 &= - A_0 - \frac{1}{4} J, \\[2mm]
E_4 &= - \omega_0 + \frac{1}{4} J
\end{aligned}
\qquad (3.24)
$$

mit

$$A_0 = + (\delta^2 + \frac{1}{4} J^2)^{1/2}$$

und die zugehörigen Eigenfunktionen

$$
\begin{aligned}
u_1 &= \alpha\,(\sigma_A)\,\alpha\,(\sigma_B) \ , \\
u_2 &= a_{22}\,\alpha\,(\sigma_A)\,\beta\,(\sigma_B) + a_{23}\,\beta\,(\sigma_A)\,\alpha\,(\sigma_B) \ , \\
u_3 &= a_{32}\,\alpha\,(\sigma_A)\,\beta\,(\sigma_B) + a_{33}\,\beta\,(\sigma_A)\,\alpha\,(\sigma_B) \ , \\
u_4 &= \beta\,(\sigma_A)\,\beta\,(\sigma_B) \ ,
\end{aligned}
\qquad (3.25)
$$

mit

$$
\begin{aligned}
a_{22} &= a_{33} = \cos \Theta \ , \\
a_{23} &= - a_{32} = \sin \Theta \ ;
\end{aligned}
$$

der Winkel Θ ist durch $A_0 \cos 2\,\Theta = \delta$ und $A_0 \sin 2\,\Theta = {}^1/_2 J$ gegeben.

Im Bereich der kernmagnetischen Resonanz ist im allgemeinen $kT \gg \hbar E_n$ und $\omega_0 \gg \delta, J$. Dann folgt für die in der Energiedarstellung ausgedrückte Kernspindichtematrix $\rho_0^{(E)}$ eines Ensembles von AB-Systemen im thermischen Gleichgewicht

$$\rho_0^{(E)} \approx \frac{1}{4} \begin{pmatrix} 1 - \dfrac{\omega_0\, \hbar}{k\,T} & 0 & 0 & 0 \\ 0 & 1 & 0 & 0 \\ 0 & 0 & 1 & 0 \\ 0 & 0 & 0 & 1 + \dfrac{\omega_0\, \hbar}{k\,T} \end{pmatrix} \qquad (3.26)$$

.

Die makroskopischen Eigenschaften des Systems, d. h. seine Ensemble-Mittelwerte berechnen sich mit Hilfe von (3.13). Die makroskopische Magnetisierung M_0 in Richtung des Magnetfeldes H_0 für eine Probe mit N AB-Systemen, die nicht miteinander in Wechselwirkung stehen, ergibt sich zu[6]

$$M_0 = -\gamma\,\hbar\,N\,\overline{\langle \mathrm{F}_z \rangle} = -\gamma\,\hbar\,N\,\mathrm{Sp}\,[\rho_0\,\mathrm{F}_z] = \frac{N}{2}\frac{(\gamma\,\hbar)^2}{k\,T}\,H_0 \,. \qquad (3.27)$$

Man findet in ähnlicher Weise

$$M_x = \gamma\,\hbar\,N\,\overline{\left\langle \sum_s \mathrm{I}_x^{(s)} \right\rangle} = M_y = \gamma\,\hbar\,N\,\overline{\left\langle \sum_s \mathrm{I}_y^{(s)} \right\rangle} = 0 \,. \qquad (3.28)$$

In Übereinstimmung mit der Erfahrung weist die makroskopische Magnetisierung im thermischen Gleichgewicht in Richtung des Feldes H_0. Man kann sich leicht davon überzeugen, daß diese Übereinstimmung nicht erreicht wird, wenn wir allen Exemplaren des Ensembles die gleiche Zustandsfunktion φ zuordnen.

3.3. Zeitabhängige Störungsrechnung mit der Kernspindichtematrix. Die Boltzmanngleichung

Im allgemeinen hat man es nicht mit Systemen im thermischen Gleichgewicht mit einem zeitlich konstanten Hamiltonoperator zu tun. Es ist aber oft möglich, den Hamiltonoperator in einen großen zeitunabhängigen Anteil $\mathcal{H}_0$ und eine kleine zeitabhängige Störung $\mathcal{H}_1(t)$ zu zerlegen. Dann lautet die Bewegungsgleichung der Dichtematrix (vgl. Slichter [11], Abragam [12], Alexandrow [13])

$$\frac{\mathrm{d}\rho}{\mathrm{d}t} = \frac{i}{\hbar}\,[\rho\,,\,\mathcal{H}_0 + \mathcal{H}_1]_- \,. \qquad (3.29)$$

---

[6] Man muß nicht von der speziellen Form der Dichtematrix (3.26) und der Annahme isolierter AB-Systeme ausgehen, um (3.27) zu erhalten. Gl. (3.27), bekanntlich das allgemeiner gültige Curiesche Gesetz, beruht auf den Annahmen, daß $(-\gamma\,\hbar\,H_z \sum_s \mathrm{I}_z^{(s)} + \mathcal{H}') \ll k\,T$ und $\mathrm{Sp}\,(\mathcal{H}' \sum_s \mathrm{I}_z^{(s)}) = 0$, wobei $\mathcal{H}'$ die dipolare Kopplung zwischen den Spins und die Quadrupolkopplung mit lokalen Feldgradienten einschließt ([12], S. 39).

Die formale Lösung dieser Gleichung ist für den Fall $\mathcal{H}_1(t) = 0$ durch

$$\rho(t) = e^{-(i/\hbar)\mathcal{H}_0 t}\,\rho(0)\,e^{(i/\hbar)\mathcal{H}_0 t} \tag{3.30}$$

gegeben. Wir definieren nun die Größe $\bar{\rho}(t)$

$$\rho(t) \equiv e^{-(i/\hbar)\mathcal{H}_0 t}\,\bar{\rho}(t)\,e^{(i/\hbar)\mathcal{H}_0 t}\;. \tag{3.31}$$

Wenn $\mathcal{H}_1$ verschwindet, ist $\bar{\rho}$ eine Konstante. Differenzieren wir (3.31) nach t und führen den erhaltenen Ausdruck in (3.29) ein, so finden wir

$$\frac{i}{\hbar}\,[\rho\,,\,\mathcal{H}_0]_- + e^{-(i/\hbar)\mathcal{H}_0 t}\,\frac{d\bar{\rho}}{dt}\,e^{(i/\hbar)\mathcal{H}_0 t} = \frac{i}{\hbar}\,[\rho\,,\,\mathcal{H}_0 + \mathcal{H}_1]_-\;. \tag{3.32}$$

Mit der Definition

$$\overline{\mathcal{H}}_1 = e^{(i/\hbar)\mathcal{H}_0 t}\,\mathcal{H}_1\,e^{-(i/\hbar)\mathcal{H}_0 t} \tag{3.33}$$

ergibt sich aus (3.32)

$$\frac{d\bar{\rho}}{dt} = \frac{i}{\hbar}\,[\bar{\rho}\,,\,\overline{\mathcal{H}}_1(t)]_-\;. \tag{3.34}$$

(3.33) beschreibt die unitäre Transformation zur sogenannten Wechselwirkungsdarstellung, die eine Separation der Bewegungen im statischen und variablen Magnetfeld gestattet.

Die formale Lösung von (3.34) erhalten wir durch Integration

$$\bar{\rho}(t) = \bar{\rho}(0) + \frac{i}{\hbar}\int_0^t [\bar{\rho}(t')\,,\,\overline{\mathcal{H}}_1(t')]_-\,dt'\;. \tag{3.35}$$

$\bar{\rho}(t')$ im Integranden ist jedoch nicht bekannt. Wir ersetzen $\bar{\rho}(t')$ durch $\bar{\rho}(0)$ und erhalten, wenn t hinreichend kurz ist, eine approximative Lösung, da sich $\bar{\rho}(t)$ bei kleinen Störungen $\mathcal{H}_1(t)$ nur langsam mit der Zeit ändert. Es gilt in 1. Näherung

$$\bar{\rho}(t) = \bar{\rho}(0) + \frac{i}{\hbar}\int_0^t [\bar{\rho}(0)\,,\,\overline{\mathcal{H}}_1(t')]_-\,dt'\;. \tag{3.36}$$

Wir gelangen zur nächst höheren Näherung, indem wir den Ausdruck (3.36) für $\bar{\rho}(t')$ in den Integranden von (3.35) einführen.

Wir untersuchen wiederum ein Ensemble von AB-Systemen, an dieser Stelle unter Bedingungen, die gewöhnlich beim Nachweis der kernmagnetischen Resonanz eingehalten werden: das Ensemble befindet sich in einem in negative z-Richtung weisenden, starken, statischen Magnetfeld H_0 und einem schwachen, variablen Magnetfeld H_1, das mit der Winkelgeschwindigkeit $+\omega_z \equiv \omega$ rotiert. Das Gesamtfeld ist dann

$$\mathbf{H}(t) = -\,(\mathbf{i}\,H_1\cos\omega t + \mathbf{j}\,H_1\sin\omega t + \mathbf{k}\,H_0)\,, \tag{3.37}$$

wo $\mathbf{i}$, $\mathbf{j}$ und $\mathbf{k}$ die Einheitsvektoren in x-, y- bzw. z-Richtung sind. Der Hamiltonoperator hat mit (3.37) die Gestalt

24*

$$\mathcal{H}'(t) = \omega_0 \left(I_z^{(A)} + I_z^{(B)}\right) + \delta \left(I_z^{(A)} - I_z^{(B)}\right) + J\left(\mathbf{I}^{(A)}, \mathbf{I}^{(B)}\right)$$
$$+ \omega_1 \left[\left(I_x^{(A)} + I_x^{(B)}\right) \cos \omega\, t + \left(I_y^{(A)} + I_y^{(B)}\right) \sin \omega\, t\right], \tag{3.38}$$

wo

$$\omega_1 = \gamma\, H_1 .$$

Die Dichtematrix ergibt sich mit Hilfe von (3.36), wobei wir

$$\overline{\rho}\,(0) = \rho_0 = \frac{\exp\left[-\hbar\,\mathcal{H}_0^{'}/(k\,T)\right]}{\mathrm{Sp}\,\exp\left[-\hbar\,\mathcal{H}_0^{'}/(k\,T)\right]} \tag{3.39}$$

und

$$\mathcal{H}_1(t)\,\hbar^{-1} = \mathcal{H}_1^{'}(t) = \omega_1 \left[\sum_s I_x^{(s)} \cos \omega\, t + \sum_s I_y^{(s)} \sin \omega\, t\right] \tag{3.40a}$$

$$= \omega_1\, e^{-i\,\omega\, t\, \sum_s I_z^{(s)}} \sum_s I_x^{(s)}\, e^{i\,\omega\, t\, \sum_s I_z^{(s)}} \tag{3.40b}$$

setzen. In bezug auf Koordinaten, die mit der Winkelgeschwindigkeit ω rotieren, ist die Störung stationär, $\mathcal{H}_1^{'(\omega)} = \omega_1 \sum_s I_x^{(s)}$. Transformieren wir auf das ruhende Koordinatensystem zurück, so erhalten wir (3.40b). Die Diagonalelemente der Dichtematrix in der Energiedarstellung sind in 1. Näherung den Boltzmannfaktoren gleich:

$$\overline{\varrho}_{nn}^{(E)}(t) = \overline{\varrho}_{nn}^{(E)}(0) + i \int_0^t \sum_l \left\{ \overline{\varrho}_{nl}^{(E)}(0)\, \overline{\mathcal{H}}_1^{'}{}_{ln}(t') - \overline{\mathcal{H}}_1^{'}{}_{nl}(t')\, \overline{\varrho}_{ln}^{(E)}(0)\right\} \mathrm{d}\,t'$$

$$= \overline{\varrho}_{nn}^{(E)}(0) = \varrho_{0\,nn}^{(E)} . \tag{3.41}$$

Für die Nichtdiagonalelemente von ρ gilt in 1. Näherung

$$\overline{\varrho}_{nm}^{(E)}(t) = i \int_0^t \sum_l \left\{ \overline{\varrho}_{nl}^{(E)}(0)\, \overline{\mathcal{H}}_1^{'}{}_{lm}(t') - \overline{\mathcal{H}}_1^{'}{}_{nl}(t')\, \overline{\varrho}_{lm}^{(E)}(0)\right\} \mathrm{d}\,t'$$

$$= i\left(\varrho_{0\,nn}^{(E)} - \varrho_{0\,mm}^{(E)}\right) \int_0^t \overline{\mathcal{H}}_1^{'}{}_{nm}(t')\, \mathrm{d}\,t' . \tag{3.42}$$

Ausgehend von (3.40b) ergibt sich mit (3.33) $\overline{\mathcal{H}}_1(t)$ in der Energiedarstellung zu

$$\overline{\mathcal{H}}_1(t) = \begin{pmatrix} 0 & h_{12} & h_{13} & 0 \\ h_{12}^{*} & 0 & 0 & h_{24} \\ h_{13}^{*} & 0 & 0 & h_{34} \\ 0 & h_{24}^{*} & h_{34}^{*} & 0 \end{pmatrix}; \tag{3.43}$$

die Elemente h_{ij} sind proportional ω_1. Die Dichtematrix des Ensembles hat dann in der Energiedarstellung in 1. Näherung die Form

$$\rho^{(E)}(t) = \begin{pmatrix} \varrho_{0\,11}^{(E)} & \varrho_{12}^{(E)} & \varrho_{13}^{(E)} & 0 \\ \varrho_{12}^{(E)*} & \varrho_{0\,22}^{(E)} & 0 & \varrho_{24}^{(E)} \\ \varrho_{13}^{(E)*} & 0 & \varrho_{0\,33}^{(E)} & \varrho_{34}^{(E)} \\ 0 & \varrho_{24}^{(E)*} & \varrho_{34}^{(E)*} & \varrho_{0\,44}^{(E)} \end{pmatrix}; \tag{3.44}$$

nach (3.42) sind die Nichtdiagonalelemente von $\rho^{(E)}$ proportional ω_1. Die Dichtematrix ρ in der I_z-Darstellung, die wir später vorwiegend benutzen werden, erhalten wir aus $\rho^{(E)}$ durch die unitäre Transformation

$$\rho = S\,\rho^{(E)}\,S^{-1}\,, \tag{3.45}$$

wo S im Hinblick auf (3.4) durch

$$u_n = S\,\psi_n \quad\text{oder}\quad u_n = \sum_m S_{mn}\,\psi_m \tag{3.46}$$

gegeben ist. Die Matrixelemente S_{mn} für das AB-System können aus den Gleichungen (3.25) abgelesen werden. Mit der Näherung (3.26) finden wir

$$\rho\,(t) = \begin{pmatrix} (1/4)\left(1 - \dfrac{\omega_0\,\hbar}{k\,T}\right) & \varrho_{12} & \varrho_{13} & 0 \\[2ex] \varrho_{12}^{*} & 1/4 & 0 & \varrho_{24} \\[1ex] \varrho_{13}^{*} & 0 & 1/4 & \varrho_{34} \\[2ex] 0 & \varrho_{24}^{*} & \varrho_{34}^{*} \left(1 + \dfrac{\omega_0\,\hbar}{k\,T}\right) 1/4 \end{pmatrix} . \tag{3.47}$$

Die Elemente der Kernspindichtematrix in (3.47) bzw. (3.44) sind mit Hilfe von (3.36), ausgehend von $\varrho_{0\,lm}^{(E)}$, errechnet worden. Damit (3.36) eine gute Näherung darstellt, muß das Zeitintervall t, über das integriert wird, hinreichend klein sein. Tatsächlich wirkt $\mathscr{H}_1\,(t')$ zur Zeit t' auf ein $\bar\rho\,(t')$, das nicht, wie in (3.36) angenommen, gleich ist $\bar\rho\,(0)$. Der relative Unterschied zwischen $\bar\rho\,(t')$ und $\bar\rho\,(0)$ ist um so kleiner, je kürzer t gegen die Larmorperiode $(\gamma|\mathscr{H}_{1e}|)^{-1} = \omega_{1e}^{-1}$ ist:

$$t \ll \omega_{1e}^{-1}\,, \quad \omega_{1e} \lesssim \omega_1\ . \tag{3.48}$$

Der Betrag des effektiven Feldes H_{1e} ist von der gleichen Größenordnung wie jener des rotierenden Feldes H_1, wenn dessen Kreisfrequenz ω bei der Resonanzfrequenz oder in deren Nachbarschaft liegt; bei den übrigen Frequenzen ist $H_{1e} < H_1$. Neben dem Effekt des rotierenden Feldes ist der Einfluß der Wechselwirkung zwischen den Kernen und ihrer Umgebung zu berücksichtigen. Die Wechselwirkung, welche die Relaxation verursacht und die Elemente der Dichtematrix gegen einen stationären Wert streben läßt, ist eine Störung 2. Ordnung, die nach einer Zeit t in Erscheinung tritt, die von der gleichen Größenordnung wie die Relaxationszeit T_1 bzw. T_2 ist. Wenn die Ungleichung

$$T_1\,,\ T_2 \ll \omega_{1e}^{-1} \tag{3.49}$$

erfüllt ist, stellt daher das im Hinblick auf die weitere Entwicklung wichtige Ergebnis (vgl. (5.6)), daß in (3.44) und (3.47) die Diagonalelemente der Dichtematrix gleich den Boltzmannfaktoren und die Elemente der Nebendiagonale Null sind, für beliebige Zeiten t, insbesondere für den stationären Zustand ($t \to \infty$), eine gute Näherung dar.

Die Relaxation stellen wir in Anlehnung an die phänomenologischen Blochschen Gleichungen, die das Verhalten von Kernen mit dem Spin $I = {}^1/_2$ in Flüssigkeiten unter den üblichen Versuchsbedingungen meist befriedigend beschreiben, durch die Terme

$$\left(\frac{\partial \rho}{\partial t}\right)_{\text{Relax}} = \frac{\rho_0 - \rho}{T_1} - \frac{\varrho_{nd}}{T_2'} \tag{3.50}$$

dar, die man als Störglieder 2. Ordnung auffassen kann; T_2' ist durch $(T_2)^{-1} = (T_2)^{-1} - (T_1)^{-1}$ festgelegt, ϱ_{nd} schließt alle Nichtdiagonalelemente von ρ (in der Energie- oder I_z-Darstellung) ein.

Nehmen wir an, daß sich die Bewegungen von ρ, die durch die Magnetfelder H_0 und H_1 sowie durch die Relaxation verursacht werden, überlagern[7], so ergibt sich (mit dem Hamiltonoperator $\mathscr{H}'$ in Kreisfrequenzeinheiten)

$$\frac{d\rho}{dt} = i\,[\rho, \mathscr{H}']_- + \frac{\rho_0 - \rho}{T_1} - \frac{\varrho_{nd}}{T_2'}\,. \tag{3.51}$$

Gleichung (3.51) wird als die Boltzmanngleichung des Kernspinsystems bezeichnet [14, 21]. Es sei noch einmal darauf hingewiesen, daß die Relaxation nur in den einfachsten Fällen durch Terme des Typs (3.50) in hinreichender Näherung dargestellt wird [15]. Eine korrekte quantitative Beschreibung der magnetischen Eigenschaften eines Ensembles von Kernen in äußeren Magnetfeldern durch Gl. (3.51) ist zu erwarten, wenn flüssige Proben mit geringer Viskosität in statischen Feldern H_0 untersucht werden, die groß gegen die Linienbreite $(\gamma\,T_2)^{-1}$ in Gauß und groß gegen die Amplitude H_1 des Radiofrequenzfeldes sind.

4. Der Umlagerungsterm in der Boltzmanngleichung

4.1. Klassifizierung chemischer Umlagerungen

Chemische Elementarprozesse, welche die Struktur und Linienform hochaufgelöster Spektren der kernmagnetischen Resonanz beeinflussen, ordnet man zweckmäßig nach folgenden Klassen:

    I. Intramolekulare Umlagerungen

        1. Symmetrische intramolekulare Umlagerungen oder intramolekulare Austauschvorgänge

        2. Asymmetrische intramolekulare Umlagerungen

---

[7] Die Annahme ist gerechtfertigt, wenn die Bedingung $\omega_1\,\tau_c \ll 1$ erfüllt ist. τ_c bezeichnet die Korrelationszeit, die in Flüssigkeiten mit geringer Viskosität im allgemeinen wesentlich kürzer als die Relaxationszeiten T_1 und T_2 ist, so daß eine lineare Bewegungsgleichung für ρ auch dann noch gültig ist, wenn die Sättigung nicht mehr vernachlässigt werden kann.

II. Intermolekulare Umlagerungen oder chemische Austauschvorgänge
   1. Monogene intermolekulare Umlagerungen
   2. Polygene intermolekulare Umlagerungen

Beispiele für intramolekulare Umlagerungen sind die Inversion und die innere Rotation. Eine intramolekulare Umlagerung ist symmetrisch, wenn die zugehörigen thermodynamischen Reaktionseffekte, wie $\Delta H_0, \Delta V_0, \Delta S_0$ usf., Null sind, also insonderheit, wenn Ausgangs- und Endzustand die gleiche Energie besitzen. Sie ist asymmetrisch, wenn die thermodynamischen Zustandsgrößen von Ausgangs- und Endzustand verschieden sind. Verläuft die Umlagerung hinreichend langsam, so besteht das Kernresonanzspektrum des einfachsten Systems, in dem eine symmetrische (asymmetrische) intramolekulare Umlagerung stattfindet, aus einem symmetrischen (asymmetrischen) Dublett. Bei intramolekularen Umlagerungen ist der Spinzustand der Kerne eines Moleküls nach der Umlagerung eindeutig durch denjenigen vor der Umlagerung bestimmt.

Dies trifft bei intermolekularen Umlagerungen im allgemeinen nicht zu. Bei letzteren tauschen zwei Moleküle mit Atomkernen in Spinzuständen, die nicht miteinander korreliert sind, Atome oder Atomgruppen miteinander aus. Der Zustand der Kernspins eines Moleküls nach dem chemischen Austausch hängt nicht nur von ihrem ursprünglichen Zustand, sondern auch vom Zustand der Kernspins des Moleküls ab, mit dem der Austausch erfolgt. Der chemische Austausch kann zwischen Molekülen einer Sorte oder zwischen Molekülen mehrerer Sorten stattfinden. Erstere Reaktion nennen wir eine monogene, letztere eine polygene intermolekulare Umlagerung.

Die bei Abwesenheit von Spin-Spin-Kopplung streng gültigen Gleichungen (2.12) von Gutowsky und Saika beschreiben den Effekt der asymmetrischen intramolekularen Umlagerung sowie den des chemischen Austausches zwischen Molekülen verschiedener Sorten. Der erwähnte Unterschied zwischen intra- und intermolekularen Umlagerungen erscheint erst dann in der Theorie, wenn die Spin-Spin-Kopplung zwischen den Kernen, die ausgetauscht werden, und den fixierten Kernen eines Moleküls zu berücksichtigen ist. Setzt man in (2.12) $\tau_A = \tau_B$ und $p_A = p_B = {}^1/_2$, so erhält man die Gleichung von Gutowsky, McCall und Slichter mit dem Effekt der symmetrischen intramolekularen Umlagerung. Der chemische Austausch zwischen verschiedenen Molekülen einer Sorte wird bei Abwesenheit von Spin-Spin-Kopplung als Effekt in der Linienform nicht beobachtet und kann deshalb in der Theorie für Systeme ohne Spin-Spin-Kopplung nicht auftreten. In Abschnitt 7.1 wird gezeigt, unter welchen Bedingungen die adiabatische Gleichung von Gutowsky, McCall und Slichter auch bei Anwesenheit von Spin-Spin-Kopplung zur Beschreibung des chemischen Austausches zwischen Molekülen einer Sorte angewandt werden kann. Im allgemeinen ist jedoch bei Gegenwart von Spin-Spin-Kopplung die Darstellung in Dichtematrixtheorien unumgänglich.

Im folgenden wird der Effekt symmetrischer intramolekularer Umlagerungen bei Anwesenheit von Spin-Spin-Kopplung behandelt (Kaplan [17], Alexander [18]). Es wird versucht, die Grundzüge der Theorie so ausführlich darzulegen, daß die Anwendung des Formalismus auf andere als die hier behandelten Systeme und der Zugang zu den Arbeiten über die Effekte der asymmetrischen intramolekularen Umlagerung (Johnson [19], Heidberg [20]) und des chemischen Austausches bei Anwesenheit von Spin-Spin-Kopplung [17, 21] erleichtert werden.

4.2. Die symmetrische intramolekulare Umlagerung. Der Diracsche Spinaustauschoperator

Um die Aufgabe, die uns gestellt ist, zu erläutern, betrachten wir zuerst ein konkretes System, nämlich eine verdünnte Lösung von N-Alkyl-2,4,6-trinitroanilin in einem inerten Lösungsmittel, an dem eine symmetrische intramolekulare Umlagerung eingehend experimentell untersucht worden ist [16]:

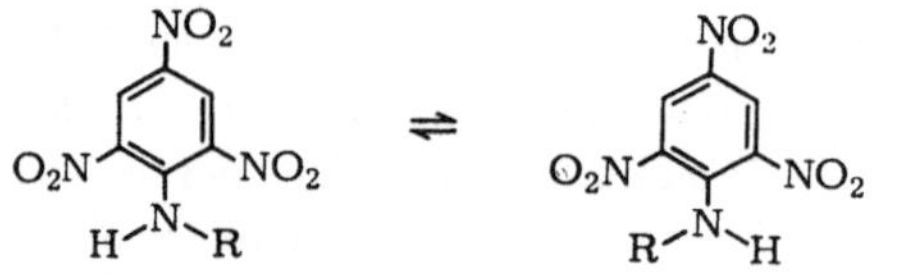

$$\text{(K)} \quad \rightleftharpoons \quad \text{(L)} \tag{4.1}$$

Die der Alkylgruppe R benachbarte Nitrogruppe in 2-Stellung ist um einen bestimmten Winkel aus der Ebene des Benzolringes herausgedreht. Die übrigen Nitrogruppen liegen in der Ringebene. Die Richtungen der Bindungen (R—N) und (N—H) sind in bezug auf die Aminostickstoff-Ringbindung (nahezu) symmetrisch. Eine derartige Konfiguration (K) besitzt bei einer vorgegebenen Temperatur eine bestimmte mittlere Lebensdauer τ. Rotiert die Aminogruppe um die Aminostickstoff-Ringbindung mit etwa 180°, so stellt sich die Nitrogruppe in 2-Stellung in die Ringebene, während sich die Nitrogruppe in 6-Stellung aus der Ringebene herausdreht. Die Konfiguration (K) geht in die Konfiguration (L) über, welche die gleiche Energie wie (K) besitzt. Bei jedem Konfigurationsübergang vertauschen die miteinander gekoppelten Protonen in 3- und 5-Stellung ihre Umgebung. Vgl. Abb. 1.

Der intramolekulare Austausch der Umgebungen zwischen Paaren von im allgemeinen gekoppelten, gleichartigen Kernspins (z. B. Protonen) kennzeichnet die symmetrische intramolekulare Umlagerung. Der Parameter, welcher die Umgebung eines herausgegriffenen Kerns (X) in einem vorgegebenen Magnetfeld beschreibt, ist seine chemische Verschiebung $\delta^{(X)}$. Der Kernspinhamiltonoperator des Moleküls nach dem Aus-

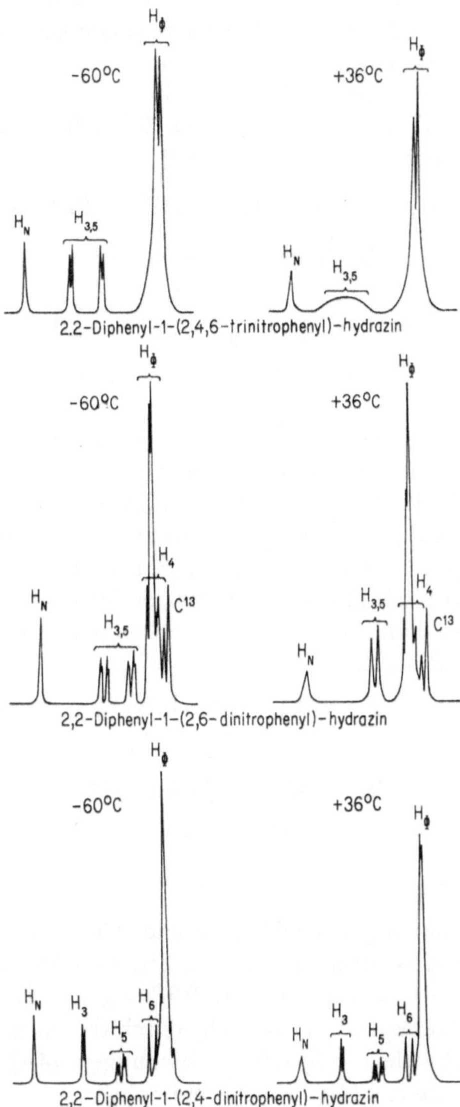

Abb. 1. Die protonmagnetischen Resonanzspektren einiger 2,2-Diphenyl-1-(polynitrophenyl)-hydrazine bei 60 MHz. Lösungsmittel: Dichlormethan. Temperatur −60°C und +36°C. Die ^{13}C-Banden des Lösungsmittels sind markiert. Die KMR-Spektren vom A_2 und AX_2−Typ der nitrosubstituierten Phenylringe, die bei 36°C beobachtet werden, spalten zu Spektren des AB- und AMX-Typs im Fall des 2,4,6-Trinitroanilin- bzw. 2,6-Dinitroanilinderivats auf. Das Spektrum des 2,4-Dinitroanilinderivats bleibt unverändert

tausch, $\mathscr{H}_L$, unterscheidet sich von jenem vor dem Austausch, $\mathscr{H}_K$, lediglich dadurch, daß ein Paar oder mehrere Paare von chemischen Verschiebungen ihre Indizes vertauscht haben.

Für die beiden gekoppelten Protonen in 3- und 5-Stellung des betrachteten Moleküls K, die wir mit A bzw. B bezeichnen, hat der Kernspinhamiltonoperator bekanntlich die Gestalt

$$\mathscr{H}'_K = \omega_0 \, (I_z^{(A)} + I_z^{(B)}) + \delta \, I_z^{(A)} - \delta \, I_z^{(B)} + J(\mathbf{I}^{(A)}, \mathbf{I}^{(B)}) \,. \tag{4.2}$$

Dann ergibt sich $\mathscr{H}'_L$, der Kernspinhamiltonoperator nach der Umlagerung, zu

$$\mathscr{H}'_L = \omega_0 \, (I_z^{(A)} + I_z^{(B)}) - \delta \, I_z^{(A)} + \delta \, I_z^{(B)} + J \, (\mathbf{I}^{(A)}, \mathbf{I}^{(B)}) \,. \tag{4.3}$$

$\mathscr{H}'_K$ und $\mathscr{H}'_L$ bestimmen die Entwicklung der Kernspinfunktion mit der Zeit t.

Wir nehmen an, daß die Umwandlung von K nach L und von L nach K sprunghaft erfolgt. Mit der Einführung dieser Annahme vernachlässigen wir den Aufenthalt des Systems in Übergangszuständen. Die Annahme dürfte im allgemeinen gerechtfertigt sein, wenn die Umlagerung in einem Zeitintervall erfolgt, das klein im Vergleich zu den mittleren Lebensdauern der Moleküle ist (vgl. Gl. (2.10) ff.). Findet der Sprung von K nach L zur Zeit $t = t'$ statt, so können wir die Zeitabhängigkeit der Spinfunktion der Kerne A und B durch

$$\varphi_K (\sigma_A, \sigma_B; t) = \exp \left[- i \, \mathscr{H}'_K t\right] \varphi_K (\sigma_A, \sigma_B; 0)$$
$$\text{für } t' \geq t \geq 0,$$
$$\varphi_L (\sigma_A, \sigma_B; t_L) = \exp \left[- i \, \mathscr{H}'_L t_L\right] \varphi_K (\sigma_A, \sigma_B; t'),$$
$$\text{für } t \geq t', \tag{4.4}$$

wo

$$t_L = t - t',$$

beschreiben, da $\mathscr{H}'_K$ und $\mathscr{H}'_L$ zeitunabhängig sind. Ähnlich kann man die Spinfunktion der Kerne darstellen, die ursprünglich zum Molekül L gehören.

Es erweist sich als zweckmäßig, die Wirkung der Umwandlung von K nach L auf die Kernspinzustände nicht durch die unmittelbar aus dem physikalischen Sachverhalt folgenden Gleichungen (4.4) auszudrücken, sondern die Änderung der Hamiltonoperatoren

$$\mathscr{H}'_K \longrightarrow \mathscr{H}'_L \tag{4.5}$$

durch die Änderung der Kernspinfunktionen

$$\varphi_K (t') \longrightarrow P^{(AB)} \varphi_K (t') \tag{4.6}$$

zu ersetzen. $P^{(AB)}$ ist der Diracsche Spinaustauschoperator [$10e$], der die Spinvariablen der Kerne A und B in den Kernspinfunktionen vertauscht, z. B.

$$P^{(AB)} \alpha \, (\sigma_A) \, \beta \, (\sigma_B) = \beta \, (\sigma_A) \, \alpha \, (\sigma_B) \,. \tag{4.7}$$

Anstelle von (4.4) treten die Gleichungen

$$\varphi_K (\sigma_A , \sigma_B ; t) = \exp [-i \mathscr{H}'_K t] \varphi_K (\sigma_A , \sigma_B ; 0)$$
$$\text{für } t' \geq t \geq 0,$$
$$P^{(AB)} \varphi_K (\sigma_A , \sigma_B ; t_L) = \exp [-i \mathscr{H}'_K t_L] P^{(AB)} \varphi_K (\sigma_A , \sigma_B ; t')$$
$$\text{für } t \geq t', \text{ wo } t_L = t - t' . \tag{4.8}$$

Es erhebt sich die Frage, ob (4.4) und (4.8) für gleiche Zeiten, insbesondere für $t > t'$, den gleichen physikalischen Zustand beschreiben. Da die Kerne innerhalb des Moleküls miteinander gekoppelt sind, kann dem einzelnen Kern kein eigener Zustand, sondern nur eine bestimmte Rolle in der Spinfunktion der gekoppelten Kerne des Moleküls zugeordnet werden. $P^{(AB)}$ vertauscht die Rollen der Kerne A und B. Im Zustand (4.4) (für $t > t'$) spielt der Kern A (B), der sich in der Umgebung $\delta^{(B)}$ $(\delta^{(A)})$ befindet, die gleiche Rolle wie im Zustand (4.8) der Kern B (A) in der Umgebung $\delta^{(B)}$ $(\delta^{(A)})$. Hieraus folgt, da A und B gleichartige Kerne, i.e. Protonen sind, daß die Funktionen (4.4) und (4.8) physikalisch äquivalente Zustände darstellen. Ähnliche Betrachtungen kann man für den Übergang $L \to K$ anstellen.

Aus Gleichung (4.6) folgt, daß mit der Einführung von $P^{(AB)}$ die Indizierung der Spinfunktionen φ mit K und L überflüssig wird. Sowohl φ als auch $P^{(AB)} \varphi$, die nicht notwendig Eigenzustände von $\mathscr{H}$ sein müssen, sind im Hilbertraum von $\mathscr{H}$ definiert. Die Eigenschaft, daß der Austauschoperator im Hilbertraum des Kernspinhamiltonoperators $\mathscr{H}$ eines Moleküls bzw. einer Konfiguration festgelegt ist, weist eine intramolekulare Umlagerung als symmetrisch aus. K und L sind Konfigurationen mit dem gleichen Kernspinhamiltonoperator.

Der Spinaustauschoperator $P^{(AB)}$ besitzt eine Reihe für seine Anwendung wichtiger Eigenschaften. Offensichtlich ist $P^{(AB)}$ linear, denn die Spinvariablen müssen in jedem der in φ enthaltenen Summanden der Form (4.7) vertauscht werden. Wenden wir $P^{(AB)}$ zweimal nacheinander an, so kommen wir zum Anfangszustand zurück:

$$P^{(AB)} P^{(AB)} = 1 \quad \text{und} \quad P^{(AB)} = (P^{(AB)})^{-1} . \tag{4.9}$$

Wenn $P^{(AB)}$ auf beide Kernspinfunktionen φ und φ' im skalaren Produkt (φ, φ') einwirkt, verändert sich das Produkt nicht:

$$(P^{(AB)} \varphi , P^{(AB)} \varphi') = (\varphi , \varphi') . \tag{4.10}$$

$P^{(AB)}$ ist mithin unitär. Es läßt sich zeigen, daß $P^{(AB)}$ als algebraische Funktion der Spinvektoroperatoren $\mathbf{I}^{(A)}$ und $\mathbf{I}^{(B)}$ nach

$$P^{(AB)} = \frac{1}{2} [1 + 4 (\mathbf{I}^{(A)} , \mathbf{I}^{(B)})] \tag{4.11}$$

ausgedrückt werden kann. Für unsere Zwecke ist es leichter, die Matrix von P auf folgendem Wege zu ermitteln. Der Kernspinzustand φ sei in einer bestimmten Darstellung, z. B. in der I_z-Darstellung durch

$$\varphi = \sum_n c_n \psi_n \qquad (4.12)$$

gegeben. Bei Anwendung von $P^{(AB)}$ erhalten wir den neuen Kernspin-zustand

$$P^{(AB)} \varphi = \sum_n c_n P^{(AB)} \psi_n \,, \qquad (4.13)$$

der im Hinblick auf die Entwicklung

$$P^{(AB)} \psi_n = \sum_m P^{(AB)}_{mn} \psi_m \qquad (4.14)$$

mit

$$P^{(AB)}_{mn} = (\psi_m \,, P^{(AB)} \psi_n)$$

in der Form

$$P^{(AB)} \varphi = \sum_m c'_m \psi_m \qquad (4.15)$$

geschrieben werden kann, wobei

$$c'_m = \sum_n P^{(AB)}_{mn} c_n \,. \qquad (4.16)$$

Es ist zweckmäßig, die Spinzustände und die Matrixelemente in der I_z-Darstellung auszudrücken, weil in dieser die Matrixelemente von $P^{(AB)}$ entweder 1 oder Null sind.

Wir untersuchen nun, wie sich die Kernspindichtematrix ρ eines Ensembles von Molekülen mit den Kernspinzuständen $\varphi^{(i)}$ transformiert, wenn jedes der Moleküle aus dem Zustand $\varphi^{(i)}$ in den Zustand $P \varphi^{(i)}$ übergeht. Die Dichtematrix vor der Transformation ist durch

$$\varrho_{kl} = \sum_i p^{(i)} R^{(i)}_{kl} \qquad (4.17)$$

mit

$$R^{(i)}_{kl} = c^{(i)*}_l c^{(i)}_k \,,$$

jene nach der Transformation durch

$$\varrho'_{kl} = \sum_i p^{(i)} R'^{(i)}_{kl} \qquad (4.18)$$

mit

$$R'^{(i)}_{kl} = c'^{(i)*}_l c'^{(i)}_k$$

definiert. Eliminieren wir $c'^{(i)*}_l$ und $c'^{(i)}_k$ mit Hilfe von (4.16), so ergibt sich wegen (4.9) und (4.10)

$$\varrho'_{kl} = \sum_i p^{(i)} (P R^{(i)} P)_{kl}$$
$$= (P \varrho P)_{kl} \,. \qquad (4.19)$$

Man beachte, daß ρ und ρ' im allgemeinen verschiedene (makroskopische) Zustände des Ensembles (und nicht verschiedene Darstellungen des gleichen Zustands) sind.

Der Erwartungswert der Kernmagnetisierung $\langle \gamma \hbar \sum_s I_z^{(s)} \rangle$ bzw. $\langle \gamma \hbar \sum_s \{ I_x^{(s)} + i\, I_y^{(s)} \} \rangle$ ist in bezug auf die Operation $P^{(AB)}$ invariant. Dagegen ändert sich der Erwartungswert der Kernspinenergie $\langle \mathscr{H} \rangle$ bei Anwendung von $P^{(AB)}$.

Eine sehr große Anzahl von Molekülen in den Konfigurationen K und L bilde ein statistisches Ensemble in einem vorgegebenen thermodynamischen Zustand, i.e. ein System mit vorgegebenen Werten für die Temperatur, die Molekülzahlen und äußeren Variablen, wie das Volumen[8]. In einem solchen System besitzen nach Voraussetzung die Konfigurationen K und L eines jeden Moleküls bei Abwesenheit eines äußeren Magnetfeldes die gleiche mittlere freie Energie bzw. freie Enthalpie und damit die gleiche mittlere Lebensdauer τ. Dann müssen die mittleren Lebensdauern der Kernspinzustände $\varphi^{(i)}$ alle einander gleich sein. Auch unter den Bedingungen, die gewöhnlich bei Kernresonanzversuchen herrschen, ist die Annahme gleicher mittlerer Lebensdauern für alle Zustände $\varphi^{(i)}$ in guter Näherung erfüllt, da die Zeeman-Energie der Kerne eines Moleküls unter diesen Bedingungen im allgemeinen sehr klein gegen kT ist. Bezeichnen wir mit $N^{(i)}$ und $N^{(j)}$ die Anzahl der Moleküle, deren Kerne sich im Zustand $\varphi^{(i)}$ bzw. $\varphi^{(j)} = P^{(AB)} \varphi^{(i)}$ befinden, so gilt

$$\frac{\mathrm{d}\, N^{(i)}}{\mathrm{d}\,t} = - \frac{N^{(i)}}{\tau} + \frac{N^{(j)}}{\tau}, \tag{4.20}$$

denn die Moleküle mit Kernspinzuständen $\varphi^{(j)}$ gehen bei der Umlagerung in Moleküle mit Kernspinzuständen $P^{(AB)} \varphi^{(j)} = \varphi^{(i)}$ über. Der Zustand $\varphi^{(j)}$ ist in der Gesamtheit der Zustände $\varphi^{(i)}$ enthalten, so daß wir für $\mathrm{d}\,N^{(j)}/\mathrm{d}\,t$ kein besonderes Zeitgesetz anzugeben haben. Dann ändern sich als Folge der Umlagerung mit den Besetzungszahlen $N^{(i)}$ der Zustände die statistischen Gewichte $p^{(i)}$ und damit die Kernspindichtematrizen nach

$$\left(\frac{\partial \varrho_{mn}}{\partial t} \right)_{\text{Umlag.}} = - \sum_{i=1} \frac{p^{(i)}}{\tau} R_{mn}^{(i)}$$
$$+ \sum_{j=1} \frac{p^{(j)}}{\tau} (P^{(AB)} R^{(j)} P^{(AB)})_{mn}, \tag{4.21}$$

$$\left(\frac{\partial \rho}{\partial t} \right)_{\text{Umlag.}} = \frac{P^{(AB)} \rho\, P^{(AB)} - \rho}{\tau}. \tag{4.22}$$

---

[8] Die folgenden Betrachtungen gelten — wie alle Gesetze der Thermodynamik — streng nur für den Grenzfall unendlich großer Systeme.

Setzen wir voraus, daß die Wirkungen der Magnetfelder, der Relaxation und der Umlagerung unabhängig voneinander sind und sich additiv verhalten (2.11), so finden wir

$$\frac{d\rho}{dt} = i\,[\rho,\mathscr{H}']_- - \frac{\rho - \rho_0}{T_1} - \frac{\varrho_{nd}}{T_2'} + \frac{P^{(AB)}\,\rho\,P^{(AB)} - \rho}{\tau} \; . \qquad (4.23)$$

Diese Gleichung stellt die generalisierte Boltzmanngleichung mit Einschluß symmetrischer intramolekularer Umlagerungen dar. Spezielle Annahmen über die Struktur des Kernspinhamiltonoperators sind in (4.23) nicht enthalten.

Wir gewinnen neue Einsicht, wenn wir den Einfluß der symmetrischen intramolekularen Umlagerung auf die Bewegung der Kernspindichtematrix von einem anderen Standpunkt ohne Anwendung des Austauschoperators P untersuchen [19]. Es sei ein Ensemble von Molekülen der Sorte (K) gegeben. Der Kernspinhamiltonoperator $\mathscr{H}'_K$ eines jeden Moleküls habe die in (4.2) festgelegte Gestalt. In jedem Molekül ist der Kern mit der chemischen Verschiebung $+\delta$ $(-\delta)$ mit A (B) bezeichnet. Das r-te Molekül befinde sich im Kernspinzustand $\varphi_K^{(i)}$:

$$\varphi_K^{(i)} = {}^Kc_1^{(i)}\,\psi_1 + {}^Kc_2^{(i)}\,\psi_2 + {}^Kc_3^{(i)}\,\psi_3 + {}^Kc_4^{(i)}\,\psi_4 \qquad (4.24a)$$

mit

$$\begin{aligned}
\psi_1 &= \alpha\,(\sigma_A)\,\alpha\,(\sigma_B)\,, & \psi_3 &= \beta\,(\sigma_A)\,\alpha\,(\sigma_B)\,, \\
\psi_2 &= \alpha\,(\sigma_A)\,\beta\,(\sigma_B)\,, & \psi_4 &= \beta\,(\sigma_A)\,\beta\,(\sigma_B)\,.
\end{aligned} \qquad (4.24b)$$

Die Anzahl der Moleküle im Zustand $\varphi_K^{(i)}$ ist ${}^KN^{(i)}$, das statistische Gewicht von $\varphi_K^{(i)}$ ist ${}^Kp^{(i)}$. Wir betrachten nun ein Ensemble von Molekülen der Sorte (L) mit dem Kernspinhamiltonoperator $\mathscr{H}'_L$ (4.3). $\mathscr{H}'_L$ geht aus $\mathscr{H}'_K$ hervor, indem die Kerne A und B ihre chemischen Verschiebungen vertauschen. Wir können den Zustand eines Moleküls der Sorte (L) in Analogie zu (4.24) in der ψ_n-Darstellung beschreiben,

$$\varphi_L^{(i)} = {}^Lc_1^{(i)}\,\psi_1 + {}^Lc_2^{(i)}\,\psi_2 + {}^Lc_3^{(i)}\,\psi_3 + {}^Lc_4^{(i)}\,\psi_4 \;, \qquad (4.25)$$

wobei die Basisfunktionen ψ_n durch (4.24 b) gegeben sind. Wir beachten, daß in (4.25) — wegen der speziellen Form von (4.3) — die Kerne mit der chemischen Verschiebung $-\delta$ mit A, jene mit der chemischen Verschiebung $+\delta$ mit B bezeichnet werden.

Die Moleküle der Sorte (K) wandeln sich in Moleküle der Sorte (L) um und umgekehrt. Die Moleküle der Sorten (K) und (L) besitzen die gleiche Lebensdauer τ. Wir können das Zeitgesetz für die Änderung der Besetzungszahlen ${}^KN^{(i)}$ bzw. ${}^LN^{(i)}$ sofort angeben, wenn wir die folgenden zwei Annahmen machen: 1. Die Umwandlung $(K) \to (L)$ und der Umkehrvorgang verlaufen sprunghaft, so daß sich der Spinzustand während der Umwandlung nicht ändert. 2. Im Umwandlungszeitpunkt $t = t'$, und nur für $t = t'$, befinden sich die $^K N^{(i)}$ Moleküle der Sorte (K) im gleichen Spinzustand wie die $^L N^{(i)}$ Moleküle der Sorte (L) [s. Gl. (4.4)]:

$$^K c_1^{(i)}(t') = {}^L c_1^{(i)}(t') \, , \quad {}^K c_2^{(i)}(t') = {}^L c_2^{(i)}(t') \, , \quad \text{usf.} \tag{4.26}$$

Dann gilt

$$\frac{\mathrm{d}\,{}^K N^{(i)}}{\mathrm{d}\,t} = \frac{{}^L N^{(i)} - {}^K N^{(i)}}{\tau} \, , \tag{4.27}$$

woraus

$$\frac{\mathrm{d}\,{}^K \rho}{\mathrm{d}\,t} = -\frac{{}^K \rho - {}^L \rho}{\tau} \tag{4.28}$$

folgt. Die generalisierte Boltzmanngleichung für $^K\rho$ lautet

$$\frac{\mathrm{d}\,{}^K\rho}{\mathrm{d}\,t} = i\,[{}^K\rho \, , \, \mathscr{H}'_K]_- - \frac{{}^K\rho - {}^K\rho_0}{T_1} - \frac{{}^K\rho_{nd}}{T'_2} + \frac{{}^L\rho - {}^K\rho}{\tau} \, . \tag{4.29}$$

Vertauschen wir (K) und (L) in (4.29), so erhalten wir die Bestimmungsgleichung für $^L\rho$. Eine Lösung der Boltzmanngleichung führen wir im nächsten Paragraphen vor. Wenn $^K\rho$ und $^L\rho$ bekannt sind, können die Ensemble-Mittelwerte $^K\langle A\rangle$ und $^L\langle\overline{A}\rangle$ der physikalischen Größe A nach (3.13) berechnet werden. Die Molenbrüche der Molekülsorten (K) und (L) sind gleich, so daß zur Bestimmung der Eigenschaften des Gesamtsystems, das aus den Ensembles der Molekülsorten (K) und (L) gebildet wird, die entsprechenden Ensemble-Mittelwerte $^K\langle\overline{A}\rangle$ und $^L\langle\overline{A}\rangle$ nur addiert werden müssen. Es ist evident, daß die beiden Verfahren zur Behandlung des Einflusses der Umlagerung physikalisch vollkommen äquivalent sind. Die Annahme (1), daß die Umwandlung sprunghaft verlaufe, liegt beiden Verfahren zugrunde. Der Annahme (2), Moleküle der Sorte (K) seien zur Zeit $t = t'$ in den gleichen Kernspinzuständen wie Moleküle der Sorte (L), entspricht die Annahme der gleichzeitigen Existenz der Kernspinzustände φ und $\mathrm{P}^{(AB)}\varphi$ zur Zeit $t = t'$. Auf den vorgezeichneten Wegen können die angegebenen Gleichungen verallgemeinert und diejenigen Beziehungen hergeleitet werden, welche die Effekte der intramolekularen Umlagerung, einschließlich der asymmetrischen, beschreiben [19, 20].

4.3. Verallgemeinerung für asymmetrische intramolekulare Umlagerungen. Hinweis auf die Behandlung des Relaxationsproblems.

Der für den allgemeinen Fall des Übergangs zwischen beliebigen Konfigurationen i eines Moleküls vorgeschlagene Ansatz [19] [20]

$$\frac{\mathrm{d}\,{}^i\rho}{\mathrm{d}\,t} = i\,[{}^i\rho, \mathscr{H}'_i]_- + \sum_j {}^j\rho\, d_{ij} \, , \tag{4.30}$$

in dem T_1- und T_2-Terme weggelassen sind, stellt die Generalisierung von Gl. (4.29) dar. Der Koeffizient d_{ij} enthält die Wahrscheinlichkeit für den

Übergang von j nach i ($d_{ii} = -1/\tau_i$). Wie in unserem Beispiel vorgeführt, kennzeichnet man die Kernspins eines Moleküls, betrachtet alle möglichen Konfigurationen i desselben, formuliert für jede den zugehörigen Spin-hamiltonoperator $\mathcal{H}_i$ und stellt die intramolekulare Reaktion $i \to j$ durch den Übergang von $\mathcal{H}_i$ nach $\mathcal{H}_j$ dar. Der Ensemble-Mittelwert der Gesamt-intensität im KMR-Spektrum, $I(\omega)$, ist gleich der Summe der Beiträge G_i aller Konfigurationen i (s. Gl. 5.4)

$$I(\omega) = \sum_i x_i\, G_i \; , \tag{4.31}$$

wobei

$$G_i = \gamma\, \hbar\, \mathrm{Sp}\, \left\{ {}^i\rho^{(\omega)} \sum_s (\mathrm{I}_x^{(is)} - i\,\mathrm{I}_y^{(is)}) \right\} \tag{4.32}$$

sowie x_i der Molenbruch der Komponente i und ${}^i\rho^{(\omega)}$ die Kernspindichte-matrix von i in dem Koordinatensystem ist, das mit der Kreisfrequenz von H_1, ω, rotiert.

Gl. (4.30) ist überdies zur Behandlung des Relaxationsproblems bei Flüssigkeiten, wie es in den Theorien von Bloch, Wangsness und Redfield gestellt wird, geeignet. Ausgangspunkt dieser Theorien ist Gl. (3.34), die in 2. Näherung gelöst wird, wobei die Bedingung $\tau_c \ll T_1$, T_2 erfüllt sein muß. Vgl. [11, 12]. Die Korrelationszeit τ_c für höher viskose Flüssigkeiten und für große, relativ langsam sich bewegende Moleküle in Lösung ist aber gewöhnlich nicht sehr kurz gegen die Relaxationszeiten T_1 und T_2. Beschreibt man die (nur in der klassischen Theorie kontinuierliche) Diffusion als einen Prozeß, der in sprungartigen Übergängen zwischen Hamilton-operatoren $\mathcal{H}_i$ besteht, wobei hier der Index i eine bestimmte Lage des Moleküls bezeichnet, so geht man bei der Behandlung des Relaxations-problems an Stelle von Gl. (3.34) zweckmäßig von Gl. (4.30) aus, wobei man die Übergangswahrscheinlichkeiten d_{ij} einer endlichen Approximation der (klassischen) Diffusionsgleichung entnimmt. Der Vorzug dieser Betrachtungsweise liegt darin, daß Gl. (4.30), die einer numerischen Lösung gut zugänglich ist, Linienformen magnetischer Resonanz von Flüssigkeiten mit beliebig langen Korrelationszeiten zu berechnen ermöglicht [R. G. Gordon et al., Veröffentl. in Vorbereitung; H. Sillescu and D. Kivelson, J. Chem. Phys. im Druck; J. H. Freed, J. Chem. Phys. im Druck].

5. Eine Lösung der generalisierten Boltzmanngleichung

Als Beispiel für eine Lösung von (4.23) berechnen wir das hochauf-gelöste Kernresonanzspektrum zweier gekoppelter identischer Kerne mit dem Spin $I = 1/2$, die in einer symmetrischen intramolekularen Umlagerung ihre Umgebungen vertauschen.

Es ist zweckmäßig, im rotierenden Koordinatensystem, in dem das rotierende Magnetfeld H_1 stationär ist, zu arbeiten. Die Dichtematrix $\rho^{(\omega)}$ im rotierenden Koordinatensystem ist nach den Gleichungen (A. 33), (A. 32b) und (3.14) durch

$$\rho^{(\omega)} = U^{-1}\, \rho\, U \tag{5.1a}$$

mit

$$U = Q^{-1} = e^{-i\,\omega\,t\,\sum_s I_z^{(s)}} \tag{5.1b}$$

festgelegt (vgl. [22]). Differentation von (5.1) nach t ergibt

$$\frac{d\rho}{dt} = U\,\frac{d\rho^{(\omega)}}{dt}\,U^{-1} - i\,\omega\,U\left[\sum_s I_z^{(s)},\rho^{(\omega)}\right]_-U^{-1}\,. \tag{5.2}$$

Eliminieren wir $d\rho/dt$ in (5.2) und (4.23), wobei $\mathscr{H}'$ in (4.23) durch (3.38) gegeben ist, und multiplizieren die erhaltene Gleichung von links mit U^{-1} und von rechts mit U, so folgt bei Beachtung der Vertauschbarkeit von P und U die Bewegungsgleichung der Dichtematrix im rotierenden Koordinatensystem

$$\frac{d\rho^{(\omega)}}{dt} = i\,(\omega - \omega_0)\left[\sum_s I_z^{(s)},\rho^{(\omega)}\right]_- - i\left[\delta\,(I_z^{(A)} - I_z^{(B)})\right.$$

$$\left. + J\,(\mathbf{I}^{(A)},\mathbf{I}^{(B)}),\rho^{(\omega)}\right]_- - i\,\omega_1\left[\sum_s I_x^{(s)},\rho^{(\omega)}\right]_- \tag{5.3}$$

$$+ \frac{\rho_0^{(\omega)} - \rho^{(\omega)}}{T_1} - \frac{\varrho_{nd}^{(\omega)}}{T_2'} + \frac{P\,\rho^{(\omega)}\,P - \rho^{(\omega)}}{\tau}\,.$$

Die Meßgröße in einem Kernresonanzexperiment ist bekanntlich die transversale Komponente G der Kernmagnetisierung in bezug auf die mit der Kreisfrequenz ω rotierenden Koordinaten. Ihr Ensemble-Mittelwert als Funktion von ω errechnet sich im Hinblick auf (3.13) nach

$$G = \gamma\,\hbar\,\mathrm{Sp}\left\{\rho^{(\omega)}\sum_s (I_x^{(s)} - i\,I_y^{(s)})\right\} \tag{5.4a}$$

in der I_z-Darstellung zu

$$G = \gamma\,\hbar\,(\varrho_{12}^{(\omega)} + \varrho_{13}^{(\omega)} + \varrho_{24}^{(\omega)} + \varrho_{34}^{(\omega)})\,. \tag{5.4b}$$

Der Imaginärteil von (5.4) als Funktion von ω stellt das Absorptionsspektrum dar, der Realteil das Dispersionsspektrum. Es genügt, (5.3) für diejenigen Elemente von $\rho^{(\omega)}$ zu lösen, die auf der rechten Seite von (5.4) erscheinen:

$$\frac{d\varrho_{12}^{(\omega)}}{dt} = -i\,(\Delta - \delta + J/2)\,\varrho_{12}^{(\omega)} + i\,\frac{1}{2}\,J\,\varrho_{13}^{(\omega)} - i\,\frac{1}{2}\,\omega_1\,(\varrho_{22}^{(\omega)}$$

$$- \varrho_{11}^{(\omega)} + \varrho_{32}^{(\omega)} - \varrho_{14}^{(\omega)}) - \frac{\varrho_{12}^{(\omega)}}{T_2} + \frac{\varrho_{13}^{(\omega)} - \varrho_{12}^{(\omega)}}{\tau}\,, \tag{5.5a}$$

$$\frac{d\varrho_{13}^{(\omega)}}{dt} = -i\left(\varDelta + \delta + J/2\right)\varrho_{13}^{(\omega)} + i\frac{1}{2}J\varrho_{12}^{(\omega)} - i\frac{1}{2}\omega_1(\varrho_{33}^{(\omega)}$$

$$-\varrho_{11}^{(\omega)} + \varrho_{23}^{(\omega)} - \varrho_{14}^{(\omega)}) - \frac{\varrho_{13}^{(\omega)}}{T_2} + \frac{\varrho_{12}^{(\omega)} - \varrho_{13}^{(\omega)}}{\tau}\ , \tag{5.5b}$$

wo $\varDelta = \omega_0 - \omega$. Die Gleichungen für $\varrho_{24}^{(\omega)}$ und $\varrho_{34}^{(\omega)}$ entsprechen jenen für $\varrho_{13}^{(\omega)}$ bzw. $\varrho_{12}^{(\omega)}$, außer, daß die Vorzeichen von $J, \varrho_{14}^{(\omega)}, \varrho_{32}^{(\omega)}$ und $\varrho_{23}^{(\omega)}$ umgekehrt sind.

Man kann die Gleichungen (5.5) vereinfachen, wenn das rotierende Magnetfeld H_1 hinreichend schwach ist, d. h. $\omega_1 T_1 \ll 1$. Da die Relaxationsterme und der Umlagerungsterm in der Boltzmanngleichung (4.23) Störeffekte höherer als 1. Ordnung beschreiben, geht (4.23) in die v. Neumanngleichung (3.29) mit den Hamiltonoperatoren (3.23) und (3.40) über, wenn wir nur die Effekte 1. Ordnung berücksichtigen. Dann folgt aus (3.47) für die Diagonalelemente in 1. Näherung

$$\varrho_{II}^{(\omega)} = \varrho_{II} \approx \varrho_{0\,II} \tag{5.6a}$$

und für die Elemente der Nebendiagonale

$$\varrho_{14}^{(\omega)} \approx \varrho_{23}^{(\omega)} \approx \varrho_{32}^{(\omega)} \approx \varrho_{41}^{(\omega)} \approx 0\ . \tag{5.6b}$$

Im allgemeinen nimmt man ein hochaufgelöstes Kernresonanzspektrum bei langsamem Durchgang (slow passage) auf. Die Kreisfrequenz ω des rotierenden Feldes H_1 (bzw. die Größe des Zeemanfeldes H_0) wird so langsam verändert, daß das System während der Messung in bezug auf das rotierende Koordinatensystem in einem stationären Zustand verharrt. Dann verschwinden in (5.3) und (5.5) die zeitlichen Ableitungen.

Mithin ergibt sich G bei vernachlässigbarer Sättigung und langsamem Durchgang zu [16]

$$G = c\left[\frac{1 - (J/2 - i/\tau)/A_-}{\varDelta + (J/2 - i/\tau) - i/T_2 + A_-} + \right.$$

$$+ \frac{1 + (J/2 - i/\tau)/A_-}{\varDelta + (J/2 - i/\tau) - i/T_2 - A_-} +$$

$$+ \frac{1 + (J/2 + i/\tau)/A_+}{\varDelta - (J/2 + i/\tau) - i/T_2 + A_+} +$$

$$\left. + \frac{1 - (J/2 + i/\tau)/A_+}{\varDelta - (J/2 + i/\tau) - i/T_2 - A_+}\right] \tag{5.7}$$

mit
$$A_{\pm} = [\delta^2 + (J/2 \pm i/\tau)^2]^{1/2}\ ;$$

$A_{\pm}$ muß überall mit dem gleichen Vorzeichen der Wurzel eingesetzt werden. Der Faktor c, der die Linienform nicht beeinflußt, ist proportional $(\omega_0\,\omega_1\,T^{-1})$. Der Imaginärteil v von G, der das Absorptionsspektrum beschreibt, ist durch

$$v = c\left[\frac{r_+ b_+ - s a_+}{a_+^2 + b_+^2} + \frac{r_- b_- - s a_-}{a_-^2 + b_-^2}\right] \tag{5.8}$$

gegeben, wo

$$a_{\pm} = [\varDelta \pm \frac{1}{2} J]^2 - (\tau^{-1} + T_2^{-1})^2 - \delta^2 - \frac{1}{4} J^2 + \tau^{-2} ,$$

$$b_{\pm} = 2 [\varDelta \pm \frac{1}{2} J] (\tau^{-1} + T_2^{-1}) \mp J/\tau ,$$

$$r_{\pm} = \varDelta \pm J ,$$
$$s = 2/\tau + 1/T_2 .$$

Die mittlere Lebensdauer τ wird bestimmt, indem man ν durch Variation von τ an das beobachtete Spektrum anpaßt (Abb. 2). Mit $\tau \to \infty$ konver-

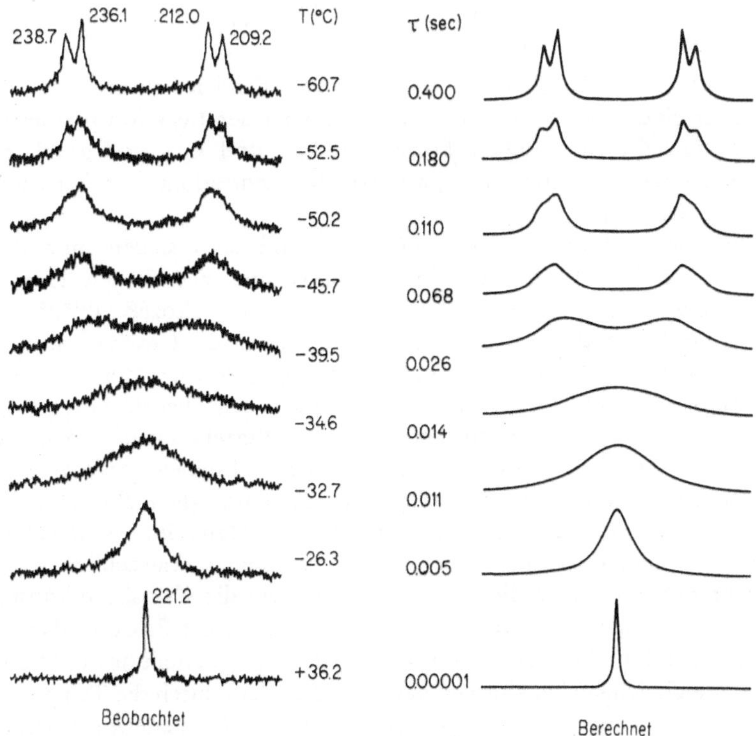

Abb. 2. 60 MHz KMR-Spektren der Picrylprotonen von N-Methyl-N-deutero-2,4,6-trinitroanilin in Dichlormethan. Linienlagen in Hz in bezug auf die Lösungsmittellinie

giert ν zu dem bekannten AB-Spektrum, das aus 4 Linien bei $\omega = \omega_0 \pm J/2 \pm (\delta^2 + J^2/4)^{1/2}$ mit den Intensitäten $1 \pm \frac{1}{2} J (\delta^2 + J^2/4)^{-1/2}$ und den Halb-

25*

wertsbreiten $2/T_2$ besteht, wobei sich im Ausdruck für die Intensität das obere und untere Vorzeichen auf die inneren bzw. äußeren Linien bezieht. Verschwindet die Spin-Spin-Kopplung, so vereinfacht sich (5.8) zu dem Ausdruck für v nach Gleichung (2.12) (mit $\tau_A = \tau_B$).

6. Kinetische Größen aus der kernmagnetischen Resonanz. Die Kinetik von Prozessen im chemischen Gleichgewicht

Das empirische Zeitgesetz für die Geschwindigkeit der Änderung der Konzentration c_i einer Spezies i lautet

$$\frac{\mathrm{d}\,c_i}{\mathrm{d}\,t} = \sum_j \left(k'_{j \to i}\, c_j - k'_{i \to j}\, c_i \right) ; \tag{6.1}$$

$k'_{j \to i}$ und $k'_{i \to j}$ sind Größen, die, wenn die betreffenden Schritte nicht unimolekular sind, noch von den Konzentrationen der Reaktionspartner abhängen. Die Geschwindigkeitskonstanten $k_{j \to i}$ und $k_{i \to j}$ sind gleich den Werten der gestrichenen Größen, wenn die Konzentrationen der Reaktionspartner Eins sind.

Die Geschwindigkeitskonstanten $k_{j \to i}$ und $k_{i \to j}$ stellen nur dann eine klare und wohldefinierte Spezifikation eines chemischen Elementarprozesses dar, wenn das System, in dem der Prozeß abläuft, im Gleichgewicht sich befindet. Nur in diesem Fall ist die Geschwindigkeitskonstante tatsächlich eine Konstante, unabhängig von der Zeit, der Konzentration und den Meßmethoden. In der statistischen Theorie wird gezeigt, daß die Geschwindigkeitskonstante von den Verteilungsfunktionen der Reaktanten für die Translation und die inneren Freiheitsgrade abhängt. Gewöhnlich kann man in makroskopischen Systemen, die nicht im Gleichgewicht sind, diese Verteilungen nicht bestimmen. Man könnte sich dadurch veranlaßt sehen, Gleichung (6.1) völlig aufzugeben. Tatsächlich ist die Gleichung aber noch nützlich, wenn nur entweder die k's nahezu konstant oder ihre Veränderungen berechenbar sind. Finden die Prozesse dagegen in Systemen im Gleichgewicht statt, so sind die Verteilungsfunktionen durch ihre Gleichgewichtsformen gegeben, die allein durch die Temperatur festgelegt sind. Die Beobachtung von chemischen Elementarprozessen in Systemen, die sich (nahezu) in Gleichgewichtszuständen befinden, ist mit Hilfe der kernmagnetischen Resonanz möglich. Ähnlich wie bei der Isotopenmarkierungsmethode sind die einzelnen Spezies von Kernen gekennzeichnet, hier durch den Spin. Die nach der Methode der magnetischen Resonanz gemessenen Geschwindigkeitskonstanten von im Gleichgewicht ablaufenden Prozessen haben eine wohldefinierte physikalische Bedeutung und sollten deshalb mit berechenbaren statistischen Geschwindigkeitskonstanten korreliert werden können.

Vorgänge in Systemen, die nicht im Gleichgewicht sind, können mit Hilfe der Methode der kernmagnetischen Resonanz natürlich auch untersucht werden.

Eine Besonderheit der Methode liegt darin, daß auch Prozesse verfolgt werden können, deren thermodynamische Reaktionseffekte Null sind. Es gelingt mit Hilfe der magnetischen Resonanz die Existenz von symmetrischen intramolekularen Umlagerungen und chemischen Austauschprozessen zwischen Molekülen einer Sorte nachzuweisen und die kinetischen Parameter dieser Reaktionen zu bestimmen. Bei den Verfahren der chemischen Relaxation hingegen stellt sich durch eine sprunghafte oder periodische Änderung geeigneter Zustandsparameter gerade wegen der endlichen Größe der Reaktionseffekte ein Nichtgleichgewichtszustand ein, dessen Übergang in den Gleichgewichtszustand verfolgt wird (EIGEN, DE MAEYER [23]).

Sorgfalt ist bei der Verknüpfung der Geschwindigkeitskonstanten mit den aus Kernresonanzexperimenten erhaltenen mittleren Lebensdauern τ geboten. Wie wir sehen, beziehen sich letztere auf Reaktionen, die als willkürliche Übergänge zwischen verschiedenen Kernspinhamiltonoperatoren bzw. Kernspinzuständen dargestellt werden können. Die mittlere Lebensdauer τ ist gleich der mittleren Zeit zwischen zwei aufeinander folgenden Übergängen. Wir kommen auf diesen Zusammenhang in Kapitel 7.1 zurück.

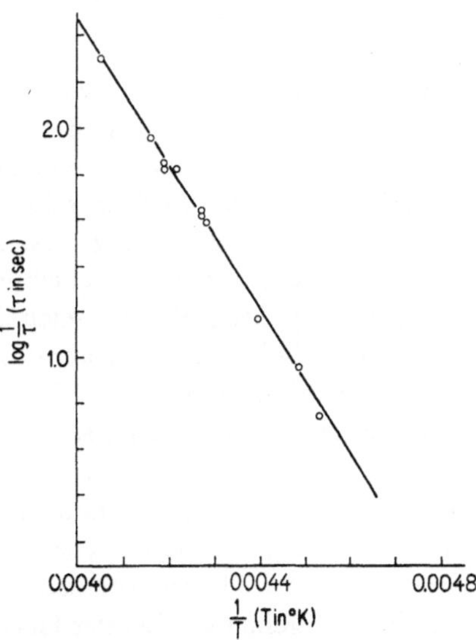

Abb. 3. Temperaturabhängigkeit der mittleren Lebensdauer τ für N-Methyl-N-deutero-2,4,6-trinitroanilin in Arrheniusscher Darstellung. Ein Beispiel für die Ermittlung kinetischer Größen von Prozessen im Gleichgewicht, deren thermodynamische Reaktionseffekte Null sind

Die Methode kann zur Messung der Geschwindigkeit von Reaktionen beliebiger Ordnung angewandt werden. Die mittlere Lebensdauer einer Spezies i, τ_i, ist mit den Geschwindigkeitskonstanten $k_{i \to j}$ durch

$$\frac{1}{\tau_i} = - \frac{1}{c_i} \frac{\mathrm{d}\,c_i}{\mathrm{d}\,t} = \sum_j k_{i \to j}\, c_i^{r_j - 1}\, c_k^{s_j}\, c_l^{t_j} \ldots \tag{6.2}$$

verknüpft; r_j, s_j, t_j, . . . sind die Ordnungen der Reaktion $i \to j$ bezüglich der Reaktanten i, k, l, . . . Um die Ordnungen zu bestimmen, mißt man τ_i als Funktion der Konzentrationen.

Die Lebensdauern, die mit dieser Methode meßbar sind, liegen im Bereich zwischen etwa 1 sec und 10^{-4} sec. Abb. 2 und 3. Es scheint erwähnenswert, daß atomare und molekulare Eigenschaften des Systems die Zeitskala liefern. Die Messung der Linienform sowie der Frequenzen δ und J ersetzt die bei kinetischen Versuchen übliche Bestimmung des Reaktionsablaufs.

7. Anwendungsbeispiele

7.1. Intermolekulare Protonenübertragung in wäßrigen Lösungen von Ammonium-ionen. Eine Anwendung der adiabatischen Theorie in Gegenwart von Kern-Spin-Spin-Kopplung

Gleichungen für adiabatische Umlagerungen vom Typ (2.12) können unter besonderen Bedingungen auch dann zur Berechnung von Kernverweilzeiten angewandt werden, wenn die betreffenden Kerne durch Spin-Spin-Kopplung mit anderen Kernen in Wechselwirkung stehen. Der Einfachheit wegen betrachten wir wieder ein AB-System, zwei Kerne A und B mit dem Spin $^1/_2$, deren chemische Verschiebung gegeneinander gleich 2δ und deren Spin-Spin-Kopplung gleich J seien. Man kann zeigen, daß der Hamiltonoperator für die Spin-Spin-Kopplung, $J\,(\mathbf{I}^{(A)}, \mathbf{I}^{(B)})$, durch das Produkt $J\,\mathrm{I}_z^{(A)}\,\mathrm{I}_z^{(B)}$ ersetzt werden kann, wenn $2\delta \gg J$. Die Eigenwerte von $J\,\mathrm{I}_z^{(A)}\,\mathrm{I}_z^{(B)}$ sind den Energien äquivalent, die der Kern B in einem Magnetfeld in z-Richtung hat, dessen Stärke proportional $J\,\mathrm{I}_z^{(A)}$ ist. Die Nachbarschaft eines Kernes in einem bestimmten Spinzustand wirkt mithin ähnlich wie die elektronische Abschirmung, welche die chemische Verschiebung verursacht. Bei der Entwicklung der Theorie für den Kollaps von Multipletts kann man daher ähnlich vorgehen wie bei der Herleitung von (2.12), wenn $2\delta \gg J$; wir betonen schon jetzt, daß diese Bedingung zwar notwendig, aber nicht hinreichend ist.

Rasche, willkürliche Übergänge des Kerns A zwischen seinen verschiedenen räumlichen Spinorientierungen können die Dublettstruktur der B-Resonanz verwischen. Wenn die Frequenz der Übergänge von A hinreichend groß ist, werden die Resonanzen von A und B nicht mehr durch die Kopplung beeinflußt, und die B-Resonanz besteht nur noch aus einer einzelnen schmalen Linie. Prozesse, die derartige Übergänge herbeiführen, sind die Spin-Gitter- und die Spin-Spin-Relaxation sowie chemische Umlagerungsprozesse. Nehmen wir an, der Kern in der Lage A ist an einem chemischen Austauschvorgang beteiligt. Seine Verweilzeit in einer AB-Konfiguration, d. h. die mittlere Zeit zwischen zwei Austauschvorgängen, sei τ_c. Wenn die genannten Relaxationen sehr viel langsamer als der Austausch ablaufen, ist die mittlere Lebensdauer τ_s eines Spinzustands in der Lage A, die gleich der mittleren Verweilzeit des Kerns B in einem gegebenen Magnetfeld ist, durch $\tau_s = 2\tau_c$ bestimmt; denn die Wahrscheinlichkeit für die Erhaltung des Spinzustands bei jedem Austausch ist für Teilchen mit dem Spin $^1/_2$ gleich $^1/_2$. Die beiden Resonanzen von B verhalten sich so, als ob sie zu zwei inäquivalenten Kernen gehörten, die ihre chemischen Verschiebungen mit der Geschwindigkeit $1/\tau_s$ vertauschen. Wir können mithin Gl. (2.12) in der Form für ein symmetrisches Dublett ($\tau_A = \tau_B = \tau$) zur näherungsweisen Beschreibung der Wirkung des geschilderten chemischen Austauschvorganges heranziehen; wir müssen nur 2δ durch J, und τ durch τ_s in (2.12) ersetzen. Von historischem Interesse ist, daß die Gleichung für die adiabatische Umlagerung bei Betrachtungen über die Koaleszenz von Multiplettstrukturen von Gutowsky, McCall und Slichter aufgestellt wurde [1].

Die Vernachlässigung der x- und y-Komponenten im skalaren Produkt ($\mathbf{I}^{(A)}$, $\mathbf{I}^{(B)}$) ist nur für den Grenzfall langsamen Austausches, $\tau_s J \gg 1$, eine hinreichende Näherung. Indem man die Bedingungen

$$J/2\delta \ll 1 \tag{7.1}$$

und

$$J\tau_s \ll 1 \tag{7.2}$$

in die vollständigen quantenstatistischen Gleichungen vom Typ (4.23) für die Linienform einführt, kann man zeigen, daß die Linienbreite $(T_2')^{-1}$ des zu einer Linie zusammengeflossenen Dubletts des fixierten Kerns B durch den Ausdruck [21]

$$\frac{1}{T_2'} = \frac{1}{T_2} + \frac{J^2}{4}\tau'\left[1 + \frac{1}{1+(2\delta)^2(\tau')^2}\right] \tag{7.3}$$

gegeben ist, während die adiabatische Theorie mit (7.1) und (7.2) nach Gl. (2.16)

$$\frac{1}{T_2'} = \frac{1}{T_2} + \frac{J^2}{4}\tau' \tag{7.4}$$

liefert ($2\tau' = \tau_s$). Man nennt das rechte Glied der Klammer in (7.3) den nichtsäkularen oder nichtadiabatischen Verbreiterungsterm oder auch die

Lebensdauerverbreiterung; Gl. (7.3) wurde von Solomon und Bloembergen [24] erstmals auf einem anderen als dem hier angegebenen Wege für flüssigen Fluorwasserstoff hergeleitet. Die chemische Umlagerung des Protons A induziert Übergänge zwischen den Energieniveaus des Spin-Systems, die, wenn die Umlagerung hinreichend rasch verläuft, nicht mehr vernachlässigt werden können.

Es gelang mit Hilfe der adiabatischen Theorie, die Geschwindigkeit der Protonenübertragung in wäßrigen Lösungen von Ammoniumionen zu messen. Überdies konnten aus den Messungen der Verweilzeiten einzelner Kerne in bestimmten Spinzuständen Aussagen über den Mechanismus der betrachteten Protonenübertragung gewonnen werden (Alexander, Connor, Grunwald, Loewenstein, Meiboom [25]).

Das protonmagnetische Resonanzspektrum des $[NH_4^+]$-Ions besteht in saurer Lösung (pH < 1) aus drei scharfen, äquidistanten Linien gleicher Intensität, deren Abstand voneinander etwa 50 Hz beträgt. Die Aufspaltung rührt von der Spin-Spin-Kopplung der Protonen mit dem ^{14}N-Kern her (I (^{14}N)=1). Offensichtlich ist der Beitrag der ^{14}N-Quadrupol-Relaxation ebenso wie jener des Protonenaustausches zur Linienbreite vernachlässigbar. Eine erhebliche Quadrupolverbreiterung der Linien ist für NH_4^+ wegen der hohen Symmetrie des Ions und des damit verbundenen geringen elektrischen Feldgradienten am Ort des ^{14}N-Kerns nicht zu erwarten. Erhöht man den pH-Wert der Lösung, so verbreitern sich die Wasserlinie und die Linien des Tripletts, bis schließlich bei weiterer pH-Steigerung alle 4 Linien in eine einzelne zusammenfließen. Die Austauschgeschwindigkeiten wurden mit Hilfe der Gleichung (2.15) für die Linienbreite $(T_2')^{-1}$ einer Komponente des Tripletts im Grenzfall sehr langsamen Austausches, $J\tau_s \gg 1$,

$$\pi\Delta = \frac{1}{T_2'} = \frac{1}{T_2} + \frac{1}{\tau_s} , \qquad (7.5)$$

ermittelt; Δ ist die Halbwertsbreite einer Komponente des NH_4^+-Tripletts in Hz, 2 $(T_2)^{-1}$ die Halbwertsbreite bei Abwesenheit von Austausch (in rad/sec), τ_s ist die mittlere Verweilzeit (in sec) eines an Stickstoff gebundenen Protons in einem gegebenen Magnetfeld (die gleich ist der mittleren Lebensdauer eines Spinzustandes in der Lage des ^{14}N).

Folgende Reaktionen dürften bei der Protolyse von Ammoniumionen eine Rolle spielen

$$NH_4^+ + NH_3 \xrightarrow{k_1} NH_3 + NH_4^+ , \qquad (7.6)$$

$$NH_4^+ + \underset{H}{OH} + NH_3 \xrightarrow{k_2} NH_3 + \underset{H}{HO} + NH_4^+ , \qquad (7.7)$$

$$NH_4^+ + OH^- \xrightarrow{k_3} NH_3 + H_2O , \qquad (7.8)$$

$$NH_4^+ + H_2O \xrightarrow{k_4} NH_3 + H_3O^+ . \qquad (7.9)$$

Obwohl die Geschwindigkeit der Reaktion zwischen NH_4^+ und OH^- sehr hoch und wahrscheinlich diffusionsbestimmt ist, spielt sie im pH-Bereich zwischen 1,5 und 3 wegen der extrem geringen OH^--Ionenkonzentration keine Rolle. Aus der Art der Abhängigkeit der beobachteten Reaktionsgeschwindigkeitskonstante von $1/[H^+]$ kann geschlossen werden, daß die Übertragung von Protonen auf H_2O relativ langsam abläuft ($k_4 < 0,6 \cdot 10^{-2}\ sec^{-1}\ Mol^{-1}\ 1$ bei 21°C). Der Zusammenhang zwischen τ_s, k_1 und k_2 ist dann im Hinblick auf

$$4\,\frac{1}{\tau_c} = -\,\frac{1}{[NH_4^+]}\,\frac{d\,[NH_4^+]}{d\,t} \tag{7.10}$$

durch den Ausdruck

$$4\,\frac{1}{\tau_s} = \left(\frac{2}{3}\,k_1 + k_2\right)\frac{[NH_4^+]}{[H^+]}\,K_A\,, \tag{7.11}$$

wo $K_A = [NH_3][H^+]/[NH_4^+]$, gegeben. Der Faktor 4 muß eingeführt werden, da τ_c als die Verweilzeit eines einzelnen Protons im NH_4^+-Ion definiert ist. Der Faktor 2/3 tritt auf, denn Protonenübergänge zwischen Stickstoffatomen im gleichen Kernspinzustand tragen nach (7.5) zur Linienverbreiterung nicht bei. Der Wert von k_2 kann aus der Verbreiterung der Wasserlinie mit Hilfe der Beziehung

$$\frac{(k_2)^{-1}}{\tau_{H_2O}} = \frac{[NH_4Cl]}{2\,[H_2O]} \tag{7.12}$$

erhalten werden; τ_{H_2O} ist die mittlere Verweilzeit eines einzelnen Protons in einem H_2O-Molekül. Die Linienbreite $\varDelta^{(H_2O)}$ (in Hz) der Wasserlinie wird durch die Gleichung

$$\varDelta^{(H_2O)}\,\pi = \frac{1}{T_2^{(H_2O)}} + \frac{1}{\tau_{H_2O}} \tag{7.13}$$

festgelegt, wenn $2\tau_{H_2O}\,\delta \gg 1$; 2δ ist die chemische Verschiebung zwischen den Signalen der H_2O- und NH_4^+-Protonen, $(T_2^{H_2O})^{-1}$ die Linienbreite der Wasserlinie in Abwesenheit des Austausches (bei niedrigen pH-Werten). Abb. 4, S. 403

Folgende Daten wurden bestimmt:

$$k_1 \cong 11 \cdot 10^8\ sec^{-1}\ Mol^{-1}\,l \quad \text{bei}\ 21\,°C$$

$$k_2 \cong 1 \cdot 10^8\ sec^{-1}\ Mol^{-1}\,l \quad \text{bei}\ 21\,°C.$$

Aus Messungen bei verschiedenen Temperaturen ergaben sich die Arrheniusschen Aktivierungsenergien für beide Reaktionen innerhalb der Fehlergrenzen zu Null. Weiter ins einzelne gehende Aussagen über den Mechanismus der Protonenübertragung sind aus diesen Messungen nicht möglich. Dies gilt insbesondere für Reaktion (7.7). Es ist nicht bekannt, wieviele H_2O-Moleküle an der Übertragung beteiligt sind. Spin-Echo-

Experimente an wäßrigen Lösungen von Trimethylammoniumionen, die mit ^{17}O angereichert waren, zeigten, daß *ein* Molekül H_2O an der Protonenübertragung zwischen dem Trimethylammoniumion und dem Trimethylaminmolekül entsprechend Gl. (7.7) beteiligt ist. Die 6 Spinzustände des ^{17}O-Kerns im H_2O-Molekül liefern die diskreten lokalen Felder für die Messung der Protonenaustauschgeschwindigkeit (Luz und Meiboom 1963) [39] [25c][9].

7.2. Ligandenaustausch in Komplexen.
Austausch von Atomen mit elektrischem Kernquadrupolmoment

Atomkerne mit einem hinreichend großen elektrischen Quadrupolmoment liefern in Flüssigkeiten nur dann ein magnetisches Kernresonanzsignal, wenn das elektrische Feld am Kernort Kugelsymmetrie besitzt. Diese ist in vielen Fällen gegeben, wenn der betreffende Kern in Gestalt seines Atomions vorliegt. Geht das betrachtete Atom dagegen eine kovalente Bindung ein, so kann die Kopplung des Quadrupolmoments an das elektrische Feld so stark sein, daß die transversale Relaxationszeit T_2 je nach der Größe des Quadrupolmoments und des Feldgradienten nur 10^{-5} bis 10^{-8} sec beträgt; in diesen Fällen ist die Resonanzlinie im allgemeinen bis zur Nichtmeßbarkeit verbreitert. So findet man, daß die Kernresonanzen von ^{79}Br und ^{81}Br in einer KBr-Lösung leicht meßbar, in einer $CdBr_2$-Lösung hingegen nicht beobachtbar sind [27].

Wenn man eine KBr-Lösung mit $CdBr_2$ versetzt, so verbreitern sich die Linien der Bromresonanzen mit zunehmender Cd-Konzentration sehr stark, bis eine so große Linienbreite erreicht ist, daß man das Signal nicht mehr erkennt. Es ist naheliegend, die Vergrößerung der Linienbreite auf die Vertauschung eines freien Br^- mit einem an Cd gebundenen Br, d. h. die Verschiebung eines Br-Kerns zwischen dem Zustand des freien Ions und dem Zustand der kovalenten Bindung, zurückzuführen. Tatsächlich konnte gezeigt werden, daß die Linienverbreiterung durch (2.12) beschrieben wird, wenn die folgenden Gleichgewichte

---

[9] Ein Beispiel für einen intramolekularen Protonenübergang, dessen Geschwindigkeit mit Hilfe der adiabatischen Theorie näherungsweise bestimmt wurde, ist die Reaktion von Protonenkomplexen substituierter Benzole in $HF-BF_3$

$$\rightleftharpoons \quad \text{[Hexamethylbenzol-Proton-Komplex, H an (1)]} \quad \rightleftharpoons \quad \text{[Hexamethylbenzol-Proton-Komplex, H an (2)]} \quad \rightleftharpoons$$

die von Mackor und MacLean zwischen -110 und $-30\,°C$ untersucht wurde [26].

$$Cd^{2+} + Br^- \underset{k_{21}}{\overset{k_{11}}{\rightleftharpoons}} CdBr^+ \tag{7.14a}$$

$$CdBr^+ + Br^- \underset{k_{22}}{\overset{k_{12}}{\rightleftharpoons}} CdBr_2 \tag{7.14b}$$

$$CdBr_2 + Br^- \underset{k_{23}}{\overset{k_{13}}{\rightleftharpoons}} CdBr_3^- \tag{7.14c}$$

$$CdBr_3^- + Br^- \underset{k_{24}}{\overset{k_{14}}{\rightleftharpoons}} CdBr_4^{2-} \tag{7.14d}$$

berücksichtigt werden und für die mittlere Verweilzeit eines Bromkerns im Ion-Zustand

$$\frac{1}{\tau_A} = - \frac{1}{[Br^-]} \frac{d[Br^-]}{dt} = k_{11}[Cd^{2+}] + k_{12}[CdBr^+] + $$
$$+ k_{13}[CdBr_2] + k_{14}[CdBr_3^-] \tag{7.15}$$

gesetzt wird (HERTZ [27]). Die einzelnen Geschwindigkeitskonstanten k_{ij} werden aus der Abhängigkeit der mittleren Verweilzeit τ_A von der Konzentration an KBr und $CdBr_2$ sowie aus den Gleichgewichtskonstanten für die Reaktionen (7.14) ermittelt. Zur Bestimmung von τ_A ist es nicht notwendig, den vollständigen Ausdruck (2.12) heranzuziehen. An dieser Stelle interessiert nur die vereinfachte Form von (2.12), die unter den Bedingungen

$$\tau_A \gg \tau_B \tag{7.16}$$

$$\frac{1}{T_{2B}} \gg |\omega_B - \omega| \tag{7.17}$$

$$T_{2A} > T_{2B} \tag{7.18}$$

gültig ist [27]. τ_B bezeichnet die mittlere Verweilzeit des Bromkerns im gebundenen Zustand. Die Bedingung (7.16) wird erfüllt, indem man die Konzentrationen geeignet einstellt ($[Br^-] \gg [CdBr^+] + [CdBr_2] + [CdBr_3^-] + [CdBr_4^{2-}]$). Die Ungleichung (7.17) besagt, daß die Linienbreite $(T_{2B})^{-1}$ des von den gebundenen Bromkernen stammenden Signals groß gegen die Ausdehnung des Spektrums und insbesondere groß gegen die chemische Verschiebung 2δ der Bromresonanzen für die beiden Zustände A und B ist. Erfahrungsgemäß gilt[10] bei schweren Kernen $10^5 \geq 2\delta \geq 10^3$ Hz und, im Hinblick auf den Wert der betreffenden Quadrupolkopplungskonstanten, $T_{2B} \approx 10^{-7}$ sec. (7.18) ist offensichtlich erfüllt. Aus Gl. (2.12) erhält man mit den Bedingungen (7.16) (7.17) (7.18) für die Linienbreite T_2^{-1} der beobachtbaren Absorptionslinie (HERTZ [27])

$$\frac{1}{T_2} = \frac{1}{T_{2A}} + \frac{1}{\tau_A (1 + T_{2B}/\tau_B)} . \tag{7.19}$$

---

[10] In einem Magnetfeld von etwa 6000 G.

Die mittlere Verweilzeit τ_A kann mithin nur bei Kenntnis von T_{2B}/τ_B ermittelt werden[11]. Diese Tatsache stellt die Grenze der Anwendbarkeit des Verfahrens dar. Eine unabhängige Bestimmung von T_{2B}/τ_B ist jedoch möglich, wenn von dem betrachteten Kern ein weiteres Isotop existiert, das ein hinreichend großes magnetisches Dipolmoment und elektrisches Quadrupolmoment besitzt, und dessen Häufigkeit hoch genug ist. Da sich die Relaxationszeiten T_{2B} für die verschiedenen Isotope umgekehrt proportional wie die Quadrate der Quadrupolmomente Q_i verhalten

$$\frac{T_{2B}^{(1)}}{T_{2B}^{(2)}} = \left(\frac{Q_2}{Q_1}\right)^2$$

und τ_A bzw. τ_B für beide Isotope praktisch den gleichen Wert haben, $\tau_A^{(1)} \approx \tau_A^{(2)}$, $\tau_B^{(1)} \approx \tau_B^{(2)}$, läßt sich aus Messungen der Verbreiterung der Resonanzlinien der beiden Isotope, ^{79}Br und ^{81}Br, $T_{2B}^{(1)}/\tau_B$ und $T_{2B}^{(2)}/\tau_B$ ermitteln. Man nimmt im Sinne der Brönstedschen Theorie für Ionenreaktionen an, daß

$$k_{11}(c) = k_{11}^{(0)} \frac{f_{Cd^{2+}} \, f_{Br^-}}{f_{CdBr^+}}$$

$$k_{12}(c) = k_{12}^{(0)} \frac{f_{CdBr^+} \, f_{Br^-}}{f_{CdBr_2}}$$

usf., wo $f_{Cd^{2+}}, f_{Br^-} \ldots$ die molaren Aktivitätskoeffizienten von Cd^{2+}, $Br^- \ldots$ sind.

Folgende Geschwindigkeitskonstanten wurden gemessen [27]; $k_{1i}^{(0)}$ in $(Mol/l)^{-1} sec^{-1}$ für die Hinreaktion, k_{2i} in sec^{-1} für die Rückreaktion.

Substanz	$k_{11}^{(0)}$	$k_{12}^{(0)}$	$k_{13}^{(0)}$	$k_{14}^{(0)}$	k_{21}	k_{22}	k_{23}	k_{24}
$CdBr_2$	$1,4\cdot10^9$	$1,4\cdot10^8$	$1,4\cdot10^7$	$1,2\cdot10^7$	$1,0\cdot10^7$	$1,4\cdot10^6$	$7,0\cdot10^6$	$6,0\cdot10^6$
$HgBr_2$	—	—	$\approx6\cdot10^9$	$\approx10^9$	—	—	$2,5\cdot10^7$	$2,5\cdot10^7$
$ZnBr_2$	$\approx5\cdot10^5$	$\approx5\cdot10^5$	$\approx5\cdot10^5$	$8\cdot10^7$	$\approx10^6$	$\approx10^6$	$\approx10^6$	$2\cdot10^7$

Die schnellsten der untersuchten Reaktionen verlaufen mit der in Lösung maximal möglichen Geschwindigkeit; sie sind mithin diffusionskontrolliert ($k \approx 10^9$ bis $10^{11} Mol^{-1} l \, sec^{-1}$). Es ist überraschend, daß das Cd^{2+}-Ion, welches wahrscheinlich von einer relativ stabilen Hydrathülle umgeben ist, welche die Annäherung von Br^- an Cd^{2+} erschweren sollte, mit Br^- mit maximaler Geschwindigkeit sich vereinigt. Bei neutralen Molekülen und Anionkomplexen dürfte nur eine labile Hydrathülle vorhanden

---

[11] Im allgemeinen existieren überdies soviele verschiedene Werte für den Quotienten T_{2B}/τ_B, wie verschiedene kovalente Bindungszustände vorhanden sind. An dieser Stelle wird der über alle Bindungszustände erstreckte Mittelwert von T_{2B}/τ_B, der konzentrationsabhängig sein kann, verwendet.

sein, so daß in diesem Fall jede durch Diffusion bedingte Begegnung zur Reaktion führen sollte. Offenbar spielen andere Faktoren, wie elektrostatische Anziehung und sterische Hinderung eine Rolle. In der Reihe der Zn-Reaktionen findet man hingegen die erwarteten Abstufungen.

Ligandenaustausch in paramagnetischen Komplexen

Versetzt man eine Lösung mit paramagnetischen Ionen, so können sich die Linienformen der ursprünglich vorhandenen Kernresonanzsignale als Folge der Verkürzung der Relaxationszeiten T_1 und T_2 verbreitern und die Resonanzfrequenzen verschieben[12]. Die Relaxation (und die Resonanzfrequenz) der Kerne in der ersten Koordinationssphäre von paramagnetischen Ionen wird naturgemäß am stärksten beeinflußt. Aber auch die Signale von Kernen, deren nähere Umgebung (A) im Innern der Lösung diamagnetisch ist, werden verändert. Dieser Einfluß wird 1. durch die direkte, mit der Entfernung rasch abnehmende Wechselwirkung der magnetischen Momente des Kerns und des Ions und 2. durch den Austausch von Kernen zwischen dem Lösungsmittel und den in der ersten Sphäre gebundenen Liganden hervorgerufen. Die Linienbreite und Höhe des Signals der Wasserprotonen der wäßrigen Lösung eines paramagnetischen Ions B, z. B. des Äthylendiaminkomplexes von Cr^{3+}, $[Cr(en)_3]^{3+}$, wird mithin nicht nur von der transversalen Relaxationszeit $T_2^{(A)}$ der Protonen des Wassers, sondern auch im Hinblick auf die Reaktion

$$Cr-NH + OH^- \longrightarrow Cr-N^- + H_2O \qquad (7.20)$$

von der mittleren Verweilzeit τ_B der Protonen in der ersten Ligandensphäre (Lage B) und der Relaxationszeit $T_2^{(B)}$ in dieser Lage (B) abhängen.

Der Einfachheit halber nehmen wir an, daß der elektronische Zustand des paramagnetischen Ions ein Dublettzustand mit den z-Komponenten des Elektronenspins, $S_z = \pm 1/2$, ist. Die Aufspaltung der Resonanz der Protonen bei B in ein Dublett, ω_{B+} und ω_{B-}, infolge der Hyperfeinwechselwirkung $J_{EP}S_zI_z$ zwischen dem Elektronenspin $\mathbf{S}$ und dem Protonspin $\mathbf{I}$ sei J_{EP}; wir vernachlässigen hiermit die nichtsäkularen Terme der Hyperfeinwechselwirkung, was bei den üblichen Feldern ($H_0 \approx 10^4$ Gauß) im allgemeinen zulässig ist. Damit können wir im Rahmen der adiabatischen Näherung für die transversale Komponente G der Protonenmagnetisierung

---

[12] Kernresonanzsignale paramagnetischer Moleküle oder Ionen in Lösung konnten bisher direkt nur dann beobachtet werden, wenn die Elektronen-Spin-Gitter-Relaxationszeiten τ_e so kurz waren, daß die Hyperfeinwechselwirkungen sich beinahe zu Null mittelten. ($\tau_e J_{EP} \ll 1$). Vgl. EATON, PHILLIPS, JOSEY, BENSON [28].

im — mit der Winkelgeschwindigkeit ω des rotierenden Feldes H_1 — sich drehenden Koordinatensystem in Analogie zu (2.11) und (2.12) ansetzen [29]:

$$\frac{\mathrm{d}\,G_A}{\mathrm{d}\,t} = -\,\alpha_A\,G_A - G_A/\tau_A + (G_{B^+} + G_{B^-})/\tau_B - i\,\gamma\,H_1\,M_z^{(A)}\,,$$

$$\frac{\mathrm{d}\,G_{B^+}}{\mathrm{d}\,t} = -\,\alpha_{B^+}\,G_{B^+} + G_A/(2\tau_A) - (G_{B^+} - G_{B^-})/\tau_e - i\,\gamma\,H_1\,M_z^{(B)}/2 - G_{B^+}/\tau_B,$$

$$\frac{\mathrm{d}\,G_{B^-}}{\mathrm{d}\,t} = -\,\alpha_{B^-}\,G_{B^-} + G_A/(2\tau_A) - (G_{B^-} - G_{B^+})/\tau_e - i\,\gamma\,H_1\,M_z^{(B)}/2 - G_{B^-}/\tau_B,$$

$$(7.21)$$

wobei　　$\alpha_A = \dfrac{1}{T_2^{(A)}} - i\,(\omega_A - \omega)\,,\ \ \alpha_{B^+} = \dfrac{1}{T_2^{(B)}} - i\,(\omega_{B^+} - \omega)\,,$

$$\alpha_{B^-} = \frac{1}{T_2^{(B)}} - i\,(\omega_{B^-} - \omega)\,;$$

der Sprung von B$^+$ nach B$^-$ (und umgekehrt) entspricht der Elektron-Spin-Relaxation; τ_e ist die Elektron-Spin-Relaxationszeit[13]. Bei langsamem Durchgang (slow passage) verschwinden die zeitlichen Ableitungen von G; bei vernachlässigbarer Sättigung ist $M_z^{(A)} \approx M_0^{(A)}$ usf., wo $M_0^{(A)}$ die Gleichgewichtsmagnetisierung von A ist. Uns interessiert die Lösung von (7.21) unter folgenden Bedingungen:

$$|\omega_A - \omega| \approx 0\,, \tag{7.22a}$$

$$\tau_A \gg \tau_B \tag{7.22b}$$

$$\frac{1}{T_2^{(B)}} + \frac{1}{\tau_B} + \frac{1}{\tau_e} \gg |J_{EP}/2|\,. \tag{7.22c}$$

Bedingung (7.22a) drückt aus, daß wir nur den Bereich des Spektrums um die beobachtbare Wasserlinie betrachten; dann ist $|\omega_{B^+} - \omega| \cong (J_{EP}/2)$ usf. Die Bedingung (7.22b) ist leicht zu verwirklichen, indem man eine verdünnte Lösung von B verwendet. Bed. (7.22c) ist hinreichend dafür, daß die Protonhyperfeinaufspaltung im Elektronspinresonanzsignal von B verschwindet, was experimentell geprüft werden kann. Bei langsamem Durchgang und vernachlässigbarer Sättigung ergibt sich aus (7.21) und (7.22) für G_A an der Stelle $\omega = \omega_A$ [29]:

$$G_A(\omega_A) = -\frac{i\,\gamma\,H_1\,M_0^{(A)}}{\dfrac{1}{T_2^{(A)}} + \dfrac{\tau_B}{(T_2^{(B)} + \tau_B)\,\tau_A}}\,. \tag{7.23}$$

---

[13] Die Unterschiede zwischen den Besetzungszahlen der Zustände B$^+$ und B$^-$ können hier vernachlässigt werden.

Die Linienbreite $(T_2)^{-1}$ des Signals der Wasserprotonen ist mithin durch (vgl. Gl. (2.7))

$$\frac{1}{T_2} = \frac{1}{T_2^{(A)}} + \frac{\tau_B}{(T_2^{(B)} + \tau_B)\,\tau_A} \approx \frac{1}{T_2^{(A)}} + \frac{p_B}{T_2^{(B)} + \tau_B} \qquad (7.24)$$

gegeben (PEARSON, PALMER, ANDERSON, ALLRED [29a]); p_A und p_B, die relativen Häufigkeiten der Protonen in den Zuständen A und B, sind mit den mittleren Lebensdauern τ_A und τ_B durch $p_A/p_B = \tau_A/\tau_B$ verknüpft. Man vergleiche (7.24) mit (7.19) und (7.22) mit (7.16)–(7.18).

Gl. (7.24) wurde benutzt, um die Geschwindigkeitskonstante k der Reaktion (7.20) für $[Cr(en)_3]^{3+}$, $[Cr(NH_3)_6]^{3+}$ und andere Komplexionen zu bestimmen. Die mittlere Lebensdauer τ_B ist nach (7.20) mit k durch die Gleichung

$$\frac{1}{\tau_B} = k\,[OH^-]$$

verbunden. In saurer Lösung ist $(T_2)^{-1}$ unabhängig vom pH-Wert, so daß $(T_2)^{-1}_{sauer} \approx (T_2^{(A)})^{-1}$. Wie erwartet, ist $T_2^{(A)}$ verschieden von der transversalen Relaxationszeit reinen Wassers. Setzt man $p_B = [M]\,6/55$, wo $[M]$ die Konzentration des paramagnetischen Ions in Mol/l ist, und nimmt man an, daß $T_2^{(B)}$ unabhängig vom pH-Wert ist, so ergeben sich folgende Daten für die Reaktion (7.20) [29]:

Verbindung	$[OH^-]$ Mol/l	T_2 sec	k Mol^{-1} l sec^{-1}
$Cr(en)_3(ClO_4)_3$	0,0044	0,024	$3{,}8 \cdot 10^6$
$Cr(en)_3(ClO_4)_3$	0,022	0,0090	$3{,}7 \cdot 10^6$
$Cr(en)_3(ClO_4)_3$	0,044	0,0052	$3{,}5 \cdot 10^6$
$Cr(en)_3(ClO_4)_3$	0,044	0,0066	$2{,}6 \cdot 10^6$

Die Konzentration des Chromkomplexes betrug stets 0,01 M, die Temperatur 25 °C; T_2 ergab sich bei pH 2,5 zu 0,04 sec.

Ebenfalls mit Hilfe von Gl. (7.24) wurden Geschwindigkeiten des Ligandenaustausches in paramagnetischen Komplexionen bei Gegenwart von HCl ermittelt [29].

$$[M(CH_3OH)_6] + CH_3OH \longrightarrow [M(CH_3OH)_6] + CH_3OH$$

$$[M(H_2O)_6] + H_2O \longrightarrow [M(H_2O)_6] + H_2O$$

$$M = Cr^{3+},\ Fe^{3+},\ Mn^{2+},\ Co^{2+},\ Ni^{2+},\ Cu^{2+},\ Ce^{3+},\ Gd^{3+}$$

Die näherungsweise Bestimmung von $T_2^{(A)}$ gelang hier durch Messung der Linienbreite des Wassersignals bzw. der Methanolsignale einer Lösung, die einen stabilen Komplex des betreffenden Ions enthielt, der frei von rasch reagierenden Protonen war, z. B. $[Cr(Ox)_3]^{3-}$, $[Ni(Dipyridyl)_3]^{2+}$. Zunächst ungelöst bleibt das Problem der Separation von $T_2^{(B)} + \tau_B$. Die Werte der

Geschwindigkeitskonstanten $(T_2^{(B)} + \tau_B)^{-1}$ für den CH_3OH-Austausch wurden durch Messung der Linienbreite sowohl der CH_3-Resonanz als auch der OH-Resonanz erhalten. Wenn die beiden Werte für ein Ion übereinstimmen, dürfte die Meßgröße τ_B^{-1} und $\tau_B \gg T_2^{(B)}$ sein. Für Ni^{2+} und Cu^{2+} ergab sich Übereinstimmung, für Fe^{3+} eine geringe Differenz, so daß für diese drei Ionen wahrscheinlich die Geschwindigkeit des Alkohol-Austausches ermittelt wurde. Auf ähnliche Weise wurde der Austausch von H_2O untersucht, indem die Linienbreite der Signale von ^1H und ^{17}O des Wassers als Funktion der Metallionenkonzentration gemessen wurde. Man fand, daß die ^1H-Signale durch $T_2^{(B)}$ und durch τ_B bestimmt werden $(T_2^{(B)} > \tau_B)$. Wahrscheinlich liefert das Sauerstoffsignal in allen Fällen die Geschwindigkeit des Austausches von H_2O; denn für ^{17}O in unmittelbarer Nachbarschaft zu dem paramagnetischen Ion sollte $T_2^{(B)} \ll \tau_B$ sein. Der Alkoholaustausch verläuft langsamer als jener von Wasser.

7.3. *Anwendung des Dichtematrixformalismus*

Analytische Lösungen zur Bestimmung von Reaktionsgeschwindigkeitskonstanten aus den Linienform-Funktionen sind nur in wenigen Fällen verfügbar. Komplexe analytische Lösungen, wie sie bereits für einfache Systeme zu erwarten sind, bieten geringe physikalische Einsicht und sind schwierig auszuwerten. Man wird deshalb unmittelbar von den generalisierten Bloch- bzw. Boltzmanngleichungen, deren physikalische Bedeutung bekannt ist, ausgehen und numerische Lösungen des Problems suchen. Berechnete und experimentelle Kurven wird man mit Hilfe einer elektronischen Rechenmaschine vergleichen und durch Variation der Verweilzeiten und unter günstigen Umständen auch anderer Parameter, wie der Linienbreite „bei Abwesenheit chemischer Umlagerungen", der Spin-Spin-Kopplungskonstanten oder der chemischen Verschiebung, einander anpassen. Insbesondere ist man zu solchen Verfahren gezwungen, wenn die Dichtematrixgleichungen anzuwenden sind. Dann tritt nämlich zu der durch die chemische Umlagerung verursachten Kopplung die nichtsäkulare Kopplung zwischen den Spins, wodurch die Zahl der gekoppelten Gleichungen sich gegenüber jener entsprechender adiabatischer Systeme noch erhöht.

Inversion von Ammoniakderivaten. In einer der wenigen Arbeiten, in der man Rechnungen dieser Art ausführte [*30, 16, 32*], wurde die Inversion von Ammoniakabkömmlingen untersucht (Saunders, Yamada [*30a*]). Die Inversion von Ammoniak kann mit der KMR-Methode nicht verfolgt werden, da sie viel zu rasch verläuft. Führt man sperrige Substituenten in das NH_3-Molekül ein, so erniedrigt sich die Inversionsrate. Diese läßt sich weiter erniedrigen, indem man Protonen an den Stickstoff anlagert. Tatsächlich erwiesen sich in $2N$ Salzsäurelösung die Protonen innerhalb jeder der beiden Methylengruppen von Dibenzylmethylammoniumchlorid als inäquivalent und lieferten im KMR-Spektrum ein *AB*-Quartett. Bei pH-Erhöhung floß das Quartett zu einer einzelnen Linie zusammen. Die Inäquivalenz der CH_2-Protonen in einem Benzylamin, dessen beide übrige N-Substituenten verschieden sind, ist zu erwarten, wenn 1. Inversion am Stickstoff hinreichend selten eintritt und 2. die Molekülasymmetrie hoch genug ist. Die chemische Verschiebung des Protons *a* (*b*) ist gleich dem Mittelwert seiner chemischen Verschiebungen in den Rotameren (1), (2) und (3). (H. S. Gutowsky, J. Chem. Phys. **37**, 2196 (1962); [*31*]). Bei einer Inversion am Stickstoff vertauschen die Protonen innerhalb einer CH_2-Gruppe ihre chemischen Verschiebungen.

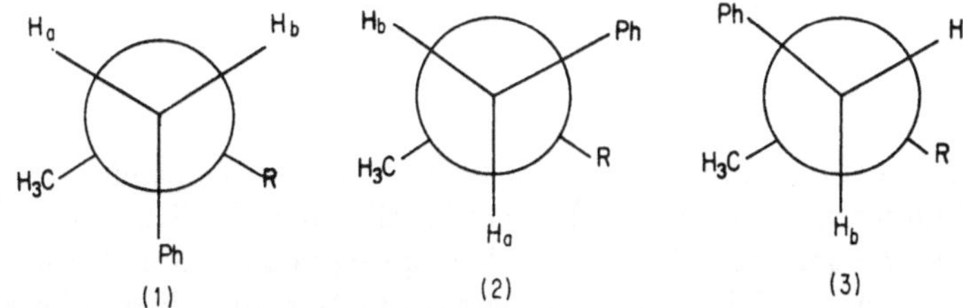

Zur Deutung der Befunde nehmen wir an, daß die Inversion nur stattfinden kann, wenn das Amin in freier nichtprotonierter Form vorliegt. Bei niedrigen pH-Werten ist die mittlere Lebensdauer des freien Amins zu kurz, um vor der Wiederanlagerung eines Protons zu invertieren. Mit steigenden pH-Werten erhöht sich die mittlere Lebensdauer des freien Amins, bis die Vertauschung der CH_2-Protonen so rasch abläuft, daß sie nach der KMR-Methode gemessen werden kann. Die mittlere Zeit zwischen zwei aufeinander folgenden Austauschvorgängen, τ, ist durch

$$\frac{1}{\tau} = -\frac{1}{[NH]}\frac{d\,[NH]}{d\,t} = k_i\,[N]/[NH]$$

mit der Inversionsrate k_i verknüpft, wenn die Konzentration von Amin gegen jene von Ammoniumsalz vernachlässigt werden kann. In dieser Gleichung bezeichnen [NH] und [N] die Konzentrationen von protoniertem bzw. freiem, nicht-invertiertem Amin. Der Quotient [N]/[NH] berechnet sich für den stationären Zustand ($d\,[N]/d\,t = 0$) zu

$$\frac{[N]}{[NH]} = \frac{k_1}{k_{-1}\,[H^+] + k_i}\,,$$

wo $[H^+]$ die Konzentration der Wasserstoffionen (bzw. des Protonendonators) sowie k_1 und k_{-1} die Geschwindigkeitskonstanten der Protonabspaltung bzw. der Protonanlagerung sind. Da $k_i \ll k_{-1}\,[H^+]$[14], gilt

$$\frac{1}{\tau} = k_i\frac{[N]_0}{[NH]_0}\,;$$

___

[14] Die Geschwindigkeit der Protonenübertragung konnte ebenfalls — mit Hilfe der KMR-Methode — direkt gemessen werden.

$[N]_0$ und $[NH]_0$ sind die Gleichgewichtskonzentrationen von Amin bzw. Ammoniumion. Durch die Bestimmung sämtlicher nach (5.4) relevanter Elemente der vollständigen Dichtematrix ergab sich für pH 3,5: $1/\tau = 21$ sec^{-1}; $k_i = (2 \pm 1)\,10^5$ sec^{-1} [30a].

Ringinversion. Die energieärmsten Konformationen von Cyclohexan und davon abgeleiteten Sauerstoff- und Schwefelheterocyclen sind die zwei Sesselformen:

Die beiden Sesselkonformationen klappen leicht ineinander um, was man Ringinversion nennt. Dabei gehen alle ursprünglich axialen Substituenten in äquatoriale über, und alle ursprünglich äquatorialen in axiale. Die Protonen von 5,5-Dimethyl-1,3-dioxan in 2-Stellung liefern im KMR-

Spektrum bei $-90\,°$C in Aceton-d$_6$-Lösung ein *AB*-Quartett (Schmid, Friebolin, Kabuss, Mecke [32]). Erhöht man die Temperatur, so verbreitern sich die 4 Linien und fließen ineinander. Bei Raumtemperatur wird nur noch eine scharfe Linie der 2-CH$_2$-Protonen beobachtet. Aus der Temperaturabhängigkeit der ermittelten τ-Werte ergaben sich die Arrheniusparameter zu $E_a = 13{,}5 \pm 0{,}8$ kcal/Mol und $\log A = 15{,}0 \pm 0{,}9$ (A in sec^{-1}).

8. Grenzen der Anwendbarkeit. Ergänzende Methoden und Entwicklungen

Die Wirkung chemischer Umlagerungen auf die Linienform in der kernmagnetischen Resonanz ist im Prinzip bekannt. Die Verfügbarkeit leistungsfähiger Rechenmaschinen ermöglicht die Berechnung der Spektren von komplexen Spinsystemen und deren Vergleich mit experimentellen Spektren. Auf diese Weise können Kenntnisse über Mechanismen und Reaktionsgeschwindigkeiten von chemischen Elementarprozessen gewonnen werden, die vielfach auf anderen Wegen unzugänglich sind. Quellen für systematische Fehler in quantitativen Untersuchungen entspringen der oft nicht zu umgehenden Annahme, daß die Parameter, welche das Spektrum im Grenzfall der langsamen Umlagerung beschreiben, unabhängig von äußeren Zustandsgrößen, wie Temperatur oder pH, sind[15].

Die wesentliche Beschränkung der Brauchbarkeit der Methode des langsamen Durchgangs dürfte darin liegen, daß ihre Anwendbarkeit auf einen relativ engen Geschwindigkeitsbereich begrenzt ist. In gewissen Fällen kann der zugängliche Geschwindigkeitsbereich durch die Anwendung der Methoden des schnellen Durchgangs (fast passage) erweitert werden. Aus dem Abfall der Nachschwingungen (wiggles) können die effektiven transversalen Relaxationszeiten und damit die Verweilzeiten in den Grenzfällen sehr langsamer und sehr rascher Umlagerungen im allgemeinen genauer bestimmt werden als aus der Linienbreite. Abb. 4.

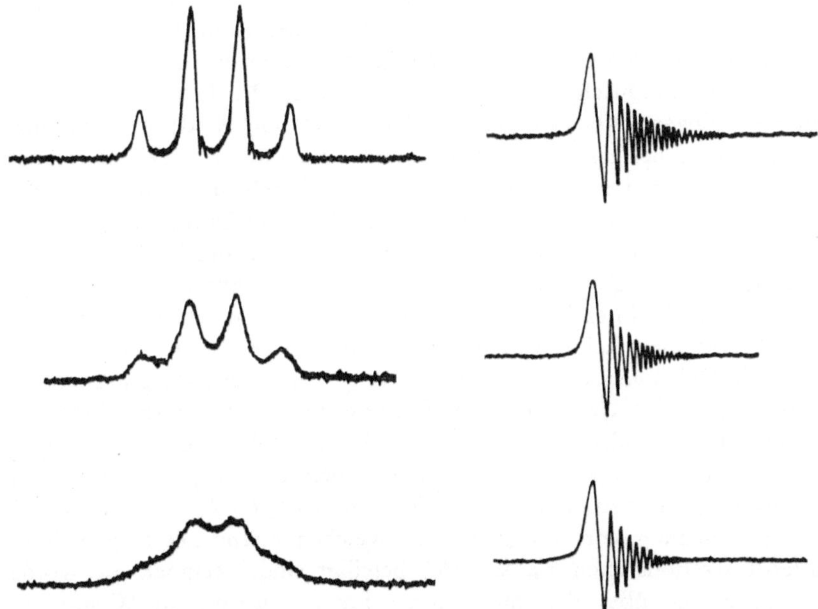

Abb. 4 Protonmagnetische Resonanzspektren von Methylammoniumion und Wasser. Die Geschwindigkeit des Protonenaustausches zwischen $[CH_3NH_3]^+$ und H_2O entsprechend Gl. (7.7) und (7.9) nimmt in der Abbildung von oben nach unten zu. Die linke Seite zeigt die CH_3-Resonanz bei langsamem Durchgang, die rechte Seite die H_2O-Resonanz der gleichen Lösung bei raschem Durchgang. Man beachte, wie mit steigender Austauschgeschwindigkeit die Nachschwingungen (wiggles) sich verkürzen, was ein Maß für die Verbreiterung der H_2O-Linie ist. [E. GRUNWALD, A. LOEWENSTEIN and S. MEIBOOM, J. Chem. Phys. 27, 630 (1957)]

---

[15] Für das System N-Alkyl-2,4,6-trinitroanilin konnte gezeigt werden, daß die Linienbreiten in den Grenzfällen sehr rascher und sehr langsamer Umlagerung nahezu identisch sind. [33] vgl. [32].

26*

Unter speziellen Bedingungen ist die Methode der „Molekularen Übertragung von Nichtgleichgewichts-Magnetisierung" [34] zur Abschätzung von Verweilzeiten zweckmäßig. Man kann ein KMR-Experiment bei schnellem Durchgang so ausführen, daß die Magnetisierung dem effektiven Magnetfeld

$$\mathbf{H}_e = \mathbf{k}\,(H_0 - \omega/\gamma) + \mathbf{i}\,H_1 \qquad (8.1)$$

folgt, was man adiabatischen Durchgang nennt; Bedingung dafür ist, daß $d\omega/dt \ll (\gamma H_1)^2$. In Gl. (8.1) sind $\mathbf{k}$ und $\mathbf{i}$ die Einheitsvektoren in z- bzw. x-Richtung im Koordinatensystem, welches mit der Winkelgeschwindigkeit ω des rotierenden Feldes H_1 sich dreht; H_0 ist das statische Magnetfeld, γ das gyromagnetische Verhältnis des beobachteten Kerns. Durchfährt man das Resonanzsignal vollständig, so kehrt sich die Richtung der Magnetisierung, die anfangs längs der z-Achse liegt, um. Die Magnetisierung kann um jeden gewünschten Winkel Θ gedreht werden, indem man das Ausmaß des Durchgangs, $\varDelta\,\omega$, geeignet wählt. Wir betrachten zwei, in gleicher Konzentration vorhandene Spezies A und B, deren Resonanzfrequenzen ω_A und ω_B seien. A und B vertauschen ihre Frequenzen mit der Geschwindigkeit $1/\tau$. Rascher adiabatischer Durchgang durch die Resonanz von A ruft eine Drehung, bei vollständigem Durchgang eine Umkehrung der Magnetisierung von A hervor. t sec später beobachtet man die Resonanz von B. Ist die mittlere Verweilzeit τ kürzer als t, so wird während des Zeitintervalls t Nichtgleichgewichts-Magnetisierung von A nach B und gleichzeitig Gleichgewichtsmagnetisierung von B nach A überführt. Dabei verkleinert sich die z-Komponente der B-Magnetisierung. Wenn $|\omega_A - \omega_B| \gg 1/\tau$, ist die Überführung von transversaler Magnetisierung zu vernachlässigen. Die Intensität der B-Resonanz wird mithin um einen von τ, (H_1, T_1 und t) abhängigen Faktor geschwächt. Der maximale Effekt wird erzielt, wenn $\Theta = \pi$ und $\tau \lesssim t < T_1$. Die Methode ist zur Untersuchung von Vorgängen, an welchen Kerne mit langen Spin-Gitter-Relaxationszeiten T_1, wie ^{13}C, beteiligt sind, geeignet. Sie wurde in einer Studie über die interessanten Fragen, warum in Wasser ein großer Anteil von gelöstem Kohlendioxid in nicht-hydratisierter Form vorliegt und warum die Hydratationsreaktion von CO_2 ein langsamer Vorgang ist, angewandt (Patterson, Ettinger [35]).

Ähnlich der Methode der Übertragung von Nichtgleichgewichts-Magnetisierung ist die Technik des „adiabatischen Halbdurchgangs". Bei dieser dreht man die Magnetisierung um 90° weg von der z-Achse, indem man bis zur Mitte der Resonanz fährt, und beobachtet den Abfall der transversalen Magnetisierung in Gegenwart des rotierenden Feldes H_1. Die Geschwindigkeit des Abfalls wird durch eine Relaxationszeit T bestimmt, die man durch Lösen von Gleichungen des Typs (2.11) erhält. T ist von der *Inhomogenität* des statischen Feldes H_0 *unabhängig*, wenn die

Veränderung von H_0 im Gebiet der Probe sehr klein gegen H_1 ist. Die Geschwindigkeit der Protonenübertragung in Wasser, dessen Protonenresonanz eine Linienbreite von weniger als 1 Hz hat, wurde nach dieser Technik ermittelt. Das in natürlichem Wasser vorkommende Isotop ^{17}O ($I = 5/2$) liefert für das Proton zusätzlich zu der einen Umgebung von ^{16}O sechs weitere und macht das Experiment möglich (MEIBOOM [36]).

Durch die Anwendung der Spin-Echo-Methode zur Messung von Geschwindigkeiten chemischer Elementarprozesse lassen sich die Grenzen der bisher erwähnten Verfahren in einer Reihe von Fällen überwinden (ALLERHAND und GUTOWSKY [37], BLOOM, REEVES und WELLS [38]). Nach dieser Methode sind höhere Geschwindigkeiten meßbar. Überdies können die chemische Verschiebung und die transversale Relaxationszeit in jedem Experiment unabhängig von der Reaktionsgeschwindigkeit bestimmt werden. Effekte der Inhomogenität des statischen Magnetfeldes werden vermieden. Ein ernster Mangel der Methode ist ihr geringes Auflösungsvermögen. Die starken Impulse des Radiofrequenzfeldes treten nicht nur mit den reagierenden Kernen, sondern auch mit den fixierten in Wechselwirkung. Die Methode wurde benutzt, um die Geschwindigkeit der Protonenübertragung in Wasser in Gegenwart von Trimethylammoniumion-Trimethylamin-Puffer zu messen und die Ordnung der Protolyse von Trimethylammoniumion in bezug auf Wasser zu ermitteln (LUZ und MEIBOOM [39]). Die innere Rotation in N,N-Dimethyltrichloracetamid und N,N-Dimethylcarbamylchlorid wurden mit ihrer Hilfe eingehend untersucht [37].

Für die Verfolgung sehr langsamer Reaktionen (auf der KMR-Zeitskala) hat sich die Doppelresonanzmethode als brauchbar erwiesen [40, 41]. Es wurde durch Anwendung dieser Technik gezeigt, daß der basenkatalysierte Austausch von Protonen zwischen den Lagen OH und $-CH=$ in Acetylaceton über die Ketoform verläuft (FORSÉN und HOFFMAN [40]).

Die angegebenen Beziehungen zwischen Linienformen und chemischen Umlagerungen gelten im allgemeinen auch im Bereich der Elektronenspinresonanz. Denn die Elektronenspinresonanz-Spektren von Radikalen in Lösung können gewöhnlich durch den Säkularanteil der isotropen Hyperfein-Wechselwirkung beschrieben werden. Schwierig von chemischen Vorgängen ist die Austauschwechselwirkung zwischen ungepaarten Elektronen zu trennen; um letztere zu vermeiden, arbeitet man zweckmäßig in verdünnter Lösung. Mitunter tragen auch anisotrope Dipolwechselwirkungen zur Linienbreite bei. Die mit Hilfe der Elektronenspinresonanz meßbaren Geschwindigkeiten sind — wegen der entsprechend größeren Energiedifferenzen — etwa um den Faktor 10^3 größer als jene mit der KMR bestimmbaren. (Vgl. JOHNSON [2a].)

9. Anhang

Der Spin und die Transformationseigenschaften von Spinfunktionen bei
räumlichen Drehungen

Der Übergang zu rotierenden Koordinaten in der quantenmechanischen
Spintheorie

In diesem Anhang stellen wir einige Elemente aus der Quantenmechanik des Spins zusammen, die wir im Hauptteil benötigen; dabei
wird vorwiegend das in elementarer Weise behandelt, was in den Darstellungen der kernmagnetischen Resonanz im allgemeinen nicht berücksichtigt werden kann.

Die Lösung der Boltzmanngleichung wird durch den Übergang zu
Koordinaten, die mit der Frequenz des Radiofrequenzfeldes rotieren,
wesentlich erleichtert. Wenn sich im Spinsystem ein stationärer Zustand
eingestellt hat, verschwinden beim Übergang zu rotierenden Koordinaten
die zeitlichen Ableitungen der Dichtematrixelemente $\varrho_{jk}^{(\varphi)}$ und man erhält
aus einem Satz von Differentialgleichungen ein leichter lösbares alge-
braisches Gleichungssystem für die $\varrho_{jk}^{(\varphi)}$. Zunächst veranlaßt uns dieser
formale Grund, die Gestalt des Operators zu bestimmen, der die Spinfunktionen und Spinoperatoren von dem stationären in ein rotierendes
Koordinatensystem transformiert. Diese Aufgabe führt uns zur Beschäf-
tigung mit einer fundamentalen Eigenschaft des Spins und des Drehimpulses im allgemeinen, nämlich der engen Verknüpfung zwischen dem
Transformationsverhalten von Wellenfunktionen bei räumlichen Drehungen und den Drehimpulsen des Systems [42].

Den inneren Drehimpuls oder Spin eines Teilchens stellen wir durch
einen linearen selbstadjungierten Vektoroperator **I** dar, dessen Komponenten
nach Definition den Vertauschungsrelationen

$$I_x I_y - I_y I_x = i I_z \qquad \text{(A. 1a)}$$

$$I_y I_z - I_z I_y = i I_x \qquad \text{(A. 1b)}$$

$$I_z I_x - I_x I_z = i I_y \qquad \text{(A. 1c)}$$

(in Einheiten von $\hbar$) genügen. Die Vertauschungsrelationen für die I_k
hatte Pauli ursprünglich einfach aus der Analogie zu denjenigen für die
Operatoren l_k der Bahndrehimpulskomponenten, die auf der Definitionsgleichung für den klassischen Bahndrehimpuls basieren, begründet.

Durch rein algebraische Entwicklung können aus den Vertauschungsrelationen die Eigenwerte der Spinoperatoren bestimmt werden. Bezeichnen wir, wie üblich, mit m und P_I die Eigenwerte von I_z bzw. I^2 und mit
ψ_{Im} die zugehörigen Eigenfunktionen, so ergibt sich

$$I^2 \psi_{Im} = P_I \psi_{Im} \qquad \text{(A. 2)}$$

mit

$$P_I = I(I+1), \qquad I = 0\,,\frac{1}{2},1,\,\frac{3}{2}\,,\,\cdots$$

und

$$I_z\,\psi_{Im} = m\,\psi_{Im}\,, \qquad m = I\,,\,I-1\,,\,\cdots\,,\,-I\,. \qquad\text{(A. 3)}$$

Wie vom Experiment her bekannt, ist für jede Teilchenart der Spin I, und damit der größte Eigenwert von I_z, eine feste Zahl, die charakteristisch für die betreffende Sorte ist. In ähnlicher Weise folgt für die Wirkung der Operatoren I_+ und I_- auf die Zustände ψ_{Im}

$$I_\pm\,\psi_{Im} = \sqrt{(I \mp m)(I \pm m + 1)}\;\psi_{Im\pm1} \qquad\text{(A. 4)}$$

wobei

$$I_\pm = I_x \pm i\,I_y\,.$$

Von der Wurzel ist nur der Absolutwert bestimmt; die Phasen werden so festgelegt, daß der Wert der Wurzel ≥ 0 ist. Für die explizite Gestalt der Matrizen der Spinoperatoren S in der Darstellung, in welcher I^2 und I_z diagonal sind, vereinbaren wir die folgende Konvention:

$$(\psi_{Im}\,,\,S\,\psi_{Im'}) \equiv \begin{pmatrix} (\psi_{II}\,,\,S\,\psi_{II}) & (\psi_{II}\,, & S\,\psi_{II-1})\,\cdots \\ (\psi_{II-1}\,,\,S\,\psi_{II}) & (\psi_{II-1}\,,\,S\,\psi_{II-1})\,\cdots \\ \cdots \end{pmatrix}$$

$$S = I_x\,,\,I_y\,,\,I_z\,,\,I^2\,,\,I_\pm\,,\,\cdots\,. \qquad\text{(A. 5)}$$

Wir interessieren uns für den einfachen Fall $I = {}^1/_2$ (Elektron, ^{1}H, ^{19}F). Dann existieren nur zwei Eigenzustände von I^2 und I_z. Der Eigenzustand $\psi_{1/2\,1/2}$ zum Eigenwert $m = {}^1/_2$ wird gewöhnlich mit α, der zum Eigenwert $m = -{}^1/_2$ mit β bezeichnet. Da der Satz der Spinfunktionen mit den beiden Eigenfunktionen α und β nach (A. 3) vollständig ist, muß sich jeder mögliche Spinzustand durch eine Linearkombination

$$\chi = a\alpha + b\beta\,, \qquad\text{(A. 6a)}$$

oder in Matrixschreibweise

$$\chi = \begin{pmatrix} a \\ b \end{pmatrix}\,, \qquad\text{(A. 6b)}$$

darstellen lassen. aa^* und bb^* sind die Wahrscheinlichkeiten dafür, daß die z-Komponente des Spins den Wert $+{}^1/_2$ bzw. $-{}^1/_2$ hat. a^* ist komplexkonjugiert zu a, b^* zu b.

Die Matrizen der Spinoperatoren für $I = {}^1/_2$ lauten nach (A. 2), (A. 3) und (A. 4) in der vereinbarten Notierung (A. 5)

$$(\psi_{1/2\,m}\,,\,I_x\,\psi_{1/2\,m'}) = \frac{1}{2}\begin{pmatrix} 0 & 1 \\ 1 & 0 \end{pmatrix}\,, \qquad (\psi_{1/2\,m}\,,\,I_y\,\psi_{1/2\,m'}) = \frac{1}{2}\begin{pmatrix} 0 & -i \\ i & 0 \end{pmatrix}\,,$$

$$(\psi_{1/2\,m}\,,\,I_z\,\psi_{1/2\,m'}) = \frac{1}{2}\begin{pmatrix} 1 & 0 \\ 0 & -1 \end{pmatrix}\,, \qquad (\psi_{1/2\,m}\,,\,I^2\,\psi_{1/2\,m'}) = \begin{pmatrix} {}^3/_4 & 0 \\ 0 & {}^3/_4 \end{pmatrix}\,.$$

Die ersten drei Matrizen sind die Paulimatrizen; in Vektorschreibweise haben sie die Gestalt

$$(\psi_{1/2 m}, \mathbf{I}\, \psi_{1/2 m'}) = \frac{1}{2}\begin{pmatrix} \mathbf{k} & \mathbf{i} - i\mathbf{j} \\ \mathbf{i} + i\mathbf{j} & -\mathbf{k} \end{pmatrix}, \tag{A. 7}$$

wo $\mathbf{i}$, $\mathbf{j}$ und $\mathbf{k}$ die Einheitsvektoren in x-, y- bzw. z-Richtung und $i = \sqrt{-1}$ sind.

Aus den Vertauschungsrelationen (A. 1) folgt unmittelbar, daß dem Spinvektor keine bestimmte Richtung im Raum zugeschrieben werden kann. Denn die Spinrichtung ist erst dann eindeutig vorgegeben, wenn alle drei Spinkomponenten exakt festgelegt sind. Es kann aber keine Zustandsfunktion existieren, die gleichzeitig Eigenfunktion von I_x, I_y und I_z ist, da nach (A. 1) $I_k I_l - I_l I_k \neq 0$. Gleichwohl ist es sinnvoll, von der durch

$$(\psi_{1/2 m}, \mathbf{I}\ \psi_{1/2 m}) = \frac{1}{2}\,\mathbf{n} \tag{A. 8a}$$

gegebenen mittleren Spinrichtung $\mathbf{n}$ oder der Richtung $\mathbf{n}$ des mittleren Spinvektors zu sprechen. (A. 8a) ist die Bestimmungsgleichung für den Erwartungswert $\langle \mathbf{I} \rangle$ des Operators $\mathbf{I}$

$$\langle \mathbf{I} \rangle = (\psi_{Im}, \mathbf{I}\, \psi_{Im}). \tag{A. 8b}$$

Befindet sich z. B. der Spin im Zustand α, so weist der mittlere Spinvektor in die positive z-Richtung. Offenbar ist die Charakterisierung eines Spinzustandes an eine bestimmte Wahl des Koordinatensystems gebunden.

Ein Beobachter beschreibe den Zustand eines physikalischen Systems durch die Wellenfunktion ψ. Wir fragen uns, welche Wellenfunktion $O_R \psi$ ein zweiter Beobachter dem gleichen Zustand zuordnet, wenn er das physikalische System in exakt gleicher Weise wie der erste Beobachter beschreibt, außer daß er ein Koordinatensystem benutzt, das in bezug auf das Koordinatensystem des 1. Beobachters gedreht ist. $O_R \psi$ kann als die Wellenfunktion des Zustandes ψ aus der Sicht des 2. Beobachters oder als die Wellenfunktion des ursprünglichen Zustandes ψ, der gedreht und dann vom 1. Beobachter beschrieben wird, definiert werden. Die Rotationen der Funktion und des Achsenkreuzes werden durch den gleichen Operator O_R bewirkt, wenn die Rotationen invers zueinander sind.

Wenn die Wellenfunktion nur von den Ortskoordinaten der Teilchen abhängt, ist die Operation O_R lediglich eine Punkttransformation P_R. Die Funktion $(P_R \psi)$ des zweiten Beobachters ordnet dem Punkt (x', y', z') die gleiche Zahl (oder Gesamtheit von Zahlen) zu wie die Funktion des 1. Beobachters dem Punkt (x, y, z). Denn der Punkt (x, y, z) hat im Bezugssystem des 2. Beobachters die Koordinaten (x', y', z').

$$(P_R \psi)(x', y', z') \equiv \psi(x, y, z). \tag{A. 9}$$

Stellt ψ eine von den Spinvariablen abhängige Funktion dar, so kann O_R nicht eine einfache Punkttransformation sein, da auf die Spinvariablen nur Spinoperatoren wirken.

Nehmen wir an, der 1. Beobachter findet den Spin im Eigenzustand α. Legt der 2. Beobachter seiner Beschreibung ein Achsenkreuz zugrunde, dessen z-Achse $\bar{z}$ mit der z-Achse des ursprünglichen Koordinatensystems den Winkel β einschließt, so kann sich der Spin bezüglich $\bar{z}$ nicht in einem Eigenzustand befinden. Die Komponente des mittleren Spinvektors in $\bar{z}$-Richtung, $\langle I_{\bar{z}} \rangle$, ist gleich der Projektion von $\langle I_z \rangle$ auf die $\bar{z}$-Achse:

$$\langle I_{\bar{z}} \rangle = \langle I_z \rangle \cos \beta \ . \tag{A. 10}$$

Im Hinblick auf (A. 6) stellt der 2. Beobachter den Zustand durch eine Linearkombination

$$\chi_+ = \overline{a}\, \psi_{1/2\ 1/2} + \overline{b}\, \psi_{1/2\ -\ 1/2} \tag{A. 11}$$

dar, in welcher $|\overline{a}|^2$ und $|\overline{b}|^2$ die Wahrscheinlichkeiten dafür sind, daß die $\bar{z}$-Komponente des Spins den Wert $+1/2$ bzw. $-1/2$ besitzt. Dann gilt außerdem

$$\langle I_{\bar{z}} \rangle = (\chi_+ , \ I_{\bar{z}} \chi_+) \tag{A. 12a}$$

oder

$$\langle I_{\bar{z}} \rangle = \frac{1}{2} \left\{ |\overline{a}|^2 - |\overline{b}|^2 \right\} \ . \tag{A. 12b}$$

Da die Überlagerungskoeffizienten normiert sein sollen,

$$|\overline{a}|^2 + |\overline{b}|^2 = 1 \ , \tag{A. 13}$$

folgt mit (A. 12b) und (A. 10)

$$|\overline{a}| = \cos \beta/2 \quad \text{und} \quad |\overline{b}| = \sin \beta/2 \ . \tag{A. 14}$$

Um das Achsenkreuz $\bar{x}, \bar{y}, \bar{z}$ des 2. Beobachters vollständig vorzugeben, muß die Rotation des Koordinatensystems durch drei unabhängige Parameter, z. B. die Eulerschen Winkel α, β, γ, spezifiziert werden. Die Eulerschen Winkel definieren wir als die Winkel dreier aufeinander folgender Drehungen. Wir beginnen die Folge mit der Drehung $R_z^{(A)}(\alpha)$ des ursprünglichen, rechtshändigen Achsenkreuzes x, y, z um die z-Achse im positiven Drehsinn um den Winkel α und bezeichnen die sich ergebenden Achsen mit x', y', z' $(= z)$. Der Drehsinn ist positiv, wenn sich beim Drehen eine rechtsgängige Schraube in Richtung der Drehachse bewegt. Dann führen wir die Drehung $R_{y'}^{(A)}(\beta)$ des Achsenkreuzes x', y', z' um die y'-Achse im positiven Drehsinn um β aus, wobei die Achsen x'', y'' $(= y')$, z'' resultieren. Schließlich drehen wir x'', y'', z'' um die z''-Achse im positiven Drehsinn um γ $[R_{z''}^{(A)}(\gamma)]$ und erhalten das neue Achsenkreuz x''', y''', z''' $(= z'')$.

                                        J. Heidberg

Es erweist sich oft als zweckmäßig, die Folge der Drehungen in bezug auf fixierte (die ursprünglichen) Achsen auszuführen. Das gleiche neue Achsenkreuz x''', y''', z''' ergibt sich, wenn die Drehungen in umgekehrter Reihenfolge wie oben um die fixierten Achsen vollzogen werden: wir rotieren zuerst um die z-Achse im positiven Drehsinn um γ, dann um die y-Achse im positiven Drehsinn um β und schließlich um die z-Achse im positiven Drehsinn um α:

$$R_{z'''}^{(A)}(\gamma)\, R_{y''}^{(A)}(\beta)\, R_z^{(A)}(\alpha) = R_z^{(A)}(\alpha)\, R_y^{(A)}(\beta)\, R_z^{(A)}(\gamma)\,. \qquad \text{(A. 15)}$$

Das Achsenkreuz $\bar{x}$, $\bar{y}$, $\bar{z}$ des 2. Beobachters gehe aus dem ursprünglichen durch die Drehungen

$$R_{y'}^{(A)}(-\beta)\, R_z^{(A)}(-\gamma) = R_{z''}^{(A)}(-\gamma)\, R_{y'}^{(A)}(-\beta) \qquad \text{(A. 16)}$$

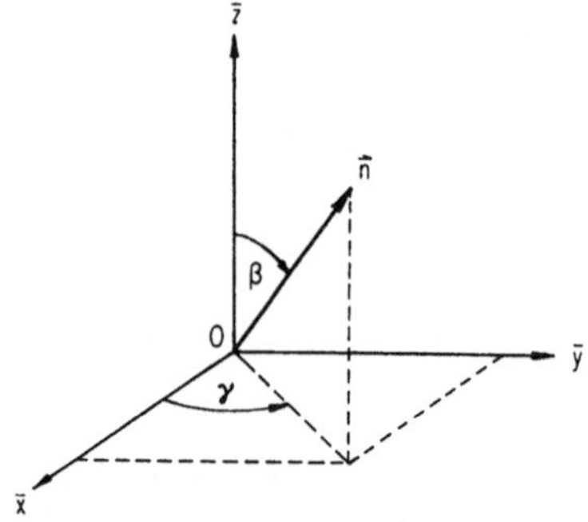

Abb. 5. Richtung $\vec{n}$ der ursprünglich in Richtung der $+z$-Achse weisenden Funktion nach den Drehungen $R_{z'}^{(F)}(\gamma)\, R_{y'}^{(F)}(\beta) = R_y^{(A)}(-\beta)\, R_z^{(A)}(-\gamma)$ S. Gl. (A. 30) und ff.

hervor; Abb. 5. Wir können $\alpha = 0$ setzen, da eine Drehung um die Achse des mittleren Spinvektors (oder, wie noch erklärt werden wird, des zugehörigen Spinors) nur eine unerhebliche Phasenverschiebung der Spinfunktion hervorruft. Offenbar wird der mittlere Spinvektor (bzw. die Richtung des zugehörigen Spinors) eindeutig bereits durch 2 Winkel, den Polarwinkel β und das Azimut γ, im Achsenkreuz des 2. Beobachters festgelegt.

Wir erhalten bei Berücksichtigung der Paulimatrizen (A. 7)

$$\langle I_{\bar{x}}\rangle = (\chi_+,\ I_{\bar{x}}\,\chi_+) = \{\bar{a}^*\,\bar{b} + \bar{a}\,\bar{b}^*\}\,\frac{1}{2}\,, \qquad \text{(A. 17a)}$$

$$\langle I_{\bar{y}}\rangle = (\chi_+,\ I_{\bar{y}}\,\chi_+) = \{-i\,\bar{a}^*\,\bar{b} + i\,\bar{a}\,\bar{b}^*\}\,\frac{1}{2}\,; \qquad \text{(A. 17b)}$$

Überdies gilt

$$\langle I_{\bar{x}}\rangle = \frac{1}{2}\sin\beta\,\cos\gamma\,, \qquad \text{(A. 18a)}$$

$$\langle I_{\bar{y}}\rangle = \frac{1}{2}\sin\beta\,\sin\gamma\,. \qquad \text{(A. 18b)}$$

Daraus folgt

$$2\,\overline{a}{}^*\,\overline{b} = \sin\beta\,e^{i\gamma} \tag{A. 19}$$

und wegen (A. 14)

$$\frac{\overline{b}}{\overline{a}} = e^{i\,\gamma}\tan\beta/2\,. \tag{A. 20}$$

Die noch willkürlichen Phasen setzen wir so fest, daß

$$\chi_+ = e^{-i\gamma/2}\cos\beta/2\ \psi_{1/2\,1/2} + e^{i\gamma/2}\sin\beta/2\ \psi_{1/2\,-1/2}\,. \tag{A. 21}$$

In ähnlicher Weise können wir die Zustandsfunktion χ_- des 2. Beobachters für den Fall bestimmen, daß sich der Spin bezüglich des Achsenkreuzes des 1. Beobachters im Zustand $\beta = \psi_{1/2\,-1/2}$ befindet. Sie lautet

$$\chi_- = -\,e^{-i\gamma/2}\sin\beta/2\ \psi_{1/2\,1/2} + e^{i\gamma/2}\cos\beta/2\ \psi_{1/2\,-1/2}\,. \tag{A. 22}$$

Wir fassen (A. 21) und (A. 22) in der Gleichung

$$Q^{(A)}\,(0,\,-\beta,\,-\gamma)\ \psi_{1/2\,m} = \sum_{m'} D^{(1/2)}_{m'm}\,(\gamma\,,\,\beta\,,\,0)\ \psi_{1/2\,m'} \tag{A. 23}$$

zusammen[16]. $Q^{(A)}\,(0,-\beta,-\gamma)$ ist der mit der Rotation $R_y^{(A)}(-\beta)\ R_z^{(A)}\,(-\gamma)$ verknüpfte Operator, welcher auf $\psi_{1/2\,m}$ angewandt die neue Funktion $Q^{(A)}\,(0,-\beta,-\gamma)\ \psi_{1/2\,m}$ ergibt. Die Funktionen $\psi_{1/2\,m}$ und $Q^{(A)}\,(0,-\beta,-\gamma)$ $\psi_{1/2\,m}$ beschreiben den gleichen Zustand, $\psi_{1/2\,m}$ aus der Sicht des ersten Beobachters, $Q^{(A)}(0,-\beta,-\gamma)\ \psi_{1/2\,m}$ aus der Sicht des zweiten. Gl. (A. 23) ist für reine Spinfunktionen das Analogon zu Gl. (A. 9). Multiplizieren wir (A. 23) von links mit $\psi_{1/2\,m'}$ und bilden das Skalarprodukt, so ergibt sich wegen der Orthonormiertheit der $\psi_{1/2\,m}$

$$(\psi_{1/2\,m'}\,,\ Q^{(A)}\,(0,-\beta,-\gamma)\ \psi_{1/2\,m}) = D^{(1/2)}_{m'm}\,(\gamma\,,\,\beta\,,\,0)$$

$$= \begin{pmatrix} e^{-i\gamma/2}\cos\beta/2 & -e^{-i\gamma/2}\sin\beta/2 \\ e^{i\gamma/2}\sin\beta/2 & e^{i\gamma/2}\cos\beta/2 \end{pmatrix}\,. \tag{A. 24}$$

Eine Zweikomponenten-Funktion, die sich bei Rotationen nach (A. 23) mit der Transformationsmatrix $D^{(1/2)}_{m'm}$ transformiert, heißt ein Spinor. Die reinen Spinfunktionen $\psi_{1/2\,1/2}$ und $\psi_{1/2\,-1/2}$ sind also Spinoren. Nach (A. 24) führt eine volle Drehung einen Spinor in sein Negatives über. Wir müssen um $4\,\pi$ drehen, um die Identität zu erhalten. Da die Eulerschen Winkel, die eine Rotation beschreiben, nur bis auf Vielfache von $2\,\pi$ bestimmt sind, entsprechen einer Rotation zwei Transformationsmatrizen, $D^{(1/2)}$ und $-D^{(1/2)}$; $D^{(1/2)}$ gehört z. B. zum Rotationswinkel φ und $-D^{(1/2)}$ zu $(\varphi + 2\,\pi)$. Diese 2 zu 1 Korrespondenz zwischen der Transformationsmatrix $D^{(1/2)}$ und räumlichen Drehungen ist für Spinoren charakteristisch.

---

[16] Die Bezeichnung $D^{(1/2)}\,(\vartheta,\varphi,\eta)$ für eine Transformationsmatrix der Gestalt (A. 24) wurde von E. WIGNER in seinem Buch „Gruppentheorie und ihre Anwendung auf die Quantenmechanik der Atomspektren", Verlag Friedrich Vieweg und Sohn, Braunschweig 1931, verwendet und ist heute gebräuchlich.

Nachdem wir die Matrixelemente des Operators $Q^{(A)}$ in der I_z-Darstellung ermittelt haben, sind wir in der Lage, die explizite Gestalt von $Q^{(A)}$ in Operatorform anzugeben. Wir nehmen zunächst an, daß $\beta = 0$ ist. Dann vereinfacht sich (A. 24) zu einer Diagonalmatrix mit den Elementen $\exp[-i\gamma/2]$ und $\exp[i\gamma/2]$ in der 1. bzw. 2. Zeile, und man erkennt, daß der Operator $Q_z^{(A)}(-\gamma)$ die Gestalt

$$Q_z^{(A)}(-\gamma) = e^{-i\gamma I_z} = \sum_{n=0}^{\infty} \frac{1}{n!}(-i\gamma I_z)^n \qquad (A.\ 25)$$

haben muß. Setzen wir in Analogie dazu

$$Q_y^{(A)}(\beta) = e^{i\beta I_y} = \sum_{n=0}^{\infty} \frac{1}{n!}(i\beta I_y)^n , \qquad (A.\ 26)$$

so folgt unmittelbar z. B. für

$$(\psi_{1/2\ 1/2},\ Q_y^{(A)}(-\beta)\ \psi_{1/2\ 1/2}) = 1 + \frac{(-i\beta/2)^2}{2!} + \frac{(-i\beta/2)^4}{4!} + \cdots = \cos\beta/2 .$$

Der Operator $Q_k^{(A)}(\vartheta)$, der mit einer Drehung des Koordinatensystems um den Winkel ϑ im positiven Drehsinn um eine Achse k, längs welcher der Spin die Komponente I_k hat, verknüpft ist, besitzt mithin die Form

$$Q_k^{(A)}(\vartheta) = e^{i\vartheta I_k} . \qquad (A.\ 27)$$

Führt man mehrere Rotationen nacheinander aus, so ist zu beachten, daß z. B.

$$(\psi_{1/2\ 1/2},\ Q_y^{(A)}(-\beta)\ Q_z^{(A)}(-\gamma)\ \psi_{1/2\ -1/2}) = -e^{-i\gamma/2}\sin\beta/2$$

$$= (\psi_{1/2\ 1/2},\ e^{-i\gamma I_z}\ e^{-i\beta I_y}\ \psi_{1/2\ -1/2})$$

usf. oder

$$Q_y^{(A)}(-\beta)\ Q_z^{(A)}(-\gamma) = Q^{(A)}(0,\ -\beta,\ -\gamma) = e^{-i\gamma I_z}\ e^{-i\beta I_y} . \qquad (A.\ 28)$$

Wir sahen bisher die Funktionen als im Raum fixiert und die Koordinatenachsen und Basisfunktionen als drehbar an. Bei der alternativen, äquivalenten Betrachtungsweise faßt man die Funktionen bezüglich der räumlich fixierten Koordinatenachsen und Basisfunktionen als drehbar auf. Unterwirft man die Achsen und Funktionen den gleichen Drehungen, so ändert sich die Lage der Funktion im Koordinatensystem nicht, und die Transformationsmatrix Q ist einfach die Einheit

$$Q^{(A)}(\gamma,\ \beta,\ \alpha)\ Q^{(F)}(\gamma,\ \beta,\ \alpha) = 1 . \qquad (A.\ 29)$$

Daher gilt

$$Q^{(A)}(-\alpha,\ -\beta,\ -\gamma) = Q^{(F)}(\gamma,\ \beta,\ \alpha) = e^{-i\gamma I_z}\ e^{-i\beta I_y}\ e^{-i\alpha I_z} . \qquad (A.\ 30)$$

Die Drehungen $R_z^{(F)}(\gamma)\ R_y^{(F)}(\beta)\ R_z^{(F)}(\alpha)$ der Funktion bezüglich fixierter Achsen sind mithin den Drehungen $R_z^{(A)}(-\alpha)\ R_y^{(A)}(-\beta)\ R_z^{(A)}(-\gamma)$ der Koordinatenachsen bei festgehaltener Funktion äquivalent.

Gleichung (A. 30) kann man — ebenso wie Gl. (A. 1) — als Grundlage zum Aufbau der Theorie des Spins wählen. Gerechtfertigt wird diese Wahl durch die Analogie von Gl. (A. 30) zu jener Beziehung, welche die Operatoren der Bahndrehimpulskomponenten mit den Rotationsoperatoren verbindet. Die Vertauschungsrelationen (A. 1) ergeben sich dann aus (A. 30) als Folge der Nichtvertauschbarkeit endlicher Drehungen des Koordinatensystems (J. v. NEUMANN und E. WIGNER 1928).

Den gesuchten Operator für die Transformation der Spinfunktionen und Kernspindichtematrizen zu rotierenden Koordinaten liefert (A. 30).

Es läßt sich zeigen, daß der Operator

$$T = \exp[i\,S] \qquad\qquad (A.\,31a)$$

unitär ist, wenn der Operator S hermitisch ist. Der Umkehrsatz ist ebenfalls gültig. Der dem hermitischen Spinoperator I_k zugeordnete Rotationsoperator

$$Q_k(\vartheta) = \exp[\,i\,\vartheta\,I_k] \qquad\qquad (A.\,31b)$$

ist mithin unitär.

Es seien ψ und F die Wellenfunktion und der Operator einer physikalischen Größe bezüglich stationärer Koordinaten, während $\psi^{(\omega)}$ und $F^{(\omega)}$ die gleichen Größen bezüglich eines Koordinatensystems seien, welches mit der Kreisfrequenz $\boldsymbol{\omega}$ rotiere. Nach (A. 27) sind ψ und $\psi^{(\omega)}$ durch die unitäre Transformation [22]

$$\psi^{(\omega)} = Q\,\psi \qquad\qquad (A.\,32a)$$
$$Q = \exp[i\,(\boldsymbol{\omega},\mathbf{I})\,t] \qquad\qquad (A.\,32b)$$

verknüpft. Substituieren wir in dem Matrixelement $(\psi_m, F\,\psi_n)$ die Funktionen ψ durch die entsprechenden Funktionen $\psi^{(\omega)}$, so erhalten wir im Hinblick auf die Unitarität von Q

$$(\psi_m, F\,\psi_n) = (Q^{-1}\,\psi_m^{(\omega)}, F\,Q^{-1}\,\psi_n^{(\omega)})$$
$$= (\psi_m^{(\omega)}, Q\,F\,Q^{-1}\,\psi_n^{(\omega)})$$

und damit

$$F^{(\omega)} = Q\,F\,Q^{-1}. \qquad\qquad (A.\,33)$$

Indem wir ψ durch $\psi^{(\omega)}$ substituieren, ersetzen wir eine Rotation des Operators F mit der Kreisfrequenz $\boldsymbol{\omega}$ bezüglich der stationären Funktion ψ durch die Rotation der Funktion $\psi^{(\omega)}$ mit der Kreisfrequenz $-\boldsymbol{\omega}$ bezüglich des stationären Operators $F^{(\omega)}$. Denn $\psi^{(\omega)}$ dreht sich relativ zu den Koordinaten, die mit der Kreisfrequenz $+\boldsymbol{\omega}$ rotieren, mit der Kreisfrequenz $-\boldsymbol{\omega}$, während $F^{(\omega)}$ relativ zu diesen rotierenden Koordinaten in Ruhe ist. Das gleiche Ergebnis (A. 33) erhält man, wenn man die Gestalt des Matrixelements von F in der Darstellung der mit der Kreisfrequenz $\boldsymbol{\omega}$ rotierenden Funktion $Q^{-1}\,\psi$ nach

$$(Q^{-1}\,\psi_m, F Q^{-1}\,\psi_n) = (\psi_m, Q F Q^{-1}\,\psi_n) \qquad\qquad (A.\,34)$$

bestimmt. Vgl. (3.14), (3.40) und (5.1).

Herrn Prof. Dr. H. Hartmann bin ich für die großzügige Förderung die er dieser Arbeit angedeihen ließ, dankbar verpflichtet. Für anregende Diskussionen und die Durchsicht des Manuskripts möchte ich auch an dieser Stelle Herrn Dr. J. A. Weil, z. Z. Gastprofessor am Department of Physics, University of Canterbury, N. Z., herzlich danken. Herrn Dr. H. Sillescu gilt mein Dank für die Überlassung einer Schrift vor ihrer Veröffentlichung und hilfreiche Hinweise, Herrn Jörn v. Jouanne für sorgfältige Mitarbeit bei den Korrekturen.

Literatur

1. a) GUTOWSKY, H. S., D. W. McCALL, and C. P. SLICHTER: J. Chem. Phys. **21**, 279 (1953).

   b) —, and A. SAIKA: J. Chem. Phys. **21**, 1688 (1953).

   c) HAHN, E. L., and D. E. MAXWELL: Phys. Rev. **88**, 1070 (1952).

2. a) JOHNSON, JR., C. S.: Advanc. Magnetic Resonance **1**, 33 (1965) [Ed. by J. S. WAUGH].

   b) REEVES, L. W.: Advanc. Phys. Org. Chem. **3**, 187 (1965).

   c) EMSLEY, J. W., J. FEENEY, and L. H. SUTCLIFFE: High Resolution Nuclear Magnetic Resonance Spectroscopy (in Two Vol.). Oxford: Pergamon Press 1965.

   d) DELPUECH, J.: Bull. Soc. Chim. France, S. 2697 (1964).

   e) LOEWENSTEIN, A., and T. M. CONNOR: Ber. Bunsenges. phys. Chem. **67**, 280 (1963).

   f) STREHLOW, H. in: Technique of Organic Chemistry, Eds. S. L. FRIESS, E. S. LEWIS, and A. WEISSBERGER, 2nd Ed., Vol. 8, Part 2, P. 865. New York—London: Interscience, 1963.

3. BLOCH, F.: Phys. Rev. **70**, 460 (1946).

4. McCONNELL, H. M.: J. Chem. Phys. **28**, 430 (1958); vergleiche die instruktiven Betrachtungen von VAN VLECK: VAN VLECK, J. H.: Ned. Tijdschr. Natuurk. **27**, 1 (1961).

5. MEIBOOM, S., Z. LUZ, and D. GILL: J. Chem. Phys. **27**, 1411 (1957).

6. ANDERSON, P. W.: J. Phys. Soc. Japan **9**, 316 (1954).

7. KUBO, R.: J. Phys. Soc. Japan **9**, 935 (1954).

8. PIETTE, L. H., and W. A. ANDERSON: J. Chem. Phys. **30**, 899 (1959).

9. a) SILLESCU, H.: Kernmagnetische Resonanz. Berlin—Heidelberg—New York: Springer-Verlag 1966.

   b) STREHLOW, H.: Magnetische Kernresonanz und chemische Struktur. Darmstadt: Steinkopff 1962.

   c) POPLE, J. A., W. G. SCHNEIDER, and H. J. BERNSTEIN: New York—Toronto—London: McGraw-Hill Book Co., Inc. 1959.

   d) SUHR, H.: Anwendungen der kernmagnetischen Resonanz in der organischen Chemie. Berlin—Heidelberg—New York: Springer-Verlag 1965.

   e) FLUCK, E.: Die kernmagnetische Resonanz und ihre Anwendung in der anorganischen Chemie. Berlin—Göttingen—Heidelberg: Springer-Verlag 1963.

   f) ROBERTS, J. D.: Nuclear Magnetic Resonance. New York—Toronto—London: McGraw-Hill Book Co., Inc. 1959.

   g) LÖSCHE, A.: Kerninduktion. Berlin: Dt. Verlag d. Wissenschaften 1957.

   h) ANDREW, E. R.: Nuclear Magnetic Resonance. Cambridge: University Press 1955.

10. a) NEUMANN, J. v.: Göttinger Nachrichten 245 und 273 (1927).

   b) MÜNSTER, A.: Statistische Thermodynamik. Berlin—Göttingen—Heidelberg: Springer-Verlag 1956.

   c) FANO, U.: Rev. Mod. Phys. **29**, 74 (1957).

   d) TER HAAR, D.: Reports Progress Phys. **24**, 304 (1961).

   e) DIRAC, P. A. M.: The Principles of Quantum Mechanics, 4th. Ed. Oxford: At the Clarendon Press 1958.

11. SLICHTER, C. P.: Principles of Magnetic Resonance. New York—Evanston—London: Harper and Row 1963.

12. ABRAGAM, A.: The Principles of Nuclear Magnetism. Oxford: At the Clarendon Press 1961.

13. ALEXANDROW, I. W.: Theorie der kernmagnetischen Resonanz. Leipzig: B. G. Teubner Verlagsges. 1966. (Übersetzung d. russischen Ausgabe von 1964).

14. WANGSNESS, R. K., and F. BLOCH: Phys. Rev. **89**, 728 (1953).

15. BLOCH, F.: Phys. Rev. **105**, 1206 (1957).

16. HEIDBERG, J., J. A. WEIL, G. A. JANUSONIS, and J. K. ANDERSON: J. Chem. Phys. **41**, 1033 (1964).

17. KAPLAN, J.: J. Chem. Phys. **28**, 278 (1958); **29**, 462 (1958).

18. ALEXANDER, S.: J. Chem. Phys. **37**, 967 (1962).

19. JOHNSON, JR., C. S.: J. Chem. Phys. **41**, 3277 (1964).

20. HEIDBERG, J.: Ber. Bunsenges. physik. Chem. **71**, 925 (1967); Veröffentl. in Vorbereitung.

21. ALEXANDER, S.: J. Chem. Phys. **37**, 974 (1962); **38**, 1787 (1963); **40**, 2741 (1964).

22. RABI, I. I., N. F. RAMSEY, and J. SCHWINGER: Rev. Mod. Phys. **26**, 167 (1954).

23. EIGEN, M., u. L. DeMAEYER: Kinetik schneller Reaktionen und chemische Relaxation, ausgearb. J. HEIDBERG u. G. KOHLMAIER, dieses Buch S. 417.

24. SOLOMON, I., and N. BLOEMBERGEN: J. Chem. Phys. **25**, 261 (1956).

25. a) CONNOR, T. M., and A. LOEWENSTEIN: J. Am. Chem. Soc. **83**, 560 (1961).

   b) MEIBOOM, S., A. LOEWENSTEIN, and S. ALEXANDER: J. Chem. Phys. **29**, 969 (1958).

   c) GRUNWALD, E.: J. Phys. Chem. **67**, 2208, 2211 (1963).

26. MACKOR, E. L., and C. MacLEAN: Pure Applied Chem. **8**, 393 (1964).

27. HERTZ, H. G.: Z. Elektrochem. Ber. Bunsenges. physik. Chem. **64**, 53 (1960); **65**, 36 (1961).

28. a) EATON, D. R., and W. D. PHILLIPS: Adv. Magnetic Resonance **1**, 103 (1965).

   b) —, A. D. JOSEY, W. D. PHILLIPS, and R. E. BENSON, Discussions Faraday Soc. **34**, 77 (1962).

29. a) PEARSON, R. G., J. PALMER, M. M. ANDERSON, and A. L. ALLRED: Z. Elektrochem. Ber. Bunsenges. physik. Chem. **64**, 110 (1960).

   b) McCONNELL, H. M., and S. B. BERGER: J. Chem. Phys. **27**, 230 (1957).

30. a) SAUNDERS, M., and F. YAMADA: J. Am. Chem. Soc. **85**, 1882 (1963).

   b) KURLAND, R. J., M. B. RUBIN, and W. B. WISE: J. Chem. Phys. **40**, 2426 (1964).

   c) NEWMARK, R. A., and C. H. SEDERHOLM: J. Chem. Phys. **43**, 602 (1965).

   d) JONÁŠ, J., A. ALLERHAND, and H. S. GUTOWSKY: J. Chem. Phys. **42**, 3396 (1965).

31. WHITESIDES, G. M., F. KAPLAN, and J. D. ROBERTS: J. Am. Chem. Soc. **85**, 2167 (1963).

32. SCHMID, H. G., H. FRIEBOLIN, S. KABUSS u. R. MECKE: Spectrochim. Acta **22**, 623 (1966).

33. HEIDBERG, J., u. J. v. JOUANNE: Veröffentl. in Vorbereitung.

34. McCONNELL, H. M., and D. D. THOMPSON: J. Chem. Phys. **31**, 85 (1959).

35. Patterson, jr., A., and R. Ettinger: Z. Elektrochem. Ber. Bunsenges. physik. Chem. **64**, 98 (1960).
36. Meiboom, S.: J. Chem. Phys. **34**, 375 (1961).
37. Allerhand, A., and H. S. Gutowsky: J. Chem. Phys. **41**, 2115 (1964).
38. Bloom, M., L. W. Reeves, and E. J. Wells: J. Chem. Phys. **42**, 1615 (1965).
39. Luz, Z., and S. Meiboom: J. Chem. Phys. **39**, 366 (1963).
40. Forsén, S., and R. A. Hoffman: J. Chem. Phys. **40**, 1189 (1964).
41. Baldeschwieler, J. D.: J. Chem. Phys. **40**, 459 (1964).
42. a) Rose, M. E.: Elementary Theory of Angular Momentum. New York: John Wiley and Sons, Inc. 1957.
    b) Tinkham, M.: Group Theory and Quantum Mechanics. New York—San Francisco—Toronto—London: McGraw-Hill Book Co. 1964.

Joachim Heidberg
Institut für physikalische
Chemie der Universität
Frankfurt/Main

Kinetik schneller Reaktionen in Lösung und chemische Relaxation

M. EIGEN und L. DE MAEYER

Ausgearbeitet von J. HEIDBERG und G. H. KOHLMAIER

Mit 3 Abbildungen

1. Einleitung

Unter schnellen Reaktionen versteht man Reaktionen, bei denen die Konzentrationsänderung eines oder mehrerer Reaktanten durch eine Halbwertszeit von weniger als einer Sekunde charakterisiert ist, so daß die konventionellen Methoden der Einleitung einer Reaktion (Mischung der Reaktionspartner) und der Beobachtung der Konzentrationsänderung versagen. Bis vor einem Jahrzehnt waren die Untersuchungen von schnellen Reaktionen im wesentlichen auf die von HARTRIDGE und ROUGHTON eingeführte Strömungsmethode beschränkt, bei der Halbwertzeiten größer als 1 msec gemessen werden können. Mit den heute verfügbaren modernen Methoden ist es möglich, komplexe chemische Reaktionsabläufe in Elementarschritte aufzulösen, deren Geschwindigkeitskonstanten erster Ordnung zwischen 1 und 10^9 sec^{-1} und deren Konstanten zweiter Ordnung zwischen 1 und 10^{11} l Mol^{-1} sec^{-1} liegen. Elementarreaktionen der Elektronen- und Protonenübertragung, Ligandensubstitutions-, Isomerisierungs- und Hydratationsreaktionen sind mit den schnellen Methoden untersucht und zum großen Teil aufgeklärt worden. Auf dieser Basis sind auch die Untersuchungen zusammengesetzter Reaktionen möglich, wie sie z. B. bei Prozessen der Wasserstoffbrückenbindung, der Enzymkatalyse, der Helix-Coil-Umwandlung, der Energieübertragung von angeregten Singulett- und Triplett-Molekülen oder der Polymerisation auftreten [1].

2. Übersicht der Meßmethoden

In dieser Zusammenfassung sollen nur die Störungsmethoden mit kleiner Störamplitude behandelt werden, also jene Methoden, bei denen die Störung eines chemischen Gleichgewichts oder eines stationären Zustands durch einen sich schnell ändernden äußeren Parameter wie Temperatur, Druck oder elektrische Feldstärke erfolgt und bei denen durch anschließende

Beobachtung der Einstellung des neuen Gleichgewichts das Relaxationsspektrum bestimmt werden kann. Die Blitzlichtphotolyse unterscheidet sich von den beschriebenen Relaxationsmethoden, indem bei ihr durch Einstrahlung von Licht Produkte entstehen, die erheblich vom Gleichgewichtszustand entfernt sein können. Als zweite Gruppe sind die Konkurrenzmethoden zu nennen, bei denen das Reaktionssystem an einen bekannten physikalischen Prozeß gekoppelt ist oder wird, der mit der chemischen Umwandlung konkurrieren kann. Ein Beispiel hierfür ist die kontinuierliche Strömungsmethode, bei der ein stationäres Reaktionsprofil entlang einer durchströmten Röhre beobachtet werden kann, das durch die Konkurrenz zwischen Massentransport und chemischer Umwandlung bestimmt ist. Ganz analoge Überlegungen gelten für die Methode des thermischen Maximums, bei der der Wärmetransport kontrollierbar ist. Bei den elektrochemischen Methoden wird die Konkurrenz zwischen Diffusion und chemischer Umwandlung, bei den photochemischen Methoden jene zwischen Bildung und chemischer Reaktion der angeregten Moleküle ausgenutzt. Die Fluoreszenzausbeute-Messungen hängen wiederum von der Konkurrenz zwischen der Fluoreszenz und der chemischen Reaktion der angeregten Singulettsysteme ab; sie ermöglichen eine Abschätzung einer chemischen Relaxation, wenn die Relaxationszeit von der Größenordnung der Lebensdauer der angeregten Moleküle ist (10^{-8} sec). Direkte Auskunft über die Lebensdauer erhält man aus Messungen der Linienbreite der Kern- und Elektronen-Spinresonanz. Mit dieser Methode lassen sich auch Reaktionssysteme untersuchen, bei denen insgesamt kein chemischer Umsatz, sondern nur ein Platzwechsel stattfindet. Interessant ist die Kombination der kontinuierlichen Strömungsmethode mit der Temperatursprungmethode; mit ihr lassen sich die schnellen vorgelagerten Gleichgewichte irreversibler Reaktionen untersuchen.

3. Theoretische Basis der Relaxationsmethoden

3.1. *Störung des chemischen Gleichgewichts*

Die Störung eines im chemischen Gleichgewicht befindlichen Systems kann durch die Änderung der äußeren Parameter, Temperatur T, Druck P oder elektrische Feldstärke E, entsprechend den thermodynamischen Beziehungen

$$(\partial \ln K_0 / \partial T)_P = \Delta H_0 / R T^2 \tag{1}$$

$$(\partial \ln K_0 / \partial P)_T = - \Delta V_0 / R T \tag{2}$$

$$(\partial \ln K_0 / \partial E)_{P,\,T} = \Delta M_0 / R T \tag{3}$$

erfolgen, wobei K_0 die auf die Aktivitäten der Molenbrüche bezogene Gleichgewichtskonstante, ΔH_0 die Reaktionsenthalpie, ΔV_0 das Reaktionsvolumen und ΔM_0 die Reaktionspolarisation ist. Bei einer Reaktionsenthalpie von 1 kcal/Mol bewirkt eine Temperaturänderung von 10 °C eine relative Änderung der Gleichgewichtskonstante von $\sim 6\%$, während bei einem Reaktionsvolumen von $\Delta V_0 = 22\ \mathrm{cm^3/Mol}$ (Reaktion $H^+ + OH^- \rightleftharpoons H_2O$) eine Druckvariation von 65 Atmosphären etwa dieselbe Änderung von K_0 hervorruft (jeweils von dem Zustand bei etwa 25 °C und 1 Atm ausgehend).

3.2. Linearisierte Geschwindigkeitsgleichungen und Relaxationszeiten

3.2.1. Einstufige Reaktionen

Entsprechend der Form der Störfunktion (und der Art der Messung des gestörten Systems) unterscheidet man Sprunganregung (Einschwing-methode) und periodische Anregung (stationäre Methode). Es läßt sich zeigen, daß bei allen einstufigen Reaktionen, gleichgültig ob sie erster, zweiter oder dritter Ordnung in der Hin- oder Rückreaktion sind, die zeitliche Änderung der Variablen

$$x_i(t) = c_i(t) - c_i^0$$

[$c_i(t)$ ist die Momentankonzentration des Reaktanten i, c_i^0 ist die entsprechende zeitunabhängige Bezugskonzentration, die sich z. B. auf den Anfangs- oder Endzustand eines Einschwingvorgangs beziehen mag] der ersten Potenz von x_i und dem Reziprokwert der Relaxationszeit τ proportional ist, unter der Voraussetzung, daß die Abweichung vom Gleichgewicht klein ist. (Siehe Anhang und vgl. [2].) Die allgemeine linearisierte Geschwindigkeitsgleichung ist gegeben durch

$$-\frac{d\,x_i(t)}{d\,t} = \frac{1}{\tau}x_i(t) - \frac{1}{\tau}\overline{x}_i(t) \tag{4}$$
$$\mathrm{mit}\ \ \overline{x}_i(t) = \overline{c}_i(t) - c_i^0$$

[$\overline{c}_i(t)$ ist die durch die, zur Zeit t herrschenden, äußeren Bedingungen bestimmte Stationäre-Zustand-Konzentration des Reaktanten i; sie heißt auch Zeitfunktion oder Zwangskraftfunktion.]

Als Beispiel sei die Reaktion

$$A + B \ \underset{k_{21}}{\overset{k_{12}}{\rightleftharpoons}}\ C \tag{5}$$

behandelt, deren ursprüngliches Gleichgewicht mit den Konzentrationen c_A^u, c_B^u und c_C^u einer zur Zeit $t = 0$ einsetzenden rechteckigen Sprungfunktion

$$t \geqslant 0 :\ \overline{c}_i(t) = c_i \tag{6}$$

unterworfen wird. Aus Gleichung (4) folgt

$$\frac{-\,d\,[c_i(t) - \overline{c}_i]}{d\,t} = \frac{1}{\tau}\left[c_i(t) - \overline{c}_i\right] \tag{7}$$

$$\text{oder} \quad -\frac{d\,z\,(t)}{d\,t} = \frac{1}{\tau}\,z\,(t) \ , \tag{7'}$$

wobei zu beachten ist, daß infolge der Massenerhaltung gelten muß

$$z\,(t) \equiv c_i\,(t) - \bar{c}_i = c_A\,(t) - \bar{c}_A = c_B\,(t) - \bar{c}_B = -\,(c_C\,(t) - \bar{c}_C) \ . \tag{8}$$

Andererseits gilt zunächst ohne Einschränkung

$$-\frac{d\,c_A\,(t)}{d\,t} = -\frac{d\,[c_A\,(t) - \bar{c}_A]}{d\,t} = k_{12}\,c_A\,c_B - k_{21}\,c_C \tag{9}$$

oder

$$-\frac{d\,z\,(t)}{d\,t} = [k_{12}\,(\bar{c}_A + \bar{c}_B) + k_{21}]\,z + k_{12}\,z^2 \ , \tag{10}$$

wenn man berücksichtigt, daß

$$k_{12}\,\bar{c}_A\,\bar{c}_B - k_{21}\,\bar{c}_C = 0 \ .$$

Nach der Linearisierung (Vernachlässigung des Gliedes $k_{12}\,z^2$) erhält man die von der Auslenkung z unabhängige Relaxationszeit:

$$\tau = [k_{12}\,(\bar{c}_A + \bar{c}_B) + k_{21}]^{-1} \tag{11}$$

mit der entsprechenden Lösung der Differentialgleichung

$$z = z_0 \exp\,[-\,t/\tau] \ . \tag{12}$$

In Tabelle 1 sind die für verschiedene einstufige Elementarreaktionen charakteristischen Relaxationszeiten zusammengestellt, die sich ganz analog zum diskutierten Beispiel herleiten lassen. Es ist beachtenswert, daß die Relaxationszeiten natürlich unabhängig von der gewählten Störfunktion sind, auf deren verschiedenen Formen später eingegangen werden soll.

Tabelle 1

Reaktion	reziproke Relaxationszeit $1/\tau$
$A \underset{k_{21}}{\overset{k_{12}}{\rightleftharpoons}} B$	$k_{12} + k_{21}$
$A + C \rightleftharpoons B + C$ (C = Katalysator)	$k_{12}\,\bar{c}_C + k_{21}\,\bar{c}_C$
$2A \rightleftharpoons A_2$	$4\,k_{12}\,\bar{c}_A + k_{21}$
$A + B \rightleftharpoons C$ (B = gepuffert $x_B \approx 0$)	$k_{12}\,\bar{c}_B + k_{21}$
$A + B \rightleftharpoons C + D$	$k_{12}\,(\bar{c}_A + \bar{c}_B) + k_{21}\,(\bar{c}_C + \bar{c}_D)$
$A + B + C \rightleftharpoons D$	$k_{12}\,(\bar{c}_A\,\bar{c}_B + \bar{c}_A\,\bar{c}_C + \bar{c}_B\,c_C) + k_{21}$

Im Gegensatz zu mehrstufigen Reaktionen existiert für einstufige Systeme eine geschlossene Lösung der vollständigen Gleichung (10).

3.2.2. Mehrstufige Relaxation. Normalvariable

Im allgemeinen läßt sich das Relaxationsverhalten von Reaktionen in Lösung nur durch ein ganzes System von gekoppelten Differentialgleichungen beschreiben, deren linearisierte Form durch (s. Anhang)

$$- \dot{x}_i(t) = \sum_k \alpha_{ik} \left[x_k(t) - \overline{x}_k(t) \right] \tag{13}$$

gegeben ist, wobei für $x_i(t)$ die Nebenbedingung

$$\sum_k \frac{1}{v_k} x_k(t) = 0 \tag{14}$$

erfüllt sein muß. v_k sind die stöchiometrischen Koeffizienten der Brutto-Reaktionsgleichung. In einem solchen Mehrstufensystem ist die zeitliche Änderung einer Komponente i, nämlich $\dot{x}_i$, nicht mehr von x_i allein abhängig. Die Definition der Relaxationszeiten im Sinne von Gl. (4) erfordert daher zunächst eine lineare Transformation der Konzentrationsvariablen x_i auf die Normalvariablen y_i, die die Beziehungen

$$- \dot{y}_i(t) = \frac{1}{\tau_i} \left[y_i(t) - \overline{y}_i(t) \right] \tag{15}$$

erfüllen. Schreibt man Gl. (13) und (15) in Matrizenschreibweise

$$- \dot{x} = A \, (x - \overline{x}) \tag{16}$$

$$- \dot{y} = D \, (y - \overline{y}) \tag{17}$$

(A ist die Koeffizientenmatrix α_{ik}, D ist die Diagonalmatrix der $1/\tau_i$), so folgt aus Gl. (16) bei der Transformation

$$x = M y \quad \text{und} \quad \overline{x} = M \overline{y} \tag{18}$$

$$- \dot{y} = M^{-1} \, A \, M \, (y - \overline{y}) \tag{19}$$

und aus Gl. (17) und (19)

$$D = M^{-1} A M \, . \tag{20}$$

Da D diagonal ist, ist M die Matrix der Eigenvektoren von A, die ihrerseits durch die Eigenwerte von A festgelegt sind. Ähnlich ergibt sich M^{-1} als die Matrix der Linkseigenvektoren von A. Die Wurzeln der Säkulardeterminante von A sind gleich den Reziprokwerten der Relaxationszeiten.

Bei Relaxation gegen einen chemischen Gleichgewichtszustand kann nach dem Prinzip der mikroskopischen Reversibilität, welches streng nur für Gleichgewichtszustände gilt, A symmetrisiert werden. Die Eigenwerte $1/\tau_i$ von A sind für diesen Fall notwendig reell, weswegen periodische Reaktionen nicht auftreten, wenn das temporäre Gleichgewicht sich nicht-periodisch ändert. Die Annäherung eines stationären Zustands, der vom chemischen Gleichgewicht weit entfernt ist, kann dagegen — wenn nicht ausschließlich Reaktionen 1. Ordnung beteiligt sind — sehr wohl über periodische Reaktionen verlaufen, die durch komplexe Terme der Gestalt $\exp[i\,\omega\,t]$ ($i = \sqrt{-1}$) dargestellt werden ([2], über einen experimentellen Befund s. z. B. B. Hess, K. Brand, and K. Pye, Biochem. Biophys. Res. Comm. **23**, 102 (1966)).

Für den schon diskutierten Fall einer stufenförmigen Anregung der Gleichgewichtslage erhält man als Lösung der Differentialgleichung

$$x_i(t) = \sum_k m_{ik}\, y_k^0\, e^{-t/\tau_k} \,, \tag{21}$$

m_{ik} sind die Elemente der Matrix $\boldsymbol{M}$, oder

$$x_i(t) = \sum_k c_{ik}\, e^{-t/\tau_k} \,, \tag{22}$$

wenn man m_{ik} und y_k^0 zu einer Konstanten c_{ik} zusammenfaßt.

Trägt man die Größe $\sum\limits_k c_{ik}\, e^{-\,t/\tau_k}$ als Funktion der Zeit auf, so lassen sich aus ihr die einzelnen Relaxationszeiten ermitteln, solange

$$\tau_i \ll \tau_{i+1}\,.$$

Da in jeder Relaxationszeit mehrere Geschwindigkeitskonstanten in Verbindung mit den vorgegebenen Gleichgewichtskonzentrationen enthalten sind, ist eine entsprechende Variation der Konzentrationen der Reaktionspartner nötig, um alle Konstanten bestimmen zu können.

Neben den hier erwähnten diskreten Relaxationsspektren treten bei sehr schnellen chemischen Reaktionen (die diffusionskontrolliert sind) und bei der dielektrischen oder mechanischen Relaxation von Polymeren kontinuierliche Relaxationsspektren auf.

3.2.3. *Quasikontinuierliche Relaxationsspektren. Mittlere Relaxationszeiten*

Die Lösung der Relaxationsgleichungen kann bei Annäherung eines konstanten stationären Zustands (z. B. bei der Äquilibrierung nach einem Temperatursprung)[1] gemäß Gl. (22) stets in der Form

$$\Phi(t) = \sum_r \beta_r\, e^{-t/\tau_r} \tag{23}$$

dargestellt werden [$\Phi(t) = x_i(t)/x_i(0)$ für $t \geqq 0$]. Die Größe $\Phi(t)$ ist mit einer Konzentrationsänderung $x_i(t)$ oder mit der Änderung einer geeigneten Meßgröße, wie der Extinktion, die eine lineare Funktion von $x_i(t)$ ist, verknüpft; in jedem Fall kann eine eindeutige Funktion von x_i in der Um-

---

[1] Bei periodischer Störung gelten andere Relaxationsfunktionen als (23).

gebung eines Bezugszustandes mit Hilfe einer Taylorentwicklung linearisiert werden. Die Koeffizienten β_r genügen der Normierungsbedingung

$$\sum_r \beta_r = 1 \, ,$$

da $\Phi(0) = 1$.

Wird ein Relaxationsvorgang durch eine sehr große Anzahl von Relaxationszeiten τ_r charakterisiert, deren Differenzen sehr klein sind, so erweist es sich als zweckmäßig, den Begriff „quasi-kontinuierliches" Relaxationsspektrum einzuführen; wir müssen uns dabei jedoch bewußt sein, daß wir es realiter nach (23) — insbesondere im Fall chemischer Reaktionen — mit einem diskreten Spektrum zu tun haben.

Zur Charakterisierung eines Spektrums hoher Spektraldichte, dessen Ausdehnung nicht zu groß ist, definieren wir die mittlere Relaxationszeit τ^* durch die Gleichung [3]

$$\frac{1}{\tau^*} = \sum_r \frac{\beta_r}{\tau_r} \, . \tag{24}$$

Aus den Gl. (23) und (24) folgt die einfache Relation

$$\frac{1}{\tau^*} = -\left(\frac{d\Phi}{dt}\right)_{t=0} \tag{25}$$

zwischen der Meßgröße $(d\Phi/dt)_{t=0}$ und τ^*. Ein Maß für die Breite des Spektrums ist das relative quadratische Mittel ϱ^* der Differenz zwischen $1/\tau_r$ und $1/\tau^*$:

$$\varrho^* = \frac{\sum_r \beta_r \, (1/\tau_r - 1/\tau^*)^2}{(1/\tau^*)^2} \, . \tag{26}$$

Diese Beziehung kann in die Gestalt

$$\varrho^* = \tau^{*2} \left\{ \left(\frac{1}{\tau^{**}}\right)^2 - \left(\frac{1}{\tau^*}\right)^2 \right\} \tag{27}$$

gebracht werden, wo

$$\left(\frac{1}{\tau^{**}}\right)^2 = \sum_r \frac{\beta_r}{\tau_r^2} = \left(\frac{d^2\Phi}{dt^2}\right)_{t=0} \, . \tag{28}$$

Auch $(d^2\Phi/dt^2)_{t=0}$ ist der Messung zugänglich. Um die Güte der Näherung

$$\Phi(t) \approx e^{-t/\tau^*} \tag{29}$$

zu ermitteln, entwickeln wir die Differenz $\Phi(t) - e^{-t/\tau^*}$ in die Taylorreihe

$$\Phi(t) - e^{-t/\tau^*} = \frac{1}{2} \varrho^* \left(\frac{t}{\tau^*}\right)^2 + \cdots \, ; \tag{30}$$

die näherungsweise Beschreibung der Relaxation durch eine Relaxationszeit ist mithin um so besser, je kleiner ϱ^* und je kürzer das Zeitintervall t ($t \ll \tau^*$) ist. Wir kommen bei der Behandlung der Relaxationsspektren der Konformationsänderungen von Biopolymeren auf diese Betrachtungen zurück.

3.3. Reaktionseffekte bezüglich Normalvariablen

In Normalvariablen y_i ausgedrückt, sind die linearisierten Relaxationsgleichungen entkoppelt und durch jeweils eine Zeitkonstante, die Relaxationszeit τ_i, charakterisiert (Kap. 3.2.2.). Auf die Reaktionseffekte $\Delta \Psi_{yi}$ (Reaktionsenthalpie, Reaktionsvolumen etc.) bezüglich der Normalvariablen y_i, die 2. Gruppe physikalisch wichtiger Größen, die mit den y_i verknüpft sind, gehen wir in Kap. 5.4.3. näher ein. Dort wird der Zusammenhang zwischen der Meßbarkeit von Relaxationszeiten mit Hilfe der Methoden der chemischen Relaxation und den auf die Normalvariablen bezogenen Reaktionseffekten an Hand eines Beispiels erläutert. Nur wenn $\Delta \Psi_{yi} \neq 0$ und die relevante Konzentrationsänderung mit der Zeit beobachtet werden kann, ist τ_i nach der (dem $\Delta \Psi_{yi}$ entsprechenden) Methode der chemischen Relaxation meßbar.

Die Meßgröße bei Relaxationsversuchen, τ_i, ist im allgemeinen eben nicht einer einzelnen Konzentrationsänderung x_i zugeordnet, sondern stellt eine Eigenschaft der Normalvariablen y_i und damit des Gesamtsystems dar.

4. Relaxationsverfahren

4.1. Sprung- und Impuls-Verfahren (Einschwingverfahren)

4.1.1. Mathematische Formulierung

Im Teil 3.2. ist die mathematische Theorie der Rechtecks-Sprungverfahren hergeleitet worden. Sie besteht im wesentlichen in der Aufstellung und Lösung der entkoppelten homogenen Differentialgleichungen der Normalvariablen y_i:

$$-\dot{y}_i = \frac{1}{\tau_i} y_i \quad ; \quad y_i = y_i^0 \, e^{-t/\tau_i}$$

und der Transformation der Variablen y_i auf die beobachtbaren Konzentrationsvariablen x_i

$$x_i = \sum_k c_{ik} \, e^{-t/\tau_k} \ .$$

Bei anderen Zeitfunktionen ist die Lösung der inhomogenen Differentialgleichung

$$-\dot{y}_i = y_i/\tau_i - \bar{y}_i(t)/\tau_i$$

durch den Ausdruck

$$y_i(t) = [e^{-t/\tau_i} \,/\, \tau_i] \int_0^t \exp[\Theta/\tau_i] f(\Theta) \, d\Theta \tag{31}$$

unter der Bedingung gegeben, daß $y_i = 0$ für $t = 0$ ist.

Man unterscheidet die Stufenreaktion mit linearem Anstieg:

$$\bar{y}_i = 0 \quad \text{für } t < 0 \;,$$
$$\bar{y}_i = (t/t_1)\,\bar{y}_{i\,\infty} \quad \text{für } 0 \leqslant t \leqslant t_1 \;, \tag{32}$$
$$y_i = \bar{y}_{i\,\infty} \quad \text{für } t_1 \leqslant t \;;$$

die Stufenfunktion mit exponentiellem Anstieg oder Abfall:

$$\bar{y}_i = \bar{y}_{i\,\infty}\,(1 - e^{-t/\tau_s}) \quad \text{für } t \geqslant 0 \;; \tag{33}$$

die rechteckige Impulsfunktion:

$$\bar{y}_i = 0 \quad \text{für } t < 0 \text{ und } t > \Theta = \text{Dauer des Impulses}$$
$$\bar{y}_i = y_{i\,R} \quad \text{für } 0 \leqslant t \leqslant \Theta \;; \tag{34}$$

die gedämpfte harmonische Funktion:

$$\bar{y}_i = 0 \quad \text{für } t < 0$$
$$\bar{y}_i = A_i\,e^{-b\,t} \sin \omega\,t \quad \text{für } t \geqslant 0 \;. \tag{35}$$

4.1.2. *Experimentelle Technik*

Bei allen Einschwingmethoden ist eine direkte Beobachtung der Konzentrationsänderung mit der Zeit erforderlich. Die optischen Eigenschaften, die von den Absorptionsspektren der einzelnen Reaktionspartner abhängen und deshalb spezifisch sind, eignen sich besonders gut zur Messung der Relaxationszeiten. Elektrische Leitfähigkeitsmessungen werden vor allem bei jenen Reaktionen angewandt, bei denen Ionen gebildet oder verbraucht werden. Bei kalorimetrischen Messungen wird die Reaktionsenthalpie des Systems ausgenutzt. In einzelnen Fällen kann durch Zusatz eines Indikators, der mit dem zu untersuchenden System gekoppelt ist, die Konzentrationsänderung beobachtet werden.

Bei der Temperatur-Sprungmethode für wäßrige Elektrolytlösungen wird ein Kondensator (0,05 bis 0,5 μF, 10 bis 100 kV, entsprechend einer Energie von 1 bis 10^3 Wattsek.) über die Elektroden einer Meßzelle von einer Länge von einigen cm in etwa 10^{-6} sec entladen, wobei eine Temperaturerhöhung von $2-10$ °C resultiert[2]. In nichtleitenden Systemen kann eine Temperaturerhöhung von einigen °C durch einen Mikrowellen-

---

[2] (Einzelheiten über die Technik der verschiedenen Relaxationsverfahren sind in A. WEISSBERGER, Technique of Organic Chemistry, 2. Aufl., Bd. 8, Teil 2, Investigation of Rates and Mechanisms of Reactions zu finden [1]).

impuls erreicht werden. Die Konzentrationsänderung wird durch den zeitlichen Verlauf der elektrischen Leitfähigkeit oder der optischen Absorption in der Meßzelle festgelegt. Messungen von Relaxationszeiten bis zu
10^{-7} sec sind möglich. Beim Drucksprungverfahren sind die über eine
Wheatstonebrücke verbundenen Meß- und Referenz-Zellen in einem Flüs-
sigkeitsautoklaven von 50 at eingebettet. Durch Bersten einer Membran
kann der Druck in 10^{-4} sec auf eine Atmosphäre reduziert werden, wobei
Relaxationszeiten im Bereich von 10^{-4} sec bis 50 sec gemessen werden
können. Mit Stoßwellen ist es möglich, Druckänderungen von 10^3 Atmosphären in 10^{-6} sec zu erreichen.

In der Feld-Sprungmethode wird ein Impuls (rechteckig oder gedämpft
harmonisch) von 10^{-7} sec bis 10^{-4} sec Dauer und einer Feldstärke bis zu
200 kV/cm benutzt, um die Dissoziation von schwachen Elektrolyten zu
untersuchen.

4.2. Stationäre Verfahren (Periodische Störungen)

4.2.1. Mathematische Formulierung

Für periodische Störungen ist das Verhältnis von Relaxationszeit zur
Frequenz der Störungsfunktion $\bar{y}\,(t)$ entscheidend[3]. Im Unterschied zu den
Einschwingvorgängen, welche von der vollständigen Lösung der Differentialgleichungen abhängen, ist bei den stationären Verfahren nur die partikuläre Lösung der Differentialgleichung erforderlich.

Läßt sich die Zeitfunktion der erzwingenden Kraft durch[3]

$$\bar{y}\,(t) = A\,e^{i\,\omega\,t} \tag{36}$$

darstellen, so ist es üblich für die partikuläre Lösung von $y\,(t)$ den Ansatz

$$y\,(t) = \gamma\,\bar{y}\,(t) \tag{37}$$

zu wählen. Man nennt die zeitunabhängige Größe γ die Übertragungsfunktion; ihr Wert ist im Falle der zu untersuchenden Differentialgleichung

$$\tau\dot{y} + y = \bar{y}$$

gegeben durch

$$\gamma = \frac{1}{1 + i\,\omega\,\tau} \tag{38}$$

oder bei einer Aufteilung in einen Realteil und Imaginärteil

$$\gamma = \gamma_{\text{re}} + i\,\gamma_{\text{im}} \tag{39}$$

durch die Komponenten

$$\gamma_{\text{re}} = \frac{1}{1 + \omega^2\,\tau^2}\,, \quad \gamma_{\text{im}} = -\,\frac{\omega\,\tau}{1 + \omega^2\,\tau^2}\,. \tag{40}$$

---

[3] Wir lassen den Index der Normalkoordinate y_l bzw. $\bar{y}_l$ in Kap. 4.2.1. der
besseren Übersicht wegen fort.

Eine ebenfalls nützliche Darstellung ist die trigonometrische Zerlegung

$$\gamma = \varrho \, (\cos \varphi - i \sin \varphi) = \varrho \, e^{-i\varphi} \, , \tag{41}$$

wobei ϱ und φ gegeben sind durch

$$\varrho = (\gamma_{re}^2 + \gamma_{im}^2)^{1/2} = \left(\frac{1}{1 + \omega^2 \tau^2} \right)^{1/2} , \tag{42}$$

$$\cos \varphi = \left(\frac{1}{1 + \omega^2 \tau^2} \right)^{1/2} , \quad \sin \varphi = \left(\frac{\omega^2 \tau^2}{1 + \omega^2 \tau^2} \right)^{1/2} , \tag{43}$$

$$\operatorname{tg} \varphi = \omega \tau \; .$$

Mit den obigen Beziehungen erhält man die partikuläre Lösung

$$y(t) = \frac{A \, e^{i\,(\omega t - \varphi)}}{(1 + \omega^2 \tau^2)^{1/2}} \; . \tag{44}$$

Ein Vergleich mit Gl. (36) zeigt, daß die Phase von y gegenüber $\bar{y}$ um den Phasenwinkel φ zurück ist, und daß die Amplitude von y gegenüber $\bar{y}$ um den Faktor $1 / (1 + \omega^2 \tau^2)^{1/2}$ reduziert ist.

Es ist zu beachten, daß bei sehr hohen Frequenzen ($\omega \gg 1/\tau$) sich φ dem Grenzwert 90° und $A / (1 + \omega^2 \tau^2)^{1/2}$ dem Grenzwert Null nähert.

Die Abhängigkeit der Geschwindigkeit c einer Schallwelle von deren Kreisfrequenz $\omega = 2\pi f$ steht in demselben funktionellen Zusammenhang wie die des Realteils γ_{re} von ω.

Aus einer Messung der Größen c (oder ε, im Falle dielektrischer Methoden) als Funktion der Frequenz (Dispersionsmethoden) können die Relaxationszeiten τ_i bestimmt werden. Analoge Beziehungen bestehen zwischen dem Imaginärteil γ_{im} und der Schallabsorption (oder dem dielektrischen Verlust).

4.2.2. Experimentelle Technik

Periodische Druck- und Temperaturschwankungen können durch einen Schwingquarz, dessen Frequenz durch einen Sender gesteuert wird, auf eine Lösung, in der die zu untersuchende Reaktion abläuft, übertragen werden. Die Relaxationszeiten der Reaktion können aus der Änderung der Schallgeschwindigkeit oder der Absorption mit der Frequenz festgelegt werden. Bei den letzteren Verfahren wird die Änderung der Amplitude A einer Schallwelle mit der Schichtdicke d

$$A = A_0 \, e^{-\alpha d} \tag{45}$$

ausgenutzt, wobei der Absorptionskoeffizient α mit μ, dem Absorptionskoeffizienten bezüglich der Wellenlänge λ, durch die Beziehung

$$\mu = \alpha \lambda \tag{46}$$

verknüpft ist. α oder μ können mit Hilfe der optischen Methode, die den Debye-Sears-Effekt ausnützt (Frequenzbereich 2–100 MHz), der Impulsmethode, bei der eine Messung des Abfalls der Schallintensität durch Sender-Empfänger-Abstimmung stattfindet (1–300 MHz), oder der akustischen Strömungsmethode (100 kHz–10 MHz) festgelegt werden. Bei den Verfahren mit stehenden Wellen, nämlich der Stimmgabelmethode ($\sim$100 Hz), der Hohlraummethode (reverberation method, 50 kHz bis 1 MHz), der Resonanzkugelmethode (5–50 kHz) und der Interferometer-Methode (50 kHz–10 MHz) wird der Absorptionskoeffizient μ aus der Abklingzeit bzw. aus der Schärfe der Resonanzkurve bestimmt.

Da der Koeffizient μ dem Imaginärteil der Übertragungsfunktion proportional ist,

$$\mu \sim \frac{\omega\, \tau_i}{1 + \omega^2\, \tau_i^2}\,, \tag{47}$$

lassen sich aus dem Wendepunkt der Kurve μ/f als Funktion von f die möglichen Relaxationszeiten bestimmen.

Die Messung der Schallgeschwindigkeit, die bei den Dispersionsverfahren notwendig ist, geschieht über die Messung der Wellenlänge (Interferometer), über die harmonischen Resonanzfrequenzen (Kugelresonator), über die Phasenverzögerung (Methoden mit kontinuierlichen Wellen) oder über die Laufzeitcharakteristik (Impulsverfahren).

Neben den akustischen Verfahren, bei denen periodische Druck- und Temperaturschwankungen durch Ultraschallwellen erzeugt werden, existiert die Methode der periodischen Änderung der elektrischen Feldstärke. Die Methode ist auf Reaktionen anwendbar, bei denen sich die Gesamtladung der Reaktion (Dissoziationsfeldeffekt) oder das Gesamtdipolmoment ändert. Im letzteren Fall können die Messungen mit kleinen periodischen Wechselfeldern und einem überlagerten Gleichfeld durchgeführt werden, damit das Reaktionsmoment ΔM genügend groß wird. Die Meßverfahren, bei denen entweder die elektrische Leitfähigkeit oder der dielektrische Verlust als Funktion des Hochfrequenzfeldes gemessen werden, leiden bis heute darunter, daß die erwarteten Effekte meistens klein sind und von anderen Prozessen (Dissoziationsfeldeffekt) überlagert werden.

5. Anwendungsbeispiele

5.1. Metallkomplexreaktionen in wäßriger Lösung

Metallionen liegen in wäßriger Lösung immer im hydratisierten Zustand vor. Die ein Ion umgebenden Wassermoleküle können verschiedenen regelmäßig angeordneten Koordinationsschalen zugeordnet werden. Die Struktur der inneren Koordinationsschalen kann auf Grund der elektrostatischen Wechselwirkung zwischen Zentralion und Ligand und zwischen den Liganden selbst (Ladung–Dipol, Dipol–Dipol) verstanden werden; für die in den äußeren Schalen gebundenen H_2O-Molekeln ist auch die Ausbildung von H-Brücken zu Wassermolekeln benachbarter Schalen von Bedeutung. Die Reaktion eines hydratisierten Metallions M^+ mit einem entsprechenden Anion X^- erfolgt stufenweise nach dem Schema:

$$M^+ + X^- \; \underset{k_{1b}}{\overset{k_{1f}}{\rightleftarrows}} \; \left[M^+ H_2O \, H_2O \, X^- \right]_{aq.}$$

$$\underset{k_{2b}}{\overset{k_{2f}}{\rightleftarrows}} \; \left[M^+ H_2O \, X^- \right]_{aq.} \quad \underset{k_{3b}}{\overset{k_{3f}}{\rightleftarrows}} \; \left[M^+ X^- \right]_{aq.}$$

Im Schritt 1 wird die Diffusion der Reaktionspartner mit ihren Hydrathüllen zum Begegnungskomplex $[M^+ H_2O \, H_2O \, X^-]_{aq.}$ beschrieben; die Begegnungsfrequenz liegt im Bereich von $10^9 - 10^{10}$ l Mol^{-1} sec^{-1}; die Lebensdauer des Komplexes, $1/k_{1b}$, ist $\sim 10^{-10} - 10^{-8}$ sec [4].

Im Schritt 2 und 3 erfolgt die Entfernung des koordinierten Wassers von dem Anion und dem Kation des betrachteten Ionenpaars. Dabei wird die kleinere Relaxationszeit des Schrittes 2 der Entfernung des Wassers vom Anion zugeordnet; k_{2f} und k_{2b} liegen bei $\sim 10^8 - 10^9$ sec^{-1} und $\sim 10^7 - 10^{10}$ sec^{-1}. Als geschwindigkeitsbestimmender Schritt ist die Entfernung eines innerkoordinierten Wassermoleküls des Kations anzusehen. Im Anschluß an INGOLD und HUGHES können die Grenzmechanismen SN_1 und SN_2 der nukleophilen Substitution (SN) unterschieden werden, wobei die Indizes 1 und 2 einen unimolekularen bzw. bimolekularen Prozeß charakterisieren. Für einen reinen SN_1-Mechanismus ist zu erwarten, daß k_{3f} unabhängig vom substituierenden Liganden ist, da der geschwindigkeitsbestimmende Primärschritt die Dissoziation von einem Metallion und einem H_2O-Molekül ist, an den sich die schnelle Reaktion der Assoziation mit dem Liganden anschließt. In der Tat erweisen sich die Geschwindigkeitskonstanten für den Substitutionsschritt für die meisten Metallionen als unabhängig von der Natur des substituierenden Liganden; mögliche Ausnahmen sind die sehr schnellen Substitutionsreaktionen ($k_{3f} > 10^7$ sec^{-1}) der Alkalimetallionen (Li^+, Na^+, K^+, Rb^+ und Cs^+) und der Erdalkalimetallionen mit großem Ionenradius (Ca^{2+}, Sr^{2+}, Ba^{2+}), bei denen eine gewisse Abhängigkeit von der Art des Liganden vorhanden ist und bei denen somit auf einen SN_2-Mechanismus geschlossen werden kann.

Für alle Metallionen mit edelgasartiger Anordnung ($d^0 s^2 p^6$ und $d^{10} s^2 p^6$) ergibt sich ein monotoner (linearer) Zusammenhang zwischen log k_{3f} und dem reziproken Ionenradius. Abweichungen von dieser Beziehung sind nur für diejenigen Ionen zu verzeichnen, bei denen ein vorgelagerter hydrolytischer Schritt die Substitution wesentlich beeinflußt (Be^{2+}, Al^{3+}, Fe^{3+}). Bei der Kinetik der Übergangsmetallionen mit d^1- bis d^9-Konfiguration sind noch die Ligandenfeldstabilisierung und der Jahn-Teller-Effekt zu berücksichtigen. So ist als Folge der Ligandenfeldstabilisierung die energetische Differenz zwischen Grund- und Übergangszustand (z. B. Oktaeder-Tetragonale Pyramide) bei d^3- und d^8-Systemen im Falle schwacher Felder und bei d^4-, d^5- und d^6-Systemen im Falle starker Felder besonders groß und dementsprechend die Geschwindigkeitskonstante k_{3f} besonders klein. Zu einer Vergrößerung von k_{3f} kommt es bei Cr^{2+}-$(d_\varepsilon^3\, d_\gamma^1)$- und Cu^{2+}-$(d_\varepsilon^6 d_\gamma^3)$-Ionen, die infolge des Jahn-Teller-Effekts eine Labilisierung der axialen Oktaeder-Positionen zeigen.

In Tabelle 2 werden die Geschwindigkeitskonstanten k_{3f} und k_{3b} einiger charakteristischer Substitutionsreaktionen von zwei- und dreiwertigen Ionen aufgeführt, die bei Zimmertemperaturen und Ionenstärke $I \approx 0,1$ gemessen wurden (vgl. [4]).

Tabelle 2

Metall-Ion	Ligand	k_{3f} (sec^{-1})	k_{3b} (sec^{-1})
Be^{2+}	SO_4^{2-}	1×10^2	$1,3 \times 10^3$
Mg^{2+}	SO_4^{2-}	1×10^5	$8 \ \times 10^5$
Mg^{2+}	$S_2O_3^{2-}$	1×10^5	$1,3 \times 10^6$
Mg^{2+}	CrO_4^{2-}	1×10^5	$1 \ \times 10^6$
Ca^{2+}	CrO_4^{2-}	$>5 \times 10^7$	$1,5 \times 10^8$
Mn^{2+}	SO_4^{2-}	3×10^6	$1,5 \times 10^7$
Fe^{2+}	SO_4^{2-}	$\sim 10^6$	$\sim 10^7$
Co^{2+}	SO_4^{2-}	3×10^5	$2,5 \times 10^6$
Ni^{2+}	SO_4^{2-}	1×10^4	1×10^5
Cu^{2+}	SO_4^{2-}	$>5 \times 10^7$	$>10^8$
Al^{3+}	SO_4^{2-}	~ 1	4
Fe^{3+}	SO_4^{2-}	5×10^2	~ 10
Zn^{2+}	SO_4^{2-}	3×10^7	$1,9 \times 10^8$

Die Reaktionsgeschwindigkeit der Umlagerung verschiedener Strukturen ist immer dann besonders groß, wenn die Struktur teilweise erhalten bleiben kann; so ist die Umwandlung des Oktaeders über die tetragonale Pyramide zur quadratisch planaren Struktur gegenüber jener über die trigonale Bipyramide zum Tetraeder bevorzugt.

5.2. Die Protolyse des Wassers. Ein diffusionsbestimmter Prozeß

Protolytische Reaktionen sind gewöhnlich schnelle, nach einem Zeitgesetz 2. Ordnung verlaufende Prozesse, deren Geschwindigkeiten bis an jene diffusionsbestimmter Vorgänge heranreichen. Die Teilnahme von Lösungsmittelionen mit besonderen Transporteigenschaften in Verbindung mit der eigentümlichen Struktur des Wassers kommt in der Kinetik protolytischer Reaktionen deutlich zum Ausdruck. Das Verständnis protolytischer Reaktionsmechanismen ist Voraussetzung für die fruchtbare Untersuchung verwickelterer chemischer Vorgänge, wie der Säure-Base-Katalyse oder der enzymatischen Hydrolyse, an welchen Protolysen beteiligt sind. Bei protolytischen Reaktionen in Wasser spielt stets eines der Ionen H_3O^+ oder OH^- eine bestimmende Rolle. Die Reaktionen zwischen diesen beiden Spezies, die Neutralisation und die Selbstdissoziation des Wassers, stellen in jedem Protolyse- oder Hydrolysemechanismus einen wichtigen Schritt dar [5].

Die obere Grenze der Geschwindigkeit von bimolekularen oder höhermolekularen Prozessen ist durch die Anzahl der Begegnungen zwischen den Reaktanten auf Grund der thermischen Molekularbewegung bestimmt. Den vollständigen Mechanismus der Assoziation von Ionen stellen wir deshalb durch ein Reaktionssystem dar, welches sowohl die Begegnung und Trennung der Reaktanten durch Diffusion als auch die chemische Transformation umfaßt:

$$
\begin{array}{ccccc}
1 & & 2 & & 3 \\[4pt]
A + B & \underset{k_{21}}{\overset{k_{12}}{\rightleftharpoons}} & A - B & \underset{k_{32}}{\overset{k_{23}}{\rightleftharpoons}} & AB
\end{array}
\qquad (48)
$$

k_{12} und k_{21} beziehen sich auf die diffusionsbestimmte Begegnung bzw. Trennung der Reaktanten A und B, k_{23} und k_{32} beschreiben die chemische Transformation, die aus mehreren Schritten zusammengesetzt sein kann.

Das Reaktionssystem (48) wird durch zwei linear unabhängige Zeitgesetze beschrieben, deren Lösungen durch die Relaxationszeiten (s. Anhang)

$$
\tau_s = 2\,(s + \sqrt{s^2 - 4\,P})^{-1} \qquad (49)
$$

$$
\tau_l = 2\,(s - \sqrt{s^2 - 4\,P})^{-1} \qquad (50)
$$

mit

$$
s = k_{12}^* + k_{21} + k_{23} + k_{32}\,,
$$

$$
P = k_{12}^*\,k_{23} + k_{21}\,k_{32} + k_{12}^*\,k_{32}
$$

gekennzeichnet sind; k_{12}^* ist durch

$$
k_{12}^* \equiv k_{12}\,(\bar{c}_A + \bar{c}_B)
$$

für konstante Ionenstärke gegeben, $\bar{c}_A$ und $\bar{c}_B$ sind die Konzentrationen von A bzw. B im Gleichgewicht. Folgende zwei spezielle Lösungen sind von Bedeutung. 1. Die Relaxationszeiten differieren um mehr als eine Größenordnung. Dann gilt $4\,P \ll s^2$ und man erhält,

$$
\tau_s = (k_{12}^* + k_{21} + k_{23} + k_{32})^{-1}, \qquad (51)
$$

$$
\tau_l = [\tau_s\,(k_{12}^*\,k_{23} + k_{12}^*\,k_{32} + k_{21}\,k_{32})]^{-1}. \qquad (52)
$$

2. Wenn die Konzentration c_{A-B} von $A-B$ klein gegen die Konzentrationen c_A, c_B oder c_{AB} von A, B bzw. AB ist, wird die Einstellung des Gleichgewichts zwischen A, B und AB durch nur eine Relaxationszeit charakterisiert:

$$\tau = \frac{k_{21} + k_{23}}{k_{12}^* k_{23} + k_{32} k_{21}} = \frac{1}{k_R(\bar{c}_A + \bar{c}_B) + k_D} \tag{53}$$

wobei

$$k_R = \frac{k_{12} k_{23}}{k_{21} + k_{23}} \quad \text{und} \quad k_D = \frac{k_{32} k_{21}}{k_{21} + k_{23}}$$

die Geschwindigkeitskonstanten der Assoziation bzw. Dissoziation sind — definiert in Analogie zu den entsprechenden Größen im Bodenstein-Lindemann-Hinshelwood-Mechanismus.

Die Reaktion ist diffusionskontrolliert, wenn die Bedingung $k_{21} \ll k_{23}$ erfüllt ist. Dann kann k_R mit dem nach der Gleichung von Debye gegebenen Ausdruck für die Geschwindigkeitskonstante k_E der diffusionsbestimmten Begegnung [6],

$$k_E = 4\pi N_L R_{ij}(D_i + D_j)\frac{\varphi_{ij}}{\exp(\varphi_{ij}) - 1}, \tag{54}$$

gleichgesetzt werden:

$$k_R = k_E. \tag{55}$$

In Gl. (54) ist N_L die Loschmidtsche Zahl, R_{ij} der Reaktionsabstand, D_i und D_j die Diffusionskoeffizienten der Reaktionspartner, φ_{ij} der Quotient aus potentieller Energie der elektrostatischen Wechselwirkung und thermischer Energie kT; für verdünnte Lösungen läßt sich die elektrostatische Wechselwirkung durch einen Coulombterm $z_i z_j e_0^2/(\varepsilon R_{ij})$ darstellen; z ist die Wertigkeit, e_0 die Elementarladung und ε die effektive Dielektrizitätskonstante. Die Größe k_D wird durch die Gleichung

$$k_D = k_s \frac{k_{32}}{k_{23}} \tag{56}$$

festgelegt, wobei k_s die Geschwindigkeitskonstante der diffusionsbedingten Trennung der Reaktanten ist:

$$k_s = (3/R_{ij}^2)(D_i + D_j)\frac{\varphi_{ij}}{1 - \exp(-\varphi_{ij})}. \tag{57}$$

Mithin können wir für die Relaxationszeit

$$\tau = \left[k_E(\bar{c}_A + \bar{c}_B) + k_s \frac{k_{32}}{k_{23}}\right]^{-1} \tag{58}$$

setzen.

Wenn die chemische Transformation im Vergleich zur diffusionsbedingten Trennung langsam abläuft, $k_{21} \geq k_{23}$, ergibt sich für die Relaxationszeit

$$\tau = [k_{23} \frac{k_E}{k_s} (\bar{c}_A + \bar{c}_B) + k_{32}]^{-1} \, . \tag{59}$$

Der Ausdruck (58) für τ ist charakteristisch für normale protolytische Reaktionen.

Für die in reinem Wasser mit dem Gleichgewicht

$$H^+ + OH^- \rightleftharpoons H_2O \tag{60}$$

verbundene Relaxationszeit ergaben Messungen nach verschiedenen Methoden den Wert

$$\tau = 3,5 \cdot 10^{-5} \sec \approx \frac{1}{k_R (\bar{c}_{H^+} + \bar{c}_{OH^-})} \tag{61}$$

bei $T = 298\ °K$, wo $\bar{c}_{H^+} = \bar{c}_{OH^-} = 10^{-7}$ M; da $[A][B]/[AB] = k_D/k_R$, folgt

$$k_R = 1,4 \cdot 10^{11} \, M^{-1} \sec^{-1}, \tag{62}$$

$$k_D = 2,5 \cdot 10^{-5} \sec^{-1}. \tag{63}$$

Der kleinstmögliche Wert für k_{12} ist also

$$k_{12} \geq 1,4 \cdot 10^{11} \, M^{-1} \sec^{-1}.$$

Mit $(D_{H^+} + D_{OH^-}) = 1,45 \cdot 10^{-4} \, cm^2 \sec^{-1}$ (Größen, die aus Messungen der elektrischen Beweglichkeit von H^+ und OH^- erhalten wurden) errechnet sich der Reaktionsabstand R_{ij} nach (54) zu 8 Å[4]. Dieser Reaktionsabstand entspricht den Querschnitten der Komplexe $H_9O_4^+$ und $H_7O_4^-$; die Komplexe entstehen bei der Assoziation von 4 Wassermolekülen mit einem „Überschußproton" bzw. einem „Defektproton". Hinweise für die hohe Stabilität der Komplexe $H_9O_4^+$ und $H_7O_4^-$ lassen sich aus einer Reihe unabhängiger Befunde, insbesondere aus solchen bei massenspektrometrischen Untersuchungen, herleiten. Zurückzuführen ist die hohe Stabilität der Proton-Wasserkomplexe auf die symmetrische Verteilung der positiven Überschußladung auf die drei Protonen im Hydroniumion, welche die Bildung dreier stabiler Wasserstoffbrücken zu benachbarten H_2O-Molekülen ermöglicht. (Die symmetrische Ladungsverteilung im H_3O^+ geht aus dem kernmagnetischen Resonanzspektrum von festem $H_3O \cdot NO_3$ hervor.) Wenn ein Reaktionsabstand von 8 Å gefunden wird, dann bedeutet das offenbar, daß die Neutralisation der Ladung „augenblicklich" geschieht, sobald die beiden Komplexe durch mindestens eine

---

[4] Für diesen großen Reaktionsabstand ist die Anwendung der Kontinuumstheorie in Gl. (54) und (57) (mit $\varepsilon_{H_2O} = 79$) gerechtfertigt.

Wasserstoffbrücke verbunden sind. Das Proton besitzt innerhalb des Molekülkomplexes eine extrem hohe Beweglichkeit. Seine Lokalisierung im zentralen H_3O^+ entspricht nur einem Maximum der Aufenthaltswahrscheinlichkeit. Die außergewöhnlich schnelle Bewegung von Protonen längs der Wasserstoffbrücken tut sich auch in anderen Systemen kund. Die Beweglichkeit von Protonen in Eis, einem mit Wasserstoffbrücken gebildeten Gitter, ist eine bis zwei Größenordnungen höher als in flüssiger Phase. Aus solchen Messungen läßt sich mit Hilfe der Nernst-Einsteinschen Beziehungen k_{23} zu etwa $10^{13}\,\text{sec}^{-1}$ abschätzen. Während der kurzen mittleren Lebensdauer eines „individuellen" H_3O^+-Ions von 10^{-13} sec kann eine Dehydratisierung, wie sie etwa bei Metallkomplexreaktionen beobachtet wird, nicht stattfinden.

Für k_{21} kann man mit $R_{ij} = 8$ Å folgenden Wert abschätzen:

$$k_{21} = 4{,}3 \cdot 10^{10}\ \text{sec}^{-1}; \tag{64}$$

aus der Gleichgewichtskonstanten und den Werten für k_{12}, k_{21} und k_{23} finden wir für die Größe k_{32}

$$k_{32} \approx 6 \cdot 10^{-3}\ \text{sec}^{-1} . \tag{65}$$

Diese Daten werden durch die folgenden Meßergebnisse ergänzt:

1. Die mit k_R verknüpfte Aktivierungsenergie wurde zu 3 kcal/Mol ermittelt. Ein Wert dieser Größe ist für eine diffusionsbestimmte Reaktion zu erwarten.

2. Für die Reaktionen von D_2O wurden die Geschwindigkeitskonstanten

$$k_R = 8{,}4 \cdot 10^{10}\ \text{M}^{-1}\ \text{sec}^{-1}\ ,$$
$$k_D = 2{,}5 \cdot 10^{-6}\ \text{sec}^{-1}$$

gefunden. Der Isotopeneffekt $k_R(H^+)/k_R(D^+) = 1{,}6$ stimmt mit der Annahme einer diffusionsbestimmten Reaktion überein.

3. Die Beweglichkeit von D^+ in D_2O-Eis ist 4 bis 7mal kleiner als die Beweglichkeit von H^+ in H_2O-Eis. Diese Größe des Isotopeneffekts — und die bereits erwähnte sehr hohe Beweglichkeit des Protons in Eis — deuten auf ein nicht-klassisches Verhalten des Protons bei der Überwindung von Potentialschwellen längs Wasserstoffbrücken hin.

Der Mechanismus der Rekombination von Hydroniumionen und Hydroxylionen schließt mithin folgende Schritte ein:

1. Begegnung der hydratisierten Ionen infolge Diffusion; Bewegung der Ionen durch elektrostatische Wechselwirkung beeinflußt.

2. Bildung einer Wasserstoffbrücke zwischen den beiden Hydraten. (Die Geschwindigkeitskonstante dieses Prozesses ist noch in k_{12} enthalten).

3. Schnelle Protonenübertragung längs der die beiden Ionen verbindenden Wasserstoffbrücken.

4. Auflösung der ursprünglichen Hydratstrukturen.

5.3. Kooperative Umwandlungen von Biopolymeren

5.3.1. Einführung

Gewisse Biopolymere, wie die Proteine und Nucleinsäuren, stellen hochpolymere Kettenmoleküle dar. Die Anzahl der verschiedenen Grundeinheiten, aus denen sich das Biopolymere aufbaut, ist relativ klein. Sowohl die Reihenfolge der Bausteine (Primärstruktur) als auch ihre räumliche Anordnung im Makromolekül (Konformation) bestimmen seine biologische Funktion. Übergänge zwischen verschiedenen Konformationen eines Biopolymeren finden im biologischen Geschehen statt. Thermodynamik, Kinetik und Mechanismus derartiger Prozesse wird man zunächst an einfacheren, wohldefinierten Systemen untersuchen. Hierzu eignen sich die Helix-Knäuel-Umwandlungen von Polypeptiden und Nucleinsäuren sowie der Übergang zwischen den Strukturen Helix I und Helix II von Poly-L-prolin. Die Strukturumwandlungen in Lösung treten bei Änderungen von Temperatur, pH-Wert, Lösungsmittelzusammensetzung und ähnlichen Zustandsänderungen ein. Die Übergänge zwischen Konformationen finden gewöhnlich in einem sehr engen Bereich der variierten Zustandsgröße statt. Der Übergang entspricht mithin eher einer Phasenumwandlung als einem chemischen Vorgang. Mit abnehmendem Molekulargewicht werden die Umwandlungskurven flacher. Aus diesen Befunden schließen wir, daß die Konformationsumwandlung offenbar einen kooperativen Prozeß darstellt [7], bei welchem die Umwandlungen der einzelnen monomeren Grundeinheiten (Segmente) des Kettenmoleküls nicht unabhängig voneinander ablaufen, sondern sich gegenseitig beeinflussen [8].

5.3.2. Thermodynamik auf der Grundlage des linearen Ising-Modells

Der Theorie kooperativer Umwandlungen legen wir die einfache Annahme zugrunde, daß die Umwandlung eines einzelnen Segments lediglich von seinen beiden nächsten Nachbarn abhängt [8]. Die Kette enthalte N Segmente. Jedes Segment kann entweder im Zustand A oder im Zustand B vorliegen. Dann gibt es 2^N verschiedene Zustände der Kette. Ein Zustand der Kette ist z. B. durch die Sequenz

$$...ABAA......BAAA... \tag{66}$$

gegeben. Die Kooperativität der Umwandlung führen wir in die Theorie ein, indem wir für die Wahrscheinlichkeit, unmittelbar benachbarte Segmente im gleichen Zustand

$$...AA... \text{ bzw. } ...BB... \tag{67}$$

zu finden, einen höheren Wert ansetzen als für jene, unmittelbare Nachbarn in verschiedenen Zuständen

28*

$$\ldots AB \ldots \text{ bzw. } \ldots BA \ldots \tag{68}$$

anzutreffen. „Keimbildungsschritte" der Art

$$\ldots AAA \ldots \to \ldots ABA \ldots, \tag{69a}$$

$$\ldots BBB \ldots \to \ldots BAB \ldots, \tag{69b}$$

d. h. die Überführung in Zustände des Gesamtsystems, in welchen unmittelbare Nachbarn in gleichen Zuständen nicht vorkommen, sollen mithin schwieriger sein als die entsprechenden „Wachstumsschritte"

$$\ldots AAB \ldots \to \ldots ABB \ldots, \tag{70a}$$

$$\ldots BBA \ldots \to \ldots BAA \ldots. \tag{70b}$$

Wir definieren folgende rela ive Wahrscheinlichkeiten [9]:

Für A: $w_{11} = s_1$ A folgt auf A (71a)

            $w_{12} = \sigma_1 s_1$ A folgt nicht auf A (71b)

Für B: $w_{22} = s_2$ B folgt auf B (71c)

            $w_{21} = \sigma_2 s_2$ B folgt nicht auf B.   (71d)

Die Größen σ_1 und σ_2 werden als Keimbildungsparameter bezeichnet. $(\sigma_1, \sigma_2 < 1)$. Der Quotient

$$s = \frac{s_2}{s_1} = \frac{[\ldots ABB \ldots]}{[\ldots AAB \ldots]} \tag{72}$$

stellt die Gleichgewichtskonstante für einen Wachstumsprozeß dar. Die Gleichgewichtskonstanten der Keimbildung an den Kettenenden lauten

$$^E K_A = \frac{[AB \ldots]}{[BB \ldots]} = \frac{\sigma_1}{s} \text{ für } A, \tag{73a}$$

$$^E K_B = \frac{[BA \ldots]}{[AA \ldots]} = \sigma_2 s \text{ für } B. \tag{73b}$$

Für die entsprechenden Größen innerhalb der Kette erhält man

$$^I K_A = \sigma_1 \sigma_2 / s \quad , \quad ^I K_B = \sigma_1 \sigma_2 s. \tag{74}$$

Die relative Wahrscheinlichkeit $W^{(\varphi)}$ eines gegebenen Zustands φ der ganzen Kette finden wir, indem wir die Wahrscheinlichkeiten w_{ij} der Zustände der Einzelsegmente miteinander multiplizieren. Wir beginnen die Multiplikation zweckmäßig am Ende der Kette (Segment N),

$$W^{(\varphi)} = w_{n_N, n_{N-1}} w_{n_{N-1}, n_{N-2}} \cdots w_{n_2, n_1} w_{n_1, n_0}, \tag{75}$$

und erhalten z. B.

$$W^{(\varphi)} = w_{22} w_{21} \cdots w_{11} w_{12}.$$

Die Zustandssumme Q_N der ganzen Kette ist der Summe der Wahrscheinlichkeiten $W^{(\varphi)}$ aller verschiedenen möglichen (Konfigurationen oder) Zustände φ gleich:

$$Q_N = \sum_{\varphi=1}^{2^N} W^{(\varphi)} \,. \tag{76}$$

Zur Ermittlung von Q_N gehen wir nach der Matrizenmethode [10] wie folgt vor[5]. Wir betrachten eine Kette von i Segmenten ($i \leqslant N$)

$$1, 2, 3, \ldots, i-r, i-r+1, \ldots, i.$$

Die letzten r Segmente seien in bezug auf ihre Zustände spezifiziert. Die Summe der Wahrscheinlichkeiten aller restlicher möglicher Konfigurationen der Kette von i Segmenten sei v_{ji}. Dann existieren so viele verschiedene Elemente v_{ji}, wie verschiedene Konfigurationen der letzten r Segmente möglich sind, nämlich 2^r. Der Index j läuft mithin von 1 bis 2^r. Die Gesamtheit der v_{ji} können wir durch den Spaltenvektor

$$v_i = \begin{pmatrix} v_{1\,i} \\ v_{2\,i} \\ \cdot \\ \cdot \\ \cdot \\ v_{2^r i} \end{pmatrix} \tag{77}$$

ausdrücken. Die Zustandssumme Q_i der Kette mit i Segmenten ist gleich der Summe aller Elemente v_{ji}. Analog zur Festlegung von v_i kann man eine Spaltenmatrix v_{i+1} definieren. Das Element $v_{j,i+1}$ stellt die Summe der Wahrscheinlichkeiten aller möglichen Konfiguration einer Kette von $i+1$ Segmenten dar, in welcher die Zustände der letzten r Segmente $i-r+2, i-r+3, \ldots, i, i+1$ vorgegeben sind. Die Gesamtheit aller $v_{j,i+1}$ bildet wiederum einen Spaltenvektor mit 2^r Elementen. Den Spaltenvektor v_{i+1} können wir nach

$$v_{i+1} = W\, v_i \tag{78}$$

aus v_i erzeugen, wo W eine quadratische Matrix der Ordnung 2^r ist. Denn die zweite Kette mit $i+1$ Segmenten läßt sich einfach herstellen, indem man an die erste Kette mit i Segmenten ein Segment anbaut. Wir wissen, wie sich die Wahrscheinlichkeiten der Konfigurationen dabei ändern.

---

[5] Bei Berücksichtigung von (71) und (75) kann (76) in die Gestalt

$$Q_N = \sum_{\substack{n_1=1 \\ n_0 \neq n_1}}^{2} \sum_{n_2=1}^{2} \ldots \sum_{n_{N\text{-}1}=1}^{2} \sum_{n_N=1}^{2} w_{n_N,\,n_{N-1}}\, w_{n_{N-1},\,n_{N-2}} \cdots w_{n_2,\,n_1}\, w_{n_1,\,n_0} \tag{76'}$$

gebracht werden. Nach den Regeln der Matrizenmultiplikation folgt sofort Gl. (87).

438   M. Eigen und L. De Maeyer

Nehmen wir an, daß $r = 3$ ist. Die folgenden Konfigurationen $K^{(i)}$ (bzw. $K^{(i+1)}$) der letzten 3 Segmente der Kette von insgesamt i (bzw. $i + 1$) Segmenten gehören zu den Elementen v_{ji} (bzw. $v_{j, i+1}$):

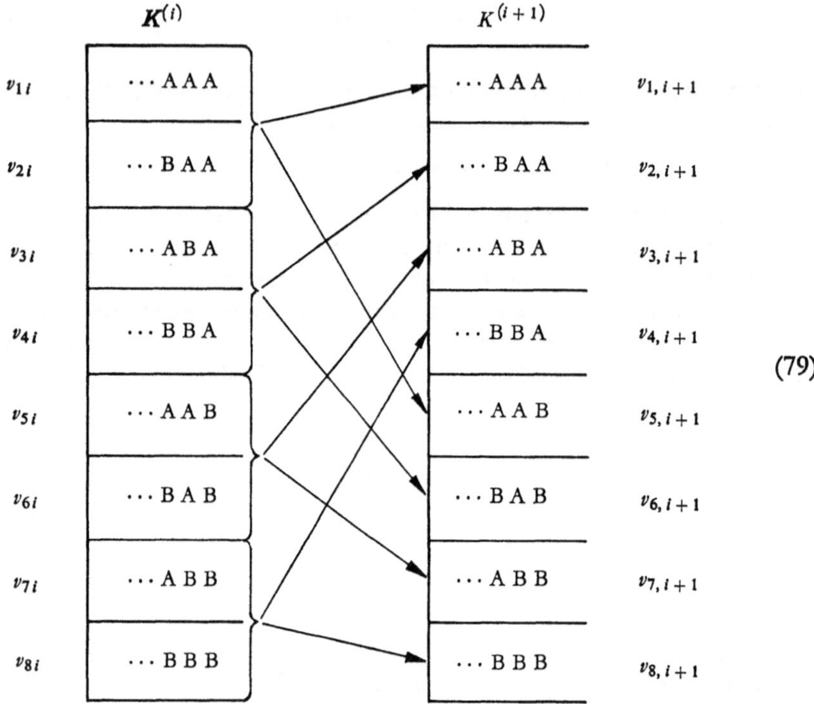

$$(79)$$

Da sich das angebaute Segment im Zustand A oder B befinden kann, sind aus jeder Konfiguration $K^{(i)}$ Übergänge zu zwei verschiedenen Konfigurationen $K^{(i+1)}$ möglich. Aus den Beziehungen (71) und (78) folgt unmittelbar für W mit $r = 3$

$$
W = \begin{pmatrix}
s_1 & s_1 & 0 & 0 & 0 & 0 & 0 & 0 \\
0 & 0 & s_1 & s_1 & 0 & 0 & 0 & 0 \\
0 & 0 & 0 & 0 & \sigma_1 \cdot s_1 & \sigma_1 \cdot s_1 & 0 & 0 \\
0 & 0 & 0 & 0 & 0 & 0 & \sigma_1 \cdot s_1 & \sigma_1 \cdot s_1 \\
\sigma_2 \cdot s_2 & \sigma_2 \cdot s_2 & 0 & 0 & 0 & 0 & 0 & 0 \\
0 & 0 & \sigma_2 \cdot s_2 & \sigma_2 \cdot s_2 & 0 & 0 & 0 & 0 \\
0 & 0 & 0 & 0 & s_2 & s_2 & 0 & 0 \\
0 & 0 & 0 & 0 & 0 & 0 & s_2 & s_2
\end{pmatrix}
\tag{80}
$$

Da W für jede Kette der Länge i mit $r \leq i \leq N$ definiert ist, ergibt sich

$$v_N = W v_{N-1} \tag{81}$$

und durch Iteration

$$v_N = W^{N-1}\, v_1, \tag{82}$$

wo mit $r = i = 1$

$$v_1 = \begin{pmatrix} \sigma_1 \cdot s_1 \\ \sigma_2 \cdot s_2 \end{pmatrix} \tag{83}$$

und

$$W = \begin{pmatrix} s_1 & \sigma_1\, s_1 \\ \sigma_2\, s_2 & s_2 \end{pmatrix}. \tag{84}$$

Die Zustandssumme Q_N der Kette mit der Länge N ist, wie bereits angegeben, gleich der Summe der Elemente der Spaltenmatrix v_N. Wir erhalten Q_N durch Multiplikation der Matrix v_N von links mit der Zeilenmatrix u_N:

$$Q_N = u_N\, v_N; \tag{85}$$

die Zeilenmatrix

$$u_N = (1 \quad 1 \quad 1 \ldots 1) \tag{86}$$

hat 2^r Elemente. Die gesuchte Zustandssumme der Kette mit N Segmenten berechnet sich somit nach

$$Q_N = u_1\, W^{N-1}\, v_1, \tag{87}$$

wobei

$$u_1 = (1 \quad 1). \tag{88}$$

Zur Bestimmung von W^{N-1} suchen wir die Matrix M, die W auf Diagonalform bringt:

$$D = M^{-1}\, W\, M = \begin{pmatrix} \lambda_1 & 0 & \ldots & 0 \\ 0 & \lambda_2 & & \\ \vdots & & \ddots & \\ 0 & & & \lambda_{2^r} \end{pmatrix}; \tag{89}$$

die Größen λ_i sind die Eigenwerte von W. Dann müssen auch die Gleichungen

$$D^{N-1} = M^{-1}\, W^{N-1}\, M = \begin{pmatrix} \lambda_1^{N-1} & 0 & & & & 0 \\ 0 & \lambda_2^{N-1} & & & & \\ \vdots & & \ddots & & & \\ 0 & & & & & \lambda_{2^r}^{N-1} \end{pmatrix} \tag{90}$$

und

$$W^{N-1} = M\, D^{N-1}\, M^{-1} \tag{91}$$

erfüllt sein. Die Diagonalmatrix D finden wir, indem wir die Eigenwertgleichung

$$W\, x_i^{(e)} = \lambda_i\, x_i^{(e)} \quad \text{bzw.} \tag{92}$$

$$(W - \lambda\, E)\, x^{(e)} = 0 \tag{93}$$

lösen, in welcher $x_i^{(e)}$ der Eigenvektor (Spaltenvektor) zum Eigenwert λ_i und E die Einheitsmatrix sind. Das lineare Gleichungssystem (93) besitzt nur dann nicht verschwindende Lösungen, wenn die aus den Koeffizienten der Komponenten des Eigenvektors $x_i^{(e)}$ gebildete Determinante Null ist:

$$|\,W - \lambda\,E\,| = 0\,. \tag{94}$$

Die Wurzeln dieser algebraischen Gleichung 2^r-ten Grades sind die Eigenwerte λ^i von W. Setzen wir eine der Wurzeln λ_i in (93) ein, so erhalten wir die Komponenten x_{ni} des Eigenvektors $x_i^{(e)}$. Wie sich leicht zeigen läßt, bilden die Komponenten der Eigenvektoren die Matrix M:

$$M = (x_1^{(e)}\,,\ x_2^{(e)}\,,\ x_3^{(e)}\,,\ \ldots\,,\ x_{2^r}^{(e)})\,. \tag{95}$$

Zweckmäßig bestimmen wir auf ähnlichem Wege M^{-1}. Ist der Zeilenvektor

$$y_i^{(e)} = (y_{i1}\,,\ y_{i2}\,,\ y_{i3}\,,\ \ldots\,,\ y_{i2^r}) \tag{96}$$

der Linkseigenvektor von W zum Eigenwert λ_i,

$$y_i^{(e)}\,W = \lambda_i\,y_i^{(e)}\,, \tag{97}$$

so lautet die Matrix

$$M^{-1} = \begin{pmatrix} y_1^{(e)} \\ y_2^{(e)} \\ \vdots \\ y_{2^r}^{(e)} \end{pmatrix}\,. \tag{98}$$

Die Zustandssumme der N-gliedrigen Kette, Q_N, läßt sich nun leicht berechnen.

Uns interessiert die Frage, wie groß der Anteil der Segmente der Kette ist, die sich bei gegebenen Parametern s, σ_1, σ_2 und N in einem bestimmten Zustand, z. B. dem Zustand B, befinden. Zunächst berechnen wir die Wahrscheinlichkeit, das Segment k im Zustand B anzutreffen. Wir definierten das Element v_{2i} des Spaltenvektors v_i als die relative Wahrscheinlichkeit dafür, das i-te Segment einer Kette, die i Segmente enthält, im Zustand B zu finden; für $i = N$ gilt nach (81)

$$v_N = \begin{pmatrix} v_{1N} \\ v_{2N} \end{pmatrix} = W^{N-k} \begin{pmatrix} v_{1k} \\ v_{2k} \end{pmatrix} = \begin{pmatrix} \varkappa\,v_{1k} + \lambda\,v_{2k} \\ \mu\,v_{1k} + \nu\,v_{2k} \end{pmatrix}\,, \tag{99}$$

wo $W^{N-k} = \begin{pmatrix} \varkappa & \lambda \\ \mu & \nu \end{pmatrix}$. Offensichtlich ist $\lambda\,v_{2k}$ die relative Wahrscheinlichkeit, daß das Segment N im Zustand A und gleichzeitig das Segment k im Zustand B ist. Ähnliche Bedeutung besitzen die übrigen Terme der rechten Seite der Gleichung (99). Die Wahrscheinlichkeit, das k-te Segment einer Kette mit N Segmenten im Zustand B anzutreffen, ist daher

$$\gamma_B^{(k)} = (\lambda + \nu)\,v_{2k}\,\frac{1}{Q_N}$$

oder

$$\gamma_B^{(k)} = (\boldsymbol{u}_1 \, \boldsymbol{W}^{N-k})_2 \, (\boldsymbol{W}^{k-1} \, \boldsymbol{v}_1)_2 \, \frac{1}{Q_N} \; ; \tag{100}$$

der Index 2 bedeutet, daß jeweils das 2. Element der eingeklammerten Vektoren einzusetzen ist. Der Anteil der Segmente im Zustand B in einer Kette mit insgesamt N Segmenten ist gleich dem Mittelwert

$$\Theta_B = \frac{1}{N} \sum_{k=1}^{N} \gamma_B^{(k)} . \tag{101}$$

Θ_B kann man als ein Maß für die Konformationsumwandlung auffassen. Wir sind nun in der Lage, den Bruchteil Θ_B explizit als Funktion von

$$\sigma = \sigma_1 \, \sigma_2, \quad \beta = \sigma_2/\sigma_1, \quad \xi = (s-1)/(2\sqrt{\sigma}), \quad N \tag{102}$$

auszudrücken. Die Eigenwerte λ_0 und λ_1 von $\boldsymbol{W} = \begin{pmatrix} s_1 & \sigma_1 \, s_1 \\ \sigma_2 \, s_2 & s_2 \end{pmatrix}$ berechnen sich nach Gleichung (94) zu

$$\lambda_{0,1} = \frac{1}{2} \, (s_1 + s_2) \pm \frac{1}{2} \, [(s_1 - s_2)^2 + 4 \, \sigma_1 \, \sigma_2 \, s_1 \, s_2]^{1/2} . \tag{103}$$

Dann ergibt sich für Θ_B [9]:

$$\Theta_B = \frac{\lambda_0 - s_1}{\lambda_0 - \lambda_1} \cdot \frac{1 + g^2 \varrho^{N-1} - (2g/N) \, (1 - \varrho^N)/(1 - \varrho)}{1 + \dfrac{\lambda_0 - s_1}{\lambda_0 - s_2} \, g^2 \, \varrho^{N-1}} \tag{104}$$

mit

$$g = \frac{1 + \sqrt{\beta} \, (\xi - \sqrt{\xi^2 + s})}{1 + \sqrt{\beta} \, (\xi + \sqrt{\xi^2 + s})}, \quad \varrho = \frac{\lambda_1}{\lambda_0} .$$

Typische Umwandlungskurven, welche die Abhängigkeit von Θ_B als Funktion von s beschreiben, sind in Abb. 1 dargestellt. Für $\beta = \sigma$, d. h. $\sigma_1 = 1$ (Abb. 1c), schneiden sich die Umwandlungskurven nicht mehr. In diesem speziellen Fall ist die Keimbildung nur für einen der Zustände, nämlich für den Zustand B, erschwert. Tatsächlich können die experimentell gefundenen Umwandlungskurven für die Helix-Knäuel-Umwandlung von Polypeptiden im Rahmen der Theorie mit einem Keimbildungsparameter (Theorie nach ZIMM und BRAGG [7]) gedeutet werden. Offenbar ist bei Umwandlungen zwischen geordneten und ungeordneten Zuständen die Keimbildung nur für die geordneten Zustände erschwert.

Dagegen wird man bei Umwandlungen zwischen zwei verschiedenen, geordneten Zuständen, z. B. zwischen den Strukturen Helix I und Helix II von Poly-L-prolin, für jeden der beiden Zustände einen Keimbildungsparameter einführen. In der Tat sind bei Versuchen mit synthetischem Poly-

L-prolin für verschiedene Molekulargewichte Umwandlungskurven beobachtet worden, die sich schneiden [*11*], ähnlich wie in Abb. 1b [*9*]; dieser Befund läßt sich im Rahmen der Zimm-Bragg-Theorie nicht deuten.

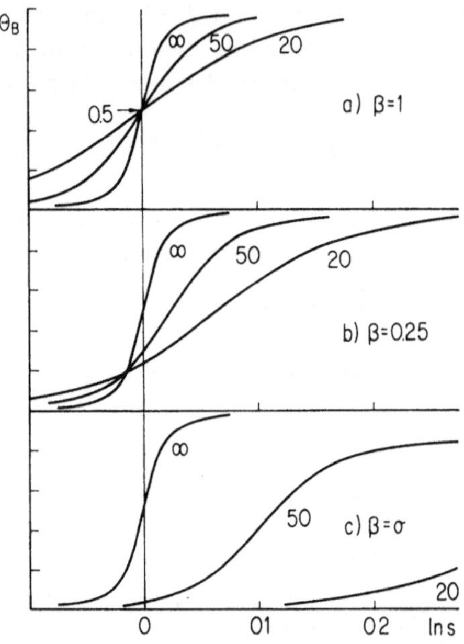

Abb. 1. Umwandlungskurven für $\sigma = 10^{-4}$ und verschiedene Werte von N (∞; 50; 20) und β (1; 0,25; σ) (G. Schwarz [*9*])

5.3.3. *Ein Modell für die Kinetik der Umwandlung. Experimentelle Ergebnisse*

Der Elementarprozeß bei der Umwandlung ist offenbar die unimolekulare Reaktion

$$A \;\rightleftharpoons\; B \tag{105}$$

irgendeines Segments i der Kette. Da der Prozeß kooperativ ist, sind seine Geschwindigkeitskonstanten vom Zustand der Nachbarsegmente abhängig. Setzen wir wieder voraus, daß nur unmittelbare Nachbarn die Umwandlung eines Segments beeinflussen, so können wir folgende Elementarprozesse unterscheiden:

$$\begin{array}{ccc} AAB & k_A & ABB \\ BAA & \overset{\longrightarrow}{\underset{k_B}{\longleftarrow}} & BBA \end{array} \tag{106}$$

Gleichung (106) beschreibt Wachstumsprozesse; der Quotient k_A/k_B ist mit der Gleichgewichtskonstanten $s = s_2/s_1$ identisch.

$$AAA \;\overset{\sigma k_A}{\underset{k_B}{\rightleftharpoons}}\; ABA \tag{107a}$$

$$BAB \; \underset{\sigma\,k_B}{\overset{k_A}{\rightleftharpoons}} \; BBB \qquad\qquad (107\,\mathrm{b})$$

In (107) ist die Annahme enthalten, daß der thermodynamische Keimbildungsparameter σ allein durch die entsprechende Verringerung der Keimbildungsgeschwindigkeit bedingt wird. Die Umwandlungen an den Kettenenden können durch die Gleichungen

$$\begin{array}{cc} BA0 & BB0 \\ & \xrightleftharpoons[\sigma_1\,k_B]{k_A} \\ 0AB & 0BB \end{array} \qquad\qquad (108\,\mathrm{a})$$

$$\begin{array}{cc} AA0 & AB0 \\ & \xrightleftharpoons[k_B]{\sigma_2\,k_A} \\ 0AA & 0BA \end{array} \qquad\qquad (108\,\mathrm{b})$$

dargestellt werden; 0 bezeichnet den Platz des fehlenden Nachbars. Im Rahmen unseres einfachen Modells läßt sich das Zeitgesetz für die Änderung des Anteils der Segmente im Zustand B sofort angeben:

$$\begin{aligned}
\frac{d\,\Theta_B}{d\,t} = \; &k_A\,[\gamma_{AAB} + \gamma_{BAA} + \gamma_{BAB} + \gamma_{BA0} + \sigma\,\gamma_{AAA} \\
&+ \gamma_{0AB} + \sigma_2\,(\gamma_{AA0} + \gamma_{0AA})] \\
&- k_B\,[\gamma_{ABB} + \gamma_{BBA} + \gamma_{ABA} + \sigma\,\gamma_{BBB} \\
&+ \sigma_1\,(\gamma_{BB0} + \gamma_{0BB}) + \gamma_{AB0} + \gamma_{0BA}] \;;
\end{aligned} \qquad (109)$$

γ_{AAB} ist der relative Anteil der Sequenz $\ldots AAB\ldots$, γ_{BAA} jener der Sequenz $\ldots BAA\ldots$ usw. Die Lösung dieses Problems setzt die Kenntnis der relativen Anteile γ als Funktion der Zeit t voraus. Ihre Bestimmung stößt jedoch wegen der großen Anzahl von Elementarreaktionen auf Schwierigkeiten. Unter besonderen Bedingungen lassen sich jedoch Aussagen über den zeitlichen Verlauf der Umwandlung gewinnen.

Stören wir das bestehende Gleichgewicht eines Systems geringfügig durch eine sprunghafte Änderung der Zustandsvariablen (z. B. der Temperatur oder der Zusammensetzung des Lösungsmittels), so läßt sich die Einstellung des neuen Gleichgewichtszustandes experimentell verfolgen. Für diesen Relaxationsvorgang ist eine kinetische Analyse möglich. Die Gleichgewichtskonstante ändert sich zur Zeit $t = 0$ sprunghaft mit den äußeren Zustandsvariablen von s auf $s + \delta s$, so daß für den Quotienten der Geschwindigkeitskonstanten k_A und k_B in (109)

$$k_A/k_B = s + \delta s \qquad\qquad (110)$$

zu setzen ist. Da im Zeitpunkt der Zustandsänderung ($t = 0$) das System in einem Gleichgewichtszustand vorliegt, kann man in die spezielle Form von Gleichung (109), mit

$$\left(\frac{d\,\Theta_B}{d\,t}\right)_{t=0} , \qquad\qquad (111)$$

444   M. Eigen und L. De Maeyer

die Gleichgewichtswerte von γ_{AAB}, γ_{BAA} usf. einsetzen. Letztere können berechnet werden. [s. Gl. (100))].

Wie bemerkt, läßt sich der Relaxationsvorgang experimentell verfolgen. Sein Ablauf wird durch die Gleichung [s. (23)]

$$\delta \Theta_B (t) = \delta \Theta_B (0) \sum_r \beta_r e^{-t/\tau_r} \qquad (112)$$

beschrieben, in welcher $\delta \Theta_B (t)$ die Abweichung der Größe Θ_B von ihrem Gleichgewichtswert zur Zeit t, β_r die Amplitudenfaktoren und τ_r die Relaxationszeiten der Elementarschritte der Umwandlung sind. Die Abweichung $\delta \Theta_B (0)$ zur Zeit $t = 0$ ist mit den Gleichgewichtswerten $\Theta_B(s + \delta s)$ und $\Theta_B(s)$ verknüpft:

$$\delta \Theta_B (0) = \Theta_B (s + \delta s) - \Theta_B (s) = \left(\frac{\partial \Theta_B}{\partial s} \right) \delta s \,. \qquad (113)$$

Die mittlere Relaxationszeit τ^* ergibt sich aus der Meßgröße $(d\Theta_B/dt)_{t=0}$ mit Hilfe der Beziehung (25) zu

$$\sum_r \frac{\beta_r}{\tau_r} = \frac{1}{\tau^*} = - \frac{1}{\delta \Theta_B (0)} \left(\frac{d \Theta_B}{d t} \right)_{t=0} \,. \qquad (114)$$

Eliminieren wir $(d\Theta_B/dt)_{t=0}$ in (111) und (114), so können wir τ^* als Funktion jener Parameter ausdrücken, welche die Umwandlung als eine kooperative charakterisieren [9]:

$$\tau^* = \frac{1}{\sigma k_A} \frac{\partial \Theta_B}{\partial s} s \frac{\lambda_0 (\lambda_0 - \lambda_1)}{s_1 (s_1 + s_2)}$$

$$\cdot \frac{1 + \dfrac{\lambda_0 - s_1}{\lambda_0 - s_2} g^2 \varrho^{N-1}}{1 - \dfrac{\lambda_0 - s_1}{\lambda_0 - s_2} g^2 \varrho^{N-2} + \dfrac{2g}{N} \dfrac{(s-1)}{(1 - \varrho s)} \dfrac{(1 - \varrho^N)}{(1 - \varrho)}} \qquad (115)$$

mit

$$\tau^*_{\max} = (4 \sigma k_A)^{-1} \,. \qquad (116)$$

Liegen lange Ketten vor, so daß man die Randeffekte vernachlässigen darf, kann (115) bei großer Kooperativität, $\sqrt{\sigma} \ll 1$, innerhalb des Umwandlungsbereichs $0 < \Theta_B < 1$ in die einfache Form

$$\tau^* \approx \frac{1}{\sigma k_A} \Theta_B (1 - \Theta_B) \qquad (117)$$

überführt werden. Der Maximalwert von τ^* liegt für $N \to \infty$ bei $\Theta_B = 0,5$ (Mittelpunkt der Umwandlungskurve). Für kleinere Kettenlängen wird τ^* kleiner. In Abb. 2 [9] sind einige nach (115) berechnete Kurven $\tau^* (\Theta_B)$ für $\sigma = 10^{-4}$ und verschiedene Werte von N und β dargestellt. Die durch Pfeile angedeuteten Maxima der Kurven verschieben sich mit fallenden Werten von N und β zu kleineren Θ_B-Werten.

Einige experimentelle Ergebnisse zur Kinetik kooperativer Umwandlungen liegen vor. Messungen der Schallabsorption durch wäßrige, $NaNO_3$-haltige Lösungen von Polyglutaminsäure zeigten, daß τ^*_{max} für die Helix-Knäuelumwandlung dieses Polypeptids bei etwa 10^{-6} bis 10^{-7} sec liegt.

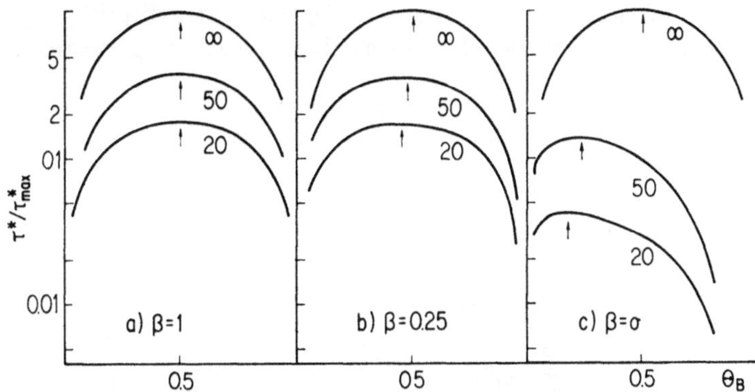

Abb. 2. Mittlere Relaxationszeiten τ^* (bezogen auf τ^*_{max}) als Funktion des Umwandlungsgrades Θ_B für $\sigma = 10^{-4}$ und verschiedene Werte von N (∞; 50; 20) und β (1; 0,25; σ) (G. Schwarz [9])

Aus thermodynamischen Daten ergibt sich für dieses System σ zu ungefähr $5 \cdot 10^{-3}$. Die Größe k_A ist in diesem Fall die Geschwindigkeitskonstante für die Bildung eines Helixsegments am Rande einer bereits vorhandenen Helix. Ein Segment der Polypeptidkette,

$$\begin{array}{c} H \\ | \\ N \quad\; H \\ \diagdown \; | \\ C \quad\;\; C \diagup \\ \| \quad\quad | \\ O \quad\quad R \end{array} \; ,$$

ist dann im Helixzustand (ein „Helixsegment"), wenn seine CO-Gruppe die richtige Wasserstoffbrücke zum NH der entsprechenden Peptidgruppe bildet; andernfalls gehört das Segment zu den Knäueleinheiten. Die Bildung einer neuen Helixeinheit sollte hier sehr schnell verlaufen. Nach experimentellen Befunden über die Bildung von Wasserstoffbrücken zwischen Peptidgruppen läßt sich k_A zu etwa 10^9 bis 10^{10} sec^{-1} abschätzen. Hieraus ergibt sich nach (116) τ^*_{max} zu etwa 10^{-6} bis 10^{-7} sec in Übereinstimmung mit dem gefundenen Wert.

Synthetisches Poly-L-prolin existiert in den zwei stabilen Konformationen I und II. Die Konformation I ist eine rechtsgängige Helix, deren Peptidgruppen in der cis-Konfiguration sind; sie ist stabil in Pyridin und aliphatischen Alkoholen. Die Konformation II ist eine linksgängige Helix mit allen Peptidgruppen in der trans-Konfiguration und stabil in Benzyl-alkohol, Wasser u. a. Lösungsmitteln[6]. Die Konformationsumwandlung wird durch Änderung der Zusammensetzung des Lösungsmittels hervorgerufen; sie kann durch Messung der optischen Drehung leicht verfolgt werden. Nach einer Reihe von Befunden ist der Elementarprozeß der Konformationsumwandlung die cis-trans-Umlagerung der Peptidgruppe. In Abb. 3 ist der Verlauf der bei Relaxationsversuchen gemessenen τ^*-

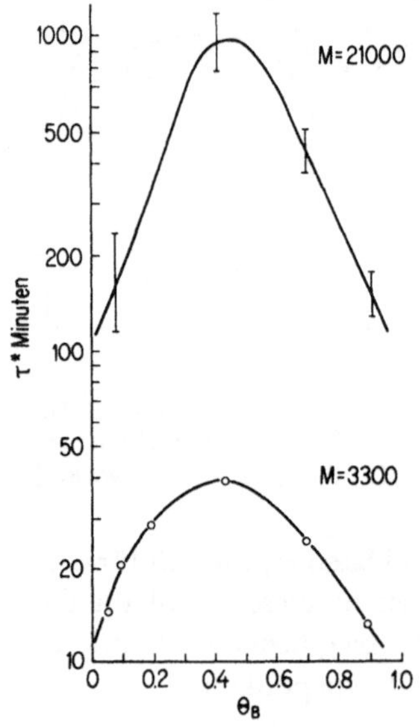

Abb. 3. Mittlere Relaxationszeit τ^* in Abhängigkeit des Umwandlungsgrades Θ_B der Umwandlung von Poly-L-prolin II in Poly-L-prolin I. Θ_B: Gehalt an der Konformation I, Temperatur: 70°C, Polymerkonzentration: 0,25 bis 0,50 g/100 ml, Lösungsmittel: n-Butanol-Benzylalkohol. Relaxation nach Störung des Gleichgewichts durch Veränderung der Zusammensetzung des Lösungsmittels
(J. Engel [11])

Werte als Funktion des Gehalts an der Konformation I, Θ_B, für zwei Molekulargewichte des Polymeren, $M = 21\,000$ und $M = 3300$, aufgezeichnet. Mit $\sigma \approx 10^{-5}$ ergibt sich nach Gl. (116) die Geschwindigkeitskonstante k_A für den Wachstumsschritt

$$(\text{trans}) \quad \overset{O}{\underset{}{\diagdown}} C\!-\!N \overset{}{\underset{H}{\diagup}} \quad \longrightarrow \quad \overset{O}{\underset{}{\diagdown}} C\!-\!N \overset{H}{\underset{}{\diagup}} \quad (\text{cis})$$

aus den gemessenen $\tau^*_{\max}$-Werten bei 70 °C zu $10^{-1}\ \mathrm{sec}^{-1}$ [11].

---

[6] Die Strukturen wurden im festen Zustand ermittelt.

5.4. Reaktionen eines allosterischen Enzyms

5.4.1. Zur Deutung sigmoider Bindungsisothermen

Bindungsisotherme von Sauerstoff an Hämoglobin zeigen einen sigmoiden Verlauf. 1925 schlug ADAIR [12] zur Deutung dieses Befundes einen Mechanismus vor, nach dem die Bindung in vier konsekutiven, bimolekularen Schritten erfolgt, wobei die Affinität des Hämoglobins zum Sauerstoff successive zunehmen soll. MONOD, WYMAN und CHANGEUX erklärten 1965 [13] das Phänomen, das inzwischen bei einer Anzahl anderer Enzymsysteme ebenfalls beobachtet wurde, mit Hilfe des folgenden Mechanismus: Es bestehe ein „allosterischer" Gleichgewichtszustand zwischen zwei Konformationen des oligomeren Proteins, von welchen jede eine verschiedene Affinität zum Liganden habe. Im Verlaufe der Bindung des Liganden verschiebe sich das allosterische Gleichgewicht. Nach der „Induced Fit"-Konzeption von KOSHLAND, NÉMETHY und FILMER [14] wird die Ligandenaffinität durch Änderungen der Sequenzen von Untereinheiten des Enzyms bei der Ligandenbindung beeinflußt.

Diese Autoren weisen nach [14], daß aufgrund von Gleichgewichtsdaten allein zwischen den einzelnen Mechanismen nicht unterschieden werden kann. Man wird für jedes einzelne System mit Hilfe kinetischer Untersuchungen sorgfältig zu prüfen haben, nach welchem Mechanismus Substrat, Aktivator und Inhibitor an das Enzym gebunden werden.

Im folgenden gehen wir auf die Gleichgewichts- und kinetischen Eigenschaften eines allosterischen Systems und schließlich auf ein spezielles Enzymsystem ein, dessen gemessene thermodynamische und kinetische Daten auf der Grundlage des allosterischen Modells diskutiert werden.

5.4.2. Gleichgewichtseigenschaften eines allosterischen Systems

Eine Substanz E (z. B. ein Enzym) möge in zwei Zuständen, T_0 und R_0, existieren. Sie besitze im Zustand R_0, und nur in diesem Zustand, n identische Bindungsstellen für je einen „Liganden" F. Die einzelnen Bindungsstellen stehen mithin untereinander nicht in Wechselwirkung. Bezeichnet R_i den Fall, daß i Bindungsstellen von R mit Liganden besetzt sind, so wird das Reaktionssystem durch die folgenden $n + 1$ Elementarprozesse [15]

$$ T_0 \underset{k'}{\overset{k}{\rightleftharpoons}} R_0 , \quad R_{i-1} + F \underset{i k_D}{\overset{(n-i+1) k_R}{\rightleftharpoons}} R_i \tag{118} $$

$$ i = 1, 2, \ldots, n $$

dargestellt. Die Gleichgewichtskonstante

$$ L = \frac{k'}{k} \tag{119} $$

nennen wir die „allosterische" Konstante. Die Geschwindigkeitskonstanten für die Bindung und Dissoziation einer einzelnen Bindungsstelle und eines Liganden sind die Größen k_R bzw. k_D. Ihr Quotient ist die Gleichgewichtskonstante K_A:

$$\frac{k_R}{k_D} = K_A \, . \tag{120}$$

Die Faktoren $(n - i + 1)$ von k_R und i von k_D müssen als statistische Faktoren eingeführt werden; es gilt z. B. für $n = 2$

$$\frac{[R_1]}{[R_0]\,[F]} = 2\,K_A \, , \; \frac{[R_2]}{[R_1]\,[F]} = \frac{K_A}{2} \, . \tag{121}$$

Die Symbole E, T_0 und R mögen auch die entsprechenden Konzentrationen bezeichnen. Dann ist $B = \sum_i i R_i$ die Konzentration der besetzten Bindungsstellen, die der Konzentration des gebundenen Liganden gleich ist. Nennen wir die Konzentrationen der freien Bindungsstellen und des freien Liganden S bzw. F, und die Gesamtkonzentration des Liganden F_{tot}, so gilt

$$B + S + n\,T_0 = nE, \tag{122a}$$
$$B + F = F_{\text{tot}} \, . \tag{122b}$$

Derjenige Bruchteil der maximalen Konzentration von Bindungsstellen, der besetzt ist, wird der Sättigungsgrad Y genannt:

$$Y = \frac{B}{nE} \, . \tag{123}$$

Gewöhnlich kann Y gemessen werden. Die Gleichgewichtswerte von Y und T_0, $\overline{Y}$ und $\overline{T}_0$, ergeben sich nach (118) und (122) zu

$$\overline{Y} = \frac{a\,(1 + a)^{n-1}}{L + (1 + a)^n} \, , \qquad \overline{T}_0 = \frac{E}{1 + \dfrac{1}{L}\,(1 + a)^n} \, . \tag{124}$$

Trägt man $\overline{Y}$ als Funktion von $a = K_A \overline{F}$ auf, wo $\overline{F}$ der Gleichgewichtswert von F ist, so erhält man eine sigmoide Kurve, die Sättigungskurve oder Bindungsisotherme.

5.4.3. *Relaxation eines allosterischen Systems. Die Meßbarkeit von Relaxationszeiten und Reaktionseffekten bezüglich Normalvariablen*

Das System von Reaktionen (118), wird durch $(n + 1)$ linear unabhängige Zeitgesetze beschrieben. Das Relaxationsspektrum des Systems wird mithin im allgemeinen $(n + 1)$ verschiedene Relaxationszeiten enthalten. Das Zeitgesetz für die Änderung der besetzten Bindungsstellen lautet:

$$\frac{dB}{dt} = k_R F \cdot S - k_D B = k_D (K_A F \cdot S - B) \, . \tag{125}$$

Wir setzen zunächst voraus, daß die unimolekulare allosterische Umwandlung $T_0 \rightleftharpoons R_0$ eingefroren ist. Dann ist von den übrigen n Relaxationszeiten nur eine meßbar, nämlich jene, welche die einstufige Relaxation von Y oder F nach Gl. (125) charakterisiert; wir nennen diese Relaxationszeit τ_B. Die restlichen $(n-1)$ Relaxationszeiten können bei Relaxationsversuchen nicht bestimmt werden, weil ihre zugehörigen thermodynamischen Reaktionseffekte, wie ΔH_{yi}, ΔV_{yl} usf. infolge der vorausgesetzten Identität der n Bindungsstellen Null sind. (Derartige Relaxationszeiten lassen sich unter gewissen Bedingungen mit Hilfe der magnetischen Resonanz messen.)

Die Änderung einer extensiven thermodynamischen Zustandsgröße Ψ, wie der Enthalpie H, des Volumens V usf. mit den Konzentrationen c_j der Komponenten j eines Systems ist durch

$$\delta \Psi = \sum_j \Psi_j \delta c_j = \sum_j \Psi_j x_j \quad \text{mit} \quad \Psi_j = \left(\frac{\partial \Psi}{\partial c_j}\right) \tag{126}$$

gegeben. Das betrachtete System enthält die Komponenten $R_1, \ldots, R_n$, R_0 und F ($j = 1, 2, \ldots, n+2$). Eliminieren wir die auf Grund der Massenerhaltung linear abhängigen Variablen,

$$x_j = \sum_k a_{jk} x_k \;, \tag{127}$$

und setzen für die verbleibenden (in unserem Fall n) Variablen

$$x_k = \sum_l m_{kl} y_l \;, \tag{128}$$

wo y_l die zur Relaxationszeit $\tau_l (\neq 0)$ gehörige Normalvariable ist, so erhalten wir

$$\delta \Psi = \sum_l \Delta \Psi_{yl} y_l \tag{129}$$

mit

$$\Delta \Psi_{yl} = \sum_k m_{kl} \Delta \Psi_{xk} \;, \quad \Delta \Psi_{xk} = \sum_j a_{jk} \Psi_j \;.$$

Wegen der Identität der n Bindungsstellen, d. h. wegen

$$\left(\frac{\partial \Psi}{\partial R_{i+1}}\right) - \left(\frac{\partial \Psi}{\partial R_i}\right) - \left(\frac{\partial \Psi}{\partial F}\right) =$$

$$= \left(\frac{\partial \Psi}{\partial R_i}\right) - \left(\frac{\partial \Psi}{\partial R_{i-1}}\right) - \left(\frac{\partial \Psi}{\partial F}\right) = -\Delta \Psi_{xi} \equiv -\Delta \Psi_{ys}$$

$$i = 1, 2, \cdots, n \tag{130}$$

ergibt sich

$$\delta \Psi = \Delta \Psi_{ys} y_s \;, \tag{131}$$

wobei $y_s = x_F = \delta F = - \sum_i i \, \delta R_i$.

$-\Delta \Psi_{ys}$ ist der Reaktionseffekt (ΔH bzw. ΔV etc.) der Reaktion $R_{i-1} + F \rightleftharpoons R_i$. Mit den übrigen $n - 1$ Normalvariablen y_k ($k \neq s$) sind keine Reaktionseffekte $\Delta \Psi_{yk}$ verknüpft. Nach den Methoden der chemischen Relaxation können daher die Normalvariablen y_k ($k \neq s$) nicht verändert und damit die zugehörigen Relaxationszeiten τ_k nicht gemessen werden. Zum Beispiel müssen bei einem Temperatursprung-Experiment die Reaktionsenthalpien ΔH_{yk} eine endliche Größe besitzen, um eine Änderung der Gleichgewichtswerte von y_k und damit die Einstellung eines neuen Gleichgewichtszustandes, die eine Messung von τ_k ermöglichte, herbeizuführen.

Dies ändert sich nicht wesentlich, solange die dem Zeitgesetz

$$\frac{dT_0}{dt} = k'R_0 - k\,T_0 = k\,(L \cdot R_0 - T_0) \tag{132}$$

folgende Konformationsumwandlung $T_0 \rightleftharpoons R_0$ sehr viel langsamer als der Bindungsprozeß verläuft. Die Konformationsumwandlung liefert die Relaxationszeit τ_T. Nehmen wir an, daß zwischen der Form R_0 und allen anderen Formen R_i stets Gleichgewicht herrscht, so kann τ_T leicht berechnet werden. Mit $\tau_B \ll \tau_T$ ergibt sich aus (125) und (132) für die reziproken Relaxationszeiten

$$\frac{1}{\tau_B} = k_R\,(\overline{S} + \overline{F}) + k_D \tag{133}$$

$$\left.\frac{1}{\tau_T} = k + \frac{k'}{(1 + a)^n}\right\} \quad \text{für } \tau_B \ll \tau_T; \tag{134}$$

$\overline{S}$ und $\overline{F}$ sind die durch die Störung bestimmten Gleichgewichtswerte von S und F. Gl. (133) ist mit dem Ausdruck für eine einstufige, bimolekulare Reaktion identisch. Wenn die allosterische Umwandlung $T_0 \rightleftharpoons R_0$ nicht hinreichend langsam sich abspielt, d. h. wenn die Bedingung $\tau_B \ll \tau_T$ nicht erfüllt ist, ändert der allosterische Prozeß R_0 während der Relaxation von Y oder F. Als Folge davon sind die Bindungsstellen in den Spezies R_i nicht mehr unabhängig und ununterscheidbar voneinander. Nun erscheint deshalb im allgemeinen ein kompliziertes Relaxationsspektrum mit ($n + 1$) Relaxationszeiten, das zweckmäßig durch mittlere Relaxationszeiten τ^* gekennzeichnet wird.

5.4.4. Die Bindung von Nikotinamid-adenin-dinucleotid an Hefe-D-Glyceraldehyd-3-phosphatdehydrogenase

Hefe-D-Glyceraldehyd-3-phosphatdehydrogenase (GAPDH) ist ein Protein mit dem Molekulargewicht 145000, das aus 4 identischen Untereinheiten aufgebaut ist. Die Zahl seiner Bindungsstellen für Nikotinamid-adenin-dinucleotid (NAD) ergab sich durch spektrophotometrische Ermittlung der Bindungsisothermen zu 4 (3,6 bis 4). Die Bindungsisotherme zeigt bei pH 8,5 und 40 °C sowie in der Umgebung dieser Werte einen schwach sigmoiden Verlauf [16]. Bei tieferen Temperaturen beobachtet man hyperbolische Kurven. Stört man den durch eine sigmoide Bindungsisotherme charakterisierten Gleichgewichtszustand des Systems bei pH 8,5 und 40 °C durch einen Temperatursprung, so beobachtet man drei auf der Zeitachse wohl separierte Relaxationsschritte, die durch die Relaxationszeiten $\tau_1 = 10^{-4}$ bis 10^{-3} sec, $\tau_2 = 10^{-3}$ bis 10^{-2} sec und $\tau_3 = 10^{-1}$ bis 10 sec gekennzeichnet sind. Die Relaxation läßt sich spektrophotometrisch verfolgen, weil das Differenzspektrum des GAPDH–NAD-Komplexes bei 366 mμ eine ausgeprägte Absorptionsbande aufweist. Die drei Relaxationszeiten hängen in charakteristischer Weise von der NAD-Konzentration ab. Der schnellste Prozeß mit der Relaxationszeit τ_1 läßt sich bei mittleren NAD-Konzentrationen durch das Zeitgesetz (133) befriedigend beschreiben[7]. Bei niedrigen NAD-Konzentrationen beobachtet man Abweichungen. Die Geschwindigkeit des langsamen Prozesses mit der Relaxationszeit τ_3 ist unabhängig von der Enzymkonzentration und fällt, wie nach Gleichung (134) zu erwarten, zunächst sehr rasch mit steigender NAD-Konzentration bis zu einem Plateauwert. Die Relaxationszeit τ_3 wird daher der allosterischen Konformationsumwandlung zugeordnet. Zur quantitativen Ermittlung von τ_3 wurde jedoch nicht Gl. (134) mit $n = 4$ herangezogen, sondern der ähnliche Ausdruck [16]

$$\tau_3^{-1} = k + k' \left[(1 + c\, a)/(1 + a) \right]^4 \,, \tag{135}$$

mit $c = 0,35$. Dieser folgt aus einem komplizierteren Reaktionsschema als in (118) angegeben, in dem insbesondere 4 identische Bindungsstellen für die T-Konformation zugelassen werden, deren Bindungsaffinität als verschieden von jener der R-Bindungsstellen angenommen wird. An dem Prozeß mit der Relaxationszeit τ_2 sind T-Bindungsstellen beteiligt, wie sich durch quantitative Anpassung der Meßpunkte an theoretische Kurven ergab.

Das in Gl. (118) angegebene allosterische Reaktionsmodell ist mit Sicherheit für eine adäquate Beschreibung des untersuchten Enzymsystems zu einfach. Gleichwohl lassen sich wesentliche Merkmale des Mechanismus der Enzym-Ligand-Wechselwirkung auf der Stufe dieses einfachen Modells qualitativ verstehen. Aus den vorgelegten Daten kann geschlossen werden, daß das Apoenzym in zwei und nur in zwei tetrameren Konformationen existiert. Die Dissoziationskonstanten der einzelnen Bindungsstellen sind unabhängig voneinander und identisch. Die Kooperativität zwischen den Untereinheiten des Enzyms scheint so hoch zu sein, daß an der Konformationsumwandlung ($T \rightleftharpoons R$) entweder alle oder keine der Untereinheiten teilnehmen. Es kann dies daraus geschlossen werden, daß 1. Relaxationsprozesse, an denen gemischte Bindungsstellen wie R_3T teilnehmen, nicht

---

[7] Dabei ist $\tau_1 = \tau_B$ zu setzen.

29*

beobachtet wurden, und 2. die gemessene Relaxationszeit τ_3 in der Weise von den Zustandsparametern abhängt, wie nach dem erweiterten allosterischen Modell Gl. (135) erwartet wird. Dies schließt nicht aus, daß genauere kinetische Untersuchungen eine weitere Verfeinerung des Modells notwendig machen.

Enzymreaktionen verlaufen nach Zeitgesetzen höher als 1. Ordnung. Der Aufklärung von Reaktionsmechanismen dient die direkte Beobachtung von intermediär, oft kurzzeitig auftretenden Spezies. Um solche Spezies in der Enzymchemie zu erkennen, muß man zu Enzymkonzentrationen übergehen, die etwa gleich hoch wie jene des Substrats, Aktivators oder Inhibitors sind. Dann reagieren die Spezies gewöhnlich sehr schnell. Überdies stellen Enzymreaktionen verwickelte Reaktionssysteme dar. Die speziellen Lösungen der zugehörigen formalen kinetischen Probleme unter den Bedingungen der Relaxation sind gewöhnlich leichter zugänglich als die allgemeinen Lösungen; in gewissen Fällen, wie bei kooperativen Konformationsumwandlungen, sind diese speziellen Lösungen die einzigen bisher erhaltenen. Relaxationsverfahren dürften mithin für die Untersuchung von Enzymreaktionen besonders geeignet sein.

Anhang:

Ein Beispiel für die mathematische Behandlung mehrstufiger Reaktionssysteme

Wir behandeln das Reaktionssystem

$$
\overset{(1)}{A} + \overset{(2)}{B} \underset{k_{21}}{\overset{k_{12}}{\rightleftharpoons}} A\,B \underset{k_{32}}{\overset{k_{23}}{\rightleftharpoons}} \overset{(3)}{C} \tag{A.1}
$$

als Beispiel, wie man bei der mathematischen Analyse eines Relaxationsproblems zweckmäßig vorgeht.

Die Abweichung der Momentankonzentration $c_i(t)$ des Reaktanten i von einer zeitunabhängigen Bezugskonzentration c_i^0 sei

$$
x_i(t) = c_i(t) - c_i^0; \tag{A.2a}
$$

überdies definieren wir die Größe $\bar{x}_i(t)$ durch

$$
\bar{x}_i(t) = \bar{c}_i(t) - c_i^0, \tag{A.2b}
$$

wobei $\bar{c}_i(t)$ die Konzentration des Reaktanten i in dem durch die augenblicklichen äußeren Bedingungen (Temperatur, Druck, etc.) festgelegten stationären Zustand ist; bei Relaxation gegen einen Gleichgewichtszustand ist $\bar{c}_i\,(t)$ die zu diesem gehörige Gleichgewichtskonzentration von i.

Bevor wir die Zeitgesetze aufstellen, eliminieren wir alle linearen Abhängigkeiten zwischen den Variablen $x_i\,(t)$ infolge der Massenerhaltung, um die Zahl der Variablen und damit der Zeitgesetze so weit wie möglich zu reduzieren. Für (A. 1) ergibt sich

$$x_1 + x_{AB} + x_2 = 0 \quad \text{oder} \quad x_{AB} = -(x_1 + x_2) \tag{A.3}$$

$$\text{mit } x_A = x_B = x_1 \quad \text{und} \quad x_C = x_2.$$

Das Reaktionssystem (A. 1) wird mithin durch zwei linear unabhängige Variablen, x_1 und x_2, vollständig charakterisiert. Zu (A. 3) analoge Beziehungen gelten für die $\bar{x}_i$. Wenn wir nicht alle linearen Abhängigkeiten beseitigen, erhalten wir bei der Lösung des Säkularproblems (A. 9) Eigenwerte λ_z, die identisch Null sind. Entsprechend (A. 22) sind die zugehörigen Normalvariablen y_z zeitunabhängig; dies heißt aber, daß diejenigen Variablen x_i, deren Linearkombination gleich einer Normalvariablen y_z ist, untereinander linear abhängig sind.

Die Zeitgesetze werden für chemische Relaxationsprozesse angesetzt. Letztere sollen durch derart kleine Störungen des Gleichgewichtszustandes oder eines anderen stationären Zustandes ausgelöst werden,

$$x_i,\ \bar{x}_i \ll c_i^0 \approx \bar{c}_i, \tag{A.4}$$

daß die in den Konzentrationsänderungen nicht-linearen Glieder in den Zeitgesetzen vernachlässigt werden können.

Für das Reaktionssystem (A. 1) gelten die Zeitgesetze

$$\frac{d\,c_A}{d\,t} = -\,k_{12}\,c_A\,c_B + k_{21}\,c_{AB}, \tag{A.5a}$$

$$\frac{d\,c_C}{d\,t} = -\,k_{32}\,c_C + k_{23}'\,c_{AB}. \tag{A.5b}$$

Sie lauten nach Linearisierung unter Beachten von $dx_i/dt = 0$ für $x_i = \bar{x}_i$

$$-\frac{d\,x_1}{d\,t} = -\dot{x}_1 = \alpha_{11}\,(x_1 - \bar{x}_1) + \alpha_{12}\,(x_2 - \bar{x}_2), \tag{A.6a}$$

$$-\frac{d\,x_2}{d\,t} = -\dot{x}_2 = \alpha_{21}\,(x_1 - \bar{x}_1) + \alpha_{22}\,(x_2 - \bar{x}_2) \tag{A.6b}$$

oder in Vektorschreibweise

$$-\dot{x} = A\,(x - \bar{x}), \tag{A.6c}$$

wo A die Matrix der α_{ij} ist; $(x - \bar{x})$ und $\dot{x}$ sind die aus den Elementen $(x_i - \bar{x}_i)$ bzw. $\dot{x}_i$ gebildeten Spaltenvektoren. Die Koeffizienten α_{ij} enthalten Geschwindigkeitskonstanten k_{ij} und die im allgemeinen zeitabhängigen Variablen $\bar{c}_i\,(t)$. Wegen (A. 4) ist die Abhängigkeit der α_{ij} von der Zeit ein Effekt 2. Ordnung, der vernachlässigt wird.

Bei Relaxationsexperimenten findet häufig die Relaxation gegen einen Gleichgewichtszustand statt. Wenn das Gesamtsystem zu einem Gleichgewichtszustand gelangt, stellt sich nach dem Prinzip der mikroskopischen Reversibilität auch für jeden Elementarprozeß ein Gleichgewichtszustand ein. Die Geschwindigkeiten von Hin- und Rückreaktion sind dann für jeden einzelnen Elementarprozeß gleich groß. Für stationäre Nicht-Gleichgewichtszustände ist das Prinzip der mikroskopischen Reversibilität nicht gültig. — Aus (A. 5) und (A. 6) folgt unmittelbar (der stationäre Zustand ist hier notwendig ein Gleichgewichtszustand)

$$
\begin{aligned}
\alpha_{11} &= [k_{12}\,(\bar{c}_A + \bar{c}_B) + k_{21}], \\
\alpha_{12} &= k_{21}, \\
\alpha_{21} &= k_{23}, \\
\alpha_{22} &= [k_{23} + k_{32}].
\end{aligned}
\tag{A.7}
$$

Wir lösen das System (A. 6) (A. 7) homogener linearer Differentialgleichungen 1. Ordnung, indem wir fordern, daß alle Variablen x_i (Abweichungen der Konzentrationen c_i von einem Bezugswert) nach dem gleichen Zeitgesetz mit der gleichen noch unbekannten Relaxationszeit $\tau = 1/\lambda$ sich ändern:

$$
-\dot{x}_1 = \lambda\,(x_1 - \bar{x}_1), \qquad -\dot{x}_2 = \lambda\,(x_2 - \bar{x}_2)
\tag{A.8a}
$$

oder in Vektorform

$$
-\dot{x} = \lambda\,(x - \bar{x}).
\tag{A.8b}
$$

Die Analogie zu dem Problem der Bestimmung der Normalschwingungen eines Moleküls, bei welchen alle Atome synchron mit der gleichen Frequenz schwingen, ist offenkundig.

Setzen wir (A. 8) in (A. 6) ein, so erhalten wir

$$
\begin{aligned}
(\alpha_{11} - \lambda)\,(x_1 - \bar{x}_1) + \alpha_{12}\,(x_2 - \bar{x}_2) &= 0 \\
\alpha_{21}\,(x_1 - \bar{x}_1) + (\alpha_{22} - \lambda)\,(x_2 - \bar{x}_2) &= 0
\end{aligned}
\tag{A.9a}
$$

oder

$$
(A - \lambda E)\,(x - \bar{x}) = 0,
\tag{A.9b}
$$

wobei E die Einheitsmatrix mit den Elementen δ_{ik} (mit $\delta_{ik} = 1$ für $i = k$ und $\delta_{ik} = 0$ für $i \neq k$) ist. Nach einem bekannten Satz der Algebra hat das lineare homogene Gleichungssystem (A. 9) nur dann von Null verschiedene Lösungen, wenn die Determinante der Koeffizienten verschwindet, d. h. wenn

$$
\begin{vmatrix} \alpha_{11} - \lambda & \alpha_{12} \\ \alpha_{21} & \alpha_{22} - \lambda \end{vmatrix} = |A - \lambda E| = 0.
\tag{A.10}
$$

Die zwei Wurzeln dieser sog. Säkulargleichung ergeben sich einfach zu

$$\lambda_{1,2} = \frac{1}{2}\,(\alpha_{11} + \alpha_{22}) \pm \sqrt{\frac{1}{4}\,(\alpha_{11} - \alpha_{22})^2 + \alpha_{12}\,\alpha_{21}}\quad. \qquad \text{(A.11)}$$

Nur für diese beiden Werte von λ ist (A. 9) lösbar. Man nennt λ_1 und λ_2 die Eigenwerte von A. Die beiden linear unabhängigen Lösungen von (A. 9), die Eigenvektoren

$$x^{(e)}_1 - \bar{x}^{(e)}_1 = \begin{pmatrix} x^{(e)}_{11} - \bar{x}^{(e)}_{11} \\ x^{(e)}_{21} - \bar{x}^{(e)}_{21} \end{pmatrix} \quad \text{zu } \lambda_1\,,$$
$$\qquad\qquad\qquad\qquad\qquad\qquad\qquad\qquad\qquad \text{(A.12)}$$
$$x^{(e)}_2 - \bar{x}^{(e)}_2 = \begin{pmatrix} x^{(e)}_{12} - \bar{x}^{(e)}_{12} \\ x^{(e)}_{22} - \bar{x}^{(e)}_{22} \end{pmatrix} \quad \text{zu } \lambda_2$$

sind, dem homogenen Charakter des Problems entsprechend, nur bis auf einen willkürlichen Faktor f_1 bzw. f_2 bestimmbar. Wir können z. B.

$$\begin{aligned} x^{(e)}_{11} - \bar{x}^{(e)}_{11} &= f_1 \\ x^{(e)}_{21} - \bar{x}^{(e)}_{21} &= f_1\,(\alpha_{12})^{-1}\,(\lambda_1 - \alpha_{11}) \end{aligned} \qquad \text{(A.13)}$$

etc. setzen. Über die Faktoren f_1, f_2 verfügen wir so, daß

$$\sum_i (x^{(e)}_{ij} - \bar{x}^{(e)}_{ij})^2 = 1 \qquad (i, j = 1, 2)\,.$$

Unter Berücksichtigung der Randbedingungen kann nun (A. 8) gelöst werden. Um die allgemeine Lösung zu erhalten, hat man die für die verschiedenen τ_k gewonnenen Partikularlösungen zu addieren. Wir erhalten z. B. für die Annäherung eines konstanten Gleichgewichtszustandes (Äquilibrierung nach einem Temperatursprung), den wir zweckmäßig als Bezugszustand festsetzen, d. h. mit $\bar{x}_i = 0$,

$$x_i(t) = \sum_k c_{ik}\,e^{-t/\tau_k}, \text{ für } t \geq 0 \qquad i = 1, 2;\; k = 1, 2, \qquad \text{(A.14)}$$

wobei $\sum_k c_{ik} = x_i(0)$. Wir stehen mithin vor der Aufgabe, den theoretischen Ausdruck (A. 14) mit den justierbaren Parametern c_{ik} und τ_k experimentellen Kurven $x_i(t)$ anzupassen. Wenn nur wenige Relaxationszeiten, wie in unserem Problem, auftreten, die um mindestens eine Größenordnung differieren, ist eine Lösung möglich. Die Relaxationsfunktionen bestehen dann aus überlagerten, aber hinreichend separierten Abfallkurven, wobei jeder nur eine Relaxationszeit zugehört.

Die Geschwindigkeitskonstanten k_{ik} erhält man aus den τ_k durch Variation der Gleichgewichtskonzentrationen der Reaktanten. In unserem Fall trägt man zweckmäßig $(1/\tau_1 + 1/\tau_2)$ sowie $(1/\tau_1 \cdot 1/\tau_2)$ gegen $(\bar{c}_A + \bar{c}_B)$ auf. Jeweils aus Steigung und Achsenabschnitt der beiden Geraden erhalten wir die vier Geschwindigkeitskonstanten des Systems (A. 1).

Besonders einfach lassen sich die Relaxationsgleichungen in den sog. Normalvariablen y ausdrücken. Diese sind durch [vgl. (18)]

$$x = My \text{ und } \bar{x} = M\bar{y} \tag{A.15}$$

festgelegt. M ist die Matrix der Eigenvektoren von A:

$$M = (x_1^{(e)} - \bar{x}_1^{(e)}, \, x_2^{(e)} - \bar{x}_2^{(e)}) \, . \tag{A.16}$$

Unter Beachtung von (A. 9) finden wir

$$AM = MD \tag{A.17}$$

mit der Diagonalmatrix D der Eigenwerte λ_i; in unserem Fall ist

$$D = \begin{pmatrix} \lambda_1 & 0 \\ 0 & \lambda_2 \end{pmatrix} . \tag{A.18}$$

Multiplizieren wir (A. 17) von rechts und links mit M^{-1}, so ergibt sich

$$M^{-1} A = D \, M^{-1} \, . \tag{A.19}$$

In Analogie zu dem Übergang von (A. 17) nach (A. 16) können wir in (A. 19) M^{-1} aus den nach

$$u_i^{(e)} A = \lambda_i u_i^{(e)} \quad \text{bzw.} \quad u \, (A - \lambda E) = 0 \tag{A.20}$$

definierten Linkseigenvektoren von A, den Zeilenvektoren $u_i^{(e)}$, bilden:

$$M^{-1} = \begin{pmatrix} u_1^{(e)} \\ u_2^{(e)} \end{pmatrix} . \tag{A.21}$$

Für den Fall einer reell symmetrischen Matrix $(A = \widetilde{A}, \, \alpha_{ik} = \alpha_{ki},$ wo $\widetilde{A}$ die zu A transponierte Matrix ist) sind die zu gleichen Eigenwerten gehörigen Rechts- und Linkseigenvektoren einander gleich. Dann ist die Transformation von A auf die Diagonalmatrix D eine Orthogonal- oder Hauptachsentransformation und M eine orthogonale Matrix $(M^{-1} = \widetilde{M})$. Vgl. Kapitel 3.2.2. Symmetrisierbarkeit von A bei Relaxation gegen einen Gleichgewichtszustand.

Aus (A. 6), (A. 15) und (A. 17) folgt, daß die Relaxationsgleichungen in Normalvariablen die einfache Form

$$-\dot{y} = D \, (y - \bar{y}) \tag{A.22}$$

oder in unserem Fall

$$-\dot{y}_1 = \lambda_1 \, (y_1 - \bar{y}_1),$$
$$-\dot{y}_2 = \lambda_2 \, (y_2 - \bar{y}_2)$$

mit $\lambda_i = 1/\tau_i$ annehmen; die Kopplung zwischen den Gleichungen ist aufgehoben.

Die allgemeine Lösung des Relaxationsproblems ist leicht zugänglich. Für einen beliebigen Satz von zeitabhängigen $\bar{y}_i \, (t)$ ergibt sich

$$y_i(t) = e^{-t/\tau_i} \left[\int \frac{1}{\tau_i} e^{t/\tau_i} \bar{y}_i(t) \, dt + c \right] ; \tag{A.23}$$

(s. z. B. E. MADELUNG, Die Mathematischen Hilfsmittel des Physikers, Springer Verlag, Berlin-Göttingen-Heidelberg, 7. Aufl. 1964, S. 269). Die Integrationskonstante c wird aus den Randbedingungen bestimmt. Mit Hilfe von (A. 15) führen wir den Übergang von den Normalvariablen y_i zu den meßbaren Konzentrationsänderungen x_i aus.

Die Verfasser sind Herrn Dr. G. Schwarz für die kritische Durchsicht des Manuskripts und für die Überlassung von Schriften vor ihrer Veröffentlichung dankbar. Herrn Dr. K. Kirschner danken wir für anregende Diskussionen, Herrn Dr. E. Holler für wertvolle Hilfe bei den Korrekturen.

Literatur

1. EIGEN, M., and L. DE MAEYER in: Technique of Organic Chemistry, Eds. FRIESS, S. L., E. S. LEWIS, and A. WEISSBERGER, 2nd Ed., Vol. 8, Part 2, p. 895. New York/London: Interscience 1963. S. a.: G. CZERLINSKI, Chemical Relaxation, M. Dekker, Inc., Newark, N. J., 1966.
2. MEIXNER, J.: Kolloid-Z. **134**, 3 (1953).
3. SCHWARZ, G.: Ber. Bunsenges. physik. Chem. **68**, 843 (1964); J. Mol. Biol. **11**, 64 (1965).
4. EIGEN, M.: Ber. Bunsenges. physik. Chem. **67**, 753 (1963); Pure and Applied Chem. **6**, 97 (1963).
5. EIGEN, M.: Angew. Chem. **75**, 489 (1963); EIGEN, M., W. KRUSE, G. MAASS, and L. DE MAEYER: Progress Reaction Kinetics **2**, 285 (1964).
6. DEBYE, P.: Trans. Electrochem. Soc. **82**, 265 (1942).
7. ZIMM, B. H., and J. K. BRAGG: J. Chem. Phys. **31**, 526 (1959).
8. ISING, E.: Z. Physik **31**, 253 (1925).
9. SCHWARZ, G. in: Physikertagung 1966 München, Plenarvorträge, Teil I, Stuttgart: B. G. Teubner 1967; Biopolymers **5**, 321 (1967).
10. KRAMERS, H. A., and G. H. WANNIER: Phys. Rev. **60**, 252, 263 (1941).
11. ENGEL, J.: Biopolymers **4**, 945 (1966).
12. ADAIR, G. S.: J. Biol. Chem. **63**, 529 (1925).
13. MONOD, J., J. WYMAN, and J.-P. CHANGEUX: J. Mol. Biol. **12**, 88 (1965).
14. KOSHLAND JR., D. E., G. NÉMETHY, and D. FILMER: Biochemistry **5**, 365 (1966).
15. SCHWARZ, G.: Rev. Mod. Phys., in press.
16. KIRSCHNER, K., M. EIGEN, R. BITTMAN, and B. VOIGT: Proc. Nat. Acad. Sci. (Washington) **56**, 1661 (1966).

M. Eigen
L. De Maeyer
Max-Planck-Institut
für physikalische Chemie
Göttingen

J. Heidberg
G. H. Kohlmaier
Institut für physikalische Chemie
der Universität Frankfurt

Sachverzeichnis

Absorption, *absorption*
  von Licht in der Blitzlichtphotolyse,
  of light in flash photolysis studies 253 bis
  256
  von Licht in Stoßwellen, *of light in
  shock waves* 225 f., 233 f., 238
Absorptionsspektrum (KMR), *absorption spectrum (NMR)* 360, 361, 385,
  386
Abstreifreaktion, *stripping reaction* 152,
  189–191, 195, 205–216
Additionsreaktionen, *addition reactions*
  263 f.
adiabatische(r), *adiabatic*
  Aufspaltung, *splitting* 61, 72
  Elektronen-Energie (Potentialflächen), *electronic energy (potential surfaces)* 43–45, 46, 60, 73, 74
  Halbdurchgang (KMR), *half-passage
  (NMR)* 404
  Kompression, *compression* 238
  Kopplung, *coupling 48*, 68
    schwache, *weak* 53–55
    starke, *strong* 50, 55–57, 67
  Näherung, *approximation* 43–45
  Potentialflächen, sich nicht kreuzende, *potential surfaces, non-crossing*
  48, *51*
  Potentialflächen, sich schneidende,
  potential surfaces, intersecting 75
  Prozesse, *processes* 349
  Reflexion, *reflection* 64
  Theorie der Linienform, *theory of
  line shape* 357
    bei Hyperfeinwechselwirkung,
    with hyperfine interaction 397,
    405
    bei Kern-Spin-Spin-Kopplung,
    with nuclear spin-spin coupling
    390
  (Elektronen)-Wellenfunktion *(electronic) wave function* 43–45, 68
Airy-Funktion, *Airy function 54*, 64, 65

Aktivierungsenergie, *activation energy*
  (Schwellenenergie), *(threshold energy)* 2, 11, 109, 157–158, 192, 228,
  235, 240, 274–275
  und Tunneln, *and tunnelling* 66
Aktivierungsenthalpie, *activation enthalpy* 274, 294–302
Aktivierungsentropie, *activation entropy*
  274, 297–302, 345
Aktivierungskomplex (s. Übergangszustand), *activated complex (see transition state)*
Aktivierungsvolumen(s), *activation volume* 297, 332–345
  Druckabhängigkeit des —, *pressure
  dependence of —* 334, 343
aliphatische(n) Moleküle(n), *aliphatic
  molecules*
  Substituenteneffekte und Reaktivität von —, *substituent effects and
  reactivity of —* 281–286
Alkali, *alkali*
  -halogenide, Potentialkurve der,
  halides, potential energy curve of
  211
  -Metallatome, Reaktionen der, *metal
  atoms, reactions of* 147–152, 157 f.,
  188–216
  -Metalle, Nachweis der, *metals, detection of* 170 f.
Alkyl, *alkyl*
  -halogenide, Reaktionen mit Alkalimetallen, *halides, reactions with alkali
  metals* 147–150, 188–191, *197–205*,
  216
Anharmonizität, Korrektur für die —,
  anharmonicity, correction for — 229
anionotropische Umlagerungen, *anionotropic rearrangements* 266
aromatisch(e, en), *aromatic*
  elektronisch angeregte — Moleküle,
  electronically excited — molecules 258 f.
  Substituenteneffekte und Reaktivi-

aromatisch(e, en)
 tät von — Molekülen, *substituent
 effects and reactivity of — molecules*
 277–281, 286
Arrhenius, *Arrhenius*
 -Aktivierungsenergie (s. a. Aktivierungsenergie), *activation energy (s. a.
 activation energy)*
  und Bindungs-Dissoziationsenergie, *and bond dissociation
  energy* 235 f.
 -Gleichung, *equation* 66, 252, 257,
 389
 -Parameter *parameters*, von Protonenübertragungsreaktionen, *of
  proton transfers* 393, 434
  einer Ringinversion, *of a ring
  inversion* 402
  bei unimolekularen Reaktionen,
  for unimolecular reactions 231 f.
Atom, *atom*
 -Abstandsänderungen, *change of—ic
 distances* 335 f.
 -Assoziation, *association* 37–39
 Quellen für —strahlen, *—ic beam
 sources* 162–167
 Reaktionen mit —en (s. a. Wasserstoffatome, Kaliumatome, Alkalimetallatome), *reactions with —s (s. a.
 hydrogen atoms, potassium atoms, alkali
 metal atoms)* 243–250
 -Rekombination, — *recombinations*
 244, 248, 256 f.
Austausch, *exchange* 360
 Druckabhängigkeit von —reaktionen, *pressure effect on — reactions*
 336
 intermolekularer —, *intermolecular —*
 375, 390–400
 intramolekularer —, *intramolecular—*
 374, 376–383, 401, 402
 langsamer — (KMR), *slow —
 (NMR)* 362, 392
 -Operator, *operator* 378
 -reaktion, *reaction* 19
 schneller — (KMR), *fast — (NMR)*
 362
 unimolekularer — (KMR), *unimolecular — (NMR)* 360

Bahndrehimpuls, *orbital angular momentum* 129

Bahnkurve, klassische, (Trajektorie),
 (s. a. Weg-Zeit-Kurve), *trajectory,
 classical* 45
Bewegungsgleichungen (s. a. Dichtematrix, Kernspindichtematrix),
 *equations of motion (s. a. density
 matrix, nuclear spin density matrix)*
 klassische —, *classical —* 124–133,
 145
bimolekulare Reaktionen, *bimolecular
 reactions*
 Berechnungen nach dem Monte-
 Carlo-Verfahren, *calculated by the
 Monte Carlo method* 144–152
 Beschreibung in verschiedenen Koordinaten, *description in different coordinate systems* 184–188
Bindung, *bond*
 —s-Dissoziationsenergie (s. Dissoziationsenergie), *dissociation energie (s.
 dissociation energy)*
  und Arrhenius-Aktivierungsenergie, *and Arrhenius activation
  energy* 235 f.
Blitzlicht, *flash*
 -Photographie, *photography* 253
 -Photolyse, *photolysis* 252–259
Blochsche Gleichungen, *Bloch equations*
 358
 erweitert für chemische Umlagerungen, *extended for chemical rearrangements* 360
Boltzmannfaktor, *Boltzmann factor* 373
Boltzmanngleichung (KMR), *Boltzmann equation (NMR)* 374
 generalisierte, *generalized* 382, 383
 eine Lösung der —, *a solution of —*
 384
 Umlagerungsterm (Reaktionsterm)
 der —, *rearrangement term (reaction term) of the —* 381–383
Born-Oppenheimer-Näherung (s. a.
 adiabatische Näherung), *Born-Oppenheimer approximation (s. a. adiabatic approximation)* 29, *131*
Breit-Wigner-Formel, *Breit-Wigner formula* 62
Brom-Molekül, *bromine molecule*
 Potentialfunktionen für —, *potential
 energy functions for —* 212
 Reaktion mit Kaliumatomen, *reaction with potassium atoms* 191, 205–216

Bromwasserstoff, Reaktion mit Kalium-atomen, *hydrogen bromide, reaction with potassium atoms* 152, 188–190, 192–197, 216
cis-Buten, *cis-butene* 124, 140–141

Carboniumion, Bildung von —, *carbonium ion, formation of* — 267f.
Chemilumineszenz, *chemiluminescence* 248 bis 250
  infrarote —, *infrared* — 151
  Nachweis von Atomen durch —, *measuring of atom concentrations by* — 244f.
chemische Aktivierung, *chemical activation* 121–124, 140–142
chemische Reaktivität, *chemical reactivity* 261–298, 290
chemische Relaxation (s. a. Relaxation), *chemical relaxation (s. a. relaxation)* 417–457
  theoretische Basis der —, *theoretical basis of* — 418–424, 452–456
chemische Verschiebung, *chemical shift* 362, 369
  Vertauschung der —, *exchange of* — 376
Coriolis-Kopplung, *Coriolis coupling* 76, 129
Cyan-Radikal, Lichtemission des —s, *cyan radical, light emission from* — 249
Cycloaddition, *cyclic addition* 293, 294, 305

Dämpfungsform der von Neumann-Gleichung, *damping form of the von Neumann equation* 88, 97
Dämpfungsterme, *damping terms* 358, *361*, 381f.
Defektproton, *defect proton* 433
Desaktivierung, *deactivation* 97, 106
  einstufige —, *single step* — 122–124
  komplexe —, *complex* — 136–138, 140–142
  mehrstufige —, *multistep* — 122–124
Detektoren bei Molekularstrahlver-suchen, *detectors in molecular beam studies* 165, 170–173
Deuterium-Isotopieeffekt, *deuterium isotope effect* 270f.
Diazoalkane, *diazoalkanes* 308–325
Dichte der Energiezustände, *density of energy states* 113, 137–138, 229
Dichtematrix (s. a. Kernspin-Dichte-matrix,), *density matrix (s. a. nuclear spin density matrix)* 78–84
  Bewegungsgleichung der —, *equation of motion of* — 79
  formale Störungsrechnung für —, *formal perturbation theory for* — 80–84
  vergröberte —, *coarse grained* — 85
  Wechselwirkungsdarstellung für —, *interaction representation for* — 80
Dielektrische Methode, *dielectric method* 427–428
Dielektrische Sättigung, *dielectric saturation* 351
diffusionskontrolliert (-bestimmt), *diffusion controlled* 334, 363
  Begegnung, Geschwindigkeit der —en, *encounter, velocity of* — 432
  Ligandenaustausch, *ligand exchange* 396
  Protolyse des Wassers, *protolysis of water* 430–434
  Trennung, Geschwindigkeit der —en, *separation, velocity of* — 432
Difluormonoxid, Dissoziation von —, *difluoromonoxide, dissociation of* — 230, 233
Dimerisierung, Druckeinfluß auf —, *dimerisation, effect of pressure on* — 336
Diracscher Spinaustauschoperator, *Dirac spin exchange operator* 378
Dispersionsspektrum (KMR), *dispersion spectrum (NMR)* 360, 385
Dissoziation, *dissociation*
  heterolytische —, *heterolytic* — 261
  homolytische —, *homolytic* — 261
  —senergie von Bindungen, *energy of bonds* 230, 255
  —skonstante von schwachen Säu-ren, *constants of weak acids* 278, 282 339
  —sreaktionen, Druckeffekte auf, *reactions, pressure effect on* 336
Distickstoffmonoxid, Zerfall von —, *nitrous oxide decomposition of* — 238 bis 240
Doppelresonanzmethode, *double resonance method* 405
Drehimpuls, *angular momentum* 125–126, 128–131
  Erhaltung des —es bei bimoleku-

Drehimpuls
laren Reaktionen, *conservation of —
in bimolecular reactions* 152, 178, 201,
203
-erhaltungssatz, *conservation of —*
124–126, 130–131
Druck, *pressure*
Einfluß des —es auf Reaktionen in
Lösung, *effect of — on reaction rates in
solution* 331–345
Düsenstrahl, *nozzle beam* 164–167
Durchgang, *passage*
schneller —, *fast* 403
langsamer —, *slow —* 360, 361, 386,
403
dynamischer ψ-Effekt, *dynamic-ψ-effect*
355
Dysonsche Integralgleichung, *Dyson
integral equation* 81, 82

elektrische Entladungen in Strömungssystemen, *discharge-flow system* 243 bis
250
elektrochemische Reaktionen, *electrochemical reactions* 348–356
Elektroden-Reaktionen, *electrode reactions* 348–356
Elektronenacceptoren, *electron acceptors*
277 f., 281
Elektronendonatoren, *electron donors*
277 f., 281
Elektronenspinresonanz und chemische
Prozesse, *electron spin resonance and
chemical processes* 405
Elektronenübertragung, *electron transfer*
Modell der — für bimolekulare
Gasreaktionen, *— model in bimolecular
gas phase reactions* 211–216
-sreaktionen, *reactions* 348–356, 357
elektronisch angeregte Moleküle, *electronically excited molecules* 244 f.,
248–250, 258 f.
elektrophile, *electrophilic*
Reaktionen, *reactions* 262–265
Substituenten (s. Elektronenacceptoren), *substituents (s. electron acceptors)*
Elektrostriktion des Lösungsmittels,
electrostriction of the solvent 276, 337 bis
344
Eliminierungsreaktionen, *elimination
reaction* 270–273

in der Gasphase, *in the gas phase* 271 f.
in Lösung, *in solution* 264 f., 269–273
E_1-Mechanismus, E_1-*mechanism* 270 bis
273
E_2-Mechanismus, E_2-*mechanism* 270 bis
273
Emission von Licht (s. a. Chemilumineszenz), *emission of light (s. a.
chemiluminescence)*
in Stoßwellen, *in shock waves* 225 f.,
234, 238
Energie, *energy*
-abhängigkeit des Ablenkungswinkels,—*dependence of deflection angle* 175,
206
-abhängigkeit von Reaktionsquerschnitten, *dependence of reaction cross
sections* 192 f., 200
-abhängigkeit der Reaktionswahrscheinlichkeit, *dependence of
reaction probabilities* 157, 201 f., 203
-abhängigkeit des unimolekularen
Zerfalls, *dependence of unimolecular
dissociation* 110 f., 153–155
-Austausch zwischen Oszillatoren,
exchange between oscillators 137–138,
140–142
Erhaltung der — und Produktanregung, *conservation of — and product
excitation* 201 f.
-erhaltungssatz, *conservation of —*
130–131
-hyperflächen (s. a. Hyperflächen,
Potentialflächen), *hypersurface (s. a.
hypersurface, potential energy surface)*
30–34, 37–41
innere — der Aktivierung, *internal
— of activation* 345
innere — der Produkte, *internal —
of products* 147 f., 150 f., 157, 160,
162, 173, 201 f., 208
Energieübertragung, *energy transfer*
intramolekulare —, *intramolecular —*
1, 156
statistische Theorie der —, *statistical
theory of —* 137–138, 140–142
bei Stößen, *collisional —* 112, 117 bis
142, 229, 236 f., 239
Theorie der —, *theory of —* 134–138,
140–142
translatorische —, *translational —* 1
Ensemble, *ensemble*

Ensemble
Mittelwert (der), *average (of)*
Gleichgewichts-Kernmagneti-
sierung, *equilibrium nuclear magnetization* 370
Gesamtintensität (KMR), *total
intensity (NMR)* 384
transversalen Kernmagnetisierung bezüglich rotierender Koordinaten, *transverse nuclear magnetization with respect to rotating
coordinates* 384, 385
Mittelwert einer physikalischen
Größe, *average of a physical quantity*
366
repräsentatives —, *representative —*
365
statistisches —, *statistical —* 381
Entartung, *degeneracy*
des Reaktionsweges, *reaction path —*
112f.
Enzym, *enzyme*
Reaktionen eines allosterischen —s,
reactions of an allosteric — 447–452
Beispiel: Bindung von NAD an
Hefe-GAPDH, *example: binding
of NAD to yeast-GAPDH* 450
bis 452
Gleichgewichtseigenschaften —,
equilibrium properties — 447
Relaxation, *relaxation* 448
Ester, *ester*
basische Verseifung von —n, *basic
saponification of —s* 282
säure katalysierte Hydrolyse von
—n, *acid catalysed hydrolysis of —s*
282, 284

Feldeffekt (s. induktiver Effekt),
field effect (s. inductive effect)
Fluoreszenz, *fluorescence*
aromatischer Moleküle, *of aromatic
molecules* 258f.
Fluoreszenzausbeute-Messungen, *fluorescence yield measurements* 418
Flusses, Erhaltung des, *flux conservation* 61
freie Enthalpie, *free energy (Gibbs)*
der Aktivierung, *of activation* 274,
290–329
lineare — Beziehungen bei organischen Reaktionen, *linear — relationships for organic reactions* 277–284,

287–289
Freiheitsgrade und unimolekulare Reaktionen, *degrees of freedom and unimolecular reactions* 89, 110

GAPDH   Hefe-D-Glyceraldehyd-3-
phosphatdehydrogenase, *GAPDH
yeast-D-glyceraldehyde-3-phosphate dehydrogenase* 450–452
Geschwindigkeit, *velocity*
—sselektion in Strahlen, *selection in
beams* 162f., 167–169, 196
—sverteilung der Produkte, *distribution of products* 195f., 207f.
Geschwindigkeitskonstante (s. Reaktionsgeschwindigkeitskonstante),
rate constant (s. reaction rate constant)
Gleichgewicht, *equilibrium*
dynamisches —, *dynamic —* 2
Prozesse im —, *processes in —* 2, 388
Störung des chemischen —s, *perturbation of chemical —* 389, 418
thermisches —, *thermal —* 368
Gleichung, *equation*
irreversible (kinetische) —, *irreversible (kinetic) —* 85, 95, 104
master —, *master —* 80, *85*, 95
reversible —, *reversible —* 80
Gloriolenstreuung, *glory scattering* 177
Greensche Funktion, *Green function* 83
Gruppenfrequenz, *group frequency* 298 bis
299

Halbwertsbreite KMR (s. a. Linienbreite), *half width NMR (s. a. line
width)* 360
Halogen, *halogen*
Reaktionen mit Alkali-Atomen,
reaction with alkali atoms, 188f., 191,
205–216
Reaktionen mit Wasserstoffatomen,
reaction with hydrogen atoms 150f.
Hamiltonsche Funktion, *Hamiltonian
function* 126–127
Hammett-Gleichung, *Hammett equation*
277–282, 287
Harpuniermechanismus, *harpooning mechanism* 211
Hauptachsentransformation, *principal
axis transformation* 128
Heisenbergdarstellung, *Heisenberg representation* 80

Helix-Knäuel-Umwandlung, *helix-coil-transition* 441, *445*

heterolytische Dissoziation, *heterolytic dissociation* 261

homolytische Dissoziation, *homolytic dissociation* 261

Hückel-Molekülorbital-(HMO)-Rechnung, *Hückel molecular orbital (HMO) calculation* 290–329

Hugoniot-Gleichung, *Hugoniot equation* 221

Hydrolyse (s. Solvolyse, Esterhydrolyse), *hydrolysis (s. solvolysis, ester hydrolysis)*

Hydroxyl-Radikal, Reaktion von —en, *hydroxyl radical, reaction of —s* 246 f.

Hyperflächen, Energie — (s. a. Potentialflächen), *hypersurface, energy — (s. a. potential surfaces)* 30—34, 37—41

Hyperkonjugation, *hyperconjugation*
Effekt der — auf die Reaktivitäten, *effect of — on reactivities* 285 f.

Impulserhaltungssatz, *conservation of linear momentum* 124–125, 130

Impulsmethoden (Sprung-), *pulse methods (jump-)*
chemische Relaxation —, *chemical relaxation —* 424–426
KMR (Spin-Echo) —, *NMR (spin echo) —* 405

induktiver Effekt und Reaktivität, *induktive effect and reactivity 275 f.*, 277–280, 282, 284–289

Infrarot, *infrared*
Chemilumineszenz im —, *chemiluminescence* 151

Intramolekulare Relaxation (s. a. Theorie unimolekularer Reaktionen, *intramolecular relaxation (s. a. unimolecular reaction rate theory)* 84–94

Inversion, *inversion*
von Ammoniakderivaten, *of ammonia derivates* 400
von Ringmolekülen, *of ring molecules* 402

Ionen, *ion*
-Molekül-Reaktion, *-molecule reaction* 12
-Neutralisierung und Molekularstrahltechnik, *neutralization and beam techniques* 165 f.

Ising Modell, lineares, *Ising model, linear* 435

isokinetische Beziehung, *isokinetic relation* 327

Isomerisierung, unimolekulare —, *isomerization, unimolecular —* 109–116

Isotopenaustauschreaktion, *isotopic exchange reaction* 353 f.

Isotopieeffekt (s. Deuteriumisotopieeffekt), *isotope effect (s. deuterium isotope effekt)*

Jod, Rekombination von — atomen, *iodine atom recombination* 256 f.

Kaliumatome, *potassium atoms*
Reaktion von —n mit Alkylhalogeniden, *reaction with alkyl halides* 147–150
Reaktion von —n mit Brom, *reaction with bromine* 205–216
Reaktion von —n mit Bromwasserstoff, *reaction with hydrogen bromide* 152, 188–190, 192–197

Kassel-Theorie, *Kassel theory* 122

kationotrope Umlagerungen, *cationotropic rearrangements* 266

Kegel-Schnittlinien (punkte) von Potentialflächen, *cone intersections of potential surfaces* 73

Kernspin-Dichtematrix (s. a. Dichtematrix), *nuclear spin density matrix (s. a. density matrix)* 365–374
Anwendungen bei Spin Umlagerungen, *applications in spin rearrangements* 400
im thermischen Gleichgewicht, *in thermal equilibrium* 368, 370
Wechselwirkungsdarstellung, *interaction representation* 371
zeitabhängige Störungsrechung, *time-dependent perturbation theory* 370
zeitliche Veränderung (Bewegungsgleichung) der —, *time dependence (equation of motion) of the —* 368

Kettenreaktionen, Elementarschritte bei—, *chain reactions, elementary steps in —* 246–248

kinematische Kopplung, *kinematic coupling*
von Elektronen und Kernen, *of electrons and nuclei* 45

Kinetik, *kinetics*

Kinetik
kooperativer Umwandlungen, *of cooperative transitions* 442
von Prozessen im Gleichgewicht, *of processes in equilibrium* 388
schneller Reaktionen in Lösung, *of fast reactions in solution* 417
kinetische Energie, *kinetic energy* 127, 130–131
kinetische Gleichung (s. Gleichung), *kinetic equation (s. equation)*
Klassifizierung chemischer Umlagerungen, *classification of chemical rearrangements* 374
organischer Reagenzien und Reaktionen, *of organic reagents and reactions* 261–266
Koaleszenz von Resonanzlinien, *coalescence (collapse) of resonance lines* 362, 387, 403
Kohlendioxid, Dissoziation von —, *carbon dioxide, decomposition of* — 232–235
Kompressibilität von gelösten Molekülen, *compressibility of solute molecules* 333 f., 344
Konfiguration, *configuration*
Änderung der — bei SN_2-Reaktionen, *change of — in SN_2-reactions* 269
Konformationumwandlungen, *conformation transitions*
eines allosterischen Enzyms, *of an allosteric enzyme* 447
von Ammoniakderivaten, *of ammonia derivates* 400
von Anilinderivaten, *of aniline derivatives* 376
von Cyclohexanabkömmlingen, *of cyclohexane derivates* 402
Elementarprozeß von kooperativen —, *elementary process of cooperative* — 442, 446
kooperative —, *cooperative* — 435
von Peptiden, *of peptides* 445
Konjugationsenergie, *Conjugation energy* 326
Konkurrenz-Methoden in der Kinetik schneller Reaktionen, *competition methods in fast reaction kinetics* 418
Konkurrenzreaktionen, *competition reactions* 121–122, 302–305
konservative Systeme, *conservative systems* 125–126
Kontinuummodell für Lösungsmittel, *continuum model for solvents* 338–344, 352, 433
kooperative Umwandlungen, *cooperative transitions*
Elementarprozeß von —, *elementary process of* — 442, 446
von Hefe-D-Glyceraldehyd-3-phosphatdehydrogenase, *of yeast-D-glyceraldehyde-3-phosphate dehydrogenase* 451
Helix-Knäuel-Umwandlung, *helix-coil-transition* 441, 445
Keimbildungsschritte bei —, *nucleation steps in* — 436, 443
von kettenförmigen Biopolymeren, *of linear biopolymers* 435, 442, 446
Kinetik, *kinetics* 442–446
mittlere Relaxationszeiten von —, *mean relaxation times of* — 444
Thermodynamik auf der Grundlage des linearen Ising Modells, *thermodynamics on the basis of the linear Ising model* 435–442
Wachstumsschritte von —, *growth steps of* — 436, 442
Koordinaten, *coordinates*
generalisierte —, *generalized* — 126 bis 127
kartesische —, *cartesian* — 126–127
nicht-kartesische —, *non-cartesian* — 126–127
krummlinige —, *curvilinear* — 23
—-systeme zur Beschreibung bimolekularer Reaktionen, *— systems for describing bimolecular reactions* 184–188
Kopplung (s. a. adiabatische, nicht-adiabatische), *coupling (s. a. adiabatic, non-adiabatic)*
Molekül an Thermostaten, *molecule to thermostat* 95–106
Korrelationsfunktion, *correlation function* 6
Kreuzungspunkt (s. Potentiale), *crossing point (s. potentials)*
Kreuzungsverbot für adiabatische Terme gleicher Symmetrie, *non-crossing rule for adiabatic terms of same symmetry* 51

Laborsystem, Beziehung zum Massenschwerpunktssystem, *laboratory*

Laborsystem
coordinate system, in relation to center of mass coordinate system 130, 160, *184–188*

Lagrangesche Funktion, *Lagrangian function* 126

Landau-Teller-Theorie, *Landau-Teller theory* 134–138

Landau-Zener-Formel, *Landau-Zener formula* 51
Anwendung bei starker Kopplung, *application in strong coupling* 66
Herleitung der —, *derivation of —* 49–52, 58–59

Landau-Zener-Modell, *Landau-Zener model*
Anwendung auf Elementarprozesse, *application to elementary processes* 75–76
Grenzen der Anwendbarkeit des —s, *limitations of the —* 52
lineares —, *linear —* 43–77
von Systemen mit vielen Zuständen, *of multistate systems* 67–70
zweidimensionale Erweiterung des —s, *two-dimensional extension of —* 70–75
Zwei-Zustand-Näherung eines eindimensionalen linearen —s, *two-state approximation of a one-dimensional linear —* 45–49, 53

Lebensdauer (s. a. Verweilzeit; unimolekulare Reaktionen), *life time (s. a. life time (residence time); unimolecular reactions)*
eines H_3^+O-Ions, *of H_3^+O-ion* 434

Lewis-Base, Reaktionen mit —n, *Lewis-base, reactions with —s* 262, 267, 270f.

Lewis-Säure, Reaktionen mit —n, *Lewis-acid, reactions with —s* 262, 313–315

Ligandenaustausch (s. a. Metallkomplexreaktionen), *ligand exchange (s. a. metal complex reactions)*
von Atomen mit elektrischem Kernquadrupolmoment, *of atoms with electrical nuclear quadrupole moment* 394
in paramagnetischen Komplexen, *in paramagnetic complexes* 397

Ligandenfeldstabilisierung, *ligand field stabilisation* 429

linearisierte Geschwindigkeitsgleichungen, *linearized rate equations* 419, *453*

Linienbreite (s. a. Halbwertsbreite) und Bestimmung von Reaktionsgeschwindigkeiten, *line width (s. a. half width) and determination of reaction rates* 360, 395, 399
bei langsamem Austausch, *with slow exchange* 362, 392
bei schnellem Austausch, *with fast exchange* 362

Linienform von KMR-Linien, *line shape of NMR lines*
Lorentzsche —, *Lorentzian —* 360

Liouville-Gleichung, quantenmechanische (s. Dichtematrix, Bewegungsgleichung), *Liouville equation, quantum mechanical (s. density matrix, equation of motion)*

Lippmann-Schwinger-Differentialgleichung, *Lippmann-Schwinger differential equation* 82, 83, 102
Chew-Goldberger-Lösung der —, *Chew-Goldberger solution of —* 82

Lösungsmittel, *solvent*
-einfluß auf organische Reaktionen, *effect on organic reaction rates* 276, 287f., 336, *338–344*
Kontinuummodell für das —, *continuum model for the —* 338–344

Lokalisierungsenergie, *localization energy* 292, 325

makroskopische Kernmagnetisierung, *macroscopic nuclear magnetization*
im thermischen Gleichgewicht, *in thermal equilibrium* 370
transversale Komponente in rotierenden Koordinaten, *transverse component in rotating coordinates* 361, 384, 385, 397

Massenschwerpunktssystem, Transformation vom — ins Laborsystem, *center of mass coordinate system, transformation from — to the laboratory coordinate system* 130, 159f., *184–188*

Massey-Parameter, *Massey parameter* 45, 74

Matrixmethode, *matrix method*
zur Berechnung der Zustandssumme von linearen Polymeren, *for the calculation of the partition function of linear polymers* 437–441

Maxwell-Boltzmann-Verteilungsfunktion, *Maxwell-Boltzmann distribution function* 10, 14

Mechanik der Stöße, *mechanics of collisions* 124–133

Mechanismus, *mechanism*
äußerer — (Stoßwechselwirkung), *external — (collision interaction)* 84, 95
innerer — (Energieumverteilung), *internal — (energy redistribution)* 84, 95
organischer Reaktionen, *of organic reactions* 261–273

Mehrquantenübergang, *multi-quantum transition* 135

Menschutkin-Reaktionen, *Menschutkin reactions* 335, 337, 340–343

mesomerer Effekt und Reaktivität, *mesomeric effect and reactivity 275f.*, 279–282, 284–289

Metallkomplexreaktionen (s. a. Ligandenaustausch), *metal complex reactions (s. a. ligand exchange)* 428–430

mikroskopische Reversibilität, Prinzip der —, *microscopic reversibility, principle of —* 363, *421*, 454

Modell der harten Kugeln, *hard sphere model* 11

Molekülstruktur und Reaktivität, *molecular structure and reactivity* 261–289

Molekularstrahl, *molecular beam*
Interpretation von —versuchen durch Monte-Carlo-Rechnungen, *studies, interpretation by Monte-Carlo calculations* 147–150, 203–205
—versuche, *studies* 157–173, 188–216

Molvolumen, partielles —, *molar volume, partial —* 333, 341

Monte-Carlo-Berechnungen, *Monte-Carlo calculations* 143–156, 203–205

Multiplizität, *multiplicity*
Änderung der — bei Reaktionen, *change of — in reactions* 235, 238f.

Nachschwingungen, *wiggles* 403

NAD (Nikotinamid-adenin-dinucleotid), *NAD (nicotinamide-adenine dinucleotide)* 450–452

von Neumann-Gleichung (s. Dichtematrix, Bewegungsgleichung), *von Neumann equation (s. density matrix, equation of motion)*

Newtonsche Gesetze, *Newton's laws* 124–125

nicht-adiabatisch(e, er, es), *non-adiabatic*
Durchgang durch einen Potentialwall, *transmission through a potential barrier* 59–64
Kopplung, *coupling* 43–45, 48, 56, 59, 63, 64, 66, 73, 74
Kopplungsparameter, *coupling parameter* 50
(nichtsäkulare) Linienverbreiterung (KMR), *(non-secular) line broadening (NMR)* 391
Reflexion von einem Potentialwall, *reflection from a potential barrier* 59–64
Relaxation der Molekülschwingungen, *relaxation, vibrational* 76
Tunneln, *tunnelling* 63
Übergänge, *transitions* 43–76
Wellenfunktion, *wave function* 47, 53, 68

Nichtgleichgewichts-, *nonequilibrium*
-bedingungen bei unimolekularen Reaktionen, *conditions in unimolecular reactions* 228
-kinetik, *kinetics* 1, 18
-magnetisierung, molekulare Übertragung von, *magnetization, molecular transfer of* 404
Relaxation gegen stationären —zustand, *relaxation towards stationary — state* 421

Normalvariable, *normal variables* 421–422, 455
thermodynamische Reaktionseffekte bezüglich —, *thermodynamic reaction quantities with respect to —* 424, 449

nukleophile, *nucleophilic*
Reaktionen, *reactions* 262–271
Substituenten (s. Elektronendonatoren), *substituents (s. electron donors)*

Olefine, *olefins* 308–325

orientierte(n) Moleküle(n), Stöße mit — *oriented molecules, collisions with —* 170, 199f.

Oszillator, anharmonischer, *oscillator, anharmonic* 135

Oxidation, *oxidation* 348–356

Pauli-Gleichung (master-), *Pauli equation (master-)* 95

Photolyse (s. Blitzlichtphotolyse), *photolysis (s. flash photolysis)*
π-Bindungsordnung, *π-bond order* 290
π-Elektronendichte, *π-electron density* 290
π-Elektronensysteme, *π-electron systems* 290–329
Plasma, *plasma* 15, 18
polarer Effekt (s. induktiver Effekt), *polar effect (s. inductive effect)*
Potential (s. a. adiabatisches P.), *potential (s. a. adiabatic p.)*
effektives —, *effective —* 178
-flächen (s. a. Potentialfunktion und Hyperflächen), *surfaces (s. a. potential energy function and hyper surfaces)*, 249 f., 348 f.
-funktion, *energy function* 146 f., 150 f., 155, 204 f., 211–213, 216
für Alkalihalogenide, *for alkali halides* 211
für Brommoleküle, *for bromine molecules* 212
intermolekulares —, *intermolecular —* 173 f., 180, 182, 193
sich kreuzende —e nullter Ordnung, *crossing zero order —s* 48, 49, 54, 69
Prädissoziation, *predissociation* 256
Präexponentieller Faktor (s. a. Arrhenius-Parameter), *pre-exponential factor (s. a. Arrhenius parameters)* 23, 228 f.
Protolyse (s. Protonenübertragung), *protolysis (s. proton transfer)*
Protonenübertragung (Protolyse), *proton transfer (protolysis)*
in reinem Wasser, *in pure water* 430–434
in wäßrigen Lösungen von Ammoniumionen, *in aqueous solutions of ammonium ions* 392
prototrope Umlagerungen, *prototropic rearrangements* 266
Pseudokreuzung, *pseudo crossing*
Energie, *energy* 65
Linie, *line* 75
von Termen (adiabatischen), *of terms (adiabatic)* 49, 53, 54, 56, 58, 68, 69

quadratische Form, *quadratic form* 127
quantenmechanische Behandlung, *quantum mechanical treatment*

der elastischen Streuung, *of elastic scattering* 179 f.
der Elementarreaktionen, *of elementary processes* 26–42
quantenstatistische Beschreibung, *quantum statistical description*
chemischer Reaktionen, *of chemical reactions* 78–108
eines Ensembles von reagierenden Kernspins, *of an ensemble of reacting nuclear spins* 364–388

Radikal, *radical*
-reaktionen, *reactions* 253, 257–259, 263–266
Spektroskopie von —en, *spectroscopy of free —s* 255 f.
Rayleigh-Gleichung, *Rayleigh equation* 221
Razemisierung, *racemization* 269
Reaktionen (s. a. bimolekulare R., unimolekulare R.), *reactions (s. a. bimolecular r., unimolecular r.)*
einstufige —, single step — 419
mehrstufige —, *multi-step* — 421, *452*
periodische —, *periodic* — 421
SN_1—, SN_1— 267–269, 429
SN_2—, SN_2— 267–269, 429
trimolekulare —, *trimolecular*—256 f.
Reaktionsabstand, *reaction distance* 335 f., 433
Reaktionsgeschwindigkeit, *reaction rate* 1–2
im Hochdruckbereich, *in the high pressure region* 94, 104, 105, 114
Reaktionsgeschwindigkeitskonstante, *reaction rate constants* 1–22
der diffusionsbestimmten Begegnung, *of the diffusion controlled encounter* 432
der diffusionsbestimmten Trennung, *of the diffusion controlled separation* 432
und die Lebensdauer von Spin-Zuständen, *and life time of spin states* 389, 393, 395, 401
und Relaxationszeiten bei chemischer Relaxation, *and relaxation times in chemical relaxation* 422, 455
schneller Reaktionen in Lösung, *of fast reactions in solution* 417
spezifische —, *specific* — 109–112
Reaktionsgruppe, *reaction group* 26

Reaktionsquerschnitt, *reaction cross section* 1–22, 149, 180f.
   differentieller —, *differential —* 1–22, 149, 162
   totaler —, *total* 1–22, *149f.*, 157, *180f.*, 190f., 194f., 199, 204f., 210, 213
Reaktionsterm in der Boltzmanngleichung (s. Umlagerungsterm), *reaction term in the Boltzmann equation (s. rearrangement term)*
Reaktionswahrscheinlichkeit, *reaction probability* 1–22, 149, 159, 180–183, 194, 198f., 209
Reaktionsweise, *reaction, mode of* 292 bis 329
Reaktivität, *reactivity*
   und Molekülstruktur, *and molecular structure* 261–289
Reduktion, *reduction* 345, 348–356
Reflexionskoeffizient, *reflection coefficient* 61
Regenbogenstreuung, *rainbow scattering* 177, 179, 207
Relaxation, *relaxation*
   chemische —, *chemical —* 417–457
   einstufige —, *single step —* 419
   elektronische —, *electronic —* 7
   Geschwindigkeits —, *velocity —* 7
   gegen einen Gleichgewichtszustand, *towards equilibrium* 421, 454
   bei innermolekularer Energieübertragung, *in intramolecular energy transfer* 156
   von Kernen mit elektrischem Quadrupolmoment, *of nuclei with electric quadrupole moment* 392, 394
   mehrstufige —, *multi-step —* 421, 452
   rotatorische —, *rotational —* 7
   Schwingungs —, *vibrational —* 7, 118–121, 227, 258
   Spektren, quasikontinuierliche, *spectra, quasi-continuous* 422
   Spin- —, *spin —* 361, 374, 384
   Spin- — in paramagnetischen Lösungen, *spin — in paramagnetic solutions* 397
   gegen einen stationären Nicht-Gleichgewichtszustand, *towards a stationary non-equilibrium state* 421
   thermische —, *thermal —* 121–124
   -sverfahren, *techniques* 424–428

Relaxationszeit (s. a. Zeit), *relaxation time (s. a. time)*
   in allosterischen Enzymsystemen, *in allosteric enzyme systems* 448
   der Bindung von NAD an GAPDH, *of binding of NAD to GAPDH* 450–452
   bei chemischer Relaxation, *in chemical relaxation* 419–420, 424, *454*
   kooperativer Umwandlungen, *of cooperative transitions* 444
   longitudinale — von Spins, *longitudinal — of spins* 359
   mehratomiger Moleküle, *of polyatomic molecules* 120–121
   Meßbarkeit bei chemischer Relaxation, *measurability in chemical relaxation* 389, 448
   mittlere —, *mean —* 422–424
   transversale — der intramolekularen Relaxation, *transverse — of intramolecular relaxation* 89
   transversale — von Spins (s. a. Linienbreite), *transverse — of spins, (s. a. line width)* 359, 360, 403
   zweiatomiger Moleküle, *of diatomic molecules* 119–120
Renormalisierung von Energieniveaus durch Störung, *renormalization of energy levels by perturbation* 92, 102
Resonanzübergang, *resonance transition* 136
Rice-Ramsperger-Kassel-Marcus, *Rice-Ramsperger-Kassel-Marcus*
   -Theorie unimolekularer Reaktionen, *theory of unimolecular reactions* 109 bis 116, 112, 154f.
Riemann-Lebesgue-Theorem, *Riemann-Lebesgue theorem* 92, 98, 101
Rotation, *rotation(al)*
   Anregung der — bei reaktiver Streuung, *excitation of — in reactive scattering* 183, 209f., 215
   Freiheitsgrade der —, Beteiligung bei unimolekularen Reaktionen, — *degrees of freedom, participation in unimolecular reactions* 110, 114, 155, 229
   innere — von Anilinderivaten, *internal — of aniline derivatives* 376
   Zustände der —, Selektion in der Molekularstrahltechnik,

Rotation
selection of — quantum states in beam studies 168
rotierend(es, en) Koordinatensystem, *rotating coordinate system* 360
Lösung der Boltzmanngleichung im —, *solution of the Boltzmann equation in —* 385
in der quantenmechanischen Spintheorie, *in quantum-mechanical spin theory* 406–413
Transformation zum —, *transformation to —* 412
Rückstoß-Reaktion, *rebound-reaction* 149, 188–190, *192–205*

Sauerstoffatome, Reaktionen von —n, *oxygen atoms, reactions of —* 244–247, 258
Schallfrequenz, *sound frequency* 119
Schrödingerdarstellung, *Schrödinger representation* 80
Schrödingergleichung, *Schrödinger equation*
der Kernbewegung, *of nuclear motion* 31
zeitabhängige —, *time dependent —* 27–31, 79, 368
zeitabhängige — als reversible Gleichung, *time-dependent — as reversible equation* 80
Schwefeldioxid, *sulfur dioxide*
elektronisch angeregtes —, *electronically excited —* 249f.
Zerfall von —, *decomposition of —* 236f.
Schwellenenergie (s. Aktivierungsenergie), *threshold energy (s. activation energy)*
Schwingung, *vibration*
-sanregung von Produkten, *—al excitation of products* 150f., 173, 183, 209f., 215, 258
Selektion, *selection*
von Quantenzuständen und Geschwindigkeiten in Strahlen, *of quantum states and velocities in beams* 162f., 167–170
SN$_1$-Mechanismus, *SN$_1$-mechanism* 267–269, 429
SN$_2$-Mechanismus, *SN$_2$-mechanism* 267–269, 429
Solvatation, *solvation* 301, 350–352, 354f.

Solvolysen, *solvolytic reactions* 267f., 279f., 282, 284f., 339, 344f.
Spin-Bahn-Kopplung, *spin-orbital coupling*
und nicht-adiabatische Übergänge, *and non-adiabatic transitions* 43, 71, 73–76
Spin-Echo-Methode, *spin-echo method* 405
Spin-Spin-Kopplung (KMR), *spin spin coupling* (NMR) 358, 369, 375, 390
spinverbotene Reaktionen, *spin-forbidden reactions* 239
Sprungmethoden (s. Impulsmethoden), *jump methods (s. pulse methods)*
stationäre Methoden, *stationary methods*
bei chemischer Relaxation, *in chemical relaxation* 426–428
bei KMR, M. des langsamen Durchgangs, *in NMR, slow passage in 360, 361, 386*
Grenzen der Anwendbarkeit, *limitations* 403
statistische, *statistical*
Matrix (s. Dichtematrix), *matrix (s. density matrix)*
Systeme, *systems* 78, 365, 381
Theorie chemischer Reaktionen, *theory of chemical reactions* 78, 202f.
—r Mittelwert (s. a. Ensemble-Mittelwert), *average (s. a. ensemble average)* 79
Stereochemie von Eliminierungen, *stereochemistry of eliminations* 270–273
sterisch, *steric*
—e Effekte bei organischen Reaktionen, *effects in organic reactions* 268, 275f., 281f., 284–289, 335, 337f.
—er Faktor bei Stößen in der Gasphase, *factor in gas phase collisions* 200f.
Stickstoff, *nitrogen*
Lichtemission von —dioxid, *light emission from — dioxide* 248
Reaktion von —atomen, *reactions of — atoms* 244f., 248f.
Stickstoffmonoxid, Reaktionen mit —, *nitric oxide, reactions with —* 244
Störung des chemischen Gleichgewichts, *perturbation of chemical equilibrium* 418
kooperativer Umwandlungen, *of cooperative transitions* 443

Störung des chemischen Gleichgewichts
periodische (oder stationäre) Methoden, *periodical (or stationary) methods* 426–428
Sprung- und Impulsmethoden, *jump- and pulse methods* 424–426
Stoß (s. a. Streuung), *collision (s. a. scattering, impact)*
elastischer—, *elastic—* 7, 132
-Kinematik, *— kinematics* 184–188
-Komplex (s. a. Übergangszustand), *— complex (s. a. transition state)* 137–138, 201f.
mit orientierten Molekülen, *with oriented molecules* 170, 199f.
reaktiver —, *reactive —* 7
Übergangswahrscheinlichkeit, *transition probability* 119–141
Übergangswahrscheinlichkeit beim harmonischen Oszillator, *transition probability for a harmonic oscillator* 120
unelastischer —, *inelastic* 7, 131–142
-Term, *term in Boltzmann equation* 103
Stoßtheorie, *impact theory* 174–183
stoßartige Wechselwirkung, *collision interaction* 84, 95, 96
Stoßparameter, *impact parameter* 146, 175
und mittlere Übergangswahrscheinlichkeit, *and mean transition probability* 65
Stoßwelle, *shock wave* 219–227
reflektierte —, *reflected —* 221–224
Stoßwellenrohr, *shock tube* 222–227
Strahlung (s. Chemilumineszenz), *radiation (s. chemiluminescence)*
Streuung, *scattering*
elastische —, *elastic —* 158–161, 174–183, 193f., 206f.
—smatrix $t^{\oplus}$, *matrix* $t^{\oplus}$ *83*, 84, 102, 104 reaktive —, *reactive —* 149, 158–161, 180–183, 194f., 208f.
—squerschnitt (s. Wirkungsquerschnitt), *cross section (see cross section)*
Theorie, quantenmechanische, *theory, quantum mechanical* 80, 179f.
Stufenleitermodell, *stepladder model*
der Energieübertragung, *of energy transfer* 123–124, 140–141, 229
Substituenten, Einfluß der — auf die Reaktivität, *substituents, influence of — on reactivity*, 268, 273–289, 292–329, 339

Substitutionsreaktionen, *Substitution reactions* 262f., 266–269

Taft-Gleichung, *Taft-equation* 282–284
thermodynamische Reaktionseffekte (Spinaustausch), *thermodynamic reaction quantities (spin exchange)* 375, 389
bezüglich Normalvariablen, *with respect to normal variables* 424, 448 bis 450
Trägheitsmoment, *moment of inertia* 128–129
Trägheitsprodukt, *product of inertia* 128
Trägheitstensor, *tensor of inertia* 128–129
Transmission, nicht-adiabatische (s. a. Tunneln) durch, über, unter einem Potentialwall, *transmission, non-adiabatic (s. a. tunnelling) through, over, under a potential barrier* 59–64
Transmissionskoeffizient, *transmission coefficient* 61, 63, 64
Tunneln, *tunnelling* 55, 59, 63, 66, 67
adiabatisches —, *adiabatic —* 63–64
nicht-adiabatisches —, *non-adiabatic* 63–64

Übergang, *transition*
über den Potentialwall, *overbarrier* 66
Übergangswahrscheinlichkeit (s. a. Landau-Zener-Formel, Verweilzeit), *transition probability (s. a. Landau-Zener formula, life time)* 239f., 248, 363, *383*
bei adiabatischer Kopplung, *at adiabatic coupling* 54, 55
mittlere —, *mean —* 64–67
mittlere zweidimensionale —, *mean two-dimensional —* 72
nicht-adiabatische —, *non-adiabatic —* 55, 74, 75
Temperatur, bei einer bestimmten, *temperature, for a given* 135
Übergangszustand, *transition state* 110, 117, 267, 341–343
Lebensdauer des —s, *lifetime of —* 149f., 200f., 210
von Prozessen in KMR, *of processes in NMR* 361, 378, 383
Theorie des —s, *theory*, 115, 274, 331f., 350
Überschußproton, *excess proton* 433

Ultraschall, *ultrasonics* 118–121

Umkehrpunkt, *turning point* 52–59, 70, 76

Umlagerung von Spins, *rearrangement of spins*
  adiabatische Theorie, *adiabatic theories* 357
  asymmetrische —, *asymmetrical —* 374, 383
  intermolekulare —, *intermolecular —* 375, 390–400
  intramolekulare —, *intramolecular —* 374
  monogene —, *monogeneous —* 375
  polygene —, *polygeneous —* 375
  quantenstatistische Behandlung (nichtadiabatische Theorie), *quantum statistical treatment (non-adiabatic theory)* 376, 391
  symmetrische —, *symmetrical —* 374, 376–383

Umlagerungen (s. a. Isomerisierung), *rearrangements (s. a. isomerizations)*
  in Lösungen, *in solutions* 265f.
  cis-trans- —der Peptidgruppe, *cis-trans— of the peptide group* 446

Umlaufen in Streuversuchen, *orbiting in scattering experiments* 178

Umwandlung, innere — von Stickstoffdioxid, *conversion, internal — of nitric dioxide* 248f.

unimolekulare Reaktionen, *unimolecular reactions*, 227–240
  Berechnungen nach dem Monte-Carlo-Verfahren, *calculated by the Monte Carlo method* 153–156
  in Flüssigkeiten, *in liquids* 106, 267, 270
  Theorie (s. a. Kassel-Theorie, Rice- Ramsperger - Kassel - Marcus - Theorie), *theory (s. a. Kassel-Theorie, Rice-Ramsperger-Kassel-Marcus-Theorie)* 95, 109–116, 227–229

Verbrennungsreaktionen, *combustion reactions* 258

Verteilung (s. a. Energieverteilung, Winkelverteilung), *distribution (s. a. energy distribution, angular distribution)*
  —sfunktion unimolekularer Zerfallszeiten, *function, for unimolecular dissociation times* 153–156

Verweilzeit, mittlere, *life time (residence time), mean* 360, 362, 387
  Geschwindigkeitskonstanten und—, *rate constants and —* 389, 393, 395, 401

Vibration-Translationsübergang, *vibration-translation transition* 134–138

Vibration-Vibrationsübergang, *vibration-vibration transition* 134–138

Wahrscheinlichkeit, *probability*
  eines Zustands (Realisierungs-), *of a state (— of realization)* 78, 367

Wasser, *water*
  Protolyse in —, *protolysis in —* 430 bis 434
  Zerfall von —, *decomposition of —* 235f.

Wasserstoff, *hydrogen*
  Konzentrationsmessung von —atomen, *concentration measurement of — atoms* 244
  -atome, Reaktionen mit Halogenen, *atoms, reactions with halogens* 150f.
  -atome, Reaktionen mit Wasserstoffmolekülen, *atoms, reactions with hydrogen molecules* 143, 152, 191
  -Sauerstoff-Reaktion, *-oxygen reaction* 246–248

Wasserstoffbrücke, *hydrogen bond*
  Beweglichkeit längs —, *mobility along —* 433–434
  Geschwindigkeit der Bildung in Peptiden, *rate of formation in peptides* 445

Wegzeitkurve, *trajectory* 45, 143–147, 203

Winkelgeschwindigkeitsvektor, *angular velocity vector* 128–129

Winkelverteilung der Produkte, *angular distribution of products* 148f., 152, 157, 190–192, 197f., 201, 205f.

Wirkungsquerschnitt, (s. a. Reaktionsquerschnitt), *cross section (s. a. reaction cross section)* 84, 103
  differentieller —, *differential —* 3–18, 159–161, 176f., 179
  totaler —, *total —* 1–22, 161

Zeit, *time*
  atomare —, *atomic* 89, 97
  transversale —, *transversal —* 89, 97

Zentralfeldtheorie der Stöße, *central field impact theory* 174–183, 193, 201, 203
Zentralkräfte, *central forces* 125
Zentrifugalpotential, *centrifugal potential* 110, 114, 178
Zerfall, *dissociation*
    Energieabhängigkeit des unimolelaren —s, *energy dependence of unimolecular —* 110f., 153–155
    Geschwindigkeit des unimolekularen —s, *rate of unimolecular —* 109 bis 116, 153–155
Zimm-Bragg-Modell für kooperative Umwandlungen, *Zimm-Bragg-model for cooperative transitions* 441
    erweitertes Modell, *extended model* 435–442

Zufälligkeit, *random*
    der Lebensdauer angeregter Moleküle, *lifetime assumption for energized molecules* 154, 156
Zufallszahl, *random number* 146
Zustand, *state*
    gemischter —, *mixed —* 365
    makroskopischer (thermodynamischer) —, *macroscopic (thermodynamic) —* 365, 381
    reiner —, *pure —* 78, *364*
Zustandssumme (Verteilungsfunktion), *partition function (state sum)*
    in der Kernspindichtematrix, *in nuclear spin density matrix* 368
    von kettenförmigen Polymeren, *of linear polymers* 437

Subject Index

absorption, *Absorption*
of light in flash photolysis studies, *von Licht in der Blitzlichtphotolyse* 253–256
of light in shock waves, *von Licht in Stoßwellen* 225f., 233f., 238

absorption spectrum (NMR), *Absorptionsspektrum (KMR)* 360, 361, 385, 386

activated complex (see transition state), *Aktivierungskomplex (s. Übergangszustand)*

activation energy, *Aktivierungsenergie* (threshold energy), *(Schwellenenergie)* 2, 11, 109, 157–158, 192, 228, 235, 240, 274–275
and tunnelling, *und Tunneln* 66

activation enthalpy, *Aktivierungsenthalpie* 274, 294–302

activation entropy, *Aktivierungsentropie* 274, 297–302, 345

activation volume, *Aktivierungsvolumen* 297, 332–345
pressure dependence of—, *Druckabhängigkeit des —s* 334, 343

addition reactions, *Additionsreaktionen* 263f.

adiabatic, *adiabatische(r)*
approximation, *Näherung* 43–45
compression, *Kompression* 238
coupling, *Kopplung* 48, 68
strong, *starke* 50, 55–57, 67
weak, *schwache* 53–55
electronic energy (potential surfaces), *Elektronen-Energie (Potentialflächen)* 43–45, 46, 60, 73, 74
half-passage (NMR), *Halbdurchgang (KMR)* 404
potential surfaces, intersecting, *Potentialflächen, sich schneidende* 75
potential surfaces, non-crossing, *Potentialflächen, sich nicht kreuzende* 48, 51

processes, *Prozesse* 349
reflection, *Reflexion* 64
splitting, *Aufspaltung* 61, 72
theory of line shape, *Theorie der Linienform* 357
with hyperfine interaction, *bei Hyperfeinwechselwirkung* 397, 405
with nuclear spin-spin coupling, *bei Kern-Spin-Spin-Kopplung* 390
(electronic) wave function, *(Elektronen)-Wellenfunktion* 43–45, 68

Airy function, *Airy-Funktion* 54, 64, 65

aliphatic molecules, *aliphatische(n) Moleküle(n)*
Substituent effects and reactivity of—, *Substituenteneffekte und Reaktivität von —* 281–286

alkali, *Alkali*
halides, potential energy curve of, *halogenide, Potentialkurve der* 211
metal atoms, reactions of, *– Metallatome, Reaktionen der* 147–152, 157f., 188–216
metals, detection of, *-Metalle, Nachweis der* 170f.

alkyl, *Alkyl*
halides, reactions with alkali metals, *-halogenide, Reaktionen mit Alkalimetallen* 147–150, 188–191, *197–205*, 216

angular distribution of products, *Winkelverteilung der Produkte 148f.*, 152, 157, 190–192, 197f., 201, 205f.

angular momentum, *Drehimpuls* 125 to 126, 128–131
conservation of—, *-erhaltungssatz* 124–126, 130–131
conservation of — in bimolecular reactions, *Erhaltung des — es bei bi-*

angular momentum
molekularen Reaktionen 152, 178, 201, 203
angular velocity vector, *Winkelgeschwindigkeitsfaktor* 128–129
anharmonicity, *Anharmonizität*
correction for, *Korrektur für die* 229
anionotropic rearrangements, *anionotropische Umlagerungen* 266
aromatic, *aromatisch(e, en)*
electronically excited — molecules, *elektronisch angeregte —e Moleküle* 258 f.
substituent effects and reactivity of — molecules, *Substituenteneffekte und Reaktivität von — Molekülen* 277–281, 286
Arrhenius, *Arrhenius*
activation energy, (s. a. activation energy), *Aktivierungsenergie (s.a. Aktivierungsenergie)*
and bond dissociation energy, *und Bindungs-Dissoziationsenergie* 235 f.
equation, *Gleichung* 66, 252, 257, 389
parameters, *-Parameter*
of proton transfers, *von Protonenübertragungsreaktionen* 393, 434
of a ring inversion, *einer Ringinversion* 402
for unimolecular reactions, *bei unimolekularen Reaktionen* 231 f.
atom, *Atom*
association, *-Assoziation* 37–39
reactions with —s (s. a. hydrogen atoms, potassium atoms, alkali metal atoms), *Reaktionen mit — en (s. a. Wasserstoffatome, Kaliumatome, Alkalimetallatome)* 243–250
recombinations, *Rekombination* 244, 248, 256 f.
— ic beam sources, *Quellen für — strahlen* 162–167
change of —ic distances, *Abstandsänderungen* 335 f.

bimolecular reactions, *bimolekulare Reaktionen*
calculated by the Monte Carlo method, *Berechnungen nach dem Monte Carlo Verfahren* 144–152

description in different coordinate systems, *Beschreibung in verschiedenen Koordinaten* 184–188
Bloch equations, *Blochsche Gleichungen* 358
extended for chemical rearrangements, *erweitert für chemische Umlagerungen* 360
Boltzmann equation (NMR), *Boltzmanngleichung (KMR)* 374
generalized, *generalisierte* 382, 383
rearrangement term (reaction term) of the —, *Umlagerungsterm (Reaktionsterm) der —* 381–383
a solution of, *eine Lösung der* 384
Boltzmann factor, *Boltzmannfaktor* 373
bond, *Bindung*
dissociation energy (s. dissociation energy), *—s-Dissoziationsenergie (s. Dissoziationsenergie)*
and Arrhenius activation energy, *und Arrhenius-Aktivierungsenergie* 235 f.
Born-Oppenheimer approximation (s. a. adiabatic approximation), *Born-Oppenheimer-Näherung (s. a. adiabatische Näherung)* 29, *131*
Breit-Wigner formula, *Breit-Wigner-Formel* 62
bromine molecule, *Brom-Molekül*
potential energy function for —, *Potentialfunktionen für —* 212
reaction with potassium atoms, *Reaktion mit Kaliumatomen* 191, 205–216
cis-butene, *cis-Buten* 124, 140–141

carbon dioxide, decomposition of —, *Kohlendioxid, Dissoziation von —* 232–235
carbonium ion, formation of —, *Carboniumion, Bildung von —* 267 f.
cationotropic rearrangements, *kationotrope Umlagerungen* 266
center of mass coordinate system, transformation from — to the laboratory coordinate system, *Massenschwerpunktssystem, Transformation vom — ins Laborsystem* 130, 159 f., *184–188*
central field impact theory, *Zentralfeldtheorie der Stöße 174–183*, 193, 201, 203

central forces, *Zentralkräfte* 125

centrifugal potential, *Zentrifugalpotential* 110, 114, 178

chain reactions, elementary steps in —, *Kettenreaktionen, Elementarschritte bei* — 246–248

chemical activation, *chemische Aktivierung* 121–124, 140–142

chemical reactivity, *chemische Reaktivität* 261–298, 290

chemical relaxation (s. a. relaxation), *chemische Relaxation (s. a. Relaxation)* 417–457
   theoretical basis of —, *theoretische Basis der* — 418–424, 452–456

chemical shift, *chemische Verschiebung* 362, 369
   exchange of —, *Vertauschung der* — 376

chemiluminescence, *Chemilumineszenz* 248–250
   infrared —, *infrarote* — 151
   measuring of atom concentrations by —, *Nachweis von Atomen durch* —244f.

classification, *Klassifizierung*
   of chemical rearrangements, *chemischer Umlagerungen* 374
   of organic reagents and reactions, *organischer Reagenzien und Reaktionen* 261–266

coalescence (collapse) of resonance lines, *Koaleszenz von Resonanzlinien* 362, 387, 403

collision (s. a. scattering, impact), *Stoß (s. a. Streuung)*
   complex (s. a. transition state), *-Komplex (s. Übergangszustand)* 137 to 138, 201f.
   elastic, *elastischer* 7, 132
   inelastic, *unelastisch* 7, 131–142
   kinematics, —*kinematik* 184–188
   with oriented molecules, *mit orientierten Molekülen* 170, 199f.
   reactive —, *reaktiver* — 7

collision, *Stoß*
   term in Boltzmann equation, *Term in Boltzmanngleichung* 103
   transition probability, *Übergangswahrscheinlichkeit* 119–141
   transition probability for a harmonic oscillater, *Übergangswahrscheinlichkeit beim harmonischen Oszillator* 120

collision interaction, *stoßartige Wechselwirkung* 84, 95, 96

combustion reactions, *Verbrennungsreaktionen* 258

competition methods in fast reaction kinetics, *Konkurrenz-Methoden in der Kinetik schneller Reaktionen* 418

competition reactions, *Konkurrenzreaktionen* 121–122, 302–305

compressibility of solute molecules, *Kompressibilität von gelösten Molekülen* 333f., 344

cone intersections of potential surfaces, *Kegel-Schnittlinien (punkte) von Potentialflächen* 73

configuration, *Konfiguration*
   change of — in SN_2-reactions, *Änderungen der* — *bei SN_2-Reaktionen* 269

conformation transitions, *Konformationumwandlungen*
   of an allosteric enzyme, *eines allosterischen Enzyms* 447
   of ammonia derivates, *von Ammoniakderivaten* 400
   of aniline derivatives, *von Anilinderivaten* 376
   cooperative —, *kooperative* — 435
   of cyclohexane derivates, *von Cyclohexanabkömmlingen* 402
   elementary process of cooperative—, *Elementarprozeß von kooperativen* — 442, 446
   of peptides, *von Peptiden* 445

Conjugation energy, *Konjugationsenergie* 326

conservation of linear momentum, *Impulserhaltungssatz* 124–125, 130

conservative systems, *konservative Systeme* 125–126

continuum model for solvents, *Kontinuummodell für Lösungsmittel* 338–344 352, 433

conversion, internal — of nitric dioxide, *Umwandlung, innere* — *von Stickstoffdioxid* 248f.

cooperative transitions, *kooperative Umwandlungen*
   elementary process of —, *Elementarprozeß von* — 442, 446
   growth steps of—, *Wachstumsschritte von* — 436, 442
   helix-coil-transition, *Helix-Knäuel-*

cooperative transition
  Umwandlung 441, 445
  kinetics, *Kinetik* 442—446
  of linear biopolymers, *von kettenförmigen Biopolymeren* 435, 442, 446
  mean relaxation times of —, *mittlere
  Relaxationszeiten von* — 444
  nucleation steps in —, *Keimbildungsschritte bei* — 436, 443
  thermodynamics on the basis of the
  linear Ising model, *Thermodynamik
  auf der Grundlage des linearen Ising
  Modells* 435–442
  of yeast-D-glyceraldehyde-3-phos-
  phate dehydrogenase, *von Hefe-D-
  Glyceraldehyd-3-phosphatdehydrogenase*
  451
coordinates, *Koordinaten*
  cartesian —, *karterische* — 126–127
  curvilinear —, *krummlinige* — 23
  generalized —, *generalisierte* — 126 bis
  127
  non-cartesian —, *nicht-kartesische*
  126–127
  systems for describing bimolecular
  reactions, —*systeme zur Beschreibung
  bimolekularer Reaktionen* 184–188
Coriolis coupling, *Coriolis-Kopplung* 76,
  129
Correlation function, *Korrelationsfunktion* 6
coupling (s. a. adiabatic, non-adiabatic),
  *Kopplung (s. a. adiabatische, nicht-
  adiabatische)*
  molecule to thermostat, *Molekül an
  Thermostaten* 95–106
cross section (s. a. reaction cross section), *Wirkungsquerschnitt (s. a. Reaktionsquerschnitt)* 84, 103
  differential —, *differentieller* — 3–18,
  159–161, 176f., 179
  total —, *totaler* — 1–22, 161
crossing point (s. potentials), *Kreuzungspunkt (s. Potentiale)*
cyano radical, light emission from —,
  Cyan-Radikal, Lichtemission des —s 249
cyclic addition, *Cycloaddition* 293, 294,
  305

damping form of the von Neumann
  equation, *Dämpfungsform der von
  Neumann-Gleichung* 88, 97

damping terms, *Dämpfungsterme* 358,
  361, 381 f.
defect proton, *Defektproton* 433
degeneracy, *Entartung*
  reaction path —, *des Reaktionsweges*
  112f.
degrees of freedom and unimolecular
  reactions, *Freiheitsgrade und unimolekulare Reaktionen* 89, 110
density of energy states, *Dichte der
  Energiezustände* 113–229, 137–138
density matrix (s. a. nuclear spin density matrix), *Dichtematrix (s. a.
  Kernspin-Dichtematrix)* 78–84
  coarse grained —, *vergröberte* — 85
  equation of motion of —, *Bewegungsgleichung der* — 79
  interaction representation for —,
  Wechselwirkungsdarstellung für — 80
  formal perturbation theory for —,
  formale Störungsrechnung für — 80–84
deactivation, *Desaktivierung* 97, 106
  complex —, *komplexe* — 136–138,
  140–142
  multistep —, *mehrstufige* — 122–124
  single step —, *einstufige* — 122–124
detectors in molecular beam studies,
  *Detektoren bei Molekularstrahlver-
  suchen* 165, 170–173
deuterium isotope effect, *Deuterium-
  Isotopieeffekt* 270f.
diazoalkanes, *Diazoalkane* 308–325
dielectric method, *Dielektrische Methode*
  427–428
dielectric saturation, *Dielektrische Sättigung* 351
diffusion controlled, *diffusionskontrolliert
  (-bestimmt)* 334, 363
  encounter, velocity of, *Begegnung,
  Geschwindigkeit der* 432
  ligand exchange, *Ligandenaustausch*
  396
  protolysis of water, *Protolyse des
  Wassers* 430–434
  separation, velocity of, *Trennung,
  Geschwindigkeit der* 432
difluoromonoxide, dissociation of —,
  Difluormonoxid, Dissoziation von —
  230, 233
dimerisation, effect of pressure on —,
  Dimerisierung, Druckeinfluß auf —
  336

Dirac spin exchange operator, *Dirac-scher Spinaustauschoperator* 378

discharge-flow System, *elektrische Entladungen in Strömungssystemen* 243–250

dispersion spectrum (NMR), *Dispersionsspektrum (KMR)* 360, 385

dissociation, *Zerfall*
energy dependence of unimolecular —, *Energieabhängigkeit des unimolekularen —s* 110f., 153–155
rate of unimolecular —, *Geschwindigkeit des unimolekularen —s* 109–116, 153–155

dissociation, *Dissoziation*
constants of weak acids, *—skonstante von schwachen Säuren* 278, 282, 339
energy of bonds, *—senergie von Bindungen* 230, 255
heterolytic —, *heterolytische —* 261
homolytic —, *homolytische —* 261
reactions pressure effect on, *—sreaktionen, Druckeffekte auf* 336

distribution (s. a. energy distribution angular distribution), *Verteilung (s. a. Energieverteilung, Winkelverteilung)*
function, for unimolecular dissociation times, *—sfunktion unimolekularer Zerfallszeiten* 153–156

double resonance method, *Doppelresonanzmethode* 405

dynamic-ψ-effect, *dynamischer ψ-Effekt*, 355

Dyson integral equation, *Dysonsche Integralgleichung* 81, 82

electrochemical reactions, *elektrochemische Reaktionen* 348–356

electrode reactions, *Elektroden-Reaktionen* 348–356

electron acceptors, *Elektronenacceptoren* 277f., 281

electron donors, *Elektronendonatoren* 277f., 281

electron spin resonance and chemical processes, *Elektronenspinresonanz und chemische Prozesse* 405

electron transfer, *Elektronenübertragung*
model in bimolecular gas phase reactions, *Modell der — für bimolekulare Gasreaktionen* 211–216
reactions, *—sreaktionen* 348–356, 357

electronically excited molecules, *elektronisch angeregte Moleküle* 244f., 248–250, 258f.

electrophilic, *elektrophile*
reactions, *Reaktionen* 262–265
substituents (s. electron acceptors), *Substituenten (s. Elektronenacceptoren)*

electrostriction of the solvent, *Elektrostriktion des Lösungsmittels* 276, 337–344

elimination reaction, *Eliminierungsreaktionen* 270–273
in the gas phase, *in der Gasphase* 271f.
in solution, *in Lösung* 264f., 269–273

E_1-mechanism, *E_1-Mechanismus* 270 to 273

E_2-mechanism, *E_2-Mechanismus* 270–273

emission of light (s. a. chemiluminescence), *Emission von Licht (s. a. Chemilumineszenz)*
in shock waves, *in Stoßwellen* 225f., 234, 238

energy, *Energie*
conservation of —, *—erhaltungssatz* 130–131
conservation of — and product excitation, *Erhaltung der — und Produktanregung* 201f.
dependence of deflection angle, *—abhängigkeit des Ablenkungswinkels* 175, 206
dependence of reaction cross sections, *—abhängigkeit von Reaktionsquerschnitten* 192f., 200
dependence of reaction probabilities, *der Reaktionswahrscheinlichkeit* 157, 201f., 203
dependence of unimolecular dissociation, *—abhängigkeit des unimolekularen Zerfalls* 110f., 153–155
exchange between oscillators, *-Austausch zwischen Oszillatoren* 137–138, 140–142
hypersurface (s. a. hypersurface, potential energy surface), *—hyperflächen (s. a. Hyperflächen, Potentialflächen)* 30–34, 37–41
internal — of activation, *innere — der Aktivierung* 345
internal — of products, *innere — der Produkte* 147f., 150f., 157, 160, 162, 173, 201f., 208

energy transfer, *Energieübertragung*
  collisional —, *bei Stößen* 112, 117 to 142, 229, 236f., 239
  intramolecular —, *intramolekulare* — 1, 156
  statistical theory of —, *statistische Theorie der* — 137–138, 140–142
  theory of —, *Theorie der* — 134–138, 140–142
  translational, *translatorische* — 1
ensemble, *Ensemble*
  average (of), *Mittelwert (der)*
    — equilibrium nuclear magnetization, *Gleichgewichts-Kernmagnetisierung* 370
      total intensity (NMR), *Gesamtintensität (KMR)* 384
      transverse nuclear magnetization with respect to rotating coordinates, *transversalen Kernmagnetisierung bezüglich rotierender Koordinaten* 384, 385
  average of a physical quantity, *Mittelwert einer physikalischen Größe* 366
  representative —, *repräsentatives* — 365
  statistical —, *statistisches* — 381
enzyme, *Enzym*
  reactions of an allosteric —, *Reaktionen eines allosterischen* —s 447–452
    equilibrium properties —, *Gleichgewichtseigenschaften* — 447
    example: binding of NAD to yeast-GAPDH, *Beispiel: Bindung von NAD an Hefe-GAPDH* 450–452
    relaxation, *Relaxation* 448
equation, *Gleichung*
  irreversible (kinetic) —, *irreversible (kinetische)* — 85, 95, 104
  master —, *master* — 80, 85, 95
  reversible —, *reversible* — 80
equations of motion, (s. a. density matrix, nuclear spin density matrix), *Bewegungsgleichungen, (s. a. Dichtematrix, Kernspindichtematrix)*
  classical —, *klassische* — 124–133, 145
equilibrium, *Gleichgewicht*
  dynamic —, *dynamisches* — 2
  perturbation of chemical —, *Störung des chemischen* — 389, 418

processes in —, *Prozesse im* — 2, 388
  thermal — *thermisches* — 368
ester, *Ester*
  acid catalysed hydrolysis of —s, *säure-katalysierte Hydrolyse von* —n 282, 284
  basic saponification of —s, *basische Verseifung von* —n 282
excess proton, *Überschußproton* 433
exchange, *Austausch* 360
  fast — (NMR), *schneller* — *(KMR)* 362
  intermolecular —, *intermolekularer* — 375, 390–400
  intramolecular —, *intramolekularer* — 374, 376–383, 401, 402
  operator, -*Operator* 378
  pressure effect on — reactions, *Druckabhängigkeit von* — *reaktionen* 336
  reaction, —*reaktion* 19
  slow — (NMR), *langsamer* — *(KMR)* 362, 392
  unimolecular — (NMR), *unimolekularer* — *(KMR)* 360

field effect (s. inductive effect), *Feldeffekt (s. induktiver Effekt)*
flash, *Blitzlicht*
  photography, -*Photographie* 253
  photolysis, -*Photolyse* 252–259
fluorescence, *Fluoreszenz*
  of aromatic molecules, *aromatischer Moleküle* 258f.
fluorescence yield measurements, *Fluoreszenzausbeute-Messungen* 418
flux conservation, *Flusses, Erhaltung des* 61
free energy (Gibbs), *freie Enthalpie*
  of activation, *der Aktivierung* 274, 290–329
  linear — relationships for organic reactions, *lineare* — *Beziehungen bei organischen Reaktionen* 277–284, 287–289

GAPDH yeast-D-glyceraldehyde-3-phosphate dehydrogenase, *GAPDH Hefe-D-Glyceraldehyd-3-phosphatdehydrogenase* 450–452
glory scattering, *Gloriolenstreuung* 177
Green function, *Greensche Funktion* 83

group frequency, *Gruppenfrequenz* 298 to 299

half width NMR (s. a. line width), *Halbwertsbreite KMR (s. a. Linienbreite)* 360
halogen, *Halogen*
    reaction with alkali atoms, *Reaktionen mit Alkali-Atomen* 188f., 191, *205–216*
    reaction with hydrogenatoms, *Reaktionen mit Wasserstoffatomen* 150f.
Hamiltonian function, *Hamiltonsche Funktion* 126–127
Hammett equation, *Hammett-Gleichung* 277–282, 287
hard sphere model, *Modell der harten Kugeln* 11
harpooning mechanism, *Harpuniermechanismus* 211
Heisenberg representation, *Heisenbergdarstellung* 80
helix-coil-transition, *Helix-Knäuel-Umwandlung* 441, *445*
heterolytic dissociation, *heterolytische Dissoziation* 261
homolytic dissociation, *homolytische Dissoziation* 261
Hückel molecular orbital (HMO) calculation, *Hückel-Molekülorbital-(HMO)-Rechnung* 290–329
Hugoniot equation, *Hugoniot-Gleichung* 221
hydrogen, *Wasserstoff*
    atoms, reactions with halogens, *—atome, Reaktionen mit Halogenen* 150f.
    atoms, reactions with hydrogen molecules, *—atome, Reaktionen mit Wasserstoffmolekülen* 143, 152, 191
    concentration measurement of —, atoms, *Konzentrationsmessung von —atome* 244
    –oxygen reaction, *–Sauerstoff-Reaktion* 246–248
hydrogen bond, *Wasserstoffbrücke*
    mobility along —, *Beweglichkeit längs —* 433–434
    rate of formation in peptides, *Geschwindigkeit der Bildung in Peptiden* 445
hydrogen bromide, reaction with pot-

assium atoms, *Bromwasserstoff, Reaktion mit Kaliumatomen* 152, 188 to 190, 192–197, 216
hydrolysis (s. Solvolysis, ester hydrolysis), *Hydrolyse (s. Solvolyse, Esterhydrolyse)*
hydroxyl radical, reaction of —s, *Hydroxyl-Radikal, Reaktion von —en* 246f.
hyperconjugation, *Hyperkonjugation*
    effect of — on reactivities, *Effekt der — auf die Reaktivitäten* 285f.
hypersurface, energy — (s. a. potential surfaces), *Hyperflächen, Energie — (s. a. Potentialflächen)* 30–34, 37–41

impact parameter, *Stoßparameter* 146, 175
    and mean transition probability, *und mittlere Übergangswahrscheinlichkeit* 65
impact theory, *Stoßtheorie* 174–183
induktive effect and reactivity, *induktiver Effekt und Reaktivität* 275f., 277–280, 282, 284–289
infrared, *Infrarot*
    chemiluminescene, *Chemilumineszenz im —* 151
intramolecular relaxation (s. a. unimolecular reaction rate theory), *Intramolekulare Relaxation (s. a. Theorie unimolekularer Reaktionen)* 84–94
inversion, *Inversion*
    of ammonia derivatives, *von Ammoniakderivaten* 400
    of ring molecules, *von Ringmolekülen* 402
iodine atom recombination, *Jod, Rekombination von — atomen* 256f.
ion, *Ionen*
    -molecule reaction, *-Molekül-Reaktion* 12
    neutralization and beam techniques, *-Neutralisierung und Molekularstrahltechnik* 165f.
Ising model, linear, *Ising Modell, lineares* 435
isokinetic relation, *isokinetische Beziehung* 327
isomerization, unimolecular —, *Isomerisierung, unimolekulare —* 109–116
isotope effect (s. deuterium isotope

isotope effect
  effect), *Isotopieeffekt (s. Deuterium-isotopieeffekt)*
isotopic exchange reaction, *Isotopen-austauschreaktion* 353f.

jump methods (s. pulse methods), *Sprungmethoden (s. Impulsmethoden)*

Kassel theory, *Kassel-Theory* 122
kinematic coupling, *kinematische Kopplung*
  of electrons and nuclei, *von Elektronen und Kernen* 45
kinetic energy, *kinetische Energie* 127, 130–131
kinetic equation (s. equation), *kinetische Gleichung (s. Gleichung)*
kinetics, *Kinetik*
  of cooperative transitions, *kooperativer Umwandlungen* 442
  of fast reactions in solution, *schneller Reaktionen in Lösung* 417
  of processes in equilibrium, *von Prozessen im Gleichgewicht* 388

laboratory coordinate system, in relation to center of mass coordinate system, *Laborsystem, Beziehung zum Massenschwerpunktsystem* 130, 160, 184–188
Lagrangian function, *Lagrangesche Funktion* 126
Landau-Teller theory, *Landau-Teller-Theorie* 134–138
Landau-Zener formula, *Landau-Zener-Formel* 51
  application in strong coupling, *Anwendung bei starker Kopplung* 66
  derivation of —, *Herleitung der —* 49–52, 58–59
Landau-Zener model, *Landau-Zener-Modell*
  application to elementary processes, *Anwendung auf Elementarprozesse* 75–76
  limitations of the —, *Grenzen der Anwendbarkeit des —s* 52
  linear —, *lineares —* 43–77
  of multistate systems, *von Systemen mit vielen Zuständen* 67–70
  two-dimensional extension of —,
    zweidimensionale Erweiterung des —s 70–75
  two-state approximation of a one-dimensional linear —, *Zwei-Zustand Näherung eines eindimensionalen linearen —s* 45–49, 53
Lewis-acid, reactions with —s, *Lewis-Säure, Reaktionen mit —n* 262, 313–315
Lewis-base, reactions with —s, *Lewis-Base, Reaktionen mit —n* 262, 267, 270f.
life time (s. a. life time (residence time); unimolecular reactions), *Lebensdauer (s. a. Verweilzeit; unimolekulare Reaktionen)*
  of H_3^+O-ion, *eines H_3^+O-Ions* 434
life time (residence time), mean, *Verweilzeit, mittlere* 360, 362, 387
  rate constants and —, *Geschwindigkeitskonstanten und —* 389, 393, 395 401
ligand exchange (s. a. metal complex reactions), *Ligandenaustausch (s. a. Metallkomplexreaktionen)*
  of atoms with electrical nuclear quadrupole moment, *von Atomen mit elektrischem Kernquadrupolmoment* 394
  in paramagnetic complexes, *in paramagnetischen Komplexen* 397
ligand field stabilisation, *Ligandenfeld-stabilisierung* 429
line shape of NMR lines, *Linienform von KMR-Linien*
  Lorentzian —, *Lorentzsche —* 360
line width (s. a. half width) and determination of reactions rates, *Linienbreite (s. a. Halbwertsbreite) und Bestimmung von Reaktionsgeschwindigkeiten* 360, 395, 399
  with fast exchange, *bei schnellem Austausch* 362
  with slow exchange, *bei langsamem Austausch* 362, 392
linearized rate equations, *linearisierte Geschwindigkeitsgleichungen* 419, *453*
Liouville equation, quantum mechanical (s. density matrix, equation of motion), *Liouville-Gleichung, quantenmechanische (s. Dichtematrix, Bewegungsgleichung)*

Lippmann-Schwinger differential equation, *Lippmann-Schwinger-Differentialgleichung* 82, 83, 102
  Chew-Goldberger solution of —, *Chew-Goldberger-Lösung der —* 82
localization energy, *Lokalisierungsenergie* 292, 325

macroscopic nuclear magnetization, *makroskopische Kernmagnetisierung*
  in thermal equilibrium, *im thermischen Gleichgewicht* 370
  transverse component in rotating coordinates, *transversale Komponente in rotierenden Koordinaten* 361, 384, 385, 397
Massey parameter, *Massey-Parameter* 45, 74
matrix method, *Matrixmethode*
  for the calculation of the partition function of linear polymers, *zur Berechnung der Zustandssumme von linearen Polymeren* 437–441
Maxwell-Boltzmann distribution function, *Maxwell-Boltzmann-Verteilungsfunktion* 10, 14
mechanics of collisions, *Mechanik der Stöße* 124–133
mechanism, *Mechanismus*
  external — (collision interaction), *äußerer — (Stoßwechselwirkung)* 84, 95
  internal — (energy redistribution) *innerer — (Energieumverteilung)* 84, 95
  of organic reactions, *organischer Reaktionen* 261–273
Menschutkin reactions, *Menschutkin-Reaktionen* 335, 337, 340–343
mesomeric effect and reactivity, *mesomerer Effekt und Reaktivität* 275f., 279–282, 284–289
metal complex reactions (s. a. ligand exchange), *Metallkomplexreaktionen (s. a. Ligandenaustausch)* 428–430
microscopic reversibility, principle of, *mikroskopische Reversibilität, Prinzip der* 363, 421, 454
molar volume, partial —, *Molvolumen, partielles —* 333, 341
molecular beam, *Molecularstrahl*
  studies, *—versuche* 157–173, 188–216
  studies, interpretation by Monte-Carlo calculations, *Interpretation von —versuchen durch Monte-Carlo-Rechnungen* 147–150, 203–205
molecular structure and reactivity, *Molekülstruktur und Reaktivität* 261 to 289
moment of inertia, *Trägheitsmoment* 128–129
Monte-Carlo calculations, *Monte-Carlo-Berechnungen* 143–156, 203–205
multiplicity, *Multiplizität*
  change of — in reactions, *Änderung der — bei Reaktionen* 235, 238f.
multi-quantum transition, *Mehrquantenübergang* 135

NAD (nicotinamide-adenine dinucleotide), *NAD (Nikotinamid-adenin-dinucleotid)* 450–452
von Neumann equation (s. density matrix, equation of motion), *von Neumann-Gleichung (s. Dichtematrix, Bewegungsgleichung)*
Newton's laws, *Newtonsche Gesetze* 124–125
nitric oxide, reactions with —, *Stickstoffmonoxid, Reaktionen mit —* 344
nitrogen, *Stickstoff*
  light emission from — dioxide, *Lichtemission von —dioxid* 248
  reactions of — atoms, *Reaktion von — atomen* 244f., 248f.
nitrous oxide, decomposition of —, *Distickstoffmonoxid, Zerfall von —* 238–240
non-adiabatic, *nicht-adiabatisch (e, er, es)*
  coupling, *Kopplung* 43–45, 48, 56, 59, 63, 64, 66, 73, 74
  coupling parameter, *Kopplungsparameter* 50
  (non-secular) line broadening (NMR), *(nichtsäkulare) Linienverbreiterung (KMR)* 391
  reflection from a potential barrier, *Reflexion von einem Potentialwall* 59–64
  relaxation, vibrational, *Relaxation der Molekülschwingungen* 76
  transitions, *Übergänge* 43–76
  transmission through a potential barrier, *Durchgang durch einen Potentialwall* 59–64
  tunnelling, *Tunneln* 63

non-adiabatic
  wave function, *Wellenfunktion* 47, 53, 68
non-crossing rule for adiabatic terms of same symmetry, *Kreuzungsverbot für adiabatische Terme gleicher Symmetrie* 51
nonequilibrium, *Nichtgleichgewichts-*
  conditions in unimolecular reactions *bedingungen bei unimolekularen Reaktionen* 228
  kinetics, *kinetik* 1, 18
  magnetization, molecular transfer of —, *magnetisierung, molekulare Übertragung von* — 404
  relaxation towards stationary — state, *Relaxation gegen stationären — zustand* 421
normal variables, *Normalvariable* 421 to 422, 455
  thermodynamic reaction quantities with respect to —, *thermodynamische Reaktionseffekte bezüglich* — 424, 449
nozzle beam, *Düsenstrahl* 164–167
nuclear spin density matrix (s. a. density matrix), *Kernspin-Dichtematrix (s. a. Dichtematrix)* 365–374
  applications in spin rearrangements, *Anwendungen bei Spin-Umlagerungen* 400
  interaction representation, *Wechselwirkungsdarstellung* 371
  in thermal equilibrium, *im thermischen Gleichgewicht* 368, 370
  time dependence (equation of motion) of the —, *zeitliche Veränderung (Bewegungsgleichung) der* — 368
  time-dependent perturbation theory, *zeitabhängige Störungsrechnung* 370
nucleophilic, *nukleophile*
  reactions, *Reaktionen* 262–271
  substituents (see electrondonors), *Substituenten (s. Elektronendonatoren)*

olefins, *Olefine* 308–325
orbital angular momentum, *Bahndrehimpuls* 129
orbiting in scattering experiments, *Umlaufen in Streuversuchen* 178
oriented molecules, collisions with —, *orientierte(n) Moleküle(n), Stöße mit* — 170, 199 f.

oscillator, anharmonic, *Oszillator, anharmonischer* 135
oxidation, *Oxidation* 348–356
oxygen atoms, reactions of —, *Sauerstoffatome, Reaktionen von* —n 244 to 247, 258

partition function (state sum), *Zustandssumme (Verteilungsfunktion)*
  of linear polymers, *von kettenförmigen Polymeren* 437
  in nuclear spin density matrix, *in der Kernspindichtematrix* 368
passage, *Durchgang*
  fast —, *schneller* — 403
  slow —, *langsamer* — 360, 361, 386, 403
Pauli equation (master —), *Pauli-Gleichung (master —)* 95
perturbation of chemical equilibrium, *Störung des chemischen Gleichgewichts* 418
  of cooperative transitions, *kooperativer Umwandlungen* 443
  jump- and pulse methods, *Sprung- und Impulsmethoden* 424–426
  periodical (or stationary) methods, *periodische (oder stationäre) Methoden* 426–428
photolysis (s. flash photolysis), *Photolyse (s. Blitzlichtphotolyse)*
π-bond order, π-*Bindungsordnung* 290
π-electron density, π-*Elektronendichte* 290
π-electron systems, π-*Elektronensysteme* 290–329
plasma, *Plasma* 15, 18
polar effect (s. inductive effect), *polarer Effekt (s. induktiver Effekt)*
potassium atoms, *Kaliumatome*
  reactions with alkyl halides, *Reaktion von* —n *mit Alkylhalogeniden* 147–150
  reaction with bromine, *Reaktion von* —n *mit Brom* 205—216
  reaction with hydrogen bromide, *Reaktion von* —n *mit Bromwasserstoff* 152, 188–190, 192–197
potential (s. a. adiabatic p.), *Potential (s. a. adiabatisches P.)*
  crossing zero order —s, *sich kreuzende* —e *nullter Ordnung* 48, 49 54, 69
  effective —, *effektives* — 178

potential
 energy function, *-funktion* 146f.,
 150f., 155, 204f., 211–213, 216
  for alkali halides, *für Alkalihalogenide* 211
  for bromine molecules, *für Brommoleküle* 212
 intermolecular —, *intermolekulares —*
 173f., 180, 182, 193
 surfaces (s. a. potential energy function and hyper surfaces), *-flächen
 (s. a. Potentialfunktion und Hyperflächen)* 249f., 384f.
predissociation, *Prädissoziation* 256
pre-exponential factor (s. a. Arrhenius
 parameters, *präexponentieller Faktor
 (s. a. Arrhenius-Parameter)* 23, 228f.
pressure, *Druck*
 effect of — on reaction rates in
 solution, *Einfluß des —es auf Reaktionen in Lösung* 331–345
principal axis transformation, *Hauptachsentransformation* 128
probability, *Wahrscheinlichkeit*
 of a state (— of realization), *eines
 Zustands (Realisierungs-)* 78, 367
product of inertia, *Trägheitsprodukt* 128
protolysis (s. proton transfer), *Protolyse (s. Protonenübertragung)*
proton transfer (protolysis), *Protonenübertragung (Protolyse)*
 in aqueous solutions of ammonium
 ions, *in wäßrigen Lösungen von Ammoniumionen* 392
 in pure water, *in reinem Wasser*
 430–434
prototropic rearrangements, *prototrope
 Umlagerungen* 266
pseudo crossing, *Pseudokreuzung*
 energy, *Energie* 65
 line, *Linie* 75
 of terms (adiabatic), *von Termen
 (adiabatischen)* 49, 53, 54, 56, 58, 68, 69
pulse methods (jump-), *Impulsmethoden
 (Sprung-)*
 chemical relaxation, *chemische Relaxation* 424–426
 NMR (spin echo) —, *KMR (Spin-
 Echo) —* 405

quadratic form, *quadratische Form* 127
quantum mechanical treatment, *quan-*

tenmechanische Behandlung
 of elastic scattering, *der elastischen
 Streuung* 179f.
 of elementary processes, *der Elementarreaktionen* 26–42
quantum statistical description, *quantenstatistische Beschreibung*
 of chemical reactions, *chemischer
 Reaktionen* 78–108
 of an ensemble of reacting nuclear
 spins, *eines Ensembles von reagierenden
 Kernspins* 364–388

racemization, *Razemisierung* 269
radiation (s. chemiluminescence), *Strahlung (s. Chemilumineszenz)*
radical, *Radikal*
 reactions, *-reaktionen* 253, 257–259,
 263–266
 spectroscopy of free —s, *Spektroskopie von —en* 255f.
rainbow scattering, *Regenbogenstreuung*
 177, 179, 207
random, *Zufälligkeit*
 lifetime assumption for energized
 molecules, *der Lebensdauer angeregter
 Moleküle* 154, 156
random number, *Zufallszahl* 146
rate constant (s. a. reaction rate constant), *Geschwindigkeitskonstante (s.
 a. Reaktionsgeschwindigkeitskonstante)*
Rayleighequation, *Rayleigh-Gleichung* 221
reactions (s. a. bimolecular r., unimolecular r.), *Reaktionen (s. a. bimolekulare R., unimolekulare R.)*
 multi-step —, *mehrstufige —* 421, *452*
 periodic —, *periodische —* 421
 single step —, *einstufige —* 419
 SN_1—, SN_1— 267–269, 429
 SN_2—, SN_2— 267–269, 429
 trimolecular, *trimolekulare* 256f.
reaction cross section, *Reaktionsquerschnitt* 1–22, 149, 180f.
 differential —, *differentieller —* 1–22,
 149, 162
 total —, *totaler —* 1–22, *149f.*, 157,
 180f., 190f., 194f., 199, 204f., 210, 213
reaction distance, *Reaktionsabstand* 335f.,
 433
reaction group, *Reaktionsgruppe* 26
reaction, mode of, *Reaktionsweise*
 292–329

reaction probability, *Reaktionswahr-scheinlichkeit* 1–22, 149, 159, 180–183, 194, 198f., 209

reaction rate, *Reaktionsgeschwindigkeit* 1–2

in the high pressure region, *im Hochdruckbereich* 94, 104, 105, 114

reaction rate constants, *Reaktionsgeschwindigkeitskonstante* 1–22

of the diffusion controlled encounter, *der diffusionsbestimmten Begegnung* 432

of the diffusion controlled separation, *der diffusionsbestimmten Trennung* 432

of fast reactions in solution, *schneller Reaktionen in Lösung* 417

and life time of spin states, *und die Lebensdauer von Spin-Zuständen* 389, 393, 395, 401

and relaxation times in chemical relaxation, *und Relaxationszeiten bei chemischer Relaxation* 422, 455

specific —, *spezifische* — 109–112

reaction term in the Boltzmann equation (s. rearrangement term), *Reaktionsterm in der Boltzmanngleichung (s. Umlagerungsterm)*

reactivity, *Reaktivität*

and molecular structure, *und Molekülstruktur* 261–289

rearrangements (s. a. isomerizations), *Umlagerungen (s. a. Isomerisierung)*

cis-trans—— of the peptide group, *cis-trans—— der Peptidgruppe* 446

in solutions, *in Lösungen* 265f.

rearrngement of spins, *Umlagerung von Spins*

adiabatic theories, *adiabatische Theorie* 357

asymmetrical —, *asymmetrische* — 374, 383

intermolecular —, *intermolekulare* — 375, 390–400

intramolecular —, *intramolekulare* — 374

monogeneous —, *monogene* — 375

polygeneous —, *polygene* — 375

quantum statistical treatment (non-adiabatic theory), *quantenstatistische Behandlung (nichtadiabatische Theorie)* 376, 391

symmetrical —, *symmetrische* — 374, 376–383

rebound-reaction, *Rückstoß-Reaktion* 149, 188–190, *192–205*

reduction, *Reduktion* 345, 348–356

reflection coefficient, *Reflexionskoeffizient* 61

relaxation, *Relaxation*

chemical —, *chemische* — 417–457

electronic —, *elektronische* — 7

in intramolecular energy transfer, *bei innermolekularer Energieübertragung* 156

multi-step—, *mehrstufige* —421, 452

of nuclei with electric quadrupole moment, *von Kernen mit elektrischem Quadrupolmoment* 392, 394

rotational —, *rotatorische* 7

single step —, *einstufige* — 419

spectra, quasi-continuous, *-Spektren, quasikontinuierliche* 422

spin —, *Spin-* — 359, 374, 384

spin — in paramagnetic solutions, *Spin- —in paramagnetischen Lösungen* 397

techniques, *-sverfahren* 424–428

thermal, *thermische* 121—124

towards equilibrium, *gegen einen Gleichgewichtszustand* 421, 454

towards a stationary non-equilibrium state, *gegen einen stationären Nicht-Gleichgewichtszustand* 421

velocity —, *Geschwindigkeits-* — 7

vibrational —, *Schwingungs-* —7, 118 to 121, 227, 258

relaxation time (s. a. time), *Relaxationszeit (s. a. Zeit)*

in allosteric enzyme systems, *in allosterischen Enzymsystemen* 448

of binding of NAD to GAPDH, *der Bindung von NAD an GAPDH* 450–452

in chemical relaxation, *bei chemischer Relaxation* 419–420, 424, *454*

of cooperative transitions, *kooperativer Umwandlungen* 444

of diatomic molecules, *zweiatomiger Moleküle* 119–120

longitudinal — of spins, *longitudinale — von Spins* 359

mean —, *mittlere* — 422–424

relaxation time
measurability in chemical relaxation *Meßbarkeit bei chemischer Relaxation* 389, *448*
of polyatomic molecules, *mehratomiger Moleküle* 120–121
transverse — of intramolecular relaxation, *transversale — der intramolekularen Relaxation* 89
transverse — of spins (s. a. line width), *transversale — von Spins (s. a. Linienbreite)* 359, 360, 403

renormalization of energy levels by perturbation, *Renormalisierung von Energieniveaus durch Störung* 92, 102

resonance transition, *Resonanzübergang* 136

Rice-Ramsperger-Kassel-Marcus, *Rice Ramsperger-Kassel-Marcus*
theory of unimolecular reactions, *-Theorie unimolekularer Reaktionen* 109–116, 122, 154 f.

Riemann-Lebesgue theorem, *Riemann-Lebesgue-Theorem 92*, 98, 101

rotating coordinate system, *rotierend (es, en) Koordinatensystem* 360
in quantum-mechanical spin theory, *in der quantenmechanischen Spintheorie* 406–413
solution of the Boltzmann equation in — *Lösung der Boltzmanngleichung im —* 385
transformation to —, *Transformation zum —* 412

rotation(al), *Rotation*
degrees of freedom, participation in unimolekular reactions, *Freiheitsgrade der — Beteiligung bei unimolekularen Reaktionen* 110, 114, 155, 229
excitation of — in reactive scattering, *Anregung der — bei reaktiver Streuung* 183, 209 f., 215
internal — of aniline derivatives, *innere — von Anilinderivaten* 376
selection of — quantum states in beam studies, *Zustände der —, Selektion in der Molekularstrahltechnik* 168

scattering, *Streuung*
cross section (see cross section), *squerschnitt (s. Wirkungsquerschnitt)*

elastic —, *elastische* — 158–161, 174 to 183, 193 f., 206 f.
matrix $t^{\oplus}$, *Matrix $t^{\oplus}$ 83*, 84, 102, 104
reactive —, *reaktive* —149, 158–161, 180–183, 194 f., 208 f.
theory, quantum mechanical, *Theorie, quantenmechanische* 80, 179 f.

Schrödinger equation, *Schrödingergleichung*
of nuclear motion, *der Kernbewegung* 31
time dependent —, zeitabhängige — 27–31, 79, 368
time-dependent — as reversible equation, *zeitabhängige — als reversible Gleichung* 80

Schrödinger representation, *Schrödingerdarstellung* 80

Selection, *Selektion*
of quantum states and velocities in beams, *von Quantenzuständen und Geschwindigkeiten in Strahlen* 162 f., 167 to 170

Shock tube, *Stoßwellenrohr* 222–227

Shock wave, *Stoßwelle* 219–227
reflected —, *reflektierte* — 221–224

SN_1-mechanism, *SN_1-Mechanismus* 267–269, 429

SN_2-mechanism, *SN_2-Mechanismus* 267–269, 429

solvation, *Solvatation* 301, 350–352, 354 f.

solvent, *Lösungsmittel*
continuum model for the —, *Kontinuummodell für das —* 338–344
effect on organic reaction rates, *—einfluß auf organische Reaktionen 276*, 287 f., 336, *338–344*

solvolytic reactions, *Solvolysen* 267 f., 279 f., 282, 284 f., 339, 344 f.

sound frequency, *Schallfrequenz* 119

spin-echo-method, *Spin-Echo-Methode* 405

spin-forbidden reactions, *spinverbotene Reaktionen* 239

spin-orbital coupling, *Spin-Bahn-Kopplung*
and non-adiabatic transitions, *und nicht-adiabatische Übergänge* 43, 71, 73–76

spin spin coupling (NMR), *Spin-Spin-Kopplung (KMR)* 358, 369, 375, 390

state, *Zustand*
  dynamic, *dynamischer* 78
  macroscopic (thermodynamic) —, *makroskopischer (thermodynamischer)* — 365, 381
  mixed —, *gemischter* — 365
  pure —, *reiner* — 78, *364*
stationary methods, *stationäre Methoden*
  in chemical relaxation, *bei chemischer Relaxation* 426–428
  in NMR, slow passage in, *bei KMR, M. des langsamen Durchgangs 360, 361,* 386
    limitations, *Grenzen der Anwendbarkeit* 403
statistical, *statistische*
  average (s. a. ensemble average), *—r Mittelwert (s. a. Ensemble-Mittelwert)* 79
  matrix (s. density matrix), *Matrix (s. Dichtematrix)*
  systems, *Systeme* 78, 365, 381
  theory of chemical reactions, *Theorie chemischer Reaktionen* 78, 202f.
stepladder model, *Stufenleitermodell*
  of energy transfer, *der Energieübertragung* 123–124, 140–141, 229
stereochemistry, *Stereochemie*
  of eliminations, *von Eliminierungen* 270–273
steric, *sterische*
  effects in organic reactions, *Effekte bei organischen Reaktionen* 268, 275f., 281f., 284–289, 335, 337f.
  factor in gas phase collisions, *—r Faktor bei Stößen in der Gasphase* 200f.
stripping reaction, *Abstreifreaktion* 152, 189–191, 195, 205–216
substituents, *Substituenten*
  influence of — on reactivity, *Einfluß der — auf die Reaktivität* 268, 273–289, 292–329, 339
substitution reactions, *Substitutionsreaktionen* 262f., 266–269
sulfur dioxide, *Schwefeldioxid*
  electronically excited —, *elektronisch angeregtes* — 249f.
  decomposition of —, *Zerfall von* — 236f.

Taftequation, *Taft-Gleichung* 282–284

tensor of inertia, *Trägheitstensor* 128–129
thermodynamic reaction quantities (spin exchange), *thermodynamische Reaktionseffekte (Spinaustausch)* 375, 389
  with respect to normal variables, *bezüglich Normalvariablen* 424, 448 to 450
threshold energy (s. activation energy), *Schwellenenergie (s. Aktivierungsenergie)*
time, *Zeit*
  atomic, *atomare* — 89, 97
  transversal —, *transversale* — 89, 97
trajectory, *Wegzeitkurve* 45, 143–147, 203
trajectory, classical, *Bahnkurve, klassische,* 45
transition, *Übergang*
  overbarrier —, *über den Potentialwall* 66
transition probability, *Übergangswahrscheinlichkeit*
  (s. a. Landau-Zener formula, life time), *(s. a. Landau-Zener-Formel, Verweilzeit)* 239f., 248, 363, *383*
  at adiabatic coupling, *bei adiabatischer Kopplung* 54, 55
  mean —, *mittlere* — 64–67
  mean two-dimensional, *mittlere zweidimensionale* 72
  non-adiabatic —, *nicht-adiabatische* — 55, 74, 75
  temperature, for a given, *Temperatur bei einer bestimmten* 135
transition state, *Übergangszustand* 110, 117, 267, 341–343
  lifetime of —, *Lebensdauer des —s* 149f., 200f., 210
  of processes in NMR, *von Prozessen in KMR* 361, 378, 383
  theory, *Theorie des —s* 115, 274, 331f., 350
transmission coefficient, *Transmissionskoeffizient* 61, 63, 64
transmission, non-adiabatic (s. a. tunnelling), *Transmission, nicht-adiabatische (s. a. Tunneln)*
  through, over, under a potential barrier, *durch, über, unter einem Potentialwall* 59–64
tunnelling, *Tunneln* 55, 59, 63, 66, 67
  adiabatic —, *adiabatisches* — 63–64

tunnelling
non-adiabatic—, *nicht-adiabatisches—* 63–64
turning point, *Umkehrpunkt* 52–59, 70, 76

ultrasonics, *Ultraschall* 118–121
unimolecular reactions, *unimolekulare Reaktionen* 227–240
calculated by the Monte Carlo method, *Berechnungen nach dem Monte Carlo Verfahren* 153–156
in liquids, *in Flüssigkeiten* 106, 267, 270
theory (s. a. Kassel-Theorie, Rice-Ramsperger-Kassel-Marcus-Theorie), *Theorie (s. a. Kassel-Theorie, Rice-Ramsperger-Kassel-Marcus-Theorie)* 95, 109–116, 227–229

velocity, *Geschwindigkeit*
distribution of products, —*sverteilung der Produkte* 195f., 207f.
selection in beams, —*sselektion in Strahlen* 162f., 167–169, 196
vibration, *Schwingung*
—al excitation of products, —*sanregung von Produkten* 150f., 173, 183, 209f., 215, 258
vibration-translation transition, *Vibration-Translationsübergang* 134–138
vibration-vibration transition, *Vibration-Vibrationsübergang* 134–138

water, *Wasser*
decomposition of —, *Zerfall von —* 235f.
protolysis in —, *Protolyse in —* 430–434
wiggles, *Nachschwingungen* 403

Zimm-Bragg-model for cooperative transitions, *Zimm-Bragg-Modell für kooperative Umwandlungen* 441
extended model, *erweitertes Modell* 435–442